S0-BLM-008

Cyclopolymerization and Cyclocopolymerization

Cyclopolymerization and Cyclocopolymerization

George B. Butler

Department of Chemistry
University of Florida
Gainesville, Florida

Marcel Dekker, Inc.   New York • Basel • Hong Kong

ISBN: 0-8247-8625-4

This book is printed on acid-free paper.

MARCEL DEKKER, INC.
270 Madison Avenue, New York, New York 10016

Current printing (last digit):
10 9 8 7 6 5 4 3 2 1

PRINTED IN THE UNITED STATES OF AMERICA

This volume is dedicated to all of those authors who believed in our early papers with sufficient conviction to undertake their own experimental programs and to publish their findings

Preface

The aim of this volume is to summarize in one reference the vast amount of information published in the chemical literature on cyclopolymerization since its inception and discovery about four decades ago. The topic is broadly defined as any type of addition polymerization that leads to the introduction of cyclic structures into the main polymer chain. Its scope is broad, including (at least conceptually) not only symmetrical diene counterparts of all known monomers capable of undergoing addition polymerization, but the unsymmetrical ones as well. Many of the latter have not been investigated. Thus, the scope of cyclopolymerization is considerably broader than that of addition polymerization itself, which includes vinyl, alkyl, carbonyl, isocyanate, nitrile, and epoxy polymerizations, topics that are covered in numerous volumes. Cyclopolymerization represents an exception to the general principle of polymerization established by Staudinger (Nobel Laureate, 1953) in 1934 that nonconjugated dienes, when polymerized, lead to cross-linked, and therefore nonsoluble and nonlinear, polymers or copolymers. It also represents an exception to the generally accepted hypothesis advanced by Flory (Nobel Laureate, 1974) in 1937, which states that the more stable intermediate controls the course of propagation in vinyl polymerization. Numerous studies have shown that, in many cases, cyclopolymerization does not conform to this hypothesis, but forms cyclic structures derived via propagation through the less stable intermediate.

Examples of monomers that cyclopolymerize in addition to both symmetrical and unsymmetrical nonconjugated dienes are diynes, dialdehydes, diisocyanates, diepoxides, dinitriles, and some organometallic monomers. Also, certain monomers that contain two different polymerizable functional groups cyclopolymerize. A variety of cyclopolymerizable monomers undergo copolymerization with their monofunctional counterparts or with other copolymerizable monomers. Reactivity ratios and other copolymer parameters have been obtained for some comonomer pairs. Certain 1,4-bis-multiple bonded structures undergo cyclopolymerization with suitable monofunctional structures to form six-membered cycles via an apparent 4+2 process. Post cyclopolymerization of suitable preformed polymers leads to polymers that contain extensive blocks of ladder structures.

Certain diene monomers undergo cyclopolymerization to yield polymers that include larger, less probable, cyclic structures in the main polymer chain. The mechanisms of both cyclopolymerization and cyclocopolymerization have been studied extensively, including kinetic measurements and derivation of suitable mathematical expressions to describe the generally accepted mechanisms. However, certain aspects of the cyclopolymerization mechanism have not been adequately explained. Cyclopolymers have a broad spectrum of applicability in commerce, often possessing enhanced properties relative to their open chain analogs. Several cyclopolymers and copolymers have now become commercial products manufactured around the world in multimillion-dollar annual quantities.

This volume is designed to be the most complete book on cyclopolymerization and cyclocopolymerization and should have a place in all science and engineering libraries. It should also be of interest to all polymer chemists, as well as to all industrial chemists having as their major objective synthesis of novel or specialty polymers. Structural and theoretical chemists will be intrigued by the potential for additional studies and justification of the unusual and unpredictable mechanistic aspects of cyclopolymerization. The majority of new references appearing in the literature during the last five years have dealt with new uses and applications of commercially available cyclopolymers and copolymers rather than with new monomers and their polymerization. Thus, applications chemists, largely from industrial laboratories, will find the content useful and challenging.

It would not have been possible to complete a project as extensive as this one without the help, understanding, and participation of numerous individuals of almost unprecedented dedication. To mention all of them would be impossible; however, they know who they are. It is with deep gratitude that I acknowledge these contributions. In addition to all of those research associates who contributed to the experimental program, I would like to specifically acknowledge the diligence and dedication of Ms. Carol Albert in typing the manuscript from a set of often undecipherable script. Mr. Nai Zheng Zhang must be gratefully recognized for his diligence and patience in preparation of the numerous figures. Also, without the patience, love, understanding, and proofreading par excellence by my wife, Josephine, this project could not have been completed. It is with love and appreciation that I sincerely acknowledge her support.

George B. Butler

Contents

1
Introduction and Historical Background

An important and well-accepted principle that had been established early in the history of modern polymer science is that nonconjugated dienes, when subjected to addition polymerization conditions, result in formation of either linear polymers containing unreacted double bonds or, more likely, crosslinked polymers. However, during a research program, the results of which were reported in a series of papers published from 1949 to 1957,[1-7] it was found that several polymers produced from diallyl quaternary ammonium salts were soluble in water, and therefore not crosslinked, and yet they did not contain residual unsaturation. It was realized that, in these cases, both allyl double bonds were involved in a reaction leading to a linear polymer, and a mechanism was proposed in which chain growth occurred *via* alternating intramolecular and intermolecular steps (1–1).[8] The presence of cyclic units in the main chain of these polymers was

Initiation

$Z\cdot$ (Radical initiator)

Intramolecular propagation

Intermolecular propagation

monomer

(1–1)

established by degradation reactions carried out on representative polymers.[9] However, these studies did not establish the size of the ring in the polymers. In this mechanism, now known as *cyclopolymerization,* cyclization occurs as a characteristic feature of the polymerization process and not as a deficiency in cross-linking and network formation.

PROOF OF STRUCTURE

This rigorous structure study[9] substantiated the cyclopolymerization hypothesis. The polymer of diallylamine hydrobromide was degraded as shown in 1–2. The structure

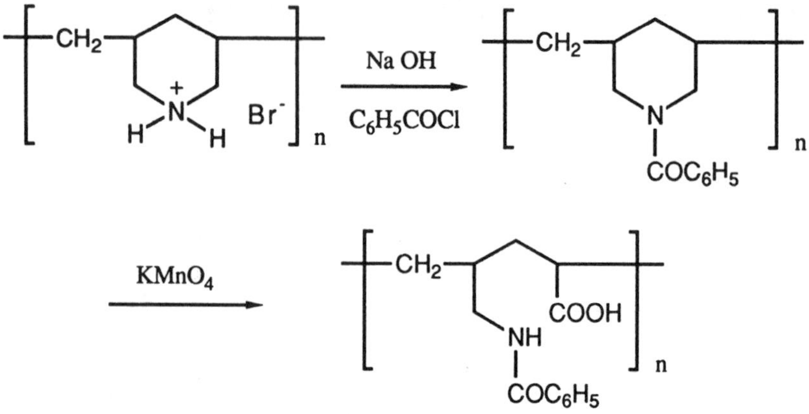

Na OH, C_6H_5COCl → ; $KMnO_4$ →

(1–2)

was found to be correct by the following: (a) The analysis was that required for this structure. (b) Potentiometric titrations in dimethylformamide solution afforded typical titration curves for carboxylic acids under these conditions. (c) The infrared spectrum showed absorption bands that correlate with such a structure. (d) On heating slightly above its melting point, the product yielded a sublimate of benzoic acid, and the residual polymer was clearly crosslinked, being insoluble and forming gels with several solvents. This product was thus shown to have resulted from the cleavage of rings present in the original poly(diallylammonium bromide).[1] Poly(diallyldimethylammonium bromide)[4] was degraded as indicated in (1–3); this degradation also gave products that can reasonably be explained only by assuming this structure for poly(diallyldimethylammonium bromide).

1. To hydroxide 2. Thermal decomposition → ; CH_3I → ; 1. To hydroxide 2. Thermal decomposition → insoluble polymer (9) + $(CH_3)_3N$

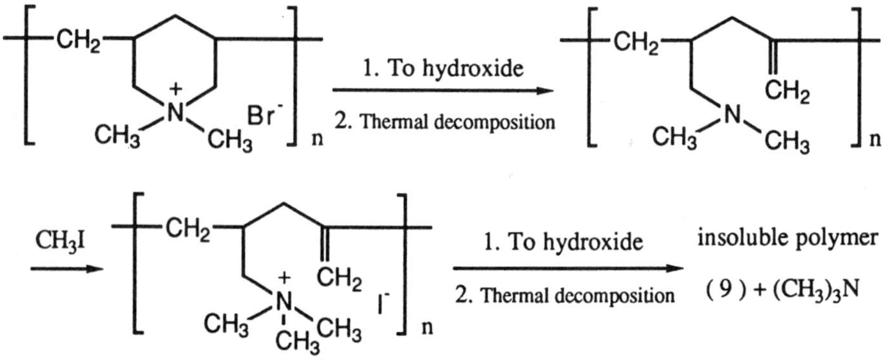

(1–3)

More recent investigations have shown that in the case of the quaternary ammonium salts, almost exclusive formation of the kinetically formed five-membered ring occurs instead of the thermodynamically favored six-membered ring (1–4). Details of these observations will be discussed later.

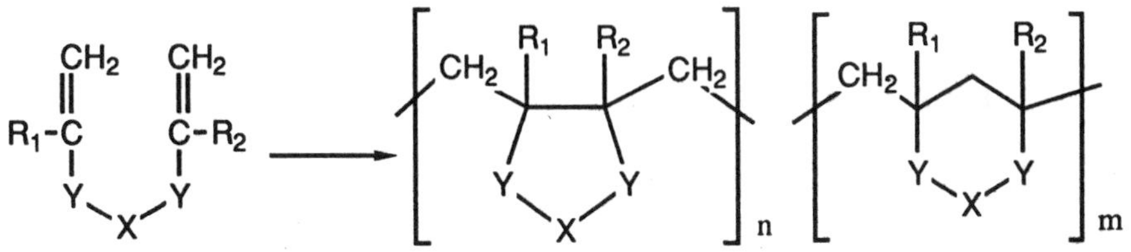

(1–4)

Since these initial investigations it has been demonstrated that by using appropriate catalysts, 1,6–dienes of the type (1–4) can be polymerized to yield soluble, saturated polymers with structures in which rings alternate with methylene groups along the linear chain.[3] Cyclic units in the polymer chain have been obtained by the polymerization of monomers containing other unsaturated systems and polymerizable groupings. It has been established that cyclic units can form by intramolecular cyclization of the monomer or after interaction of two or more monomer or comonomer units. Cyclopolymerization is, thus, defined as any type of addition polymerization which leads to the introduction of cyclic structures into the main chain of the polymer.

HISTORICAL

Before the principles of cyclopolymerizations were established, it had been generally assumed that polymerization and copolymerization would lead to only crosslinked polymers. Although the observed extents of reaction at the gel point for addition of certain crosslinkable systems such as methyl methacrylate-ethylene dimethacrylate and vinylacetate-divinyl adipate[10] were always greater than statistical predictions[10,11], the formation of cyclic structures was considered of minor importance.

In polymerization[12] studies of diallyl phthalate, it was found that 40% of the units were cyclized in the polymer. Statistical calculations[13] predicted that 31% of converted monomer should undergo some kind of cyclization in the polymerization of diallyl phthalate. Earlier results were in fair agreement with the prediction and further work has been pursued in this direction.[14] It was[15] subsequently found that diallyl phthalate would yield a polymer, by solution polymerization, which was 81% cyclized.

CYCLOPOLYMERIZATION

After the cyclopolymerization mechanism had been discovered, it became clear that a similar explanation could be offered for an unusual copolymerization reaction. In 1951, a soluble 1:2 copolymer of divinyl ether and maleic anhydride had been synthesized which contained no carbon-carbon unsaturation. A mechanism (1–5) now known as *cyclocopolymerization,* leading to polymers containing bicyclic units, was proposed and supported by experimental evidence.[16]

Following these investigations, a vast amount of work has been carried out on cyclopolymerization and cyclocopolymerization. Much of the earlier work went into the

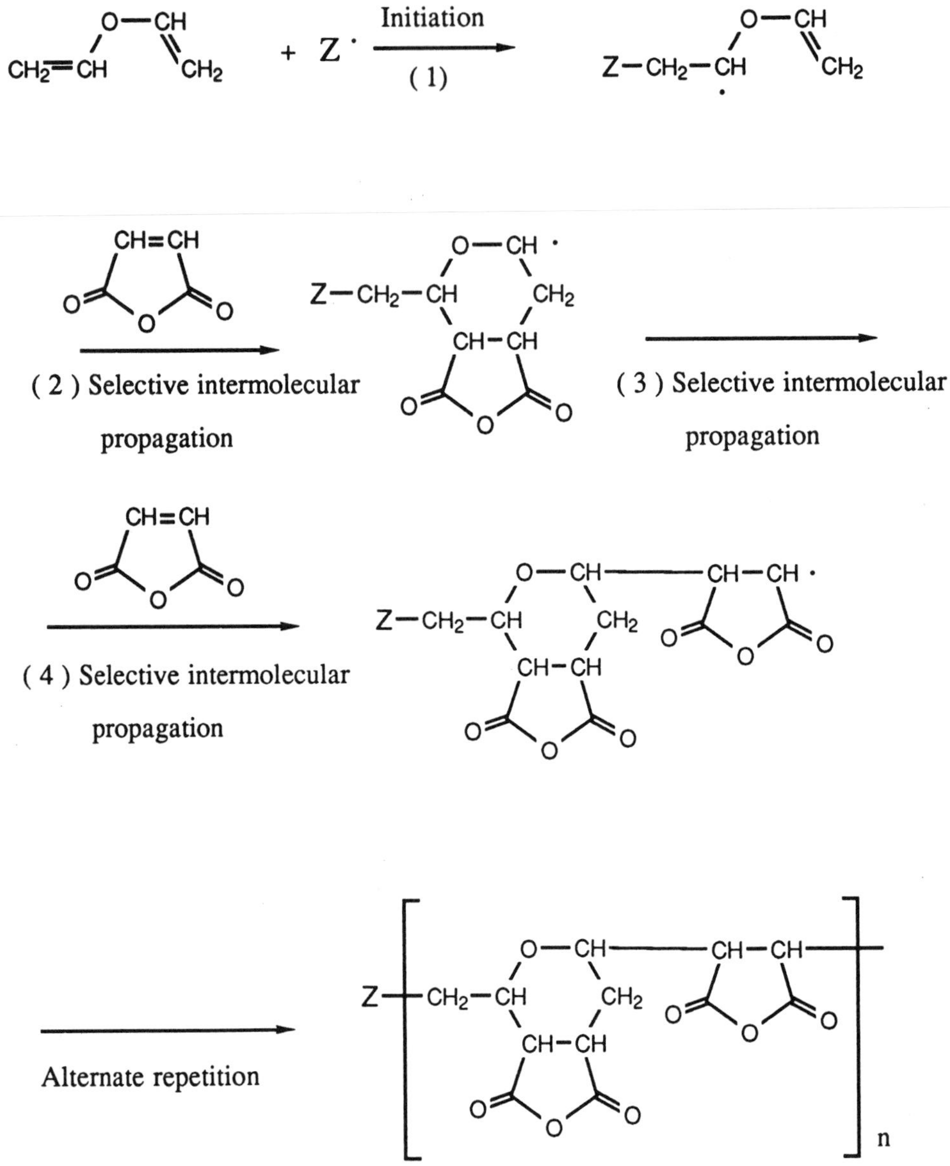

Initiation

(1)

(2) Selective intermolecular propagation

(3) Selective intermolecular propagation

(4) Selective intermolecular propagation

Alternate repetition

(1-5)

synthesis of new monomers and polymers to discover the range of structures which could be achieved and the properties that they possessed. It is the purpose of this volume to summarize the reactions which are encompassed within the terms cyclopolymerization and cyclocopolymerization and discuss the more recent investigations of new monomers or combinations of monomers which lead, by a chain-growth polymerization mechanism, to polymers having cyclic units as the dominant feature of the polymer structure. Useful polymers have resulted from research in this field and a brief review of the properties of polymers of industrial and medical importance is included. However, particular emphasis will be placed upon those studies which have provided an analysis of the microstructure of the polymer chain and work which has attempted to explain the mechanism of the cyclization reactions.

RING SIZE

Although it has long been understood that properties are closely related to the structure of polymers, early workers in this field were often satisfied when the linear nature of the polymers and the presence of cyclic units had been established. Thermodynamic factors were considered most important and hence, six-membered rings were assigned to the products obtained, based on an earlier well-accepted principle of polymerization theory that the more stable of two competing reactive species will predominate in the reaction. The corresponding five-membered structures are feasible (1–6), and ring–size has been the subject of extensive investigations during recent years.

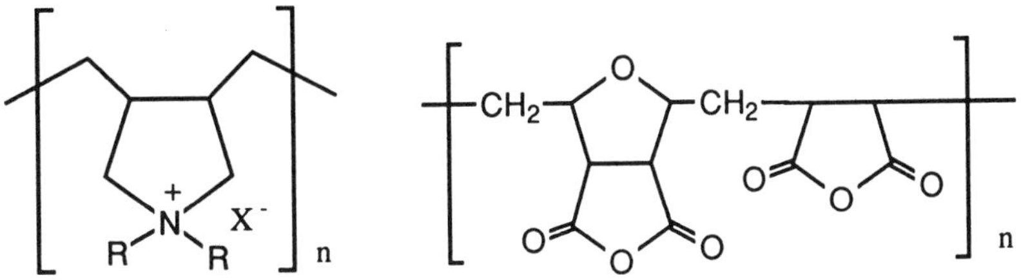

(1–6)

The six-membered rings follow from the expectation that the reaction would proceed *via* head-to-tail propagation, involving secondary radicals as intermediates, whereas the five-membered rings require head-to-head cyclization steps with formation of a primary radical, commonly accepted as a less stable intermediate, which leads to the more strained five-membered ring. Recent studies using a combination of chemical and spectroscopic methods (particularly ^{13}C NMR) have yielded significant information on the structures of cyclopolymers. Other approaches to a study of the ring sizes of cyclic units in these polymers are to detect or trap the initially formed reactive species. Such structural studies are developing our understanding of the driving force for the cyclization reaction.

MECHANISTIC PROPOSALS

Various attempts have been made to provide a general explanation for the characteristic features of the cyclopolymerization and cyclocopolymerization reactions. As a fundamental explanation of cyclopolymerization, it was proposed[17] that an electronic interaction occurs between the unconjugated double bonds of 1,6 dienes, or between the intramolecular double bond and the reactive center which provides a favorable pathway for cyclization. Alternatively, others[18] considered steric effects to be more important. The presence of the large pendant group was regarded as a hindrance to intermolecular reaction and conformational interconversions would bring the unconjugated double bond to a favorable position for reaction. Later, unsatisfactory correlations between polymer structures and telomerization reactions[19] led to concern over the relative importance of kinetic and thermodynamic factors in the cyclization step.[20] In the cyclocopolymerization process, participation of charge-transfer complexes has been proposed.[21] Proposals have also been made, based on molecular orbital theory, to account for the preference for five-membered rings in certain of these systems.[22] Studies of these ideas have provided information of importance not just to the mechanisms of cyclopolymerization and cyclocopolymerization but also to chain-growth polymerization in general. Detailed discussion of each of these proposals will follow in appropriate sections of this volume.

THE FIRST TEN YEARS

The experimental observations which led to postulation of the alternating intramolecular-intermolecular chain propagation (now referred to generally as "cyclopolymerization") were made in 1949.[1] However, the details of the proposal were not published until more convincing evidence was available.[23] Following this publication, the potential scope of the concept became apparent to synthetic polymer chemists, which, stated simply, would include all suitable monomers capable of undergoing addition polymerization in presence of suitable catalysts or initiators; also an almost unlimited group of hetero-monomers, i.e. those having different multiple-bonded, suitably distributed functions within the molecule could be postulated. The limitations implied in the original postulate applied to ring size, only stable rings such as five- or six-membered rings being immediately obvious. It is not surprising then that within the first ten years of its history the scope of cyclopolymerization was well established. Because of the unusual nature of the required monomers, many had to be synthesized and characterized for the first time.

Radical Initiated Polymerization

Dimethyl α,α-dimethylenepimelate, a monomer closely related to methyl methacrylate, was synthesized and polymerized by use of various free-radical initiators.[24] Other monomers of this general type prepared and studied are shown in (1–7).[25] Dimethyl

$$CH_2=\overset{\overset{R}{|}}{C}-(CH_2)_3-\overset{\overset{R}{|}}{C}=CH_2$$

(a) $R = COOH$ (d) $R = COCl$

(b) $R = COOCH_3$ (e) $R = CONH_2$

(c) $R = COOC_2H_5$ (f) $R = CN$

(1–7)

α,α-dimethyleneadipate was also prepared and investigated to determine the ease of formation of a five-membered ring. The five-membered ring was found to form less readily than the six-membered ring; however, a quantitative infrared analysis indicated that more than 90% of the monomer units were incorporated to produce a polymer containing five-membered ring structures as the predominating recurring unit. The polymer had an inherent viscosity of 0.21 dl/g (0.34 g/100 ml, benzene). α,α'-Dimethylenepimelonitrile (7f) polymerized in solution and in emulsion to produce a soluble polymer having an inherent viscosity of 0.65 dl/g (0.235 g/100 ml, dimethylformamide). The spectral and solubility data were in agreement with the cyclic structure. This polymer was found to be much more thermally stable than the monoolefinic counterpart, polymethacrylonitrile. The latter undergoes thermal discoloration at 140°C, whereas the poly-α,α'-dimethylenepimelonitrile did not change color until heated to temperatures above 275°C.

Dimethyl α,α'-dimethylenepimelate copolymerized with acrylonitrile to produce a soluble polymer with no residual unsaturation. The analysis indicated a 65% incorporation of acrylonitrile in the copolymer, a ratio that did not change on repeated purification. It was also shown that monomers of the above type would undergo copolymerization with acrylonitrile to produce copolymers having fiber-forming properties superior to those of polyacrylonitrile.[26] For example, a copolymer of 15% 2,6-dicyano-1,6-heptadiene with 85% acrylonitrile was spun and drawn fourfold to strong filaments having a wet initial modulus of 10 g/den at 90°C and a recovery from 3% elongation in 50°C water of 56%. Typical values for fiber prepared in a similar manner from acrylonitrile homopolymer are 3 g/den and 40% respectively. The preparation of other polymers from these dienes are summarized in Table 1-1.

Two groups[27–29] independently described polymerization of acrylic anhydride under a variety of conditions to soluble polymers. Molecular weights were as high as 95,000. Organic peroxides were found to initiate the polymerization. On bulk polymerization, brittle white polymers were obtained which swelled considerably in dimethylformamide and dimethyl sulfoxide but dissolved completely on standing for several hours. This behavior, along with the observation of a definite softening range, suggested that these polymers were linear and of high molecular weight. The molecular weights were calculated from determinations of intrinsic viscosities (ν) in 2N sodium hydroxide solution, the equation $(\nu) = 4.27 \times 10^{-3} dp^{0.69}$ for poly(acrylic acid)[30] under these conditions being used to calculate the values for the degree of polymerization (DP). Molecular weights for the poly(acrylic anhydrides) of from 95,000 to < 5,000 depending on the particular polymerizing conditions were obtained. From these results, it is evident that acrylic anhydride polymerizes in solution exclusively by the cyclopolymerization route (1–8), being the structure of the products.

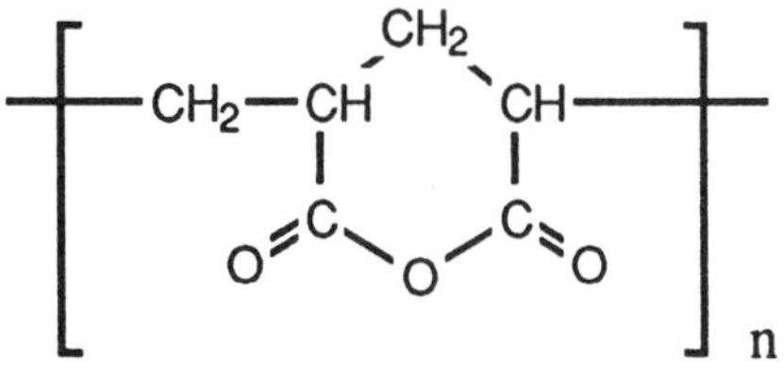

CH_2 — CH — CH_2 — CH — (cyclic anhydride: C(=O)—O—C(=O))]$_n$

(1–8)

An effort to synthesize diacrylylmethane by a Claisen condensation of methyl vinyl ketone with ethyl acrylate[31], gave instead poly(diacrylylmethane), the production of which was attributed to an anionic polymerization of the monomeric compound as it formed. Also described was a cyclopolymer from alloocimene (2,6-dimethyl-2,4,6-octatriene).[32]

Polymers[33] containing phosphine oxide units in the chain have been prepared through cyclopolymerization. Diallylphenylphosphine oxide and dimethallylphenylphosphine oxide were polymerized to polymers of apparent low molecular weights in the

Table 1–1 Polymers from 2,6-Disubstituted 1-6-Heptadienes[a]

X in monomer	Monomer, g	Solvent	Initiator	Temp, °C	Time, hr	% conversion	Intrinsic viscosity[b]
$-COOCH_3$	3.0	Emulsion (25 ml)	Metabisulfite (0.06g)-peroxydisulfate (0.01 g)	55	3	43	<0.10
$-COOCH_3$	2.0	Benzene (15 ml)	Benzoyl peroxide	75	24	22	0.21
$-COOCH_3$	2.0	t-Butyl alcohol (15 ml)	Azobisisobutyronitrile (0.01 g)	65	24	65	0.47
$-COOCH_3$	1.0	γ-Butyrolactone (3 ml)	UV, benzoin (0.005g)	20	16	92	0.60
$-COOCH_3$	1.0	Tetramethylene sulfone (3 ml)	UV, benzoin (0.002 g)	20	4	80	2.03
-CN	1.0	γ-Butyrolactone (3 ml)	UV, benzoin (0.003 g)	20	4	25	1.00
-CN	1.0	Tetramethylene sulfone (4 ml)	UV, benzoin (0.003 g)	20	4	74	1.22

[a] Formula: $CH_2=C(X)-(CH_2)_3-C(X)=CH_2$

[b] dl/g in dimethylformamide.

presence of free-radical initiators. Poly(diallylphenylphosphine oxide) was soluble in ethanol, dimethylformamide, and acetic acid. An infrared examination of the polymer showed only a small peak for carbon-carbon double bond. A thermal gravimetric analysis of the polymer showed that it retained 95% of its original weight at 325°C, 90% at 375° and 60% at 400°.

Soluble copolymers of diallylalkylamine oxides with acrylonitrile and other monomers have been described.[34] These copolymers were assumed to possess one unreacted double bond for each molecule of diallylamine oxide entering the chain; however, copolymerization by an alternating intramolecular intermolecular mechanism had probably occurred, producing saturated copolymers. No subsequent reexamination of these copolymers has been reported. A number of soluble copolymers involving several diallyl monomers and several conventional monomers have been described.[35] These copolymerizations were postulated to follow the cyclopolymerization mechanism. Cyclization of allyl ether and allyl sulfide to substituted tetrahydropyrans and tetrahydrothiopyrans, respectively, occurred when attempts were made to add various reagents under free-radical conditions.[36] The activity of the reagent as a chain-transfer reagent was found to be important in determining the degree to which addition was accompanied by cyclization.

Diallyl- and dimethallylphosphonium salts have been polymerized to soluble cyclopolymers through use of free-radical initiators.[37] These monomers appear to lend themselves to higher molecular weight formation than the corresponding diallylphosphine oxides. For example, poly(diallyldiphenylphosphonium chloride) was converted to poly(diallyldiphenylphosphine oxide) by treatment with sodium hydroxide. The intrinsic viscosity of this polymer was much higher than that of poly(diallylphenylphosphine oxide) prepared by direct polymerization of the corresponding monomer.[33]

Although the literature reveals that many other symmetrical diene monomers have been shown to undergo cyclopolymerization by free-radical initiators, those discussed here are typical members. In every case the experimental results offer strong evidence that cyclopolymerization is the predominant mode of polymer formation. These results show that the extent of ring closure is somewhat concentration dependent; however, in general, it can be stated that, in most cases, cyclization occurred to the extent of 95–100% of all monomer units entering the chain.

Cationic Initiated Polymerization

The number of monomers of the 1,5 or 1,6-heptadiene type which had been subjected to cationic initiation was somewhat limited. The synthesis and polymerization of 2,6-diphenyl-1,6-heptadiene had been described.[38] It was found that all the known general types of initiation—cationic, conventional anionic, free-radical, Ziegler-type, and thermal—resulted in polymerization only by cyclopolymerization leading to the same polymer in all cases. The polymers obtained were soluble in a number of solvents, and spectral studies supported the earlier conclusions.

Thermal polymerizations without initiator were run at 100 and 180°C. Only the latter temperature afforded any appreciable polymer. The inherent viscosity in benzene was 0.23 dl/g (0.5%, 30.0°C). A polymer with an inherent viscosity of 0.35 dl/g (0.5%, 30.0°C) had a polymer-melt temperature of 265°C.

Cationic initiated polymerization of this monomer using boron trifluoride in methylene chloride gave polymers having polymer-melt temperatures of 250°. *Polymerization with a Ziegler-type catalyst* from triisobutylaluminum and titanium tetrachloride gave a polymer of similar structure.

The best method for obtaining both high-conversion and high-molecular-weight polymer was *anionic polymerization.* The lithium complex of naphthalene ("Lithium naphthalene") was used as the initiator and tetrahydrofuran as the solvent. An 80% yield of polymer with an inherent viscosity in benzene of 0.49% dl/g(0.5%, 30.0°C) and a polymer-melt temperature of 300°C was obtained.

All the polymers thus prepared from 2,6-diphenyl-1,6-heptadiene were soluble in tetrahydrofuran, chloroform, and benzene. Clear films were cast from these solvents and fibers could be drawn from the polymer melts. Melts could be held at 300°C open to the atmosphere for at least one minute without any sign of discoloration, gelation, or depolymerization—in marked contrast to the noncyclic analog poly-α-methylstyrene, which readily depolymerized under these conditions. By extrapolation of inherent viscosities, the limiting polymer-melt temperature of the polymers appeared to be about 300°C. This high value is interesting in that the polymers do not contain polar groups and presumably results from a stiff chain structure. The conditions used for synthesis of some homopolymers of this monomer are summarized in Table 1–2.

1-Vinyl-4-methylenecyclohexane has been polymerized to a polymer containing 1,3-methylenebicyclo[2.2.2]octane units by a cationic mechanism.[39] The polymer had an intrinsic viscosity of 0.08 dl/g in benzene and the infrared and nuclear magnetic resonance spectra of the polymer indicated residual unsaturation equivalent to one double bond per two monomer units remained. Polymer structure (1–9) was supported by

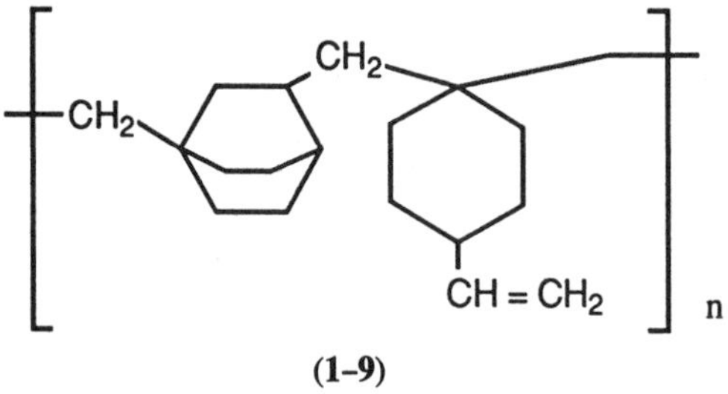

$$\left[-CH_2- \text{(bicyclo[2.2.2]octane)} -CH_2- \text{(cyclohexane, } CH=CH_2\text{)} - \right]_n$$

(1–9)

the evidence. Aliphatic divinyl formals and acetals have been found to undergo cationic polymerization[40] to produce cyclic polymers.

Anionic Initiated Polymerization

The earliest record of an anionic cyclic polymerization was reported in 1958.[31] A polymer was inadvertently obtained in an effort to synthesize diacrylmethane by a Claisen condensation of methyl vinyl ketone with ethyl acrylate. More recent work[41], has shown the polymer to be a copolymer of diacrylmethane and methyl vinyl ketone. The anionic polymerization[42] of pure diacrylmethane, prepared by a known method, was reported in 1963.[43] Sodium methoxide at −75°C gave 90% yield of a bright yellow soluble polymer. Evidence supports the cyclic nature of the polymer. Dimethacrylmethane was also cyclopolymerized anionically. 2,6-Diphenyl-1,6-heptadiene was also shown to undergo anionic initiated cyclopolymerization.

Table 1–2 Polymers from 2,6-Diphenyl-1,6-heptadiene

Monomer, ml	Solvent	Initiator	Temp, °C	Time, hr	% Yield	Inherent viscosity, dl/g[a]	Polymer-melt temp, °C
1.0	None	Heat	180	96	41	0.23	280
1.0	None	Cumene hydroperoxide (0.004 g)	100	120	36	0.35	265
1.0	Methylene chloride (5 ml)	Boron trifluoride	−78		94	0.19	255
1.0	Tetrahydrofuran (10 ml)	Lithium naphthaline	26–28		80	0.49	300
1.5	Decalin (7 ml)	Titanium tetrachloride (6 ml of 0.001 M soln)	26–28	96	71		140
		Triisobutylaluminum (1 ml of 0.001 M soln)					

[a] 0.5%, benzene, 30°C

Polymerization of trimethylenediisocyanate to a predominantly cyclic polymer by use of anionic initiators such as sodium cyanide in dimethylformamide and butyllithium in heptane has been described.[44] The polymers were characterized by solubility determinations, differential thermal analysis, thermogravimetric analysis, infrared spectroscopy, x-ray diffraction, and viscosity measurements. The polymers were linear, crystalline, thermally stable materials having inherent viscosities up to 0.12 dl/g. The observed properties are consistent with (1–10).

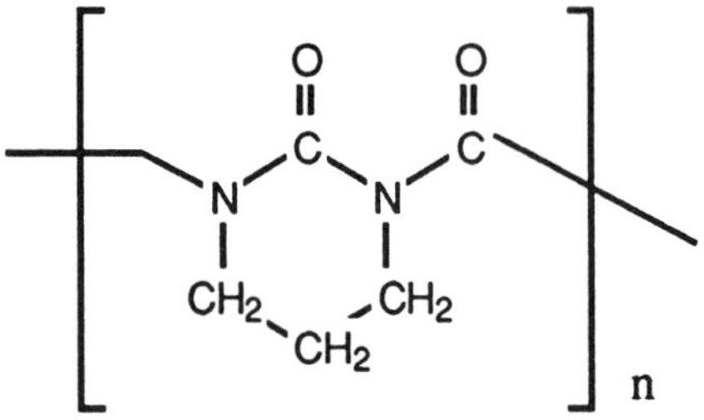

(1–10)

In a similar type of polymerization, cyclopolymerization of 1,2-diisocyanates and 1,2,3,-triisocyanates to cyclic polymers by an anionic mechanism has been described. The structure obtained for propane–1,2,3,–triisocyanate[45] is shown (1–11). These polymers were found to have number-average molecular weights as high as 45,000.

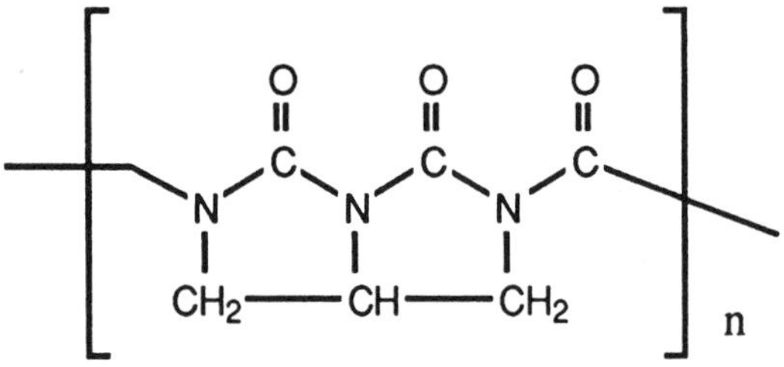

(1–11)

Other anionic polymerization of monomers functionally capable of undergoing cyclopolymerization have been reported. Several publications have appeared describing the cyclic polymerization of glutaraldehyde.[46–49] All these investigators stated that the properties of the polymers were consistent with those predicted from an alternating intramolecular-intermolecular chain propagation (1–12).

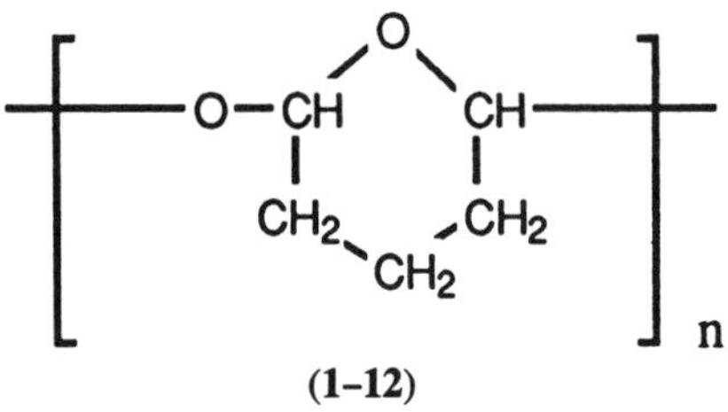

(1–12)

By use of tertiary phosphines having pKa values greater than 8.0, crotonaldehyde[50] was converted to a polymer with a structure produced from a vinyl-type polymerization of crotonaldehyde with a simultaneous cyclization of some vicinal, pendant aldehyde groups to appropriately substituted tetrahydropyran units. This type of cyclization was previously postulated to explain the structure of various polyacroleins.[51] Polycrotonaldehyde showed that over 80% of the conjugated carbonyls were singly conjugated, indicating that only a small amount of aldol condensation side-chain branches were present.

Coordinate Polymerization

The earliest reported example of cyclopolymerization by use of Ziegler catalysts was that of 1,5-hexadiene and 1,6-heptadiene.[52] Both monomers led to soluble, predominately saturated, hydrocarbon polymers. These properties are consistent with the cyclopolymerization mechanism leading to the formation of cyclopentane and cyclohexane rings, respectively. More recent work[53] adds more definitive evidence for the nature of the repeating unit in poly-1,5-hexadiene. The polymers were prepared with a number of modified alkyl metal coordinate catalysts. The properties varied with the catalyst modification used and the mode of preparation, but every polymer prepared was crystalline, had a density exceeding 1.0 g/cc and a melting point in excess of 100°C, and was very flexible. One particular poly-1,5-hexadiene had a tensile strength of 5400 psi, a melting point of 139°C, and a density of 1.122 g/cc, and yet was very flexible, having an apparent modulus of elasticity at −50°C of 200,000 psi.

The x-ray fiber repeat distance of poly-1,5-hexadiene was found to be 4.80 Å, which corresponded to a chain structure consisting of a linear, zigzag array of 1-methylene-3-cyclopentyl groups. Extensive calculations on model systems and other considerations ruled out four- and six- membered rings as recurring units in the polymer chain. Although a repeat distance of 4.80 Å was calculated if the methylene substituents in the 1,3 positions of the cyclopentane ring were placed *trans* on a planar ring, this system did not take into account known facts about 1,3-disubstituted cyclopentanes. *cis*-1,3-Dimethylcyclopentane is known to be more stable than the *trans* isomer by about 0.5 kcal/mole, and this is most likely the result of ring puckering, a proven phenomenon in cyclopentane and its derivatives. If the methylene substituents of poly-1,5-hexadiene were placed *cis* and the number 2 carbon atom of the ring puckered toward the methylene substituents to a perpendicular distance of 0.46 Å above the plane of planar cyclopentane, the distance between the methylene groups (fiber repeat distance) would be 4.80 Å. This system (See Structure 2–7), would[12] account for all the known facts about cyclopentane and poly-1,5-hexadiene.

Diallyldimethylsilane was polymerized[54] using a Ziegler-type catalyst of triisobutylaluminum and titanium tetrachloride in heptane at 30°C. The polymers ranged from moderately viscous oils to high-melting white solids. The high-melting solid obtained in one experiment shrank slightly between 100° and 300°C, above which temperature it began to turn yellow, and at 376° it formed a clear amber melt. This sample had an inherent viscosity of 0.22 dl/g (0.25 g/100 ml, benzene, at 25°C). A quantitative comparison of the polymers with diallyldiphenylsilane showed that there were 6.0% of the monomer units incorporated in the polymer chain which still retained one double bond.

The infrared studies are thoroughly consistent with a polymer structure in which at least 94% of the monomer units have undergone cyclization during the polymerization process.

Distillable liquids[55] had already been obtained as early as 1959, presumably cyclic trimers and tetramers, by the action of triethylaluminum-titanium tetrachloride catalysts on diallyldiethylsilane and diallyldimethylsilane. The liquids were oily and viscous and were distilled at 200–300°C (4–5 mm). The solid polymers obtained were stable to temperatures above 300°C.

Polymers of both diallyldimethylsilane and diallydiphenylsilane having structures consistent with the cyclopolymerization mechanism by use of triethylaluminum-titanium tetrachloride catalysts were obtained.[56] The poly(diallyl-diphenylsilane) was found to have a melting range of 80–110°C and an intrinsic viscosity in benzene of 0.13 dl/g. The poly(diallyldiphenylsilane) softened at 125–155°C, and was found to have a weight-average molecular weight, obtained by light-scattering measurements, of 6.0×10^4. Polymers obtained from diallyldiphenylsilane using di-t-butyl peroxide possessed similar properties. The infrared spectrum of this polymer was identical to that of the Ziegler-catalyzed polymer. A thermal gravimetric analysis of poly-(diallyldiphenylsilane) showed that it retained 100% of its original weight to 375°C, 95% to 400°, and 50% to 425°.

COPOLYMERIZATION OF 1,6-DIENES WITH VINYL MONOMERS

Numerous examples of copolymerization of 1,6-dienes with common vinyl monomers and other comonomers such as sulfur dioxide have been recorded in the literature. In every case, the predominant role of the diene was that of cyclization, although at certain concentrations of diene in certain instances marked tendencies toward crosslinking were observed. A limited number of copolymerizations of this type will be discussed.

Copolymers of bis-ethylenically unsaturated amines such as diallylamine, dimethylallylamine, and diallylalkylamines with olefins such as styrene, alkyl acrylates, acrylamide, acrylonitrile, methacrylontrile, and vinyl acetate have been described.[57] Copolymerization of these amines with acrylamide produces copolymers of extremely high molecular weight. For example, a linear copolymer prepared from a mixture of 98.5 mole % of acrylamide and 1.5 mole % of diallylbenzylamine had a viscosity of over 100,000 Cp at a concentration of 11.1% in water. Terpolymers of acrylic acid, acrylamide, and diallylethylamine, or acrylonitrile, vinyl acetate and diallylamine, were also prepared. In all cases the copolymers and terpolymers were low in residual unsaturation.

The copolymerization of divinyl acetals with vinyl acetate has been studied.[58] Copolymers that, for 25–91% by weight conversion, had the same composition as the initial monomer compositions were obtained. Unsaturation was practically absent in the copolymers and all were soluble. Also studied was the copolymerization of divinyl acetals with styrene; the reaction proceeded according to the cyclopolymerization mechanism, with m-dioxane rings forming in the main chain of the polymer.[59]

Crystalline copolymers[60] of ethylene and 1,5-hexadiene over a wide range of compositions have been prepared, using a preformed coordination catalyst. Copolymer compositions ranging from 14.9 to 92.8 mole % of ethylene were obtained. X-ray diffraction patterns of the copolymer series showed that crystalline units were present over the entire composition range. The copolymers were interpreted as having crystallized in blocks of poly-1,5-hexadiene and polyethylene units. The melting-point curve resembled that of a eutectic composition.

Copolymers of a wide variety of 1,6-dienes with sulfur dioxide have been prepared.[61] These copolymers were found to have inherent viscosities up to 1.67 dl/g (1.11 g/100 ml, dimethylformamide), and degrees of polymerization up to 1000. The molar ratio of diene to sulfur dioxide was 1:1. The copolymers were soluble and essentially saturated.

Substantially linear copolymers[62] of diallylbarbituric acid with unsaturated compounds such as maleic anhydride, vinyl acetate, vinylene carbonate, methyl acrylate, acrylonitrile, styrene, and vinylidene chloride were synthesized in the presence of conventional free-radical initiators.

The homopolymerization and copolymerization[63] of bicyclo-[2.2.1]hepta-2,5-diene were studied, and it was found that homopolymerization led to a low-molecular-weight soluble polymer containing both 3,5-disubstituted nortricyclene and 5,6-disubstituted bicyclo[2.2.1]hepta-2-ene units in the polymer backbone. When the bicycloheptadiene was copolymerized with vinyl chloride, vinylidene chloride, acrylonitrile, ethyl acrylate, and methyl methacrylate, only the cyclized monomer unit, 3,5-disubstituted nortricyclene, was found in the high-molecular-weight copolymers. Polymerization of this monomer[64,65] was also studied by another group who reported the structure of the resulting polymer to be solely that formed *via* polymerization through only one double bond of the bicyclohepta-2,5-diene.

Copolymerizations of methacrylic anhydride[66] and a variety of common types of vinyl monomers were extensively studied. The results showed that both soluble and insoluble copolymers were obtained, depending on the comonomer used and the experimental conditions under which the copolymerization was carried out. In general, soluble copolymers were formed under the following conditions: (a) the less reactive the comonomer in free-radical copolymerizations, (b) the greater the dilution, (c) the greater the difference in the number of moles of the two components in the charge, and (d) the lower the conversion.

The monomer reactivity ratios for the styrene-o-divinylbenzene system have been determined;[67] cyclopolymerization was observed to some extent. The copolymers were soluble and the copolymerization showed no signs of crosslinking. Subsequent studies on homopolymerization of o-divinylbenzene showed that only cyclopolymerization occurred.

The reactivity ratios for the copolymerization of lauryl methacrylate with diallyl phenylphosphonate, diallyl butylphosphonate, and diethyl allylphosphonate, respectively, have been reported.[68]

CYCLOPOLYMERIZATION INVOLVING OTHER DOUBLE BONDS

Numerous examples of monomers, having double bonds between carbon and heteroatoms, have been polymerized and their polymers studied. For example, numerous aldehydes have been polymerized through the carbon-oxygen double bond and alkyl isocyanates have been polymerized through the carbon-nitrogen double bond to linear polymers designated as 1-nylons. Difunctional monomers of both the above types of doubly bonded structures have now been prepared and their polymerizations studied. Examples of these polymerizations studied during this early period will be discussed.

Polymerization of epoxy compounds has been under study for quite some time. In an application of the cyclopolymerization concept to this type of monomer, bifunctional epoxides capable of forming a stable ring system during the propagation step have been synthesized and their polymerization studied.

1,2,5,6-Diepoxyhexane[69] has been polymerized by a variety of catalysts to produce a polymer that has tetrahydropyran recurring units as the result of the cyclopolymerization mechanism. A phosphorus pentafluoride-water catalyst system gave soluble polymer with inherent viscosity as high as 0.37 dl/g (0.25 g/ 100 ml, 66:100 tetrachloroethane:phenol). In contrast, 1,2,4,5-diepoxypentane polymerized to a completely insoluble polymer by all catalyst systems employed.

Cyclopolymerization of glutaraldehyde[46–49] has been reported and the properties of the polymer were interpreted in terms of a structure containing tetrahydropyran rings bonded together at the 2,6 positions by an oxygen atom (See Structure 1–12).

Polymerization of suberaldehyde has also been studied.[70] By use of carbonyl polymerization catalysts, a polyacetal-type polymer containing free pendant carbonyl groups was obtained. Evidence of some cyclopolymerization in this compound, even though this necessitates formation of nine-membered rings (1–13) along the chain was obtained.

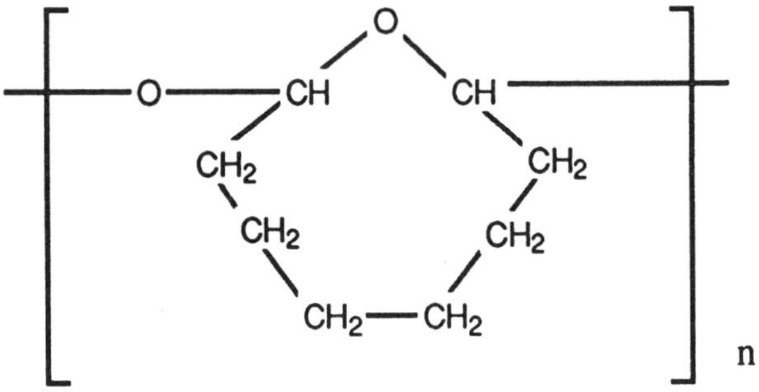

(1–13)

Previous mention has been made of the cyclic polymerization of trimethylenediisocyanate,[44] 1,2-diisocyanates,[45] and 1,2,3-triisocyanate.[45] These heterobonded monomers led to cyclic structures having alternating carbon and nitrogen atoms along the polymer backbone. (See Structures 1–10 and 1–11).

Glycidyl methacrylate and glycidyl acrylate[71] undergo cyclic copolymerization of the methacrylate double bond and the epoxy group by initiation with boron trifluoride etherate, to produce cyclic lactone–ether recurring units in the polymer (1–14).

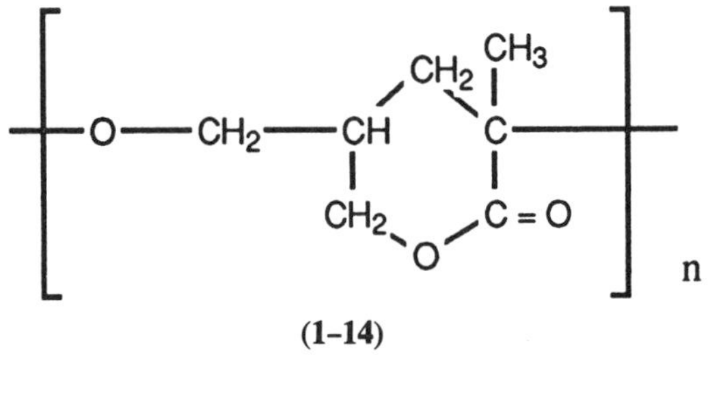

(1–14)

Free-radical polymerization of glycidyl methacrylate and acrylate yields linear polymers, reaction taking place at the vinyl bonds without participation of the glycidyl group.[72]

CYCLOPOLYMERIZATION OF UNSYMMETRICAL MONOMERS

Although the greater part of the work done in the area of cyclopolymerization has been limited to symmetrical monomers, unsymmetrical 1,6-dienes should be capable of homopolymerization *via* cyclization. Particularly interesting and promising monomers for cyclic polymerization are the monomers containing double bonds of comparable reactivity toward copolymerization. Homopolymerizations of such unsymmetrical 1,6-dienes as allyl acrylate,[73] allyl methacrylate,[74,75] and substituted allyl methacrylates were reported before the proposal of the cyclopolymerization hypothesis; consequently, no previous study directed specifically toward verification of the presence of cyclic units in the polymers derived from such dienes has been made.

The homopolymerization of unsaturated esters of maleic and fumaric acids[76] have been studied. These monomers were considered particularly suited for such a study because of the marked preference of the maleic and fumaric double bonds to undergo copolymerization, and were probably the first example of unsymmetrically substituted dienes to be studied in cyclopolymerization. Also, a qualitative comparison of the relative reactivities of various unsaturated ester groups was made possible, as well as a comparison of the degree of cyclization of the *cis* and the *trans* isomers. Although, in general, these unsymmetrical monomers ultimately led to crosslinked polymers, the extents of cyclization in the monoallyl, crotyl, and 3-butenyl esters of monomethylmaleic acid were found to be 49, 60, and 38% respectively. The extents of cyclization in the corresponding monoesters of monomethylfumaric acid were found to be 63, 23, and 43% respectively. Thus, it was shown that, although ultimately crosslinking occurs, cyclization constitutes a significant mode of propagation in these unsymmetrical monomers which are functionally capable of forming six-membered rings, even though the relative reactivities toward copolymerization are far apart. A typical structure obtainable from methyl allyl fumarate is shown (1–15).

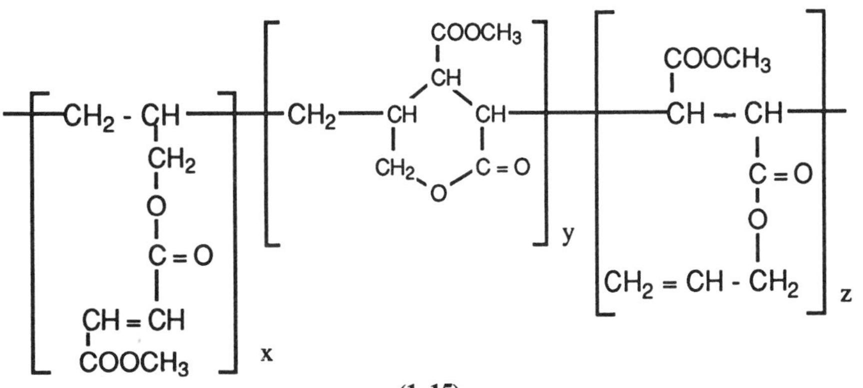

(1–15)

It was recognized quite early that less-favored ring sizes may be formed in cyclopolymerization and that the less-stable radical may predominate in the propagation.[77] In an extension of the study of unsymmetrical monomers to the allyl, crotyl, and methallyl esters of both *trans*-crotonic acid and vinylacetic acid, and to the propargyl ester of *trans*-crotonic acid and the monoallylamide of the same acid, it was shown that the less-stable radical predominated in the cyclization step during polymerization of allyl and methallyl crotonates (1–16).

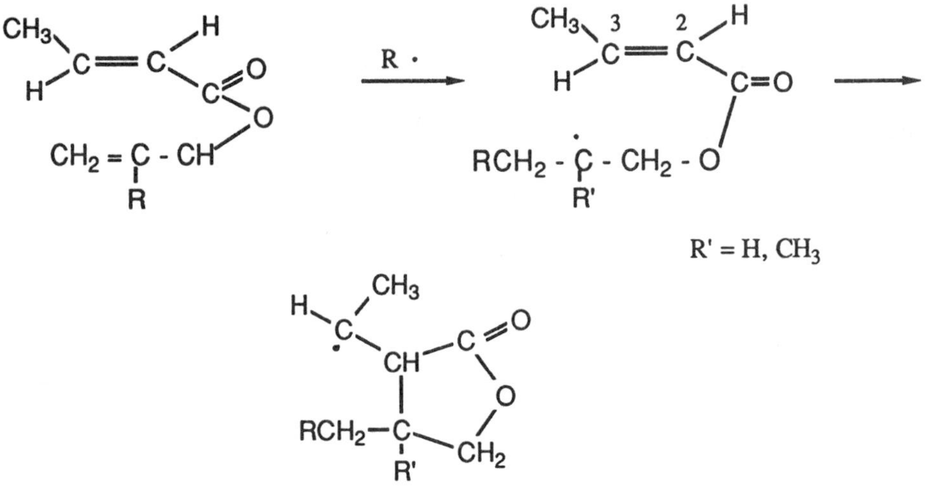

R' = H, CH_3

(1–16)

More recent work with allyl acrylate[78] has shown that polymerization to 60% conversion leads to crosslinked polymers. Determination of residual unsaturation showed that 24.2% of the structural units were linked through the acrylate group, 17.2% were linked through the allyl group, and 58.6% of the units had undergone cyclization in accordance with the cyclopolymerization mechanism, leading to lactone groups (1–17).

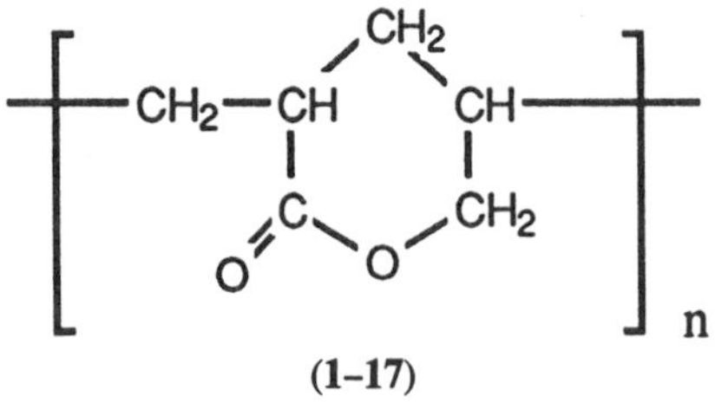

(1–17)

The syntheses of linear cyclopolymers from unsaturated acid monoallylic esters[79] have been reported. On bulk polymerization, linear polymers of monoallyl maleates and citraconates, soluble in organic solvents, were obtained. On the basis of residual unsaturation determinations, it was concluded that the polymers contained 63–65% cyclic units.

Allyl crotonate and allyl cinnamate formed copolymers with vinyl chloride that were soluble and free of gels.[80] In another example of copolymers of vinyl chloride with unsymmetrical dienes, soluble copolymers free of gel were obtained with allyl β-allyloxypropionate.[81] Other unsymmetrical monomers used in forming similar copolymers were allyl allyloxyacetate, allyl *o*-allyloxybenzoate, allyl *p*-allyloxybenzoate, and allyl γ-allyloxybutanoate.

CYCLOPOLYMERIZATION LEADING TO POLYCYCLIC SYSTEMS

As was previously mentioned, polymerization of open-chain 1,5- or 1,6-dienes leads to polymers having five- or six-membered rings alternating in the backbone of the polymer with methylene links. Numerous examples of polymerization of cyclic dienes to polymers containing repeating units of polycyclic structures were reported during this early period.

The formation of bicyclic and tricyclic structures have been reported by polymerization of open-chain but branched (star-shaped) trienes and tetraenes, the bicyclic and tricyclic structure being formed by a succession of ring closures within a single monomer molecule before propagation to a neighboring monomer occurred. For example, triallylethyl- and tetraallylammonium bromides were converted to soluble polymers containing low degrees of residual unsaturation.[82] A typical polymer of triallylethylammonium bromide had an intrinsic viscosity of 0.14 dl/g and a molecular weight estimated to be in excess of 5000. A triolefinic monomer, 3-vinyl-1,5-hexadiene, with a Ziegler catalyst, was converted to a soluble polymer having an intrinsic viscosity of 0.46 dl/g and low residual unsaturation.[82] These results indicate an extensive chain structure of 2,6-linked bicylco[2.2.1]heptyl rings (1–18). Triallylmethylsilane was also

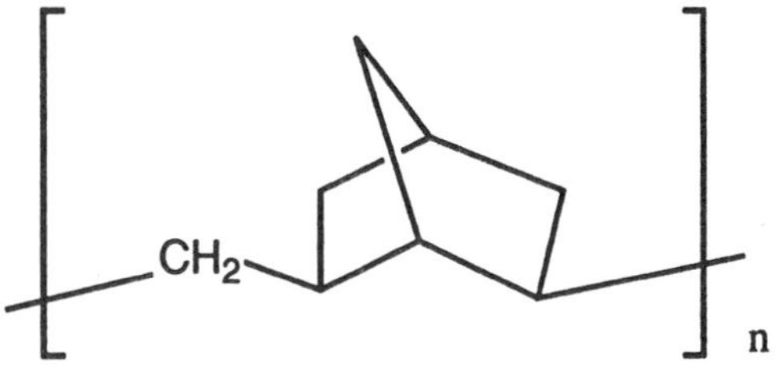

$$\left[-CH_2- \text{(bicyclo[2.2.1]heptane-2,6-diyl)} - \right]_n$$

(1–18)

polymerized by a Ziegler catalyst to soluble, solid polymers showing low residual unsaturation, properties which are consistent with a large [3.3.1]bicyclic ring content. Thus it appears that monomers functionally capable of closing a second and even a third ring before propagation to a neighboring monomer tend to do so.

2-Carbethoxybicyclo[2.2.1]-hepta-2,5-diene was polymerized to high-molecular-weight, soluble, essentially saturated polymers reported to contain nortricyclene repeating units.[83] Polymerization studies of other similar structures suggested that polymerization of bicycloheptadienes requires activation of the double bonds and resonance stabilization of radicals formed during the polymerization. The polymers had inherent viscosities from 0.45 to 0.71 dl/g (0.1% benzene, 25°C) corresponding to molecular weights of 66,000 to 123,000, respectively.

By use of coordinate catalysts, cyclopolymerization of *cis,cis*-1,5-cyclooctadiene to a bicyclopolymer has been described. The spectral and other physical properties of the polymer were consistent with a poly(2,6-bicyclo[3.3.0]octane) structure.[84] Thus it is seen that even though this monomer contains only internal double bonds, a marked driving force exists toward polymerization by the alternating intra-intermolecular mechanism.

SOLID-STATE CYCLOPOLYMERIZATION

Apparently, during this early period, only one example of solid-state polymerization of monomers capable of proceeding by the cyclic polymerization mechanism was reported. Through cobalt-60 irradiation of N,N-diallylmelamine, a cyclic polymer with an intrinsic viscosity of 0.20 dl/g and in 91.7% conversion was obtained.[85] The polymers were soluble in acids, indicating no crosslinking. Both monomer double bonds disappeared during the polymerization without leading to crosslinking, strongly suggesting cyclopolymerization as the predominate mode of chain growth.

CYCLOPOLYMERIZATION LEADING TO LARGER RINGS

Although most of the work dealing with cyclopolymerization has been done with monomers functionally capable of forming the sterically favored five- or six-membered rings, a limited amount of work was done during this early period with systems that can lead to larger rings. Varying ratios of soluble polymers were obtained from higher α-diolefins in which the terminal double bonds were separated from four to eighteen methylene groups.[86] The results obtained and the percentage of cyclization reported corresponded roughly to those which have been realized in other cyclization reactions. Seven-membered rings form fairly readily, the intermediate ring sizes form with greater difficulty, the larger rings (fourteen- and fifteen-membered) are easier to obtain, and finally, the yields decrease again with still higher rings sizes.

CYCLOCOPOLYMERIZATION

Previously discussed examples of cyclopolymerizations and copolymerizations involving monomers which undergo cyclopolymerization have involved ring closures in such a manner that all members of the cyclic structure were derived from a single monomer. In 1958,[16,87] an example of a bimolecular alternating intramolecular-intermolecular copolymerization was reported. It was shown that divinyl ether undergoes copolymerization with maleic anhydride in such a manner that a six-membered cyclic structure is formed in which the divinyl ether contributes four members and the maleic anhydride contributes the remaining two (1–5). Because of the reluctance of either monomer to undergo free-radical homopolymerization, and the marked tendency of both to undergo copolymerization with each other, the cyclic copolymer appeared to be the preferred product. The structure of the copolymer was shown to be that proposed as follows: (a) the yield of copolymer based upon either monomer was nearly quantitative; (b) the copolymer was soluble in a number of solvents; (c) the infrared absorption was that predicted by the proposed structure; (d) the polyanhydride was readily hydrolyzed to a water-soluble polyacid that possessed infrared and other properties consistent with the proposed structure; and (e) the cyclic ether units could be opened with hydriodic acid to produce a polyacid containing iodine. Subsequent studies have shown the cyclic structures to also contain five-membered rings.

A number of other examples of cyclocopolymerization have been reported[88] involving a wide variety of 1,4-dienes and olefins, indicating that this type of copolymerization is general for 1,4-dienes and polymerizable olefins. The copolymerization of N,N-divinylaniline[89] with a wide variety of monomers has been studied, and the copolymerization parameters for divinylaniline, divinyl sulfone, and divinyl ether were determined.

In a somewhat modified cyclocopolymerization process, it has been[90] shown that 1,5-dienes such as 1,5-hexadiene undergo cyclocopolymerization with sulfur dioxide in a manner quite analogous to the divinyl ether-maleic anhydride system with the exception that five members of the ring system are derived from the 1,5-diene and the remaining member is derived from the sulfur dioxide molecule (1–19). This type of cyclocopolymerization has been extended to the formation of bicyclic repeating units from bicyclocopolymerization of sulfur dioxide and *cis,cis*–1,5-cyclooctadiene.[91]

SO_2 + → $-[SO_2-$ ... SO_2 ...$]-$

(1–19)

STEREOCHEMICAL ASPECTS OF CYCLOPOLYMERIZATION

The strong interest of polymer chemists in recent years in stereoregular polymers[92] has led to investigations concerned with whether or not the cyclopolymerization process leads to steric control during propagation. Two papers[93,94] have demonstrated clearly that homogeneous free-radical polymerization of vinyl monomers to chains containing asymmetric atoms is stereospecific to an extent dependent on the temperature and other environmental factors. The polymers obtained by free-radical polymerizations conducted at low temperature have been identified[95,96] as the predominantly syndiotactic form. In a study of stereoregulating influences during the free-radical polymerization of methacrylic anhydride[97] it was observed that, indeed, stereochemical control does occur, and the temperature and solvent effects on this control can be determined.

The isotactic character of the poly(methyl methacrylate) obtained by methylation of the poly(methacrylic anhydride) prepared at various temperatures was found to increase with increasing temperature, while the syndiotactic and random-placement characteristics were found to decrease in an approximately parallel manner. These results were explained[98] on the basis of the strained conformation of the free radical formed after the ring-closure step. Even though the cyclic free radical would have a marked tendency to assume planarity, complete planarity at this position would place considerable strain on the remaining bonds of the cyclic structure in which the energetically favorable chair conformation would ordinarily predominate.

Retention of some pyramidal character would thus be expected in this cyclic structure, and would favor an axial propagation. The axial propagations would be more favored at low temperatures. It was shown that the favored axial propagation would lead to a new type of stereoregularity termed "syndioduotactic" and consisting of ddll sequences throughout the polymer chain. In support of this hypothesis, a polymer prepared at 15°C showed 7% isotactic, 48% syndiotactic, and 44% heterotactic character, whereas one prepared at 80°C showed 67% isotactic, 13% syndiotactic, and 20% heterotactic character.[97]

These results have been confirmed and extended;[99] methacrylic anhydride was polymerized to 46% conversion by ultraviolet irradiation in 50% toluene solution at

$-50°C$, and the polyanhydride was converted to its methyl and ethyl esters. Both were found to exhibit considerable crystalline character. The poly-(methyl methacrylate) was found to possess less than 2% isotactic character and equal syndiotactic and heterotactic character, each greater than 49%.

The unique nature of the above polymer is further substantiated by a study of monomolecular films of monolayers of the polymer.[100] The values obtained in this study were different from those of the other forms of poly-(methyl methacrylate) studied, and in agreement with the "syndioduotactic" nature of this polymer.

At least two instances of stereoregulation in Ziegler—catalyzed polymerization of nonconjugated dienes have been reported. A systematic study of the polymer obtained from 1,5-hexadiene[53] was carried out and the polymer was shown to possess an all-*cis* structure. Copolymers of 1,5-hexadiene and ethylene[60] over a wide range of compositions exhibit high degrees of crystallinity, with the patterns of the copolymers also showing the presence of both poly-1,5-hexadiene and polyethylene diffractions. The copolymers were interpreted as consisting of blocks of poly-1,5-hexadiene and polyethylene units.

CYCLOPOLYMERIZATION LEADING TO LADDER POLYMERS

An extension of the cyclopolymerization concept involves formation of ladder or double-strand polymers through a second-stage cyclopolymerization of a suitably substituted linear polymer. For example, 1,2-polybutadiene[101,102] and 3,4-polyisoprene,[103] are examples of such suitably substituted linear polymer. 1,2–Polybutadiene can lead to a double–strand polymer (1–20). Other structures which are believed to possess largely

(1–20)

the double-strand polymer structure are "black Orlon" (1–21) and polyacrolein (1–22). Both structures (1–21) and (1–22) may be considered special cases of 1,6-polyenes, as are 1,2-polybutadiene and 3,4-polyisoprene, and their conversion to double-strand polymers represents a variation of cyclopolymerization.

N≡C N≡C N≡C → N=C N=C N=C

(1–21)

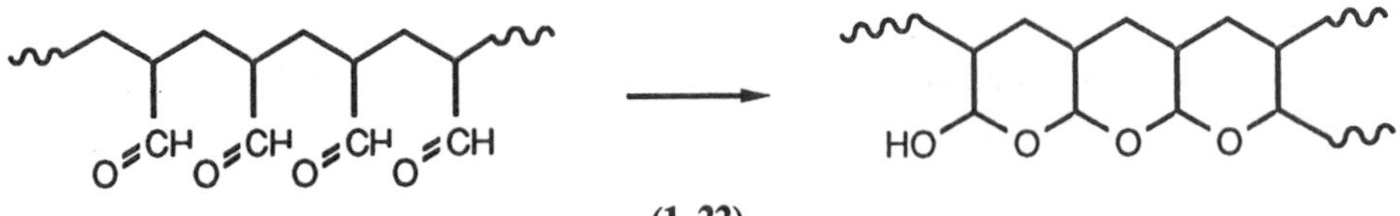

O=CH O=CH O=CH O=CH → HO O O O

(1–22)

Work in this special type of cyclopolymerization during this early period led to some interesting and unique structures and was a stimulus for extensive subsequent studies.[104–110] Several papers reported evidence for ladder or double-strand segments in polymers obtained by treating either various polyisoprenes or the monomer itself[93–95] with cationic initiators. Cationic cyclization[104] of the polymer led to crosslinking unless dilute solutions were used. The soluble polymers were obtained by using boron trifluoride or phosphorus oxychloride in polymer concentrations of about 0.25% in benzene. The resulting material consisted of segments of fused structure, although chain scission also resulted, reducing the molecular weight to about one-fourth of the original polyisoprene.[111,112]

Isoprene, 1,3-butadiene, and 1-chloro-1,3-butadiene were polymerized[110] with complex metal alkyl catalysts to obtain powdery, insoluble, probably crosslinked polymers with high density and high heat resistance. Evidence was presented that these polymer chains consist of fused six-membered saturated rings in the form of a linear ladder or a spiral ladder structure.

Structural evidence indicates that these polymers contain residual linear segments with 1,4 units which can be isomerized to a cyclic form by the action of sulfuric acid. It was proposed that the cyclic structures were formed during the polymerization itself from intermediate 1,2-polymeric structures rather than as the result of initiator on first-formed linear chains. Two alternative mechanisms were proposed to account for this single-step cyclopolymerization of conjugated diene monomers to ladder polymers. Since the metal halide was present in considerable excess, the polymerization was considered to be cationic. The first of the proposed mechanisms involved a reversal of the direction of polymerization, resulting in an intramolecular cyclopolymerization to yield a sequence of fused six-membered saturated rings, as shown in 1–23 and 1–24. The

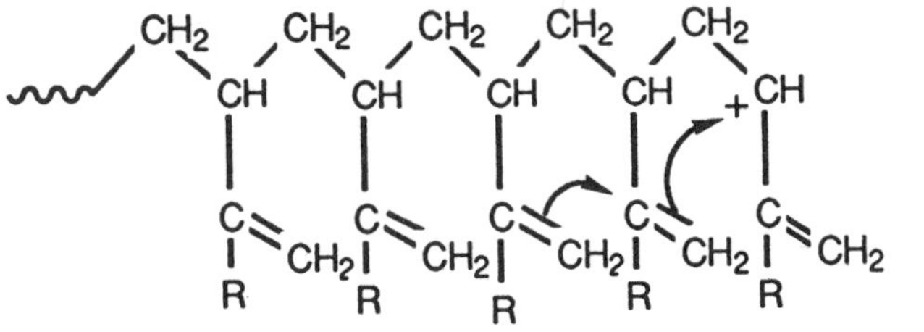

(1–23)

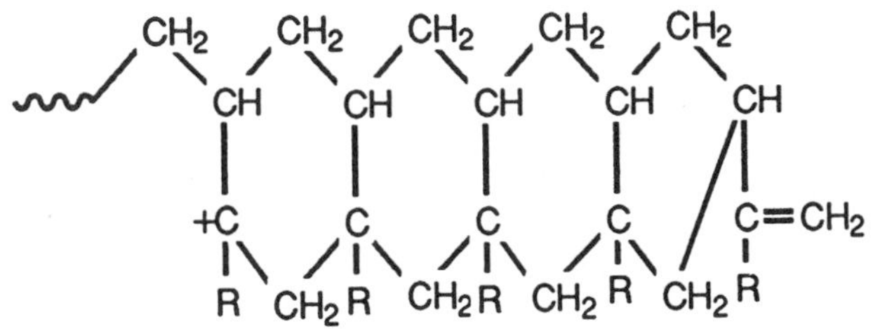

(1–24)

second proposed mechanism involved a cyclization reaction initiated by the attack of an active center of a monomer unit or of a growing chain on the pendant double bonds of a polymer molecule, as indicated in 1–25 and 1–26. It was shown that the catalyst components individually or together failed to cyclize either 3,4-polyisoprene or *cis*-1,4-

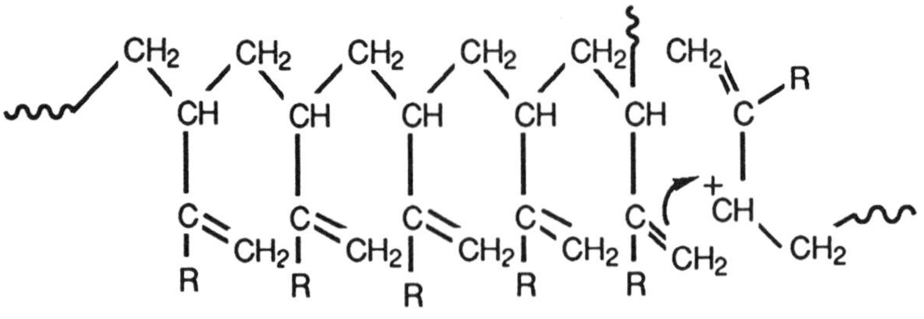

(1–25)

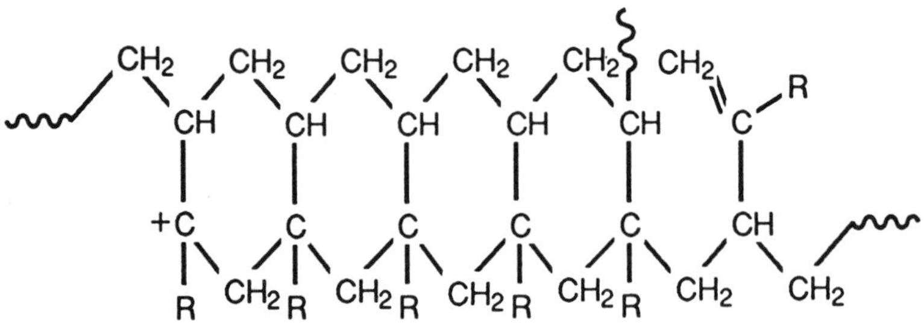

(1–26)

polyisoprene. Both proposed mechanisms for the extensive degree of cyclization are consistent with these observations.

An extension of this work,[113] reported cyclizations of *cis*-1,4-; *trans*-1,4- and 3,4-polyisoprene with sulfuric acid and comparison of the spectra with those of the cyclopolymers of isoprene, butadiene, and chloroprene prepared directly from the monomers by use of organometallic catalysts. The similarity of all the spectra indicate that both types of polymer have the same polycyclic structure. However, it was concluded that even though these similarities in infrared spectra exist, further evidence is required to show that the independently derived polymers possess identical structures.

Since both isotactic and syndiotactic 1,2-polybutadiene have been reported, second-stage cyclization experiments on each of these stereoregular polymers should lead to interesting results. As the structures 1–27 and 1–28 indicate, there should be

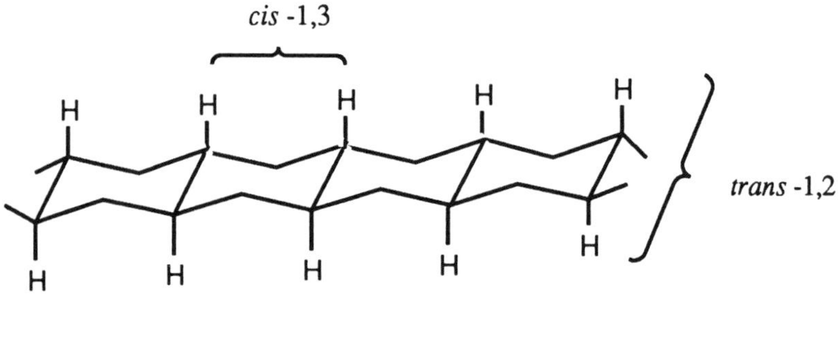

cis -1,3

trans -1,2

conformation of *cis* -1,3 , *trans* -1,2 polycyclohexane

(1–27)

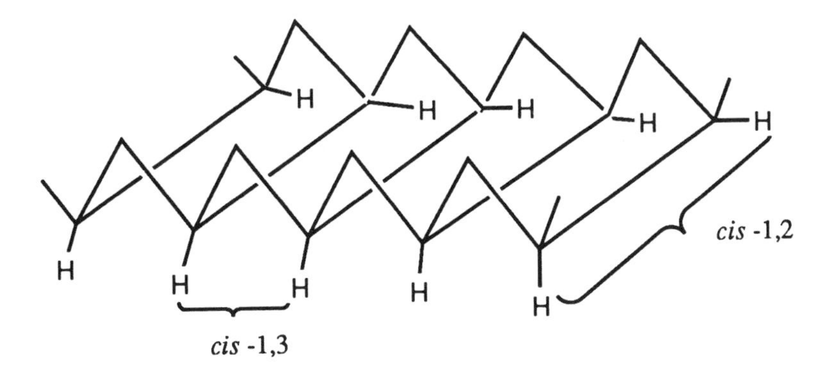

conformation of *cis* -1,3, *cis*- 1,2 polycyclohexane

(1–28)

large energetic factors to favor one of the two stereoregular cyclized structures in each case over the other possibility.[98] For example, from isotactic 1,2-polybutadiene, the *cis*-1,3 *trans*-1,2 conformation can readily assume the conformationally favored chair form of the derived cyclohexane rings (1–27) while the other stereoregular form, the *cis*-1,3 *cis*-1,2 conformation, can only assume the higher energy boat form of the derived cyclohexane rings (1–28). The difference in strain energy between the chair and boat forms of a cyclohexane ring is usually considered to be 6 kcal/mole.

From syndiotactic 1,2-polybutadiene, the *trans*-1,3 *cis*-1,2 conformation can readily assume the conformationally favored chair form (1–29) of the derived cyclohexane

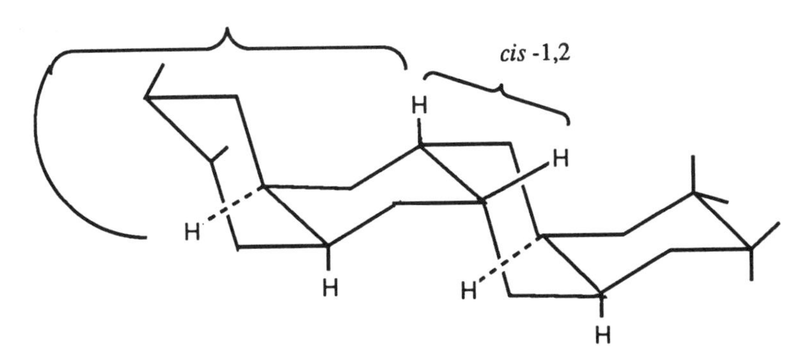

conformation of *trans* -1,3 , *cis* -1,2 polycyclohexane

(1–29)

rings whereas the other stereoregular form, the *trans*-1,3 *trans*-1,2 conformation, can only assume the higher energy boat form of the cyclohexane rings (1–30). Thus, from each stereoregular form of 1,2-polybutadiene, one of the two stereoregular forms derivable from each should be energetically favored over the other form by approximately 6n kcal/mole of polymer of average degree of polymerization n.

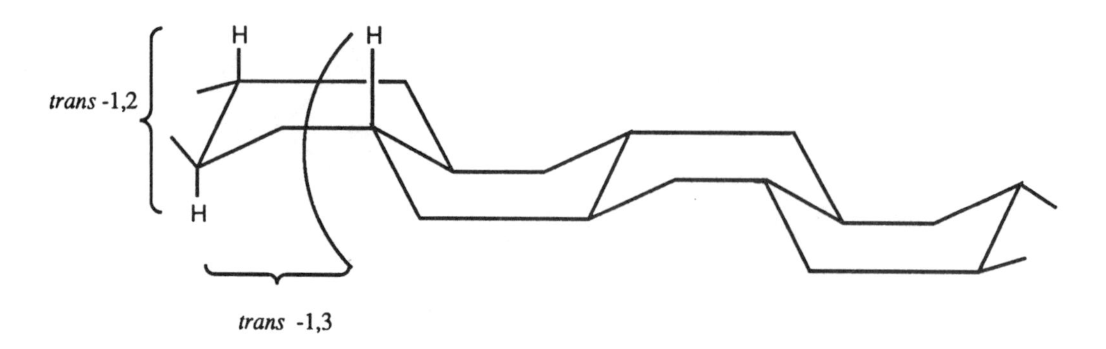

conformation of *trans* -1,3 , *trans* -1,2 polycyclohexane

(1–30)

Another example of formation of a double-strand polymer reported during this early period was that obtained through a two-stage polymerization of vinylisocyanate[114,115]. It was shown that polymers containing a large portion of ladder polymer structure could be obtained by either of the following routes: (a) Polymerization of vinylisocyanate through the vinyl double bond followed by cobalt-60 irradiation. The resulting polymer was found to contain 85% ladder structure although it was insoluble. (b) Polymerization of vinylisocyanate to N-vinyl-1-nylon by several routes, followed by treatment of the vinyl nylon with azobisisobutyronitrile. The resulting polymer (1–31) was found to contain 90% ladder structure and was soluble in several solvents.

(1–31)

The synthesis of N-vinyl-1-nylon by anionic polymerization of vinylisocyanate has been reported as well as the preparation of several copolymers of this monomer.[116]

THEORETICAL INTERPRETATION OF CYCLOPOLYMERIZATION—EARLY PROPOSALS

During polymerization of a 1,6-diene by free-radical initiation, the radical produced by the initial attack of the initiator fragment appears to have the choice of reacting with the internal double bond to close the ring or reacting with a neighboring molecule to form a structure that has a pendant double bond. Each such event leading to a polymer chain with pendant double bonds offers the possibility of crosslinking. Since the many examples of cyclic polymerization reported appear to proceed in this manner without appreciable crosslinking and without appreciable residual unsaturation, an explanation having some basis other than the probability of ring closure is necessary.

A possible explanation has been offered[17] which involves an electronic interaction between the double bonds of the 1,6-diene, or between the reactive species generated by initiator attack and the intramolecular double bond. Such an interaction can be represented in its simplest form (1–32). The only quantitative evidence found in the

(1–32)

literature to support such a proposal is that derived from a study[117] of the radical-initiated polymerization of allylsilanes. The results showed that the total activation energy of radical polymerization for diallyldimethylsilane is about 9 kcal/mole of double bond less than that for allyltrimethylsilane. These results suggested an energetically more favorable pathway from monomer to polymer in the diallyl derivative than in the corresponding monoallyl derivative. An electronic interaction of the type proposed, either in the ground state or in the excited state of the molecule, would be expected to provide such a pathway.

Several papers[34,35,52] support the proposed interaction, or at least a more favorable energy pathway from monomer to polymer in a variety of dienes. Furthermore, other work[38] has shown that the second double bond of 2,6-diphenyl-1,6-heptadiene reacts many times faster with a free radical, a carbonium ion, or a carbanion generated at the site of the first double bond than with the double bond of a neighboring molecule. These results provide convincing evidence that a stabilizing influence on the excited state of the molecule has been exerted through some means, thus providing an energetically favorable path from diene to cyclic product.

The ultraviolet absorption properties of a series of compounds which are functionally capable of undergoing cyclic polymerization have been investigated[118] and a departure from the predicted properties based upon the monoolefinic counterparts was found. For example, the *cis* and *trans* isomers of 1,3,8-nonatriene exhibited bathochromic shifts of 3.5 and 5.0 nm, respectively, which could be interpreted as resulting from the stabilizing influence of interspatial resonance.

In an extension of this work, the probability of ring closure in cyclopolymerizing systems was calculated.[119,120] The results indicated that far greater ratios of cyclization are realized in all systems reported than predicted on the basis of probability alone. Thus, this strong tendency toward cyclization can be explained satisfactorily only on the basis of some driving force other than the probability factor. In this work, a series of tetraenes having two pairs of conjugated dienes situated in such manner that the proposed electronic interaction could occur were synthesized and their ultraviolet absorption properties determined. A bathochromic shift of 10.6 nm was observed when 1,3,6,8-nonatetraene was compared with 1,3-nonadiene.

Although many additional references could be cited which support the proposed interaction as a driving force in cyclopolymerization, only a few of the more significant ones are referred to. In the polymerization of 2-carbethoxybicyclo [2.2.1]hepta-2,5-diene[83] to polymers believed to contain nortricyclene repeating units, the evidence suggests that this type of polymerization requires activation of the double bonds and a resonance stabilization of radicals formed during the polymerization. In a study of acetolysis of 5-hexenyl p-nitrobenzenesulfonates,[121] it was found that a 71% increase in acetolysis rate occurred when these compounds were compared with the corresponding hexyl esters, and 34% yield of cyclohexylacetate was obtained. These results are quite analogous to the cationic polymerization of 2,6-diphenyl-1,6-heptadiene.[38] However, in this case, because of the increased resonance stabilization of the intermediate benzyl carbonium ion, this cyclopolymerization results in exclusive ring closure. In a similar instance involving free-radical intermediates, 6-heptanoyl peroxide[122] was decomposed thermally and the 5-hexenyl radicals produced cyclized to form cyclohexane radicals which produced both cyclohexane and cyclohexene in the products. However, no group participation by the double bond in the peroxide decomposition was observed.

In retrospect, this proposal while accounting for many observations, does have weaknesses. Few interactions of the type suggested have been documented. However, support can still be found. Other proposals not totally dissimilar, have been offered and will be discussed in detail in subsequent sections of this volume.

It is of considerable interest, however, that many aspects of the originally projected scope of cyclopolymerization have been realized within this first ten year period. However, many new systems have been studied, and the scope and utility of this mode of polymer synthesis has been broadened considerably.[29]

LITERATURE SURVEYS

The literature of cyclopolymerization has been surveyed a number of times during the period 1960–1989. It is of interest to summarize each of these reviews, and to characterize them according to major topic covered, extent of coverage and the scope and limitations of each. These summaries are presented in this Introductory Chapter, and are presented chronologically by years.

The principles of cyclization in condensation polymerization were discussed in 1946.[123] It was pointed out that competition between chain polymerization and ring formation in condensation polymerization is readily accounted for in bifunctional monomers with less than five and more than six members through application of statistical probability and ring stability considerations. However, these factors do not completely account for the large tendency of nonconjugated diene monomers capable of forming five- and six-membered rings to cyclize, as in the case of cyclopolymerization.

A review of polymerization of unconjugated dienes with 54 references through 1959 was published.[124] A review of the literature dealing with cyclopolymerization or intramolecular-intermolecular polymerization of nonconjugated dienes was presented in 1960.[125] Carbocyclic and heterocyclic rings containing 0, N, P, Si, and Sn were discussed. The synthesis of new polymers containing rings in the chain, and having superior thermal stability was reviewed in 1961.[126] A brief textbook of polymer chemistry including an introduction to cyclopolymerization was published in 1962.[127]

A review of cyclopolymerization or ring formation during polymerization with 56 references was published in 1962.[128] A review of synthetic methods for heat-resistant polymers with rings in the chain was published in 1962 with 47 references.[129] Polymers containing benzene rings and heterocyclic rings in the chain were discussed. A review of acrolein polymers, with reference to the diverse nature of the monomer, including its tendency to undergo cyclopolymerization to ladder polymers was published in 1964 with 131 references.[130]

A 1965 review, of ring-forming polymerization, with 65 references, covered the topics of divinyl compounds, methacrylic anhydride, 1,5-hexadiene polymers, 1,4-dienes, cyclic dienes, dialdehydes, diisocyanates and diepoxides.[131] Ionic polymerization of acrolein was reviewed with 36 references in 1965.[132] A review of strained polycyclic systems with three- or four-membered ring units with 54 references was published in 1965.[133] A variety of cyclopolymerizable monomers was discussed.

A review of the literature dealing with cyclopolymerization through 1964 with 170 references has been published.[134] A review including a discussion of the concept of cyclopolymerization, proof of structure of cyclopolymers, and the scope of the reaction, with 106 references was published in 1966.[135] Cyclopolymerization of non-conjugated dienes was reviewed in 1967 with 69 references.[136]

Allyl polymers and their properties were reviewed in 1967 with 29 references.[137] Conditions for cyclization reactions of 1,4-, 3,4–, or 1,2–polydienes and cyclopolymerization of conjugated dienes, their mechanism, and side reactions were discussed in a review in 1967.[138] A review of the antitumor, antiviral, and antibacterial action of heparin, heperinoids, anionic dyes and synthetic polymers including cyclocopolymers with 557 references was published in 1968.[139]

A theoretical discussion of the growth-regulating activity of polyanions including those derived from hydrolysis and neutralization of the 1:2 alternating copolymer of divinyl ether and maleic anhydride, was published with 660 references in 1968.[140] The place of polyanions in the intercellular environment, and their role in cell physiology were also discussed. Intramolecular cyclopolymerization and photochemical dimerization have been discussed in a 1969 review with ten references.[141] The role of charge-transfer complexes in cyclocopolymerization was reviewed in 1969 with 40 references.[142]

Cyclopolymerization of bifunctional monomers was reviewed in 1969 with 44 references.[143] A discussion, including experimental results on a series of unsymmetrical nonconjugated dienes, e.g. N-allylacrylamide, N-allylmethacrylamide, allylacrylate, and allylmethacrylate, was published in 1969 with 11 references.[144] Polymerization of allyl ethers, esters, amines, alcohols, and halides, allylaromatic compounds, and other allyl compounds, the mechanisms of allyl polymerization by radical, radical-complex, radiation and ionic initiation, and cyclopolymerization of gem-diallyl compounds were reviewed with 386 references in 1970.[145]

In a 1970 review, it was shown that the addition of a radical to an alkene would occur most readily when the orbitals bearing the three electrons which were redistributed remain in the same plane throughout the reaction.[146] A review of cyclopolymerization of divinyl and dialdehyde monomers with 111 references was published in 1971.[147] It was shown in a 1971 review that in cyclopolymerization of certain monomers *via* cationic initiation, the ratio of the rate constants for cyclization and intermolecular propagation is strongly dependent on the nature of the catalyst and on the polarity of the medium.[148]

A review with 74 references on the homopolymers of diallyl phthalate and diallylisophthalate was published in 1972.[149] A review of radical and anionic initiated polymerization of allyl acrylates and N-allylacrylamides with 25 references was published in 1973.[150] A review of certain aspects of polycyclization with reference to application of the various types of ring forming reactions, including cyclopolymerization with 221 references was published in 1973.[151]

A review of polyelectrolytes, ionomers and recent advances in ion-containing polymers with 139 references was published in 1975.[152] Polyelectrolytes synthesized *via* cyclopolymerization were discussed. The subject of cyclopolymerization with emphasis on the mechanisms and the structure of the resulting polymers was reviewed with 81 references in 1975.[153] A review of cyclopolymerization with 37 references was published in 1975.[154]

The mechanism of cyclopolymerization of nonconjugated dienes and supporting evidence, as well as the scope of cyclopolymerization, were reviewed with 98 references in 1975.[155] The topic of cyclopolymerization, including the mechanism and scope with numerous examples, structure studies, and supporting evidence was reviewed with 509 references in 1975.[156] A review of cyclopolymerization which concentrated on diallylamine polymers because of their importance in areas such as fiber formation, flocculants, dispersants, thickening agents, ion exchange resins, antistatics, and the ability of amphoteric copolymers to convert chemical energy into mechanical energy was published in 1975.[157]

A review with 49 reference on the mechanism of the cyclocopolymerization of divinyl ether with maleic anhydride and on the properties of the copolymer was published in 1975.[158] Experimental and theoretical studies of the cyclic population of polymers prepared by ring-chain equilibration reactions and the conformation of the polymer chains were reviewed with 224 references in 1976.[159] The practical applications of polymers derived from diallylamines as well as theoretical studies related to the mechanism of cyclopolymerization were reviewed in 1976.[160]

A review with 27 references which deals with the placement of monomers in vinyl polymerization and more specifically with the repeat units in cyclopolymerization was published in 1976.[161] The concept that certain traditional methods used to predict and establish structure in polymers may be insufficient in certain cases was also discussed. Recent topics on cyclopolymerization with 27 references was reviewed in 1977.[162] The feasibility of formation of five-membered rings compared with six-membered rings during cyclopolymerization, including a discussion of cyclopolymerization of dimethacrylic amides, and cyclopolymerization in the presence of Lewis acids were discussed. A review of cyclopolymerization and cyclocopolymerization with 101 references was published in 1978.[163] The principles of cyclopolymerization were critically reviewed, conflicting aspects of the mechanism were discussed, and the effects of the stability of propagating species on ring size was presented.

A review with 124 references of the biological activity of polycarboxylic acids, including those derived from hydrolysis and neutralization of the 1:2 alternating cyclocopolymer of divinyl ether and maleic anhydride, was published in 1978.[164] Water-soluble organic polymers, including cyclopolymers, and their uses with 14 references was reviewed in 1974.[165] An introduction to polymers in biology and medicine with reference to cyclopolymerization, with 64 references was published in 1980.[166]

The structure and biological activities of some anionic polymers including those derived *via* cyclopolymerization were reviewed with 99 references in 1980.[167] A review with 22 references dealing with the synthesis of polyamines, with emphasis on poly(triallylamine) resins, and their role in thermally regenerable ion exchange processes was published in 1980.[168] The formation of cyclic structures during cyclopolymerization of diallyl monomers, and the possibility that the reaction is kinetically controlled to preferentially produce five-membered rings was discussed in a review with 90 references in 1980.[169]

The chemistry, properties, and applications of poly(diallylammonium salts) were reviewed with 56 references in 1980.[170] Emphasis was placed on the polymerization mechanism, polymer structure, and flocculation properties of the polymers. The synthesis, polymerization mechanism, characterizations, and biological activity of DIVEMA (the 1:2 alternating cyclocopolymer of divinyl ether and maleic anhydride) were discussed in a review with 162 references in 1980.[171] Synthetic polyelectrolytes, including poly(diallyldimethylammonium chloride), useful for water and wastewater treatment were reviewed with 118 references in 1981.[172] The theory of regioselectivity and stereoselectivity in radical reactions was discussed in detail in a review with 132 references in 1981.[173] The theory of ring-closure of alkenyl radicals and supporting evidence for ring size control was emphasized. Polyelectrolytes and coagulants for the dissolved and induced air and vacuum flotation processes, their applications, and future trends were reviewed with 19 references in 1981.[174] Intramolecular addition modes in radical-initiated cyclopolymerization of some non-conjugated dienes were discussed in a review with 39 references in 1982.[175] The antitumor and antiviral effects of carboxylic acid polymers, including cyclopolymers, were reviewed with 60 references in 1982.[176] A discussion of the scope of cyclopolymerization published in 1982, was extended to include large rings.[177] By use of a strongly-electron-donating 1,13-nonconjugated diene and a strongly-electron-accepting alkene, cyclocopolymerization occurred to yield copolymers consisting of large rings, the most probable of which would contain fifteen members.

The role of difunctional olefinic monomers in vinyl polymerization, the theories of cyclopolymerization and cyclocopolymerization, and the structures of the variety of polymers obtained from nonconjugated dienes were emphasized in a 1982 review, with 99 references.[178] A review with 163 references in which the theory, polymerization mechanisms and polymer structures derived *via* cyclopolymerization and cyclocopolymerization were emphasized, was published in 1982.[179] The synthetic procedures and biological activities of the 1:2 cyclocopolymer of divinyl ether and maleic anhydride were discussed in a review with 152 references in 1983.[180] The term "pyran copolymer" is a misnomer, and should not be used further to represent this copolymer.

Synthetic anionic polymers having interferon-inducing and antitumor activity including those synthesized *via* cyclopolymerization were reviewed with 161 references in 1984.[181] Under the general subject of drugs and pharmaceutical sciences, the synthesis and properties of anticancer and interferon inducing agents were reviewed in 1984 in a volume with 325 pages.[182]

The scope of cyclopolymerization, with emphasis on cyclopolymerization of dienes, diynes, dialdehydes, diisocyanates, diepoxides, dinitriles and organo-metallic compounds, was discussed in a 1985 review with 303 references.[183] Cyclic polymers including those synthesized *via* cyclopolymerization were reviewed in a recently (1986) published volume with 22 tables and 152 illustrations.[184] A review of the syntheses and uses of cation-type water-soluble polyelectrolytes derived *via* cyclopolymerization of dimethyldiallylammonium chloride was published with six references in 1987.[185]

A review of the preparation, properties and uses of polymers and copolymers based on dimethyldiallylammonium chloride with 80 references was published in 1987.[186] A review of the preparation and biological properties of DIVEMA (the 1:2 alternating cyclopolymer of divinyl ether and maleic anhydride) and Metton (crosslinked polydicyclopentadiene) with eight references was published in 1988.[187] A review of polymers of quaternary ammonium salts as clay swelling inhibitors and fine migration control agents for drilling fluids and completion and workover fluids for petroleum wells, with 12 references, was published in 1988.[188]

A review with 10 references on the preparation, cyclization and uses of poly(dimethyldiallylammoniumchloride) was published in 1988.[189] A review of preparation, properties and uses of water soluble polymers with 237 references was published in 1989.[190] Polyelectrolytes obtained *via* cyclopolymerization were included. A comprehensive review of the topic of cyclopolymerization and cyclocopolymerization with details of the current understanding of the mechanism, and the scope of cyclopolymerization with 340 references was published in 1989.[191]

The incorporation of hydrophobic side chains into a polyelectrolyte can produce profound changes in the physical chemical behavior of the parent macromolecule.[192] This topic was reviewed with 27 references in 1989. A review of cyclopolymerization with 28 references was published in 1989.[193] A discussion of synthetic procedures and properties of polymers with crown ether rings, developed *via* cyclopolymerization, was included. Also, some polymers obtained *via* asymmetric cyclopolymerization were discussed.

The discovery of cyclopolymerization has been classified as the eighth major structural feature of synthetic high polymers to be established.[194] Synthesis of a variety of new polymers including those prepared by cyclization was reviewed in 1969 with 77 references.[195] Ring-forming polymerizations including cyclopolymerizations, have been reviewed in 1972 in a 568 page volume on Heterocyclic Rings.[196] Cyclization polymerization was reviewed in 1976 with 27 references.[197]

REFERENCES

1. Butler, G. B. and Bunch, R. L., J. Am. Chem. Soc., 1949, 71, 3120; CA 44, 1988f, (1950).
2. Butler, G. B. and Ingley, F. L., J. Am. Chem. Soc., 1951, 73, 895; CA 45, 7516b, (1951).
3. Butler, G. B. and Goette, R. L., J. Am. Chem. Soc., 1952, 74, 1939; CA 48, 1947a, (1954).
4. Butler, G. B., Bunch, R. L. and Ingley, F. L., J. Am. Chem. Soc. 1952, 74, 2543; CA 48, 11299h, (1952).
5. Butler, G. B. and Johnson, R. A., J. Am. Chem. Soc., 1954, 76, 713; CA 49, 2997b, (1954).
6. Butler, G. B. and Goette, R. L., J. Am. Chem. Soc., 1954, 76, 2418; CA 49, 8801e, (1954).
7. Butler, G. B. and Angelo, R. J., J. Am. Chem. Soc., 1956, 78, 4797; CA 51, 2547h (1956).
8. Butler, G. B. and Angelo, R. J., J. Am. Chem. Soc., 1957, 79, 3128; CA 51, 15514h (1957).
9. Butler, G. B., Crawshaw, A. and Miller, W. L., J. Am. Chem. Soc., 1958, 80, 3615; CA 53, 2224e, (1959).

10. Walling, C., J. Am. Chem. Soc., 1945, 67, 441; CA 39, 1796(6) (1945).
11. Stockmayer, W. H., J. Chem. Phys., 1944, 12, 125: CA 38, 2546(5) (1944).
12. Simpson, W., Holt, T., and Zeite, R. J., J. Polym. Sci., 1953, 10, 489; CA 47, 7817d (1953).
13. Haward, R. N., J. Polym. Sci., 1954, 14, 535; CA 49, 3575y (1955).
14. Haward, R. N. and Simpson, W., J. Polym. Sci., 1955, 18, 440; CA 50, 16175e (1955).
15. Oiwa, M. and Ogata, Y., Nippon kogaku zasshi, 1958, 79, 1506; CA 54, 4488b (1960).
16. Butler, G. B., J. Macromol. Sci. (Chem.) A, 1971, 5, 219; CA 73, 120933k (1970).
17a. Butler, G. B., J. Polym. Sci., 1960, 48, 279; CA 55, 17070b (1961).
17b. Butler, G. B., U.S. Pat. US 3,239,488, 1966, (Nov. 8).
18. Gibbs, W. E. and Barton, J. M., Vinyl Polymeriz., G. E. Ham, Ed. Vol. 1, 1967, 1, Pt. 1, 59; CA 70, 88271t (1969).
19. Corfield, G. C., Chem. Soc. Rev., 1972, 1, 523; CA 78, 98030t (1973).
20. Solomon, D. H. and Hawthorne, D. G., J. Macromol. Sci., Rev. Macromol. Chem., 1976, 15, 143.
21. Butler, G. B., Pure Appl. Chem., 1970, 23, 255; CA 75, 36729g (1971).
22. Beckwith, A. L. J., Gream, G. E. and Struble, D. S., Aust. J. Chem., 1972, 25, 1081; CA 77, 61313r (1972).
23. Butler, G. B., Science (Gordon Res. Conf., Polym. Prog.), 1955, 121, 574.
24. Marvel, C. S. and Vest, R. D., J. Am. Chem. Soc., 1957, 79, 5771; CA 52, 7219i, (1958).
25. Marvel, C. S. and Vest, R. D., J. Am. Chem. Soc., 1959, 81, 984; CA 53, 13997i (1959).
26. Milford, G. N. and Wall, F. T., U.S. Pat. 2,976,268, 1961 (Mar. 21); CA 55, 16016a (1961).
27. Crawshaw, A. and Butler, G. B., J. Am. Chem. Soc., 1958, 80, 5464; CA 53, 21055a (1959).
28. Jones, J. F., J. Polym. Sci., 1958, 33, 15; CA 53, 4809b (1959).
29. Jones, J. F., Ital. Pat. 563,941, 1957 (June 7).
30. Kagawa, I. and Fuoss, R. M., J. Polym. Sci., 1955, 18, 535.
31. Jones, J. F., J. Polym. Sci., 1958, 33, 7; CA 53, 6988d (1959).
32. Horio, T., Oshima, A. and Sakaguchi, E., Japan Pat. JP 13,611, 1967 (Aug. 2); CA 68, 50350p (1968).
33. Berlin, K. D. and Butler, G. B., J. Am. Chem. Soc., 1960, 82, 2712; CA 55, 431a (1961).
34. Price, J., U.S. Pat. 2,871,229, 1959 (Jan. 27); CA 53, 8651f (1959).
35. Schuller, W. H., Price, J. A., Moore, S. T., and Thomas, W. M., J. Chem. Eng. Data, 1959, 4, 273; CA 54, 5156d (1960).
36. Friedlander, W. S., Abstr. 133rd ACS Meet., San Francisco, 1958, p. 18N.
37. Butler, G. B. and Skinner, D. L., Conf. High Temp. Polym. and Fluid Res., 1959 (May).
38. Field, N. D., J. Org. Chem., 1960, 25, 1006; CA 54, 20980d, (1960).
39. Butler, G. B. and Miles, M. L., Polym. Preprints, 1962 (Sept.), 3 (2), 288; CA 61, 5768a (1964).
40. Matsoyan, S. G., Avetyan, M. G., and Voskayan, M. G., Vysokomol. Soedin., 1961, 3, 562; CA 56, 7498g (1961).
41. Otsu, T., Mulvaney, J. E. and Marvel, C. S., J. Polym. Sci., 1960, 46, 546; CA 55, 12275c (1961).
42. De Winter, W., Marvel, C. S. and Abdul-Karim, A., J. Polym. Sci. A, 1963, 1, 3261; CA 60, 672b (1964).
43. Bloomfield, J. J., J. Org. Chem., 1962, 27, 3327; CA 58, 1338d (1963).
44. Miller, W. L. and Black, W. B., Polym. Repr., 1962, 3, 345; CA 61, 5775e (1964).
45. Beaman, R. G., U.S. Pat. US 3,048,566, 1962 (Aug. 7); CA 59, 782b (1962).
46. Moyer, W. W., Jr. and Grev, D. A., J. Polym. Sci. B, 1963, 1, 29; CA 58, 8047b (1963).

47. Overberger, C. G., Ishida, S. and Ringsdorf, H., J. Polym. Sci., 1962, 62, S1; CA 58, 10306f (1963).
48. Myerson, K., Schulz, R. C. and Kern, W., Makromol. Chem., 1962, 58, 204; CA 58, 5792b (1963).
49. Aso, C. and Aito, Y., Bull. Chem. Soc. Japan, 1962, 35, 1426; CA 57, 13967d (1962).
50. Koral, J. N., J. Polym. Sci., 1962, 61, S37; CA 59, 4045h (1963).
51. Schulz, R. C., Kunststoffe, 1957, 47, 303; CA 51, 14317a (1957).
52. Marvel, C. S. and Stille, J. K., J. Am. Chem. Soc., 1958, 80, 1740; CA 52, 13608e, (1958).
53. Makowsky, H. S. and Shim, B. K. C., Polym. Prepr., 1960, 1 (1), 101; CA 57, 4847c (1962).
54. Marvel, C. S. and Woolford, R. G., J. Org. Chem., 1960, 25, 1641; CA 55, 8279h (1961).
55. Topchiev, A. V., Nametkin, N. S., Durgar'yan, S. G. and Dyankov, S. S., Chem. & Pract. Apply. Organosilicon Comp 1958, 2, 118; CA 53, 8686h (1959).
56. Butler, G. B. and Stackman, R. W., J. Org. Chem., 1960, 25, 1643; CA 55, 9262e (1961).
57. Schuller, W. H. and Price, J. A., U.S. Pat. 3,032,539, 1962 (May 1); CA 57, 6151d (1962).
58. Matsoyan, S. G., Avetyan, M. I. and Voskanyan, M. G., Vysokomol. Soedin., 1961, 3, 1140; CA 57, 8726a (1962).
59. Matsoyan, S. G., Avetyan, M. G. and Voskanyan, M. G., Vysokomol. Soed., 1962, 4, 882; CA 58, 8045d (1963).
60. Shim, B. K. C., Makowski, H. S. and Wilchinsky, Z. W., Polym. Prepr., Div. Polym. Chem., ACS, 1963, 4, (1), 43; CA 62, 645e (1965).
61. Wright, C. D. and Friedlander, W. S., U.S. Pat. 3,072,616, 1963 (Jan. 8); CA 58, 11491e (1963).
62. Wright, C. D., U.S. Pat. 3,057,829, 1962 (Oct. 9); CA 59, 7671b (1963).
63. Zutty, N. L., J. Polym. Sci. A, 1963, 1, 2231; CA 59, 14116f (1963).
64. Polyakova, A., Plate, A., Pyranishnikova, M. and Pilatnikov, N., Neftekhimiya, 1961, 1, 521; CA 57, 10026g (1962).
65. Kargin, V. A., Plate, N. A. and Dudnik, L. A., Vysokomol. Soedin., 1959, 1, 420; CA 54, 5151a (1960).
66. Hwa, J. C. H. and Miller, L., J. Polym. Sci., 1961, 55, 197; CA 56, 8918g (1962).
67. Wiley, R. H. and Davis, B., J. Polym. Sci. B, 1963, 1, 463; CA 59, 14116h (1963).
68. Beynon, K. I., J. Polym. Sci. A, 1963, 1, 3343; CA 59, 15387c (1963).
69. Stille, J. K. and Culbertson, B. M., J. Polym. Sci. A, 1964, 2, 405; CA 60, 9365h (1964).
70. Koran, J. N. and Smolin, E. M., J. Polym. Sci. A, 1963, 1, 2831; CA 59, 10242h, (1963).
71. Arbuzova and Efremova, V. N., Vysokomol. Soedin., 1959, 1, 455; CA 54, 5151h (1960).
72. Arbuzova, I. A. and Efremova, V. N., Vysokomol. Soedin., 1960, 2, 1586.
73. Gindin, L., Medvedev, S. and Flesher, E., J. Gen. Chem. USSR, 1949, 19, a127; CA 44, 6387d (1950).
74. Blout, E. and Ostberg, B., J. Polym. Sci., 1946, 1, 230; CA 40, 4912(1) (1946).
75. Cohen, S. G., Ostberg, B. E., Sparrow, D. B. and Blout, E. R., J. Polym. Sci., 1946, 1, 230; CA 42, 5256h (1948).
76. Barnett, M. D., Crawshaw, A., and Butler, G. B., J. Am. Chem. Soc., 1959, 81, 5946; CA 54, 9755b (1960).
77. Barnett, M. D. and Butler, G. B., J. Org. Chem., 1960, 25, 309; CA 54, 16380d (1960).
78. Schulz, R. C., Marx, M. and Hartmann, H., Makromol. Chem., 1961, 44–46, 281; CA 55, 22110d (1961).

79. Arbuzova, I. A., Plotkina, S. A., and Sokolova, O. V., Vysokomol. Soedin., 1962, 4, 843; CA 59, 6523h (1963).
80. Martin,Jr., R. N., U.S. Pat. 3,043,814, 1962 (July 10); CA 57, 11341h (1962).
81. Martin, Jr. R. H., U.S. Pat. 3,025,272, 1962 (Mar. 13); CA 56, 14485b (1962).
82. Trifan, D. S. and Hoglen, J. J., J. Am. Chem. Soc., 1961, 83, 2021; CA 55, 19760a (1961).
83. Graham, P. J., Buhle, E. L., and Pappas, N., J. Org. Chem., 1961, 26, 4658; CA 57, 1047f (1962).
84. Reichel, B. Marvel, C. S. and Greenley, R. Z., J, Polym. Sci. A, 1963, 1, 2935; CA 59, 11668c (1963).
85. Gibbs, W. E. and Van Deusen, J. Polym. Sci., 1961, 54, S1, CA 56, 4946h (1962).
86. Marvel, C. S. and Garrison, W. E., Jr., J. Am. Chem. Soc., 1959, 81, 4737; CA 54, 3171d (1960).
87. Butler, G. B., Abstr., 133rd ACS Meet., San Francisco, 1958, p. 6R.
88. Butler, G. B., 134th ACS Meet. Abstr., Chicago, 1958, p. 32T.
89. Chang, E. Y. C. and Price, C. C., J. Am. Chem. Soc., 1961, 83, 4650; CA 56, 14465c (1962).
90. Stackman, R. W., J. Macromol. Sci. (Chem.) A, 1971, 5, 251; CA 73, 120934m (1970).
91. Frazer, A. H. and O'Neil, W. P., ACS Div. Polym. Chem., Polym. Preprints, 1963, 4, (1), 21.
92. Gaylord, N. G. and Mark, H. F., eds., Linear and Stereoreg. Add. Polym., 1959, 581; CA 53, 18543e (1959).
93. Fox, T. G., Garrett, B. S., Goode, W. E., Gratch, S., Kincaid, J. F., Spell, A. and Stroupe, J. D., J. Am. Chem. Soc., 1958, 80, 1768; CA 52, 10635c (1958).
94. Fox, T. G., Goode, W. E., Gratch, S., Huggett, C. M., Kincaid, J. E., Spell, A. and Stroupe, J. D., J. Polym. Sci., 1958, 31, 173; CA 55, 7966f (1961).
95. Stroupe, J. D. and Hughes, R. E., J. Am. Chem. Soc., 1958, 80, 2341; CA 52, 12505f (1958).
96. Bovey, F. A. and Tiers, G. V. D., J. Polym. Sci., 1960, 44, 173; CA 54, 1808e (1960).
97. Miller, W. L., Brey, W. S., Jr., and Butler, G. B., J. Polym. Sci., 1961, 54, 329; CA 56, 6815f (1962).
98. Butler, G. B., Pure Appl. Chem. 1962, 4, 299; CA 57, 8718i (1962).
99. Hwa, J. C. H., J. Polym. Sci., 1962, 60, 512; CA 57, 11375i (1962).
100. Hwa, J. C. H. and Ries, H. E., Jr., J. Polym. Sci. B, 1964, 2, 389; CA 60, 13333f (1964).
101. Natta, G., Belg. Pat. 549,554.
102. Natta, G. and Porri, L., Fr. Pat. 1,154,938.
103. Natta, G., Porri, L., Carbonaro, A. and Stoppa, G., Makromol. Chem., 1964, 77, 126; CA 61, 10783a (1964).
104. Angelo, R. J., Polym. Preprints, 1963, 4, 32; CA 62, 645g (1965).
105. Tocker, S., U.S. Pat. 3,168,501, 1962; CA 57, 3635c (1962).
106. Tocker, S., J. Am. Chem. Soc., 1963, 85, 640; CA 58, 12700a (1963).
107. Kossler, I., Stolka, M. and Mach K., Paper 36, IUPAC, Paris, 1963, (Pub. 1964); CA 60, 7005h (1964).
108. Gaylord, N. G., Kossler, I., Stolka, M. and Vodehnal, J., Paper 8, ACS Div, Pol. Chem. Polym. Prepr., 1963, 4, 69.
109. Gaylord, N. G., Kossler, I., Stolka, M., and Vodehnal, J., J. Am. Chem. Soc., 1963, 85, 641; CA 58, 12678a (1963).
110. Gaylord, N. G., Kossler, I., Stolka, M. and Vodehnal, J., J. Polym. Sci., A, 1964, 2, 3969; CA 61, 14788g (1964).
111. Anon., Chem. Eng. News, 1963, 41, 37.
112. Anon., Chem. Eng. News, 1963, 41, 43.

113. Stolka, M., Vodehnal, J. and Koessler, I., J. Polym. Sci., A, 1964, 2, 3987; CA 61, 14788h (1964).
114. Anon., Chem. Eng. News, 1963, 41, 41.
115. Anon., Chem. Eng. News, 1963, 41, 33.
116. Schulz, R. C. and Stenner, R., Makromolek. Chem., 1964, 72, 202; CA 60, 15991f (1964).
117. Mikulasova, D. and Hrivik, A., Chemicke Zvesti, 1957, 11, 641; CA 52, 9028c (1958).
118. Butler, G. B., and Brooks, T. W., J. Org. Chem., 1963, 28, 2699; CA 59, 11670a (1963).
119. Butler, G. B. and Raymond, M. A., ACS, Div. Org. Chem., Denver, CO, 1964, 95, Paper 51.
120. Butler, G. B. and Raymond, M. A., J. Org. Chem., 1965, 30, 2410; CA 63, 5485g (1965).
121. Bartlett, P. D., Ann., 1962, 653, 45; CA 57, 7080i (1962).
122. Lamb, R. C., Ayers, P. W. and Toney, M. K., J. Am. Chem. Soc., 1963, 85, 3483; CA 59, 15167g (1963).
123. Flory, P. J., Chem. Rev., 1946, 39, 137.
124. Kolesnikov, G. S. and Davydova, S. L., Uspekhi Khim., 1960, 29, 679; CA 55, 7268i (1960).
125. Marvel, C. S., J. Polym. Sci., 1960, 48, 101; CA 55, 15989c (1960).
126. Koton, M. M., J. Polym. Sci., 1961, 52, 97; CA 55, 26517c (1961).
127. Stille, J. K., "Introd. to Polym. Chem.", Wiley, 1962, Ch. 12.
128. Chiang, T.-C., Hua Hsueh T'ung Pao, 1962, 3, 1, 51; CA 57, 7459i (1962).
129. Koton, M. M., Usp. Khim., 1962, 31, 153; CA 57, 1025e (1962).
130. Schulz, R. C., Encycl. Polym. Sci. Technol., 1964, 1, 160; CA 64, 19778e (1966).
131. Aso C., Kogyo-kwagaku zasshi, 1965, 20, 29; CA 64, 12812b (1965).
132. Schulz, R. C., Chimia, 1965, 19, 143; CA 63, 3053e (1965).
133. Seebach, D., Angew. Chem., 1965, 77, 119; CA 62, 10305a (1965).
134. Matsoyan, S. G., Rus. Chem. Rev.; IUPAC, USSR, 1960, Sec. 1, p. 189, 1966, 35, 32; CA 64, 12799f (1966).
135. Butler, G. B., Encycl. Polym. Sci. Technol., 1966, 4, 568; CA 65, 17052d (1966).
136. Schulz, R. C., J. Kolloid-Z. Z. Polym., 1967, 216/217, 309; CA 67, 54480h (1967).
137. Wakayoshi, T., Purasuchikkusu, 1967, 18, 11; CA 68, 115227a (1968).
138. Koessler, I., Plast. Hmoly. Kauc., 1967, 4, 234; CA 67, 100456e (1967).
139. Regelson, W., Advan. Chemother., 1968, 3, 303; CA 69, 17741g (1968).
140. Regelson, W., Advan. Cancer Res., 1968, 11, 223; CA 72, 117517b (1970).
141. DeSchryver, F.C. and Smets, G., Meded. kon. Vlaam. Acad. Wetensch., Lett 1969, 31, 14; CA 74, 64436g (1971).
142. Butler, G. B., IUPAC Symposium, Prague, 1969, 255.
143. Aso, C., IUPAC Spmp., Prague, 1969, 287.
144. Trossarelli, L. Guaita, M. and Priola, A., J. Polym. Sci. C, 1969, 16, 4713; CA 71, 22388a (1969).
145. Volodina, V. I., Tarasov, A. I. and Spaskii, S. S., Usp. Khim., 1970, 39, 276; CA 73, 4193v (1970).
146. Beckwith, A. L. J., Special Publication 24 (Lon. Chem. Soc.) 1970, 239.
147. Aso, C., Kunitake, T. and Tagami, S., Progr. Polym. Sci., Japan, 1971, 1, 149; CA 78, 30232e (1973).
148. Aso, C., Kunitake, T. and Tagami, S., Prog. Polym. Sci., Japan, 1971, 1, 170; CA 78, 30232e (1973).
149. Yudkin, B., Dudina, A. and Sheptun, N., Plast. Massy, 1972, 16; CA 77, 35303t (1972).
150. Madruga, E. I., Fontan, J. and San Roman, J., Rev. Plast. Mod., 1973, 26, 581; CA 80, 48424v (1974).
151. Krongauz, E. S., Usp. Khim., 1973, 42, 1854; CA 80, 37468m (1974).
152. Hoover, M. F. and Butler, G. B., Polymer Symposia, 1974, 45, 1; CA 82, 17137r (1975).

153. Solomon, D. H., J. Macromol. Sci. (Chem) A, 1975, 9, 97; CA 82, 58133e (1975).
154. Solomon, D. H., J. Polym. Sci., Polym. Symp., 1975, 49, 175; CA 83, 79629g (1975).
155. Butler, G. B., Proc. Int. Symp. Macromol., (Mano, Ed.) 1975, 57; CA 85, 6058b (1976).
156. Butler, G. B., Corfield, G. C. and Aso, C., Prog. Polym. Sci., 1975, 4, 71; CA 88, 153021z (1978).
157. Solomon, D. H., J. Macromol. Sci. (Chem), A, 1975, 9, 95.
158. Butler, G. B., Reidel Dordrecht (Rembaum & Seligny,Eds) 1975, 97; CA 84, 90592f (1976).
159. Semlyen, J. A., Adv. Polym. Sci., 1976, 21, 41; CA 85, 124391c (1976).
160. Solomon, D. H., J. Macromol. Sci. (Chem), A, 1976, 10, 855.
161. Hawthorne, D. G. and Solomon, D. H., J. Polym. Sci., Polym. Symp., 1976, 55, 211.
162. Otsu, T. and Yamada, B., Kagaku (Kyoto), 1977, 32, 166; CA 87, 6373s (1977).
163. Butler, G. B., J. Polym. Sci.-Polym. Symp., 1978, 64, 71; CA 90, 169064t (1979).
164. Ottenbrite, R. M, Regelson, W., Kaplan, A., Carchman, R., Morahan, P. and Munson, A., Polymer Drugs, (Donaruma & Vogl, Eds.), 1978, 263; CA 91, 83997h (1979).
165. Reinisch, G., Mitteilungsbl. Chem. Ges. D. D. R., 1979, 26, 79; CA 91, 158121d (1979).
166. Ottenbrite, R. M., Polym. Biol. Med., (Anion. Polym. Drugs) 1980, 1, 1; CA 95, 12619h (1981).
167. Ottenbrite, R. M., Polym. Biol. Med., (Anion. Polym. Drugs) 1980, 1, 21; CA 95, 34955s (1981).
168. Bolto, B. A., Polym. Amines & Ammon. Salts, (Goethals, Ed., 1980, 365; CA 94, 16403k (1981).
169. Ottenbrite, R. M. and Ryan, W. S, Jr., Industrial & Eng. Chem, Prod. Res. & Dev. 1980, 19, 528; CA 93, 205045k (1980).
170. Butler, G. B., Polym. Amines & Ammon. Salts, (Goethals, Ed., 1981, 125; CA 94, 4257c (1981).
171. Butler, G. B., Anionic Polymeric Drugs., (Donaruma, et al, Eds.), 1980, 49; CA 95, 15656t (1981).
172. Vorchheimer, N., Polyelectrolytes Water Wastewater Treat., 1981, 1; CA 95, 220336r (1982).
173. Beckwith, A. L. J., Tetrahedron, 1981, 37, 3073; CA 96, 103268e (1982).
174. Walzer, J. G., Polyelectrolytes Water Wastewater Treat., 1981, 145; CA 96, 37451f (1982).
175. Matsumoto, A., Iwanami, K., Kitamura, T., Oiwa, M. and Butler, G. B., ACS Symp. Series, 1982, 195, 29; CA 97, 145293k (1982).
176. Ottenbrite, R. M. ACS Symp. Ser., 1982, 186, 205; CA 96, 210263a (1982).
177. Butler, G. B. and Kresta, J. E., Eds., ACS Symp, Ser. 1982, 195, 1–442.
178. Butler, G. B., Acct. Chem. Res., 1982, 15, 370; CA 97, 216742u (1982).
179. Corfield, G. C. and Butler, G. B., Develop. in Polymeriz. 3, (Haward, R. N., Ed.) 1982, 3, 1; CA 97, 110406x (1982).
180. Butler, G. B., J. Macromol. Sci.-Rev. Mac. Chem. Phys., 1982, 22, 89; CA 98, 64952t (1983).
181. Ottenbrite, R. M. and Butler, G. B., Drugs Pharm. Sci.: Anticancer Agents., 1984, 24, 247; CA 101, 163081v (1984).
182. Ottenbrite, R. M. and Butler, G. B., Drugs and Pharmaceutical Sci. (Dekker), 1984, 24, 325 pp; CA 101, B163692v (1984).
183. Butler, G. B., Encycl. Polym. Sci. Eng., 1985, 4, 543; CA 105, 43356z (1986).
184. Semlyen, J. A., Cyclic Polymers (Elsevier Sci. Pub. Co.) 1986.
185. Kryuchkov, V. V., Amburg, L. A., Parkhamovich, E. S. and Boyarkina, N. M., Plast. Massy, 1987, 22; CA 107, 176492d (1987).
186. Boyarkina, N. M., Kryuchkov, V. V., Parkhamovich, E. S., Amburg, L.A., Topchiev, D. A. Kabanov, V. A., Plast. Massy, 1987, 17; CA 107, 198941b (1987).
187. Breslow, D. S., Polym. Mater. Sci. Eng., 1988, 58, 223; CA 109, 93643u (1988).

188. Borchardt, J. K., Prepr. A.C.S., Div. Pet. Chem., 1988, 33, 60; CA 109, 76254a (1988).
189. Jaeger, W., Hahn,M. and Wandrey, C., Mitteilungsbl.- Chem. Ges. DDR, 1988, 35, 151; CA 109, 170942 (1988).
190. McCormick, C. L., Encycl. Polym. Sci. and Eng., 1989, 17, 730.
191. Butler, G. B., Comp. Polym. Sci. (Pergamon Press), 1989, 4, 1423.
192. Strauss, U. P., Adv. Chem. Ser.(Polym. Aqueous Media), 1989, 223, 317; CA 112, 21501k (1990).
193. Yokota, K., Kobunshi, 1989, 38, 976; CA 112, 119466c (1990).
194. McGrew, F. C., J. Chem. Ed., 1958, 35, 178.
195. Furukawa, J., Kobunshi Kako, 1969, 18, 73; CA 71, 39399c (1969).
196. Cotter, R. J. and Matzner, M., Heterocyclic Rings: A Ser. of Monogrph. 1972, 13B, 568 pp; CA 78, 30455e (1973).
197. Fukuda, W. and Kakiuchi, H., Kagaku (kyoto), 1976, 31, 652; CA 86, 5864v (1977).

2

Nonconjugated Symmetrical Dienes

CYCLOPOLYMERIZATION OF NONCONJUGATED SYMMETRICAL DIENES

The initial observation that 1,6-hexadienes do not cross-link, but undergo cyclopolymerization was made during an attempt to synthesize strongly basic ion exchange resins. According to the Staudinger theory of cross-linking,[1] dialkyldiallylammonium halides were predicted to produce a highly cross-linked network. However, highly soluble products of reasonable molecular weights were obtained when such monomers were subjected to radical-initiated polymerization.[1-1] The concept of cyclopolymerization otherwise described as an alternating intramolecular–intermolecular chain propagation was advanced to account for these observations.[1-23]

SYNTHESIS OF POLYMERS CONTAINING CARBOCYCLIC RINGS

The discovery of cyclopolymerization was extended to the formation of six-membered carbon rings *via* an alternating intramolecular–intermolecular polymerization reaction.[1-24, 1-25] (2–1) 1,6–Heptadiene–2,6–dicarboxylic acid, its methyl and ethyl esters, and nitrile (2–1a–d) were polymerized in free radical systems to produce soluble polymers which were essentially free of unsaturation. Two polymers [from (2–1a) and (2–1c)] were partially dehydrogenated with potassium perchlorate. Infrared and ultraviolet spectral data indicated the presence of aromatic rings in the dehydrogenated material. Thus the predominant structural unit of these polymers must be cyclic (2–1e).

(a) R = COOH (b) R = $COOCH_3$

(c) R = $COOC_2H_5$ (d) R = CN

(2–1)

$$CH_2{=}\overset{R}{\overset{|}{C}}{-}(CH_2)_3{-}\overset{R}{\overset{|}{C}}{=}CH_2 \longrightarrow \left[CH_2 - \underset{\text{(cyclohexane ring)}}{\overset{R \quad R}{\bigcirc}} \right]_n$$

(e)

(2-1, cont'd)

In order to obtain high molecular weight polymers, the polymerization of 2–1b and 2–1d was investigated in detail.[2] Low molecular weight products were obtained from all variations tried in solution or emulsion systems. Polymerization, under the influence of ultraviolet light, gave soluble, high molecular weight products at high conversions (Table 2–1). The polymers exhibited good thermal stability and contained little, if any, unsaturation.

It was found that 1,6-heptadiene polymerized in the presence of Ziegler type catalysts, yielding soluble predominantly saturated polymers.[1-52] Again, partial dehydrogenation with potassium perchlorate resulted in a polymer showing the spectral characteristics of meta-substituted aromatic rings. Cyclopolymerization of 1,6-heptadiene gave material with a slightly higher thermal stability and density than polyethylene.[3]

The polymerization of 2,6-diphenyl-1,6-heptadiene was described and it was found that thermal, free radical, cationic, anionic, and Ziegler type initiators all gave soluble polymers of essentially the same infrared spectrum with little or no residual unsaturation.[1-38] Details of these polymerizations as well as some of the polymer properties are given in Table 2–2. These results were independently confirmed.[4]

Polymers of highest molecular weight were obtained from monomers which led to ring sizes of six. Anionic polymerization with phenyllithium in these systems gave the best yields (85%). Yields of 50–70% were obtained for cationic, Ziegler, and free-radical initiation. Inherent viscosities ranged rom 0.25 with phenyllithium to 0.128 and 0.145 with cumene hydroperoxide and boron trifluoride, respectively. For 2,7-diphenyloctadiene-1,7, only low molecular weights and low to moderate yields were obtained by cationic initiation. 2,5-Diphenylhexadiene-1,5 gave low molecular weights and moderate yields with cationic and Ziegler initiators (58–83%). Cumene hydroperoxide gave only a small yield of polymer.

The polymerization of a number of substituted (positions two, four, and six) 1,6–heptadienes was studied.[5-8] 1,6-Heptadienes having an electron acceptor substituent in the 4-position could be polymerized in the presence of radical initiators to yield linear polymers containing cyclohexane units in the chain. The cyclic structure of the polymers was demonstrated by chemical and spectroscopic methods. The presence of methyl groups in positions two and six retarded polymerization, while the presence of chlorine atoms sharply accelerated it. Under the conditions of the polymerization, 2,6-dichloro–4–substituted 1,6–heptadienes underwent up to 70% dehydrochlorination to polymers containing 1,4–cyclohexadiene rings in the main chain (2–2).

Table 2–1 Cyclopolymers from 2,6-disubstituted 1,6-Heptadienes

Str. 2-1	Monomer (g)	Polymerization conditions and/or solvent	Initiator	Temp. (°C)	Time (h)	Conversion (%)	Intrinsic viscosity[a]
b	3.0	Emulsion (25 ml)	Metabisulfite (0.06 g)-peroxydisulfate (0.01 g)	55	3	43	0.10
b	2.0	Benzene (15 ml)	Benzoyl peroxide	75	24	22	0.21
b	2.0	γ-Butyrolactone (15 ml)	Benzoyl peroxide	75	24	28	0.28
b	2.0	t-Butyl alcohol (15 ml)	Azobisisobutyronitrile (0.01 g)	65	24	65	0.47
b	1.0	γ-Butyrolactone (3 ml)	UV, benzoin (0.005 g)	20	16	92	0.60
b	1.0	Formic acid (3 ml)	UV, benzoin (0.005 g)	20	3	25	1.00
b	1.0	Tetramethylene sulfone (3 ml)	UV, benzoin (0.005 g)	20	4	80	2.03
d	1.0	γ-Butyrolactone (3 ml)	UV, benzoin (0.005 g)	20	4	25	1.00
d	1.0	Tetramethylene sulfone (4 ml)	UV, benzoin (0.005 g)	20	4	74	1.22

[a] dl/g in dimethylformamide

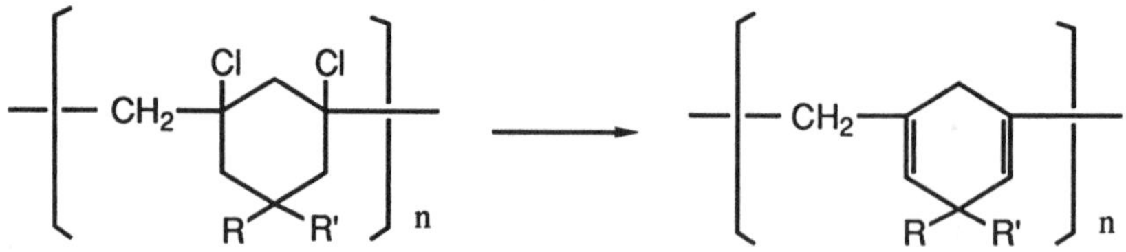

(2–2)

The cyclopolymerization of some of the 4-substituted 1,6-heptadienes discussed above has also been studied by other workers.[9,10]

In an effort to synthesize diacrylylmethane by a Claisen condensation of methyl vinyl ketone with ethyl acrylate, a soluble, fusible polymer was isolated.[1-31] It was proposed that this was a cyclopolymer of diacrylylmethane formed by initiation of the monomer, as rapidly as produced, by the anions present in the reaction mixture. After studying some reactions of this polymer, it was suggested[1-41] that the product was, in fact, a copolymer of diacrylylmethane and methyl vinyl ketone. Pure diacrylylmethane was ultimately prepared,[1-43] followed by anionic initiation of the pure monomer with sodium methoxide at $-75°$, to give a 90% yield of a bright yellow polymer which was soluble in water, dimethyl sulfoxide (DMSO), dimethylacetamide, and dimethylformamide (DMF).[1-42] This polymer was a sodium salt of polydiacrylylmethane. Infrared evidence supported a cyclic structure as predicted by the cyclopolymerization mechanism. The molecular weight was low and the polymer, formed on acidification of the sodium salt, was insoluble in all solvents tested. Dimethyldiacrylylmethane was also synthesized and polymerized[11] to yield a soluble polymer with a cyclic 1,3-diketone structure. A mono-oxime of this polyketone with 90% oximated structure was obtained. This mono-oxime underwent the Knoevenagel reaction to yield a partial ladder polymer (2–3), which had good thermal stability.

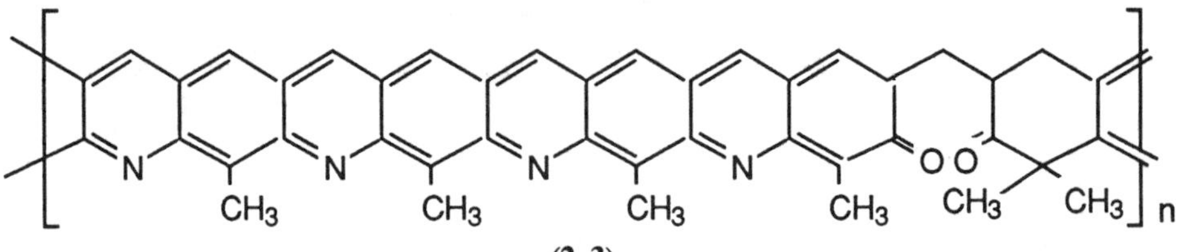

(2–3)

cis-1,3-Divinylcyclohexane and cis-1,3-divinylcyclopentane have been polymerized[12] to yield soluble, virtually saturated polymers which are clearly cyclopolymers containing bicyclo[3.3.1]nonane (2–4) and bicyclo[3.2.1]octane (2–5) structures respectively. 1,8-Divinylnaphthalene has been polymerized to give low molecular weight polymers. The tricyclic unit (2–6) was assigned to the polymers on the basis of (a) the absence of any maxima in the infrared spectrum which could be ascribed to vinyl unsaturation, (b) the complete solubility of the polymers in common organic solvents, and (c) the elemental analysis.[13]

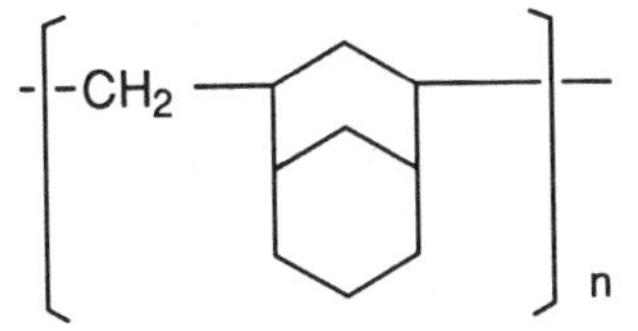

(2–4)

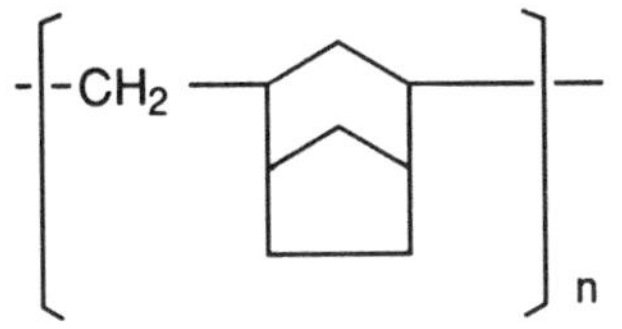

(2–5)

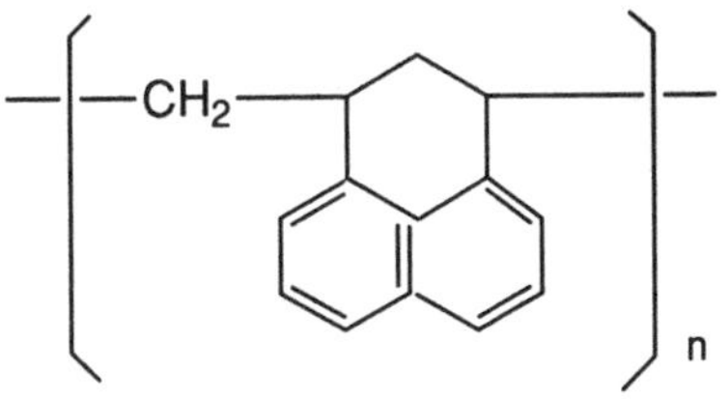

(2–6)

Perfluoro-1,6-heptadiene and 4-chloroperfluoro-1,6-hptadiene have been polymerized and their glass-transition temperatures determined.[14] The highly saturated polymers were prepared by exposing monomer to radiation. The ratio of combined monomer units per perfluorovinyl group was about 500. Soluble materials were prepared by stopping the polymerization before the conversion became high. Ring structures were formed. It was shown that polymers of the perfluoro-α,ω-diene had much lower glass transition temperatures than did polymers of perfluoro-α-olefins. The presence of the chlorine atom raised the glass transition temperature by about 45°C in comparison to the cyclopolymer of perfluoro-1,6-heptadiene.

The preparation of hydrocarbon soluble polymers from the polymerization of 1,5-hexadiene with the triisobutylaluminum–titanium tetrachloride catalyst system was reported in 1958.[1-52] These polymers were found to contain only 5–8 % residual unsaturation. Based on this information, the principal recurring unit in these polymers was proposed to be the 1-methylene-3-cyclopentyl group. The two bands in the high resolution infrared spectrum of poly(1,5-hexadiene), obtained with typical Ziegler catalysts, could be interpreted by the presence of units containing cyclopentane rings.[15] Later, the structure and properties of poly(1,5-hexadiene) prepared with triethylaluminum–titanium tetrachloride catalyst combinations were studied.[1-53, 16] Although the yields of the polymers and their respective properties varied with the catalyst modification and the mode of preparation, every poly(1,5-hexadiene) prepared was crystalline, had a density exceeding 1.0 g/cc, a melting point in excess of 100°C, and was very flexible. The density of poly(1,5-hexadiene) was higher than polyethylene or any aliphatic polyolefin. It had about the same strength and melting point as high density polyethylene yet it was about as flexible as low density polyethylene. This combination of properties was offered as proof that the polymers contain carbocyclic rings, resulting from a cyclopolymerization mechanism.

The x-ray fiber repeat distance of poly-1,5-hexadiene was 4.80 Å, which corresponded to a chain structure consisting of a linear, zigzag array of 1-methylene-3-cyclopentyl groups. Extensive calculations on model systems and other considerations ruled out four- and six-membered rings as recurring units in the polymer chain. Although a repeat distance of 4.80 Å was calculated if the methylene substituents in the 1,3 positions of the cyclopentane ring are placed trans on a planar ring, this system did not take into account known facts about 1,3-disubstituted cyclopentanes. *cis*-1,3-Dimethylcyclopentane is more stable than the trans isomer by about 0.5 kcal/mole, and this is most likely the result of ring puckering, a proven phenomenon in cyclopentane and its derivatives. If the methylene substituents of poly-1,5-hexadiene are placed cis and the number 2 carbon atom of the ring is puckered toward the methylene substituents to a perpendicular distance of 0.46 Å above the plane of planar cyclopentane, the distance between the methylene groups (fiber repeat distance) would be 4.80 Å. This system (2–7) accounted for all the known facts about cyclopentane and poly-1,5-hexadiene.

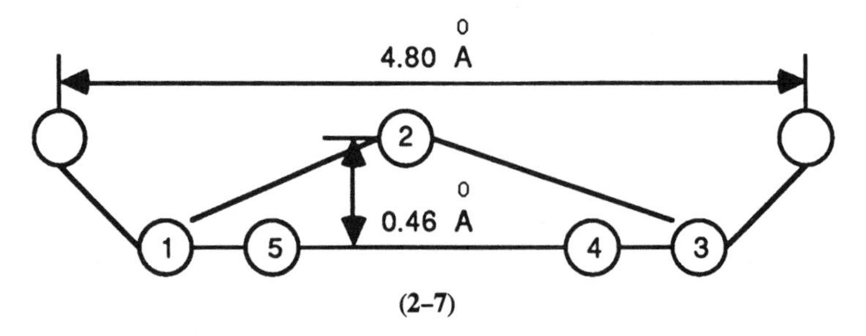

(2–7)

Later work[1-53] added more definitive evidence for the nature of the repeating unit in poly-1,5-hexadiene. One particular sample had a tensile strength of 5400 psi, a melting point of 139°C, and a density of 1.122 g/cc, and yet was very flexible, having an apparent modulus of elasticity at −50°C of 200,000 psi.

The cyclopolymerization of 1,5-hexadiene with ethylaluminum dichloride[17] and diethylaluminum chloride-vanadylbis-(acetylacetonate)[18] catalysts has been described more recently. The thermal stability of a 1,5-hexadiene cyclopolymer has been investigated.[3]

The polymerization of 2,5-dimethyl-1,5-hexadiene with Ziegler catalysts[1-52] and boron trifluoride and titanium tetrachloride[19] has been studied. Soluble polymers were obtained showing only a low content of double bonds. The properties of the polymers prepared with Ziegler and the cationic initiators appear to be identical. In both cases it was assumed that propagation involved a cyclopolymerization mechanism with the formation of five-membered rings in the chain. It was observed[1-52] that 2-methyl-1-pentene would not polymerize beyond a pentamer with Ziegler catalysts. This seemed to indicate that the cyclopolymerization process constituted a driving force to effect polymerization through an otherwise poorly polymerizable structure.

The polymerization of o-divinylbenzene has been investigated in detail. Soluble polymers containing a low pendant double bond content were obtained with free radical,[20] cationic,[21-23] and anionic[24] initiators under a variety of conditions. It was concluded, therefore, that o-divinylbenzene was converted to a linear polymer having five-membered ring units (2–8) in the main chain, formed through a cyclopolymerization mechanism. However, the polymerization of o-divinylbenzene has been reinvestigated,[25,26] and on the basis of kinetic studies and the energetics of the process, the authors have proposed that the polymers are mainly comprised of seven-membered rather than the originally proposed five-membered rings (2–9).

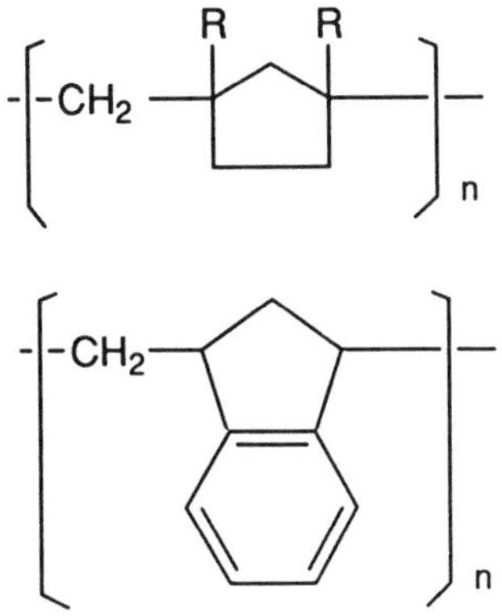

(2–8) 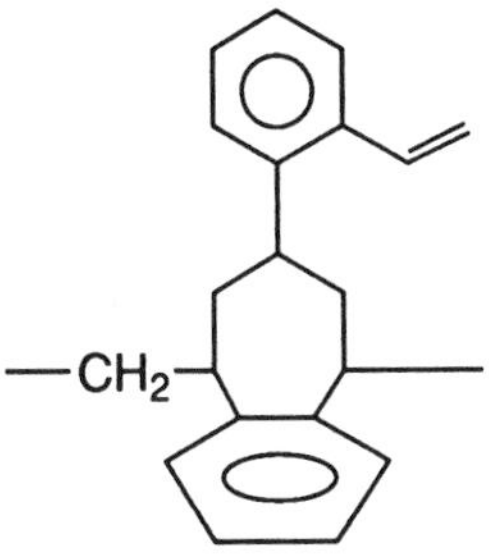 (2–9)

A comparison between the most probable propagation mechanism of cyclopolymerization of o-divinylbenzene, *cis*–1,2–divinylcyclohexane, o–phthalaldehyde, *cis*–cyclohexene–4,5–dicarboxaldehyde and o–vinylbenzaldehyde by cationic initiators has been made.[27] o-Divinylbenzene was proposed to propagate by a stepwise process and an intermediate, whereas the divinyl cyclohexane and the aldehydes were proposed to proceed by a concerted process. o-Vinylbenzaldehyde showed the existence of both reaction schemes, depending upon the type of cation produced.

The cyclization constant r_c, defined as the ratio of rate constants for the intramolecular cyclization and intermolecular propagation, respectively, in anionic polymerization was compared with that of free radical and cationic polymerizations for o-divinylbenzene.[24] In free radical polymerizations, r_c ranged from 2 to 4 mole/l while in cationic polymerization, r_c varied from 0.1 to 5 mole/l depending on polymerization conditions. In anionic systems, r_c values were less than 1 mole/l.

Interpretation of the data suggested that the uncyclized species interacts with the π orbital of the neighboring double bond in cationic polymerization to facilitate cyclization (2–10), whereas in the anionic system, a similar interaction would be less important. An electronic interaction is the basis for one of the two proposed mechanisms for cyclopolymerization which will be discussed later.

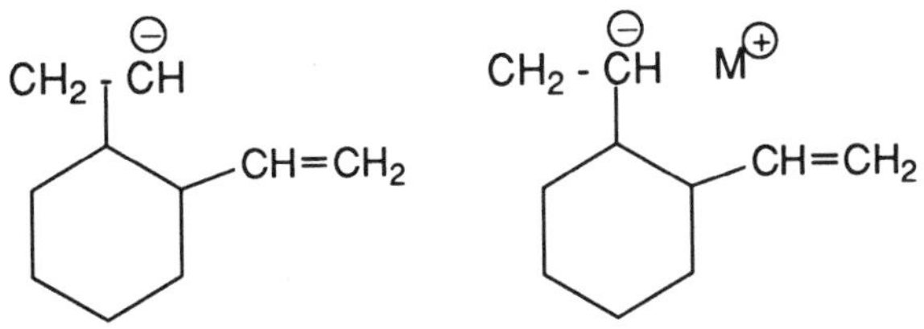

FREE ION ION PAIR

(2–10)

The uncyclized anion of o-divinylbenzene could exist as a free anion or an ion pair. The free ion should be more reactive than the ion pair and the reactivity of the ion pair should be influenced by solvent and counterion. The free anion should give higher r_c values. Addition of $NaB(C_6H_5)_2$ converted the free anion to the ion pair and thus lowered r_c.

Polymerization of the conjugated monomer 1,3-pentadiene,[28,29] to linear and cyclic polymers, by using alkali metals as initiators, has been reported. The mechanism postulated for this cyclopolymerization is not clear.

Aromatization of cyclized 1,2-polybutadiene has been claimed.[30] Polybutadiene was prepared by using butyllithium as an initiator. Addition of sulfuric acid caused cyclization. Chloranil in xylene was added to the polymer with aromatization resulting after refluxing for 72 hr.

It has been shown[1–19, 1–156] that a wide variety of 1,6–dienes can be polymerized to soluble–saturated polymers, and the structures having five– and or six–membered rings have been proposed (2–11).

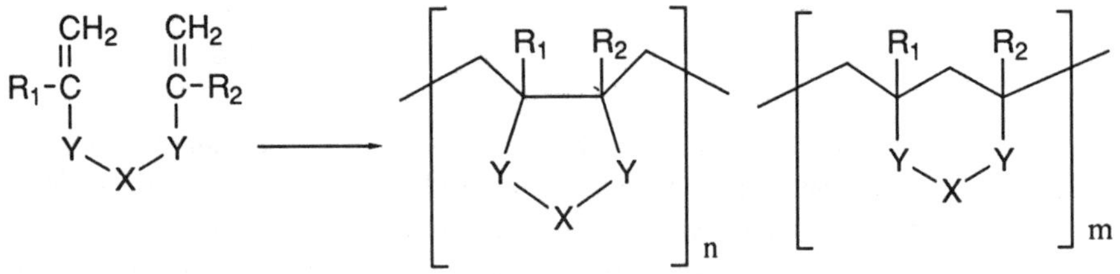

(2–11)

Cyclocopolymerization and telomerization reactions of symmetrical divinyl compounds were reviewed in 1968.[31]

cis–1,3-Divinylcyclohexane[32] was prepared from *cis*–1,3–diformylcyclohexane *via* the Wittig synthesis and its cyclopolymerization was studied by use of Ziegler catalysts. A cyclopolymer containing bicyclo[3.3.1]nonane units with little residual unsaturation was obtained.

The influence of solvents and catalysts on the formation of soluble and insoluble polymer during cationic polymerization of 2,5-dimethyl-1,5-hexadiene has been investigated.[33] In nonpolar medium, the amount of the insoluble part was dependent on the catalyst used and increased as follows: $BF_3 << AlBr_3 < TiCl_4$. Especially soluble polymers were obtained in a solution of CH_2Cl_2 or in a mixture of CH_2Cl_2-nitrobenzene. The temperature, $-30°$ to $-78°C$, did not show any important influence on the polymer compound. The bulky anion of the initiator and the polarity of the medium considerably influenced the extent of cyclization.

The formation of certain carbocyclic and metallorganic rings by cyclopolymerization mechanisms[34] have been reviewed.

Cyclopolymerization of *cis*–1,2–divinylcyclohexane[35] by Friedel–Crafts and other cationic initiators have been reported. $C_2H_5AlCl_2$-alkylhalide systems were most effective, but $(C_2H_5)_3AlTiCl_4$, $(C_2H_5)_2AlCl/TiCl_4$, and similar systems were also effective. The polymers were composed almost entirely of the cyclized unit but lacked the pendant vinyl group. The monomer had a rather small cationic reactivity.

cis–1,2–Divinylcyclohexane[36] was polymerized with a titanium tetrachloride–triethylaluminum initiating system to a thermally and chemically resistant polymer of 1100–3000 molecular weight, 100–55°C softening point, in 35–92% yield and of the following structure (2–12):

A large number of cyclopolymerizable monomers have subsequently been synthesized, and a variety of polymerization processes studied to further define the scope

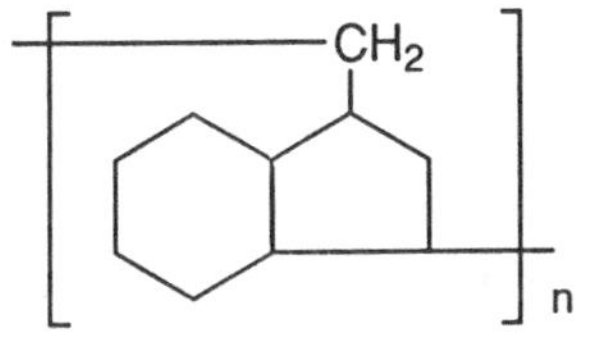

(2–12)

and versatility of the cyclopolymerization procedure. A variety of these will be described in this chapter under each sub-chapter heading.

A hydrocarbon polymer of bicyclo[2.2.1]heptadiene was obtained by the diene synthesis.[1-63, 1-65] The polymer was insoluble in common organic solvents but has a softening point above 350°. It is stable up to 400° where slow decomposition begins. A linear structure for the polymer was assumed. It was proposed that the structure of the polymer consisted of both 3,5-disubstituted nortricyclene and 5,6-disubstituted bicyclo[2.2.1]heptene-2 repeat units. 1,5,9-Cyclododecatriene was synthesized in 71% yield by treating butadiene in benzene solution with a catalyst derived from di-(isobutyl)-aluminum hydride with titanium trichloride.[37]

General copolymer composition equations, relating copolymer composition, in terms of the possible structural units, to monomer concentration have been derived by applying a statistical stationery process for cyclopolymerization of nonconjugated dienes.[38]

Polymeric diacryloylmethane was claimed by reaction of ethyl acrylate with methyl vinyl ketone in presence of sodium methoxide.[39, 40] It was postulated that the monomer was formed by the reaction followed by methoxide initiated cyclopolymerization.

Evidence was presented that cyclopolymerization involving cumulative 1,2- and 1,4-addition could be accomplished.[41] For example, *trans*–1,3,8–nonatriene was polymerized by use of a complex catalyst consisting of triethylaluminum and titanium tetrachloride to the extent of 23%, half of which was soluble in benzene. Symmetrical 1,6-diolefins were polymerized to cyclic polymers *via* the cyclopolymerization mechanism using radical initiation.[42] Other modes of initiation could also be used.

Copolymers of diallylbarbituric acids and vinyl monomers were prepared by use of radical initiators.[1-62] The copolymers were substantially linear as indicated by their solubility, and suggested that cyclopolymerization of the diallyl compound had occurred.

Copolymers of 1,5 or 1,6-diolefins and α-olefins, prepared *via* metal-coordination catalysts, were described.[43] A copolymer containing 20 mole % ethylene units and 80 mole% diene units had no unsaturation detectable by IR spectroscopy. In a study of the effect of chain length on the reaction product, the reactions between diisobutylaluminum hydride and α,ω-dienes, having numbers of carbons from five to eleven were investigated.[44] The results showed that 1,5-hexadiene gave 97.6% methylcyclopentane, whereas, 1,6-heptadiene gave only 0.9% methylcyclohexane. None of the other dienes studied gave any cyclic product. The mechanistic aspects of the reaction were discussed.

Cyclic hydrocarbon polymers were obtained by polymerization of α,ω-diolefinic monomers in the presence of triethylaluminum-titanium tetrachloride.[45] The resulting polymers were soluble, high-molecular-weight materials. Copolymers were also described. Typical polymers were obtained from 1,5-hexadiene, 1,6-heptadiene, and 3-

methyl-1,5-hexadiene. Typical copolymers were obtained from the diene monomers with ethylene, 1-hexene, and 1-octadecene, as well as with each other.

It has been reported that *cis*–1,2-divinylcyclopentane isomerizes reversibly at >300° to the corresponding *trans*-isomer.[46] It was reported in this paper that at 220°, this compound gave an equilibrium mixture containing approximately 5% *cis,cis*-cyclonona-1,5-diene, produced by the Cope rearrangement. Evidence was obtained in support of the six-centered transition state mechanism for the rearrangement.

Perfluro-1,7-octadiene could be polymerized under pressure to give samples of polymer which showed a high degree of cyclization, as indicated by infrared studies.[47]

The cyclopolymerization behavior of multiolefinic monomers such as 1,4-pentadiene and triallylmethane was described and cyclopolymer structures discussed.[48] A homopolymerization study of bicyclo[2.2.1]hepta-2,5-diene (norbornadiene) showed that, by radical initiation, a pre-gel polymer of molecular weight 17,000 showed thermosetting characteristics but was soluble in many solvents.[49] Reactivity ratios of this monomer with vinyl acetate and p-chlorostyrene were reported.

Perfluoro-1.5-hexadiene, perfluoro-1,6-heptadiene, and perfluoro-1,7-octadiene were polymerized under high pressure.[50] The polymers, however, were often insoluble and brittle. Trace impurities were believed to cause interference with the polymerization process, causing crosslinking.

Evidence has been presented that molecules containing a pair of conjugated diene systems, suitably positioned in the molecule to permit 1,4–1,4–cumulative cyclopolymerization, do proceed by such a mechanism.[51] 1,3,6,8–Nonatetraene and 3,6-dimethylene-1,7-octadiene were synthesized and their cyclopolymerization studied. The proposed structures of the two cyclopolymers are shown (2–13):

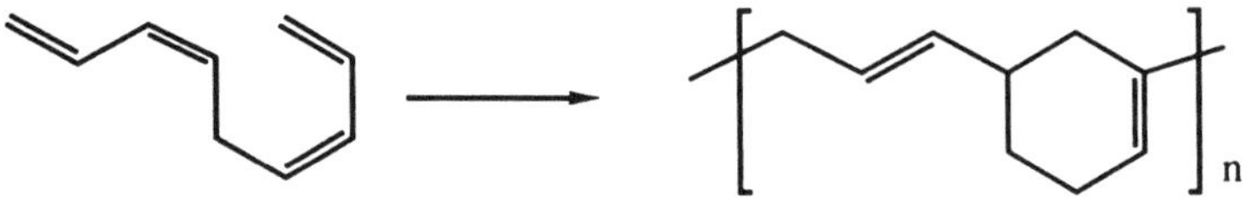

1,3,6,8-nonatetraene

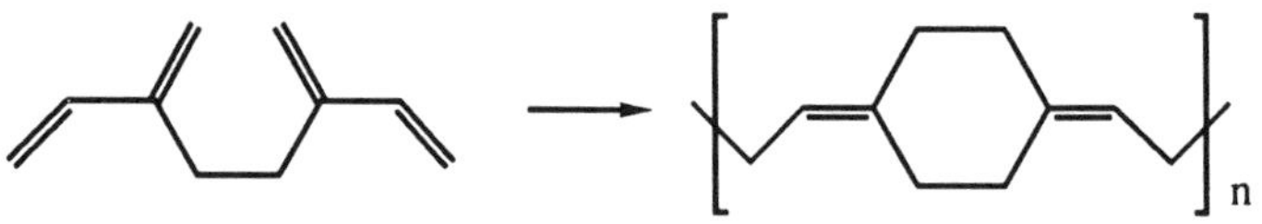

3,6-dimethylene-1,7-octadiene

(2–13)

In a novel study of cyclooxidation of non-conjugated dienes such as biallyl, it was shown that in presence of suitable oxidizing agents, e.g. lead tetraacetate, a two step addition–cyclization took place (2–14)[52]. Diacetoxycyclohexane was also obtained. Diallyl ether and 1,6-heptadiene were found to undergo cyclization to five-membered ring products when treated with radicals derived from iodoperfluoroalkanes.[53]

In a study of 1,5-hexadiene oligomerization equilibrium, a polymerization of the monomer in benzene in presence of titanium tetrachloride gave two oligomers, a solid, isolated by precipitation with methanol, and a liquid, isolated by vacuum distillation of

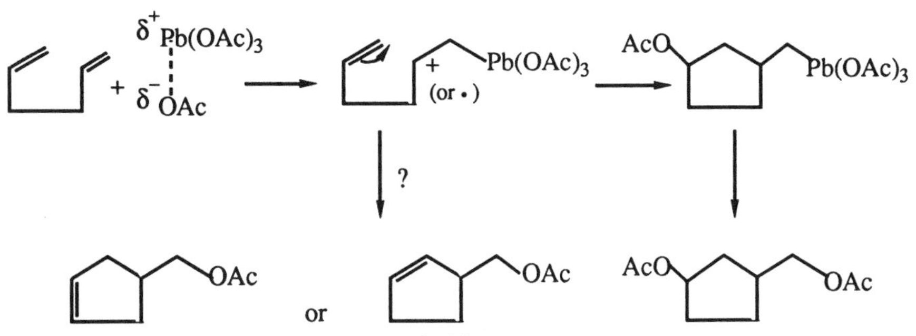

(2–14)

the residue.[54] Structures of the oligomers were not elucidated. However, since the solid polymer was soluble in benzene, it was not crosslinked.

Crystalline polymers of α,ω-hexadienes and heptadienes were described.[55] 1,5-Hexadiene was polymerized in heptane in the presence of tri-(isobutyl)aluminum and either titanium tetrachloride or vanadium trichloride. The poly-1,5 hexadiene and poly-1,6-heptadiene were substantially saturated, highly crystalline film and fiber forming polymers which apparently have the structures, poly(methylene-1,3-cyclopentylidene) and poly(methylene-1,3-cyclohexylidene), respectively, and are not crosslinked.

Cyclization of ethyl diallylacetate or ethyl diallylmalonate to cyclopentane derivatives, during the addition of perfluoroalkyl and trichloromethyl radicals, was shown to occur.[56] No six-membered rings were obtained. A pair of oxygen-bridged cyclooctyl compounds, 2,5-diiodo-9-oxabicyclo[4.2.1]nonane and isomeric 2,6-diiodo-9-oxabicyclo[3.3.1]nonane, were obtained by a transannular reaction of *cis,cis*–1,5–cyclooctadiene with mercuric oxide and iodine.[57]

Allene was converted to a mixture of cyclic trimers, a tetramer, a pentamer, isomeric hexamers, waxes and polyallene *via* treatment with *cis*(triphenylphosphine)nickel.[58] The mechanism of the photochemical cyclization of non-conjugated dienes was discussed. Dimethyl 2,7-nonadiene-1,9-dioates were studied, and the structures of the products were given.[59]

During a fundamental study of the basis for cyclopolymerization, a UV spectroscopic investigation of a series of dimethacrylimides was carried out.[60] The spectra of the dimethacrylimides were compared with those of the N-isobutyrylmethacrylimides. However, it was concluded that interactions of the carbonyls, through the nitrogen atom, precluded attachment of any significance to the spectral differences observed.

Divinyl benzene, butadiene and diallyl maleate have been used in conjunction with γ- or UV-irradiation to produce dimensionally stable, cross-linked polyvinyl chloride films.[61] Dimethyl-*cis*-divinylcyclobutane and *cis*-divinylcyclobutane were prepared from piperylene and butadiene, respectively, in the presence of nickel acetylacetonate, a triarylphosphine and diethylaluminum ethoxide.[62]

Highly substituted sebacic and azelaic acids have been synthesized by free radical addition of isobutyric acid to 1,5-hexadiene and 1,4-pentadiene, respectively, initiated by di-(tert.-butyl) peroxides.[63] Both *cis*- and *trans*-isomers were obtained.

Terminal alkadienes from 1,4-pentadiene through 1,7-decadiene gave mono- and bis adducts from iodoperfluoroalkanes with radical initiators.[64] 1,6-Heptadiene was unusual in preferentially cyclizing under conditions which favored linear adducts from the other dienes. Analogous cyclizations of 1,6-heptadienes had been shown previously to occur in cyclopolymerization reactions.

Diels-Alder polymerization of (9-anthryl)methyl methacrylate or (9-anthryl)methyl propionate at room temperature, in presence of titanium tetrachloride, gave polymers in good yield.[65] Spectral data indicated the presence of cyclic units in the polymers.

Cationic polymerization of 1,4-dimethylenecyclohexane yielded relatively low molecular weight polymers containing appreciable amounts of endocyclic double bonds.[66] The monomer did not undergo appreciable cyclopolymerization, although, structural aspects suggest that it should, except for the steric limitations imposed on the postulated cyclopolymer, accounted for by Bredts rule.

An experiment on cyclopolymerization has been included in a laboratory manual published in Japan.[67] Cyclopolymerization of [E,E-[6.2]-paracyclophane-1,5-diene proceeded by an intra-intermolecular mechanism to give a polymer containing [3.2]-paracyclophane in the repeat unit.[68] Thermal analysis of the polymer indicated a complex thermooxidative behavior in the presence of oxygen, while depolymerization occurred above about 400° in an inert atmosphere.

SYNTHESIS OF POLYMERS CONTAINING HETEROCYCLIC RINGS

A review of cyclopolymerization and telomerization reactions of hetero-unsaturated compounds has been published.[31]

In the early studies on radical initiated polymerization of unsaturated quaternary ammonium salts,[1-1] it was found that monomers containing three allyl double bonds generally produced crosslinked, water-insoluble polymers, that monomers containing two allyl double bonds produced linear water-soluble polymers, and that monomers containing only one allyl double bond failed to polymerize, or polymerized very slowly. The production of linear, soluble polymers from monomers containing two allyl double bonds was explained[1-7] by the cyclopolymerization mechanism. Two of the polymers were degraded by well-known methods.[1-9] Both poly(diallylammonium bromide) and poly(diallyldimethylammonium bromide) gave products resulting from the cleavage of nitrogen-containing rings present in the structures. These results confirmed the cyclopolymerization mechanism, but not the size of the ring.

A wide variety of diallyl monomers, for example, diallyl ether, diallylamine, diallyl-substituted malonic esters and acetoacetic esters, etc. immediately presented themselves as candidates for such a ring forming polymerization process. It is not surprising, then, that after the initial observation was publicized, a flood of novel monomers and their cyclopolymerization occurred.

The following section describes synthesis of heterocyclic polymers *via* cyclopolymerization. As early as 1959, however, it was shown that formation of five- or six-membered rings could also be accomplished from preformed polymers.

During partial decomposition of vinyl polymers, such as polymethylmethacrylamide and polymethacrylic acid, volatile products were given off and cyclic units with primary valences, were formed in the macromolecular chains.[69] The formation of these cyclic structures was shown to increase the thermostability of the polymers.

Electron spin resonance (ESR) spectra were recorded for a number of radicals by interaction of diallylmalonic acid, diallyl ether, diallylamine, and related compounds with amino, hydroxyl and phenyl radicals.[70] The results indicated that the radicals, initially formed by addition at one of the double bonds, rapidly cyclize to produce radicals containing five-membered rings.

Radical induced reactions of perfluoroalkyl iodides with diallyl ether, N-methyldiallylamine, and 1-heptene were described.[71] The diallyl compounds cyclized to produce heterocyclic products.

Polymerization of Nitrogen-Containing Compounds

In a series of studies on the radical initiated polymerization of unsaturated quaternary ammonium salts,[1-2 thru 1-9] it was reported that monomers containing two allyl double bonds produced linear water-soluble polymers. The production of linear, soluble polymers from monomers containing two allyl double bonds was explained[1-9] by the cyclopolymerization mechanism. The polymerization experiments were carried out in aqueous solution, using t-butylhydroperoxide as catalyst. Ammonium persulfate-induced cyclopolymerization[72] of diallyl quaternary ammonium chlorides, bromides, and N-alkyldiallylammonium chlorides in dimethyl sulfoxide (DMSO) has been described. The dependencies of yield and molecular weight of the polymers on the polymerization conditions were examined and quaternary ammonium chlorides were found to have better polymerizability than bromides. The cause was attributed to the ease of oxidation of the bromide ion. Bromide ion is probably oxidized more easily by catalytic persulfate to generate bromine, which can then act as an inhibitor (2–15). Thus, bromide ion inhibits the polymerization, both by consumption of the starting radical and by termination of the growing chain.

$$R\cdot + Br^- \longrightarrow R^- + Br\cdot$$

$$Br\cdot + Br\cdot \longrightarrow Br_2$$

$$R\cdot + CH_2{=}CHCH_2X \longrightarrow RCH_2\dot{C}HCH_2X$$

$$RCH_2\dot{C}HCH_2X + Br_2 \longrightarrow Br\cdot + RCH_2CH(Br)CH_2X$$

(2–15)

Viscosity measurements on the polymers were carried out in 1N sodium chloride aqueous solutions. From the limiting viscosity numbers, it was concluded that some of the poly(diallylquaternaryammonium chlorides) obtained with the ammonium persulfate-dimethylsulfoxide system were of quite high molecular weight.

The solid state polymerization of *N,N*-diallylmelamine has been reported[1-85] to proceed by the cyclopolymerization mechanism (2–16). Samples were irradiated at room temperature in a ^{60}Co source and then thermostatted at higher temperatures to promote the postirradiation reaction. The polymers were soluble in acids, but the intrinsic viscosities were all of a low order. Microcatalytic hydrogenation indicated a residual unsaturation level less than 1% and vinyl absorption peaks, present in the monomer, were absent in the infrared spectra of the polymers. A cyclopolymer, with an intrinsic

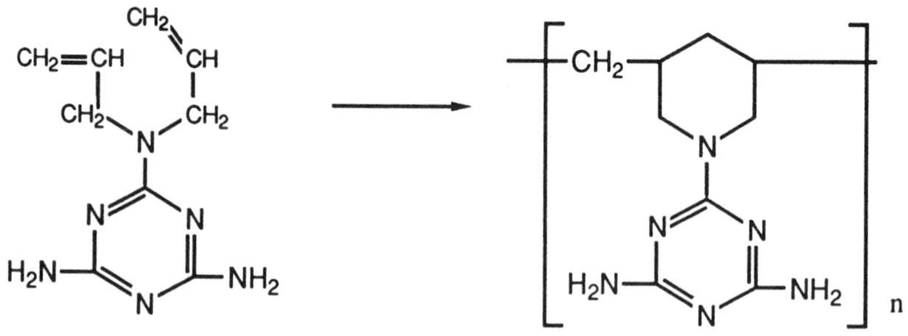

(2–16)

viscosity of 0.20 dl/g, was obtained in 91.7% conversion from a sample of monomer that had received a radiation dose of 360×10^{-5} rad at room temperature and had been given a postirradiation treatment of 96 h at 60°C.

Free radical telomerization of diallylcyanamide, in presence of perfluoroalkyl iodides, has been reported[73] to give *cis*- and *trans*-isomers of 1-cyano-3-iodomethyl-4-(perfluoroalkyl)-methylpyrrolidine.

It was observed[74] that during the radical cyclopolymerization of diallylcyanamide (2–17), the presence of zinc chloride increased the polymerization rate and polymer yield.

The ability of diallyl derivatives of quaternary ammonium salts to undergo cyclopolymerization may be due to the displacement of the electron density of the double bond under the influence of the nitrogen cation. On the basis of this, it was shown[75,76] that the introduction of electron-accepting groups onto the nitrogen atom of diallylamine increased the tendency of the N-substituted diallylamines to polymerize in the presence of radical initiators. Increasing the electron-accepting ability of the substituent promoted polymerization. For instance, for N,N-diallylamides, the influence of the substituent was $ClCH_2CO$- > CH_3CO- > CH_3CH_2-.

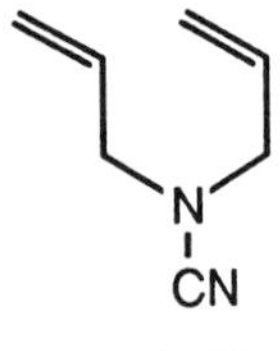

(2–17)

The solubility and fusibility of the resulting polymers and also, the absence of residual unsaturation indicated that polymerization had occurred through both double bonds of the N-substituted diallylamines with the formation of linear cyclopolymers (2–18).

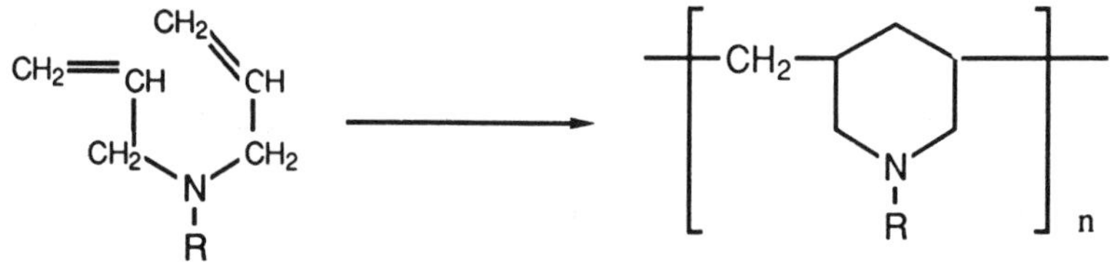

(2–18)

It was established that the capacity of N-substituted derivatives of dimethallylamine for cyclopolymerization is much lower than that of the corresponding N-substituted diallylamines. The cyclopolymerization of some similar compounds has been reported.[77]

Diacrylamide has been polymerized in solution, using a radical initiator.[78] The polymer was soluble in hot dimethylformamide and contained only a small amount of double bond. The six-membered ring was presumed to be present since the polymer showed an infrared (IR) peak characteristic of the carbonyl group in glutarimide (2–19) and was considered to proceed *via* intramolecular-intermolecular propagation.

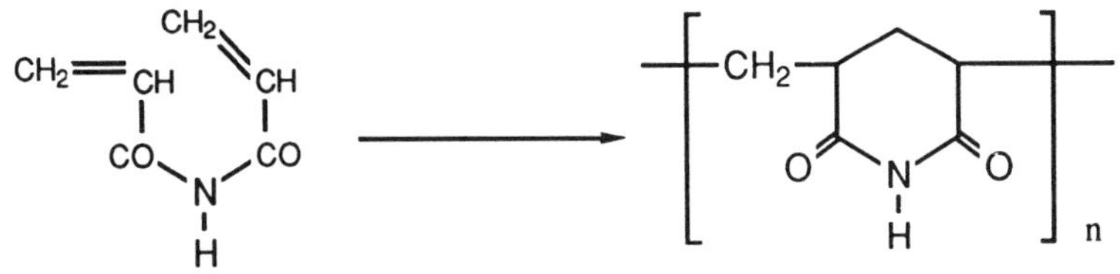

(2–19)

Dimethacrylamide and N-substituted dimethacrylamides have been polymerized as well using free radical initiators. Soluble and fusible polymers, which contain practically no unsaturation,[79] were obtained. These polymers were expected to contain six-membered ring structures. However, the formation of two different cyclic structures is possible (2–20); a five- or six-membered ring, depending on the orientation of the second vinyl group at the intramolecular propagation step, head to tail (2–20a) or head to head (2–20b).

Polymers of similar structure were obtained by partial deamination of polymethacrylamides. This reaction can only give six-membered rings (2–21) from a pure head-to-tail polymer chain.

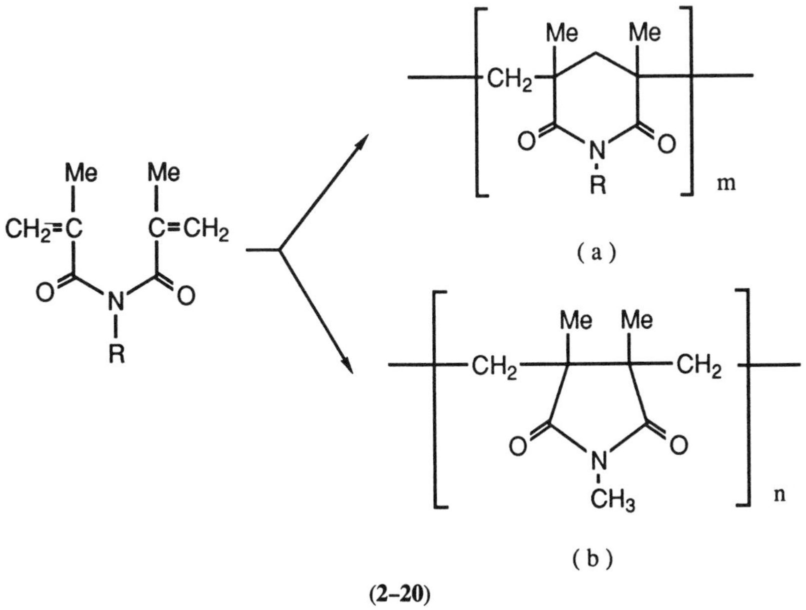

(a)

(b)

(2–20)

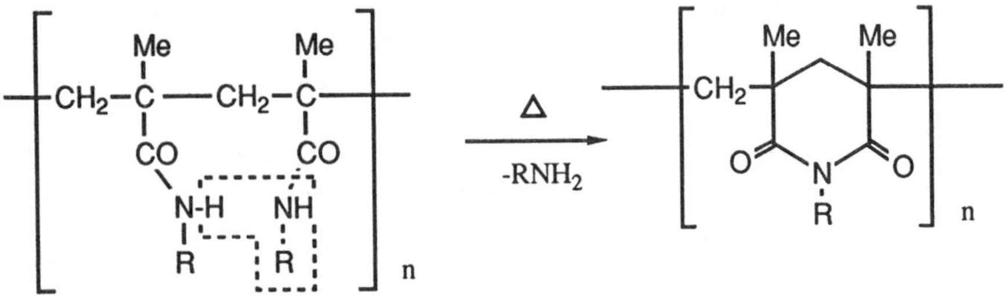

(2–21)

A comparison of the spectra of the polymers, obtained by the polymerization of the monomers of the dimethacrylic series with those of the polymers formed by partial thermal deamination of the corresponding polymethacrylamides, indicated that important differences existed in the carbonyl absorption region. This absorption region for five- and six-membered cyclic imides was investigated. The difference in the absorption for the five-membered cyclic imides and the six-membered ones, which were shifted towards lower frequencies, was 68–76 cm^{-1}. This agreed closely with the difference in the position of the absorption bands of poly(*N*-substituted dimethacrylamides). The difference between the absorption bands in six-membered cyclic imides was 52–55 cm^{-1}, which agreed with the difference in the position of the bands for the polymers obtained by deamination of the corresponding polymethacrylamides. Thus, the polymerization of N-substituted dimethacrylamides yielded cyclopolymers containing predominantly five-membered rings, while the deamination of the corresponding polymethacrylamides gave polymers with six-membered cyclic units.

The infrared spectrum of the sample of polydimethacrylamide, obtained by the polymerization of a 20% solution of dimethacrylamide in DMSO (Table 2–2), had absorption bands in the carbonyl region at 1700 and 1725 cm^{-1} and a band of low intensity at 1775 cm^{-1}. The spectrum of deaminated polymethacrylamide, which consisted almost exclusively of six-membered rings, did not exhibit the absorption band at

Table 2–2 Infrared Spectra of Polydimethacrylamide

Polymerization Conditions or Polymer Source	Carbonyl Absorption (cm^{-1}) 1	2	3
Deaminated polymethacrylamide	1695	1725	
40% solution of dimethacrylamide in dimethyl sulfoxide	1695	1725	1775
20% solution of dimethacrylamide in dimethyl sulfoxide	1700	1725	1775
16% solution of dimethacrylamide in benzene	1698	1725	1762
3% solution of dimethacrylamide in benzene	1700	1725	1775

1775 cm^{-1}. This band was assigned to the presence of some five-membered rings in the polydimethacrylamide chain. The IR spectrum of a mixture of poly(N-phenyldimethacrylamide) and deaminated poly(N-phenylmethacrylamide) had absorption bands at 1675, 1702, and 1767 cm^{-1} due to overlapping of the spectra of the two polymers. Thus, the absorption band at 1775 cm^{-1} was taken as evidence for the presence of a certain amount of five-membered ring structures in the polydimethacrylamide chain.

In contrast to the radical cyclopolymerization, the anionic cyclopolymerization of *N*-methyldimethacrylamide yielded a polymer which consists exclusively of glutarimide units.[80]

It has been reported that an increase in polarity of polymerization medium for cyclopolymerization of dimethacrylamides[81] increases the ratio of five- to six-membered rings in the polymer. The probable conformations of the imide group in dimethacrylamides and their cyclic polymers (See 2–20a and b) were determined by IR spectroscopy. A planar conformation of the imide groups with a transoid orientation of the carbonyl group was observed in solutions of acyclic imides in nonpolar solvents. In polar solvents, conformations with cisoid orientation could also be detected. In dimethacrylamide molecules, where steric factors have a marked effect on the conformation of the imide group, a nonplanar conformation of this group was observed. Increasing the solvent polarity favored the formation of the six-membered over the five-membered ring. The relation between the conformation of the imide group in the monomers and the five:six membered ring obtained from this was discussed.

The radiation–induced cyclopolymerization of *N*–substituted dimethacrylamides (2–22), in the liquid, supercooled–liquid, glassy, and crystalline states has been studied.[82–84] For 2–22a–d, cyclopolymerization occurred completely, predominantly to five-membered rings, in all states except the glassy state, where no polymerization was observed. The proportion of six-membered rings was a little higher in the crystalline state than in liquid or supercooled-liquid states. The cyclopolymerization of 2–22e showed some differences, which were attributed to the ordering effect of the octadecyl group.

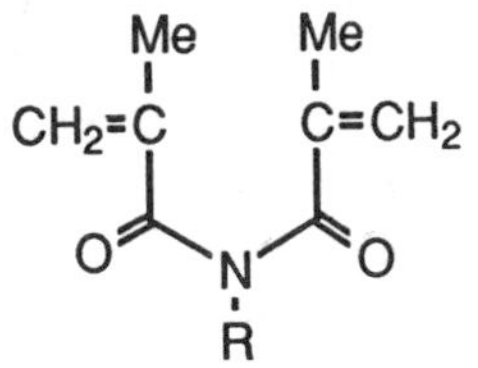

Me Me
CH_2=C C=CH_2
O N O
R

a) R = CH_3
b) R = C_6H_5
c) R = $CH_3CH_2CH_2$
d) R = $C_6H_5CH_2$
e) R = $CH_3(CH_2)_{16}CH_2$

(2–22)

It has been reported that radical cyclopolymerization of *N,N*-dimethyl-*N′*, *N′*-dimethacryloylhydrazine yielded a polymer containing five–membered ring units (2–23) by a head–to–head mechanism.[85]

A cyclopolymerization study of divinylurea[86] to obtain heterocyclic-ring-containing polymers has been published.

Radical cyclopolymerization of N-propyldimethacrylamie has been reinvestigated. Kinetic orders of the polymerization rate with respect to the monomer and the initiator were 1.0 and 0.5, respectively, at concentrations greater than 1.6 M. Absolute values of

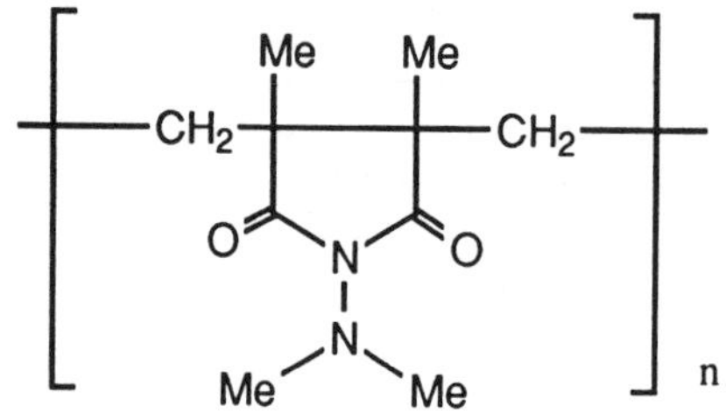

(2–23)

rate constants of propagation and termination reactions were determined by the rotating sector method in bulk and in 50% acetonitrile solution at 30°C. The polymer consisted of five-membered cyclic repeating units almost exclusively. The rate constants for the addition of polymer radical to vinyl monomers and those for the addition of polymer radicals to N-propylmethacrylamide were also evaluated. On the basis of the rate constants obtained, the rate-determining step of propagation was the intermolecular addition of the growing radical to the monomer. The growing radical, formed by intramolecular addition to C:C of the same molecule was as reactive as polymer radicals from resonance-stabilized monomers. The five-membered ring formation was highly favored by the conformation of the monomer, in which coplanarity of the newly formed bond and the semi-occupied orbital of the new radical center was readily achieved. Relatively low reactivity of the polymer radical was accounted for by delocalization of the unpaired electron on the coplanar three carbon atoms. Radical addition of 2-methyl-2-propanethiol to the monomer gave only the substituted succinimide. The five-membered ring formation in the addition reaction was due to the favorable conformation of the monomers[87]

Other studies on dimethacrylamide and N-substituted dimethacrylamides have provided evidence that, in the solid state, the monomers preexist in a conformation favorable for cyclization.[88,89] The thermal or photochemical polymerization of $RN(COCMe:CH_2)_2$ ($R-CH_3$ or C_6H_5) in benzene gave polymers containing five-membered rings and some unsaturation. The polymerization of the monomer ($R = H$) gave polymers containing five and six-membered rings and linear imides. Polymers, prepared by the photochemical polymerization of the three monomers in a solid matrix (PVC film), had structures analogous to those of polymers prepared in benzene. The polymers and the IR spectra of the monomers indicate that the monomers have a conformation favorable for ring formation during polymerization. Polymerization of N-phenyldimethacrylamide and N-methyldimethacrylamide in poly(vinyl chloride) film gave polymers having five-membered ring recurring units without unsaturated side groups. The results suggested that dimethacrylamides preexist in a conformation favorable for cyclization and the polymerization medium exerted no effect on the structure of the polymers.

A performance-structure relations study of electroconductive polymers, including the cyclopolymer of dimethyldiallylammonium chloride[90] has been published. The three main classes of water-soluble polymers (nonionic, anionic, and cationic) were compared for their electroconductive response on paper. Cationic quaternary ammonium polymers were the most electroconductive, particularly at low relative humidities. Of the eight structural types of quaternary ammonium polymers studied, the cyclopolymer of dimethyldiallylammonium chloride had optimum functional properties for application to electrographic papers.

Some of the experimental procedures for making homopolymers of dimethyldiallylammonium chloride,[91] along with a discussion of the experimental variables, such as effects of catalyst ratio, monomer concentration, and impurities, have been summarized. Also, some of the procedures used for evaluation of the polymers in flocculation and sludge dewatering were discussed.

Synthesis of diallylamine derivatives[92] with the general formula (2–24), in which R_1 and R_2– are H or CH_3, R_3 and R_4 are H, C_{1-8} alkyl, C_6H_5, substituted phenyl, benzyl, substituted benzyl, or cyclohexyl groups or $R_3 \ldots R_4$ is piperazine, morpholine, or N-alkylpiperazine, and X is Cl, Br, or iodine has been reported. The monomers were polymerized in DMSO to give H_2O-soluble, linear quaternized polyamines (2–24) useful as water-soluble paints, adhesives, dye additives, surfactants, flocculation accelerators, etc.

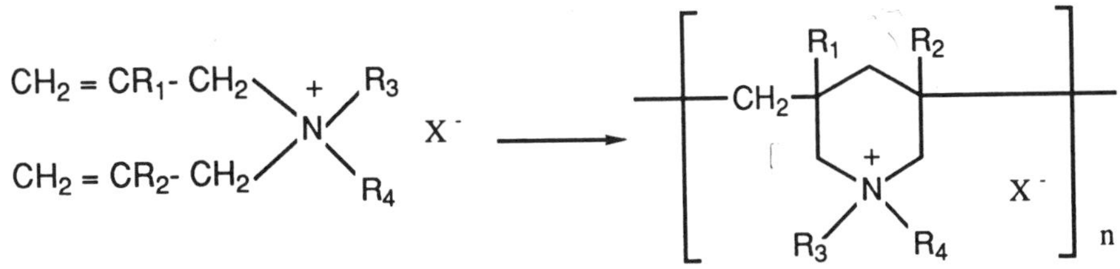

$CH_2 = CR_1\text{-}CH_2$, $CH_2 = CR_2\text{-}CH_2$, $\overset{+}{N}$, R_3, R_4, X^- → CH_2, R_1, R_2, $\overset{+}{N}$, R_3, R_4, X^-, n

(2–24)

The cyclopolymerization of various diallylamines[93–95] has been investigated and thermally regenerable ion-exchange resins were prepared. Reports on methods for the determination of residual unsaturation were also published. The cyclopolymerization of dialkyldiallylammonium halides has also been studied by others.[96–98]

The preparation and cyclopolymerization of *cis*- and trans-1,3,5-triisocyanato-cyclohexane[99] have been described. The initiator used was NaCN in dimethylformamide (DMF). Both monomers led to polymers of predominantly bicyclic structures as shown below (2–25):

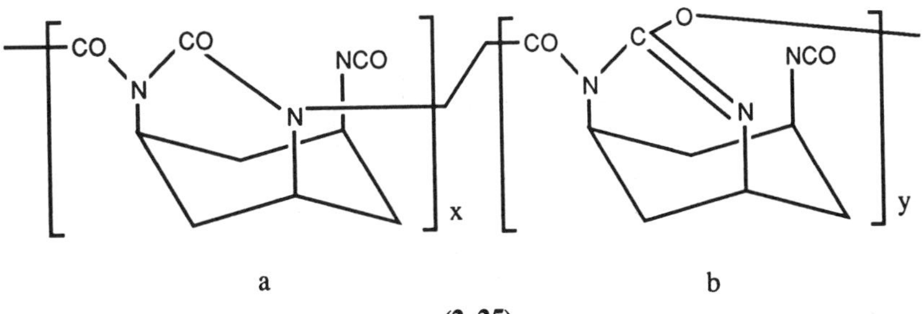

CO, CO, N, N, NCO, x; CO, C, O, N, N, NCO, y

a b

(2–25)

Cyclic polymers[100] having functional side chains that undergo various chemical reactions, including chelate formation, prepared by the free radical polymerization of quaternary ammonium monomers, have been disclosed. The functional side chains react to modify the structure and the chemical and physical properties of the polymer fundamentally. Thus, diallyl(carbethoxymethyl)–methylammonium bromide yielded solid polymer of structure (2–26):

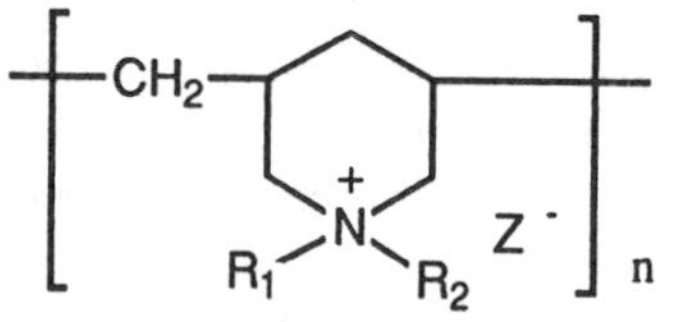

(2–26)

($R_1 = CH_3$, $R_2 = CH_2CO_2C_2H_5$, $Z = Br$). These polymers formed cupric chelates after heating in an aqueous suspension of $CuCO_3$. The use of methyl 3-(diallyl-amino)propionate hydrochloride, diallyl-[ω-(carbomethoxy)undecyl]methylammonium bromide, diallyl-[1-(carbethoxy)undecyl]methylammonium bromide, diallyl-(cyano-methyl)methylammonium chloride, and N,N-diallyl-α-aminobutyric acid for this purpose was demonstrated.

Amphoteric polyelectrolytes of high molecular weight of structure 2–27 were

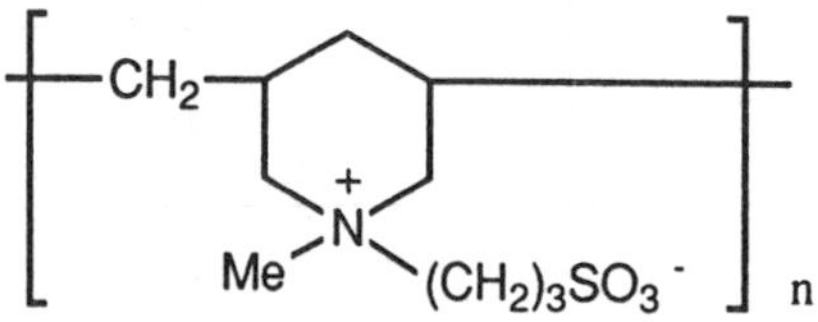

(2–27)

obtained by radical–initiated polymerization of the monomer obtained by reaction of methyldiallylamine with propane sultone.[101] Other similar monomers and polymers were also reported.

It has been reported that in radical polymerization of $C_6H_5CH_2NHCH_2CH = CH_2$, $C_6H_5CH_2N(CH_2CH = CH_2)_2$ and their HC1 salts, $(CH_2 = CHCH_2)_2(C_6H_5CH_2)C_2H_5N^+Br^-$, and $(CH_2{:}CHCH_2)_2(C_6H_5CH_2)CH_3N^+I^-$ in the presence of azodiisobutyronitrile,[102] the lone pair of electrons on the secondary and tertiary N atoms readily form complexes with radicals to promote α-H abstraction and conjugate with the resultant allyl radicals. On the other hand, the lone pair of electrons in the ammonium derivatives do not conjugate with allyl radicals and increase the Coulomb integral of α-C atoms. The polymerization rate of diallylmethylbenzyl ammonium iodide in methanol and acetone suggested the formation of intermediate cyclopolymers, leading to degradative chain transfer. Synthesis and cyclopolymerization of the N,N-diallyl- and N,N-dimethallyl derivatives of methanesulfonamide and ethanesulfonamide[103] have been reported. These monomers gave low molecular weight cyclopolymers containing predominantly six-membered rings and no residual double bonds, as evidenced by their comparison with N-(alkylsulfonyl)piperidines as model compounds.

Since the initial investigations, it has been shown[1-19, 1-156] that a wide variety of 1,6–dienes can be polymerized to soluble, saturated polymers, and structures having five- and/or six-membered rings have been proposed.

Cyclopolymerization of diallylamines has been reviewed by several authors[1-20, 1-169, 1-170] and the papers presented at two symposia on this topic have been published.[1-157] The effect of β–substituents on the cyclopolymerization of diallylamines has been studied.[104] Electron spin resonance studies and analysis of products[105-107] showed that the piperidine ring content increased with the bulk of the β-substituent, which caused a small decrease in basicity of the polymers (2–28).

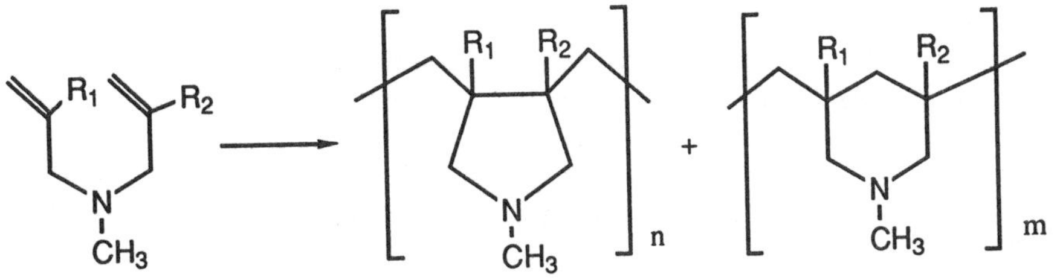

(2–28)

A large number of diallylamines, with a variety of polar and potential metal chelating groups as substituents on the nitrogen atom, have been prepared.[108–111] Their ability to form cyclopolymers and the structures and some properties of these polymers were investigated. For cross-linking these polyamines, it has been shown[112] that use of 1,6-bis(diallylamino)hexane (2–29) allowed optimum control of cross-linking, PK_a values, and ion-exchange capacity.

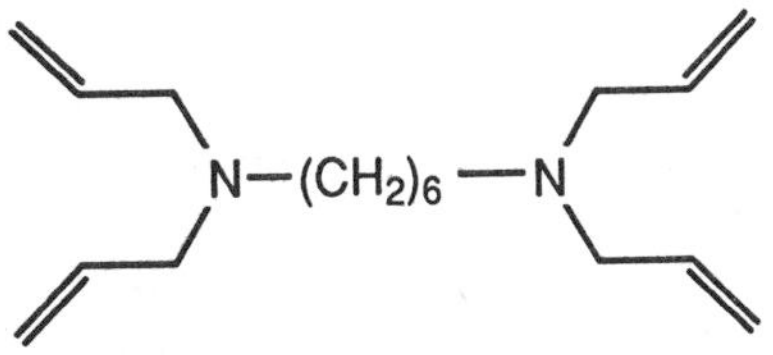

N—$(CH_2)_6$—N

(2–29)

A novel monomer, 4-(diallylamino)pyridine, has recently been reported and its cyclopolymerization studied. The polymers were water-soluble. The absence of NMR peaks due to vinyl groups in the spectrum of the polymer indicated a high degree of intramolecular reaction during polymerization. The monomer was also copolymerized with diallylamine hydrochloride and dimethyldiallylammonium chloride.[113]

Another novel monomer, N,N-diacryloylmetribuzin, synthesized by treating the herbicide, metribuzin, and acryloyl chloride in the presence of $C_2H_5)_3N$ was cyclopolymerized using free-radical initiators to form soluble polymers containing no residual unsaturation measurable by IR.[114]

Numerous recent references deal with the synthesis of the commercially important monomer, diallyldimethylammonium chloride (DADMAC) or other salts,[115-118] its polymers or copolymers,[98, 119–131] kinetic and mechanistic studies on the polymerization process,[132–134] and analytical procedures.[135] Also, a study of the polymerization process for monoallyl and diallylamine derivatives has been reported.[136] A variety of unsaturated quaternary ammonium compounds, functionally capable of undergoing free-radical initiated cyclopolymerization or thermosetting polymerization to give crosslinked polymers suitable as ion exchangers, were described.[137]

In an attempt to synthesize 1,4-bis(diallylamino)butane and 1,5-bis(diallylamino)pentane in which 1,4- and 1,5-dihalo- and di(arylsulfonoxy)alkanes were treated with diallylamine, the predicted products were not obtained.[138] Instead the products of the reactions were allylpyrrolidine and allylpiperidine, respectively. The products of the reaction were explained on the basis of an intramolecular cyclization to the five- or six-membered cyclic quaternary ammonium salt, followed by allylation of excess diallylamine to produce triallylamine and the respective allyl-substituted heterocyclic amine.

A process for polymerizing vinyl compounds containing a basic nitrogen atom was described. Diallylamines were claimed as comonomers, although no examples utilizing such monomers were given.[139] Linear homopolymers of diallylamines were obtained by radical initiation of a wide variety of substituted diallylamines.[140] A cyclopolymer of diacrylimide was obtained by radical initiated polymerization of the monomer.[141]

The cyclopolymerizable monomer, N-methyldimethacrylamide, was synthesized for polymerization studies by heating methacrylic anhydride or methacryloyl chloride with N-methylmethacrylamide in presence of a polymerization inhibitor.[142] Linear, thermoplastic poly(diallylamines) were obtained by radical initiation of aqueous solutions of the hydrochloride salts of the amines.[143] The syntheses of the cyclopolymerizable monomers, 1,3-divinylparabanic acid and 1,3-divinylparabanic acid 4-imide were described.[144] A variety of unsaturated quaternary ammonium salts was synthesized and characterized.[145] An unsaturated quaternary ammonium salt, containing an asymmetric nitrogen atom, and one containing an asymmetric carbon atom, were resolved to obtain the detrorotatary isomer in each case.

Polymerization of N-substituted dimethacrylamides results in cyclic polymers containing predominantly five-membered rings, while deamination of the corresponding polymethacrylamide forms repeating six-membered cyclic units.[146] Diallylamine was complexed with zinc chloride and polymerized and copolymerized with other polar monomers *via* radical initiation.[147] Polymerization occurred rapidly at low temperatures, and to high conversion, in contrast to a similar experiment without zinc chloride.

N,N-Dimethacrylmethacrylamide was synthesized for the purpose of photolysis to the adamantane-like structure shown in 2–30.[148] However, the products apparently were

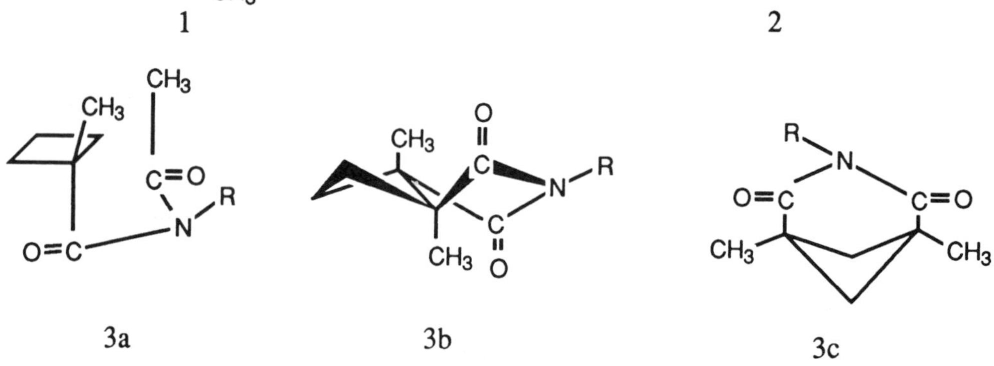

1 2

3a 3b 3c

(2–30)

a mixture of two or more of the cyclobutane structures shown, rather than the adamantane-like structure predicted. Cyclopolymerization of diallylamine and vinylpyridines *via* radical initiation is reported to yield a copolymer with recurring piperidene ring units.[149] This copolymer, when blended with polypropylene, gave a dyeable fiber with good receptivity to acid, cationic and disperse dyes.

Poly(N-phenyldimethacrylamide) was synthesized by several methods, including free-radical initiation.[150] All polymers had superimposable IR spectra. Evidence was presented to support the following structure for the polymer (2–31).

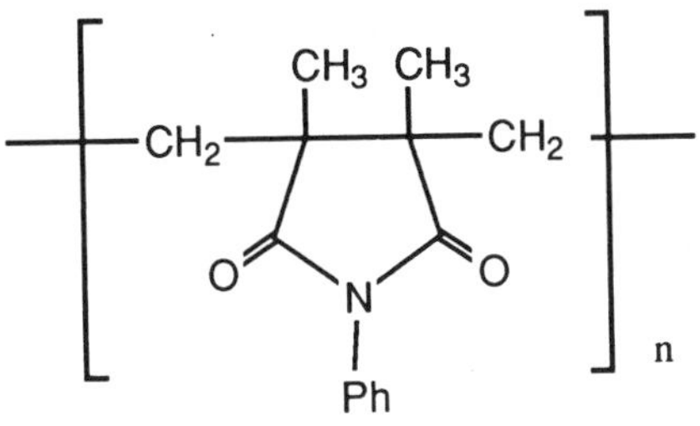

(2–31)

Anionic cyclopolymerization of N-methyldiacrylamide was studied.[151] When toluene was used as solvent, precipitation polymerization occurred. However, in tetrahydrofuran at low temperatures, solution polymerization occurred.

The polymerization of diallyldimethylammonium chloride as a flocculation agent was described.[152] Copolymers of this monomer with acrylamide were also described. Water-soluble linear polymers were obtained by polymerization of monomers containing two olefinic groups separated by three atoms, one of which was a quaternary ammonium nitrogen atom.[153] A variety of cyclopolymers from monomers of this type was described.

A variety of commercial products has been developed utilizing cyclopolymerization either in the form of the homopolymers of diallyldimethylammonium chloride or its copolymers with acrylamide.[154] Polymers of diallyldimethylphosphonium chloride, diallyldiphenylphosphonium chloride, benzyldiallylsulfonium chloride and a copolymer of diallyldimethylammonium chloride and diallyldimethylammonium bromide were prepared by radical initiation.[155] A study of radical-initiated polymerization of N,N-divinylureas led to the conclusion that the predicted cyclopolymer structure was not obtained, probably, since evidence shows that such monomers react only through their tautomeric forms.[156]

A method of purification of quaternary ammonium monomers, such as dimethyldiallylammonium chloride, which involves a vacuum stripping step, a steam stripping step at a regulated pH, and an activated carbon treatment was described. Improved properties of the polymers obtained from monomers purified by this procedure were observed.[157] Triallylamine polymers, suitable as weakly basic quasispherical ion exchange resins of 0.5–10 particle size, were prepared by polymerization of triallylamine salts of common inorganic acids, either alone or with such comonomers as methyldiallylamine salts or methyltriallylammonium salts; polymerization was induced by ionizing radiation of 5–15-megarads at 10°–40°.[158]

The influence of the polymerization conditions on the polymer structure of 1,1′-divinylferrocene, when cationic and radical initiators were used, has been investigated.[159] The polymers obtained with cationic initiators contained 70–80% cyclized units as well as uncyclized units. Radical initiators led to polymer having more than 96% cyclized units.

A variety of polymers and copolymers of dimethyldiallylammonium chloride, along with a discussion of methods of synthesis, properties and applications were discussed in a 1970 review, which included 375 references.[160] Linear N,N-diallylurea polymers were synthesized from the monomer by radical initiation; copolymers with other vinyl monomers, containing >3% of this monomer, were also obtained. All yielded soluble products.[161]

A polymerization study of 1,3-divinylimidazolid-2-one and 1,3-divinylhexahydropyrimide-2-one led to the conclusion that no cyclopolymerization occurred.[162] N,N-diphenyl-N,N-divinylurea could not be polymerized under a variety of conditions and initiators. In a study of the imidization reaction in poly(vinylamides), it was shown that the products varied in their nitrogen contents from the theoretical value to considerably lower values, depending on the method used.[163] The low values were ascribed to loss of ammonia *via* intramolecular imide formation as shown in the structures below (2–32). Such structures were analogous to those obtainable by cyclopolymerization of the diacrylimides.

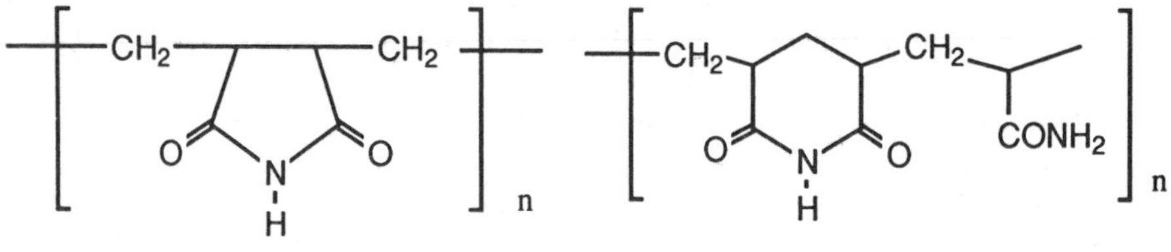

(2–32)

Linear poly(N,N-diallylureas) were treated with paraformaldehyde to give a resin, which would undergo crosslinking with acid.[164] Linear poly(N,N-diallylcyanamide), prepared by bulk polymerization with x-rays, was treated with hexamethylenediamine to give a thermostable resin.[165] Cyclization of N-substituted diallylamines to pyrrolidine derivatives during the radical addition of perfluoroalkyl iodides was shown to occur.[166]

Cyclopolymers, analogous to those derived *via* cyclopolymerization of N-alkylbismethacrylimides with radical initiation, were synthesized by different procedures.[167] A variety of other synthetic procedures were claimed in the patent; also, the cyclopolymers were crosslinkable. Polymers, containing N-substituted acrylamide and methacrylamide units useful as flocculants, were prepared by reaction of polyacrylimide, polymethacrylimide or their copolymers with primary amino compounds containing other functionality.[168] An electron spin resonance study of irradiated N-methyldimethacrylamide was carried out, and the spectra of the propagating radicals showed an intramolecular cyclization mechanism which proceeded stepwise.[169]

Salts of diallylamine derivatives, e.g. dimethyldiallylammonium chloride, were irradiated in presence of photosensitizers by a fluorescent lamp to yield high conversion to water-soluble polymers.[170] N-Propyl-, N-methyl, and N-phenyldimethacrylamides were converted to the respective cyclopolymers in presence of AIBN.[171] The relative rates of polymerization were measured, and the reactivity ratios with methyl methacrylate were determined.

Diallylhydroxylamine hydrochloride was cyclopolymerized in presence of ammonium persulfate to yield a polymer postulated to contain N-hydroxypiperidine rings.[172] The monomer was also copolymerized with the hydrochlorides of methyldiallylamine, benzyldiallylamine, triallylamine, and o-allyldiallylhydroxylamine. Phosphoric acid amides containing at least two allyl or methallyl groups were cyclopolymerized by radical initiation with sulfur dioxide to give polysulfones (2–33).[173]

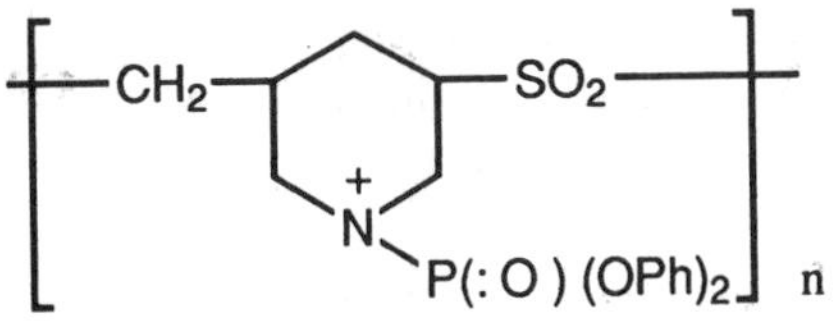

CH_2 SO_2 $\overset{+}{N}$ $P(:O)(OPh)_2$ n

(2–33)

The hypothesis of a highly cyclized polymer formed from symmetric unconjugated dienes has been confirmed using sym-dimethacryldimethylhydrazine and sym-dimethacryloylhydrazine.[174] The IR spectra of the polymer showed no presence of C:C bonds, and 94.5% cyclization was suggested from NMR spectra.

Polymers containing cyclo-1,1-diallylguanidine linkages have been studied.[175] The polymerization of 1,1-diallylguanidine salts, in the presence of peroxides and glyoxalation, gave H_2O soluble cationic polymers containing cyclic linkage for use in wet strengthening of paper and as flocculant for solids suspended in aqueous medium.

Poly(triallylamine hydrochloride) and similar polymers were obtained by polymerizing the monomers in suspension, solution or bulk with radical or redox initiators to produce polymers useful in anion exchanger manufacture.[176] The ESR spectra of diallylamines and related compounds treated with hydroxyl or amino radicals were assigned to five-membered ring radicals as the major radical species present; however, the dimethallylamine series gave both five- and six-membered cyclic radicals.[177]

Based on ^{13}C and lanthamide induced physical shift data, the aminoketone formed by treating N-methyl-N,-N-diallylamine with cyanoisopropyl radicals was 2,2,6-trimethyl-*cis*-perhydroisoindolid-5-one. The ^{13}C NMR spectrum in the presence of acid indicated the existence of two protonated forms of the compound.[178]

The effects of ionic strength, molecular weight and temperature were examined for the Huggins constant of poly(dimethyldiallylammonium chloride) in aqueous solutions containing sodium chloride.[179] Aqueous solutions of diallyldimethylammonium fluoride were polymerized in presence of free-radical initiation to yield high molecular weight polymers, with intrinsic viscosity of 0.57 dl/g.[180]

The unperturbed dimensions of poly(dimethyldiallylammonium chloride) were determined by light scattering on aqueous solutions of the polymer containing different concentrations of sodium chloride.[181] The results were in agreement with those calculated using standard literature methods.

The synthesis and polymerization of N-allylamines has recently been studied.[182] A mixture of two equivalents of allylchloride with methylamine, when heated with hydrated silica, was postulated to produce a copolymer which could be converted to a solid residue by heating with epichlorohydrin.[183] The dilute solution properties of poly(dimethyldiallylammonium chloride) in water and methanol containing tetraethylammonium chloride were further investigated by light scattering.[184] The characteristic dimensions and conformation of the polymer chain were calculated.

The reactivity of allyl monomers ($CH_2{=}CHCH_2R_1$, where R = NH_2, $N(CH_3)_2$, $NHCH_2CH{=}CH_2$, CH_3CO_2, OH, or $OCH_2CH{=}CH_2$ was studied by irradiation with γ-rays at $-196°$, in order to determine radical behavior during gradual thawing, which controls the postpolymerization of the monomer.[185] Reactivity was found to be dependent upon the nature of the functional group, both in the presence and absence of protonic acids. Protonation of monomer functional groups, having a high pK value ($R{=}NH_2$, $N(CH_3)_2$, $NHCH_2CH{=}CH_2$), changed the polarizing action so that degradative chain transfer was suppressed and low temperature postpolymerization was possible.

A method of purification of aqueous solutions of dialkyldiallylammonium chlorides involved pH adjustment to 10.5–11.5 after steam distillation at 110°, and finally, the residue was passed through an activated carbon column.[186] Improved polymerization properties were observed. Conformational effects on the cyclopolymerization of N-(p-bromophenyl)dimethacrylamide were studied by x-ray diffraction techniques.[187]

An improved poly(diallyldimethylammonium) chloride was obtained when the allyl chloride used in monomer synthesis was purified by filtering and washing with water prior to use.[188] The pH dependent changes in poly(diallyldimethylammonium chloride) aqueous solutions were attributed to conversion of sulfate ion impurities to bisulfate ions at low pH and resulted in a different electrophoretic mobility and intrinsic viscosity at low ionic strengths.[189] The relation between electrophoretic mobility and molecular weight of poly(diallyldimethylammonium chloride) was investigated using electrophoretic light scattering. The electrophoretic mobility of the polymer was heterogeneous, indicating a distribution of mobilities.[190]

The homopolymerization and copolymerization of 1,1′-bis(methoxycarbonyl)divinylamine were studied in a variety of solvents using AIBN or boron trifluoride-etherate as initiator.[191] The monomer polymerized with both initiators but did not polymerize in presence of butyl lithium. It also copolymerized with styrene by radical initiation but not with methyl methacrylate or butyl vinyl ether. No gelation occurred in the homopolymers, and a nine-membered bicyclic ring structure was proposed for the polymer.

A method for separating sodium ions from solutions of quaternary diallylammonium salts involved precipitating the sodium ion as sodium carbonate.[192] The method included addition of a saturated solution of a controlled mixture of ammonium bicarbonate, potassium bicarbonate and potassium carbonate, resulting in removal of up to 63% of the sodium ion as a precipitate of sodium carbonate.

A continuous polymerization process for dialkyldiallylammonium salts has been described.[193] The process involves use of 30–70% solutions of the monomer in presence of buffer salts to maintain the pH at 6.7–10.3 so that product yields are greater than 90%.

A polyelectrolyte complex membrane from a polyanion and poly(diallyldimethylammonium) chloride has been prepared and found to be useful in biological systems.[194]

Cyclopolymerization of α,α'-dimethoxycarbonyldivinylamine using AIBN as an initiator gave a polymer consisting essentially of structure 2–34.[195]

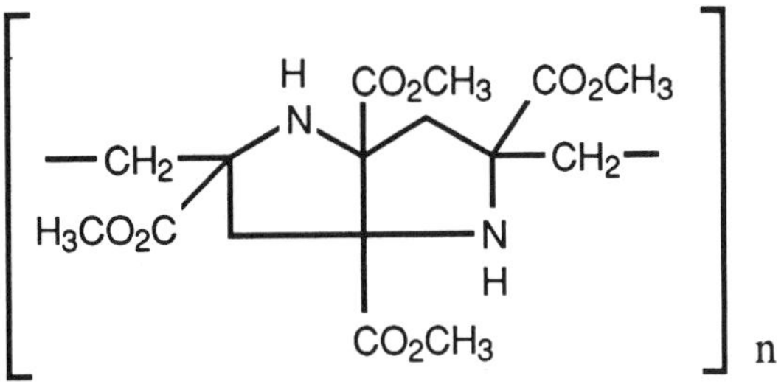

H CO2CH3 CO2CH3
N
—CH2 CH2—
H3CO2C N
H
CO2CH3 n

(2–34)

The kinetics of polymerization of diallyldimethylammonium chloride was studied.[196] The polymerization, initiated by ammonium persulfate, followed kinetics of 2.3 order in monomer and 0.47 order in initiator with overall activation energy of 15.4 kcal/mole. Comparison of the experimental kinetic equation with the theoretical equation for polymerization of unconjugated symmetrical dienes suggested that polymerization occurred with partial cyclization and chain termination of cyclized radicals. The K and a constants of the Mark-Houwink equation for the polymer were found to be 12.6×10^{-4} and 0.51, respectively.

A study of polymerization of N,N-dimethyl-3,4-dimethylenepyrrolidinium bromide, both in presence and absence of initiators has been published.[197] In aqueous solutions of 10–50% monomer, polymerization at 60° showed that conversion time decreased with increased monomer concentration; the molecular weight of the polymer increased with monomer concentration.

Diallylamine was treated with epichlorohydrin to give an intermediate which could be converted to diallylazetidinium salts.[198] These salts (2–35) were converted to homo-

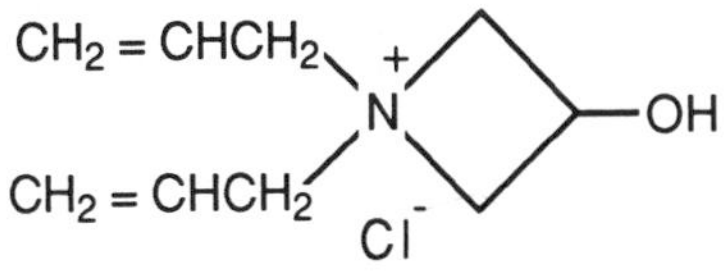

CH2 = CHCH2 +
N —OH
CH2 = CHCH2
Cl⁻

(2–35)

polymers and copolymers of acrylamide, useful in oil field technology. A discussion of the preparation, structure and flocculation capability of poly(N,N-dimethyl-3,4-dimethylenepyrrolidinium bromide) has been published.[199]

A colloid-titration system was examined in which a cationic polyelectrolyte [poly(diallyldimethylammonium chloride), molecular weight 50,000] was titrated with an anionic polyelectrolyte, [poly(vinyl alc.) sulfate K salt] under various conditions, and the characteristic of this system was analyzed in terms of the process of polyion-complex formation.[200] Deviation of the interaction from stoichiometry (excess cation concentration) provides a positively charged polyion-complex.

A single crystal x-ray diffraction and solid-state polymerization study of N-(p-bromophenyl)dimethacrylamide has led to the conclusion that the monomer crystallized in a conformation which slightly favors the intramolecular over the intermolecular polymerization mode.[201]

Poly(diallylamine hydrochloride) was synthesized as an agent for improving dye fastness in textile fibers.[202] A discussion of the synthesis of dimethyldiallylammonium chloride as a prospective monomer for synthesis of water soluble polyelectrolytes has been published.[203] Studies of the kinetics of radical polymerization of diallyldimethylammonium chloride has been continued.[204] The overall rate of polymerization was found to be proportional to the 0.75 and third power of initiator and monomer, respectively.

The effect of micellar aggregation on polymerization and the polymer products of allyldimethyldodecylammonium bromide was studied by polymerizing the monomer under both micellar and isotropic conditions.[205] The monomer was effectively polymerized by X-ray irradiation in the micellor state to yield a polymer of mean molecular weight of 11,000 $\pm$1,000 consisting of a mixture of head-to-tail and head-to-head configuration in the ratio of 85:15.

Mossbauer spectra of potassium ferrocyanide complexes with poly(N.N-diallyldimethylammonium chloride) were studied to aid in interpretation of the interaction between the two.[206] High-molecular-weight polymers of diallyldimethylammonium chloride have been synthesized by continuous addition of methanol solutions of azo compounds to aqueous solutions of the monomer.[207] Poly(N,N-diallylethylammonium phosphate) has been prepared as an antistatic agent for textiles.[208]

A polymeric pyrrolidinium methanesulfonate viscosifier was prepared from diallyl-3-bromopropylamine. A 1 (wt) % solution of the polymer containing 10% or 20–25% sulfonate in brine (7–60 wt. % salt) had viscosity = 1.21–2.79 and 2.07–3.82 Cp, respectively.[209]

The thermal and photoinduced polymerizations of 1,3-bis(p-vinylphenyl)propane were carried out by using tetracyanoethylene as a strong electron acceptor in several solvents to initiate charge-transfer cyclopolymerization.[210] The cyclic polymer contained [3.3]paracyclophane units in the main chain. The rate of polymerization increased with increasing polarity of the solvent, consistent with charge-transfer initiation. All available evidence indicated that both polymerizations proceeded by a cationic mechanism.

Diallyldialkylammonium bromides were prepared and their polymerizability as oriented molecular assemblies in aqueous solution was studied.[211] Films of the salts on a glass wall were incubated by adding a phosphate buffer soluion to give giant spherical vesicles with diameters ranging from 10 to 100 M. The vesicles were polymerized by UV irradiation.

4-(Diallylamino)pyridine was cyclopolymerized for 36 hours in presence of 2.2′-azobisamidinopropane hydrochloride.[212] The new monomer was also cyclocopolymerized with diallylamines or dimethyldiallylammonium chloride. The copolymers were random, and a high degree of intramolecular cyclization occurred in both the homopolymer and the copolymer.

Synthesis of the vesicle-forming quaternary ammonium salts, allyldidodecylmethylammonium bromide and diallyldidodecylammonium bromide was reported, and topochemical polymerization of the monomers was accomplished by x-ray irradiation.[213] Only polymerized vesicles, which resulted from the diallyl monomer, retained the structure of the monomer vesicles and also, exhibited higher stability.

Radical polymerization of N,N-diallyl-N-methyl-N-(carbisopropyloxymethyl) ammonium chloride was studied in aqueous and methanolic solutions at 35°-60°.[214] The polymerization was 0.5 order with respect to initiator and the rate of polymerization increased sharply and nonlinearly with increasing initial concentration of monomer. The initial rate of polymerization was independent of the pH and concentration of sodium chloride in the reaction media. Bimolecular termination without degradative transfer to monomer was observed.

The kinetics of radical polymerization of N,N–dimethyldiallylammonium chloride was studied in aqueous solutions of 2–5 mol./L at 60° and low monomer conversion values. The reaction order was 0.5 with respect to initiator and showed no degradative chain transfer to monomer.[215] The initial viscosity of monomer solutions had a significant effect on the rate constant of bimolecular chain termination.

Vesicles of diallyldidodecylammonium bromide were prepared using a ^{60}Co source.[216] The polymerization rate was compared to that of allyldidodecylmethylammonium bromide. The conclusion was drawn that the diallyl compound polymerized *via* the cyclopolymerization mechanism whereas the monoallyl compound polymerized *via* an addition mechanism.

Polymers containing 4-N-pyrrolidinopyridine were prepared by cyclopolymerization of N-4-pyridylobis-methacrylamide in aqueous solution followed by lithium aluminum hydride reduction.[217] Cyclopolymerization of 4-(diallylamine)pyridine hydrochloride *via* free radical initiation gave water-soluble polymers.[218] The palladium-catalyzed oxidation of poly(N-phenyl-3,4-dimethylenepyrrolidine) which was prepared by AIBN-initiated polymerization of N-phenyl-3,4-dimethylene-pyrrolidine yielded thermally stable poly(N-phenyl-3,4-dimethylenepyrrole).[219] Hydrogenation and reduction of the latter polymer by hydrogen-platinum oxide and diimide, respectively, failed to yield poly(N-phenyl-3,4-dimethylenepyrrolidine).

Cyclopolymerization of N,N–diacryloyl metribuzin (2–36) and copolymerization of this monomer with acrylamide and acrylic acid were studied.[220] An aqueous buffered system to release metribuzin at a controlled rate was developed.

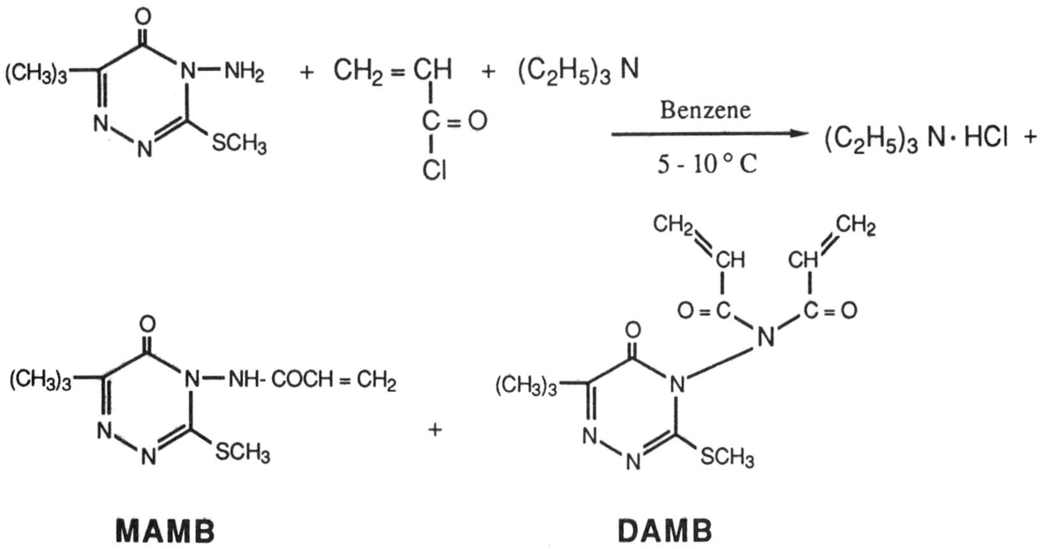

$$\text{Metribuzin} + CH_2{=}CH{-}C({=}O){-}Cl + (C_2H_5)_3N \xrightarrow[5\text{-}10\,^\circ C]{\text{Benzene}} (C_2H_5)_3N\cdot HCl + \text{MAMB} + \text{DAMB}$$

MAMB

DAMB

(2–36)

The solution properties of poly(dimethyldiallylammonium chloride) were studied.[221] The Mark-Houwink preexponential parameters and exponents in 1 M sodium chloride were 4.61×10^{-5} and 0.81, respectively, at pH 5.5 and were 7.2×10^{-4} and 0.5, respectively under theta conditions. The radius of gyration, expansion factor, and second viral coefficients in the molecular weight range $(1\text{–}5) \times 10^5$ were determined.

The reactivity ratios in cyclopolymerization of 4-(diallylamino)pyridine with N,N-dimethacrylamide, dimethyldiallylammonium chloride or acrylic acid showed that the pyridine monomer reacted with comonomers faster than with itself while the comonomers displayed the opposite behavior.[222] ^{13}C NMR spectra showed evidence of both cis- and trans-isomers in the five-membered ring of the repeat unit of the pyridine monomer.

A process for the manufacture of high solids, free-flowing granular poly(dimethyldiallylammonium chloride) has been described.[223] A polymer having molecular weight of 2.9×10^6 and containing less than 10% residual monomer could be obtained. The content of poly(dimethyldiallylammonium chloride) in aqueous dispersions was determined by adding alkali metal hydroxide followed by potentiometric titration with an aqueous solution of potassium ferricyanide solution at pH 0.5–12 to end-point.[224]

A laboratory experiment has been described which utilizes two major types of polymerization processes involving heterocyclic monomers or repeat units.[225] The first type involves polymer synthesis involving ring-opening to yield a polyamide from oxazoline, and the second type involves ring-closure *via* cyclopolymerization of a diallylamine derivative to give a polymer containing pyrrolidine units.

Complete conversion of polydiallylamine to the acetamide derivative was attained under mild conditions. Formation of the formamide polymer required harsher conditions and did not give complete functionalization.[226] Complexation with iodine and transition metal cations by the polymers was strong. A highly branched polymer was obtained by copolymerizing diallyldimethylammonium chloride with controlled amounts of methyltriallylammonium chloride.[227] The polymer was useful in paper manufacture.

The kinetics of polymerization of aqueous solutions of diallyldimethylammonium chloride in inverse emulsion was studied and the mechanism discussed.[228] The polymerization mechanism conformed to the Smith-Ewart mechanism. The influences of ion strength and partition equilibria on the rate of polymerization were studied also, and were considered to be the causes of the high order with respect to the monomer concentration.

Polymerization of dimethyldiallylammonium chloride by gamma irradiation of a film of the monomer was shown to have a polypyrrolidine structure.[229] Polymerization occurred at all radiation dosages, with 98% conversion at 1 Mrad. Further studies on the polymerization and vesicular properties of allyl and diallyl based monomeric and polymerized quaternary ammonium salts have been carried out.[230] In contrast to vesicles, based on the monoallyl monomers which were not stable, simple polymerized vesicles resulting from the diallyl monomer, showed excellent long-term stability, up to two years.

Higher molecular-weight polymers of diallyldialkyl ammonium compounds were obtained when polymerized in presence of fluoride ion.[231] Also, higher conversions to polymer were obtained in presence of fluoride ion than in its absence. A copolymer of dimethyldiallylammonium chloride and acrylic acid of high molecular weight and low acrylic acid content was effective in improving the performance of hair conditioners.[232]

Vesicles were formed from aqueous solutions of diallyldodecylammonium bromide by treatment with ultrasound, and the vesicles were polymerized by γ irridation or by free radical initiators.[233] The vesicles were stable, and freeze fracture electron micrographs showed that the products were single-compartment dilayer membrane vesicles.

A water-soluble cationic poly(diallylamine)-epichlorohydrin resin was heated with a carboxymethyl hydroxyethyl cellulose solution in aqueous sodium hydroxide to give crosslinked structures with $<1\%$ solubility in 10% aqueous sodium hydroxide.[234] The products were useful as ion exchangers. A study of the kinetics of radical polymerization of diallyldimethylammonium chloride in aqueous solution under high monomer-to-polymer conversion showed that the polymerization was accompanied by autoacceleration.[235]

Head-to-head and tail-to-tail additions in anionic polymerization were confirmed for the first time in cyclopolymerization of N-methyldiacrylamide.[236] The effect of the polymerization temperature on the polymers from N-propyldimethacrylamide, dimethacrylamide, and N-methyldiacryamide was studied.[237] The n-propyl monomer gave polymer with a five-membered ring as the main repeat unit. Dimethacrylamide gave mainly a six-membered ring, while the methyl monomer gave exclusively five-membered rings. The observed structures were independent of temperature over the range of $-78°$ to $180°$.

Homopolymers of decyl- and hexadecyldiallylamines and copolymers of these monomers with diallyldimethylammonium chloride were synthesized.[238] The copolymer of the hexadecyl monomer exhibited side-chain crystallization of the long alkyl chain, and the copolymer of the decyl monomer showed formation of polymeric micelles. Aqueous dispersions of polymer vesicles were prepared with increasing concentration by radical polymerization of ultrasonically dispersed unsaturated quaternary ammonium bromides, having at least two allyl groups, one alkyl group of 6–10 carbon atoms, and one group, a carboxymethyl ester of an alcohol, of 12–16 carbons.[239]

2,6-Bis[(dimethylamino)formyl]-4-(diallylamino)pyridine was cyclopolymerized by radical initiation in aqueous acid to give a linear, water-soluble, homopolymer. This polymer exhibited strong complexing ability which was ascribed to the pyridine nitrogen, along with the amide functions.[240]

Poly(dimethyldiallylammonium chloride), among other polyelectrolytes, was analyzed by size-exclusion chromatography, using Superose 6 gel packing.[241] The elution curves showed no adsorptive affects for the polymers under appropriate ionic strength conditions.

In a continuing study of the structural chemistry of polymerizable monomers, the effects of the crystal and molecular structure of N-methyldimethacrylamide on the initial stage of solid-state cyclopolymerization were studied.[242] The shortest spacing between vinyl carbon atoms was 2.908 Å; the planes of the two intramolecular vinyl groups arranged a dihedral angle of 40.6°. The location of the vinyl carbon atoms allowed an intramolecular closure in the initiation process, i.e., tail-to-tail cyclization was observed.

Polymerization of Oxygen-Containing Compounds

Acrylic anhydride[1-27, 1-28] and methacrylic anhydride[1-97] were early examples of monomers which were found to undergo cyclopolymerization. The polymers, obtained by radical polymerization, were soluble in certain polar organic solvents and were believed to consist of recurring six-membered anhydride rings. Later, some apparently conflicting data was reported by others. Both acrylic[243] and methacrylic anhydride[244, 245] were shown to give insoluble crosslinked polymers, under certain conditions.

A systematic study[246] of the polymerization of acrylic and methacrylic anhydride was undertaken. Both were polymerized by radical initiators in bulk or solution at − 50° to 80°C, and 50–99% conversions were achieved (Table 2–3). Depending on the conditions used, the polymers obtained were either soluble (linear) or insoluble (crosslinked) in dimethyl sulfoxide or dimethylformamide. These findings were in reasonable agreement with those reported by most other workers.

Methacrylic anhydride was readily polymerized to give soluble, linear cyclopolymers either in bulk or in nonpolar solvents.[1-97] In a polar solvent, crosslinked polymers were obtained at high monomer concentrations and at high conversions.[1-96] Acrylic anhydride appeared to be more easily crosslinked; soluble polymers were only obtained by solution polymerization. Thus, linear poly(acrylic anhydrides), having cyclic recurring units, could be obtained by the cyclopolymerization mechanism. The extent of the cyclization step, however, depended on the monomer and its environment. Occasional pendant vinyl groups resulted in some of the soluble polyanhydrides. This was less severe with methacrylic than acrylic anhydride; the former did not gel under bulk polymerization conditions. It had been reported earlier, perhaps erroneously, that crosslinked poly(methacrylic anhydride) was obtained in bulk and solution polymerization.

The stereochemical configuration of poly(methyl methacrylate) derived from the cyclopolymer poly(methacrylic anhydride) has been the subject of some interest. It has been demonstrated[1-27] that there may be some stereoregulating influence in the cyclopolymerization mechanism. Poly(acrylic acid) was obtained by hydrolysis of poly(acrylic anhydride) which was substantially more crystalline than conventional poly(acrylic acid).

It had been shown[1-96] that the stereochemical configuration of poly(methyl methacrylate) could be determined from its nuclear magnetic resonance spectrum. The three observed α-methyl proton peaks were assigned to isotactic, heterotactic, and syndiotactic configurations respectively.

Through other work,[1-93, 1-95] it became possible to assign the various peaks observed in the NMR spectrum to the α-methyl protons in the three types of configuration. Samples of isotactic and syndiotactic poly(methyl methacrylate) of nearly pure stereoregular configuration were prepared and the identity of each determined by X-ray analysis. The NMR spectra of two samples of poly(methyl methacrylate) derived from crosslinked poly(methacrylic anhydride) were examined.[244] It was concluded that while there is, in fact, a distinct tendency to favor formation of heterotactic sequences compared to what is observed in methyl methacrylate polymerization, this is, by no means, the only sequence formed. The spectrum of a sample of poly(methyl methacrylate) derived from poly(methacrylic anhydride) prepared at −50°C (UV in toluene) was studied[1-99] and it was calculated that the stereochemical configuration of this polymer was composed of a random run of syndiotactic and heterotactic blocks of various lengths. A study[1-100] of the monomolecular films (monolayers) of this polymer, using the film balance, gave results which appeared to be consistent with this structure. The fraction of polymer present, in each of the three possible stereoconfigurations, was

Table 2–3 Cyclopolymerization of Certain Unsaturated Anhydrides

Anhydride	Solvent	Monomer concentration (%)	Initiator[a]	Temperature (°C)	Time (hrs.)	Yield (%)	Solubility
Methacrylic	—	—	1% BP	50	17[b]	99.5	+
Methacrylic	—	—	BP/DMA/T	25	0.5	>50	+
Methacrylic	—	—	UV	−50	7	60	+
Methacrylic	benzene	10	1% BP	80	5	88	+
Methacrylic	DMSO	15	0.5% AIBN	65	64	58	+
Methacrylic	DMSO	30	0.5% AIBN	65	64	64	+
Methacrylic	DMSO	50	0.5% AIBN	65	64	>50	−
Methacrylic	DMSO	75	0.5% AIBN	65	64	>50	−
Methacrylic	toluene	50	UV	−50	7	46	+
Acrylic	—	—	0.5% AIBN	65	64[c]	~95	−
Acrylic	—	—	UV	−50	3	—	—
Acrylic	benzene	15	0.5% BP	80	4	~90	+
Acrylic	toluene	50	UV	−50	5	95	+

[a] AIBN = azobisisobutyronitrile; BP = benzoyl peroxide; BP/DMA/TP = benzoyl peroxide/dimethylaniline/thiophenol; UV = ultraviolet irridation.

[b] Heated at 125°C for 2 hrs and at 140°C for 3 hrs.

[c] Heated at 120°C for 2 hrs.

also determined[1-97] by an investigation of the NMR spectrum of samples of poly(methyl methacrylate) derived from poly(methacrylic anhydride). The results indicated that the stereoconfiguration was influenced by the cyclopolymerization mechanism. Some conformational aspects of polymers obtained through the intra–intermolecular mechanism have been discussed.[1-98]

The cyclopolymerization of acrylic anhydride[247-249] and methacrylic anhydride[248, 250-252] has been investigated extensively by other groups.

Some practical applications of polyanhydrides have been made. Optical fibers were prepared by melt spinning together a polymer containing polyacrylic or polymethacrylic anhydride units, unsaturated carboxylic acid units, aromatic vinyl compound units, and methyl methacrylate units as core along with a polymer having lower refractive index as sheath. Thus, such a copolymer as core and poly(methyl methacrylate) as sheath were melt spun together and drawn to give flexible optical fibers with good resistance to heat and cracking.[253]

The possibility of obtaining acetals of poly(vinyl alcohol) directly from monomers by the cyclopolymerization of divinyl acetals was first demonstrated in 1960.[254] Treatment with radical initiators produced soluble, saturated polymers, which upon hydrolysis gave poly(vinyl alcohol). This reaction was extended[255-263] to the preparation of a wide variety of poly(vinyl acetals). Bulk polymerization, using radical initiators, proceeded exclusively by a cyclopolymerization mechanism. Using ionic catalysts, aromatic divinyl acetals, under all conditions studied, gave linear cyclopolymers containing no detectable residual unsaturation. The polymerization of aliphatic divinyl acetals, however, gave soluble polymers which, from their infrared spectra, contained some units with pendant vinyl groups. The ionic polymerization of divinyl formal[259, 264] gave only crosslinked polymer, regardless of the reaction conditions. The cyclic mechanism of divinyl acetal polymerization has been confirmed by a measurement of the effective dipole moments of the polymers.[265, 266] The polymerization of divinyl acetals has been extensively studied.[267-271] Five- and six-membered rings were formed.[271] To determine the average concentrations of each type of cyclic unit, the poly(divinyl acetals) were hydrolyzed and the resultant polymers containing 1,2- and 1,3-glycol units derived from five- and six-membered rings respectively, were analyzed. According to the analytical data, the average poly(divinyl acetal) structure was 2–37, where x:y = 77:23.

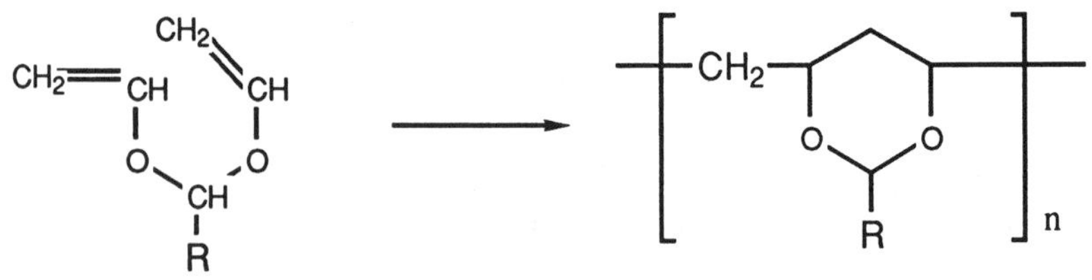

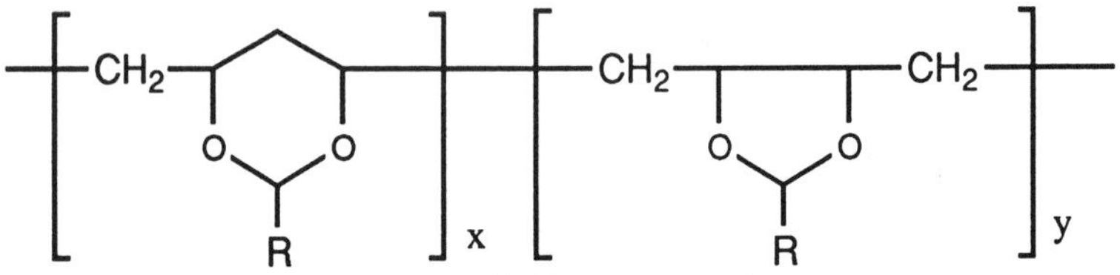

(2–37)

The x:y ratio does not depend on R. Divinyl ketals also polymerize[272] under the influence of radical initiators to give linear cyclopolymers containing 1,3-dioxane rings.

Poly(divinyl butyral) and poly(divinyl formal) with molecular weights of 5000–500,000 have been claimed.[273] Polymerization was initiated by an oxygen-trialkyl boron system.

Radical polymerization of divinylformal,[274] with azobisisobutyronitrile initiator in several solvents, has been investigated. In many solvents (benzene, cyclohexane, acetonitrile, dimethyl sulfoxide, etc.), the polymers consisted of the cyclized monomer unit and 5–8% of the pendant formate group. The amounts of the residual vinyl group were quite small in these solvents. The formate group was formed by hydrogen migration and subsequent ring scission of the cyclic propagating radical. However, a polymer obtained in carbon disulfide contained ~30% of the pendant vinyl group but no formate group. The carbon and hydrogen contents of this polymer were lower than expected and sulfur was detected. The polymerization in benzene carbon disulfide indicated a much stronger influence of carbon disulfide than benzene. Carbon disulfide molecules interact strongly with the propagating radical and are incorporated into the polymer.

The structure of poly(divinyl butyral)[275] has been reported to depend on the polymerization temperature and to possess a high degree of branching and to have large cyclic structures as well as the six-membered ring structure, predicted by the originally proposed cyclopolymerization mechanism.

Cyclopolymerization of divinyl butyral[276] by free radical initiation was studied and it was reported that initiation occurred by hydrogen atom abstractions. Also isomerizations of the growing chain occurred to yield ester and ketal functions, along with some crosslinking.

Gamma-ray initiated polymerization of divinylformal at 0°C in the liquid state, or at −78°C by solution polymerization, gave polymers which contained cyclic units. ^{13}C NMR spectroscopy was used to show that the polymers contained a *cis*-4,5-disubstituted-1,3-dioxolane ring as the predominant unit. Other divinyl acetals were also studied.[1-183]

Cyclopolymerization of the dialdehyde, *cis*1,3-diformylcyclohexane,[32] was investigated and a cyclopolymer consisting of 3-oxabicyclo[3.3.1]nonane rings was obtained. The synthesis of cyclopolymers containing heterocyclic rings by cyclopolymerization of dialdehydes[86] has been published.

The polymerization of divinyl carbonate was studied under various reaction conditions,[277] and it was found that cyclopolymerization was not the predominant mode of propagation. The polymer structure was analyzed and found to contain a five-membered cyclic carbonate structure, a six-membered cyclic carbonate structure, and a large number of pendant unsaturated carbonate units. Gelation usually occurred at low conversions.

The chemical structure of poly(vinyl alcohol), derived from poly(divinyl carbonate) or divinyl carbonate-vinyl acetate copolymers,[278] has been extensively investigated. The poly(vinyl alcohol) thus derived contained an unusually large amount of 1,2-glycol structure, which was considered to be due to the 5-membered ring structure in poly(divinyl carbonate) or divinyl carbonate-vinyl acetate copolymers.

Preparation of divinyl carbonates,[279] by reaction of a mercury(II) aldehyde or ketone complex with phosgene in presence of a tertiary amine in an inert solvent, has been accomplished. Divinyl and diisopropenyl carbonates were prepared by this method. These monomers were converted to their polymers by heating with azobisisobutyronitrile.

Poly(divinyl carbonate)[280] was used to prepare poly(vinyl alcohol) by alkaline hydrolysis. The obtained poly(vinyl alcohol) was reported to possess infrared spectral and differential thermal analytical properties essentially the same as those of poly(vinyl alcohol) prepared from poly(vinyl acetate). These results suggested formation of six-membered rings in the poly(divinyl carbonate). Diisopropenyl carbonate[281] was also prepared and polymerized, *via* a free radical mechanism.

Phthalaldehyde[282] was polymerized with anionic and coordination catalysts at $-78°C$. All polymers were composed only of cyclized units (1,3-dialkoxyphthalan rings). The stereochemical structure, *cis, and trans,* of the ether ring on the polymer was proposed. Coordination catalysts gave high contents of the *trans* configuration. The amount of cis-structure decreased in the following order: γ-ray irradiation > cationic catalysts ⩾ anionic catalysts > coordination catalysts. Cyclopolymers of phthaldehyde, obtained by cationic polymerization, had high *cis* content.

Polymerization of 3,4-dihydro-2H-pyran-2-carboxyaldehyde (acrolein dimer)[283] under several conditions has been described. Poly(acrolein dimer) obtained at room temperature was found to have structural units (2–38) of 6,8-dioxabicyclo[3.2.1]octane

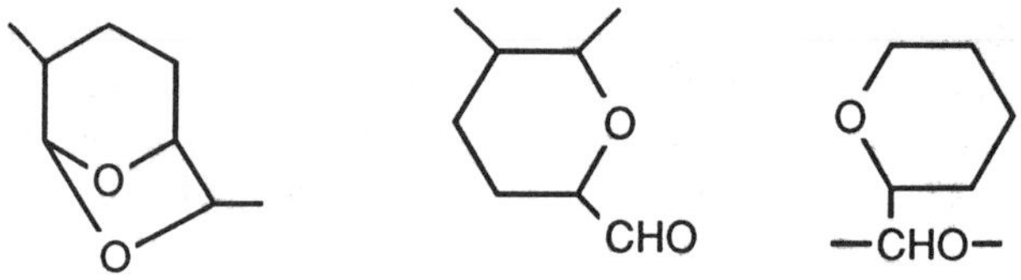

O O CHO O O —CHO—

(2–38)

produced by intra-intermolecular reaction, in addition to the ordinary structures produced by the unsaturated ether group participation and those produced by the aldehyde group participation. The lower the polymerization temperature the higher the selectivity for carbonyl polymerization. The stronger the acidic character of initiators, the higher the selectivity for the open chain structures. With anionic catalysts, such as triethyl aluminum and dibenzene chromium, acrolein dimer gave a polymer containing only recurring structural units derived solely from the C=C double bond at $-78°C$, while it did not give polymer above $-20°C$. There appears to be a ceiling temperature in the polymerization with anionic catalysts. The solubility of the polymers increased with increasing amounts of the bicyclic structure.

The cyclopolymerization of divinylformal has been investigated further.[284] Gamma-ray bulk polymerization in the solid state at $-190°C$ and $-78°C$ gave polymers having a linear structure with pendant vinyl groups, while polymers obtained at $0°C$, in the liquid state, or at $-78°C$ by solution polmerization, contained cyclic units. ^{13}C NMR spectroscopy has been used[285] to show that polymers of divinylformal contain a *cis*–4,5–disubstituted–1,3–dioxolane ring as the predominant unit, both in the main chain and as a pendant group.

Cyclopolymerization of 1-cyclohexene-4,5-dicarboxaldehyde and 1-methyl-1-cyclohexene-4,5-dicarboxaldehyde[286] with cationic, anionic and coordination catalysts, e.g. BF_3-$O(C_2H_5)_2$, $SnCl_4$, $Al(C_2H_5)_3$, and tert–BuOLi has been investigated yielding cyclopolymers [(2–39), R = H or CH_3] soluble in common organic solvents (benzene,

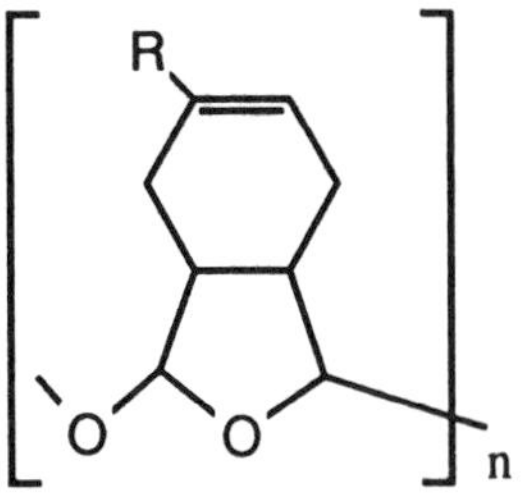

(2–39)

zene, THF, CCl_4, etc.), with all the catalysts used. The extent of cyclization was >98%. The cyclopolymers were obtained at 30°C, suggesting unusually high ceiling temperatures for such dialdehydes. The enhanced reactivity of these monomers relative to 1-cyclohexene-4-carboxaldehyde and the high tendency of cyclization apparently were caused by fixation of the adjacent aldehyde groups in the *cis* positions.

Based upon a telomerization study, it was concluded that diallyl ether, divinyl ether, and divinyl formal[287] propagated through the -CH· radical, inconsistent with earlier predictions.

Cyclopolymerization of a number of bifunctional monomers[27] has been discussed. Propagation schemes, depending on the nature of the monomer, were proposed in the cyclopolymerization of a number of bifunctional monomers where the existence of the functional groups in the vicinal position led to a steric advantage for cyclopolymerization. The cationic polymerization of o-divinylbenzene proceeded by a stepwise propagation and an intermediate; a concerted scheme was proposed for cis-1,2-divinylcyclohexane, o-phthaldehyde, or cis-cyclohexane-4,5-dicarboxaldehyde. Both reaction schemes coexisted in the cationic polymerization of o-vinylbenzaldehyde, depending on the type of cation produced. In the stepwise reaction, the influence of polymerization conditions on cyclization gave information on the polymerization mechanism of related monofunctional monomers. Cyclopolymerization of diallyl carbonate[288] with benzoyl peroxide, as initiator, has been extensively studied. Several dimethacrylates have been synthesized and cyclopolymerized, including dialkylsilyl dimethacrylates,[289–291] divinyl oxalate[292] and divinyl esters of other dibasic acids,[293] to give polymers with cyclic structures. Hydrolysis to poly(vinyl alcohol) yields information on the propagation mechanism.

Carbon suboxide C_3O_2, has been reported to undergo an unusual type of cyclopolymerization. A recent review has been published with forty-five references dealing with current studies of its cyclopolymerization in which a mechanism involving a zwitterionic intermediate with competing propagation or termination steps was proposed to explain the details of the mechanistic pathway (2–40).[294]

A cyclopolymerization study and regioselective synthesis of vinyl itaconates has been published. A selective synthesis for both 1- and 4-alkyl vinyl itaconate regioisomers was presented. Polymerization experiments with these two monomers revealed that only 4-vinyl itaconic acid and 1-alkyl-4-vinyl itaconates lead to cyclopolymers. A synthesis of the γ-lactone repeat unit found in the cyclopolymer confirmed the structure and stereochemistry of the polymer.[295]

A wide variety of additional monomers, capable of cyclopolymerizing to oxygen-containing rings, have been synthesized and their polymerization characteristics studied. A number of these studies are presented briefly in a chronological fashion in this section.

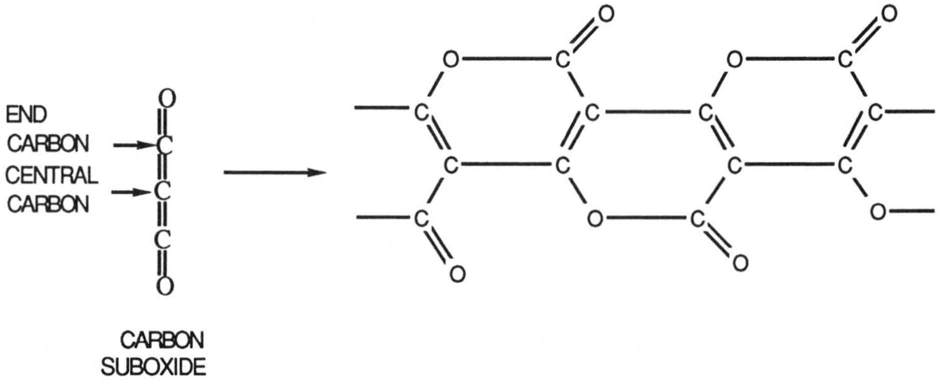

END CARBON
CENTRAL CARBON
CARBON SUBOXIDE

Poly(carbon suboxide): a polycyclic chain of six-membered lactone rings with polyconjugated double bonds

(2–40)

Polymers containing amide and carboxyl groups were prepared by reaction of poly(acrylic anhydride) or poly(methacrylic anhydride) with amines.[296] Polymers containing ester and carboxyl groups were prepared by reaction of poly(acrylic anhydride) or poly(methacrylic anhydride) with alcohols.[297] Exclusive formation of six-membered ring was reported in free radical additions to 1,6-dienes, such as diallyl ether with bromotrichloromethane.[298] More recent data tends to refute this claim.

A process for quantitatively producing polyacrylic anhydride by treating an acrylic acid solution with ketene followed by radical-initiated polymerization of the reaction product,[299] as well as other synthetic procedures have been reported. An alternate procedure for producing poly(acrylic anhydride) or poly(methacrylic anhydride) to direct polymerization of the respective monomers has been described.[300] The procedure involved polymerization of the alkenic acids in the presence of an anhydride forming reagent, e.g. acetic anhydride. Linear polymers containing cyclic anhydride units were obtained.

Cyclopolymerization of symmetrical and unsymmetrical 1,5- and 1,6-diolefin monomers has been reported.[301] Among the compounds studied were divinylacetals and vinyl crotonate. Evidence was obtained that cyclization occurred in both types of monomers. Poly(vinylacetals) were prepared in presence of cationic initiators.[302] The polymers were linear and soluble cyclopolymers.

A linear, cyclic polyacrylic anhydride, when blended with a synthetic fiber-forming hydrophobic polymer, produced a more dye receptive fiber.[303] Grafted polymers were prepared by copolymerizing, in appropriate monomer ratios, vinyl monomers such as styrene with diene monomers such as allyl acrylate.[304] Allyl methacrylate, methallyl methacrylate, allylstyrene, and several symmetrical divinyl compounds were used.

Novel vinyl monomers, having polymerizable double bonds positioned in such manner to allow cyclopolymerization were synthesized.[305] The compounds were obtained by the reaction of anhydrides of acrylic or methacrylic acid and saturated carboxylic acids with acrolein or methacrolein, in presence of ferric chloride. Compounds of the general formula: $RCO_2CH(R')OCOR'$ where R is alkyl, vinyl or isopropenyl and R′ is vinyl or isopropenyl were obtained. The resulting cyclopolymers were shown to possess unusual dimensional stability and improved strength characteristics.

A reliable procedure for synthesizing poly(acrylic anhydride) *via* cyclopolymerization of the corresponding monomer has been provided in Macromolecular Syntheses.[306]

The divinylacetal of (R)(+)-3,7-dimethyloctanal was polymerized by radical and cationic initiators.[307] The former gave only viscous liquids, while the latter gave solid products. Optical rotatory dispersions of these polymers fit simple Drude plots. The polymer prepared by boron trifluoride etherate exhibited negative optical rotations, while both the polymer obtained by AIBN and the model compound showed positive rotations. From these observations, it was suggested that the polymers might differ from each other in stereoregularity.

In a paper on cyclopolymerization, a study of the temperature's effect on the polymerization mechanism of methacrylic anhydride was reported.[308] For the monomolecular and bimolecular mechanisms of cyclopolymerization, an equation was derived, $-d[M]/d[m] = A + B/[M]$, where [M] is the monomer concentration, [m] is the pendant double bond concentration and A and B are constants. This equation was verified by polymerizing methacrylic anhydride under different temperatures. In all cases, -d[M]/d[m] gave a linear relation with 1/[M] as expected, and the intercepts of the plots obtained were 2.9, 2.15, and 1.7 at 60°, 40°, and 25° respectively The values of the ratio of the bimolecular rate to the monomolecular rate were 2.7[M], 2.0[M], and 1.4[M] at 60°, 40°, and 25°, respectively, and thus decrease as temperature decreases. The energy of activation of bimolecular cyclopolymerization was 3.7 kcal/mole higher than that of the monomolecular reaction.

Cyclopolymerization of acrylic anhydride has been studied by radiation initiation in toluene at 30°C.[309] The cyclization ratio, k_c/k_i was shown to be 11.1 mole/L and $E_c - E_i$ was 2.3 kcal/mole.

The structures of the poly(vinyl alcohol) derived form poly(divinyl *n*-butyral) and from poly(vinyl formate) were compared with ordinary poly(vinyl alcohol).[310] That derived from the poly(divinyl *n*-butyral) was different from the ordinary sample; however, that derived from poly(vinyl formate) showed nearly the same spectra as the ordinary atactic polymers.

Divinyl malonate was homopolymerized and copolymerized with styrene, methyl methacrylate, acrylonitrile, and vinyl acetate *via* radical initiation.[311] The homopolymer apparently was soluble only in certain solvents. However, the copolymers with all monomers except vinyl acetate were completely soluble with no gel.

A high-resolution proton magnetic resonance (PMR) and infrared (IR) spectral study of poly(vinylformal) and its model compounds was carried out.[312] The data from the polymer were compared with those of model formals obtained from stereoisomers of pentane-2,4-diol and heptane-2,4,6-triol. A highly isotactic polymer of allyl vinyl ether, with boron trifluoride-etherate as initiator, at −78° in toluene was obtained.[313] The reaction proceeded through the vinyl double bond, and the polymer contained allyl ether side chains. An infrared study of poly(vinylformal), in highly stretched films, favored a symmetrical puckered m–dioxane backbone ring structure (2–41):[314]

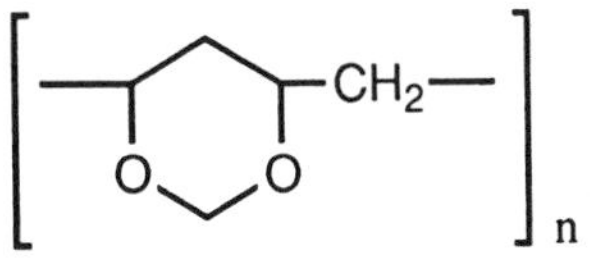

(2–41)

The monomer reactivity ratio of diallyl oxalate in copolymerization with allyl benzoate at 60° was 0.89. Corresponding ratios for divinyl succinate and diallyl adipate were 0.92 and 0.99, respectively.[315] The monomer reactivities for cyclization were 1.089, 1.38 and 1.39, respectively for the oxalate, succinate and adipate.

The *cis*- and *trans*- isomers of 2,6-divinyltetrahydropyran were prepared and characterized.[316] Conformational analysis of the isomers yielded the activation free energy of ring inversion of the *trans* isomer. In support of the electronic interaction theory, bathochromic shifts have been reported for a number of monomers; however, others failed to show such shifts.[317] Also, the bathochromic shifts exhibited by certain monomers have an alternative explanation.

The high tendency for cyclization in cyclopolymerization has been explained as being due to a smaller decrease in entropy for such a reaction compared to intermolecular addition.[318] Therefore, the Gibbs Free Energy of activation would be less for cyclopolymerization than for the intermolecular propagation. A mathematical, statistical analysis of the free radical polymerization of methacrylic anhydride in $CHCl_3$, at various temperatures, was conducted. The results showed that the dependence of the mole fraction of the cyclic structural units on the monomer concentration at which the polymer was obtained, agreed with free radical cyclopolymerization kinetics for symmetrical unconjugated dienes.[319]

In the cyclopolymerization of allylalkylmaleates, increasing the size of the alkyl substituent led to steric hindrance, decreased the polymerization rate, and increased the selectivity of addition of growing radical to the allylic double bond.[320] Free radical addition of perfluoroalkyl iodides to alkenes, for example, diallyl ether or 1-heptene, was initiated by amines.[321]

In a study of the structure of some allyl and epoxy monomers, it was observed that no reaction between the double bonds in diallyl ether and between the double bonds and the oxirane ring in allyl glycidyl ether occurred; therefore, such possible reactions are the driving forces in the cyclopolymerization of these monomers.[322] UV spectra of all monomers showed a bathochromic shift with an increase in concentration, probably due to molecular association.

It has been shown that the ratio of five- to six-membered ring formation in cyclopolymerization of acrylic anhydride at 50°C varies with the dipole moment (M) of the solvent from 5% five-membered ring at $M = 0$ to 30% at $M = 4$.[323] A pronounced temperature effect was also observed. At 90°C, the five-membered ring content was 20% at $M = 0$ and increased to 75% at $M = 4$. Under all conditions described, methacrylic anhydride led only to six-membered rings.

The five-membered ring content of poly(acrylic anhydride) has been shown to vary with both conversion at constant monomer concentration, and with monomer concentration at constant conversion.[324] In benzene, the five-membered ring increased from 5% at 40% conversion to 10% at 90% conversion; however, in τ-butyrolactone, the five-membered ring content increased from 28% at 15% conversion to 75% at 50% conversion, a much greater change than in the non-polar medium.

A study of cyclopolymerization of divinyl formal confirmed that the polymer varied with polymerization conditions.[325] The polymer contained the *cis*-dioxolane ring as the major structure, along with the *trans* ring and the branched structure. The unsaturated unit and the six-membered unit were not formed under any conditions.

Cyclopolymerization of acrylic and methacrylic anhydrides led to head-to-head and head-to-tail linkages in the poly(methylacrylate) and poly(methylmethacrylate), derived from the respective polyanhydrides.[326]

The ether of methyl α-(hydroxymethyl)acrylate was prepared as a nonhydrolyzable crosslinking agent for acrylate esters.[327] However, cyclopolymerization of the ether led to the cyclic polymer, postulated to contain the six-membered tetrahydropyran ring. Partial hydrolysis to the corresponding polycarboxylic acid was reported.[328]

Polymers, which contain the anhydride unit derived from cyclopolymerization of methacrylic anhydride, have been claimed in a patent.[329] The units were introduced by heating copolymers of styrene and methacrylic acid at 234°.

Polymerization of Compounds Containing Sulfur, Silicon, and Other Elements

Several S,S-divinylmercaptals[330,331] were polymerized and soluble, rubber-like polymers were obtained, containing little residual unsaturation. On hydrolysis, these cyclopolymers were converted to poly(vinyl mercaptan) and on oxidation, to the corresponding sulfones. It has also been reported[332] that the cyclopolymerization of formaldehyde-S,S′-divinylmercaptal occurred. Also obtained was a soluble cyclopolymer from the solution polymerization of S,S′-divinyl dithiocarbonate in benzene.[333]

In contrast to diallyl sulfide and diallyl sulfone, di-2-chloroallyl sulfide and di-2-chloroallyl sulfone[8] were found to be capable of polymerizing in the presence of radical initiators with the formation of cyclic sulfur-containing polymers, which were extensively dehydrochlorinated under the reaction conditions.

The polymerization of a number of substituted diallylsilanes has been investigated[334-337] under a variety of conditions.[337] In general, the polymers obtained by radical initiation were viscous oils or low melting solids having structures consistent with the cyclopolymerization mechanism. However, metal-coordination initiation has led to high polymers.

Divinyloxydimethylsilane[338,339] was polymerized using both radical and cationic initiators. Radical polymerization gave polymers containing five-membered rings, six-membered rings, and pendant vinyl groups. The five-membered ring content was determined from the amount of 1,2-glycol structures in the derived poly(vinyl alcohol). Compared to the divinyl acetals, its capacity for cyclization was rather low. The unreacted double bond caused crosslinking at all except low conversions or low monomer concentrations. Polymerization in toluene or nitroethane, using cationic initiators, proceeded without cyclization, and only a small amount of cyclopolymerization occurred in methylene chloride.

N,N-Diallylmethanesulfonamide, N,N-diallylethanesulfonamide, N,N-dimethallylmethanesulfonamide, and N,N-dimethallylethanesulfonamide[103] were cyclopolymerized to low molecular weight cyclopolymers.

The synthesis and polymerization of a number of unsaturated silanes[340] which are functionally capable of polymerizing by the cyclopolymerization mechanism were studied. Allyl substituted silanes studied included diallylmethylphenylsilane, diallylcyclotetramethylenesilane, and diallylcyclopentamethylenesilane. Methallyl substituted silanes studied included dimethallyldimethylsilane, dimethallyldiphenylsilane, dimethallylmethylphenylsilane, and dimethallylcyclopentamethylenesilane. All of the diallylsilanes yielded polymers by use of coordination type initiators. The dimethallylsilanes failed to yield polymer *via* coordination initiators, but yielded low polymers by a free radical mechanism.

A wide variety of diunsaturated phosphonium salts were synthesized and their cyclopolymerization characteristics studied.[341] Monomers which were functionally capable of leading to five-, six-, or seven-membered rings were generally concluded to cyclopolymerize as predicted by the proposed mechanism. Divinylphenylphosphine copolymerized with acrylonitrile in accord with the cyclocopolymerization mechanism.

A report on the influence of the polymerization conditions on the polymer structure of 1,1′-divinylferrocene,[342, 1-148] when cationic and free radical initiation were used, has been published. The polymers obtained with cationic initiators were shown to contain 70–80% cyclized units and uncyclized units, whereas the radically initiated polymer consisted of more than 96% cyclized units.

The polymerization of 1,1′-divinylferrocene has been investigated in several independent research centers.[343-345] It was reported that soluble polymers could be obtained, using radical and cationic initiators, with the formation of three-carbon bridged ferrocene units in the chain (2–42). Recent investigations of these polymers by

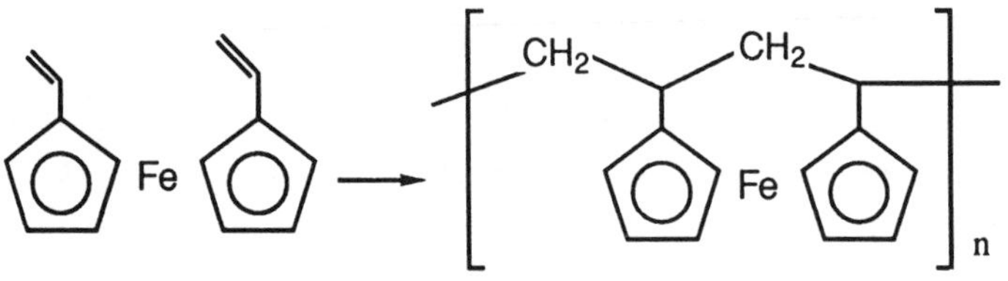

(2–42)

Mossbauer spectroscopy[346] have shown that the radical- and cationic initiated polymers have significantly different parameters (Table 2–4). Using model compounds, it has been shown that radical initiation does yield cyclopolymers with the three-carbon bridged ferrocene unit but that polymers produced by cationic initiation do not have this structure.

Perfluorinated C_5-C_8, 4-chloroperfluoro-1,6-heptadiene, and perfluoro-1,11-dodecadiene[347] have been synthesized by telomerization, and the dienes were polymerized by γ-radiation at high pressure. The cyclization/noncyclization ratio was computed for polymerization of the dienes.

Table 2–4 Mossbauer Parameters of Polymers of 1,1′-Divinylferrocene

Initiator	Isomer shift[a] δ (mms^{-1})	Quadrupole splitting ΔE_Q (MMS^{-1})
Radical	0.23 (2)	2.29 (2)
Cationic	0.27 (2)	2.40 (2)

[a] Relative to Fe in Pd.

The polymerization of methylvinyloxygermanes,[348] using radical and cationic initiators, has been studied. Di- and trivinyloxygermane radical polymerizations involved the formation of five- and six-membered rings. A number of diallylsilanes have been prepared[349] and their cyclopolymerization reported.

In contrast to earlier work, soluble, fusible, linear cyclopolymers were obtained by radical polymerization of divinyl phosphonates in dilute solutions, especially if the polymerizations were carried out at low temperatures and the initiator was decomposed photolytically.[350]

Soluble copolymers of diallylalkylamine oxides with acrylonitrile[1-34] and other monomers have been described. These copolymers were assumed to possess one unreacted double bond for each molecule of diallylamine oxide entering the chain; however, copolymerization by an alternating intramolecular intermolecular mechanism had probably occurred, producing saturated copolymers. A number of soluble copolymers, involving several diallyl monomers[1-35] and several conventional monomers,[1-36] have been reported and it has been postulated that these copolymerizations follow the mechanism under discussion. Cyclizations of allyl ether and allyl sulfide to substituted tetrahydropyrans and tetrahydrothiopyrans, respectively, occurred when attempts were made to add various reagents to these dienes under free-radical conditions.

Diallyl- and dimethallylphosphonium salts have been polymerized to soluble cyclopolymers through use of free-radical initiators.[1-37] These monomers appear to lend themselves to higher molecular weight formation than the corresponding diallylphosphine oxides. For example, poly(diallyldiphenylphosphonium chloride) was converted to poly-(diallyldiphenylphosphine oxide) by treatment with sodium hydroxide. The intrinsic viscosity of this polymer was much higher than that of poly(diallylphenylphosphine oxide) prepared by direct polymerization of the corresponding monomer. Both poly-(diallylphenylphosphine oxide) and poly-(dimethallylphenylphosphine oxide) were synthesized by cyclopolymerization of the respective monomers *via* a radical mechanism.[351]

Dimethyl- and diethyldiallylsilanes were subjected to metal coordination catalysts and found to give products consisting of liquids and solids.[1-55] The liquids could be distilled under vacuum and were identified as trimers, tetramers and pentamers. The solids were insoluble but could be swollen in heptane.

Allyldiethylsilane and allylethylphenylsilane led to viscous liquids *via* radiation.[352] The polymers of diallyl ethylsilane and triallylsilane *via* radiation were infusible, insoluble polymers. The silicon-hydrogen bond was retained in the radiation and radical-initiated polymers; however, when platinum on carbon was used as catalyst, this bond disappeared, which suggest platinum catalyzed addition of the silicon-hydrogen bond across the allyl double bond.

Divinyl acetals were cyclopolymerized *via* radical initiation to yield polymers hydrolyzable to poly(vinyl alcohol).[353] Copolymers of unsaturated polyesters with ethylenically unsaturated phosphine oxides, which contain three vinyl, allyl, or methallyl groups were synthesized.[354] The products were crosslinked; however, considerable cyclopolymerization and successive ring closures probably occurred in certain instances.

B-Triallyl-N-triphenylborazine and B-trivinyl-N-triphenylborazine were studied as monomers for homopolymerization and copolymerization with a variety of vinyl monomers, including styrene and methyl methacrylate, *via* radical initiation.[355] Homopolymerization did not occur, and only low molecular weight soluble copolymers were obtained. These results were attributed to steric hindrance by the phenyl groups. Reactivity ratios were reported.

Methacrylic anhydride was used as a comonomer in the preparation of plasticlenses.[356] The possibility of cyclopolymerization was not considered, although it most certainly occurred.

A study of the preparation and polymerization of unsaturated phosphorus-containing cyclopolymerizable monomers has been carried out.[357] Divinylsiloxane oligomers were polymerized *via* free radical initiation to give a soluble polymer at 30% conversion; crosslinking occurred at higher conversions. The oligomers were also copolymerized with isoprene.[358] Reactivity ratios were reported. Heterocyclic polymers, containing phosphorus acid ester bonds in the main chain, were synthesized *via* repetition of the 1,3-dipolar cycloaddition of terephthalonitrile to a series of diallyl esters of organophosphorus acids.[359] Examples are diallyl phenylphosphonate, diallyl phenylthiophosphonate, diallyl ethyl phosphate, and diallyl phenylphosphate. 1:1 Alternating polymers were proposed.

A variety of unsaturated phosphonium salts were synthesized.[360] These salts could be characterized as 1,5- 1,6- 1,7- or 1,8-dienes having a diphenylphosphonium or alkylphenylphosphonium substituent in the chain. The salts were cyclopolymerized to yield cyclopolymers postulated to contain 5-,6-,7-, and 8-membered ring structures, respectively. The participation of the β-styrylphosphonic group in cyclopolymerization has been studied.[361] The polymerization of diallyl styrylphosphonate involves all its double bonds. In the case of diethylstyrylphosphonate, the polymerization did not occur at all, while the homopolymerization of diallyl phenylphosphonate proceeded slowly. The relative ease of polymerization of the styryl monomer was attributed to the formation of the intermediate cyclic free radical structure; the formation of such structures in the other monomer is not possible.

Radical bulk polymerization of methyl, ethyl, propyl, octyl, 2–chloroethyl, and 2–bromoethyl divinylphosphinate at 50–100° gave crosslinked polymers; bulk polymerization of phenyldivinylphosphinate or $(CH_2{=}CH)_2P(O)(C_2H_5)_2$ gave rubbery polymers.[362] Poly(ethyldivinylphosphinate) obtained in benzene solution was crosslinked, but that obtained in C_2H_5OH solution was linear and oligomeric, apparently having been cyclopolymerization. Radical polymerization of phosphoric acid derivatives has also been studied.[363] The Mannich reaction product of phosphoric acid with diallylamine and paraformaldehyde gave diallylaminomethyl phosphonic acid, which was cyclopolymerized to the structure shown (2–43).

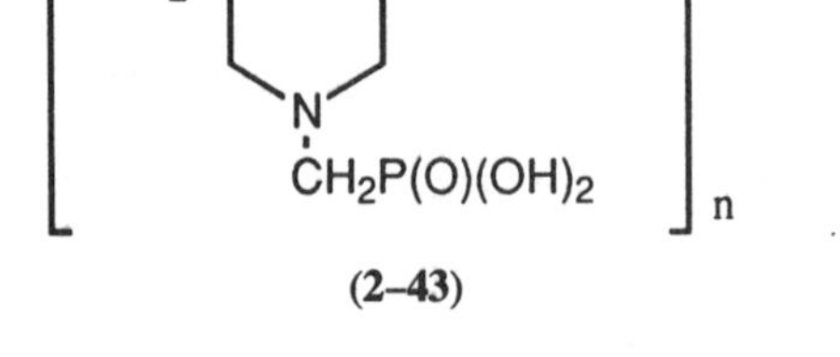

$$\left[-CH_2-\text{(piperidine-3,5-diyl, N-}CH_2P(O)(OH)_2\text{)}- \right]_n$$

(2–43)

Reaction of tungsten hexachloride with tetraallylsilane gave a product which catalyzed ring opening polymerization of cycloolefins.[364] Polyoctadienamers containing 80–90% *cis* units were obtained in up to 100% yield. Similar catalysts from diallyldimethylsilane and phenyltriallysilane were less effective.

Allyl halides have been polymerized by using alkali metal silicates and alkali earth metal silicates to react chemically with the halide to produce a silicon acid which acts as a catalyst to polymerize the allyl halide, giving allyl halide-allyl alcohol copolymers.[365]

Cyclopolymerization has been used as a new method of constructing polymers with defined arrangement of functional groups in the chain.[366] Water-soluble acrylamide polymer samples of various molecular weight, containing salicylaldehyde and lysine moieties, were prepared by copolymerization of (N″-5-methacryloylaminosalicylidene-N′-methacryloyl-(S)-lysinato)(pyridine)copper (II) with acrylamide, followed by destruction of the copper complex by the disodium salt of EDTA. At pH > 8, the formation of "internal" aldimine occurred.

Cyclopolymerization of triallylphenylsilane *via* radical initiation was carried out.[367] The polymer was reported to contain both five- and six-membered rings, as well as one allyl side chain in its structure. The polymer softened at 127–30° and had molecular weight of 47,000. A new negative photoresist for a bilayer resist system was developed, using this polymer as a component.

RINGS LARGER THAN FIVE OR SIX MEMBERS

Prior to the discovery of cyclopolymerization, the polymerization and copolymerization of nonconjugated dienes were predicted to yield only insoluble crosslinked products. Although the observed extents of reaction at the gel point for the addition polymerization of the systems methyl methacrylate–ethylene dimethacrylate and vinyl acetate–divinyl adipate[1-10] were always greater than theoretical predictions,[1-11] the formation of cyclic structures was considered of minor importance. However, evidence had been provided[1-12] of the formation of cyclic structures to the extent of approximately 41% in the polymers of diallyl phthalate. It was calculated[1-13] that in the polymerization of diallyl phthalate the probability of intramolecular reaction would be appreciable. In a study of fourteen diallyl esters, it was observed[368] that during the formation of crosslinked polymers intramolecular reaction occurred in all cases. The lack of agreement between the theory of gelation and experiment was due to the intramolecular reaction.[369-373]

After the principle of cyclopolymerization had been established, attempts to obtain soluble polymers containing ring structures with more than six atoms were made.[1-86] The polymerization, by Ziegler catalysts, of the hydrocarbons $CH_2{=}CH(CH_2)_n CH{=}CH_2$ in which n was equal to 4, 5, 6, 7, 8, 9, 11, 12, 14, and 18 was studied. In every case, the evidence indicated that both cyclic and acyclic recurring units were produced. The percentages of cyclization reported corresponded roughly to those which had been obtained in other cyclization reactions for forming rings of the sizes in question.[374, 375] Thus, seven-membered rings form fairly readily; the intermediate ring sizes form with greater difficulty; the larger rings (fourteen- and fifteen-membered) are easier to obtain, and, finally, the yields decrease again with still higher ring sizes.

In a similar study using the monomers $CH_2{=}C(C_6H_5)\ (CH_2)_n\ C(C_6H_5){=}CH_2$, where n = 2, 3, and 4, it was reported[4] that the ease of cyclopolymerization with ring size was $6 > 5 \cong 7$.

A series of bis-N-vinyl compounds by reaction of vinyl isocyanate with ethylene glycol, 1,4-butanediol, 1,6-hexanediol, and 1,8-octanediol were synthesized.[376] These monomers [(2–44) $R = (CH_2)_n$, where n = 2, 4, 6, and 8] were polymerized free radically under identical conditions. The monomer and catalyst concentrations were chosen in such a manner that only soluble polymers were produced. Table 2–5 indicates the

$$2\,CH_2=CH-N=C=O \;+\; HO-R-OH \longrightarrow$$

$$CH_2=CH-NH-CO-O-R-O-CO-NH-CH=CH_2$$

(2–44)

influence of solvent and the number of CH_2 groups present on the rate of polymerization. The degree of cyclization was calculated from the residual double bond content (extrapolated to 0% conversion). The percentage of pendant N-vinyl groups was determined by quantitative analysis of the acetaldehyde, formed upon acidic hydrolysis of the polymer (2–45) using a very sensitive colorimetric method.

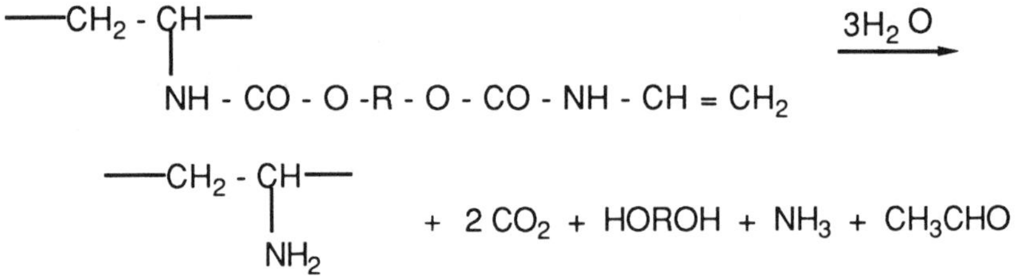

$$-CH_2-\underset{\displaystyle NH-CO-O-R-O-CO-NH-CH=CH_2}{\underset{|}{CH}}- \xrightarrow{3H_2O}$$

$$-CH_2-\underset{\displaystyle NH_2}{\underset{|}{CH}}- \;+\; 2\,CO_2 + HOROH + NH_3 + CH_3CHO$$

(2–45)

From Table 2–5 it can be seen that, at the beginning of the polymerization, the polymers contain about 83% cyclization which is relatively independent of the number of atoms in the ring. In the cyclopolymerization of α,ω-diolefins, the degrees of cyclization were considerably lower.[1-86]

However, due to the differences in monomer concentrations and the presence of insoluble fractions in the diolefin polymers, the results are not easily comparable.

Corresponding bis-N-vinyl compounds were prepared in good yield by the reaction of vinyl isocyanate with catechol, resorcinol, and hydroquinone. It is known that to bridge the para-position, a minimum of ten elements is necessary; for the meta-position,

Table 2–5 Polymerization of Bis-Vinylcarbamates

$CH_2=CH\text{-}NH\text{-}CO\text{-}O\text{-}(CH_2)_n\text{-}O\text{-}CO\text{-}NH\text{-}CH=CH_2$
Monomer concentration: 025 mole/1;
Catalyst concentration: 1.2×10^{-2} mole/1
AIBN; temp. 50°C

n	Projected ring size	Polymerization time for 40% conversion (h) In DMF	In DMSO	Extent of cyclization[a] (%)
2	11	8.0	1.9	83
4	13	11.2	2.6	85
6	15	13.5	3.1	83
8	17	14.5	2.8	82

[a] For polymerization in DMF, extrapolated to 0% conversion.

only nine are needed. The bis-N-vinyl compounds under discussion offer a maximum of nine elements, apart from the benzene ring carbons, toward ring formation. Intra-intermolecular chain propagation is therefore only possible with the catechol and resorcinol derivatives. Indeed, both N-vinyl groups of the hydroquinone derivative polymerized almost independently and eventually, crosslinking occurred. On the other hand, the catechol derivative, under the same conditions, gave a soluble polymer with approximately 70% cyclization. These results agree qualitatively with those detailed earlier on the allyl esters of the three isomeric phthalic acids.[1-12, 377]

Several unsymmetrical silane monomers,[378] capable of forming seven- or eight-membered rings, were synthesized and cyclopolymerized in an effort to determine the feasibility of formation of larger rings through use of coordination type initiators. It was concluded that the predicted cyclizations occurred. Allyldimethyl-(3-buten-1-yl)silane and allyldiphenyl(3-buten-1-yl)silane are capable of producing polymers containing a seven-membered ring.

Dimethylbis(3-buten-1-yl)silane and diphenylbis(3-buten-lyl)silane are capable of producing polymers containing an eight-membered ring. The diphenyl monomers gave benzene-soluble, solid polymers containing a low degree of residual unsaturation, evidence interpreted that the predicted cyclization had occurred. The dimethyl derivatives gave only viscous oils or semisolid products, which were not characterized.

Radical initiated homopolymerization of triallidene sorbitol (2–46) and tricro-

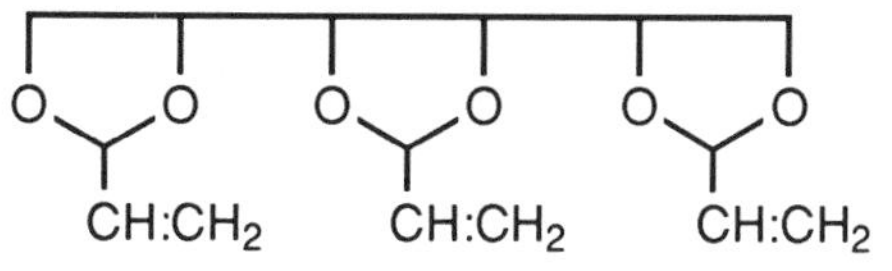

O O O O O O

$CH{:}CH_2$ $CH{:}CH_2$ $CH{:}CH_2$

(2–46)

tylidene sorbitol,[379] monomers which are functionally capable of producing ring sizes of nine members by the cyclopolymerization mechanism that was described. The ratios of the unimolecular cyclization rate to the total rate of bimolecular propagation and uncyclized radical chain transfer for the homopolymerizations were determined.

The polymerization of o-divinylbenzene has been reinvestigated.[25, 26] The kinetics and energetics of the process strongly support a proposal that the polymers are mainly comprised of seven-membered rings.

Large rings have been synthesized from a range of other monomers. Work on the polymerization of 1,3-bis(4-vinylphenyl)propane and similar compounds has been reviewed.[380] Cyclopolymers having [3.3]paracyclophane repeating units are only obtained by a cationic mechanism. Ethylene glycol divinyl ether has been cyclopolymerized,[381] by radical and cationic initiators, to polymers having six- and seven-membered rings, respectively, with some residual unsaturation. Diethylene glycol divinyl ether,[382] by cationic initiation, yields a soluble polymer with ten-membered rings.

Tetraethylene glycol divinyl ether was reported to yield a polymer containing 16-crown-5 units that was effective as a phase-transfer catalyst.[383, 384] The divinyl monomer, 1,2-bis(2-ethenyloxyethoxy)benzene was shown[385] to cyclopolymerize *via* cationic initiation to yield the 13-membered ring polymer.

Di-3-butenyldiphenylphosphonium bromide, a monomer functionally capable of forming a cyclic polymer containing an eight-membered ring, showed little tendency to polymerize by free radical initiation.[341] Macrocyclic rings have been formed from a number of other monomers by cyclopolymerization and cyclocopolymerization.[384,385]

A number of dimethacrylates have been synthesized and cyclopolymerized, including dialkylsilyl dimethacrylates.[289-291] Divinyl oxalate[292] and divinyl esters of other dibasic acids[293] have given polymers with cyclic structures. Hydrolysis to poly(vinyl alcohol) yields information on the propagation mechanism. Detailed studies of the polymerization of diallyl compounds, particularly diallyl dicarboxylates have been continued.[386-388] Although poly(allyl propyl phthalate) contains about 15% of the head-to-head structure, on radical initiation at 80°C, polymers of dialyl phthalate contained lesser amounts, which were dependent upon the extent of cyclization. By extrapolation of results, completely cyclized poly(diallyl phthalate) was found to consist exclusively of an eleven membered ring (2–47) from head-to-tail propagation. This is not the case

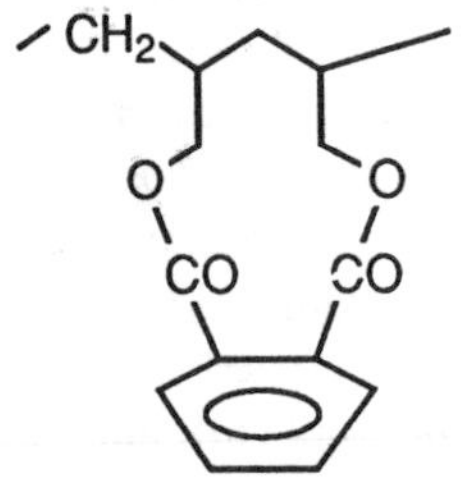

(2–47)

with polymers from diallyl cis-cyclohexane-1,2-dicarboxylate or diallyl succinate, where at first a decrease in head-to-head structures was noted but then an increase occurred at high extents of cyclization. An explanation based on steric factors was provided. Allyl esters of unsaturated acids have also been studied in detail.[389]

The possibility of some cyclization occurring during the polymerization of a non-conjugated diene had been recognized earlier. The fact that observed gel points for some dienes occurred later than calculated by Stockmayer's equation was explained on the basis that the equation failed to account for occasional cyclization. Earlier reference was made to these studies.[1-10 thru 1-13, 368-373]

To clarify the possibility of preparing a polymer that contained head-to-head (HH) methyl methacrylate units, radical cyclopolymerization of o-dimethacryloyloxybenzene was investigated. The intramolecular propagation proceeded mainly by a head-to-tail (HT) mechanism.[390] Preparation, thermal properties, and reaction of HH cyclopolymers by polymerization of N-substituted dimethyacrylamide derivatives to soluble HH cyclopolymers have been studied.[391] A recent study included preparation and characterization of poly(methylacrylate) and poly(methyl methacrylate) consisting of HH and HT units through cyclopolymerization of acrylic and methacrylic anhydrides. The content of HH units of these HH/HT polymers was determined by ^{1}H-NMR and ^{13}C-NMR spectra.[326]

The kinetics of polymerization of N,N′-methylenebisacrylamide were studied using Ce(IV)-thiourea redox system as initiator. The rate of polymerization was proportional to $[Ce(IV)]^{1/2}$, $[thiourea]^{1/2}$ and $[monomer]^{3/2}$. A cyclopolymerization mechanism fits in with the experimental results.[392]

During studies of cyclopolymerization in the presence of alkylaluminum chlorides, cyclopolymerizations of 2-(o-allylphenoxy)ethyl methacrylate and the higher homologs in the 11–19-membered-ring region were investigated. The addition of alkylaluminum chlorides to these monomers was effective for increasing the extent of cyclization. The effect of alkylaluminum chlorides was remarkable in formation of the 11-membered ring, but was less so for lower-membered rings.[393]

Direct observation of the propagating species in the cationic polymerization of 1,3-bis(p-vinylphenyl)propane in 1,2-dichloroethane was accomplished. Stopped-flow spectroscopy was used to study the cationic propagating species in the cyclopolymerization of 1,3-bis(p-vinylphenyl)propane and 1,2-bis(p-vinylphenyl)ethane by CF_3SO_3H in $C_2H_4Cl_2$. The cation generated from 1,3-bis(p-vinylphenyl)propane showed a bathochromic shift at 360 nm, exceeding those of cations from 1,2-bis(p-vinylphenyl)ethane (345 nm) and p-methylstyrene (332 nm). This shift was ascribed to the stabilization of the cationic center by an intramolecular interaction through space with the adjacent terminal styryl group, which was also suggested by molecular orbital calculations. This interaction facilitated cyclopolymerization.[394] Cationic cyclopolymerization of 1,3-bis(4-vinylnaphthyl)propane and the polymer structure have been studied. The cyclopolymers contained predominantly syn-[3,3](1,4)naphthaleneophane units in the main chain.[395]

Cyclopolymerization of ethylene glycol divinyl ether and photocrosslinking reaction of its polymer has been studied. Soluble polymers with methylene-1,4-dioxane-2,3-diylmethylene and 2-[2-(vinyloxy)ethoxy]ethylene structural units were obtained.[396]

Polymerization of N,N′-methylene bisacrylamide was carried out with a redox initiator. A polymerization mechanism involving cyclopolymerization in the propagation step was suggested.[397]

Cyclopolymerization of the divinyl ethers, (-)-(S)-2,2′-bis(2-vinyloxyethoxy)-1,1′-binaphthyl or (-)-(S)-2,2′-bis[2-(2-vinyloxyethoxy)ethoxy]-1,1′-binaphthyl produced optically active polymeric products, having a crown ether structure.[398]

Poly(triethylene glycol dimethyacrylate) obtained by microhetereogeneous, three dimensional polymerization at 293°K in the presence of 2,2,6,6-tetramethylpiperidine-1-oxyl inhibitor, introduced at various stages of the polymerization, exhibited higher compression elasticity modulus (E) than uninhibited samples at the same conversion.[399] Polymerization of triethylene glycol dimethacrylate gave a soluble polymer of cyclolinear structure, containing $\leqslant$1–2 long–chain branches per macromolecule. Gel formation was observed on polymerization of other dimethacrylate compounds.[400]

Electron spin resonance studies of solid-state polymerization of irradiated N-(p-bromophenyl)dimethacrylamide and sym-dimethacryloyldimethylhydrazine have been published. The intermediate radicals, formed in the cyclopolymerization of both monomers, were studied. An initiation radical could be observed in each case.[401] A similar study of the cyclopolymerizability of N,N′-dimethacryloylmethylhydrazine and solvent effects on its polymerization have been published. Cyclopolymerizability of the monomer was explained by the difference in the reactivities of the two methacryloyl groups. Solvents influenced not only the polymerization rate of N,N′-dimethacryloylmethylhydrazine but also the structure of the resulting poly(N,N′-dimethacryloylmethylhydrazine) to a small extent. The polymerization of sym-dimethacryloyldimethylhydrazine was not affected by solvents.[402]

Thermal and photoinduced polymerization of 1,3-bis(p-vinylphenyl)propane were carried out by using tetracyanoethylene as a strong electron acceptor.[125] The cyclopolymerizations produced polymer containing [3.3] paracyclophane units in the main chain. It was suggested that both polymerizations had proceeded by a cationic mechanism.

Cyclopolymerization for 2,2′-dimethyl-5,5′-diphenylmethane dimethacrylate *via* free radical and anionic initiation gave mainly heterotactic polymers, except for butyllithium which gave mainly isotactic polymers.[289]

Reactions of phenylene, naphthylene, biphenylene and methylenediphenylene were polymerized by radical and ionic initiators to give cyclopolymers as well as crosslinked polymers.[116]

Poly(vinyl alcohol) derived from divinyl oxalate has been studied.[292] Bulk polymerization of the monomer gave organic solvent-insoluble poly(divinyl oxalate), whereas, solution polymerization in benzene, ethyl acetate, or acetonitrile gave organic solvent-soluble polymers. Further studies indicated that poly(vinyl alcohol) prepared from the insoluble polymer contained more head-to-head linkages than the usual polymer obtained prepared from vinyl acetate.

The divinyl esters of the symmetrical aliphatic dibasic acids of 4–12 carbon atoms were polymerized *via* radical initiation, and the polyvinyl alcohol derived was studied.[293] The extent of cyclopolymerization varied from 15–65%. Monomers capable of yielding even-membered rings cyclopolymerized more easily than those of odd-membered rings. The poly(vinyl alcohol) derived showed a similar structure with respect to 1,2-glycol content and stereoregularity to that from poly(vinyl acetate).

Copolymerization of vinyl n-butyl ether with divinyltetramethylene glycol and divinylethylene glycol gave soluble copolymers with I_2 as initiator.[382] The divinyl ethers also gave homopolymers having 9- and 10-membered rings, respectively.

The reaction kinetics of polymerization reactions involving cyclization processes was studied theoretically and formulas were derived with which the degree of cyclization could be calculated from experimentally obtainable amounts.[1-15]

Copolymerization of diethylene glycol dimethacrylate with acrylonitrile was studied, as well as homopolymerization of the acrylate monomer.[403] The mechanism of cyclopolymerization was discussed in terms of a cyclization constant. The extent of cyclization varied in presence and absence of a diluent.

The fraction of residual unsaturation was studied for polymers of diallyl terephthalate and diallyl isophthalate prepared at 80° and 100° with benzoyl peroxide as initiator.[404] The results confirmed the postulate of the formation of primary linear chains by opening of the double bond in the monomer as well as confirming ring formation in diallyl phthalate polymerization. The fraction of residual unsaturation for poly(diallyl phthalate) was 10% as compared to 30% for the other polymers under similar conditions.

Several di- and triolefins of the octadecene series were polymerized with metal coordination catalysts. The dienes yielded soluble homopolymers.[405] The trienes gave only crosslinked polymers. In a further study of cyclic polymerization reactions, poly[2,2′-methylenebis(N-methylacrylamide)] was obtained *via* radical initiation.[406] On the basis of all the structural evidence, a linear structure with rings in the chain was ascribed to the polymer obtained.

N,N′-Methylene bis-acrylamide was used in a process for shrinkproofing proteinous fibers, e.g., wool, by treating the fibers with the monomer in presence of a redox system of radical initiation of polymerization.[407]

A polarographic study of the polymerization rates of cyclopolymerizable monomers such as divinyl and diallyl adipates in presence of radical initiators was carried out.[408] The diallyl monomers polymerized relatively much slower than the corresponding divinyl monomers. In a continuing study of synthesis of N-substituted methacrylamides,

N,N′-alkylenebismethacrylamides were synthesized by reaction of methacrylic anhydride or methacryloyl chloride with suitable diamines.[409]

Allyl silanes were polymerized and copolymerized with propylene by use of the trialkyl aluminum-titanium tetrachloride catalytic system.[410] Cyclopolymerization of bis-N-vinylcarbamates *via* free radical initiation was studied.[411] *The results related to the degree of cyclization were consistent with other reported results dealing with formation of highly probable vs. less probable ring closures.*

Head-to-head polymers isomeric with the corresponding head-to-tail structures obtained by conventional polymerization of vinyl monomers were claimed.[412] The polymers were obtained by cyclopolymerization of monomers which contain two double bonds in such positions that during a free radical initiated addition polymerization, a ring of six atoms would be formed. The two double bonds must be separated by four atoms to result in the head–to–head polymers after being hydrolyzed (2–48).

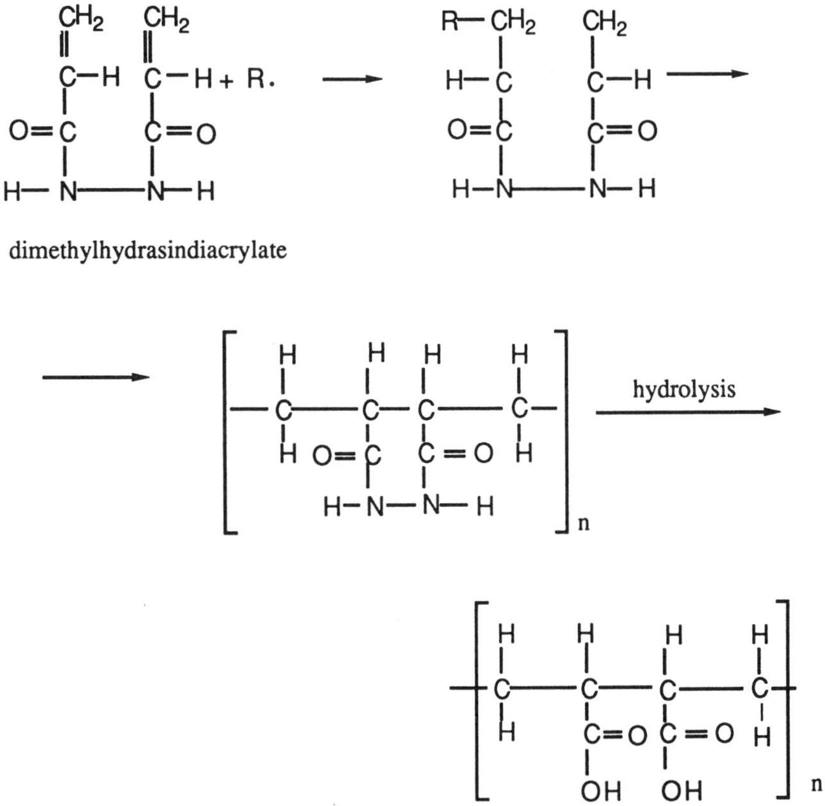

dimethylhydrasindiacrylate

hydrolysis

(2–48)

Perfluorinated divinyl ether polymers, described as thermosetting, have been synthesized from monomer structures that fit the pattern of cyclopolymerizable monomers.[413] The monomers studied were the 1,7- to 1,10-diene structures, more specifically, di(perfluorovinyl)ethers of the hypothetical perfluoro-ethylene glycol, tri-(perfluoromethylene)glycol, tetraperfluoromethylene glycol and pentaperfluoromethylene glycol.

In a study of photopolymerization of dimethacrylate esters of glycols, the rate of reaction decreased sharply with time in the initial stages.[414] This was ascribed to the existence of ordered regions among the monomers, confirmed by lowered activation energy of the process with declining temperature. This was not correlated with possible cyclization reactions during the process.

A prepolymer of diallyl phthalate was prepared by polymerizing monoallyl phthalate followed by esterification with allyl alcohol.[415] A process for synthesis of a polymerizable prepolymer of diallyl phthalate which involves heating the monomer in presence of a radical inhibitor has been described.[416] The prepolymer had a relatively low iodine value, indicating that considerable cyclopolymerization had occurred. Solid prepolymers of diallyl phthalate, diallyl isophthalate, and diallyl terephthalate were obtained by using chain transfer agents, e.g., carbon tetrachloride to control gelation.[1-32]

Polymeric N-vinyllactams, vinyl esters and acrylate esters were crosslinked with α,ω-diolefins when heated with di-(tert-butyl)peroxide.[417] 1,7-Octadiene was specifically cited as an example. The relationship of this and closely related α,ω-dienes raised a question regarding the efficiency of these compounds as crosslinking agents.

The cyclopolymerizable monomers, triallyl- and trimethallyl isocyuranate have been synthesized in high yields.[418]

Diallidenepentaerythritol was copolymerized with styrene, acrylonitrile, vinyl acetate, methyl methacrylate and methyl acrylate, using a free radical initiator.[419] The monomer reactivities were determined, as well as the Q and e values for the diallidene monomer. The intrinsic viscosities of the copolymers were determined, and were found to decrease rapidly with increasing ratio of the allidene monomer in the monomer mixtures. The polymers were soluble and therefore not crosslinked; however, cyclopolymerization was not considered. Since the allidene monomer is a 1,10-nonconjugated diene, and is a rather inflexible molecule, cyclopolymerization may be highly unlikely.

The kinetics of high-intensity electron beam-initiated polymerization of bis(2-methacryloyloxyethyl)-4-methyl-m-phenylenediurethane was studied during network formation up to complete gelation and up to 56% conversion of unsaturation.[420] The extent of cyclization was not considered, although literature references were given which concurred that deviations from gel point theory lead to a low estimate of the kinetic chain length.

The polymerization of diallyl isophthalate and diallyl terephthalate was studied by free radical initiation.[421] The ratio of the rate constants of monomolecular propagation to that of bimolecular propagation of the uncyclized radical was determined as 1.6 mole/L for the isophthalate ester and 0.6 mol/l for the terephthalate ester. Diallyl phthalate was also polymerized *via* radical initiation and studied kinetically under various conditions.[422] The ratio of the cyclization rate to propagation rate was discussed on the basis of the degree of polymerization and the degree of unsaturation. The activation energy of cyclization was larger than that of propagation by 1.6 kcal/mole.

Diallidenepentaerythritol was polymerized *via* free radical initiators to give polymer yields of less than 10% in most cases.[423] The structure of the polymers was studied as well as the polymerization process, and it was concluded that both degradative chain transfer and cyclization occurred. The ratio of the cyclization velocity to the sum of the velocities of chain propagation and chain transfer was calculated to be 0.276. Copolymerization of diallyl phthalate with allyl benzoate *via* free radical initiation was studied.[424] The monomer reactivity ratios were determined and the ratio of the rate constant for intramolecular cyclization to that of propagation of the uncyclized radical was determined to be 6.85 mole/L.

The solid state photopolymerization of trans,trans-2,5-distyrylpyrazine was investigated, and the polymerization observed through a polarizing microscope and *via* x-ray diffraction patterns.[425] The polymer was highly crystalline and the direction of polymer axes was simply related to that of the monomer crystal. The four-center type of polymerization was shown to proceed topochemically by a photochemically induced stepwise mechanism.

A series of diallylethers of the tetrachloroxylenes were prepared, characterized, epoxidized, homopolymerized and copolymerized with maleic anhydride.[426] The homopolymers formed with peroxides at $>130°$ were soluble and did not cross-link at temperatures as high as 150°. Although cyclopolymerization was not suggested, the reported evidence indicated that it may have occurred to a considerable extent, since bromine numbers reported were far below theory. The diepoxides described were not studied further.

Polymerization of diallylcarbonate, a 1,8-diene, was studied with the objective of producing a useful product, rather than to study its structure.[427] However, a soluble prepolymer could be obtained at 80° *via* free-radical initiation. Cyclopolymerization was not considered; however, the observations reported were not inconsistent with a reasonable high degree of cyclization.

A copolymerization study of 5-allyl-1,3-diallyl isocyanurates with n-butyl methacrylate led to the conclusion that improved properties of the copolymers were obtained relative to poly(n-butyl methacrylate).[428] The best monomer ratio was 95:5 of the acrylate-isocyanurate. The possibility that cyclopolymerization of the comonomer may have contributed, although not considered by the authors, is apparent.

Polymerization studies in dimethallylidene-pentaerythritol and dicrotylidenepentaerythritol were carried out by free radical initiation, and the results compared with those previously reported on the corresponding diallidene monomer.[429] Cyclization during polymerization was less in these cases because of increased steric hindrance and degradative chain transfer occurred more readily than on the allidene group.

Dynamic mechanical properties and glass transition temperatures were measured for crosslinked polymers derived from diallyl succinate monomers.[430] The mobility of the diester having an eleven-membered ring and of homologous structures which were introduced into the crosslinked polymer system was discussed on the basis of the parameter for cyclopolymerization of a monomer, dynamic mechanical properties and glass transition temperatures.

A comparative study of the polymerization properties of N,N′ (disubstituted methylene) bisacrylamides with those of N,N′-methylenebisacrylamide was carried out.[431] The monomers were shown to readily undergo copolymerization with a variety of vinyl monomers.

The study of four-centered type photopolymerization was extended to α,α'-dicyano-p-benzenediacrylic acid and its ester and amide derivatives.[432] The polymerization was found to proceed in a manner analogous to similar monomers studied earlier. The polymers obtained were highly crystalline with high melting points and limited solubility.

Copolymerization of diallidenepentaerythritol with maleic anhydride *via* free radical initiation was studied. The resulting polymer always contained a molar ratio of vinyl group to maleic anhydride of 1:1 and the formation of a molecular complex during the copolymerization was postulated.[433] It was kinetically estimated that the copolymerization may have proceeded *via* a pseudo-ter-polymerization of the vinyl group, maleic anhydride and the charge-transfer complex.

Dicarboxylic acids were prepared by free radical initiated addition of acids to both ends of α,ω-diolefin molecules.[434] 2,2,9,9-Tetramethylsebacic and 2,2,8,8-tetramethylazelaic acids were prepared by addition of 2-methylpropanoic acid to 1,5-hexadiene and 1,4-pentadiene, respectively. However, 1,7-octadiene led to substituted cyclic monocarboxylic acids, resulting from addition of the carboxylic acid radical to one end of the diene, followed by cyclization of the free radical formed. A six or seven-membered ring structure was probably formed.

A method of preparation of allyl esters of phthalic, isophthalic, trimellic, or maleic acids, which involved reaction of the sodium salts of the acids with allyl chloride was described.[435] Diallyl phthalate was cyclopolymerized with vinyl acetate and methyl acrylate *via* radical-initiation.[436] The copolymerization parameters were reported. Cyclic, linear poly(dimethylsilylene dimethacrylate) and poly[α,β-bis(methacryloyloxy)tetramethyldisiloxane] were obtained by free radical initiated cyclopolymerization of the corresponding monomers.[437] The following structures were postulated (2–49).

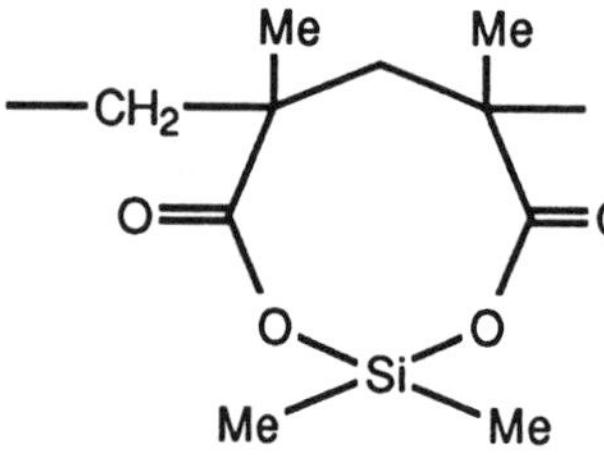

I   II

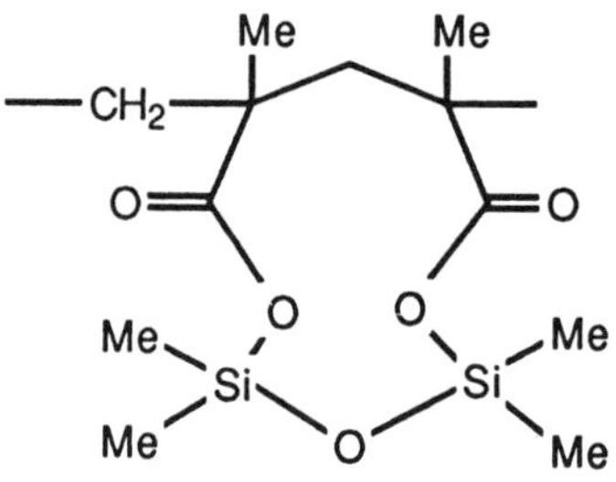

(2–49)

The rate of cure of diallyl phthalate, using differential scanning calorimetry, was determined.[438] Cyclopolymerization of diallyl phthalate with diallyl isophthalate, diallyl terephthalate and diallyl oxalate was studied and the monomer reactivity and cyclization reactivity ratios determined.[439]

Photopolymerization of platelike crystals of 2,5-distyrylpyrazine gave poly(2,5-distyrylpyrazine) crystals which approximately duplicated the molecular arrangement in the monomer crystals.[440] Recrystalization of the polymer crystals resulted in a different x-ray pattern, indicating that the original crystal form was not the most stable. The term "lattice-duplicating polymerization" was proposed for this rearrangement of monomer crystals to polymer crystals.

Photopolymerization and photocopolymerization of o-, m-, and p-phenylene diacrylates were studied.[441] The p-isomer polymerized most rapidly but gave a crosslinked polymer. The o-isomer was reported to give a ladder polymer *via* cyclization. Quite likely, the structure of the cyclized polymer was as shown below (2–50). A soluble copolymer was formed with styrene and the o-isomer, but insoluble copolymers were obtained with the m- and p-isomers.

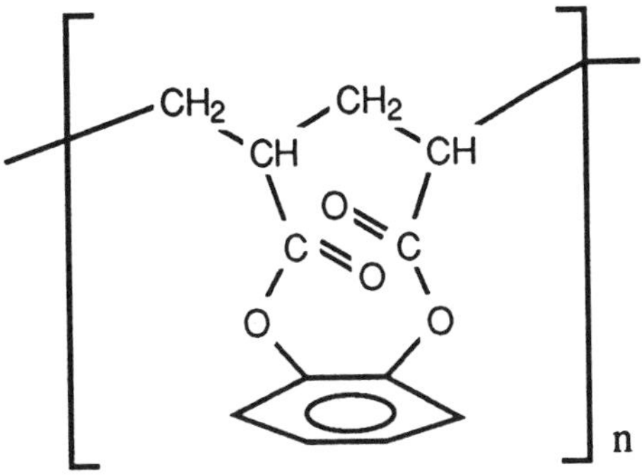

(2–50)

In the copolymerization of diallyl phthalate with methallyl benzoate, diallyl isophthalate, and diallyl terephthalate by radical initiation, residual unsaturation in the copolymers varied.[442] Intramolecular cyclization and steric hindrance differences accounted for these results. Monomer reactivity ratios were calculated for cyclized and uncyclized radicals. Diallyl phthalate was polymerized in aqueous emulsion to 30–60% conversion and a film of the prepolymer was further reacted to an insoluble polymer by heating in presence of a radical initiator.[443] Diallyl phthalate was cyclocopolymerized with allyl benzoate, vinyl acetate, methyl acrylate and vinyl chloride by free radical initiation.[444] The reactivity ratio of the cyclized radical was smaller than that of the uncyclized radical, owing to steric hindrance in the addition of the cyclized radical to the diallyl monomer.

The free radical copolymerizability of some vinyl monomers decreased as follows: diallyl phthalate, vinyl acetate, and allyl benzoate, with essentially no differences between diallyl tetrachlorophthalate and diallyl tetrabromophthalate.[445] Cyclization during polymerization decreased as follows: Diallyl tetrabromophthalate, diallyl tetrachlorophthalate, and diallyl phthalate. Diallylphenylphosphonate was bulk polymerized by free radical initiation to give a crosslinked polymer.[446] Cyclic structures were not considered. Diallyl phthalate-styrene copolymer was obtained by bulk polymerization by free radical initiation.[447] The monomer reactivity ratios of the monomer pair were $r_1 = 0.145 \pm 0.099$ and $r_2 = 25.5 \pm 2.6$, respectively. The ratio of rate constants $k_c/k_1 = 0.27$ mole/dm^3.

Poly[1,3-bis(4-vinylphenyl)propane] was prepared by polymerizing the monomer by free radical initiation.[448] Cationic initiation led to polymer having no pendant vinylphenyl groups.

Four α,ω-bis(4-vinylphenyl)alkanes $CH_2 = CH\text{-p-}C_6H_4\text{-}(CH_2)_nC_6H_4\text{-p-}CH = CH_2$, n = 1–4, were prepared and polymerized for n = 3 and n = 4 to give [3,n] paracyclophane containing polymers by cationic initiation, whereas monomers in which n = 1 and n = 2 gave polymers formed by intermolecular coupling of two styryl double bonds followed by H transfer.[449]

Thermally stable imide-diallyl resins were prepared as raw materials for multilayer printed circuit boards.[450]

A study of branched polymers, based on diallyl isophthalate, showed that a linear increase in intrinsic viscosity occurred and a nonlinear dependence on molecular weight was observed with increased time.[451] The degree of residual unsaturation decreased from 40 to 10% with conversion. Cyclopolymerization of diallyl tartrate diacetate has been studied.[452] Its rate of polymerization was less than that for diallyl tartrate but greater than that of diallyl succinate. The cyclization constant k_1/k_2, was found to be 2.5 mol/L compared to 2.1 and 2.8 mol/L for diallyl tartrate and diallyl succinate, respectively.

The cyclopolymerizability of 1,7-dienes, for example, sym-dimethacrylodimethylhydrazine was inversely related to the polymerizability of their monofunctional counterparts, for example, isobutyroyl-2-methacryloyl-1,2-dimethylhydrazine.[453] The difference was not related to their ability to form intermolecular hydrogen bonds.

Diallyl formal, diallyl acetal, and diallyl butyral were polymerized at 80°C in presence of benzoyl peroxide to degrees of polymerization of 13.4, 6.3, and 4.9, respectively. The ratios of the unimolecular cyclization rate constant to that of bimolecular propagation were 8.1, 9.3, and 12.1, respectively, compared with 32 and 8.6 M, respectively, for diallyl phthalate.[454]

Enhanced polymerizability was observed in the polymerization of diallyl hydroxydicarboxylates.[455] In radical polymerization, the polymerization rate increased in the order diallyl succinate < diallyl malate < diallyl tartrate. The effect of intermolecular hydrogen bonding on the polymerization was discussed.

Synthesis and cyclopolymerization of sym-dimethacryloyldimethylethylenediamine and sym-dimethacryloyldimethylhexamethylenediamine were studied.[456] Cyclization occurred to a large extent in both monomers.

Diallyl isophthalate was used as a crosslinking agent in blends with polyesters such as poly(ethylene terephthalate).[457] The extent of cyclopolymerization was not determined.

Radiation-cured methacrylate-dimethacrylate copolymers gave optimum strength results at 5-10% dimethacrylate contents.[458] Results varied with the connection number (molecular bridge length) between the methacrylate groups, indicating that the extent of cyclization was a critical factor.

In an attempt to determine the characteristic hydroxyl group effect observed previously in the cyclopolymerization of diallyl tartrate, copolymerizations of several monomers with and without hydroxyl groups with styrene and vinyl acetate were studied.[459] Reactivity ratios were determined for the two systems, and the results were discussed in terms of hydrogen bonding in the hydroxyl group containing monomers.

Polymerizations of m- and p-divinylbenzenes were carried out without crosslinking with acylium salts and trifluoromethyl sulfonic acid.[460] A stepwise polymerization mechanism was proposed which led to an unsaturated but linear polymer backbone.

Head-to-head addition for the cyclopolymerization of diallyl and dimethallyldicarboxylates was enhanced by an increase in acid-group bulkiness and an increase in temperature above the ceiling temperature for head-to-tail addition.[461] In a study of branched polymers, it was indicated that branched molecules of poly(diallyl isophthalate) consisted of a set of linear branch chains of degree of polymerization, approximately 20, each of which contained randomly distributed cyclic structures and unreacted double bonds, capable of further polymerization.[462]

Radical cyclopolymerization of diallyl dicarboxylates in presence of carbon tetrachloride gave telomers, except for diallyl oxalate which gave a gel.[463] An explanation was proposed.

Macrocyclic ring-containing polymers were synthesized and characterized.[464] The monomers studied were divinyloxy compounds capable of undergoing cyclopolymerization to yield seven- or 13-membered rings, all containing two or more ether oxygens. The resulting large ring structures were capable of crown ether activity (2–51).

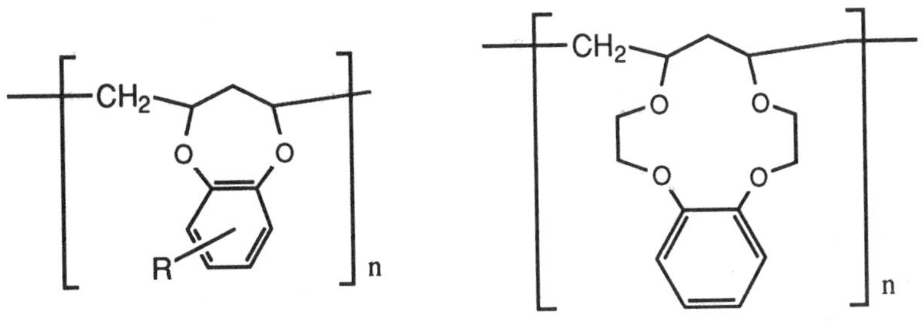

(2–51)

A differential scanning calorimetric study of the polymerization of diallyl dicarboxylates was conducted at elevated temperatures.[465] Two exothermic peaks were observed, the second of which was attributed to postpolymerization of unreacted monomer.

Divinyl ethers of $HO(CH_2CH_2O)_4H$ were cyclopolymerized with boron trifluoride etherate to give polymers which contained mainly the cyclic units shown (See 2–52) along with random crosslinks.[466] The polymer exhibited a phase-transfer catalytic ability comparable to that of 18-crown-6.

Radical copolymerization of acryloyl chloride with 2-hydroxylpropyl methacrylate occurred in dioxane at 60° under nitrogen in the presence of azobisisobutyronitrile.[467] Conversion increased while acryloyl chloride mol fraction was unchanged with increasing reaction time. It is quite probable that the actual monomer was the derived acrylic ester of 2-hydroxypropyl methacrylate, followed by cyclopolymerization.

The rate of cyclopolymerization of diallyl phthalate in bulk at 80° in presence of CBr_4 or CCl_4 decreased with decreasing chain length.[468] The effect of remote units on the formation of cyclized units may be considerable, and the steric suppression effect of bulky side chains on the intermolecular propagation of uncyclized radicals was enhanced with increasing chain length with promotion of intramolecular cyclization of uncyclized radicals. In a study of cyclopolymerization of reactive oligomers, for example, glycol bis(allylphthalates) and glycol bis(allyl succinates), it was shown that the rates of the glycol phthalates were less than that of diallyl phthalate but those for the glycol succinates were greater.[469] The results were explained on the basis of excluded volume effects caused by crosslinking.

Radical telomerization of diallyl phthalate and diallyl terephthalate with carbon tetrabromide was extensively investigated.[470] Telomerization kinetics was discussed and chain-transfer constants were estimated to be 57.9 and 53.8, respectively, for the two esters.

Radical cyclopolymerization of N,N′-dimethyl-N,N″-dimethacryloylureas was investigated in connection with a program designed to prepare and study head-to-head vinyl polymers.[471] The polymer was completely soluble in benzene and was postulated to consist of a seven-membered cyclic unit formed *via* alternating intra-intermolecular propagation. Heating gave a head-to-head cyclopolymer of N-methyl-N,N-dimethacrylamide.

1,2-bis[2-(2-Vinyloxyethoxy)ethoxy]benzene was cationically cyclopolymerized to give a poly(crown ether) which could be used to extract Li, Na, K, Rb, and Cs picrates into CH_2Cl_2.[472]

Ethylene glycol bis(methyl fumarate) was prepared as a new type of divinyl compound and reactive oligomer. The cyclization constant was 1.64 mol/L, being rather low compared with diallyl oxalate.[473] Copolymerization with diallyl phthalate was explored and a remarkable rate enhancement was observed for copolymerization. The monomer reactivity ratios were $r_1 = 0.96$ and $r_2 = 0.025$, with the r_1 value being reduced compared with the copolymerization system. The results were discussed from the standpoint of steric effects on the polymerization of the fumarate as an internal olefin. Thermal cyclization of monoesters of maleic anhydride copolymers, for example, the propylene copolymer monomethyl ester, provided information about their structure. Solid-phase cyclization was used.[474]

The quantum yields of generation and growth of chains in photoinitiated solid-state cyclopolymerization of di-ethyl-p-phenylenediacrylate at 150–300° K were determined.[475] The effective chain growth in the solid phase at low temperatures was related to partial transformation of the energy of chemical reactions into the energy of stresses decreasing the growth barrier. A review of cationic macrocyclopolymerization, a novel synthesis of crown ether polymers, has been published. Synthesis and characterization of the monomer and conversion to the polymer were described (2–52).[476]

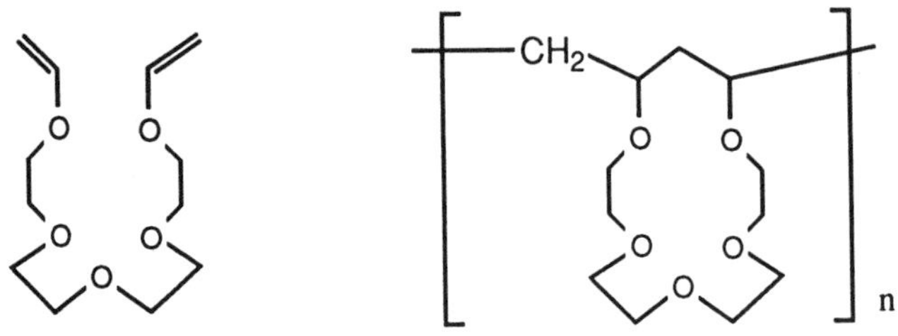

(a) (b)

(2–52)

N,N′-Methylenebisacrylamide was cyclopolymerized by a radical mechanism.[477] The cyclization constant was determined. Cyclopolymerization kinetics of this monomer, initiated by a redox couple, was investigated.[478] Rate equations were derived. Also, a kinetic study of photosensitized cyclopolymerization of N,N′-methylenebisacrylamide, using uranyl ion, was made in an effort to establish the mechanism.[479] Evidence for cyclization of the radical prior to propagation was obtained.

Cationic cyclopolymerization of 1,2-bis[2-(2-vinyl-oxyethoxy)ethoxy]benzene was achieved by use of the HI/iodine initiator to give a polymer containing benzo-19-crown-6-units.[480] The molecular weight distribution (MWD) of the polymer was 1.4. Other studies were reported.

The quaternary salt of N,N-dimethylamine-ethyl methacrylate underwent spontaneous radical polymerization with allyl bromide in presence of persulfate at ambient temperature.[481] The allyl group was reported to be retained in the polymer. Activation energy of polymerization was 53.7 kJ/mol.

In continuing studies of polymerization of diallyl compounds, it was observed that no substantial differences in the gel points for the three isomeric diallyl phthalates existed.[482] Discussion of the unusual results in terms of the correlation between gelation and the difference in cyclization modes, and the differences in reactivity between the uncyclized and cyclized radicals for crosslinking was included.

Cyclopolymerization of divinyl and diepoxide monomers, for example, 1,2-bis(2-ethenyloxyethoxy)benzene and 1,2-bis(2,3-epoxypropyl)benzene, was carried out, producing polymers having benzo-13-crown-4 and benzo-9-crown-3 units, respectively, in the polymer chain.[483]

THEORETICAL INTERPRETATION

Cyclopolymerization of symmetrical 1,6-dienes has now been well established. It has been applied to a wide variety of monomer systems utilizing all known types of initiator systems. In general, the resultant polymers have contained five- or six-membered rings as the predominant structural feature with little residual unsaturation or crosslinking present. In an effort to provide a fundamental explanation of cyclopolymerization, it was proposed that an electronic interaction may occur between the unconjugated double bonds of 1,6-dienes, or between the intramolecular double bond and the reactive species after initiation.[1-17] Such an interaction can be represented in its simplest form by Structure 1-32. An electronic interaction of the type proposed in the ground state of a 1,6-diene would decrease the energy of the pseudocyclic conformation of the molecule. The molecule would then require a smaller entropy change in going from the ground state to the activated state required for the intramolecular propagation reaction. Similarly, the energy of the activated state would be reduced by an interaction between the intramolecular double bond and the reactive species after initiation. A ground state-activated state interaction would suggest an energetically favorable pathway for cyclopolymerization.

A statistical approach to cyclopolymerization has been taken[484] and the results compared with experimental observations. Probability calculations indicated that, under normal polymerization conditions, the intramolecular step leading to cyclopolymerization is not favored, on a purely statistical basis, compared to the competitive intermolecular step leading to branching and eventual crosslinking. The model (2-53) for this determination was based on a stepwise mechanism for the polymerization of 1,6-heptadiene. It was assumed that a reaction site was formed at the second carbon atom of the chain, C,

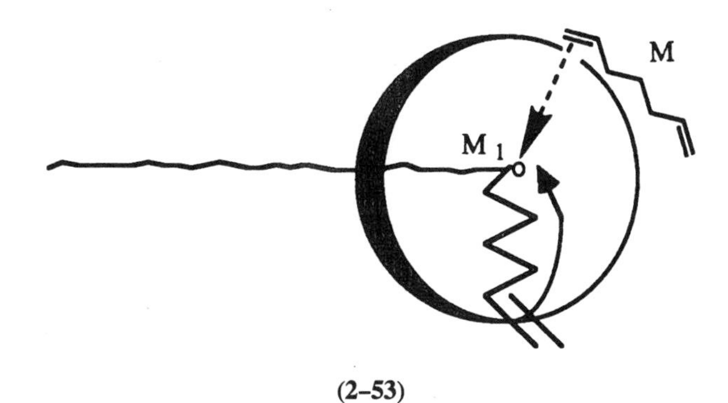

(2-53)

and the probability of C, being in a volume element surrounding this position was then calculated. The probability of any other double bond approaching within this same volume element was calculated merely by determining the density of double bonds in the system. The results could be used to explain the high degree of cyclization in dilute solutions, but they show that, statistically, at high monomer concentrations, the extent of cyclization should be less than 50%. The results, summarized in Table 2–6, which show that complete cyclization may occur at much higher monomer concentrations, were interpreted to mean that a more favorable pathway from 1,6-diene to cyclic polymer exists than would be predicted on a purely statistical basis.

However, steric effects were considered to be important in explaining the difference between the statistical and observed values of cyclization.[1-18] The rate of intermolecular reaction of an initiated monomer molecule would be expected to be lower than that of a similar monomer with less than five atoms in the pendant group. The presence of the large pendant group would tend to prevent intermolecular reaction from occurring at a rate that might be expected simply on the basis of the activation energy for the step. Further, the pendant double bond will frequently be presented to the reactive species in a conformation which is favorable for reaction.

The strong interest of polymer chemists in recent years in stereoregular polymers[1-92] has led to investigations concerning whether or not the cyclopolymerization process leads to any steric control during propagtion. Several publications[1-93,1-94] have demonstrated clearly that homogeneous free-radical polymerization of vinyl monomers to chains containing asymmetric atoms is stereospecific to an extent dependent on the temperature and other environmental factors. The polymers obtained by free-radical polymerizations conducted at low temperatures have been identified[1-95,1-96] as the predominantly syndiotactic form. In a study of stereoregulating influences during the free-radical polymerization of methacrylic anhydride, it was observed[1-97] that, indeed, stereochemical control does occur, and that temperature and solvent effects on this control are observable. By use of a nuclear magnetic resonance technique, the isotactic character of the poly(methyl methacrylate) obtained by methylation of the poly(methacrylic anhydride) prepared at various temperatures was found to increase with increasing temperature, while the syndiotactic and random-placement characters were found to decrease in an approximately parallel manner. These results were explained[1-98] on the basis of the strained conformation of the free radical formed after the ring-closure step. Even though the cyclic free radical would have a marked tendency

Table 2–6 Extent of Cyclization in Polymerization of 1,6-Heptadienes

Monomer	Concentration (mole/1)	Soluble polymer (%)	Estimated cyclization	Reference
Diallyl quaternary ammonium salts	>5.0	100	96–100	1-9
Diallyl silanes	1.2–2.3	92–100	>95	1-54, 1-56
1,6-Heptadiene	1.9–6.2	100	90–96	1-52
N,N-Diallylmelamine	Solid state	100	99	1-85
Acrylic anhydride	~8.0	100	98–100	1-27, 1-28
Mathacrylic anhydride	~5.0	a100	96–100	317

to assume planarity, complete planarity at this position would place considerable strain on the remaining bonds of the cyclic structure in which the energetically favorable chair conformation would ordinarily predominate.

Retention of some pyramidal character would thus be expected in this cyclic structure, and would favor an axial propagation. The axial propagation would be favored at low temperatures. It was shown that the favored axial propagation would lead to a new type of stereoregularity termed "syndioduotactic" and consisting of *ddll* sequences throughout the polymer chain. In support of this hypothesis, a polymer prepared at 15°C showed 7% isotactic, 48% syndiotactic, and 44% heterotactic character, whereas one prepared at 80°C showed 67% isotactic, 13% syndiotactic, and 20% heterotactic character.[1-97]

These results have been confirmed and extended[1-99] in a polymerization study of methacrylic anhydride to 46% conversion by ultraviolet irradiation in 50% toluene solution at −50°C. The polyanhydride was converted to the methyl and ethyl esters, and both were found to exhibit considerable crystalline character by x-ray diffraction measurements. By the nuclear magnetic resonance technique, the poly-(methyl methacrylate) was found to possess less than 2% isotactic character and equal syndiotactic and heterotactic character, each greater than 49%.

The unique nature of the above polymer is further substantiated by a study of monomolecular films or monolayers of the polymer.[1-100] The values obtained in this study were different from those of the other forms of poly-(methyl methacrylate) studied, and were in agreement with the "syndioduotactic" nature of this polymer.

At least two instances of stereoregulation in Ziegler-catalyzed polymerizations of nonconjugated dienes have been reported. A systematic study of the polymer obtained from 1,5-hexadiene has been made, and it was found that it possesses an all-*cis* structure (2-7).

The mechanism of free-radical cyclopolymerization of acryloyl- and methacryloylmethylallylamines has been studied. Cyclopolymerizability of N-allyl-N-methylacrylamide was investigated and compared with that of N-allyl-N-methylmethacrylamide which is known to yield a highly cyclized polymer with a small amount of pendant methacryloyl groups. The difference in cyclopolymerizabilities between the two monomers could be interpreted on the basis of the hypothesis that the lower the polymerizabilities of the monofunctional counterparts of unconjugated dienes, the higher their cyclopolymerizabilities.[485]

Fully identified cyclic structures formed *via* radically initiated cyclizations and cyclopolymerizations were explained, using frontier orbital theory. This theory, briefly introduced for general application to organic polymer chemistry, worked very well in explaining the ring size in cyclopolymerization except when steric hindrance was predominant in controlling the course of cyclization.[486]

MINDO/3 and Carbon-13 NMR results have been used to predict the free radical polymerizability of allyl monomers. Polymerizable monoallyl and diallyl monomers exhibited high negative eigenvalues for the C-H bond alpha to the allyl group and small delta theta values (distance between the beta- and gamma-carbon NMR peaks) as correlated by MINDO/3 calculations and ^{13}C NMR results, respectively. Monomers with low negative eigenvalues (corresponding to weak alpha-C-H bonds) underwent degradative chain transfer instead of polymerization. Monomers with intermediate eigenvalues or delta theta values polymerized with difficulty.[487]

Electrophoretic mobilities were determined of $(CH_2{:}CHCH_2)_2N^+(CH_3)_2Cl^-$, $CH_2{:}C(CH_3)CO_2CH_2CH_2CH_2N(CH_3)_2$, $CH_2{:}C(CH_3)CO_2CH_2CH_2N^+(CH_3)_2C_2H_5Cl^-$, $C(CH_3)CO_2CH_2-CH_2N^+(CH_3)_2CH_2CH_2CO_2^-$, and their corresponding polymers. Low electrophoretic mobility of the polymers of $(CH_2{:}CHCH_2)_2N^+(CH_3)_2Cl^-$, $CH_2{:}C(CH_3)CO_2CH_2CH_2N(CH_3)_2$ and $CH_2{:}C(CH_3)-CO_2CH_2CH_2N^+(CH_3)_2C_2H_5Cl^-$ supported the contention that insensitivity of the polymerization kinetics of $(CH_2{:}CHCH_2)_2N^+(CH_3)_2Cl^-$, $CH_2{:}C(CH_3)CO_2CH_2CH_2N(CH_3)_2$, and $CH_2{:}C(CH_3)CO_2CH_2CH_2N^+(CH_3)_2C_2H_5Cl^-$ in aqueous solutions to pH, ionic strength of the medium, and the nature of counterions reflects the formation of intramolecular complexes (*via* N^+ and ether O) having high affinity to counterions and giving stable ion pairs. These intramolecular complexes exist as growing macroradicals and polymers, and render them effectively neutral in spite of formal electric charges.[488]

The Monte Carlo method was used to model radical polymerization of polyfunctional monomers with regard to intramolecular cyclization, i.e., conformational changes in the growing chain which allow the reaction between two radicals or between a double bond and one radical of the same chain.[489]

In a continuing study of head-to-head polymers, N-phenyldimethacrylamide was cyclopolymerized to poly(N-phenyldimethacrylamide) containing 94% five-membered rings, but more importantly containing 6% six-membered rings in the cyclopolymer.[490]

Poly(diallyldimethylammonium chloride) was prepared by $Na_2S_2O_8$-initiated polymerization of the monomer in the presence of disodium versenate, and the effect of the concentration of monomer, $K_2S_2O_8$, disodium versenate, temperature, and pH of the reaction mixture on the yield and intrinsic viscosity of the polymer was determined.[120]

A paper relevant to this discussion of cyclopolymerizations deals with computer simulation of gelation. The influence of cyclization on the extent of gelation of telechelic linear prepolymers with tetrafunctional crosslinking agents at the critical point was investigated by computer simulation, using the upper-lower bound methods and scaling methods.[491]

Binding of methyl orange and its homologs by polyion complexes in aqueous solution has been studied. Polyion complexes of Na polymethacrylate and poly(diallyldimethylammonium chloride) which are insoluble in water and have an equal number of positive and negative charges, bind organic anions (methyl orange, ethyl orange, propyl orange, butyl orange, and pentyl orange) in aqueous solution. The significance of the hydrophobic and electrostatic interactions which accompany the binding were described.[492]

Complexation of poly(diallyldimethylammonium chloride) and 1-anilino-8-naphthalenesulfonic acid in water, aqueous NaCl, and methanol-water mixture was studied by measurements of fluorescence, UV spectroscopy, and potentiometric titration.[493] The maxima in the absorption and fluorescence spectra of 1-anilino-8-naphthalenesulfonic acid were shifted toward long-wave and short-wave spectral regions, respectively, in the presence of the polymer. Simultaneously, the intensity of fluorescence increased appreciably. The spectral change depended on the molar ratio of the interacting components, nature of solvent (water and methanol and aqueous KOH solutions were used) and ionic strength of the solutions. The electrostatic interaction of the components results in the formation of a 1:32 stoichiometric complex between the reactions.[494]

Interaction of poly(diallyldimethylammonium chloride) with ferro- and ferricyanide anions has been studied, and the conclusion was drawn that the binding of the anions proceeded by an anionic mechanism and was accompanied by a spectral shift of the cyanide bands.[495]

Complexing of eosine in analysis of cationic flocculants has received some attention. The spectrophotometric determination of cationic flocculants such as poly(diallyldimethylammonium chloride) in the presence of eosine was affected by the pH of the aqueous solutions used, with pH 3.4–3.6 being optimal for the determination.[496]

Association of polyelectrolytes with oppositely charged mixed micelles has been studied. Phase diagrams were obtained from mixtures of a strong polycation, poly(diallyldimethylammonium chloride), with a nonionic surfactant (Triton X-100) and an anionic surfactant, sodium dodecyl sulfate (SDS) in the presence of NaCl. The results were consistent with intrapolymer condensation of micelles.[497]

In a contiuing study of polyelectrolyte-micelle complexes, solutions of poly(diallyldimethylammonium chloride) in NaCl and in Triton X-100 in the presence of SDS were characterized by quasi-elastic light scattering technique. The effect of solution variables on the phase boundary was discussed.[498]

The effect of the salt content on formation and structure of polyelectrolyte complexes (symplexes) has been studied. The particle size of complexes of poly(diallyldimethylammonium chloride) with Na cellulose sulfate in H_2O was shown by light scattering to increase with increasing NaCl concentration. Further salt addition resulted in limited aggregation.[499] The adsorption of an aqueous saline solution of the polymer on colloidal silica was investigated as a function of pH, sodium chloride concentration, and molecular weight. These adsorption characteristics were discussed in connection with the charge and conformation of the polymer in water or salt solution and also with the surface charge of colloidal silica.[500]

Characteristics of the behavior of N-cetyl-N,N-diallyl(dodecyloxycarbonylmethyl)ammonium bromide in water were studied. The ultrasonic dispersion of the monomer in water gave vesicular colloids, which could be used for immobilization of biological objects and modeling membrane processes.[501]

Salt linkage formation of poly(diallyldimethylammonium chloride) with acidic groups in the polyion complex between human carboxyhemoglobin and potassium poly(vinyl alcohol) sulfate has been observed.[502]

Polysalts were prepared from poly(acrylic acid), poly(itaconic acid), maleic anhydride-α-methylstyrene copolymer, or maleic anhydride-propene copolymer with poly(diallyldimethylammonium chloride), polyethylenimine, or two quaternized acrylic polymers. The results were discussed in terms of the acid strength and distribution of the anionic groups, both of which affect the composition of the polysalts.[503] Symplex polysalts were also prepared by treating sodium polyphosphates with poly(diallyldimethylammonium chloride) polyethylenimine, and cationically modified polyacrylamides. Morphological studies indicated that the ratio of the components in the complexes was controlled more strongly by electrostatic charge than in the case of symplexes involving cellulose derivatives as anionic components.[504]

Sparingly soluble metal salts of poly(diallyldimethylammonium chloride) or its copolymer were obtained instantaneously at ambient temperature at pH 1–8 by treating with a dilute aqueous solution of metal–containing anions. This reaction was utilized for detoxication and purification of solutions and wastewater, metal concentration, and dressing or preservation of wood surfaces.[505]

Further discussion of the polyelectrolytic titration, method, information and limits, has been published. In the determination of polyacrylamide, anionic and cationic starch, and polyethyleneimine in pulp suspensions or their filtrates by titration with potassium poly(vinyl sulfate) or poly(diallyldimethylammonium chloride) as anionic or cationic standard reagent, respectively, the end point of titration was traced with high sensitivity using a streaming current detector.[506]

The conditions of formation and some properties of compositions from polyelectrolyte complexes and latexes have been studied. Films were formed by mixing polyelectrolyte solutions, e.g., poly(diallyldimethylammonium chloride) in aqueous NH_3 or 50% HCO_2H with 15% acrylate latexes, pouring on a polyethylene substrate, and drying at ambient temperature. The physical properties of the composites depended on composition.[507]

Thermal degradation of poly(diallyldimethylammonium chloride) and 1:1 diallyldimethylammonium chloride-methacrylic acid copolymer gave products containing methyl chloride (main component), allyl chloride, trimethylamine, allyldimethylamine, and (from diallyldimethylammonium chloride-methacrylic acid copolymer) HCl and CO_2. The phase transition temperatures were determined from elongation as a function of temperature.[508]

Problems of interfacial phenomena associated with application to determining the amount of cationic polyelectrolyte (poly-diallyldimethylammonium chloride) adsorbed on cellulose fibers have been studied. Poly(vinyl alcohol potassium sulfate) was used as an anionic polymer.[509] A theoretical model for the electric double layer of fibers with an adsorbed layer of cationic polyelectrolyte has been proposed to account for the structural changes by the adsorbed monolayer. This model could also be applied to general cases, in which the reversal of polarity of the charge occurs by an adsorbed monolayer of cationic polyelectrolyte, in equilibrium with inorganic salts having counter ions with relatively low polarization energies.[510]

In studies of electric conduction properties of cellulose fibers with adsorbed layers of cationic polyelectrolyte, the effect of multilayer formation of poly(diallyldimethylammonium chloride) on the electric conduction behavior of carboxymethyl cellulose (CMC) fiber was investigated. The excluded volume effect of protonic transition mechanism accounted for the conduction behavior of CMC fiber with absorbed multilayers.[511] Electric conduction properties of cellulose fibers with adsorbed layers of cationic polyelectrolyte have been determined. The electric conduction of a poly(diallyldimethylammonium chloride), containing dissolving pulp pad, was determined as a function of the amount of adsorbed poly(diallyldimethylammonium chloride), KCl concentration, and pH, and the conduction data were analyzed. The results suggested that mobile ions in the electric double layer of the pad were replaced by immobile poly(diallyldimethylammonium chloride) ions as adsorption of poly(diallyldimethylammonium chloride) proceeded.[512] In a continuing study of electric conduction properties of cellulose fibers with adsorbed layers of cationic polyelectrolyte and to provide theoretical interpretations for the empirical relation $R = GL + W$ (R = total pad resistance, L = pad length, G and W = constants), a mathematical model was proposed for the electrical conduction of dissolving pulp, cotton linters, and CMC containing poly(diallyldimethylammonium chloride) as polyelectrolyte. The model correlated the electrical conduction with the hydrodynamic properties of the cellulosic pad and permitted estimation of specific conductance values of fiber surfaces using the hydrodynamic data obtained by the streaming current method.[513]

Electron spin-resonance spectroscopy (ESR) has been used[70,177,514] to study the radical addition reactions of non-conjugated dienes. Hydroxyl, amino, methyl or phenyl radicals were allowed to react with diallylmalonic acid and its derivatives), diallyl ether, diallylamine and variously substituted diallylamines in aqueous solutions. Using a flow technique, the products of the reaction were generated in the cavity of an electron spin resonance spectrometer (2–54). The results were interpreted as clear evidence that the

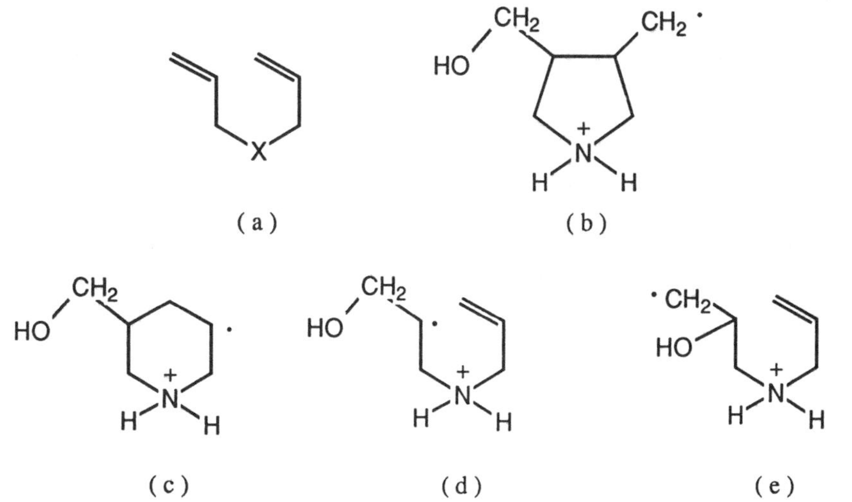

(a) (b)

(c) (d) (e)

(2–54)

radicals present in highest concentration in the flow cell when 1,6-dienes of general type 2–54a underwent reaction were five-membered ring radicals. For example, the spectrum obtained from the reaction of diallylamine with hydroxyl radicals consisted of a triplet (splitting due to α protons) of doublets (splitting due to β protons). This spectrum was assigned to the five-membered ring radical (2–54b), rather than the six-membered ring radical (2–54c) or the uncyclized radicals (2–54d and 2–54e), since it requires that the unpaired electron interact with three protons, two of which are equivalent. Only a and e have such a structure. The uncyclized radical (e) was rejected since this constitutes an abnormal mode of addition to an allylic double bond.

The sensitivity of this cyclization to steric and resonance effects was shown by investigations on a number of substituted diallylamines. The ESR spectrum of radicals from N,N-bis(2-isopropylallyl)methylamine and $NH_3^+\cdot$ showed a multiplet spectrum which was assigned to six-membered ring radicals. This was in accord with other results on cyclization of radicals[515] and showed that a bulky substituent on the internal position of the double bond disfavors attack at that position.

Certain of the diallyl amines and quaternary ammonium salts yield cyclopolymers consisting largely of five-membered rings. A ^{13}C NMR study of poly(N-methyl-N,N-diallylamine)[105] showed formation of five-membered rings; however, when the allyl groups were substituted in the 2-position, mixtures of both five- and six-membered rings formed. Also, it has been shown by ^{13}C NMR and model-compound studies that poly(diallyldimethylammonium chloride)[516] consists mostly of five-membered rings linked mainly in a 3,4-cis configuration.

Poly(triallylamine) has been used as a component of an ion exchange process which includes both cationic and anionic exchangers. All remaining unsaturation in this polymer was removed by a rapid reaction with sodium bisulfite at pH = 5.[95] The percent sulfur in the product provided a convenient way to measure the degree of unsaturation in cyclopolymerization studies.

Cationic polyelectrolytes based on diallyldialkylammonium salts have been synthesized and their properties determined.[96] Poly(diallyldimethylammonium chloride) flocculates TiO_2 suspended in water. Polymer having intrinsic viscosity [n] = 0.9 was prepared in aqueous solutions containing 0.25% $(NH_4)_2S_2O_8$ for 72 hours at 30° and then for 48 hours at 60°. Polymerization rates could be increased by substituting DMSO for H_2O but the polymer prepared in DMSO had low flocculating ability.

The conversion of monomer during radical cyclopolymerization of diallyldimethylammonium chloride in water has been continuously determined by measuring the change of density of the solution with an apparatus which detected changes in the frequency of electronically stimulated vibrations of the ends of a glass polymerization tube containing the solution.[97] The frequency of vibration was controlled by the solution density which changed with increasing conversion of monomer. The apparatus detected density changes of $\pm$ 4 $\mu g/cm^3$ for water.

The polymerization of diallyldimethylammonium chloride initiated in aqueous solutions by AIBN had perceptible and nearly equal rates at pH 6–10; $(NH_4)_2S_2O_8$ was also an effective initiator at pH 3.[98] The polymerization had a small induction period, and was 0.5-order in initiator. The rates of the polymerization increased rapidly and nonlinearly with increasing monomer concentration, apparently as a result of intermolecular association. The rates of polymerization of diallyldimethylammonium bromide, and diallyldiethylammonium bromide, and the molecular weights of the polymers, were considerably lower than those of the dimethyl chloride derivative.

The azobisisobutyronitrile-initiated polymerization of various N-substituted diallylammonium, N-ethyldimethallylammonium, and triallylammonium chlorides yielded as the respective major by-products, 2-substituted 6,6-dimethylperhydro-5-isoindolones, 2-ethyl-3a,6,6,7a-tetramethylperhydro-5-isoindolone, and 2-allyl-6,6-dimethylperhydro-5-isoindolone.[104] These isoindolones were formed from pyrrolidylmethylene radical precursors and indicated that, in agreement with recent ESR studies, radical addition to diallylamines resulted in the formation of pyrrolidine derivatives and not piperidines as previously supposed.

A ^{13}CNMR spectral study of cyclopolymers obtained from N,N-diallylamines was carried out.[107] The structures of polymers formed by the radical-induced cyclopolymerization of N-substituted-N,N-diallylamines and N-methyl-N,N-bis(2-alkylallyl)amines were determined by pulsed Fourier transform ^{13}C natural abundance NMR spectroscopy. The polymers of N,N-diallylamines all contained cis- and the trans-substituted pyrrolidine rings in a 5:1 ratio. The polymers of N-methyl-N,N-bis(2-alkylallyl)amines gave complex spectra due to the presence of both cis- and trans-pyrrolidine and piperidine rings. The differences in chemical shifts of the N-methyl signals from the different types permitted structural analysis of the spectra.

Allylamino monomers have been synthesized by the Mannich reaction and their cyclopolymerization studied.[108] Diallylamino derivatives of phenol acetates, aniline, and pyrrole underwent cyclopolymerization as their HCl salts at pH 1–2 in the presence of $TiCl_3$ and H_2O_2. Molecular weight could not be detected by viscometry because of the anomalous viscosity behavior of these polymers. The nature of the amino groups in the polymers was indicated by potentiometric titration curves.

Cyclopolymerization of N-substituted diallylamine monomers with a variety of polar groups attached to the N atom, were studied and their cyclopolymerization to multifunctional polypyrrolidines evaluated.[111] Cyclopolymerization of some monomers with small amounts of tetraallyldiamines gave crosslinked polymers suitable for ion exchange testing. The structures of both types of polymers were determined by ^{13}C NMR.

Poly(diallyldimethylammonium chloride) exhibited some peculiar behavior in aqueous solution.[184] This peculiar behavior was attributed to the influence of the simple electrolyte counterion by hydrophobic interaction.

From previously known facts and the results of a study of the Cope rearrangement of 1,4 dienes, it was concluded that the four-centered, chair-like transition state is favored in this rearrangement over the six-centered, boat-like structure.[517] The Cope rearrangement has some mechanistic implications related to cyclopolymerization.

The polymerization of methacrylic esters by use of ^{14}C labeled benzoyl peroxide made it possible to distinguish between initiation by benzoyloxy and phenyl radicals derived from the peroxide.[518] The proposal that a secondary propagating radical should be predominant over a primary one is consistent with "head-to-tail" enchainment in radical-initiated polymerization of vinyl monomers, an arrangement which has been unquestionably established as the major mode of enchainment in a wide variety of vinyl monomers.[519]

Kinetics and rate constants for the reduction of alkyl halides by organotin hydrides have been determined. Rate constants for halogen abstraction by tri-n-butyltin radical varied from $8 \times 10^2\ M^{-1}\ sec^{-1}$ for 1-cyclopentane to $2.5 \times 10^9\ M^{-1}\ sec^{-1}$ for methyl iodide. These data have been used to advantage in studying radical cyclizations.[520] The frequency of intramolecular collisions between two groups linked to a polymer chain was studied and an ESR technique devised to permit determination of this frequency.[521]

The He photoelectron spectra of α,ω -diolefins, containing from five to nine carbon atoms were determined.[522] The π-level splittings were 0.34, 0.46 (or 0.31), 0.41, 0.21, and 0.14 eV, respectively. The observed fine structure of the first two bands suggested that there was no crossing of the π-levels. Extended Hueckel MO calculations on different conformations of 1,4-pentadiene and 1,5-hexadiene indicated that a through-space dominated interaction throughout the series was being observed.

Free radical cyclizations in small molecules have been studied extensively.[523] Deuteration studies showed that in the cyclization of 1-naphthylbutyl radicals two different mechanisms were involved. The results have been used as a mechanistic tool.

It was shown that during methacrylic anhydride cyclopolymerization, the intramolecular propagation step had a higher activation energy (2.6 ± 0.3 kcal/mole) than the intermolecular step, but steric factors favored cyclization.[317] Conditions favoring cyclization were high temperature and conversion, low monomer concentration, and poor solvent systems. The increase in cyclization caused by heterogeneous conditions was due to tight polymer chain coiling and slow diffusion. The effect of decreasing temperature and conversion was low, but increased the heterotacticity and decreased the syndiotacticity by equivalent amounts. The presence of noncyclic anhydride units increased the syndiotacticity.

The effect of β-substituents on the cyclopolymerization of diallylamines has been studied.[524] The radical-induced cyclization of the diallylamines, $CH_3N(CH_2CR{:}CH_2)_2$($R = H, CH_3, C_2H_5, i-C_3H_7, t-C_4H_9, CO_2C_2H_5, C_6H_5$), or of $(CH_3)_2NCH_2C(t-C_4H_9){:}CH_2$, did not necessarily involve a concerted reaction as acyclic radical intermediates were observed in reactions of sterically-hindered monomers.

The direction of ring closure of the acyclic radical was susceptible both to steric and conjugative effects of β-substituents, and to a lesser extent, to increases in reaction temperature, all of which favored an increase in the proportion of piperidines in the product, reflecting increased cyclization to the thermodynamically preferred piperidyl radical. Increased piperidine content in the polymers increased interaction between neighboring ionized amino groups and a decrease in polymer basicity. An increase in piperidine content of the polymer in which $R = CH_3$ was obtained by increasing the polymerization temperature, but this increase was accompanied by reduction in degree of polymerization and conversion.

In the free radical polymerization of methacrylic anhydride in $CHCl_3$, the cyclization propagation reaction had an entropy decrease much more than that of intermolecular cyclization.[318] The supported hypothesis that the large cyclic unit content of the products of free radical polymerization of symmetrical 1,6-dienes was caused by the entropy effect favoring cyclization, and overruled the energy effect favoring intermolecular additions.

A simple, rapid, reproducible, nonhazardous procedure for the preparation of porous supports for the gel permeation chromatography of cationic polymers has been reported which involves treating a controlled porosity support, having a siliceous surface with a silane and a quaternizing agent.[525] When the support was used as packing in a gel permeation chromatography column, high plate counts were achieved and no adsorption of cationic polymers occurred.

It has been shown that the reactivity of allylic monomers such as CH_2:$CHCH_2R$($R = CO_2H$, CH_3CO_2, OCH_2CH:CH_2, OH, NH_2, $N(CH_3)_2$, $NHCH_2CH$:CH_2) in polymerization is also influenced by polar effects of the functional group.[526] A correlation between the rate of polymerization and the Taft constant was observed. The influence of protonation on the kinetics of chain transfer and polymerization was discussed.

The difference in the polymerizability of a series of allyl monomers, e.g., vinylacetic acid, allyl acetate, dimethylvinylcarbinol, diallyl ether, allyl alcohol, dimethylallylamine, diallylamine, and allylamine, has been shown to be related to the polar effects of the functional groups, whose influence determines the relative stability of the C-H bonds at the position and thus the rate of chain transfer to monomer (degradative chain transfer).[527] The initial polymerization rates obey the Hammett-Taft equation. The polymerization rates were increased in the presence of acids, e.g., CH_3CO_2H, HCl, H_2SO_4, H_3PO_4, polyphosphoric acid, and methylphosphoric acid. The nature of the monomers, the acid-monomer molar ratio, and the nature, functionality, strength, and concentration of the acid determined the increase in the polymerization rate. A kinetic scheme for the polymerization was prcposed.

The reactivity of allyl monomers in low-temperature postpolymerization reactions has been studied.[528] In CH_2:$CHCH_2R$ ($R = NH_2$, $NHCH_2CH$:CH_2, CH_3CO_2, OCH_2CH:CH_2, OH) irradiated with γ-rays at $-196°$, radical behavior during gradual thawing, which determined the postpolymerization reactivity of the monomer, depended on the nature of the functional group, both in the presence and absence of protonic acids. Protonation of monomer functional groups having a high pK value ($R = NH_2$, $N(CH_3)_2$, $NHCH_2CH$:CH_2 changed the polarizing action so that degradative chain transfer was suppressed and low-temperature postpolymerization was possible. With such monomers, the concentration of active centers did not change significantly during thawing. The retention of alkyl radicals in the final stages of thawing indicated suppression of degradative chain transfer in such systems. For monomers with low or moderate pK values, degradative chain transfer was not suppressed, and polymer formation did not occur in low-temperature postpolymerization conditions.

The trajectories of colliding 2.6- and 4.0-μm diameter polystyrene latex spheres, previously studied in aqueous glycerol containing potassium chloride, undergoing Poiseville flow were observed in presence of poly(diallyldimethylammonium chloride).[529] The cationic polymer adsorbed on the latex and caused a reversal in the surface charge on going from low polymer concentrations to high polymer concentrations.

The formation of stoichiometric polyelectrolyte complexes between poly (diallyldimethylammonium chloride) and its alternating copolymer with SO_2 and oppositely charged polyelectrolytes has been studied.[530] The molecular weight distribution for poly(diallyldimethylammonium chloride) was determined by fractional precipitation from methanol-dioxane.[531] Distribution curves prepared by various means showed no substantial differences in the high-molecular-weight regions. Reproducibility was good at intrinsic viscosity > 0.3 dL/g.

A thermodynamic model of equilibrium polymerization was developed which is capable of treating all physically important effects such as monomer activation, growth of linear and cyclic polymers, excluded volume, (bi)functionality, polymer rigidity, and solution effects.[532] During a spectrophotometric study of reaction rates, determination of cleaning rate equation was accomplished.[533] A reaction rate equation that can be applied for systems for which spectrophotometric measurement is possible was theoretically derived, and its validity discussed with the results from determination of the rate of cyclopolymerization initiation of hydrobenzamide and spectrophotometric measurement of laundering rate.

Methyl methacrylate and styrene were copolymerized with dimethylallyl (16-carbomethoxycetyl)ammonium chloride in presence of radical initiators to obtain antistatic copolymers.[534] The surface electrical resistivity of the copolymer was 2.1×10^{10} to 3.1×10^{12} at 65% relative humidity.

Continuing studies on the mechanism of cyclopolymerization led to the inclusion of charge transfer complexes in the monomer in order to influence the ring size.[535] It was anticipated that if a donor and acceptor group were substituted at the C_2 and C_2' positions, respectively, of an allyl ether, an intramolecular complex would be formed. Monomers selected for the study were 2-chloroallyl 2′-phenylallyl ether, 2-carboethoxyallyl 2′-phenylallyl ether, 2-cyanoallyl 2′-phenylallyl ether and 2-carboethoxyallyl 2′-methoxyallyl ether. Intramolecular charge transfer complexation was proved by omitting the point of unsaturation having the donor group. For this study 2-carboethoxyallyl 2′-phenylpropyl ether, 2-cyanoallyl 2′-phenylpropyl ether and 2-carboethoxyallyl 2′-methoxypropyl ether were synthesized and the ^{13}C and ^1H NMR spectra were compared with those of the corresponding monomers in order to determine "charge transfer." The monomers were polymerized with 2,2′-azobisisobutyronitrile in benzene; only 2-carboethoxyallyl 2′-phenylallyl ether gave a linear cyclopolymer soluble in most organic solvents. 2-Chloroallyl 2′-phenylallyl ether did not afford any polymer and 2-cyanoallyl 2′-phenylallyl ether and 2-carboethoxyallyl 2′-methoxyallyl ether afforded branched polymer when the percentage conversion and percent monomer concentration were kept low. This was determined *via* gel permeation chromatography. The polymer of 2-carboethoxyallyl 2′-phenylallyl ether consisted largely of five-membered rings at 40°C and six membered rings at 60°C. This ring distribution supported intramolecular "charge transfer" complexation at lower temperatures and normal cyclopolymerization domination at higher temperatures, corresponding to dissociation of the charge transfer complex. Theoretical calculations such as geometry optimizations were carried out on N-methylmaleimide and methyl vinyl ether and possible complexes involving them in order to support charge transfer between these two molecules.

It has been shown that bis(α-substituted-vinyl)phenylene oxides (n = 1, 2, 3, R = $(CH_3)_2$, C_6H_5) can be polymerized with simultaneous ring closure to give thermally stable polyindanyl polymers (2–55).[536]

$$CH_2{=}CR{-}[C_6H_4{-}O]_n{-}C_6H_4{-}CR{=}CH_2 \xrightarrow[CH_2Cl_2]{SnCl_4}$$

(2–55)

1,3-Bis(p-vinylphenyl)propane has been polymerized in presence of radical, anionic or coordination polymerization initiators.[537] Radical polymerization yielded linear polymer with pendant styryl groups without gelation. Anionic polymerization in presence of BuLi or Na naphthalene gave insoluble polymer containing no cyclized units. Polymerization in presence of PhMgBr gave soluble polymer. Bis(π-allylnickel trifluoroacetate) also gave soluble polymer.

One of the earliest telomerization experiments conducted after the cyclopolymerization mechanism had become known involved radical-initiated cyclization of 1,6-heptadiene with SO_2ClF.[538] The six-membered ring structure, 1-chloro-3-(fluorosulfonylmethyl)cyclohexane was reported to be formed. Five-membered rings were not considered in this early work for obvious reasons; however, in light of present knowledge, five-membered rings were quite likely formed as well.

Cyclopolymerization of N-phenyldimethacrylamide by the group transfer polymerization technique led to polymer consisting predominantly of six-membered imide rings in contrast to the predominantly five-membered ring structure obtained by radical initiation.[539]

REFERENCES

1. Staudinger, H. and Heuer, W., Ber. 1934, 67, 1159; CA 28, 6119(8) (1934).
2. Milford, G. N., J. Polym. Sci., 1959, 41, 295; CA 54, 14774d (1960).
3. Anderson, W. S., Proc. Batt. Symp. Therm. Stabil. Polym., 1963, 1242; CA 60, 9374c, (1964).
4. Marvel, C. S. and Gall, E. J., J. Org. Chem., 1960, 25, 1784; CA 55, 2525f, (1960).
5. Matsoyan, S. G., Pogosyan, G. M., and Skripnikova, R. K., Vysokomolek. Soedin., 1962, 4, 1142; CA 58, 14107g, (1963).

6. Matsoyan, S. G., Pogosyan, G. M., Skripnikova, R. K, and Nikogosyan, L. L., Izv. Akad. Nauk armyan. SSR, Khim Nauki 1962, 15, 541; CA 59, 7655f, (1963).
7. Matsoyan, S. G., Pogosyan, , G. M., Skripnikova, R. K., and Mushegyan, A. V., Vysokomolek. Soedin., 1963, 5, 183; CA 59, 7654b, (1963).
8. Matsoyan, S. G., Pogosyan, G. M., and Cholakyan, A. A., Izv. Akad. Nauk armyan. SSR, Khim. Nauki 1965, 18, 178; CA 63, 14693g, (1965).
9. Chang, H.-C., Tsao, W.-H., and Feng, H.-T., K'o Hsueh T'ung Pao, (6), 1963, 40; CA 60, 1844f, (1964).
10. Feng, H.-T., Chang, H.-C., Tsao, W.-T., Sheng, P., Hsi, F. and Chi, H.-C., K'o Hsueh T'ung Pao, 1963 (5), 53; CA 60, 3114c (1964).
11. De Winter, W. and Marvel, C. S., J. Polym. Sci. A, 1964, 2, 5123; CA 62, 6456(1965).
12. Corfield, G. C.,A. Crawshaw, G. B. Butler and Miles, M. L., Chem. Commun., 1966, 238; CA 65, 5537c (1966).
13. Stille, J. K. and Foster, R. T., J. Org. Chem. 1963, 28, 2703; CA 59, 12719(1963).
14. Brown, D. W. and Wall, L. A., Polym. Preprts., 1968, 9, 1401; CA 71, 50602j (1969).
15. Valvassori, A., Sartori, G. and Ciampelli, F., Chimica Ind. Milan, 1962, 44, 1095; CA 58, 2554d (1963).
16. Makowski, H. S., Shim, B. K. C. and Wilchinsky, Z. W., J. Polym. Sci. A, 1964, 2, 1549; CA 61, 3209a (1964).
17. Yuki, H., Hateda, K. and Sakano, I., Kogyo-kagaku Zasshi, 1967, 70, 1998; CA 68, 115014d (1968).
18. Angelescn, E. and Raducu, A., Studil Cerc. Chim., 1968, 16, 389; CA 69, 67782z (1968).
19. Marek, M., Roosova, M., and Doskocilova, D., J. Polym. Sci. C, 1967, 16, 971; CA 67, 64756s (1967).
20. Aso, C., Nawata, T., and Kamao, H., Makromol. Chem., 1963, 68, 1; CA60, 3104a (1964).
21. Aso, C. and Kita, R., Kogyo-kagaku zasshi, 1965, 68, 707; CA 63, 8485c (1965).
22. Aso, C., Kunitake, T., and Kita, R., Malromol. Chem., 1966, 97, 31: CA 65, 18687e (1966).
23. Aso, C., Kunitake, T., Matsugama, Y., and Imaizumi, Y., J. Polym. Sci. A1, 1968, 6, 3049; CA 69, 107169p (1968).
24. Aso, C., Kunitake, T., and Imaizumi, Y., Makromol Chem., 1968, 116, 14; CA 69, 67769a (1968).
25. Costa, L., Chiantore, O. and Guaita, M., Polymer, 1978, 19, 197; CA 89, 44296t (1978).
26. Costa, L., Chiantore, O. and Guaita, M., Polymer, 1978, 19, 202; CA 89, 44297u (1978).
27. Aso, C., Pure Appl. Chem., 1970, 23, 287; CA 74, 142453b (1971).
28. Friedmann, G., Bull. Soc. chim. Fr., 1967, 2, 698; CA 66, 105314w (1967).
29. Friedmann, G., Peint.-Pigm.-Vern., 1967, 43, 457; CA 67, 73930w (1967).
30. Kiji, J. and Iwamoto, M., J. Polym. Sci. B, 1968, 6, 53; CA 68, 50430q (1968).
31. Aso, C., Kogyo-kwagaku zasshi, 1967, 70(11), 1920; CA 68, 105529m (1968).
32. Corfield, G. C. and Crawshaw, A., J. Macromolec. Sci. (Chem.), 1971, 5, 1873; CA 73, 131387m (1970).
33. Marek, M. and Pecka, I., J. Macromolec. Sci. (Chem.), 1971, 5(1), 2063; CA 73, 120968a (1970).
34. Cotter, R. J. and Matzner, M., Ring-Form. Polymeriz.(Acad. Press) N.Y., 1969; CA 72, 22132h (1970).
35. Aso, C., Kunitake, T., Khattak, R. K. and Sugi, N., Makromolek. Chem., 1970, 134, 147; CA 73, 25952g (1970).
36. Aso, C., Kunitake, T. and Uchio, H., Ger. Offen. 1,955,519, 1970 (Dec. 3); CA 74, 54411f (1971).
37. Stamicarbon N. V.,Brit. Pat. 856,858, 1960 (Dec. 21), CA 55, 13913h (1961).

38. Trossarelli, L. and Guaita, M., J. Macromolec. Chem., 1966, 1, 471; CA 66, 2792w (1967).
39. Jones, J. F., U.S. Pat. US 2,978,436, 1961, (Apr. 4);CA 55, 18194e (1961).
40. Jones, J. F., Ger. Pat. 1,076,376, 1960, (Feb. 25); CA 55, 19342c (1961).
41. Butler, G. B. and Brooks, T. W., Polym. Preprints, 1962, 3 (1), 168; CA 59, 12931e (1963).
42. Butler, G. B., U.S. Pat. 3,044,986; Brit. Pat. 912401, 1962, (July 17); CA 57, 15356e (1962).
43. Angelo, R. J., Belg. Pat. 632,633 ;Fr. Pat. 1,357,995, 1963 (Nov. 21); CA 61, 799e (1964).
44. Hata, G. and Miyake, A., J. Org. Chem., 1963, 28, 3237; CA 60, 4166e (1964).
45. Shell Internationale, Brit. 930,985; U.S. 3,223,638; Ger. 1,165,865 1963 (July 10); CA 59, 10259b (1963).
46. Vogel, E., Grimine, W. and Dinne, E., Argew. Chem., 1963, 75, 1103; CA 60, 5351c (1964).
47. Fearn, J. E. and Wall, L., U.S. Govt. Research Report, 1964, 39, 19; CA 63, 18272a (1965).
48. Shelden, R. A., Dissert. Abstr., 1964, 25, 2404; CA 62, 6568g (1965).
49. Pellon, J., Kugel, R. L., Marens, R. and Rabinowitz, R., J. Polym. Sci., A, 1964, 2, 4105; CA 61, 12097f (1964).
50. Fearn, J. E. and Wall, L., U.S. Govt. Res. Develop. Rept., 1965, 40, 53; CA 64, 8321b (1966).
51. Butler, G. B. and Raymond, M. A., J. Macromol. Sci. (Chem), 1966, 1, 201; CA 65, 10677e (1966).
52. Tabushi, I. and Oda, R., Tet. Letters, 1966, 2487; CA 65, 5322h (1966).
53. Brace, N. O., J. Org. Chem., 1966, 31, 2879; CA 65, 15213d (1966).
54. Naumova, S. F., Erofeev, B. V. and Maksimova, T. P., Vestsi Akad. Navuk Belar. SSR, Ser. Khim 1969, 11; CA 72, 66277p (1970).
55. Olson, S. G., U.S. Pat. 3,435,020, 1969 (Mar. 25); CA 70, 97398s (1969).
56. Brace, N. O., J. Org. Chem., 1969, 34, 2441; CA 71, 70157w (1969).
57. Cope, A., McKervey, M. and Weinshenker, N., J. Org. Chem. 1969, 34, 2229; CA 71, 38725u (1969).
58. De Pasquale, R. J., J. Organometalic Chem., 1971, 32, 381; CA 76, 72849a (1972).
59. Scheffer, J. R. and Wostradowski, R.A., J. Chem. Soc., D, 1971, 1217; CA 75, 151073a (1971).
60. Butler, G. B. and Myers, G. R., J. Macromol. Sci. (Chem), A, 1971, 5, 105; CA 73, 120935n (1970).
61. Ito, S., Ohnishi, S., Yamauchi, A. and Yamamoto, M., Jap. Pat. JP 72 45,590, 1972 (Nov. 17); CA 80, 71694s (1974).
62. Wilke, G., Heimbach, P. and Brenner, W., Ger. Pat. 1,643,063, 1972 (Mar. 16); CA 77, 20339k (1972).
63. Bradney, M. A. M., Forbes, A. D. and Wood, J., J. Chem. Soc., Perkin, 1973, 11, 1655; CA 80, 26715p (1974).
64. Brace, N. O., J. Org. Chem., 1973, 38, 3167; CA 79, 104494m (1973).
65. Dumitrescu, S., Grigoras, M. and Natansohn, A., J. Polym. Sci., Polym. Lett. Ed., 1979, 17, 553; CA 91, 158141k (1979).
66. Ball, L. E., Sebenik, A. and Harwood, H. J., ACS Symp. Ser., 1982, 195, 207; CA 97, 182962y (1982).
67. Kanbara, M. E., Ed., Kobun. Jikken., 4:Fukaj., Kaikan., 1983, 1–520; CA 101, 131267u (1984).
68. Longone, D. T. and Glatzhofer, D. T., J. Polym. Sci., Part A: Polym. Chem., 1986, 24, 1725; CA 105, 191651g (1986).

69. Bresler, S., Koton, M., Osmingkaya, A., Papov, A. and SavitsKaya,M., Vysokomol. Soedin, 1959, 1, 1070; CA 54, 15998c (1960).
70. Beckwith, A. L. J.,Ong, A. K. and Solomon, D. H., J. Macromol. Sci. (Chem) A, 1975, 9, 115; CA 82, 86678e (1975).
71. Brace, N. O., J. Org. Chem., 1979, 44, 212; CA 90, 86351h (1979).
72. Negi, Y., Harada, S. and Ishizuka, O., J. Polym. Sci. A, 1967, 5, 1951; CA 67, 82428r (1967).
73. Brace, N. O., J. Polym. Sci. A1, 1970, 8, 2091; CA 73, 76964v (1970).
74. Danielyan, V. A., Karslyan, S. V. and Matsoyan, S. G., Arm. Khim. Zh., 1979, 32, 970; CA 93, 72381t (1980).
75. Matsoyan, S. G., Pogosyan, G. M., and Zhamkochyan, G. A. Izv. Akad. Nauk armyan. SSR, Khim. Nauki 1964, 17, 62; CA 61, 4489a (1964).
76. Ostroverkhov, V. G., Brunovskaya, L. A., and Korneinko, A. A., Vysokomol. Soedin., 1964, 6, 925; CA 61, 5777d (1964).
77. Crawshaw, A. and Jones, A. G., Chemy. Ind., 1966, 2013; CA 66, 38264q (1967).
78. Miyake, T., Kogyo-kagaku zasshi, 1961, 64, 359; CA 57, 6113f (1962).
79. Sokolova, T. A. and Rudkovskaya, G. D., J. Polym. Sci C, 1967, 16, 1157; CA 67, 64772h (1967).
80. Gotzen, F. and Schroder, G., Makromol. Chem., 1965, 88, 133; CA 64, 3694e (1966).
81. Boyarchuk, Y. M., Vysokomolek. Soedin. Ser. A., 1969, 11, 2161; CA 72, 32400h (1970).
82. Kodaira, T. and Aoyama, F., J. Polym. Sci., Polym. Chem. Ed., 1974, 12, 897; CA 81, 121139h (1974).
83. Kodaira, T., Ni-imoto, M. and Aoyama, F., Makromol. Chem., 1978, 179, 1791; CA 89, 75502m (1978).
84. Kodaira, T. and Sakai, M., Polym. J., 1979, 11, 595; CA 91, 193671e (1979).
85. Sokolova, T. A. and Osipova, I. N., Vysokomol. Soedin., Ser. B, 1968, 10, 384; CA 69, 27924r, (1968).
86. Ishida, S., Dissertation, Poly. Insti. of Brooklyn, 1967; CA 68, 69372u (1968).
87. Yamada, B., Saya, T. and Otsu, T., Makromol. Chem., 1982, 183, 627; CA 96, 163289x (1982).
88. Jakus, C. and Smets, G., Proc. IUPAC, Macromol. Symp., 28th., 1982, 171; CA 99, 105814r (1983).
89. Jakus, C., Smets, G. and Zeegers-Huyskens, T., Bull. Soc. Chim. Belg., 1984, 93, 175; CA 101, 24023q (1984).
90. Hoover, M. F. and Carr, H. E., Tappi, 1968, 51(12), 552; CA 70, 47997r (1969).
91. Boothe, J. E., Flock, H. G. and Hoover, M. F., J. Macromolec. Sci. (Chem.), 1970, 4, 1419; CA 73, 45870g (1970).
92. Negi, Y., Harada, T. and Ishizuka, O., Japan Pat. 70 01,457, 1970 (Jan. 19); CA 72, 91003d (1970).
93. Jackson, M. B., J. Macromol. Sci. (Chem), A, 1976, 10, 959; CA 85, 109023r (1976).
94. Jackson, M. B., J. Macromol. Sci. (Chem), A, 1978, 12, 853; CA 90, 39790r (1979).
95. Eppinger, K. H. and Jackson, M. B., J. Macromol. Sci. (Chem), A, 1980, 14, 121; CA 92, 77114x (1980).
96. Wyroba, A., Polimery (Warsaw), 1978, 23, 86; CA 90, 24035z (1979).
97. Jaeger, W., Wandrey, C., Reinisch, G. and Linow, K. J., Faserforsch Textiltech., 1978, 29, 647; CA 90, 39294g (1979).
98. Topchiev, D. A., Bikasheva, G. T., Martynenko, A. I., Kaptsov, N. N., Gudkova, L. A. and Kabanov, V. A., Vysokomol. Soedin., Ser. B, 1980, 22, 269; CA 93, 72371q (1980).
99. Butler, G. B. and Corfield, G. C., J. Macromolec. Sci. (Chem.), A, 1971, 5, 37; CA 73, 131385j (1970).
100. Allied Chemical Corp., Brit. Pat. 1,178,371, 1970 (Jan. 21); CA 72, 67495v (1970).

101. Harada, T., Jap. Pat. JP 70 35,793, 1970 (Nov. 14); CA 74, 54408k (1971).
102. Tsuchida, E., Tomono, T. and Sano, H., Kogyo-kwagaku zasshi, 1970, 73(9), 2024; CA 74, 32043k (1971).
103. Crawshaw, A. and Jones, A. G., Macromolec. Sci. (Chem.), 1972, 6, 65; CA 77, 20066u (1972).
104. Hawthorne, D. G. and Solomon, D. H., J. Macromol. Sci. (Chem), A, 1975, 9, 149; CA 82, 98441k (1975).
105. Hawthorne, D. G., Johns, S. R., Solomon, D. H. and Willing, R. I., J. Chem. Soc., Chem. Commun., 1975,982; CA 84, 74657a (1976).
106. Hawthorne, D. G., Johns, S. R., Solomon, D. H. and Willing, R. I., Austr. J. Chem., 1976, 29, 1955; CA 86, 17017j (1977).
107. Johns, S. R., Willing, R. I., Middleton, S. and Ong, A. K., J. Macromol. Sci. (Chem), A, 1976, 10, 875; CA 85, 124479n (1976).
108. Hodgkin, J. H. and Solomon, D. H., J. Macromol. Sci. (Chem), A, 1976, 10, 893; CA 85, 124406m (1976).
109. Hodgkin, J. H. and Allan, R. J., J. Macromol. Sci. (Chem), A, 1977, 11, 937; CA 87, 68697j (1977).
110. Hodgkin, J. H.,Polym. Preprnts., ACS, Div. Polym. Chem. 1978, 19, 420; CA 93, 205067u (1980).
111. Hodgkin. J. H. and Demerac, S., ACS Adv. Chem. Ser., 1980, 187, 211; CA 93, 186852c, 239980b.
112. McLean, C. D., Ong, A. K. and Solomon, D. H., J. Macromol. Sci. (Chem) A, 1976, 10, 857; CA 85, 178242y (1976).
113. Mathias, l. J., Vaidya, R. A. and Bloodworth, R. H., J. Polym. Sci., Polym. Ltrs.,, 1985, 23, 289; CA 103, 71727y (1985).
114. McCormick, C. L., Zhang, Z. R. and Anderson, K. W., Polymer Prepr., (ACS Div. Polym. Chem.), 1983, 24, 364; CA 100, 210487y (1984).
115. Nippon Kayaku Co., Ltd., Japan Pat. 81 16,448, 1981 (Feb. 17); CA 95, 43911b (1981).
116. Kartashevskii, A.I., Varfolomeev, D.F., Sementsova, L.G., Kirillov, T.S., Skundina, L.Y. Izmailov, I. E., Khim. Tekhnol. Topl. Masel., 1981, 6, 37; CA 95, 208941b (1981).
117. Saeki, S., Suehiro, M. and Fukui, Y., Japan Pat. 60 184,052, 1985 (Sept. 19); CA 104, 69296i (1986).
118. Bakhitov, M. I., Davlethaev, I. G., Prokop'ev, V. P. and Kuznetsov, E. V., Vysokomol. Soedin., Ser. A, 1984, 26, 1153; CA 101, 73181y (1984).
119. Wyroba, A., Polimery (Warsaw), 1981, 26, 50; CA 95, 63036y (1981).
120. Wyroba, A., Przem. Chem., 1981, 60, 531; CA 97, 39481k (1982).
121. Topchiev, D. A., Nazhmetdinova, G. T., Krapivin, A. M., Shreider, V. A. and Kahanov, V. A., Vysokomol. Soedin., Ser. B., 1982, 24, 473; CA 97, 128368w (1982).
122. Hahn, M., Jaeger, W., Seehaus, F., Wandrey, C. and Reinisch, G., Ger. (East) Pat. DD 156,979, 1982 (Oct. 6); CA 98, 107993b (1983).
123. Neigel, D. and Szymanski, C. D., Brit. UK Pat. Appl. GB 2,125,051; 1984 (Feb. 29); CA 101, 7798d (1984).
124. Topchiev, D. A., Nazhmetdinova, G, T., Kartashevskii, A. I., Nechaeva, A. V. and Kabanov, V. A., Izv. Akad. Nauk SSSR, Ser. Khim., 1983, 10, 2232; CA 100, 34887t (1984).
125. Nazpmetdinova, G. T., Shreider, V. A., Topchiev, D. A. and Kabanov, V. A., Izv. Akad. Nauk SSSR, Ser. Khim., 1984, 5, 1024; CA 101, 73146r (1984).
126. Neigel, D., Eur. Pat. Appl. EP 103,698, 1984 (Mar. 28); CA 101, 73258d (1984).
127. Schaper, J., U.S. Pat. 4,439,580, 1984 (Mar. 27); CA 100, 210998r (1984).
128. Szymanski, C. D. and Neigel, D., U.S. Pat. 4,517,351, 1985 (May 14); CA 103, 54890d (1985).

129. Golubkova, N.A., Martynenko, A.I., Babaev, N.A., Nechaeva, A.V., Efendiev, A.A., Topchiev, D.A. and Kabanov, V.A., Izv. Akad. Nauk SSSR, Ser. Khim., 1986, 2, 485; CA 105, 134367v (1986).
130. Ishimi, K., Japan Pat. JP 61 126,114 [86 126,114], 1986 (Jun. 13); CA 105, 173268h (1986).
131. Dumitriu, E., Oprea, S. and Dima, M., Mater. Plast. (Bucharest), 1981, 18, 202; CA 96, 163281p (1982).
132. Wandrey, C., Jaeger, W. and Reinisch, G., Acta Polym., 1982, 33, 156; CA 96, 200490d (1982).
133. Berlin, K. D. and Butler, G. B., J. Org. Chem., 1960, 25, 2006; CA 55, 15333f (1961).
134. Hahn, M., Jaeger, W., Wandrey, C. and Reinisch, G., Acta Polym., 1984, 35, 350; CA 100, 210526k (1984).
135. Klyachko, Y.A., Shnaider, M.A., Korshunova, M.L., Kolganova, I.V. and Topchiev, D.A., Zh. Vses, Khim. O-va., 1984, 29, 111; CA 100, 176887s (1984).
136. Kageno, K., Ueda, T. and Harada, S., Eur. Pat. EP 140,309, 1985 (May 8); CA 103, 105434t (1985).
137. Butler, G. B. and Bunch, R. L., U.S. Pat. US 2,611,768, 1952 (Sept. 23); CA 47, 6446e (1953).
138. Butler, G. B. and Angelo, R. J., J. Amer. Chem. Soc., 1955, 77, 1767; CA 50, 2562f (1956).
139. Price, J. and Thomas, W., U.S. Pat. 2,840,550, 1958 (June 24); CA 53, 1827f (1959).
140. Butler, G. B., Angelo, R. J. and Crawshaw, A., U.S. Pat. US 2,926,161, 1960 (Feb. 23); CA 54, 13743g (1960).
141. Miyake, V. and Matsumoto, M., Jap. Pat. 61 10,842, 1962 (July 15); CA 56, 2574e (1962).
142. Sokolova, T. A. and Rudkovskaya, G. D., Zhur. Obshch. Khim., 1961, 31, 2224; CA 56, 2327h (1962).
143. Peninsular ChemResearch, Inc.,Brit. Pat. 905,831, 1962 (Sept. 12), CA 58, 4662c (1963).
144. Schulz, R. C. and Hartmann, H., Angew. Chem., 1962, 74, 250; CA 57, 8559f (1962).
145. Butler, G. B. and Squibb, S. D., J. Chem. and Eng. Data, 1965, 10, 404; CA 64, 3336f (1966).
146. Rudkovskaya, G. D., Sokolova, T. A. and Koton, M. M., Dokl. Akad. Nauk SSSR, 1965, 164, 1069; CA 64, 3693c (1966).
147. Katayama, M., Harada, A., Seno, T. and Miyamichi, K., Japan Pat. 65 13,674, 1965 (July 1); CA 63, 18300h (1965).
148. LaLonde, R, T. and Aksentijevich, R. I., Tet. Letters, 1965, 23; CA 62, 9005c (1965).
149. Katayama, M., Miyamichi, K., Seno, T., Ueki, S. and Masuda, Y., Japan Pat. 65 17,139, 1965 (Aug. 4); CA 65, 12335g (1966).
150. Azori, M., Plate, N., Rudkovskaya, G., Sokolova, I and Kargin, V. , Vysokomol. Soedin., 1966, 8, 759; CA 65, 10676h (1966).
151. Feng, H-T. and Hsiao, Y. H., K'o Hsueh Tung Pao, 1966, 17, 163; CA 67, 32998f (1967).
152. Peninsular ChemResearch, Inc., Neth. Appl. 6,514,783, 1966 (May 18); CA 68, 3330x (1968).
153. Peninsular ChemResearch, Inc., Brit. Pat. 1,037,028, 1966, (July 20); CA 65, 15611e (1966).
154. Butler, G. B., U.S. Pat. US 3 288 770, 1966 (Nov. 29).
155. Peninsular ChemResearch, Inc., Brit. Amend. Pat. 1,084,089, 1967, (Sept. 27); CA 69, 107432u (1968).
156. Overberger, C. G., Montaudo, G. and Ishida, S., J. Polym. Sci., A1, 1969, 7, 35; CA 71, 3775h (1969).

157. Boothe, J. E., U.S. Pat. US 3,472,740, 1969 (Oct. 14); CA 71, 123530w (1969).
158. Battaerd, H. A. J., Ger. Offen. 1,964,174, 1970 (July 9); CA 73, 56737v (1970).
159. Kunitake, T., Nakashima, T. and Aso, C., J. Polym. Sci. A1, 1970, 8(10), 2853; CA 73, 131365c (1970).
160. Hoover, M. F., J. Macromol. Sci. (Chem), A, 1970, 4, 1327; CA 73, 56636m (1970).
161. Iwakura, Y., Uno, K. and Tsuruoka, K., Japan Pat. JP 70 31,309, 1971 (Oct. 9); CA 74, 32357r (1971).
162. Crawshaw, A. and Jones, A. G., J. Macromol. Sci. (Chem), A, 1971, 5, 1903; CA 73, 131624m (1971).
163. Haas, H., J. Polym. Sci. A1, 1971, 9, 3583; CA 76, 113820e (1972).
164. Iwakura, Y., Uno, K. and Tsuruska, K., Japan Pat. JP 71 03,177, 1971 (Jan. 26); CA 74, 112780p (1971).
165. Iwakura, Y., Uno, K. and Tsuruoka, K., Japan Pat. JP 71 03,176, 1971 (Jan. 26); CA 74, 112781q (1971).
166. Brace, N. O., J. Org. Chem., 1971, 36, 3187; CA 75, 151623e (1971).
167. Gaenzler, W. and Schroeder, G., Ger. Offen. 2,041,736, 1972 (Feb. 24); CA 77, 6439m (1972).
168. Voelker, T. and Hugener, H., Ger. Offen. 2,134,717, 1972 (Jan. 20); CA 76, 141727y (1972).
169. Kodaira, T., Morishita, K., Aida, H. and Yamaoka, H., J. Polym. Sci., Polym. Lett. Ed., 1973, 11, 347; CA 79, 105616q (1973).
170. Shimizu, K., Harada, S., Ueda, T. and Endo, T., Ger. Offen. 2,416,675, 1974 (Oct. 31); CA 83, 59819r (1975).
171. Kodaira, T., Aoyama, F., Morishita, K. Tsuchida, M. and Nogi, S., Kobunshi Ronbunshu, 1974, 31, 682; CA 82, 98504h (1975).
172. Nagasawa, K., Kuroiwa, K. and Narita, K., Jap. Pat. 74 14,351, 1974 (Apr. 6); CA 82, 58510a (1975).
173. Hashimoto, T., Ueda, T. and Harada, S., Japan Pat. JP 74,122,600, 1974 (Nov. 22); CA 82, 125837c (1975).
174. Kodaira, T., Sakai, M. and Yamazaki, K., J. Polym. Sci., Polym. Lett., 1975, 13, 521; CA 84, 74669f (1976).
175. Panzer, H. P. and McAdams, L. V., U.S. Pat. 3,878,170, 1975, April 15; CA 83, 45048B (1975).
176. Solomon, D. H. and McLean, C. D., Ger. Offen. 2,428,096, 1975 (Jan. 23); CA 83, 206969v (1975).
177. Beckwith, A. L. J., Ong, A. K. and Solomon, D. H., J. Macromol. Sci. (Chem) A, 1975, 9, 125; CA 82, 58211d (1975),
178. Johns, S. R. and Willing, R. I., J. Macromol. Sci. (Chem), A, 1975, 9, 169; CA 82, 98422m (1975).
179. Maxim, S., Dumitriu, E., Ioan, S. and Carpov, A., Eur. Polym. J., 1977, 13, 105; CA 87, 39929f (1977).
180. Rabinowitz, R. and Welcher, R. P., U.S. Pat. 4,064,333, 1977 (Dec. 20); CA 88, 90255f (1978).
181. Ioan, S., Dumitriu, E., Maxim, S. and Carpov, A., Eur. Polym. J., 1977, 13, 109; CA 87, 39930z (1977).
182. Sirotkina, E. E., Kogan, R. M., Kovalevich, L. I. and Galkina, G. F., Mineral'n. Syr'e i Neftekhimiya, Tomsk, 1979, 148; CA 93, 205057r (1980).
183. Blount, D. H., U.S. Pat. US 4,157,438, 1979 (Jun. 5);CA 91, 108487g (1979).
184. Ioan, S. and Maxim, S., Eur. Polym. J., 1979, 15, 161; CA 91, 92107v (1979).
185. Kumar M. V., Masterova, M. N., Goluveb, V. B., Zubov, V. P. and Kabanov, V. A., Vestn. Mosk. Univ. Ser. 2: Khim., 1979, 20, 490; CA 92, 42445d (1980).

186. Calgon Corp.,Jap. Pat. JP 79 01,282, 1979 (Jan. 23); CA 90, 205125z (1979).
187. Stone, S. A., Diss. Abstr. Int. B, 1979, 39, 3356; CA 90, 104446z, (1979).
188. Hunter, W.E. and Sieder, T. P., U.S. Pat. US 4,151,202, 1979 (Apr. 24); CA 91, 21433n (1979).
189. Bowman, L. M. and Cha, C. Y., J. Polym. Sci., Polym. Lett. Ed., 1979, 17, 167; CA 90, 187483h (1979).
190. Cha, C. Y., Folger, R. L. and Ware, B. R., J. Polym. Sci., Polym. Phys. Ed., 1980, 18, 1853; CA 93, 150779v (1980).
191. Zaima, T., Nishikubo, T. and Mitsuhashi, K., Kobunshi Ronbunshu, 1980, 37, 173; CA 92, 215824g (1980).
192. Ohme, R., Ruscha, J. and Ballschuh, D., Ger. Pat.(East) 141,921, 1980 (May 28); CA 94, 191675x (1981).
193. Ohme, R., Ballschuh, D. and Rusche, J., East Ger. Pat.141,028, 1980 (Apr. 9); CA 94, 122269f (1981).
194. Ballschuh, D., Ohme, R. and Rusche, J., Ger. (East) Pat. 141,029, 1980 (Apr. 9); CA 94, 122273c (1981).
195. Zaima, T., Nishikubo, T. and Mitsuhashi, K., J. Polym. Sci., Polym. Lett., 1980, 18, 1; CA 92, 129389n (1980).
196. Wyroba, A., Polimery (Warsaw), 1981, 26, 139; CA 95, 220422r (1981).
197. Ottenbrite, R. M.,Polymer Bull. (Berlin), 1981, 6, 225; CA 96, 123426h (1982).
198. Buriks, R. S. and Lovett, E. G., U.S. Pat. US 4,341,887, 1982 (July 27); CA 97, 163699b (1982).
199. Ottenbrite, R. M., ACS Symp. Ser., 1982, 195, 61; CA 97, 163597s (1982).
200. Onabe, F., Makurai Gakkaishi, 1982, 28, 437; CA 97, 163623x (1982).
201. Stone-Elander, S. A., Butler, G. B., Davis, J. H. and Palenik, G. J., Macromolecules, 1982, 15, 45; CA 96, 69482h (1982).
202. Nippon Senka Kogyo Co., Ltd., Japan Pat. 57 82,591, 1982 (May 24); CA 98, 36017y (1983).
203. Kryuchkov, V. V., Parkhamovich, E. S. and Amburg. L. A., Iss., v Ob. Pol. i Pr. Ter. Smol. i Ion. 1983,32; CA 101, 91880e (1984).
204. Hahn, M., Jaeger, W. and Reinisch, G., Acta Polym., 1983, 34, 322; CA 99, 158920d (1983).
205. Paleos, C. M., Dais, P. and Malliaria, A., J. Polym. Sci., (Chem.), 1984, 22, 3383; CA 102, 79391j (1985).
206. Bekturov, E. A., Kudaibergenov, S. E., Saltybaeva, S. S., Khlystov, A. S. and Yaskevich, V. I., Makromol. Chem., Rapid Commun., 1984, 5, 763; CA 102, 62790r (1985).
207. Hahn, M., Jaeger, W., Wandrey, C., Seehaus, F. and Reinisch, G., Ger. (East) Pat. DD 205,690, 1984 (Jan. 4); CA 101, 39006k (1984).
208. Daiichi Kogya Seiyaku Co., Ltd., Japan Pat. JP 59 216,981 [84 216,981], 1984 (Dec.7); CA 102, 115118m (1985).
209. Schmitt, K. D., U.S. Pat. 4,504,622, 1985 (Mar. 12);CA 103, 90176j (1985).
210. Nishimura, J., Mimura, M., Nakazawa, N. and Yamashita, S. J., J. Polym. Sci., Polym. Chem. Ed., 1980, 18, 2071; CA 93, 168692w (1980).
211. Iino, Y., Ogata, Y., Shigehara, K. and Tsuchida, E., Makromol. Chem., 1985, 186, 923; CA 103, 54519q (1985).
212. Mathias, L. J., Vaidya, R. A. and Bloodworth, R. H., Polymer Prepr. (ACS Div. Polym. Chem.), 1985, 26, 182; CA 104, 88999g (1986).
213. Babilis, D., Dais, P., Margaritis, L. H. and Paleos, C. M., J. Polym. Sci., Polym. Chem. Ed., 1985, 23, 1089; CA 102, 221261n (1985).
214. Babaev, N. A., Martynenko, A. I., Topchiev, D. A. and Kabanov, V. A., Acta Polym., 1985, 36, 396; CA 103, 123962t (1985).

215. Topchiev, D. A., Malkanduev, Yu. A., Korshak, Yu. V., Mikitaev, A. K. and Kabanov, V. A., Acta Polym., 1985, 36, 372; CA 103, 123960r (1985).
216. Babilis, D., Dais, P. and Paleos, C., Polymer Preprints, 1985, 26 (1), 206; CA 103, 37761v (1985).
217. Narang, S. C., Ventura, S. and Ramharack, R., Polym. Prepr. (ACS, Div. Polym. Chem.), 1986, 27, 67; CA 105, 191691v (1986).
218. Mathias, L. J., U.S. Pat. 4,591,625, 1986, (May 27);CA 105, 98178y (1986).
219. Ottenbrite, R. M. and Chen, H., Polym. Preprints, 1986, 27, 13; CA 105, 191688z (1986).
220. McCormick, C. L., Zhang, Z. B. and Anderson, K. W., J. Controlled Release, 1986, 4, 97; CA 105, 185753a (1986).
221. Burkhardt, C. W., McCarthy, K. J. and Parazak, D. P., J. Polym. Sci., Part C: Polym. Lett., 1987, 25, 209; CA 107, 59848p (1987).
222. Mathias, L. J. and Cei, G., Macromolecules, 1987, 20, 2645; CA 107, 218122r (1987).
223. Hunter, W. E. and Craun, G. P., U.S. Pat. US 4,654,378, 1987 (Mar. 31); CA 107, 7834m (1987).
224. Skripchuk, V. G. and Blekher, T. A., USSR SU 1,346,640, 1987 (Oct. 23); CA 108, 76268n (1988).
225. Mathias, L. J. and Viswanathan, T., J. Chem. Educ., 1987, 64, 639; CA 107, 95994s (1987).
226. Mathias, L. J., Vaidya, R. A. and Halley, R. J., J. Appl. Polym. Sci., 1987, 33, 1157; CA 106, 156974p (1987).
227. Jaeger, W., Wandrey, C., Hahn, M., Nicke, R., Pensold, S., Borchers, B. and Tappe, M., Ger. Offen. DE 3,733,587, 1988 (Apr. 14); CA 109, 151790c (1988).
228. Huang, P. C. and Reichert, K. H., Angew. Makromol. Chem., 1988, 162, 19; CA 109, 231625t (1988).
229. Huber, E. W. and Heineman, W. R., J. Polym. Sci., Part C: Polym. Lett., 1988, 26, 333; CA 109, 110989f (1988).
230. Babilis, D., Paleos, C. M. and Dais, P., J. Polym. Sci., Chem. A, 1988, 26, 2141; CA 109, 190943z (1988).
231. Butler, G. B. and Pledger, H., Jr., U.S. Pat. US 4,742,134, 1988 (May 3),CA 109, 38430x (1988).
232. Boothe, J. E., Morse, L. D. and Klein, W. L., Eur. Pat. EP 269,243, 1988 (Jun. 1); CA 110, 198905e (1989).
233. Yu, Z. and Zhao, D., Gongneng Gaofenzi Xuebao, 1988, 1, 39; CA 112, 78461b (1990).
234. Lukach, C. A. and Sau, A. C., U.S. Pat. 4,853,437, 1989 (Aug.1); CA 112, 57821d (1990).
235. Topchiev, D. A., Malkanduev, Yu. A., Yanovskii, Yu. G., Oppengeim, V. D. and Kabanov, V. A., Eur. Polym. J., 1989, 25, 1095; CA 112, 78020g (1990).
236. Kodaira, T. and Tanahashi, H., Macromolecules, 1989, 22, 4643; CA 111, 233750v (1989).
237. Kodaira, T., Kitagawa, N. and Aoyagi, K., Kobunshi Ronbunshu, 1989, 46, 507; CA 112, 78011e (1990).
238. Do, C. H. and Butler, G. B., Polym. Prepr. (ACS, Div. Polym. Chem.), 1989, 30, 352; CA 112, 179961v (1990).
239. Zubov, V. P., Egorov, V. V., Batrakova, E. V. and Ksenofontova, O. B., USSR SU 1,523,550, 1989 (Nov. 23); CA 112, 159191r (1990).
240. Mathias, L. J., Kloske, T. and Cei, G., Macromolecules, 1989, 22, 4615; CA 111, 233782g (1989).
241. Strege, M. A. and Dubin, P. L., J. Chromatogr., 1989, 463, 165; CA 111, 234048j (1989).

242. Higuchi, T. and Kodaira, T., Makromol. Chem., 1989, 190, 2885; CA 112, 21305z (1990).
243. Mercier, J. and Smets, G., J. Polym. Sci., 1962, 57, 763; CA 57, 8715a (1962).
244. Tiers, G. V. D. and Bovey, F. A., J. Polym. Sci., 1960, 47, 479; CA 55, 25344i (1961).
245. Gibbs, W. E. and Murray, J. T., J. Polym. Sci., 1962, 58, 1211; CA 57, 4819a (1962).
246. Hwa, J. C. H., Fleming, W. A., and Miller, L., J. Polym. Sci. A, 1964, 2, 2385; CA 61, 3209c (1964).
247. Mercier, J. and Smets, G., J. Polym. Sci. A, 1963, 1, 1491; CA 59, 1761b (1963).
248. Miyake, T., Kogyo-kagaku zasshi, 1961, 64, 710; CA 57, 6113g (1962).
249. Okada, M., Hayashi, Kanae, Hayashi, Koichiro and Okamura, S., Kobunshi kagaku, 1965, 22, 441; CA 63, 16477b (1965).
250. Aso, C., Kogyo-kagaku zasshi, 1960, 63, 363; CA 56, 3629a (1962).
251. Smets, G., Hous, P. and Deval, N., J. Polym. Sci. A, 1964, 2, 4825; CA 62, 6557e, (1965).
252. Sakurada, I, Iwagaki, T. and Sakaguchi, Y., Kobunshi Kagaku, 1964, 21, 270; CA 62, 6583c (1965).
253. Asahi Chemical Industry Co., Ltd.,Japan Pat. JP 60 115,902 [85 115,902], 1985 (Jun. 22); CA 103, 179368w (1985).
254. Matsoyan, S. G. and Avetyan, M. G., Zh. Obsch. Khim., 1960, 30, 697; CA 54, 24358b (1960).
255. Matsoyan, S. G., J. Polym. Sci., 1961, 52, 189; CA 55, 7900e (1961).
256. Matsoyan, S. G., Avetyan, M. G., Akopyan, L. M. Voskanyan, Morlyan, N. M. and Eliazyan, M. A., Vysokomol. Soedin., 1961, 3, 1010; CA 55,13081b (1961).
257. Matsoyan, S. G., Eliazyan, M. A., and Gervokyan, E. Ts., Vysokomol. Soedin., 1962, 4, 1515; CA 59, 1764g (1963).
258. Matsoyan, S. G. and Voskanyan, M. G., Izv. Akad. Nauk armyan. SSR, Khim Nauki 1963, 16, 151; CA 59, 11668f (1963).
259. Matsoyan, S. G., Voskanyan, M. G., and Saakyan, A. A., Izv. Akad. Nauk armyan. SSR, Khim Nauki 1963, 16, 455; CA 60, 10796b (1964).
260. Matsoyan, S. G. and Akopyan, L. M., Vysokomol. Soedin., 1963, 3, 1311; CA 57, 12704c (1962).
261. Matsoyan, S. G. and Akopyan, L. M., Vysokomol. Soedin., 1963, 5, 1329; CA 60, 665f (1964).
262. Matsoyan, S. G., Pogosyan, G. M., and Saakyan, A. A., Vysokomol. Soedin., 1963, 5, 1334; CA 60, 666a (1964).
263. Matsoyan, S. G. and Akopyan, L. M., Izv. Akad. Nauk armyan. SSR, Khim Nauki 1963, 16, 51; CA 59, 11367f (1963).
264. Miyake, T.,Kogyo-kagaku zasshi, 1961, 64, 1272; CA 57, 4850f (1962).
265. Kocharyan, N. M., Barsamyan, S. T., Matsoyan, S. G., Pikalova, V. N., Tolapchyan, L. S. and Voskanyan, M. G., Izv. Akad. Nauk armyan. SSR, Khim. Nauki 1965, 18, 441; CA 64. 12799g (1966).
266. Barsamyan, S. T.,Vysokomol. Soedin. Ser. A, 1967, 9, 749; CA 67, 33111y (1967).
267. Arbuzova, I. A. and Sultanov, K., Vysokomol. Soedin. 1960, 2, 1077; CA 55, 8921i (1961).
268. Sultanov, K. and Arbuzova, I. A., Usbek. khim. Zh., 1963, 7, 57; CA 59, 4043c (1963).
269. Arbuzova, I. A., Kostikov, R. R., and Propp, L. N., Vysokomol. Soedin, 1960, 2, 1402; CA 55, 19318d (1961).
270. Sultanov, K. and Arbuzova, I. A., Uzbek. khim. Zh., 1963, 7, 58; CA 60, 5648h (1964).
271. Sultanov, K. and Arbuzova, I. A., Uzbek. khim. Zh., 1965, 9, 38; CA 64, 12799h (1966).
272. Matsoyan, S. G. and Saakyan, A. A., Vysokomol. Soedin., 1961, 3, 1317; CA 57, 12704e (1962).

273. Ardis, A. E., Dietrich, H. J., Raymond, M. A.and Urs, V., U.S. Pat. US 3,514,435, 1970 (May26); CA 73, 15718w (1970).
274. Aso, C., Kunitake, T. and Ando, S., J. Macromolec. Sci. (Chem.), 1971, 5(1), 2019; CA 73, 120931h (1970).
275. Raymond, M. A. and Dietrich, H. J., J. Macromolec. Sci. (Chem.), 1972, 6, 207; CA 77, 75721j (1972).
276. Dietrich, H. J. and Raymond, M. A., J. Macromolec. Sci. (Chem.), 1972, 6, 191; CA 77, 75579u (1972).
277. Murahashi, S., Nozakura, S. , Fuji, S., and Kikukawa, K., Bull. Chem. Soc. Japan, 1965, 38, 1905; CA 64, 6765h (1966).
278. Kikukawa, K., Nozakura, S. and Murahashi, S., Kobunshi kagaku, 1968, 25, 19; CA 69, 19684x (1968).
279. Rhum, D. and Moore, G. L., Brit. Pat. 1,129,229, 1968 (Oct. 2),CA 69, 105926j (1968).
280. Air Reduction Co. Ltd.,Brit. Pat. 1,129,230, 1968 (Oct. 2); CA 69, 107270q (1968).
281. Rhum, D. and Weintraub, L., Brit. Pat. 1,129,228, 1968, (Oct. 2); CA 69, 107255p (1968).
282. Aso, C. and Tagami, S., Macromolecules, 1969, 2, 414; CA 71, 71019q (1969).
283. Kitahama, Y., Ohama, H., and Kobayashi, H., J. Polym. Sci. C, 1966 (Pub. 1969), 23, 785; CA 71, 13440j (1969).
284. Yamakita, H. and Hayakawa, K., Nippon Kagaku Kaishi, 1977, 5, 706; CA 87, 53631j (1977).
285. Tsukino, M. and Kunitake, T., Polym J., 1979, 11, 437; CA 91, 124101g (1979).
286. Aso, C., Kunitake, T., Sasaki, M. and Koyama, K., Kobunshi kagaku, 1970, 27, 260; CA 73, 99266h (1970).
287. Aso, C. and Kunitake, T., Mem. Fac. Engng. Kyushu Univ., 1969, 29, 31; CA 72, 79533u (1970).
288. Matsumoto, A., Takashima, K. and Oiwa, M., Bull. Chem. Soc. Japan, 1969, 42(7), 1959; CA 71, 70965h (1969).
289. Gueniffey, H., Kaemmerer, H. and Pinazzi, C., Makromol. Chem., 1973, 165, 73; CA 79, 42898k (1973).
290. Kaemmerer, H., Steiner, V., Gueniffey, H. and Pinazzi, C., Makromol. Chem., 1976, 177, 1665; CA 85, 78395g (1976).
291. Aminovskaya, D. N., Andreev, D. N. and Davidyuk, L. N., Vysokomol. Soedin., Ser. A, 1977, 19, 1325; CA 87, 53634n (1977).
292. Nishino, J., Yoshida, T., Makino, I., Nanya, S.,Tamaki, K.,and Sakaguchi, Y., Kobunshi Ronbunshu, 1974, 31, 177; CA 81, 91983j (1974).
293. Kikukawa, K., Nozakura, S. I. and Murahashi, S., J. Polym. Sci. , A1, 1972, 10, 139; CA 76, 127505t (1972).
294. Yang, N., Haubenstock, H. and Snow, A., ACS Symp. Ser., 1982, 195, 417; CA 97, 163520m (1982).
295. Bjorkquist, D. W., Bush, R. D., Ezra, F. S. and Keough, T. W., J. Org. Chem., 1986, 51, 3192; CA 105, 60962m (1986).
296. Jones, J. F. and Summers, R. M., U.S. Pat. US 3,005,784, 1961 (Oct. 24); CA 56, 13101c (1962).
297. Jones, J. F., Tucker, H. and Arnold, L. F., U.S. Pat. US 3,005,785, 1962; CA 56, 13101f (1962).
298. Friedlander, W. S. and Tiers, G. V. D., Ger. Pat. 1,098,942, 1958; CA 56, 5810c (1962).
299. Semon, W. and Jones, J. F., Ger. Pat. 1,086,892; Can. Pat. 575,854, 1959 (May 12); CA 55, 22913e (1961).
300. Jones, J. F., Ger. Pat. 1,079,836, 1960 (Apr. 14); CA 55, 22923g (1961).

301. Arbuzova, I. A. and Rostovskii, E., J. Polym. Sci.; IUPAC, USSR (1960), 1961, 52, 325; CA 55, 4038d (1961).
302. Arbuzova, I. A. and Sultanov, K., USSR Pat. SU 140,988, 1961 (Sept. 20); CA 56, 11818e (1962).
303. Arnold, L. F.; U.S. Pat. US 3,066,113, 1962 (Nov. 27); CA 59, 8085c (1963).
304. Kreidl, W., Aust. 223,820; Brit. 924,460, 1962 (Oct. 10); CA 58, 1555c (1963).
305. Voelker, T., Ross, K. and Hosch, L., Ger. Pat. 1,125,177, 1962 (Mar. 8); CA 69, 4887d (1962).
306. Butler, G. B., Crawshaw, A. and Miller, W. L., Macromol. Synth., 1963, 1, 38; CA 60, 12109c (1964).
307. Abe, A. and Goodman, M., J. Polym. Sci. A, 1964, 2, 3491; CA 61, 12096h (1964).
308. Feng, H-T., Chang, H. C. and Tsao, W., Ko Pen-T'ung Hsun, 1965, 2, 120; CA 64, 3694d (1966).
309. Okada, M., Hayashi, Kanae, Hayashi, Koichiro and Okamura, S., Nipp. Hosha. Kobun. Kenkyu Kyokai Nenpo, 1963, 5, 95; CA 63, 7112f (1965).
310. Sakaguchi, Y., Nishino, J., Itori, K. and Yato, T., Kobunshi kagaku, 1966, 23, 759; CA 66, 65938u (1967).
311. Yoshizumi, N., Kinochita, M. Masayoshi, I. and Imoto, M., Mem. Fac. Eng., Osaka City Univ., 1967, 9, 81; CA 70, 12024u (1969).
312. Shibatani, K., Fujiwara, Y. and Fujii, K., J. Polym. Sci., A1, 1970, 8, 1693; CA 73, 40287n (1970).
313. Yuki, H., Hatada, K., Ota, K. and Sasaki, T., Bull. Chem. Soc. Jap., 1970, 43, 890; CA 72, 121954j (1970).
314. Padhye, M. R. and Iyer, B. B., Curr. Sci., 1972, 41, 528; CA 77, 127163a (1972).
315. Matsumoto, A. and Oiwa, M., Nippon Kagaku Kaishi, 1973, 2035; CA 80, 83720k (1974).
316. Corfield, G. C., Crawshaw, A., Thompson, S. J. and Jones, A. G., J. Chem,. Soc., Perkin Trans. 2, 1973,1549; CA 80, 2993b (1974).
317. Gray, T. F. and Butler, G. B., J. Macromol. Sci. (Chem), A, 1975, 9, 45; CA 82, 58274b (1975).
318. Chiantore, O., Camino, G., Chiorino, A. and Guaita, M., Makromol. Chem., 1977, 178, 125; CA 86, 55735z (1977).
319. Chiantore, O. Camino, G., Chiorino, A. and Guaita, M., Makromol. Chem., 1977, 178, 119; CA 86, 55734y (1977).
320. Matsumoto, A., Ueoka, T. and Oiwa, M., J. Polym. Sci., Polym. Chem. Ed., 1978, 16, 2695; CA 90, 55339f (1979).
321. Brace, N. O., Colloq. Int. C. N. R. S. 1977, 1978, 278, 387; CA 91, 174493g (1979).
322. Donescu, D., Carp, N. and Gosa, K., Rev. Roumaine Chim., 1979, 24, 501; CA 92, 76955k (1980).
323. Butler, G. B. and Matsumoto, A., J. Polym. Sci., Polym. Ltrs. Ed., 1981, 19, 167; CA 94, 157390r (1981).
324. Matsumoto, A., Kitamura, T., Oiwa, M. and Butler, G. B., J. Polym. Sci., Polym. Chem. Ed., 1981, 19, 2531.
325. Tsukino, M. and Kunitake, T., ACS Symp. Ser., 1982, 195, 73; CA 97, 183092b (1982).
326. Ohya, T. and Otsu, T., J. Polym. Sci., Polym. Chem. Ed., 1983, 21, 3503; CA 100, 68833n (1984).
327. Mathias, L. J. and Kusefoglu, S. H., J. Polym. Sci.:Part C:Polym. Letters, 1987, 25, 451; CA 108, 6939h (1988).
328. Mathias, L. J., Kusefoglu, S. H. and Ingram, J. E., Macromolecules, 1988, 21, 545; CA 108, 95019y (1988).
329. Otani, I., Hosoya, K. and Ouchi, K., Japan Pat. JP 01,279,911 [89,279,911], 1989 (Nov. 10); CA 112. 180106b (1990).

330. Matsoyan, S. G. and Saakyan, A. A., Vysokomol. Soedin., 1961, 3, 1755; CA 56, 14443g (1962).
331. Matsoyan, S. G. and Saakyan, A. A., Izv. Akad. Nauk armyan. SSR, Khim. Nauki 1962, 15, 463; CA 59, 7655d (1963).
332. Ringsdorf, H., and Overberger, C. G., J. Polym. Sci., 1962, 61, 511; CA 57, 15344e (1962).
333. Ringsdorf, H. and Overberger, C. G., Makromol. Chem., 1961, 44, 418; CA 55, 22322d (1961).
334. Bogomol'nyi, G. S.,Vysokomol. Soedin., 1959, 1, 1469; CA 54, 14753h (1960).
335. Kolesnikov, G. S., Davydova, S. L., and Ermolaeva, T. N., Vysokomol. Soedin., 1959, 1, 1493; CA 54, 17940c (1960).
336. Chang, H.-C., Feng, H.-P., and Feng, H.-T., Ko fen tzu t'ing hsun, 1964, 6, 487; CA 63, 16474h (1965). 337. Gusel'nikov, L. E., Nametkin, N. S., Polay, L. S., and Chemysheva, T. I., Vysokomol. Soedin., 1964, 6, 2002; CA 62, 6560e, (1965).
338. Furue, M., Nozakura, S. and Murahashi, S., Kobunshi kagaku, 1967, 24, 522; CA 68, 69390y (1968).
339. Sumi, M., Nozakura, S. and Murahashi, S., Kobunshi Kagaku, 1967, 24, 512; CA 68, 69389c (1968).
340. Butler, G. B. and Stackman, R. W., J. Macromol. Sci. (Chem.), 1969, 3(5), 821; CA 71, 50559a (1969).
341. Butler, G. B., Skinner, D. L., Bond, W. C.,Jr. and Rogers, C. L., J. Macromolec. Sci. (Chem.), 1970, 4, 1437; CA 73, 45900s (1970).
342. Kunitake, T., Nakashima, T. and Aso, C., Makromol. Chem., 1971, 146, 79; CA 75, 110628s (1971).
343. Sosin, S. L., Dzhashi, L., Antipova, V. A. and Korshak, V. V., Vysokomol. Soedin., Ser. B, 1970, 12, 699; CA 74, 13698m (1971).
344. Sosin, S. L., Dzhashi, L., Antipova, V. A. and Korshak, V. V., Vysokomol. Soedin., Ser. B, 1974, 16, 347; CA 81, 78323z (1974).
345. Korshak, V. V. and Sosin, S. L., Organometal. Polym., Academic Press, NY, 1977, 26.
346. Corfield, G. C., Brooks, J. S. and Plimley, S., Polym. Prepr., ACS, Polym. Div., 1981, 22, 3.
347. Fearn, J. E., J. Res. Natn. Bur. Stand., Sect. A, 1971, 75, 41; CA 74, 126097h (1971).
348. Kida, S., Nozakura, S. and Murahashi, S., Polym. J., 1972, 3, 234.
349. Billingham, N. C., Jenkins, A. D.,Kromfli, E. B. and Walton, D. R. M., Polym. Sci., Polym. Chem., 1975, 15, 675; CA 86, 121870t (1977).
350. Corfield, G. C. and Monks, H. H., J. Macromol. Sci. (Chem), A, 1975, 9, 1113; CA 84, 17894b (1976).
351. Berlin, K. D. and Butler, G. B., Polym. Prepr., 1960, 1 (1), 93.
352. Topchiev, A. V., Nametkin, N. S. and Polak, L., J. Polym. Sci., 1962, 58, 1349; CA 57, 11377h (1962).
353. Matsoyan, S. G. and Avetyan, M. G., USSR Pat. 126,264, 1960 (Feb. 10); CA 54, 16024c (1960).
354. Ludington, R.,U.S. Pat. 3,009,897; Ger. Pat. 1115464, 1961 (Nov, 21); CA 56, 11801b (1962).
355. Pellon, J., Deichertand, W. and Thomas, W., J. Polym. Sci., 1961, 55, 153; CA 56, 8922i (1962).
356. Hungerford, A. and Mullane, P. J., U.S. Pat. US 3,038,210, 1962 (June 12); CA 57, 12745d (1962).
357. Bond, Jr., W. C., Dissertation Abstr., 1964, 25, 3260; CA 62, 13250d (1965).
358. Andrianov, K. A., Gavrikova, L. and Rodionova, E., Izv. Akad. Nauk armyan. SSR, Khim. Nauki 1968, 1786; CA 69, 97223k (1968).

359. Shizunobu, H., Isao, F. and Akio, Y., Kobunshi Kagaku, 1968, 25, 6; CA 70, 47979m (1969).
360. Butler, G. B., Skinner, D. S., Bond, Jr., W. C. and Rogers, C. L., Polym. Prepr., 1969, 10 (2), 923; CA 74, 14243t (1971).
361. Galeev, V. S., Levin, Y. A. and Kovalenko, V. I., Inst. Fiz. Khim. Akad. Nauk USSR, 1972,103; CA 78, 84875k (1973).
362. Levin, Y. A., Pyrkin, R. I., Yagfarova, T. A. and Usol'tseva, A. A., Vysokomol. Soedin., Ser. A, 1973, 15, 2070; CA 80, 15256p (1974).
363. Haruta, M., Kageno, K., Soeta, M. and Harada, S., Jap. Pat. JP 75 72,987, 1975, June 16; CA 83, 164855x (1975).
364. Oreshkin, I. A., Red'kina, L. I., Kershenbaum, I. L., Chernenko, G. M., Makovetskii, K. L., Tinyakova, E. I. and Dolgoplosk, B. A., Izu. Akad. Nauk SSSR, Ser. Khim, 1977, 2566; CA 88, 62646r (1978).
365. Blount, D. H., U.S. Pat. US 4,115,635, 1978 (Sep. 19); CA 90, 55503e (1979),
366. Belokon, Y. N., Tararov, V. I., Savel'eva, T. F., Lependina, O. L., Timofeeva, G. I. and Belikov, V. M., Makromol. Chem., 1982, 183, 1921; CA 97, 128194m (1982).
367. Saigo, K., Ohnishi, Y., Suzuki, M. and Gokan, H., J. Vac. Sci. Technol. B, 1985, 3, 331; CA 102, 157849f (1985).
368. Holt, T. and Simpson, W., Proc. R. Soc. Lond. A, 1956, 238, 154; CA 51, 15169e (1957).
369. Gordon, M., J. Chem. Phys. 1954, 22, 610; CA 48, 8582e (1954).
370. Gordon, M. and R.-J. Roe, J. Polym. Sci. 1956, 21, 27; CA 50, 14318a (1956).
371. Gordon, M. and Roe, R.-J., J. Polym. Sci. 1956, 21, 39; CA 50, 14318d (1956).
372. Gordon, M. and Roe, R.-J., J. Polym. Sci., 1956, 21, 57; CA 50, 14318f (1956).
373. Gordon, M. and Roe, R.-J., J. Polym. Sci., 1956, 21, 75; CA 50, 14318i (1956).
374. Stoll, M., Rowe, A., and Stoll-Comte, G., Helv. Chim. Acta, 1934, 17, 1289; CA 29, 1064(4) (1935).
375. Stoll, M. and Rowe, A., Helv. Chim. Acta, 1935, 18, 1087; CA 30, 1361(2) (1936).
376. Schulz, R. C. and Stenner, R., Makromol. Chem., 1966, 91, 10; CA 64, 14286b (1966).
377. Simpson, W. and Holt, T., J. Polym. Sci., 1955, 18, 335; CA 50, 15175d (1956).
378. Butler, G. B. and Iachia, B., J. Macromolec. Sci. (Chem.), 1969, 3(8), 1485; CA 71, 124979m (1969).
379. Ouchi, T., Yaguchi, Y. and Oiwa, M., Bull. Chem. Soc. Japan, 1971, 44(6), 1623; CA 75, 110641r (1971).
380. Nishimura, J. and Yamashita, S., Polym. Prepr., ACS, Div. Polym. Chem., 1981, 22, 46.
381. Nishikubo, T., Iizawa, T. and Yoshinaga, A., Makromol. Chem., 1979, 180, 2793; CA 92, 7031u (1980).
382. Seung, S. L. N. and Young, R. N., J. Polym. Sci., Polym. Lett., 1978, 16, 367; CA 89, 90314v (1978).
383. Mathias, L. J., and Canterberry, J. B., Polym. Prepr., ACS, Div. Polym. Chem., 1981, 22, 38.
384. Chu, S. C. and Butler, G. B., Polym. Lettr., 1977, 15, 277; CA 87, 23769t (1977).
385. Butler, G. B. and Lien, Q. S., Polym. Prepr.(ACS-Div. of Polym. Chem.), 1981, 22, 54.
386. Matsumoto, A., Iwanami, K. and Oiwa, M., J. Polym. Sci., Polym. Lett., 1980, 18, 307; CA 92, 198870z (1980).
387. Matsumoto, A., Iwanami, K. and Oiwa, M., J. Polym. Sci., Polym. Chem. Ed., 1981, 19, 213; CA 94, 122029c (1981).
388. Matsumoto, A., Iwanami, K., Kitamura, T., Oiwa, M. and Butler, G. B., Polym. Preprints, ACS Div. Polym. Chem., 1981, 22, 36.
389. Urushido, K., Matsumoto, A. and Oiwa, M., J. Polym. Sci., Polym. Lett., 1981, 19, 59.

390. Ohya, T. and Otsu, T., J. Polym. Sci., Polym. Chem. Ed., 1983, 21, 3169; CA 100, 23062u (1984).
391. Otsu, T. and Ohya, T., J. Macromol. Sci. (Chem.), A, 1984, 21, 1; CA 100, 68836r (1984).
392. Paulrajan, S., Gopalan, A., Subbaratnam, N. R. and Venkatarao, K., Polymer, 1983, 24, 906; CA 99, 88620e (1983).
393. Yokota, K., Kakuchi, T., Nanasawa, A., Iwata, J. and Takada, Y., Polym. J. (Tokyo), 1982, 14, 509; CA 97, 110464q (1982).
394. Nishimura, J., Ishida, Y. Yamashita, S., Hasegawa, H., Sawamoto, M. and Higashimura, T., Polym. J. (Tokyo), 1983, 15, 303; CA 98, 216046r (1983).
395. Nishimura, J., Furukawa, M., Yamashita, S., Inazu, T. and Yoshino, T., J., J. Polym. Sci., Polym. Chem. Ed., 1981, 19, 3257; CA 96, 86049q (1982).
396. Nishikubo, T., Iizawa, T., Yoshinaga, A. and Nitta, M., Makromol. Chem., 1982, 183, 789; CA 97, 6840y (1982).
397. Gopalan, A., Paulrajan, S., Subbaratnam, N. R. and Rao, K. V., J. Polym. Sci., Polym. Chem. Ed., 1985, 23, 1861; CA 103, 71748f (1985).
398. Kakuchi, T. and Yokota, K., Makromol. Chem., Rapid Commun., 1985, 6, 551; CA 103, 160933e (1985).
399. Lagunov, V. M., Golikov, I. V. and Korolev, G. V., Vysokomol. Soedin., Ser. A, 1982, 24, 131; CA 96, 163299a (1982).
400. Bol'bit, N. M., Izyumnikov, A. L., Rogozhkina, E. D., Faizi, N. K. and Chikin, Y. A., Vysokomol. Soedin., Ser. A, 1985, 27, 1621; CA 104, 19892e (1986).
401. Kodaira, T. and Butler, G. B., J. Macromol. Sci. (Chem), 1985, A22, 213; CA 102, 114016q (1985).
402. Kodaira, T., Tsuji, S., Okuyama, K. and Hosoki, Y., Polym. J. (Tokyo), 1984, 16, 901; CA 102, 132511b (1985).
403. Aso, C.and Sodakata, K., Kogyo-kwagaku zasshi, 1960, 63, 188; CA 56, 156678 (1962).
404. Lalan-Keraly, F., Compt. Rend., 1961, 253, 2975; CA 56, 15663d (1962).
405. Marvel, C. S. and Rogers, J. R., J. Polym. Sci., 1961, 49, 335; CA 55, 19749d (1961).
406. Sokolova, T. A. and Rudkovskaya, G. D., Vysokomol. Soedin., 1961, 3, 706; CA 55, 26505a (1961).
407. Pardo, Jr., C. E. and Lundgren, H. P., U.S. Pat. 3,005,730, 1961 (Oct. 24); CA 56, 3690e (1962).
408. Shur, A., Filimonova, B. and Filimonova, M., Vysokomol. Soedin., 1961, 3, 1661; CA 56, 10385h (1962).
409. Sokolova, T. A. and Tikhodeeva, I., Zhur. Obshch. Khim., 1961, 31, 2222; CA 56, 2327g (1962).
410. Nametkin, N. S., Topchiev. A. V. and Durgar'yan, S. G., Akad. Nauk. SSSR, Inst. Neftek. Sint. SS 1962, 168; CA 58, 8047c (1963).
411. Schulz, R. C. and Stenner, R., Angew. Chem., 1963, 75, 379.
412. Mark, H.,U.S. Pat. 3,081,282; Br. 856861; Ger. 1098205, 1963 (Mar. 12); CA. 58, 14145g (1963).
413. Fritz, C., Brit. Pat. 1,106,343, 1968 (Mar. 13); CA 68, 96496u (1968).
414. Berlin, A. A., Tvorogov, N. and Korolev, G., Izv. Akad. Nauk armyan. SSR, Khim. Nauki 1966, 193; CA 64, 12799a (1966).
415. Umezu, Y. and Kobayashi,B., Japan Pat. 26,698, 1967 (Dec. 18); CA 68, 96364z (1968).
416. Fukui, K., Kagiya, T., Aimi, M., Nose, S., Futsugan, Y. and Tanaka, Y., Japan Pat. JP 26,699, 1967 (Dec. 18); CA 68, 96521y (1968).
417. Merijan, A., U.S. Pat. 3,350,366, 1967 (Oct. 31); CA 68, 3517p (1968).
418. Francis, W. C. and Burdick, D. L., U.S. Pat. US 3,376,301, 1968 (Apr. 2); CA 68, 105255u (1968).
419. Ouichi, T., Yamamoto, S., Akao, Y., Nagaoka, Y., and Oiwa, M., Kogyo-kwagaku zasshi, 1968, 71, 1078; CA 69, 97225n (1968).

420. Labana, S. S., J. Polym. Sci., A1, 1968, 6, 3283; CA 70, 38274a (1969).
421. Matsumoto, A. and Oiwa, M., Nippon Kagaku Zasshi, 1969, 90, 1278; CA 72, 67320j (1970).
422. Matsumoto, A., Asano, K. and Oiwa, M., Nippon Kagku Zasshi, 1969, 90, 290; CA 71, 3737x (1969).
423. Ouchi, T. and Oiwa, M., Kogyo-kwagaku zasshi, 1969, 72, 746; CA 71, 39517q (1969).
424. Matsumoto, A. and Oiwa, M., Kogyo-kagaku zasshi, 1969, 72, 2127; CA 72, 21960q (1970).
425. Nakanishi, H., Suzuki, Y., Suzuki, F. and Hasegawa, M., J. Polym. Sci., A1, 1969, 7, 753; CA 71, 13397a (1969).
426. Stallings, J. P., J. Polym. Sci. A1, 1970, 8, 1557; CA 73, 15325r (1970).
427. Berry, D. and Gynn, G., Ger. Offen. 2,017,545, 1971;CA 74, 43040h (1971).
428. Chovnik, L., Khomenkova, K., Pazenko, Z. and Kovnev, K., Khim Pron. Ukv., 1970, 9; CA 72, 133239t (1970).
429. Ouchi, T., Tatsuno, S., Nakayama, T. and Oiwa, M., Bull. Chem. Soc. Jap., 1970, 43, 2241; CA 73, 77708b (1970).
430. Isaoka, S., Mori, M., Mori, A. and Kumanotani, J., J. Polym. Sci., A1, 1970, 8, 3009; CA 74, 42762h (1971).
431. Murfin, D. L., Hayashi, K. and Miller, L. E., J. Polym. Sci. , A1, 1970, 8, 1967; CA 73, 77692s (1970).
432. Nakanishi, F. and Hasagawa, M., J. Polym. Sci., A1, 1970, 8, 2151.
433. Ouchi, T. and Oiwa, M., Kogyo kagaku zasshi, 1970, 73, 1713; CA 73, 120944q (1970).
434. Bradney, M. A. M., Forbes, A. D. and Wood, J., Am. Soc., Div. Petr. Chem., Prepr., 1971, 16, 20; CA 78, 147295u (1973).
435. Linden, G. B. and Brooks, W., U.S. Pat. 3,574,705, 1971 (Apr. 12); CA 75, 6687z (1971).
436. Matsumoto, A., Kogyo-kwagaku zasshi, 1971, 74, 1913; CA 76, 46562g (1972).
437. Andreev, D. N. and Usacheva, N. T., Khim. Geterotsikl. Soedin. (10), 1972, 1436; CA 78, 30273u (1973).
438. Willard, P., Polym. Eng. Sci., 1972, 12, 120; CA 76, 154677g (1972).
439. Matsumoto, A. and Oiwa, M., Nippon Kagaku Kaishi, 1972, 1934; CA 78, 84847c (1973).
440. Nakanishi, H., Hasegawa, M. and Sasada, Y., J. Polym. Sci., A-2, 1972, 10, 1537; CA 77, 114937f (1972).
441. Azuma, C. and Ogata, N., Polym. J., 1973, 4, 628; CA 80, 15255n (1974).
442. Matsumoto, A. and Oiwa, M., Makromol. Chem., 1973, 166, 179; CA 79, 79260c (1973).
443. Fujio, Y., Yokoyama, K., Matsumoto, K., Nose, S. and Kodama, T., Jap. Pat. JP 73 92,483, 1973 (Nov. 30); CA 80, 121901e (1974).
444. Matsumoto, A., Aso, T., Tanaka, S. and Oiwa, M., J. Polym. Sci., Polym. Chem. Ed., 1973, 11, 2357; CA 80, 15279y (1974).
445. Matsumoto, A., Kitagami, J. and Oiwa, M., J. Polym. Sci., Polym. Chem. Ed., 1973, 11, 2365; CA 80, 15252j (1974).
446. Toy, A. D. F., Macromol. Synth., 1974, 5, 77; CA 83, 28641x (1975).
447. Bledzki, A., Krolikowski, W., Spychaj, S. and Spychaj, T., Pr. Nauk. Inst. Technol. Org., 1975, 16, 193; CA 83, 43986g (1975).
448. Furukawa, J. and Nishimura, J., J. Polym. Sci., Polym. Lett. Ed., 1976, 14, 85; CA 84, 151027z (1976).
449. Furukawa, J. and Nishimura, J., J. Polym. Sci., Polym. Symp., 1976, 56, 437; CA 89, 180417k (1978).
450. Tamamura, T., Rev. Electr. Commun. Lab., 1978, 26, 1686; CA 90, 153138j (1979).
451. Pavlova, O. V. and Kireeva, S, M., Fiz.-Khim Protsessy Gazov. Kondens. Faz. 1979, 91; CA 93, 47372h (1980).

452. Ohata, T., Kasahara, K., Matsumoto, A. and Oiwa, M., J. Polym. Sci., Polym. Chem. Ed., 1979, 17, 4107; CA 93, 8576h (1980).
453. Kodaira, T., Yamazaki, K. and Kitoh, T., Polym. J., 1979, 11, 377; CA 91, 108277p (1979).
454. Matsumoto, A. and Oiwa, M., J. Polym. Sci., Polym. Lett. Ed., 1980, 18, 421; CA 93, 95711q (1980).
455. Ohata, T., Matsumoto, A. and Oiwa, M., J. Polym. Sci., Polym. Chem. Ed., 1980, 18, 467; CA 92, 181734k (1980).
456. Kodaira, T, Murato, O. and Edo, Y., J. Polym. Sci., Polym. Lett. Ed., 1980, 18, 737; CA 94, 103879p (1981).
457. Teijin Ltd., Japan Pat. 80 115,429, 1980 (Sep. 5); CA 94, 31609a (1981).
458. Micko, M. M. and Paszner, L., J. Radiat. Curing, 1980, 7, 1; CA 93, 47653a (1980).
459. Ohata, T., Fukumori, E., Matsumoto, A. and Oiwa, M., J. Polym. Sci., Polym. Chem. Ed., 1980, 18, 1011; CA 93, 8594n (1980).
460. Hasegawa, H. and Higashimura, T., Macromolecules, 1980, 13, 1350; CA 93, 221147a (1980).
461. Matsumoto, A., Iwanami, K. and Oiwa, M., J. Polym. Sci., Polym. Lett. Ed., 1981, 19, 497; CA 95, 220404m (1981).
462. Kireera, S. M., Pavlova, O. V., Berlin, A. A. and Sivergin, Y. M., Vysokomol. Soedin., Ser. A, 1981, 23, 1791; CA 96, 35963a (1982).
463. Matsumoto, A., Matsui, Y., Tamura, I., Yamawaki, M. and Oiwa, M., Technol. Rep. Kamsai Univ., 1982, 23, 129; CA 98, 72768c (1983).
464. Butler, G. B. and Lien, Q. S., ACS Symp. Ser. No. 195, (Butler and Kresta, Eds.) 1982, 149; CA 97, 182961x (1982).
465. Yamawaki, M., Miyagawa, M., Matsumoto, A. and Oiwa, M., Technol. Rep. Kansai Univ., 1982, 23, 115; CA 98, 72812n (1983).
466. Mathias, L. J. and Canterberry, J. B., ACS Symp. Ser., 1982, 195, 139; CA 97, 182960w (1982).
467. Constantinescu, A. C., Dumitriu, S. and Simionescu, C. I., Acta Polym., 1983, 34, 60; CA 98, 89987w (1983).
468. Matsumoto, A., Nakane, T. and Oiwa, M., J. Polym. Sci., Polym. Lett. Ed., 1983, 21, 699; CA 99, 140469d (1983).
469. Matsumoto, A., Kukimoto, Y., Aoki, K. and Oiwa, M., J. Polym. Sci., Polym. Chem. Ed., 1983, 21, 3493; CA 100, 68790w (1984).
470. Matsumoto, A., Nakane, T. and Oiwa, M., J. Appl. Polym. Sci., 1983, 28, 1105; CA 98, 143889f (1983).
471. Otsu, T. and Ohya, T., Polym. Bull. (Berlin), 1983, 9, 355; CA 98, 161234t (1983).
472. Yokota, K., Matsumura, M., Yamaguchi, K. and Takada, Y., Makromo Chem., Rapid Commun., 1983, 4, 721; CA 100, 7277n (1984).
473. Matsumoto, A., Jiang, G. and Oiwa, M., J. Polym. Sci., Polym. Chem. Ed., 1983, 21, 3191; CA 100, 23025j (1984).
474. Raetzsch, J. C., Steinert, V., Seiler, S. and Buettner, B., Acta Polym., 1984, 35, 373; CA 101, 7762n (1984).
475. Gerasimov, G. N., Mikova, O. B. and Abkin, A. D., Vysokomol. Soedin., Ser. A, 1985, 27, 1280; CA 103, 123967y (1985).
476. Mathias, L. J. and Canterberry, J. B., Polym. Prepr., 1985, 26, 58; CA 103, 6728k (1985).
477. Subbaratnam, N. R., Manickam, S. P., Venuvanalingam, P. and Gopalan, A., J. Macromol. Sci. (Chem.), A, 1986, 23, 117; CA 104, 6231p (1986).
478. Rathnasabapathy, S., Marisami, N., Manickam, S. P., Venkatarao, K. and Subbaratnam, N. R., J. Macromol. Sci. (Chem.), A, 1988, 25, 83; CA 108, 38478y (1988).
479. Babu, B. S., Rao, K. N., Sethuram, B. and Rao, T. N., J. Macromol. Sci. (Chem.), A, 1988, 25, 109; CA 108, 38480t (1988).

480. Kakuchi, T., Kobayashi, O., Nakaya, D. and Yokota, K., Polym. J. (Tokyo), 1989, 21, 649; CA 112, 36564g (1990).
481. Maksumova, A. S. and Mirzhalolova, M. A., Dokl. Akad. Nauk UzSSR, (5), 1989, 36; CA 112, 99289s (1990).
482. Matsumoto, A., Ogasawara, Y., Nishikawa, S., Aso, T. and Oiwa, M., J. of Polym. Sci.: Part A: Polym. Chem., 1989, 27, 839; CA 110, 213479q (1989).
483. Bartulin,J., Parra, M., Ramirez, A. and Zunza, H., Polym. Bull. (Berlin), 1989, 22, 33; CA 112, 36608z (1990).
484. Butler, G. B. and Raymond, M. A., J. Polym. Sci. A, 1965, 3, 3413; CA 64, 2168e (1966).
485. Kodaira, T. and Sumiya, Y., Makromol. Chem.(Eng.), 1986, 187, 933; CA 104, 207777v (1986).
486. Lai, Y. C. and Butler, G. B., J. Macromol. Sci. (Chem), 1984, 21(A), 1547; CA 101, 131182n (1984).
487. Vaidya, R. A. and Mathias, L. J., J. Polym. Sci., Polym Symp., 1986, 74, 243; CA 105, 153597m (1986).
488. Martynenko, A.I., Krapivin, A.M., Zezin, A.B., Topchiev, D.A. and Kabanov, V. A., Vysokomol. Soedin., Ser. B., 1982, 24, 580; CA 97, 216762a (1983).
489. Romantsova, I. I., Pavlova, O. V., Kireeva, S. M. and Sivergin, Yu. M., Dokl. Akad. Nauk SSSR, 1986, 289, 422; CA 105, 191669u (1986).
490. Xi, F. and Vogl, O., J. Macromol. Sci. (Chem.),A, 1983, 20, 321; CA 99, 123026q (1983).
491. Shy, L. J., Leung, Y. K. and Eichinger, B. E., Polym. Preprints, 1984, 25, 278; CA 101, 7748n (1984).
492. Takagishi, T., Kozuka, H. and Kuroki, N., J. Polym. Sci., (Chem.), 1981, 19, 3237; CA 96, 86288s (1982).
493. Bekturov, E. A., Khamzamulina, R. E., Reinisch, G., Jaeger,W. and Belgibaeva, Z. K., Acta Polym., 1984, 35, 521; CA 101, 73210g (1984).
494. Khamzamulina, R.E., Bel'gibaeva, Z.K., Reinisch, G., Bekturov, E.A. and Jaeger, W., Izv. Akad. Nauk Kaz. SSR, Ser. Khim., 1985, 3, 33; CA 103, 160956q (1985).
495. Bekturov, E. A., Kudaibergenov, S. E., Ushanov, V. Z. and Saltybaeva, S. S., Makromol. Chem., 1985, 186, 71; CA 102, 96197t (1985).
496. Klyachko, Y.A., Shnaider, M.A., Korshunova, M.L., Kolganova, I.V. and Topchiev, D.A., Zh. Vses, Khim. O-va. im. D.I.Mendeleeva 1985, 30, 573; CA 104, 170570u (1986).
497. Dubin, P. L. and Oteri, R., J. Colloid Interface Sci., 1983, 95, 453; CA 99, 200981b (1983).
498. Dubin, P. L., McQuigg, D. W. and Rigsbee, D. R., Polym. Prepr. (ACS, Div. Polym. Chem.), 1986, 27, 420; CA 104, 208114p (1986).
499. Dautzenberg, Horst and Dautzenberg, Herbert, Acta Polym., 1984, 36, 102; CA 102, 79418y (1985).
500. Kokufuta, E. and Takahashi, K., Macromolecules, 1986, 19, 351; CA 104, 89413y (1986).
501. Egorov, V. V., Batrakova, E. V., Titkova, L. V., Demin, V. V., Zubov, V. P. and Barnakov, A. N., Vysokomol. Soedin. , Ser.8, 1982, 24, 370; CA 97, 72806t (1982).
502. Kokufuta, E., Shimizu, H. and Nakamura, I., Macromolecules, 1981, 14, 1178; CA 95, 164118g (1981).
503. Philipp, B., Hong, L. T., Linow, K. J. and Cowie, J. M. G., Eur. Polym. J., 1981, 17, 615; CA 95, 151300k (1981).
504. Philipp, B., Hong, L. T., Dawydoff, W., Linow, K. J. and Schuelke, U., Z. Anorg. Allg. Chem., 1981, 479, 219; CA 95, 187851e (1981).
505. Rusche, J., Ballachuh, D. and Ohme, R., Ger. (East) 147,949, 1981 (Apr. 29); CA 95, 187949t (1981).

506. Schempp, W. and Tran, H. T., Wochenbl. Papierfabr., 1981, 109, 726; CA 95, 221563z (1982).
507. Shalbaeva, G. B., Deposited Doc. USSR, 1982, VINITI, 3676, 20; CA 101, 193433d (1984).
508. Vasil'eva, O. V., Davydov, A. V., Kolganova, I. V., Mikaya, A. I., Zaikin, V. G., Yanovakii, Y. G., Topchiev, D. A. and Vinogradov, G. V., Dokl. Akad. Nauk SSSR, 1984, 276(3), 621; CA 101, 152462u (1984).
509. Onabe, F., Mokuzai Gakkaishi, 1982, 28, 445; CA 97, 111466d (1982).
510. Onabe,F., Mokuzai Gakkaishi, 1983, 29, 593; CA 99, 196809g (1983).
511. Onabe, F., Mokuzai Gakkaishi, 1983, 29, 513; CA 99, 196798c (1983).
512. Onabe, F., Mokuzai Gakkaishi, 1983, 29, 467; CA 99, 124286m (1983).
513. Onabe, F., Mokuzai Gakkaishi, 1983, 29, 459; CA 99, 124285k (1983).
514. Beckwith, A. L. J., Hawthorne, D. G. and Solomon, D. H., Aust. J. Chem., 1976, 29, 995; CA 85, 122976y (1976).
515. Beckwith, A. L. J., Blair, I. A. and Phillipou, G., Tetrahedron Letters, 1974, 2251; CA 82, 72397a (1975).
516. Lancaster, J. E., Baccei, L. and Panzer, H. P., Polym. Lett., 1976, 14, 549; CA 85, 143666n (1976).
517. Doering, W. and Roth, W. R., Tetrahedron, 1962, 18, 67; CA 57, 2044g (1962).
518. Bevington, J. C. and Malpass, B. W., J. Polym. Sci. A, 1964, 2, 1893; CA 60, 15708d (1964).
519. Lenz, R. W., Org. Chem. of Synth. High Polym., Wiley, 1967, 346.
520. Carlsson, D. J. and Ingold, K. U., J. Am. Chem. Soc., 1968, 90, 7047; CA 70, 19359r (1969).
521. Connor, H., Shimada, K. and Szwarc, M., Macromolecules, 1970, 5, 80; CA 78, 58916e (1973).
522. Buenzli, J. C., Burak, A. J. and Frost, D. C., Tetrahedron, 1973, 29, 3735; CA 81, 3229q (1974).
523. Julia, M., Pure & Appl. Chem., 1974, 40, 553; CA 83, 177813g (1975).
524. Hawthorne, D. G. and Solomon, D. H., J. Macromol. Sci., Chem., A, 1976, 10, 923; CA 85, 109052z (1976).
525. Talley, C. P., Bradley, G. M. and Guliana, R, T., U.S. Pat. 4,118,316, 1978, Oct. 3; CA 90, 55533q (1979).
526. Kumar, M. V. and Lomonosova, M. V., Deposited Doc., VINITI 1805–78, 1978, 168; CA 91, 211882f (1979).
527. Zubov, V. P., Kumar, M. V., Masterova, M. N. and Kabanov, V. A., J. Macromol. Sci. (Chem.), A, 1979, 13, 111; CA 90, 169102d (1979).
528. Kumar, M. V., Masterova, M. N., Golubev, V. B., Zubov, V. P. and Kabanov, V. A., Vestn. Mosk. Univ., Ser. 2, 1979, 20, 490; CA 92, 42445d (1980).
529. Takamura, K., Goldsmith, H. L. and Mason, S. G., J. Colloid Interface Sci., 1981, 82, 190: CA 95, 50033v (1982).
530. Kurokawa, Y., Shirakawa, N., Terada, M. and Yui, N., J. Appl. Polym. Sci., 1980, 25, 1645; CA 93, 186911w (1980).
531. Wandrey, C., Jaeger, W. and Reinisch, G., Acta Polym., 1982, 33, 442; CA 97, 163761r (1982).
532. Jaric, M. V. and Bennemann, K. H., Phys. Rev. A, 1983, 27, 1228; CA 98, 126671m (1983).
533. Yamada, I., Kenkyu Kiyo-Konan Joshi Daigaku, 1984 (Pub.1985), 651; CA 104, 110256m (1986).
534. Khazryatova, L.K., Kuznetsov, E.V., Serzhanina, V.A., Kacherovskaya, F.B., Bikmullina, L. A., Mikhalevskaya, S. V., Kushkova, T. M., Gerasimov, S. S., Vasilenok, Y. I. and Lagunova, V.N., Dokl. Akad. Nauk BSSR, 1985, 29, 622; CA 103, 216159b (1986).

535. Vaz, R. J. N., Diss. Abstr. Int. B, (PhD Thesis UF 1985) 1986, 47(1), 218; CA 105, 173115f (1986).
536. Neville, R. G., U.S. Pat. 3,725,346, 1973 (Apr. 3); CA 79, 19380b (1973).
537. Nishimura, J., Ishida, Y., Mimura, M., and Nakazawa, N. and Yamashita, S., J. Polym. Sci., Chem., 1980, 18, 2061; CA 93, 132890v (1980).
538. Friedlander, W. S. and Tiers, G. van D., Ger. Pat. 1,098,942, 1962; CA 56, 5810c (1962).
539. Kozakiewicz, J.J., Kurose, N. S, Draney, D. R., Huang, S. Y. and Falzone, J., Polym. Preprints, 1987, 28 (2), 347.

3
Nonconjugated Unsymmetrical Dienes

This chapter deals predominantly with unsymmetrical nonconjugated dienes, and the effects of the different reactivities of the double bonds in the molecule on the extent of cyclization as well as on the mode of initiation, and selectivity of initiators. Certain dienes have been shown to permit initiator attack and propagation through only one double bond of the diene, whereas the same molecule may undergo extensive cyclization when initiation according to a different mechanism is chosen. Unsymmetrical monomers capable of leading to five- or six-membered carbocyclic rings will be discussed initially followed by those monomers capable of forming heterocyclic rings. This group will be discussed from the standpoint of the heteroatom beginning with nitrogen as the heteroatom, to oxygen, and to other elements. Rings larger than those containing five- or six-members will also be discussed as well as theoretical interpretations of the results.

Methyl allyl maleate and methyl allyl fumarate were perhaps the earliest examples of unsymmetrical dienes to be studied,[1-77] but numerous other such monomers have now been studied and discussed in detail in an earlier review.[1-156] Cyclopolymerization[1-77] of allyl and methallyl crotonates appear to represent the first examples in which it was shown that the less-stable radical predominated in the cyclization step in contrast to prediction by the Flory hypothesis.[1]

Although the greater part of the work done in the area of cyclopolymerization has been limited to symmetrical monomers, unsymmetrical 1,6-dienes should be capable of homopolymerization *via* cyclization. Particularly interesting and promising monomers for cyclic polymerization are the monomers containing double bonds of comparable reactivity toward copolymerization. Homopolymerizations of such unsymmetrical 1,6-dienes as allyl acrylate,[1-73] allyl methacrylate,[1-74, 1-75] and substituted allyl methacrylates were reported before the proposal of the cyclopolymerization hypothesis; consequently no previous study directed specifically toward verification of the presence of cyclic units in the polymers derived from such dienes has been made. Crosslinking was assumed in most cases.

CYCLOPOLYMERS CONTAINING CARBOCYCLIC RINGS

Cyclopolymerization of o-allylstyrene has been studied by free radical and Friedel-Crafts initiators; all of the vinyl groups participated in polymerization, and the allyl groups partly remained as residual double bonds.[2] The allyl groups participated only *via* the cyclopolymerization mechanism. Spectral studies indicated that the cyclic structural units in the radical polymerization were six-membered rings which were formed by addition of the propagating radicals to the vinyl groups followed by cyclization of the allyl groups. In cationic polymerization, the allyl groups predominately isomerized prior to cyclization or addition, and the polymer contained considerable five-membered rings. In the polymerization of o-allyl-α-methylstyrene by Friedal-Crafts catalysts, all the unsaturated groups in the iso-propenyl and allyl groups participated in the polymerization and the polymer obtained had only the cyclic structural units. The cyclic structural units of the polymer obtained with $BF_3C_2H_5O$ catalyst in various solvents were six-membered rings, whereas those obtained with $AlBr_3$ catalyst in nonpolar solvents were six-membered rings resulting from isomerization of the allyl groups.

1-Allylindene and 1-allylindan were prepared and polymerized cationically.[3] 1-Allylindene was polymerized in the presence of $TiCl_4$, $TaCl_5$, $AlCl_3$, and $AlBr_3$. C_2H_5Cl and CH_2Cl_2 were used as solvents. The polymers were studied by nuclear magnetic resonance and infrared spectra showing that the allyl double bonds participated in the polymerization.

A study of cyclopolymerization of 3-allylcyclopentene and 3-vinylcyclopentene has been reported.[4] Vinylcyclopentene showed higher reactivity than 3-allylcyclopentene in cationic homopolymerization. Poly(3-allylcyclopentene) (>76% bicyclic structure), possibly containing the units shown (3–1), was soluble in common organic solvents,

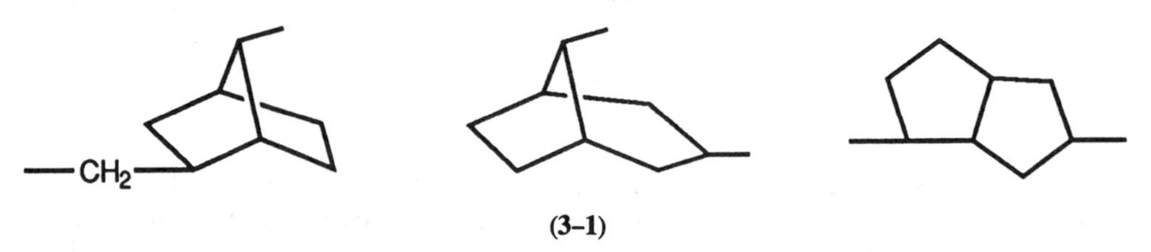

(3–1)

while poly(3-vinylcyclopentene) (>72% bicyclic structure) contained insoluble fractions. Maleic anhydride copolymerized with vinylcyclopentene and 3-allylcyclopentene in the presence of azobisisobutyronitrile to give 2:1 copolymers.

Cationic polymerization of some polycyclic dienes has been investigated.[5] exo-Dicyclopentadiene was polymerized in the presence of boron trifluoride etherate or palladium dichloride dibenzonitrile without rearrangement to form a 1,2-polymer. endo-Dicyclopentadiene rearranged to the exo configuration during polymerization in the presence of BF_3Et_2O but not with $PdCl_2(PhCN)_2$, and formed a 1,3-polymer. In the presence of $PdCl_2(PhCN)_2$, endo-dicyclopentadiene formed a 1,2-polymer. Diendomethylenehexahydronaphthalene was polymerized in the presence of the boron catalyst to form a completely saturated polymer, but in the presence of the palladium catalyst, the polymer had 38.9% residual unsaturation. A half-cage structure (3–2) was proposed as the repeating unit of the saturated polymer.

1-Vinyl-4-methylenecyclohexane has been polymerized to a polymer containing 1,3-methylenebicyclo[2.2.2]octane units by a cationic mechanism.[1-39] By use of gaseous boron trifluoride at $-78°C$, 79% yield of polymer was obtained. The polymer had an

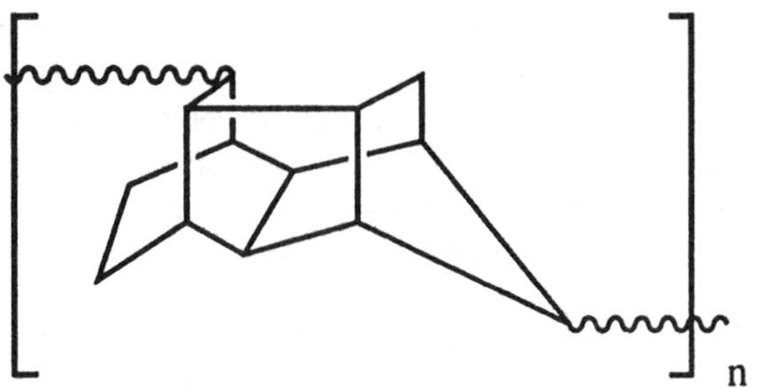

(3–2)

intrinsic viscosity of 0.08 dl/g in benzene, a softening point of 150°C, and a melting range of 150–190°, with decomposition. The infrared spectrum of the polymer indicated the presence of residual terminal unsaturation. A nuclear magnetic resonance spectrum of the polymer gave an integrated peak-area ratio for sp^2 to sp^3 hydrogen atoms of 1:7.5, indicating that the residual unsaturation is equivalent to one double bond per two monomer units. The nuclear magnetic resonance analysis confirmed the infrared evidence that the unsaturation was composed of terminal methylene, and that it was due to the presence of the vinyl groups of the monomer rather than to the methylene. This evidence led to the conclusion that two structural units were present in the polymer chain in approximately equal amounts; the polymer could be represented as in (3–3); no evidence was available to indicate that the units are present in the polymer chain in an alternating manner.

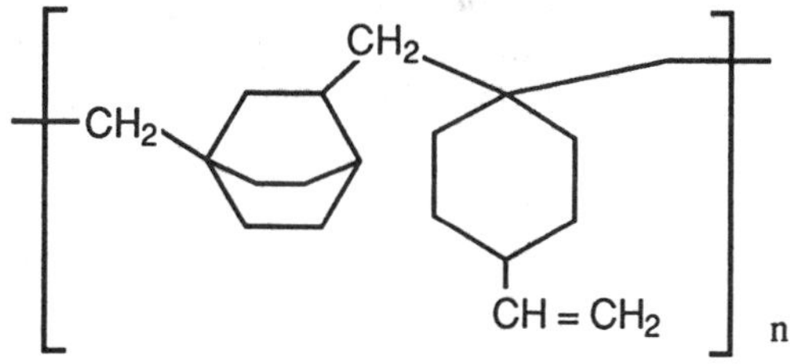

(3–3)

The ultraviolet absorption properties of a series of compounds[1-118] which are functionally capable of undergoing cyclic polymerization were investigated and a departure from the predicted properties based upon the monoolefinic counterparts was found. For example, the cis and trans isomers of 1,3,8-nonatriene exhibited bathochromic shifts of 3.5 and 5.0 nm, respectively, which could be interpreted as resulting from the stabilizing influence of interspatial resonance.

In an extension of this work,[1-119, 1-120] the probability of ring closure in cyclopolymerizing systems was calculated. The results indicated that far greater ratios of cyclization were realized in all systems reported than predicted on the basis of probability alone. Thus, this strong tendency toward cyclization could be explained satisfactorily only on the basis of some driving force other than the probability factor. In this work, a series of tetraenes having two pairs of conjugated dienes situated in such manner that the proposed electronic interaction could occur were synthesized and their ultraviolet absorption properties determined. A bathochromic shift of 10.6 nm was observed when 1,3,6,8-nonatatraene was compared with 1,3-nonadiene.

Although many additional references could be cited which support the proposed interaction as a driving force in cyclopolymerization, only a few of the more significant ones will be referred to. It has been pointed out[1-83] that, in the polymerization of 2-carbethoxybicyclo[2.2.1]hepta-2,5-diene to polymers believed to contain nortricyclene repeating units, evidence suggests that this type of transannular polymerization requires activation of the double bonds and resonance stabilization of radicals formed during the polymerization.

In a study of acetolysis of 5-hexenyl *p*-nitrobenzene sulfonates, it was found that[1-121] a 71% increase in acetolysis rate occurred when these compounds were compared with the corresponding hexyl esters, and 34% yield of cyclohexylacetate was obtained. These results are quite analogous to the cationic polymerization of 2,6-diphenyl-1,6-heptadiene reported earlier[1-38] however, in this case, because of the increased resonance stabilization of the intermediate benzyl carbocation, this cyclopolymerization results in exclusive ring closure. In a similar instance involving free-radical intermediates, 6-heptanoyl peroxide was thermally decomposed and it was shown[1-122] that the 5-hexenyl radicals produced resulted in formation of cyclohexane radicals, as both cyclohexane and cyclohexene were observed in the products. However, no group participation by the double bond in the peroxide decomposition was observed.

It has been proposed that the strong driving force for cyclic polymerization has its basis in an electronic interaction between an intramolecular neighboring double bond and the activated species (either a radical, carbanion, or carbocation).

A variety of allyl ethers of 2-allylphenol were synthesized and polymerized by a cationic mechanism to yield thermoplastic polymers containing residual unsaturation,[6] indicating that the allyl side-chains of the ring were not involved in the polymerization. A reinterpretation of these results would be in order, since residual unsaturation was apparently not determined quantitatively, and if cyclopolymerization had occurred, the same result could have been obtained.

4-Vinylcyclohexene has been converted partially to an oligomer containing the structural units shown (3–4) on treating with butyl lithium in tris(tri-

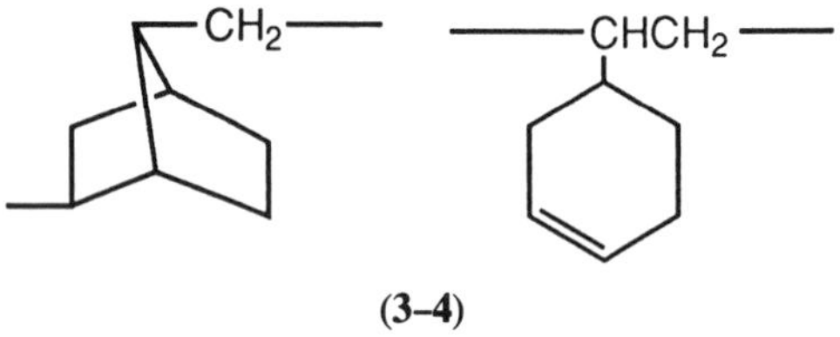

$-CH_2-$ $-CHCH_2-$

(3–4)

methylamino)phosphine oxide solution.[7] Also, it has been shown that the polymerization of 4-vinyl-1-cyclohexene in the presence of complex coordination catalytic systems involved primarily endo- and exocyclic double bonds and gave a bicyclic structure.[8]

Cationic polymerization of 4-vinyl-1-cyclohexene proceeded *via* three different carbocation intermediates leading to the different structural units shown in Structure 3–5. The presence of the structural units was demonstrated by NMR spectroscopy.[9]

Cyclic dimers of butadiene, piperylene, and isoprene, e.g. 1-vinyl-3-cyclohexene were polymerized in presence of an aluminum halide catalyst to yield polymers soluble in benzene.[10] By controlling conditions a dimer of vinylcyclohexene could be obtained in 35% yield by cationic initiation rather than the usual homopolymer.[11]

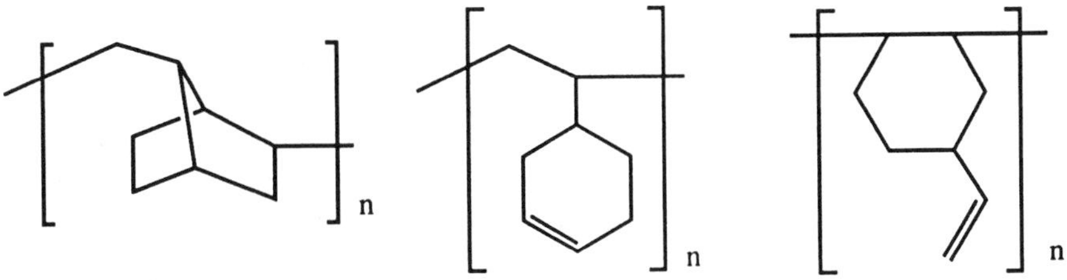

(3–5)

Other investigations provided evidence that 4-vinylcyclohexene polymerized *via* cationic initiation could yield a viscous oil.[12]

Studies on polymerization of 4-vinyl-1-cyclohexene have been continued. The monomer was converted to dimer, trimer and higher polymers by heating at elevated temperatures with phosphorus pentoxide.[13]

Crystalline poly-4-vinyl-1-cyclohexene was obtained by using anionic-coordinated catalytic systems composed of lithium aluminum organo-metallic complexes and titanium tetrahalides. The polymer is soluble in aromatic solvents and contains double bond unsaturation in the ring.[14]

Poly(4-vinylcycloxene) was obtained by treating a hydrocarbon solution of 4-vinyl-1-cyclohexene with ethylaluminum dichloride and stannic chloride.[15] The soluble polymer was isolated by precipitation with methanol.

4-Vinyl-1-cyclohexene was also cyclopolymerized with ethylaluminum dichloride and the hydrocarbon soluble polymer isolated by precipitation with methanol.[16]

Crystalline polymers of 1-vinyl-3-cyclohexene have also been obtained *via* a variety of coordination catalysts.[17] Presumably, the internal double bond did not participate as in a cyclopolymerization mechanism, although the internal double bond of 1,4-hexadiene is known to become involved in a cyclopolymerization process.

2-Methylene-5-norbornene has been polymerized by use of a variety of cationic initiators to yield soluble saturated polymers containing the polycyclic repeat units shown in Structure 3–6. Copolymerization and terpolymerization studies led to the conclusion that the monomer entered the chain only through the norbornene double bond.[18]

A solid homopolymer of 5-methylene[2.2.1]hept-2-ene with repeating unit shown in Structure 3–7 was prepared by a cationic mechanism.[19] The polymer was soluble and possessed no residual unsaturation, indicating a product of cyclopolymerization leading to the nortricyclene ring structure.

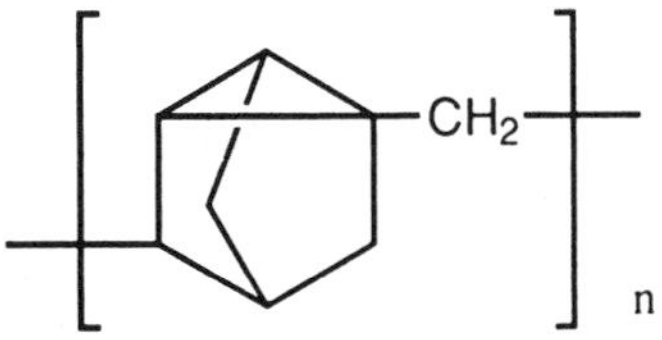

CH_2 n

(3–6)

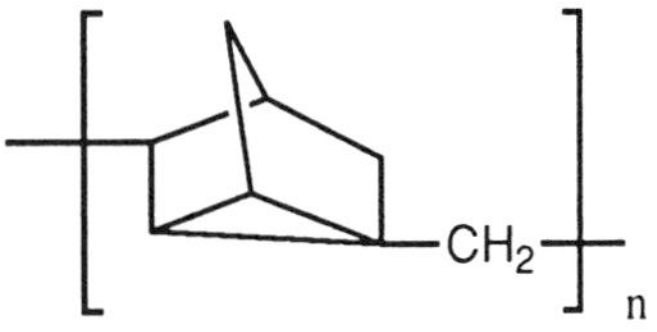

CH_2 n

(3–7)

In support of the electronic interaction theory of cyclopolymerization, the ultraviolet spectra of 1,4-dimethylenecyclohexane and methylenecyclohexane were compared.[20] The diene showed a bathochromic spectral shift in accord with the predictions of the theory.

Also, it has been shown that conjugated diene polymers having an unsaturated bond in the side chain are cyclized in the presence of cationic initiators to give heat-resistant polymers.[21]

The cyclopolymerization characteristics of 3-allylcyclopentene, 4-allylcyclopentene, 3-allylcyclohexene, and 4-allylcyclohexene have been studied and the extent of their cyclopolymerization by cationic initiation was compared with solvolysis data for the same intermediate carbocations.[22] The allylcycloalkenes were also polymerized using metal alkyl coordination catalysts. Calculations based on nuclear magnetic resonance spectra showed that the cationic polymers were 68–95% cyclized. Coordination polymerization gave 54–80% cyclized products. Maleic anhydride-diene copolymers (2:1 molar) were also prepared.

3- and 4-Allylcyclopentenes as well as 3- and 4-allylcyclohexenes were synthesized as cyclopolymerizable monomers.[23] A comparison between their cyclopolymerization behavior and their ultraviolet (UV) absorption characteristics was made. Neither UV nor NMR spectral studies showed definite ground state interactions between neighboring double bonds of the monomers.

In a study of the reactions of diolefins at high temperatures, the kinetics of the cyclization of 3,7-dimethyl-1,6-octadiene was studied.[24] Radical initiation of polymerization yielded 51% polymer with 14% cyclization to 1,2-dimethyl-3-isopropenedicyclopentane. Although the structure of the polymer was not discussed, it was postulated that the intramolecular cyclization to the isomer did not occur *via* a radical mechanism.

In a continuing study of the reactions of diolefins at high temperatures, the thermal cyclization of 1,6-octadiene and 7-methyl-1,6-octadiene was studied. Unfortunately no polymerization reactions were described.[25]

5,7-Dimethyl-1,6-octadiene was employed as a termonomer in a terpolymerization with ethylene and propylene *via* metal-coordination catalysts to introduce a residual double bond into the terpolymer.[26] The residual double bond content is desirable in order to provide a vulcanizable site for application of the terpolymer as an elastomer. Results obtained indicated that the terminal double bond participated in the polymerization whereas the trisubstituted double bond remained inactive. No evidence was presented to show that cyclization of the diene did not occur.

By use of Ziegler-type coordination catalysts, 1,5- and 1,6-dienes were cyclopolymerized and cyclocopolymerized to produce polymers capable of forming clear films.[27]

Free radical additions of polyhalomethanes and other addends were found to undergo transannular cycloadditions to cis,cis-1,5-cyclooctadiene to give [4.4.0]bicyclooctane products.[28]

d-Limonene was polymerized by use of triisobutyl-aluminum or diisobutylaluminum chloride to give low molecular weight polymer which is identical in structure to that prepared by cationic initiators.[29] The evidence indicates that the polymer has more than one type of recurring unit, at least one of which is bicyclic. α- and β-Pinenes were also polymerized by these catalysts. Data on optical activity and NMR spectra are reported.

1,2-Divinyl-3-methylcyclobutane was synthesized as a candidate for terpolymerization with ethylene and propylene.[30] Since this compound is a 1,5-diene, its relationship to cyclopolymerization is apparent.

When 1,4-hexadiene was used as a termonomer in EPDM elastomer synthesis, there appeared to be a dramatic effect on propylene incorporation.[31] This may be related to the tendency of this termonomer to undergo cyclopolymerization. Also, when this diene was used as a termonomer in EPDM elastomer synthesis, loss of the desirable residual unsaturation necessary for satisfactory vulcanization occurred *via* cyclopolymerization.[32]

Allylcyclopentadiene was polymerized by cationic initiation.[33] The soluble polymers obtained contained most of the allyl groups unchanged. The polymerization mechanism involved equal amounts of addition *via* 1,4- and 3,4-polymerization. Metal-coordination catalysts resulted in involvement of the allyl groups to some extent to give insoluble polymers. Cyclopolymers derived *via* a cumulative 1,4–1,2-cyclopolymerization would produce polymers which violate Bredt's rule. Polymerization of other monomers related to allylcyclopentadiene; e.g., methallylcyclopentadiene and allylmethylcyclopentadiene, was carried out with boron trifluoride as catalyst. All three polymers were soluble in organic solvents.[34] It was shown that the proportion of 1,4-structure increased with increasing steric hindrance in the monomer. No mention of bicyclic structures was made although allylcyclopentadiene is a structurally ideal monomer for cyclopolymerization.

The cationic initiated polymerization of 1,2-dihydro-, 9,10-dihydro- and endo-dicyclopentadiene was studied under various conditions.[35] Linear polymers were obtained from the 1,2-dihydro- and endo-dicyclopentadiene, but only oligomers from the 9,10-dihydro-monomer. It was shown that both the cyclopentenic unsaturation of the monomers as well as the norborene double bond reacted.

A copolymerization study of dicyclopentadiene with isobutylene in the presence of a Ziegler-Natta catalyst was carried out.[36] An unsaturated liquid copolymer of molecular weight up to 6000 was obtained. The copolymerization constants were determined; however, the structure of the copolymer was not discussed further.

The concept of cyclopolymerization was applied to alloocimene, (2,6-dimethyl-2,4,6-octatriene), by use of cationic initiators.[37] The poly(alloocimene) obtained was soluble and gave a definite softening point of from 68° to 70°C. The mechanism postulated required the loss of two double bonds, which is consistent with a cyclopolymerization mechanism. Alloocimene was also polymerized to a linear soluble polymer by use of triisobutylaluminum-titanium tetrachloride catalyst.[38] The structural units present in the polymer chain arose *via* 1,2-polymerization involving the 2,3-double bond and 1,4-polymerization involving the 2,5-conjugated diene structure. No other modes of polymerization were indicated.

Alloocimene was polymerized by a variety of catalysts including a number of coordination type combinations.[39] Free radical initiators apparently were not used. All initiators produced polymer having the same recurring structural units, and only a slight amount, if any, cyclopolymerization occurred. It was suggested that possibly all of the catalysts used were acting as cationic initiators. This work failed to confirm the work of an earlier paper. 2,6-Dimethyl-2,4,6-octatriene (alloocimene) was polymerized over activated clay, aluminum chloride, activated carbon, benzil or sulfuric-acetic acid mixtures.[40] The structure of the polymer was not discussed, although both liquid and solid polymer were obtained.

Polymyrcenes were synthesized by treating the monomer with cationic, anionic, radical, and metal-coordination type initiators.[41] Polymers derived from anionic, radical and metal-coordination initiation retained two double bonds in the 1,4-arrangement, while those obtained *via* cationic initiation retained only one double bond. These results in conjunction with the absence of vinyl, vinylidene, and isopropylidene structures are consistent with a cyclopolymerization mechanism for the cationic initiated polymerization.

Cyclization of ϵ,δ-alkenyl radicals was studied to show that radicals designed to form six-membered rings *via* thermodynamic control led to increasing amounts of five membered rings with decreasing radical stabilizing β-substituents on the alkene double bond.[42]

The 5-hexenyl radical and its five 1- and 5-methyl substituted derivatives were generated from the appropriate bromide and tributylstannane.[43] Cyclization to the corresponding cyclopentylmethyl and cyclohexyl radicals occurred. Ratios of six- to five-membered rings increased with temperature and with 1-substitution, while 5-substitution increased the five- to six-membered ring ratio. 1-Substitution increased the rate of cyclization to six-membered rings by a factor of at least twenty.

The carbon-carbon double bond has been discussed with regard to its nucleophilic reactivity in solvolytic ring closure reactions.[44] The double bond in 2-(Δ^3-cyclopentyl)-ethyl tosylate and p-nitrobenzenesulfonate participates directly in the acetolysis of these esters, as shown by direct formation of exo-norbornyl acetate.

The compounds shown exhibit acetolysis rates 600–3300 times greater than those for the corresponding saturated compounds (3–8) to confirm participation of the double

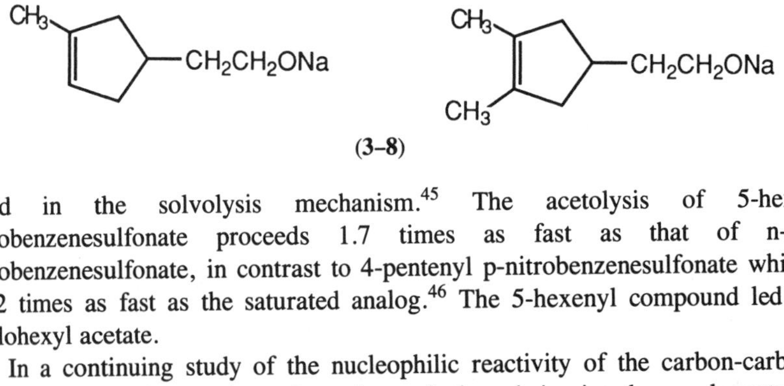

CH_3 — CH_2CH_2ONa $\qquad$ CH_3, CH_3 — CH_2CH_2ONa

(3–8)

bond in the solvolysis mechanism.[45] The acetolysis of 5-hexenyl p-nitrobenzenesulfonate proceeds 1.7 times as fast as that of n-hexyl p-nitrobenzenesulfonate, in contrast to 4-pentenyl p-nitrobenzenesulfonate which is only 0.72 times as fast as the saturated analog.[46] The 5-hexenyl compound led to 11.6% cyclohexyl acetate.

In a continuing study of the nucleophilic reactivity of the carbon-carbon double bond, it was shown that elongation of the chain in the cyclopentyl-ethyl p-nitrobenzenesulfonates from ethyl to n-propyl brings about an acetolysis rate decrease by a factor of 600 of the anchimerically assisted solvolysis.[47] The sources of the observed difference lie partly in entropy but mostly in energy factors.

In a continuing study of radical cyclization, cyclization of diversely substituted δ,ϵ-unsaturated free radicals was studied. Structures, where R and R′ are H or CH_3 were studied.[48] When X = $CO_2C_2H_5$ and y = $CO_2C_2H_5$ or x = Cl and y = $CO_2C_2H_5$ or x = CH_3CO and y = H, the five-membered ring was the main product. When x = CN and y = $CO_2C_2H_5$, the two ring products were obtained in equal amounts. When x = CN and y = $CO_2C_2H_5$, the six-membered ring was the main product (3–9).

A variety of unsaturated ethers of vinyl alcohols was prepared and polymerized *via* a cationic mechanism.[49] Linear polymers having residual unsaturation were postulated. If cyclopolymerization had occurred to some degree, the results could be interpreted in this manner, particularly since residual unsaturation was apparently not determined

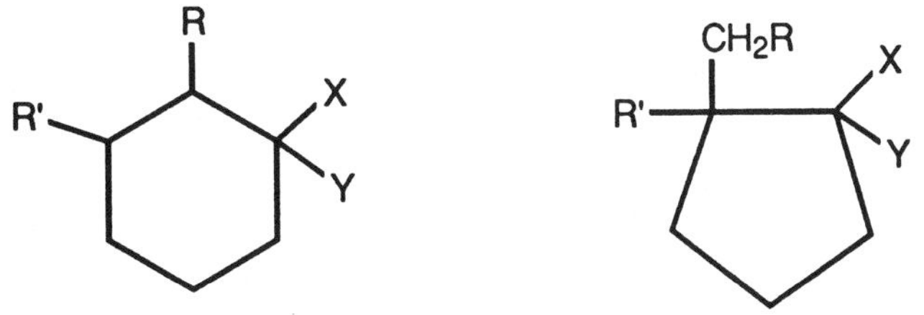

(3–9)

quantitatively. A variety of vinyloxyethylethers of 2-allyl and 2-methallylphenols was subjected to cationic initiated polymerization to yield soluble copolymers containing considerable unsaturation.[50] The extent of cyclization if any was not determined.

Thermal degradation of poly(methacrylic acid) in vacuo at 60–400°C yielded volatile material (monomer, H_2, CO_2, CO) and a cyclized product. The apparent activation energy of the cyclization stage was 19.5 kcal/mol.[51] The cyclized product was apparently poly(methacrylic anhydride).

A proton-NMR study of the esterification of syndiotactic poly(methacrylic acid) with carbodiimides has been conducted.[52] The esterification by CH_3OH in the presence of dicyclohexyl-carbodiimide was examined by ^1H NMR. The reaction proceeded *via* the cyclic anhydride intermediate leading to a limited conversion and a tendency to alternation in the sequence of esterified and nonesterified repeat units on the chain. In a continued proton-NMR study of the esterification of syndiotactic poly(methacrylic acid) with carbodiimides, the partial esterification of syndiotactic poly(methacrylic acid) by $C_6H_5CH_2OH$ and F_3CCH_2OH in the presence of dicyclohexylcarbodiimide was examined by evaluation of triad and pentad probabilities from ^1H-NMR measurements.[53] The reaction mechanism tends to give an alternating chain sequence of esterified and nonesterified acid groups with limited conversion suggesting the cyclic anhydride as intermediate. Esterification by F_3CCH_2OH in concentrated H_2SO_4 gave a random distribution of esterified chain units.

The 1:1 cyclopentadiene-maleic anhydride copolymer of structure 3–10 was

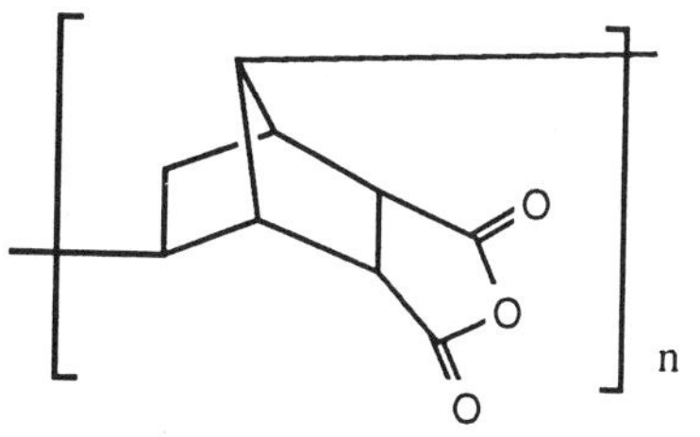

(3–10)

claimed. The copolymer could be formed either by di(tert-butyl)peroxide initiated polymerization of the Diels-Alder adducts of the monomer pair or from the free monomers.[54]

Three isomeric non-conjugated dienes, o- m-, and p-(2-vinyloxyethoxy)styrenes were selectively polymerized *via* radical or anionic initiators through the styryl double bond while leaving the vinyl ether function unchanged.[55] The polymers were subsequently cross-linked *via* cationic initiation through the vinyl ether linkages. No cyclocopolymerization was reported.

Polymers containing 1,1,3,3-tetraphenyl-1-butene and/or 3-methyl-1,1,3-triphenylindene structures were prepared by acid-catalyzed polymerization of bis[p-(1-phenylvinyl)]methane.[56] The indane units apparently arose *via* a cyclization step in the mechanism.

Vinyl radicals derivable from reaction of vinyl halides with trialkylstannanes undergo cyclizations to form five- or six-membered rings containing double bonds in predictable positions.[57]

The cyclization of squalene to steroids was investigated with homogenates of rat and hog liver.[58] Both converted squalene to lanosterol; however, rat liver homogenate also produced cholesterol. The process is aerobic and requires reduced pyridine nucleotide. This process appears to be a biological analog of cyclization of 5-hexenyl radicals, a process studied extensively in connection with mechanistic investigations of cyclopolymerization. The cationic or oxygen initiated cyclization of squalene to lanosterol was proposed to occur by an intramolecular series of four successive cyclization reactions with the neighboring double bond reacting with the active species generated.[59] This process is analogous to cyclopolymerization with the exception that no intermolecular interactions are proposed.

Linear polymers containing bicyclic structures formed *via* cyclopolymerization were described in a Ph.D. dissertation.[60] Radical polymerization of 2,6-diallylphenyl acrylate was studied. Its rate of polymerization was lower than that for 2-allylphenyl acrylate.[61] Anionic polymerization of the monomer gave crosslinkable linear polymers.

Intramolecular charge-transfer interaction and its potential influence on cyclopolymerization led to the first practical synthesis of 3,3-dimethoxycyclopropene and a study of its chemical properties.[62] Also, cyclopolymerization of monomers with diene- or cyclopropane-like structures has been observed.[63] A cationic or Ziegler-Natta initiated polymerization study of 2,3-dimethyl-1,3-butadiene, bicyclopropyl and spiropentane showed that cyclic structures were formed in all cases. Spiropentane polymer consisted wholly of cyclized units. Bicyclopropyl gave polymer containing cyclohexane units.

Polymerization of spiro[2.4]hepta-4,6-diene in presence of cationic initiators proceeded at the cyclopentadiene portion without involving the cyclopropyl ring.[64] The spiro structure contributed both to the cationic reactivity of the monomer and the enhanced stability of the polymer. Cationic homo- and copolymerization of spiro[4.4]-nona-1,3-diene was accomplished to give polymer of structure 3–11.[65] Apparently the

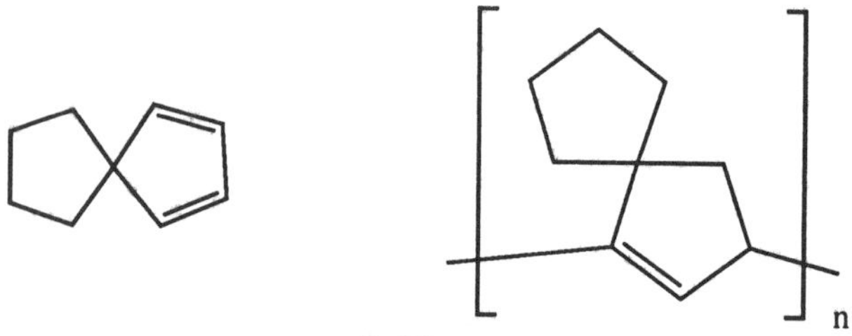

n

(3–11)

severe steric crowding suppressed the usual 1,2-and 1,4-addition. Reactivity ratios for copolymerization of this monomer with a-methylstyrene were given.

Poly(o-isopropenylstyrene) with a high thermal stability and good processability was manufactured by polymerizing the monomer with anionic initiators, e.g. butyllithium.[66] The polymers were soluble in organic solvents and had molecular weights up to 120,000. Poly(o-isopropenylstyrene) was manufactured by polymerizing the monomer first with anionic initiators, then by cationic initiators to give heat and chemically resistant polymers containing the repeat unit shown in Structure 3–12.[67]

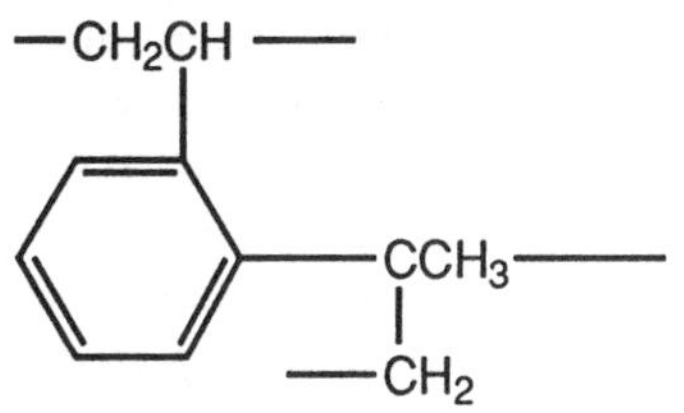

—CH_2CH—

—CCH_3—

—CH_2

(3–12)

HETEROCYCLIC RINGS

A wide variety of unsymmetrical nonconjugated dienes capable of undergoing cyclopolymerization to produce polymers containing heterocyclic units in the main chain has been studied. This section will deal with the details of this topic.

Nitrogen-Containing Rings

Vinyl methacrylate, allyl methacrylate, N-allylacrylamide, and N-allylmethacrylamide were polymerized in toluene using α,α'-azobisisobutyronitrile as initiator.[68] N-Allylacrylamide and N-allylmethacrylamide did not produce soluble polymer, even at a low conversion. It was assumed that these polymers crosslinked as a result of degradative chain transfer reactions.

However, it had been previously shown,[1-144, 69–71] that N-allylacrylamide and N-allylmethacrylamide, when polymerized under suitable conditions, yielded soluble polymers containing appreciable amounts of cyclic units. Free radical polymerization of N-allylmethacrylamide in benzene or toluene solutions gave only insoluble crosslinked polymers even when the conversions were kept lower than 1%. However, at low enough conversions, by polymerization in methanol at temperatures between 35° and 70°C using α,α'-azobisisobutyronitrile as initiator, only soluble polymers containing considerable amounts of γ-lactam type structural units were obtained. In highly diluted solutions, conversions up to 50% were attained without any formation of crosslinked products. N-Allylacrylamide[1-144, 72] showed a greater tendency to form cyclic units. The polymerization of N-allylacrylamide has also been studied more extensively.[73] Polymerization at 60–80°C in solution in the presence of α,α'-azobisisobutyronitrile or dibenzoyl peroxide gave soluble cyclopolymers containing only 2–5% residual unsaturation. The presence of the γ-lactam ring was established.

A radiation polymerization study of N-allylmaleamic acid in the solid state[74] led to the conclusion that this monomer must proceed by a different mechanistic pathway than N-allylsuccinamic acid and maleamic acid. A comparison of irradiation dosage versus percent yield curves revealed that N-allylsuccinamic acid polymerized at a constant rate without an induction period, while N-allylmaleamic acid had an induction period and also exhibited a postpolymerization effect. Intermittent irradiation of N-allylmaleamic acid greatly increased its yield in a given time; the effect was explained as due to the accelerating action of the polymerized phase. Electron paramagnetic resonance spectroscopy showed that polymerized N-allylmaleamic acid contained free radicals; however, their concentration changed only a little during postpolymerization. Maleamic acid gave only approximately 3% conversion on γ-irradiation.

Radical polymerization of unconjugated, unsymmetrical dienes[75] such as N-allylmethacrylamide, N-allylacrylamide, allyl acrylate, and allyl methacrylate was reviewed in 1968. It was concluded that the cyclic structure imparts rigidity and thermal stability to the polymer.

Cyclopolymerization experiments on 1-vinyluracil[76] have been described. The polymer was prepared by free radical initiated polymerization of the monomer and contained substituted dihydrouracil rings believed to have been formed by a cyclopolymerization mechanism.

The synthesis of polyampholytes[77] by cyclopolymerization of N-2-phenylallylacrylamide and N-ethyl-2-phenylacrylamide followed by hydrolysis of the cyclic amides has been published.

Cyclopolymerization of N-allylmaleimide[78] has been described. The monomer was polymerized by free radical and anionic initiation. The content of bicyclic units in poly(N-allylmaleimide) increased as monomer concentration in the polymerization solution decreased (3–13).

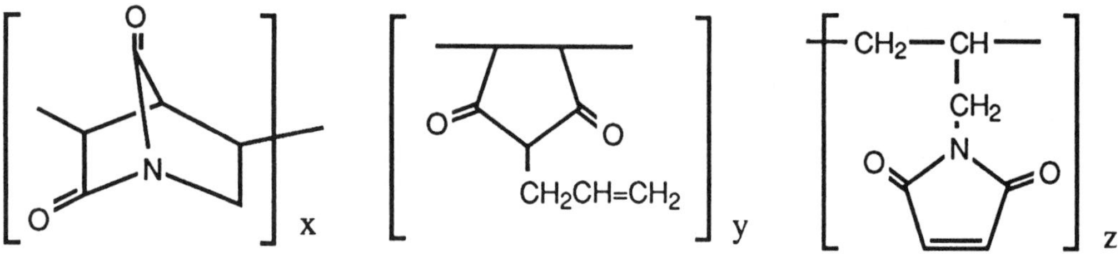

(3–13)

Cyclopolymerization of N-allyl-N-methylmethacrylamide has been investigated[79, 80] Radical initiation over the temperature range −78° to 120°C resulted in polymers with 88–93% cyclic units, mainly five-membered rings with some six-membered rings and pendant methacryl groups. However, the monofunctional counterpart, N-(n-propyl)-N-methylmethacrylimide could not be polymerized. It was shown that by studying the polymerization of the monofunctional analogs (N-methyl-N-propylmethacrylamide, N-allyl-methylisobutyramide[80] by γ-rays at low temperature, the propagating radical involved only the methacryloyl group and not the allyl group. The additional unsaturation in polymer obtained at −78° was due to the allyl group.

Polyampholytes have been prepared by cyclopolymerization of N-(2-phenylallyl)acrylamide and the *N*-ethyl derivative,[81] followed by hydrolysis of the cyclopolymer (3–14). Cyclization was not complete since cross-linking readily occurred. The polymerization was compared with the copolymerization reaction of α-methylstyrene and acrylonitrile which gave a 1:1 copolymer, independent of monomer charge in the presence of $ZnCl_2$ and azobisisobutyronitrile. Lewis acids gave high molecular weight polymers from these complex monomers.

Cyclopolymerization of N-methacryloyl-N-methylcrotonamide has been studied. While N,N-disubstituted methacrylamide and crotonamide, which can be considered as

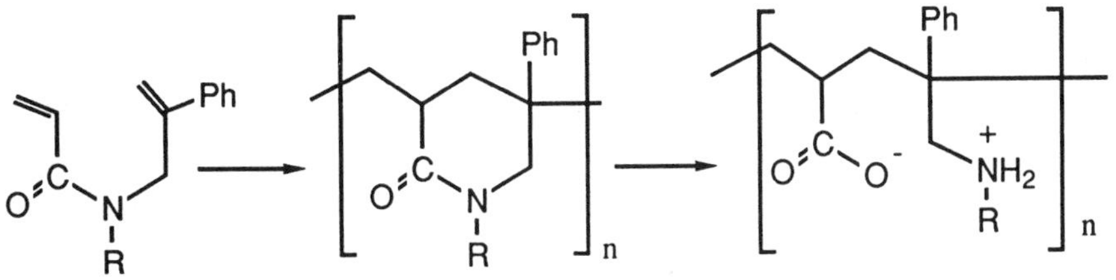

R = H, C_2H_5

(3–14)

structural moieties of N-methacryloyl-N-methylcrotonamide do not homopolymerize, N-methacryloyl-N-methylcrotonamide nevertheless yielded a homopolymer consisting of five-membered cyclic repeating units. The coplanarity of the three carbon atoms and reluctance of the reacting double bonds to homopolymerize seem to be the minimum requirements for forming a polymer consisting of solely five-membered cyclic repeating units.[82] These results offer further support that cyclopolymerizable monomers possess an internal driving force for cyclopolymerization.

Symmetrical unconjugated dienes, for example sym-dimethacryldimethylhydrazine, gave polymer cyclized to the extent of 95%.[2-174] Radical-induced cyclization and cyclopolymerization of N-methyl-N-allyl-N-(2-alkylallyl)amines and N-methyl-N,N′-bis(2-alkylallyl)-amines have been studied.[2-105] The structures of the products were shown by ^{13}C NMR spectroscopy to be pyrrolidines and piperidines, the proportion of each depending on the bulk of the 2-alkyl substituent. The cyclopolymerization of N-allyl-N-methyl(2-substituted allyl)amines has been studied.[2-106] The structures of the poly(N-methylpyrrolidines), poly(N-methylpiperidines), and low-molecular-weight products obtained by radical-induced cyclizations were determined by mass and ^{13}C NMR spectra. Radical polymerization of 1-vinylindole and 3-methyl-1-vinylindole gave cyclic indoline structures as determined by ^{13}C NMR spectra.[83] Cationic polymerization of 1-vinylindole also gave cyclic structures.

Poly(methylamine-allyl halide) copolymers and their reaction products have been claimed.[84] The monomers were synthesized and polymerized, apparently, in a one-step process, using hydrated silica as a catalyst. The synthesis and free radical polymerization of N-substituted citraconimides have been reviewed.[85] N-Phenyl and N-cyclopropylcitraconimide did not polymerize with azobisisobutyronitrile as initiator. N-Allylcitraconimide readily polymerized to produce polymers which contained $\leqslant 85\%$ cyclic repeating units.

Cyclopolymerization of methyl α-(methacrylamido)acrylate gave the cyclopolymer containing the structural unit shown in Structure 3–15 to the extent of $\sim 90\%$.[86]

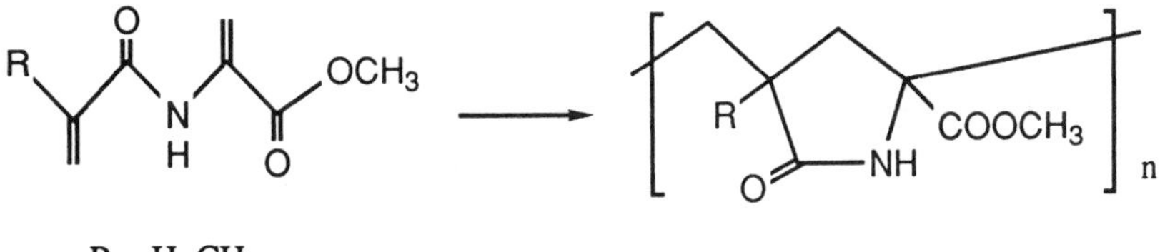

R = H, CH_3

(3–15)

The cyclopolymerizability of N-allyl-N-methylacrylamide was investigated and compared with that of N-allyl-N-methylmethacrylamide which had been shown to yield a highly cyclized polymer.[2–485] The structural study showed that the acrylamide monomer gave a high retention of the allyl group and little cyclization. Only the acryloyl group underwent propagation. The different relative rates of the competing propagation reactions in the two systems were responsible for the major differences in behavior.

Free-radical polymerization of N-allyl-N-alkylacrylamides was also studied.[87] The degree of cyclization increased either with a decrease of the monomer concentration or an increase in the length of the N-alkyl group. Cyclopolymerization of N-tert-butyl-N-allylacrylamide led to exclusively cyclic polymer, in contrast to the rapid linear polymerization by the analogous monoene, N-tert-butyl-N-propylacrylamide.[88] The high degree of cyclization in the allyl monomer was attributed to hindrance of intermolecular propagation of the acryloyl radical.

N-Allylmethacrylamide and N-allylacrylamide were cyclopolymerized over the temperature range below and above the ceiling temperature of the monofunctional counterpart, N-propylmethacrylamide.[89] Increasing the temperature enhances the cyclopolymerizability of the acrylamide monomer, while that for the methacrylamide monomer increased only slightly. The differences were explained by the hypothesis that the lower the polymerizability of the corresponding monoene counterpart of the unconjugated dienes, the higher their cyclopolymerizabilities.

Linear polymers were obtained from N,N′-methylene-bisacrylamide with an energy of activation of 32 ±2 kJ mol^{-1} which supported the concept of cyclization prior to polymerization.[90] Copolymers containing the structural unit shown in Structure 3–16

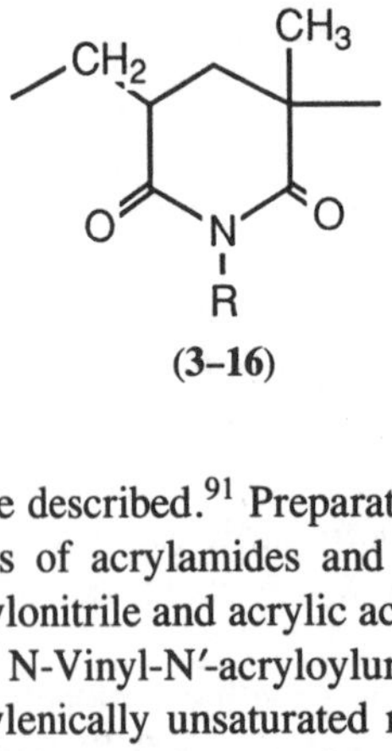

CH_3 CH_2 O N O R

(3–16)

were described.[91] Preparation of such copolymers was accomplished by heating copolymers of acrylamides and acrylate esters in presence of water. Also, copolymers of acrylonitrile and acrylic acid were shown to yield the described copolymers.

N-Vinyl-N′-acryloylurea was prepared from acrylamide as a cross-linking agent for ethylenically unsaturated monomers.[92] This compound is a 1,7-diene and may be susceptible to a degree of cyclopolymerization, which would reduce its effectiveness as a crosslinking agent.

Oxygen-Containing Rings

A large number of unsymmetrical nonconjugated dienes capable of leading to five- or six-membered oxygen-containing heterocyclic rings has been studied. This section will be devoted to a discussion of such monomers and their polymers.

Polymerization of allyl acrylate *via* radical initiation led to soluble polymer in conversion up to 60%.[1–78] A study of the polymer structure led to the conclusion that

24% of the structural units were linked through the acrylate group, 17% through the allyl group and the saturated 60% contained lactone groups (3–17) formed by cyclopolymerization.

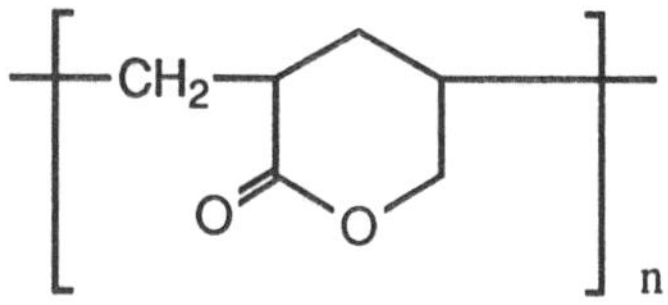

(3–17)

A free radical polymerization and copolymerization study of allyl acrylate has been carried out.[93] Polymerization of allyl acrylate gave partially cyclized, glassy, brittle, soluble poly(allyl acrylate). The cyclization constant (ratio of propagation rate to cyclization rate) was 2.42. The heat of activation was 15.4 kcal/mole and the difference between the activation energies of inter- and intramolecular propagation was 1.4–1.8 kcal/mole. Reactivity ratios in the preparation of copolymers with styrene, methyl acrylate, vinyl acetate, and allyl chloride were determined. The molecular weight of allyl acrylate-allyl chloride polymer was low because of strong chain transfer.

A study of the free radical cyclopolymerization of vinyl methacrylate and comparison with the results obtained from the cyclopolymerization of allyl methacrylate supported the hypothesis that degradative chain transfers are responsible for the discrepancy in the copolymerization behavior of methacrylic and allylic unsaturations as derived from copolymerizations and from cyclopolymerizations.[94]

Free radical and anionic homopolymerization of vinyl acrylate and vinyl methacrylate gave soluble polymers containing vinyl group residual unsaturation.[95] Poly(vinyl methacrylate) prepared with radical catalysts at monomer concentrations 1.8–0.5 mole/L contained 10–20% cyclic units and poly(vinyl acrylate) prepared similarly at monomer concentrations 0.9–0.3 mole/l contained 50–60% cyclic units.

Vinyl methacrylate and allyl methacrylate have been cyclopolymerized *via* radical initiation.[68] The vinyl ester led to a polymer in which up to 60% of the monomer units had cyclized. In contrast the allyl ester gave only crosslinked polymer. However, the allyl ester was further investigated[82] and the conclusion was reached that, under suitable conditions, this monomer yielded soluble polymers containing appreciable amounts of cyclic units.

Cyclopolymerization of o-isopropenylstyrene with radical, cationic, anionic, and Ziegler type initiators has been reported.[96] These studies showed that in the free radical polymerization all the vinyl groups participate in the reaction, and the isopropenyl groups partly remain as residual double bonds. The cyclization constant for o-isopropenylstyrene is higher than that for o-allylstyrene, and cyclized more readily, probably due to the steric effects. In anionic polymerization with BuLi, the vinyl groups selectively participate in the reaction and all isopropenyl groups remain free, while when Friedel-Crafts and Ziegler type catalysts are used, both the vinyl and isopropenyl groups participate, and the polymer prepared has only cyclic structural units, probably due to carbonium ion stabilization.

The synthesis and cyclopolymerization of cinnamyl methacrylate have been studied.[97] Poly(cinnamyl methacrylate) prepared in the presence of Bz_2O_2 contained cyclic structural units [other than benzene ring] (3–18); the ratio of vinyl propagation rate con-

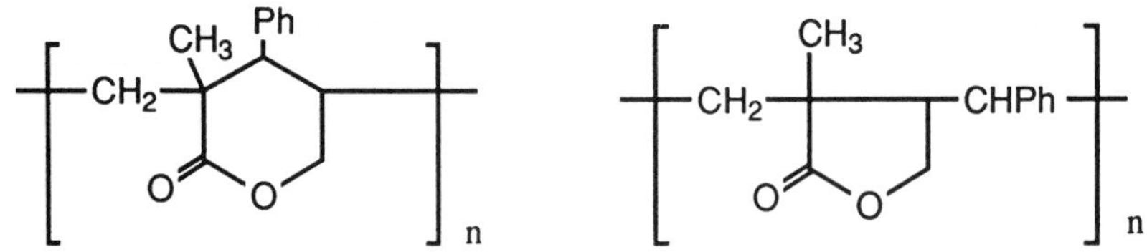

(3–18)

stant to the cyclization rate constant was ~9 L/mole. Poly(cinnamyl methacrylate) prepared in the presence of BuLi did not contain the cyclic structures, and the monomer reactivity ratios for the anionic copolymerization of methyl methacrylate (M_1) and cinnamyl methacrylate (M_2) in toluene were: $r_1 = 0.80 \pm 0.13$, $r_2 = 1.10 \pm 0.15$.

The radical cyclopolymerization of γ-chlorocrotyl acrylates has been reported.[98] γ-Chlorocrotyl acrylate and γ-chlorocrotyl methacrylate gave white soluble poly(γ-chlorocrotyl acrylate) and poly(γ-chlorocrotyl methacrylate) containing 13–17% double bonds. The polymerization rate was first order with respect to the monomer and 0.5 order with respect to initiator.

Cyclocopolymerization with intramolecular lactone ring formation has been described.[99] Copolymerizations in the presence of azobisisobutyronitrile gave a 1:1 allyl alcohol-maleic anhydride copolymer, and a 1:1 maleic anhydride-methallyl alcohol copolymer which were linear and soluble and contained lactone structures and carboxyl groups. Copolymers of maleic anhydride, allyl, or methallyl alcohol and another vinyl monomer, contain cyclic anhydride groups and lactone structures. A maleic anhydride-2-phenylallyl alcohol copolymer, a maleic anhydride-2-methylene-1,3-propanediol copolymer, and a 1,1-dimethallyl alcohol-maleic anhydride copolymer were also prepared.

o-Isopropenylphenyl vinyl ether has been polymerized by cationic and radical initiation to soluble cyclopolymers (3–19).[100] The relative reactivities of the two functional

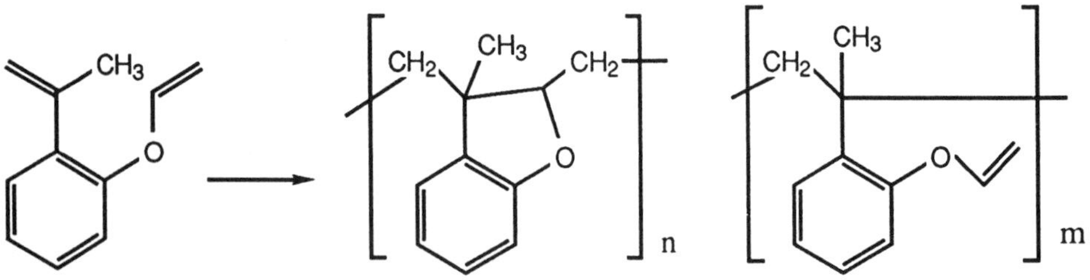

(3–19)

groups in cationic polymerization were estimated from the copolymerization of phenyl vinyl ether and α-methylstyrene. During cyclopolymerization, inter-molecular addition of a vinyl ether group was always followed by cyclization, but addition to an isopropenyl group led to cyclization or further inter-molecular reactions. Hence, only vinyl ether groups are present in any residual unsaturation in the polymers. The cyclization reaction was suppressed by cross propagation with p-chlorostyrene and acrylonitrile, and the vinyl ether groups partly remained as double bonds in the copolymer. Polymers produced by radical initiation in bulk are completely cyclized. Infrared spectroscopy showed that cyclization occurred *via* head-to-head addition, leading to five-membered rings.

The cyclopolymerization of a number of other unsymmetrical monomers among which are vinyl *o*-isopropenyl benzoate, *o*-allylphenyl acrylate, *o*-vinylphenyl acrylate and 2-(*o*-allylphenoxy)ethyl acrylate have also been studied.[101-106]

Vinyl o-isopropenylbenzoate *via* radical polymerization gave a polymer with a small amount of residual isopropenyl groups.[101] The diene copolymerized with acrylonitrile to give a polymer containing some residual vinyl ester groups. The degree of cyclization in the copolymer decreased with increasing diene concentration and with increasing polymerization temperature.

Cyclopolymerization of o-allylphenyl acrylate was studied in the presence of alkylaluminum chlorides.[102] The presence of the complexing agents increased the extent of the cyclization in the polymer. Much faster polymerization rates and higher extent of cyclization were obtained in the presence of $Al_2Cl_3(C_2H_5)_3$ or $C_2H_5AlCl_2$ than with $(C_2H_5)_2AlCl$. At an Al-monomer ratio $\geqslant 1$, the extent of cyclization increased to 93–7%. The cyclic structure of the polymer consisted of eight-membered rings.

The study of cyclopolymerization in the presence of alkylaluminum chlorides has been continued.[103] The addition of alkylaluminum chlorides to 2-(o-allylphenoxy)ethyl acrylate and 4-(o-allylphenoxy)butyl acrylate increased the copolymerization tendency and produced polymers containing eleven and thirteen-membered rings, respectively. At Al:monomer molar ratios of $\geqslant 1.0$, the extent of cyclization increased to 90–93%.

Cyclopolymerization studies in the presence of alkylaluminum chlorides have been extended to copolymerization.[104] The cyclocopolymerizations of o-allylphenyl acrylate and 2-(o-allylphenoxy)ethyl acrylate with a donor molecule, p-chlorostyrene, in the presence of $Al(C_2H_5)_{1.5}Cl_{1.5}$ showed that the electron-withdrawing double bonds were consumed in an equal proportion to the electron-releasing double bonds, irrespective of the initial composition of monomer feed. The character of the cyclopolymerization was explained by a molecular complex mechanism, rather than by a complexed radical mechanism, and the nature of intramolecular cyclization was similar to that for propagation in the alternating copolymerization.

The cyclopolymerization of o-vinylphenyl acrylate with a conventional radical initiator and with alkylaluminum chlorides gave a polymer containing residual acrylic and vinyl double bonds.[105] The polymerization proceeded through head-to-tail additions with formation of a cyclic seven-membered ring, the sequence of addition being random for the radical polymerization and alternating for polymerization in the presence of the aluminum compounds. The latter cyclopolymerization was explained by a molecular-complex mechanism.

The cyclopolymerization of o-allylphenyl methacrylate in presence of these complexing agents showed that rapid polymerization in the presence of $(C_2H_5)_2AlCl$ was induced.[106] The extent of cyclization was extremely small compared with that of o-allylphenyl acrylate. In the presence of $C_2H_5AlCl_2$, the extent of cyclization was far larger and increased to 92% at Al/monomer molar ratio 1.24. The value of the ratio of rate constants for linear propagation and cyclization, for this monomer capable of forming eight-membered rings, was comparable to those values of the analogous acrylic monomers capable of forming eleven- and thirteen-membered rings. There were indications that the effects of alkylaluminum chlorides on the cyclopolymerization should be smaller for the methacrylic monomers than for the acrylic monomers. The polymerization of unsaturated esters of maleic and fumaric acids were studied.[1-76] Although, in general, these unsymmetrical monomers ultimately led to crosslinked polymers, the extents of cyclization of the soluble polymers obtained by stopping the polymerization just short of the gel time, varied from 23 to 63 % (Table 3–1). It was noted that those polymers containing the greatest degree of cyclization did possess double bonds of comparable reactivity (allyl and fumarate bonds in (4), 2-butenyl and maleate bonds in (2). The structure 3–20 was proposed for poly(methyl allyl fumarate), where x:y:z = 1:3:1.

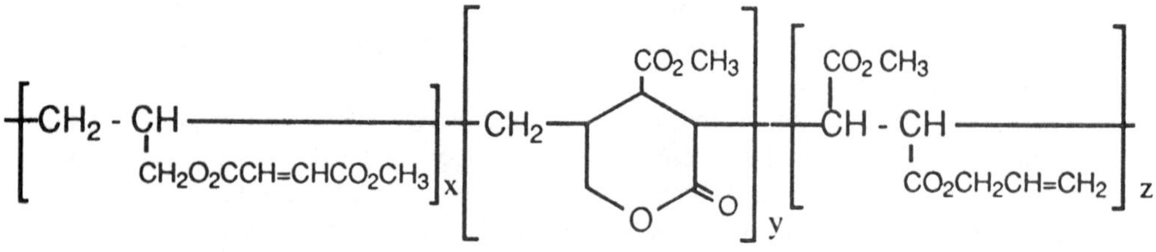

$$\left[CH_2-\underset{CH_2O_2CCH=CHCO_2CH_3}{CH}\right]_x\left[CH_2-(\text{ring: } CO_2CH_3,\ O,\ C{=}O)\right]_y\left[\underset{CO_2CH_3}{CH}-\underset{CO_2CH_2CH=CH_2}{CH}\right]_z$$

(3–20)

Linear polymers from the radical polymerization of the monoallyl esters of maleic and citraconic acids were obtained.[1-79] Bulk polymerization at 50–60°C in the presence of benzoyl peroxide yielded soluble polymers of molecular weights of 15,000–40,000. On the basis of residual unsaturation determinations it was concluded that the polymers contained 63–65% cyclic units. Similarly, the polymerization of diallyl maleate yielded a cyclopolymer containing 33.4% unsaturation (one double bond per monomer unit) of which 31.4% was due to allyl double bonds and 2 % to maleate double bonds.[107] Thus very little allyl-allyl intramolecular reaction had taken place. The presence of both five- and six-membered rings was established in the polymers of the monoallyl and diallyl esters of maleic acid.[108]

Vinyl trans-cinnamate has been polymerized to linear, soluble polymers containing γ-lactone rings having 10–30% residual unsaturation. This monomer which has two double bonds of widely different chemical nature, undergoes radical-initiated polymerization in accord with a cyclization rate constant: vinyl propagation rate constant ratio, $K_c/k_p = 13$.[109, 110]

It was found that by polymerizing a 0.65 m solution of allyl methacrylate in benzene at 60°C using 6.1×10^{-3} moles/l of α,α′-azobisisobutyronitrile as initiator, soluble polymers containing about 20% of γ-lactone type structural units were obtained at conversions lower than 10%.[1-144, 111] Other reports of the cyclopolymerization of unsymmetrical nonconjugated dienes can be found in the literature.[1-77, 112-122] Soluble polymers have also been obtained by polymerization of allyl methacrylate in carbon

Table 3–1 Degree of Cyclization in Poly(Methyl Alkenyl Maleates and Fumarates)

Polymer	Monomer	Per cent cyclization
1	Methyl allyl maleate	49
2	Methyl 2-butenyl maleate	60
3	Methyl 3-butenyl maleate	38
4	Methyl allyl fumarate	63
5	Methyl 2-butenyl fumarate	23
6	Methyl 3-butenyl fumarate	43

tetrachloride, dioxane, and diallyl ether solutions.[123] It was suggested that the propagation and cyclization reactions proceed only *via* addition to the methacrylyl groups of the monomer. Some degradative chain transfer occurred with the allyl groups, and it was considered that the solvents may ensure the production of soluble polymers by reactions in which allyl-radical side chains were terminated without crosslinking.

A more recent study of cyclopolymerization of methallyl methacrylate has been published. Cyclization is depressed in comparison with the cyclization of allyl methacrylate, but cyclization of methallyl methacrylate is highly enhanced at elevated temperatures. The formation of five-membered rings occurred at higher polymerization temperatures.[124]

The mechanism of cyclopolymerization of vinyl acrylate, methacrylate, and α-chloroacrylate has been studied more recently.[125] The residual double bond in the polymers was confirmed to be mainly a vinyl group by IR spectroscopy and ICl titrimetry. The residual double bonds could be reduced by lowering the monomer concentrations.

Controlled ring closure has been shown possible in the radical cyclopolymerization of allyl methacrylate.[126] The cyclopolymerization was examined, especially above the ceiling temperature for head-to-tail propagation in methyl methacrylate polymerization. Above this ceiling temperature, cyclization and formation of five-membered rings were enhanced. The polymer contained no unreacted pendant methacryloyl groups. The cyclopolymerization kinetics and mechanism were discussed. Radical initiated cyclopolymerization of isopropenyl acrylate and methacrylate has been studied at 60° in benzene. The residual double bond in the polymers was determined to be mainly isopropenyl groups by spectroscopy and titrimetry.[127]

o-Isopropenylphenyl vinyl ether has been polymerized by cationic catalysts to give soluble poly(o-isopropenylphenyl vinyl ethers) which contained a part of vinyl ether groups as residual double bonds.[128] The variation in the relative reactivities of two functional groups with the polymerization conditions was estimated from the copolymerization of phenyl vinyl ether and α-methylstyrene. Two kinds of intramolecular cyclization processes participated in the polymerization reactions. The intermolecular addition of the vinyl ether group was always followed by the cyclization of the isopropenyl group, which accounted for the absence of the isopropenyl groups in the resulting polymers. On the other hand, the intermolecular addition of isopropenyl group led to the cyclization by the vinyl ether group in competition with other intermolecular additions. Because of the very different reactivity of these cyclization processes, the relative reactivity of the two functional groups in the intermolecular addition reaction controlled the cyclopolymerizability of the monomer. The content of the cyclized monomer unit was affected by polymerization conditions such as catalyst, solvent and temperature.

Cyclopolymerization of methyl 2-methallyl maleate has been studied and has been shown to lead preferentially to five-membered lactone ring formation. This monomer has a stronger tendency to cyclopolymerize than allyl methyl maleate.[129]

The kinetics of radical-initiated cyclopolymerization of methyl 2-methallyl fumarate was investigated and compared with the polymerization results of allyl methyl fumarate.[130] In the former, the extent of cyclization and five-membered ring formation were enhanced. The steric effect was considered important.

Although an earlier study of polymerization of allyl acrylate had been carried out, based on residual unsaturation in the polymer obtained in this study, it was concluded that polymerization could proceed solely through the acrylic double bond under the conditions used.[131]

Cationic polymerization of vinyl p-(vinyloxy)benzoate led to a linear polymer having the vinyl ester group intact.[132] Radical postpolymerization was reported to lead to a ladder polymer of a molecular weight of 2000.

A polarographic study of the homopolymerization and copolymerization of ethyl methacrylate and vinyl isobutyrate was carried out in order to evaluate the reactivity of the vinyl and methacrylate double bonds in vinyl methacrylate.[133] Complex ester vinyl groups were detected in the polymer.

Complex formation in radical-initiated polymerization and mixed polymerization has been studied.[134] The unexpectedly low degree of cyclization in the preparation of poly(allyl methacrylate) by cyclopolymerization in the presence of zinc chloride resulted from complexation of $ZnCl_2$ by the allyl group as well as by the acrylic double bond. Although maleic anhydride-norbornene copolymer was of 1:1 composition, polymerization rate was maximum at anhydride-norbornene ratio 2:1, indicating formation of a highly reactive 2:1 complex but propagation only by the 1:1 complex. Previous work was reviewed with 40 references.

It has been shown that the photopolymerization of vinyl methacrylate in the presence of zinc chloride proceeded *via* a radical mechanism and was not accompanied by a carbocation ion mechanism; the accelerating effect of the $ZnCl_2$ was due to an increase in the rate constant for the propagation reaction.[135] Polymerization in the absence of the $ZnCl_2$ was propagated by methacrylic double bond; the vinyl double bonds were only involved in the intramolecular cyclization.

Diallyl maleate was used as a crosslinking agent for polyethylene when molded into mixed disks with the polymer and irradiated.[136] The extent of loss of crosslinking effectiveness as the result of intramolecular cyclization was not considered nor discussed.

Unsymmetrical 1,6-heptadienes which may form intramolecular charge-transfer complexes have been studied by cyclopolymerizing 2-phenylallyl methacrylate, 2-phenylallyl 2′-(carboethoxy)allyl ether, methallyl methacrylate, allyl-2-phenylallyl ether and 2-phenylallyl methallyl ether.[137] The extent of cyclization in the latter monomer was greater than in the former, although the presence of charge-transfer complexes could not be detected in either monomer.

Model compounds which may form intramolecular charge-transfer complexes and thus enhance their cyclopolymerization *via* free-radical initiation were prepared and their spectra analyzed to aid in the study of the mechanism of cyclopolymerization.[138] 2-Phenylallyl methacrylate, 2-phenylallyl 2′-carboethoxyallyl ether, methallyl methacrylate, 2-phenylallyl ether and 2-phenylallyl methallyl ether and their corresponding monoenes were prepared. IR, NMR and UV spectra of the dienes and their corresponding monoenes were studied. No ground state electronic interactions were observed.

Maleic anhydride was reported to copolymerize with an unsaturated alcohol to yield a water-soluble polymer having a lactone structure.[139] Quite possibly, the monoester formed before copolymerization occurred and the actual monomer was the monoallyl fumarate which then underwent cyclopolymerization to yield the obtained product.

Spectroscopic studies of the relation of the structure of poly(allyl α-methylcinnamate) to the polymerization temperature have been conducted.[140] The polymer obtained in radical polymerization of the monomer at 80–140° contained units resulting from intermolecular reactions of cinnamic and allylic double bonds, and five-membered lactone rings produced in chain-propagating intramolecular cyclization.

Cationic polymerization of vinyl acrylate was reported to proceed only *via* the double bond of the acrylic moiety.[141] Cyclopolymerization was not reported, although other reports have presented strong evidence that such does occur.

Cyclopolymerization of methyl allyl maleate has been studied.[142] The ratios of the rate constants of the unimolecular cyclization reaction to those of the bimolecular propagation reaction of the uncyclized allyl and vinyl radicals were estimated to be 9.7 and 1.35 mol/L, respectively.

Radical polymerization of allyl methacrylate in the presence of Lewis acids has been studied.[143] Cyclopropagation was accelerated by complex formation with $SnCl_4$. The rate of polymerization was increased by the presence of a small amount of $SnCl_4$. A linear polymer of allyl cinnamate is shown in Structure 3–21.[144] The process was temperature dependent and indicated higher reactivity of the cinnamate double bonds. The copolymers retained varying percentages of each of the original double bond types but cyclization was a major mode for the polymerization mechanism.

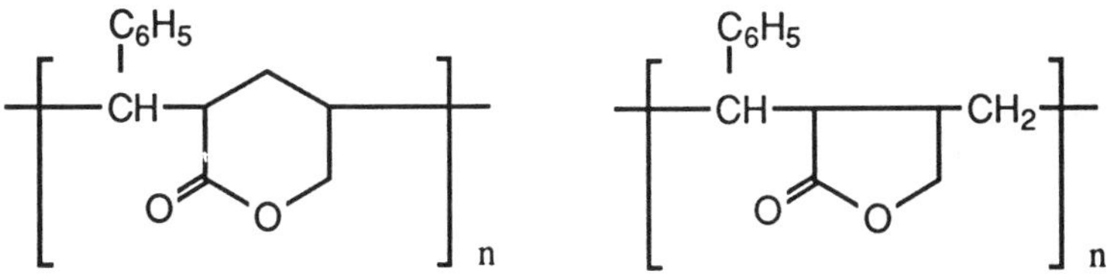

(3–21)

It has been shown that in the cyclopolymerization of the allyl alkyl maleates increasing the size of the alkyl substituent led to steric hindrance, decreased the polymerization rate, and increased the selectivity of addition of the growing radical to the allylic double bond.[2-320] Polymer-analog conversions of poly(allyl cinnamate) have been studied. Heating poly(allyl cinnamate) resulted in crosslinking and formation of new five-membered rings.[145]

It has been shown that the bulk polymerization of allyl crotonate was relatively slow, and was 0.97-order in AIBN, suggesting termination solely by degradative chain transfer.[146] In polymerization in C_6H_6, the extent of cyclization increased with decreasing monomer concentration and increasing temperature. The ratios of the rate constants of unimolecular cyclization to those of bimolecular propagation for uncyclized allyl and crotyl radicals were established to be 1.4 and 0.31 mol/L, respectively.

A large decrease in the degree of polymerization, an increase in residual saturation of the polymers, and evolution of CO_2 with increasing temperature has been observed during the polymerization of diallyl oxalate.[147] These findings are interpreted by considering the dismutation of the uncyclized growing radical to yield the allyl radical, CO_2, and polymer containing a terminal double bond. The polymerization kinetics were studied. Allyl groups were introduced into poly(methyl acrylate) by transesterification.[148] The products could be crosslinked by electron beams and UV irradiation. Similar experiments were carried out with maleic anhydride copolymers.[149] Allyl alcohol was used to convert the copolymers to their esters followed by crosslinking by use of electron beams or UV light.

1,4-Dienes, e.g., 1,4-pentadiene, 1,1-divinylcyclohexane, etc., were converted to cyclic peroxides by treating with a mixture of thiophenol and oxygen in presence of free radical initiators.[150] Polymerization of allyl vinyl ether to isotactic poly(allyl vinyl ether) has been studied and a mechanism for its stereocontrol was proposed.[151] No participation by the allyl group was indicated.

The kinetics of radical polymerization of allyl methyl fumarate was studied and compared with the corresponding maleate.[152] The fumarate ester cyclization was much higher, and gelation occurred much later. The cyclization constants were 2.73 and 1.48 mol/L, respectively. In an effort to improve the utility of methyl allyl maleate[175] as a monomer, studies were carried out to isomerize the maleate to the more reactive fumarate by polymerizing in presence of morpholine an isomerization catalyst for the maleate.[2-389]

Negative electron beam resists with high gel rigidity have been prepared from allyl methacrylate-styrene copolymers.[153]

The reduced viscosity of acrylic acid-allyl alcohol copolymer polyelectrolyte goes through a maximum as a function of pH in neutral and acidic media, depending on the nature of the anion.[154] Formation of water resistant aggregates was correlated with increasing viscosity.

Triethylene glycol dimethacrylate was polymerized in the presence of an inhibitor introduced at various stages of the polymerization. Higher conversions were obtained.[2-399]

Polymerization of vinyl *trans*-cinnamate, an unsymmetrical monomer having two double bonds of widely different chemical nature has been further studied.[155] Solvent effects in radical polymerization of this monomer have been investigated. The effects of polarity of solvent on the degree of cyclization was insignificant, however, the overall polymerization rate in homogeneous media was dependent on solvent polarity.

An allyl methacrylate-methacrylic acid copolymer was shown to be useful in light-sensitive compositions and recording materials.[156] The extent of cyclopolymerization and/or crosslinking was not discussed.

A selective synthesis for both 1- and 4-alkyl vinyl itaconate regioisomers was developed.[2-295] Polymerization experiments showed that only 4-vinyl itaconic acid and 1-alkyl vinyl itaconates led to cyclopolymers.

The effect of the stability of the cyclized radical on the rate of cyclopolymerization of unsymmetrical monomers such as 2-phenylallyl methacrylate and 2-phenylallyl 2-carboethoxyallyl ether has been studied.[157] The relative rates of the dienes were greater than their monoolefin counterparts, and less cyclization occurred in the acrylate than in the ether.

Allyl methacrylate was copolymerized with di-(tri-n-butyltin) itaconate, and the reactivity ratios reported.[158]

Allyl methacrylate-butyl acrylate-styrene graft copolymers were effective as components of impact-resistant vinyl polymer molding compositions.[159]

The use of allyl methacrylate as a component of radiation grafting and crosslinking of poly(vinyl alcohol) was described.[160] Low dose-radiation induced grafting of the monomer and crosslinking of the polymer, whereas, in the absence of allyl methacrylate no crosslinking occurred.

Methyl and ethyl 2-acryloyloxyacrylate and ethyl 2-methacryloxyacrylate were polymerized to linear cyclopolymers.[161] Solubility studies, chemical analysis for residual double bonds, and IR spectra of the polymers indicate that these monomers polymerized *via* the cyclopolymerization mechanism. Reactivity of the residual allyl groups in poly(allyl methacrylate), prepared by lithium-initiated polymerization of the monomer was studied.[162] The allyl double bond was shown to effect crosslinking *via* radical initiation, to provide vulcanization sites, and to undergo epoxidation. All reactions proceeded smoothly.

Allyl acrylate and methacrylate were polymerized *via* anionic initiation.[163] The allyl double bonds were not involved in reactions during polymerization but were presumably responsible for the crosslinking that occurred when the polymers were exposed to air.

Allyl acrylates were polymerized by anionic initiation to linear, soluble polymers having pendant unsaturated groups.[164] Cross-linking was accomplished by copolymerizing with other monomers. The corresponding methallyl esters and methacrylates were also used.

Vinyl chloride-methyl methacrylate copolymers and their intramolecular cyclization to a lactone structure were studied.[165] The kinetic study yielded rate constants and activation energies which indicated that lactonization was first order with respect to the number of cyclizable pairs.

In a study of synthesis of polymers with polar side groups, 1,5-diene monomers of Structure 3–22,[166] in which x and y were polar groups, e.g. cyano, were synthesized.

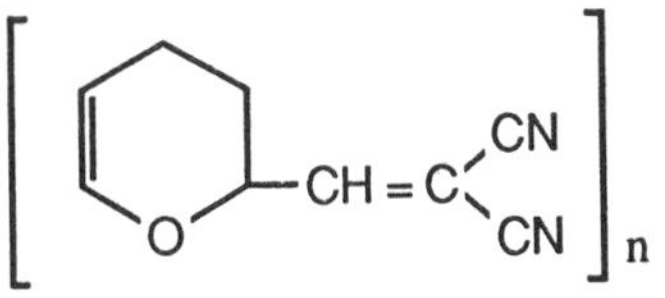

CH = C(CN)(CN)

n

(3–22)

The monomers were polymerized cationically, to form polymers *via* the vinyl double bond. Some properties of the polymers were reported. No other types of initiation were employed.

In a study of the properties of structurally homogeneous poly(1-vinyluracils) and their interactions with polynucleotides, the hypochromism and fluoresence emission intensity of the polymer increased with its syndiotacticity.[167] Highly syndiotactic polymer formed a triple stranded complex with poly(adenylic acid). Complexes with polymers of lower syndiotacticity were also studied.

Rings Containing Other Elements

Only a limited number of studies have been conducted on unsymmetrical monomers containing elements other than carbon, nitrogen and oxygen. The synthesis and a polymerization study of an unsymmetrical silane monomer, allylvinyldimethylsilane have been reported.[2-340] This monomer yielded a low molecular weight, noncyclic polymer *via* coordination type initiation in which only the allyl double bond participated.

Telomerization studies with allyl ethylsulfonate and allyl allylsulfonate have been carried out.[168] Telomerization of allyl ethylsulfonate in the presence of butyl mercaptan yielded a mixture of a five-membered ring sultone containing a sulfide group and a five-membered ring sulfonium salt formed by reaction of the sultone with its own sulfide function. These results indicate that the cyclic units of poly(allyl ethylsulfonate) are five-membered rings. Telomerization of allyl allylsulfonate in the presence of butyl mercaptan yielded a mixture of two products formed by addition of butyl mercaptan to one of the two allyl functions. Telomerization of allyl allylsulfonate in $BrCCl_3$ yielded a small amount of 1,1,1,5,5,5-hexachloro-3-bromo-pentane formed *via* fission of an O-C bond and a rearranged adduct. The rearrangement of the allyl group to a propenyl group in the case of allyl allylsulfonate was not observed when allyl ethylsulfonate or propyl allylsulfonate were telomerized under the same conditions. The rearrangement of the allyl double bond in allyl allylsulfonate is due to the presence of a second double bond in the same molecule. Poly(allyl allylsulfonate) might have a more complicated structure than expected from a simple cyclopolymerization mechanism.

A radical initiated polymerization and copolymerization study of allyl allylsulfonate was also carried out.[169] The monomer was polymerized to low molecular weight polymers containing cyclic structures in contrast to the behavior of allyl ethylsulfonate and of propyl allylsulfonate which did not polymerize under the same conditions. Allyl allylsulfonate was copolymerized with styrene, methyl acrylate, and vinyl acetate. Copolymerization reactivity factors indicate that allyl allylsulfonate has a higher reactivity than allyl ethylsulfonate or propyl allylsulfonate.

RINGS LARGER THAN FIVE- OR SIX-MEMBERS

Detailed studies of the polymerization of diallyl compounds, particularly diallyl dicarboxylates, have been continued.[2-386, 2-387, 2-388] Although poly(allyl propyl phthalate) contains approximately 15% of the head-to-head structure, polymers of diallyl phthalate (on radical initiation at 80°C) contained lesser amounts, which were dependent upon the extent of cyclization. By extrapolation of results, completely cyclized poly(diallyl phthalate) was found to consist exclusively of an eleven-membered ring from head-to-tail propagation. This is not the case with polymers from diallyl *cis*-cyclohexane-1,2-dicarboxylate or diallyl succinate, where at first a decrease in head-to-head structures was noted, but then an increase occurred at high extents of cyclization. An explanation based on steric factors is provided.

Cyclopolymerization reactions leading to larger ring sizes has dealt with conformationally controlled unsymmetrical dienes such as 2-(o-vinyphenoxy)ethyl acrylate and methacrylate[2-384] and their cyclopolymerization, utilizing both radical and photoinitiated polymerization. The polymers obtained (3–23) were essentially completely cyclized.

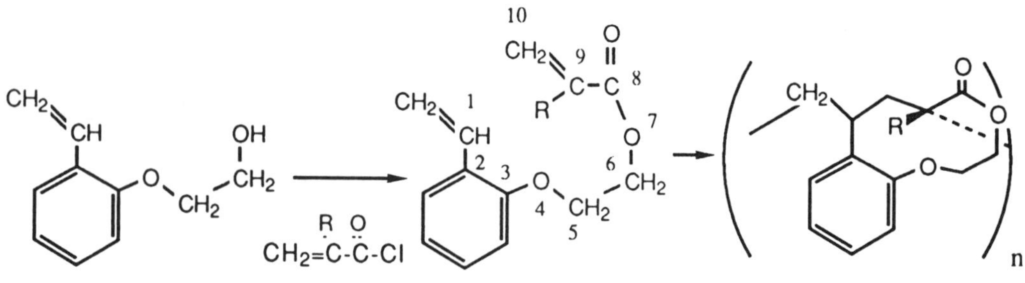

R = H or CH_3

(3–23)

This facile formation of the large ring sizes was attributed to two structural features of these monomers which may contribute to this unusual extent of cyclization: (a) a weak intramolecular charge-transfer complex may exist between the weakly accepting (meth)acrylate portion of the molecule and the electron-donor styrene portion, and (b) the geometry of the molecule, the fact that atoms 1–4 and 7–10, respectively, are inclined to be coplanar with atoms 1 and 4 *cis*, may exert a strong conformational control, thus aiding in preorientation of the molecule to favor cyclization.

A method of synthesis of polymers containing crown lactone units *via* cyclopolymerization in the presence of alkylaluminum chlorides has been developed. Polymerization of the acrylates CH_2:$CHCH_2(OCH_2CH_2)_yO$-o$C_6C_4O(CH_2CH_2O)_XCH$:CH_2 (x = 1–2, y = 1–2) or o-CH_2:$CHCH_2C_6C_4O(CH_2CH_2O)_xCOCH$:$CH_2$ (x = 2–5) by aluminum chloroalkyls gave cyclopolymers containing crown ether lactone units. $(C_2H_5)_3Al_2Cl_3$ and $C_2H_5AlCl_2$ were more effective than C_2H_5AlCl, but often gave insoluble polymers. The polymerizations were faster than radical cyclopolymerization. The cation binding ability of the polymers was demonstrated.[170]

Studies on cyclopolymerization in the presence of alkylaluminum chlorides have been extended to acrylates and methacrylates containing oligooxyethylene units in the 11–20-membered-ring region. 2-[2-(o-Allylphenoxy)ethyloxy]ethyl acrylate and higher homologs containing oligooxyethylene units capable of forming 14-, 17- and 20-membered rings, respectively, were studied. Although the effect of alkylaluminum chlorides was gradually reduced with increasing ring size, it was remarkable in the formation of 14-membered rings for methacrylates[171] and even the 20-membered rings for acrylates.

Diallyl styrylphosphonate, allyl allylstyrylphosphinate, allyl vinylstyryl phosphonate, diallyl styrylphosphonite and bis[2-(2-methacryloyloxyethoxy) ethyl] styrylphosphonate were polymerized to give products which varied from viscous liquids to rubbery polymers or crosslinked polymers, depending on the monomer and polymerization conditions.[172] Cyclopolymerization was indicated in the case of diallylstyrylphosphonate. Extensive intramolecular cyclization occurred in p-divinylbenzene-styrene copolymers prepared by thermal copolymerization in solution as shown by the reduction of the intrinsic viscosity and radius of gyration of the polymer as the concentration of p-divinylbenzene increased.[173]

Studies of the effect of temperature on the polymerization of diallyl phthalate have been carried out.[174] The degree of polymerization increased slightly with temperature from 80–100% and then decreased from 100–50%; the tendency for cyclization became more marked with increasing temperature. Cinnamic acid derivatives were synthesized from 4-vinylbenzaldehyde and active hydrogen compounds such as diethyl malonate as potential photocrosslinkable monomers.[175]

The copolymerization of N-acetylallylamine with monomethyl maleate in the presence of gamma radiation has been studied.[176] The reactivity ratio for the polymerization was $r_I = 0.34$ and $r_{II} = 0.05$, respectively, indicating a tendency toward alternation copolymerization by a radical process. Lactam ring formation was observed on acid titration of the saponified copolymer. The mechanism of radical polymerization of ethylene glycol dimethacrylate in the absence of oxygen and under conditions of increasing temperature was established.[177] Soluble poly[3-(crotonoyloxy)-2-hydroxypropyl methacrylate] was prepared by radical solution polymerization of the monomer.[178] The polymer retained ~95% of the crotonyl double bonds and only a small amount of cyclization product was detected.

2-Allyloxyethyl methacrylate was polymerized to yield 100% of soluble polymer without gel formation.[179] Films of the polymer could be crosslinked by heating, indicating that the cyclopolymer contained some unsaturation. The synthesis of macrocyclic ring-containing polymers *via* cyclopolymerization and cyclocopolymerization has been studied.[180] 2-Allyloxyethyl methacrylate in C_6H_6 has been polymerized in solution in 100% yield without gel formation to give polymer having intrinsic viscosity 1.37 dl/g (25°, in C_6H_6).[181]

It has been shown that kinetic and relative reactivity data for the radical copolymerization of diethylene glycol divinyl ether with alkyl acrylates showed that the predominant influence on the rate of polymerization was steric and not electronic factors.[182] Steric hindrance in the copolymerization increased with increasing length of the alkyl chain. Kinetic equations were derived to describe the rate of copolymerization. It has been shown that polymerization of the α-methylstyrene derivatives shown (n = 1, n = 2, n = 3) by $SnCl_4$ in dilute solution gave soluble polymers.[183] No cyclopolymerization occurred, but addition polymerization was accompanied by transfer, giving indan units (3–24).

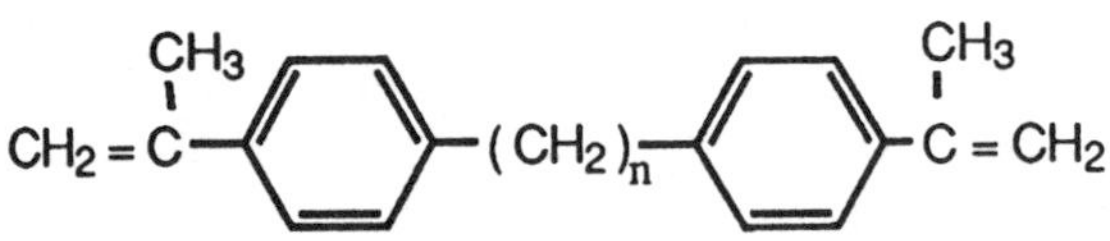

$CH_2=C(CH_3)-C_6H_4-(CH_2)_n-C_6H_4-C(CH_3)=CH_2$

(3–24)

Ethylene dimethacrylate underwent predominantly cyclization rather than crosslinking when copolymerized with styrene.[184] Procedures for determining copolymerization parameters of monovinyl-divinyl pairs of monomers which usually neglect cyclization were revised. Cyclization dynamics and thermodynamics of end-to-end cyclization of polymers were studied by utilizing pyrene-terminated polystyrene.[185] The equilibrium constant could be calculated for both long and short chains. Polymerization of diethylene glycol bis(allyl carbonate) has been studied.[186] The polymer was unsaturated, and cyclization occurred to give a 16-membered ring, only to a small extent.

The thermal stability of polymers containing indanic units in the main chain was determined.[187] The polymers were obtained by an acid-catalyzed stepwise polymerization of 1,1-diphenylethylene derivatives which selectively introduced indan type cyclic units into the polymer backbone (3–25).

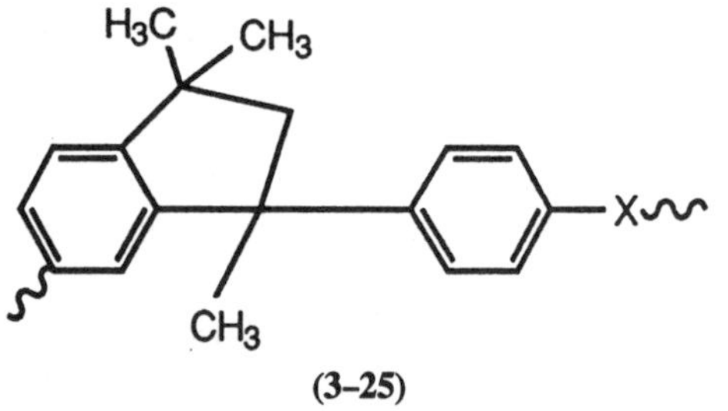

H3C CH3
CH3
X

(3–25)

Unconjugated dienes such as ω(2-allylphenoxy)alkyl acrylates and ω(2-allylphenyl)oligooxyethylene acrylates were polymerized to give polymers containing seven- to twenty-membered rings.[188]

1,3-Bis(4-vinylnaphthyl)propane was polymerized cationically to cyclopolymers containing predominantly syn-[3.3](1,4)-naphthalenophane units in the main chain.[2-395]

A soluble polymer with 1,4-dioxepane-5,7-diylmethylene units was obtained from cationic polymerization of ethylene glycol divinyl ether, by using iodine as catalyst.[2-396] Radical initiation led to polymers having the cyclic unit up to 75% as well as the 2-[2-(vinyloxy)ethoxy]-ethylene structural units.

An investigation of the effects of ethylaluminum chlorides on cyclopolymerization was extended to acrylate and methacrylic esters of ethylene oxide oligomers of o-allylphenol, in which n (no of ethylene oxide units) values were 2, 4, 6 and 10.[2-393] The dienes are capable of forming 11-, 13-, 15-, and 19-membered rings, respectively. The ethylaluminum chloride addition was effective in increasing the extent of cyclization, particularly in formation of the 11-membered ring. The cyclization constant was similar for ring sizes 8–15 for both the acrylate and methacrylate series.

The kinetics of cyclopolymerization of N,N′-methylenebisacrylamide initiated by the ceric ion-thiourea redox system was studied.[2-392] The rate of polymerization was proportional to $[Ce(IV)]^{1/2}$, $[\text{thiourea}]^{1/2}$ and $[\text{monomer}]^{3/2}$, consistent with the cyclopolymerization mechanism. In a continuing study of head-to-head vinyl polymers, radical cyclopolymerization of o- and p-dimethacryloyloxybenzene was investigated.[2-390] It was concluded that only a minor amount of head-to-head units was formed in the soluble, cyclic polymers obtained.

A variety of diunsaturated phosphonium salts functionally capable of undergoing cyclopolymerization to produce seven- or eight-membered rings were synthesized and their radical initiated polymerization studied.[2-341] Among these were allyl-3-butenyldiphenylphosphonium bromide and 3-butenylmethallyldiphenylphosphonium bromide, both of which produced polymers consistent with the cyclopolymerization mechanism. Di-3-(butenyl)diphenylphosphonium bromide, a monomer functionally capable of yielding a polymer having an eight-membered ring structure did not polymerize, or polymerized only slowly.

Diethylene glycol divinyl ether underwent cyclopolymerization in presence of boron trifluoride etherates.[189] The polymer contained seven- and nine-membered rings and was soluble in several solvents. Crosslinking occurred after 90% conversion. Ethylene glycol divinyl ether also underwent cyclopolymerization to give polymer containing seven-membered rings. Cyclopolymerization of N,N′-methylenebisacrylamide by persulfate ion-ferrous ion redox system was carried out.[2-397] The rate of polymerization was dependent on $[\text{monomer}]^{1.5}$, and $[\text{persulfate ion}]^{0.5}$. The evidence was consistent with the cyclopolymerization mechanism. Polymerization and copolymerization of triethylene glycol dimethacrylate in cyclohexane or methyl ethyl ketone gave a soluble polymer of cyclolinear structure containing a few long chain branches per macromolecule.[2-400] Small amounts of ethylene glycol dimethacrylate were copolymerized with methyl methacrylate.[190] Monomers and pendant vinyl conversion as a function of time were measured up to the gel point. Evidence showed a tendency to cyclize during formation of a primary chain; this tendency increased with dilution. Pendant reactivity led to subsequent cyclic structures as well as crosslinking.

The effects of cinnamoyl content in copolymers of 2-cinnamoyloxyethyl methacrylate with methyl methacrylate on the competitive intra- and intermolecular photo crosslinking were studied.[191] The quantum yields for intramolecular crosslink formation increased linearly with increasing dimethacrylate in the copolymer. The parameters for intermolecular crosslink formation, on the contrary, decreased with the extent of reaction since the increase of intramolecular crosslinks prevents the interpenetration of polymer coils. The polymerization behavior of N-(p-vinylphenyl)acrylamide was studied.[192] Soluble polymers were obtained by both cationic and anionic initiation. The residual double bond *via* cationic initiation was the acrylamide moiety while that from anionic initiation was the styrene moiety. The residual double bonds could be used for crosslinking. Free radical initiation was not reported.

Formation of cyclic polymers during a chain-transfer-with-cleavage reaction was analyzed theoretically.[193] The probability (ϵ) of linear/cyclic polymer was calculated. At low ϵ values ($\epsilon = mN/n_0$; where m = minimum number of monomer units/ rings, N = total number of linear polymer mole, N_0 = maximum number of monomers units in rings); the average degree of polymerization of the cyclic compounds is $\sim$3m. Radical initiated polymerizations of mixed allyl isopropenyl alkyl or aryl phosphonates were studied.[194] By IR spectroscopy, the polymerization was found to proceed with formation of polymer containing residual double bonds.

The rates of polymerization, *via* radical initiation, of several allyl isopropenyl alkyl or aryl phosphonates were compared.[195] The rate of polymerization of the ethyl phosphonate was found to be four times that of the corresponding tolylphosphonate. The reaction rate of the ethyl monomer increased approximately two and one-half times for each 10° of temperature rise. Solution polymerizations of allyl (o-vinylphenyl)ether and allyl (p-vinylphenyl) ethers with cationic and radical initiators were investigated.[196] Generally, soluble polymers were obtained by both initiator types, however, crosslinking was dependent on conditions. No comparison was made between the o- and p-isomers, unfortunately, as the extent of cyclopolymerization should vary.

N,N′-Diallyldiimides of 1,2,3,4-cyclopentanetetracarboxylic acid were polymerized *via* free radical initiation.[197] No discussion of the polymer structure was included. Although this monomer is a 1,11-diene (3–26) certain conformations of the molecule

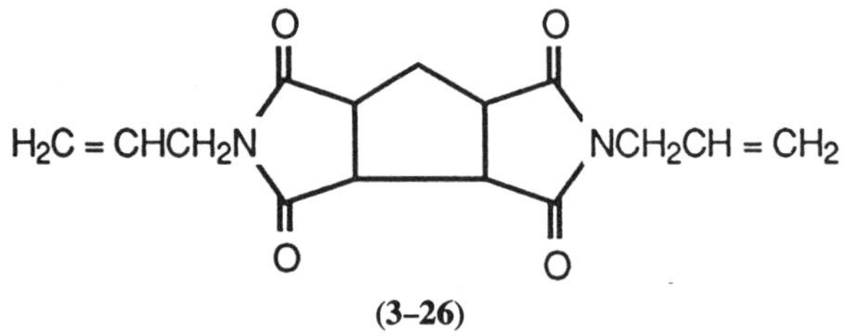

(3–26)

would place the double bonds of the allyl groups quite close together. Linear, soluble polymers were obtained *via* radical initiation of allyl isopropenyl esters of alkyl- and arylphosphonic acids, e.g. allyl isopropenyl ethyl phosphonate, allyl isopropenyl phenylphosphonate, allyl isopropenyl p-chlorophenylphosphonate, and allyl isopropenyl p-tolylphosphonate.[198] It was shown that diallyl esters of 2-substituted succinic acids undergo cyclopolymerization to a high degree *via* radical initiation, and that the extent of cyclization was affected by the structure of the 2-substituent.[199]

Bis[β-(methacryloxy)propionaldehyde pentaerythritol acetal] was polymerized and copolymerized with styrene and acrylonitrile *via* free radical initiation.[200] The ratio of the rate constant of the unimolecular cyclization reaction to that of the bimolecular propagation reaction of the uncyclized radical was 22.0 moles/L. The copolymerization of the monomer proceeded primarily *via* cyclocopolymerization. The reactivity ratios for the two systems were reported. The effect of the molecular size of the dimethacrylate in a copolymerization study of dimethacrylates with vinyl monomers was investigated.[201]

The glass transition of the copolymers was independent of their crosslinking density. Apparently, no evaluation of the extent of cyclopolymerization in any case was done.

Radical initiated cyclopolymerization of dimethallyl phthalate has been studied, and it was shown that the ratio of the rate constant for unimolecular cyclization to that of bimolecular propagation of the uncyclized radical equals 7.6 mol/L, which approaches the corresponding value for diallyl phthalate.[202] Cyclopolymerization of the diallyl esters of oxalic, malonic, succinic, adipic, and sebacic acids has been studied kinetically and rate data, as well as overall activation energies of polymerization, were reported.[203] A radical initiated copolymerization study of diethylene glycol divinyl ether with some akyl esters of acrylic acid was carried out.[204] It was found that the copolymerization rate decreased with the bulkiness of the alkyl group of the ester, e.g. from C_1-C_{12}. No examination of the extent of cyclopolymerization of the divinyl ether was reported.

2-Vinyl-4-(acryloyloxymethyl)-1,3-dioxolane and 2-vinyl-4-(methacryloxymethyl)-1,3-dioxolane were polymerized and copolymerized with styrene and acrylonitrile by free radical initiation.[205] The polymerizations proceeded mainly on the acryl or methacryl group; however, a small fraction of the allidene group participated in a cyclopolymerization process. The rate constant ratio of the unimolecular cyclization to that of the bimolecular propagation for the uncyclized radical was 0.065 moles/L for the acryl monomer and 0.114 mole/L for the methacryl monomer. Monomer reactivity ratios and Q and e values were calculated for these systems.

THEORETICAL CONSIDERATION

The major theoretical implications in predicting whether cyclization may occur in nonconjugated unsymmetrical dienes have been given some study. Among these are the relative-reactivities of the double bonds toward the initiating system and toward each other, as well as temperature and solvent effects. A major consideration, however, is the requirement of ring size, five- and six-membered rings forming most effectively, of course, and the extent to which head-to-head propagation may occur, producing a mixture of ring sizes. The possibility of intramolecular donor-acceptor complex formation exerting an influence on ring size has been given some consideration.

This section includes studies directed toward better understanding of these concepts. A more detailed discussion of the mechanism of cyclopolymerization, including kinetic and ring size studies and further interpretations of mechanistic proposals appears in Chapter 10.

A study of radical cyclizations[26] designed to shed light on the propensity of certain 1,6-dienes to lead to predominantly five-membered rings rather than the predicted six-membered rings was reported. The 5-hexenyl radical and its five 1- and 5-methyl substituted derivatives were generated and the ratios of the ring sizes of the cyclized products were determined. The results were that 1-substitution and increasing temperature led to larger ratios for six- to five-membered rings, while 5-substitution led to an increased 5:6-membered ratio. 1-Substitution increased the rate of cyclization to six-membered rings by a factor of at least twenty.

A study of the cyclopolymerization characteristics of a number of allyl-substituted cycloalkenes *via* cationic and metal alkyl coordination initiation was undertaken.[22] The results were related to the classical studies on anchimerically assisted reactions of cycloalkene double bonds with neighboring carbocations.[46,47] Cyclizations to the extent of 95% in the case of cation initiation was observed whereas a maximum of 80% cyclization was observed in the case of metal alkyl coordination initiation. These results were in accord with the earlier work, and support the "electronic interaction theory" of cyclopolymerization.

The effect of temperature on radical-initiated cyclopolymerization of N-allyl-N-methylmethacrylamide has been studied.[79,80] Initiation over the range of $-78°C$ to $120°C$ resulted in polymers with 88–93% cyclic units, mainly five-membered rings with some six-membered rings, and pendant methacryl groups. A cyclopolymerization study of N-allyl-N-methyl(2-substituted allyl)amines has been referred to earlier.[2-106] The structures of the poly(N-methylpyrrolidines), poly(N-methylpiperidines), and low-molecular-weight products obtained by radical-induced cyclizations were determined. Steric interactions induced by β-substituents favored attack at the unsubstituted allyl group, but conjugated substituents favored attack at the substituted group with formation of conjugation-stabilized radicals. Kinetically controlled cyclization to the pyrrolidines occurred in all cases except that of the tert-butyl derivative, in which steric interaction induced cyclization to both piperidine and pyrrolidine products.

The use of alkylaluminum chlorides as complexing agents for cyclopolymerizable monomers has been investigated. During the period 1976–79, four papers were published in which it was demonstrated that the complexing agents increased the extent of cyclization in the polymers. Unsymmetrical monomers such as o-allylphenyl acrylate, the corresponding methacrylates, 2-(o-allylphenoxy)ethyl acrylate, and 4-(o-allylphenoxy)butyl acrylate were studied both in homopolymerization and copolymerization with conventional vinyl monomers.[102-106] Cyclopolymers having eight-, eleven-, and thirteen-membered rings, and extents of cyclization ranging from 90–93% were obtained. The aluminum:monomer ratio was also critical, cyclization increasing when this ratio was $\geqslant 1.0$. The results indicated that the increase in cyclopolymerization tendency in presence of the complexing agent was due to an intramolecular interaction between the two double bonds of different reactivities. In later papers, it was demonstrated that diacrylate and dimethacrylate esters of ethylene oxide oligomers capable of forming fifteen-, and nineteen- and twenty-membered rings responded equally well to the complexing agents.[171, 2-393] In another paper, cyclopolymerization of N,N′-methylenebisacrylamide occurred with an energy of activation of 32 ± 2 $kJmol^{-1}$, a result which supports the concept of intramolecular interaction between the double bonds prior to polymerization.[90]

The importance of polymer ceiling temperature in controlling both degree of cyclization and ring size in cyclopolymerization has been demonstrated.[126] In a study of radical cyclopolymerization of allyl methacrylate both below and above the ceiling temperature for head-to-tail propagation in methyl methacrylate, it was shown that above this ceiling temperature, cyclization and formation of five-membered rings were enhanced. The polymer contained no unreacted pendant methacryloyl groups.

In two papers published in 1974, the possibility that intramolecular charge transfer (donor-acceptor) complexes may play a role in cyclopolymerization was studied.[137, 138] In an analogous manner, certain highly alternating copolymerizations can be explained. Compounds such as 2-phenylallyl methacrylate which in cyclopolymerization would simulate an α-methyl styrene-methyl methacrylate copolymerization, was synthesized and its polymerization studied. No physical evidence for an intramolecular donor-acceptor complex was obtained, however, cyclopolymerization occurred to a considerable extent. Another monomer which would also simulate α-methyl styrene-methyl methacrylate copolymerization, 2-phenylallyl 2′-(carboethoxy)allyl ether, was also synthesized and studied. A cyclopolymer having $[\eta] = 0.39$ dl/g, and only a small amount of allyl unsaturation was obtained.

Large ring-containing polymers produced from photoinitiation of 2-(o-vinylphenoxy)ethyl acrylate and the corresponding methacrylate were attributed to the preorientation of the monomers as the result of steric effects which permitted the double bonds of the monomers and the intermediate atoms to assume essential coplanarity which promoted the preorientation.[2-384] Cyclization was essentially quantitative in these cases.

A growing interest in pure head-to-tail vinyl polymers has prompted a number of investigations to employ cyclopolymerization of suitable monomers, particularly those 1,6-dienes such as the N-substituted bis-acrylamides and methacrylamides which produce predominantly five-membered rings as a source of head-to-head polymers. In an effort to synthesize head-to-head polymers, o-and p-dimethacryloyloxybenzenes were studied.[2-390] However, only a small amount of head-to-head units was observed in this case, which required larger ring formation, particularly in the p-isomer.

The propensity for cinnamic acid derivatives to undergo photoinitiated reactions led to a study of the effects of cinnamoyl content in copolymers of 2-cinnamoyloxyethyl methacrylate with methyl methacrylate on the competitive intra- and intermolecular crosslinking.[191] The quantum yields for intramolecular crosslink formation increased linearly with increasing cinnamoyl content in the copolymer. However, the parameters for intermolecular crosslink formation decreased with the extent of reaction. This decrease was attributed to prevention of interpenetration of polymer coils as a result of the increase of intramolecular crosslinks.

The relative reactivities and structural aspects of the cationic initiated polymerization of mono and diolefinic norbornanes has been explained on the basis of well-documented organic chemical principles.[206] A variety of monomers, including 2-methylenenorbornane, 2-methylene-5-norbornene, norbornene, 2-vinyl-5-norbornene, 2-isopropenyl-5-norbornene, 2,5-norbornadiene, 2-vinylnorbornane, and 2-isopropenylnorbornane were studied. The structures of the polymers were explained on the basis of hydride shifts and steric compression during propagation. It had been shown as early as 1961 that anchimerically assisted solvolysis of 2-(3-cyclopentenyl)ethyl tosylate by a neighboring double bond could lead to ring closure to the norbornyl system.[207] The ratio of the solvolytic rate constants, a measure of the driving force of the neighboring double bond, was 5.8, 50, and 1900 in solvents of different nucleophilic activity. This study is directly related to and supports the "electronic interaction theory" as an explanation for the propensity of cyclopolymerization.

In a theoretical approach to simulation of degradation processes, some relations for random crosslinking without cyclization were tested. The only conclusions drawn were that the problem is complex.[208] 4-(Cyclohex-1-enyl)butyl radical was shown to readily undergo intramolecular homolytic addition.[1-22] The rate constants for the 1,5- and 1,6-modes of cyclization were found to be 3.8×10^4 and 3.2×10^4 sec^{-1}, respectively. The direction and relative rates of cyclization of fifteen hex-5-enyl radicals were determined.[2-515] Substituents on the olefinic bond at C-5 had a profound effect on the rate of 1,5-cyclizations. Substituents at C-6 retarded the 1,6-cyclization rate. Favorable entropy and enthalpy contributions were responsible for preferential cyclization of hex-5-enyl, hept-6-enyl, and oct-7-enyl radicals to the cycloalkylcarbinyl radical.[209] The rate constant for the 1,5-H transfer in the hept-6-enyl radical was estimated from the results.

The ratio of isomers of 1,2-dimethylcyclopentane formed, together with 1-heptene and cyclohexane from cyclization of the hept-6-enyl radical was cis:trans = 2.3:1.[210] A similar preference for the cis cyclization was exhibited by the radical generated from 6-chloronon-1-ene and allyl 2-chloropropyl ether. The formation of the cis transition state is favored by a secondary attractive interaction between the olefinic π^* orbital and the half-filled orbital delocalized by hyperconjugation.

In a study of cyclization of 3-allylhex-5-enyl radical as part of an overall study of the mechanism and structures of cyclopolymers, cis- and trans-1-allyl-3-methylcyclopentane, bicyclo[4.2.1]nonane, and exo-3-methylbicyclo[3.2.1]octane were formed.[211] The formation of bicyclic systems from this radical involved two discrete steps, the kinetic parameters of which were determined. Homoconjugative interaction between the double bonds affects the rate and stereoselectivity but not the regioselectivity of the first step. No bicyclo[3.3.1]nonane was formed, which suggests that structures previously assigned to cyclopolymers derived from triallyl monomers may be incorrect.

In a study of the mechanism of cyclization of aryl radicals containing unsaturated ortho-substituents, for example, the radical derived *via* reduction of o-iodophenylallyl ether by tributyltin hydride, the aryl radicals initially produced cyclized regiospecifically to give products in which the newly formed radical center was exocyclic to the new ring.[212] The results were consistent with a transition state formed by primary interaction of the partly filled orbital with one lobe of the π^* orbital.

2-(o-Vinylphenoxy)ethyl methacrylate polymerized by radical initiation or photoinitiation in tetrahydrofuran at 60° or ambient temperature, respectively.[2-384] Large-ring rigid polymer structures formed *via* a cyclopolymerization mechanism, were proposed.

In a continuing study of thermal decomposition of quaternary ammonium hydroxides, the importance of conformational factors in β-elimination from quaternary hydroxides derived from piperidines, morpholines, and decahydroquinolines was studied.[213] The products of thermal decomposition of 3-oxa-6-azoniaspiro[5.5]undecane hydroxide show that β-elimination occurs several times faster in a morpholine than in a piperidine ring. The requirement for easy elimination in 6-membered rings is the anti-coplanarity of H_β, C_β, C_α and N^+.

A spectroscopic method for a separation determination of allyl and acrylate double bonds in insoluble polyacrylates based on measuring the absorbance of allyl and acrylate groups of 1630 and 1622 nm, respectively, has been described.[214] The content of acrylate and allyl double bonds in allyl methacrylate-methyl methacrylate copolymer was determined. The relative measurement error was 4–5 and 10–30% at monomer contents 5–20 and <5%, respectively. The minimum double bond contents which could be determined by the method were 0.5 and 0.3% for allyl and methacrylate, respectively.

In a study of stereoselectivity of ring closure of substituted hex-5-enyl radicals, the stereoselectivity and kinetics of ring closure of 2-substituted hex-5-enyl radicals in which the substituent was methyl or vinyl, as well as the 3- and 4-methyl-5-hexenyl radicals, were examined.[215] All of the radicals underwent regiospecific 1,5-ring closure with a rate constant greater than that for hex-6-enyl radicals. The 3-substituted radicals gave cis products preferentially. The 2- and 4-substituted radicals gave mainly trans products. The results were rationalized in terms of conformational effects in the chair-like transition state (3–27).

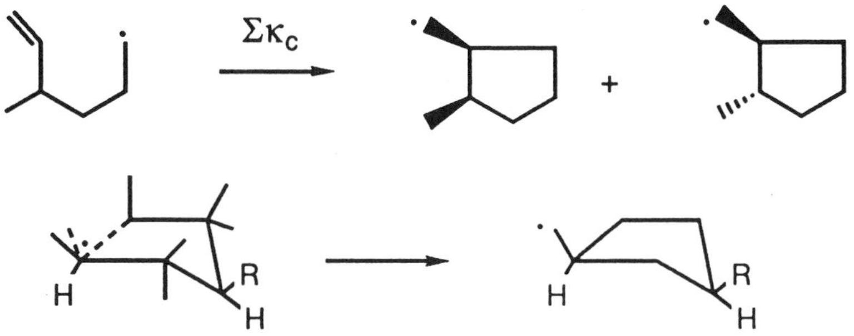

(3–27)

The mechanism of ring closure of alkenyl peroxy radicals generated by oxygenation of benzenethiol-1,3,6-triene mixtures was studied.[216] It was shown that 1,2-dioxolans were obtained by regio- and stereospecific ring closure of the alkenylperoxy radical intermediate. It has been shown that bicyclic systems such as those shown (3–28) can be formed by radical cyclization of appropriate radicals.[217] The rates and stereochemistry of the ring closures were determined and rationalized.

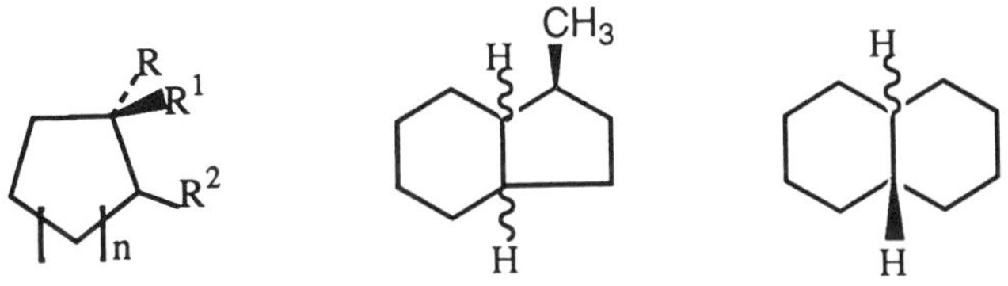

(3–28)

Stopped-flow spectroscopy was used to study the propagating species in the cationic initiated polymerization of 1,3-bis(p-vinylphenyl)propane and the corresponding ethane.[2-394] The cation generated showed a bathochromic shift at 360 nm, exceeding that of p-methylstyrene (332 nm), which was attributed to stabilization of the cationic center by a through space intramolecular interaction with the adjacent terminal styryl group, which facilitated the cyclopolymerization (3–29).

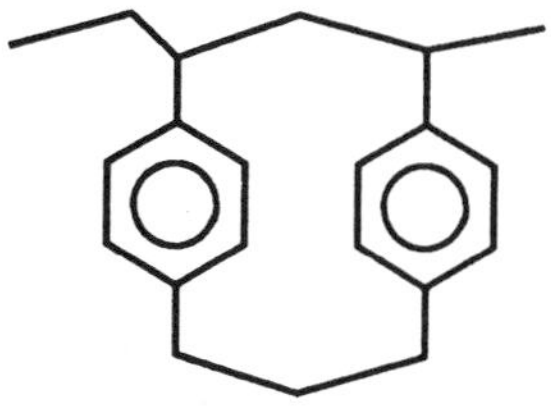

(3–29)

A theoretical study was undertaken of ring closure of a variety of alkenyl, alkenylaryl, alkenylvinyl and similar radicals.[218] The method involved the application of MM2 force-field calculations to model transition structures for which the dimensions of the arrays of reactive centers have been obtained by MNDO-UHF techniques. The results, which generally accord with guidelines based on stereochemical considerations, show excellent qualitative and satisfactory quantitative agreement with experimental data. The method was successfully applied to complex systems including ring closure of alkylperoxy radicals, and formation of the triquinane system by three consecutive cyclizations.

Alkenoxyoxymethyl iodides or selenides were converted to lactones by treatment with tributyltin or tributylgermane.[219] The reaction involved highly regio and stereoselective ring closure of alkenoyloxymethyl radicals. For example, the phenylselenomethyl ester of 2,2-dimethyl-3-butenoic acid when treated with tributyltin gave a mixture of the methyl ester of the acid and the lactone.

Cyclopolymerization has some similarities, particularly pertaining to the "electronic interaction theory," to the so-called "no mechanism" reactions of organic chemistry, namely, the Diels-Alder reaction and the Claisen and Cope rearrangements.[2-517] An elegant study of the pathway most probably taken in the Cope rearrangement supported the conclusion that a four-centered, chair-like transition state is favored over the six-centered, boat-like structure. The behavior of the meso- and rac-3,4-dimethylhexa-1,5-diene in the Cope rearrangement was examined. The results of this study are shown (3–30).

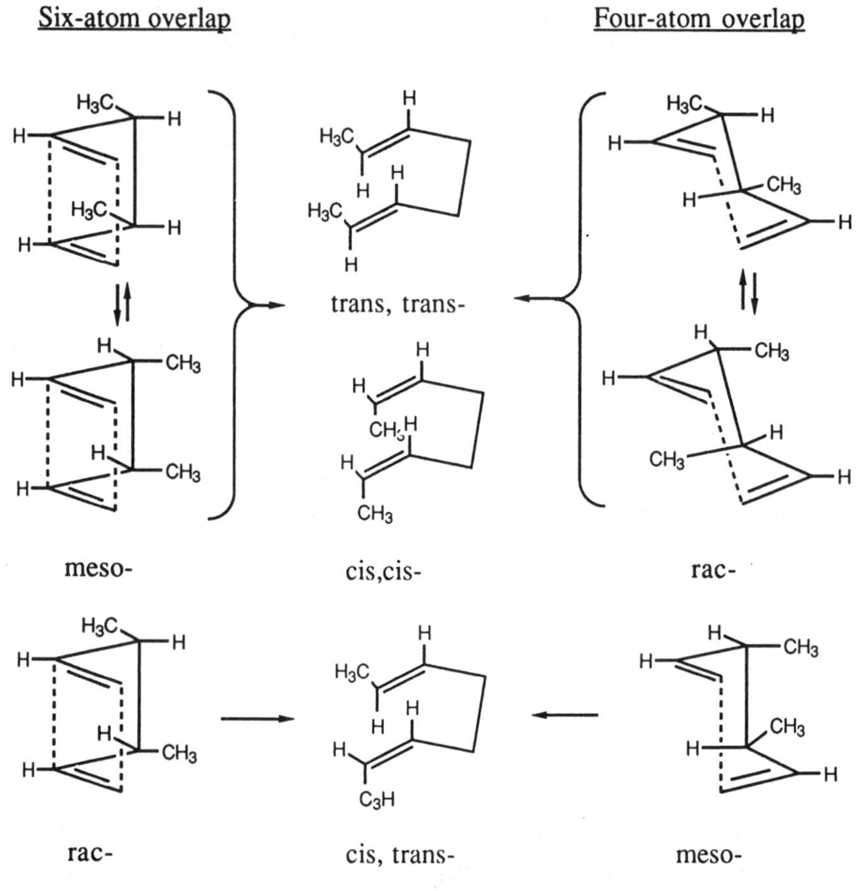

Six-atom overlap | Four-atom overlap

meso- | trans, trans- | cis,cis- | rac-

rac- | cis, trans- | meso-

(3–30)

REFERENCES

1. Flory, P. J., J. Am. Chem. Soc., 1937, 59, 241.
2. Yokota, K. and Takada, Y., Kobunski kagaku, 1969, 26(288), 317; CA 71, 22369v (1969).
3. Quere, J. and Marechal, E., Bull. Soc. Chim. Fr., 1969, 11, 4087; CA 72, 44218t (1970).
4. Ohara, O., Kobunshi kagaku, 1971, 28, 285; CA 75, 98863e (1971).
5. Foster, R. J. and Hepworth, P., Kinet. Mech. Polyreactions, Int. Symp., 1969, 1, 339; CA 75, 152127h (1971).
6. Butler, G. B. and Ingley, F. L., J. Am. Chem. Soc., 1951, 73, 1512; CA 45, 7372g (1951).
7. Alieva, A. G., Stotskaya, L. L. and Krentsel, B. A., Zh. Vses. Khim. Obshchest, 1972, 17, 588; CA 78, 84844z (1973).
8. Alieve, A. G., Stotskaya, L. L. and Krentsel, B. A., Vysokomol. Soedin., Ser. A, 1973, 15, 1005; CA 79, 79269g (1973).
9. Heublein, G. and Henblein, B., Plaste Kautsch., 1976, 23, 95; CA 84, 136125m (1976).
10. Leary, R. F. and Garber, J. D., U.S. Pat. 2,513,243, 1950 (June 27); CA 44, 9192a (1950).
11. Bondhus, F. J. and Johnson, H. L., U.S. Pat. US 2,543,092, 1951 (Feb. 2); CA 45, 4486d (1954).
12. Johnson, H. L. and Stuart, A. P, U.S. Pat. US 2,576,515, 1951 (Nov. 27); CA 46, 1807i (1952).
13. Edmonds, Jr., J. T., U.S. Pat. US 2,913,443, 1960 (Nov. 17); CA 54, 6191f (1960).
14. Marconi, W., Cesca, S. and Fortuna, G. D., J. Polym. Sci., B, 1964, 2, 301; CA 60, 12124h (1964).
15. Toshima, K., Ichiki, E. and Fujita, Y., Japan Pat. 65 21,436, 1965 (Sept. 22); CA 64, 2193c (1966).
16. Toshima, K., Ichiki, E. and Fujita, Y., Japan Pat. 66 5,396, 1965 (Nov. 11); CA 65, 9052b (1966).
17. Marconi, W., Cesca, S., Fortuna, G. D. and Cesari, M., Ital. Pat. 699,021, 1965 (Dec. 2); CA 68, 96362x (1968).
18. Sartori, G., Valvassori, A, Turba, V. and Lachi, M. P., Chimica Ind. Milan, 1963, 45, 1529; CA 60, 13327d (1964).
19. Pledger, H., Jr., U.S. Pat. 3,252,957, 1966 (May 24); CA 65, 7305h (1966).
20. Van Heiningen, J. J., Diss. Abstr. B, (Ph.D. Diss., U. of FL), 1967, (August, 1966) 27, 3471; CA 67, 99467a (1967).
21. Ichikawa, M., Harita, Y. and Tashiro, M., Japan Pat. JP 73 29,880, 1973, Sept. 13; CA 81, 38653a (1974).
22. Van Heiningen, J. J. and Butler, G. B., J. Macromol. Sci. (Chem) A, 1974, 8, 1175; CA 82, 43799c (1975).
23. Butler, G. B. and Van Heiningen, J. J., J. Macromol. Sci. (Chem), A, 1974, 8, 1139; CA 83, 59316z (1975).
24. Huntsman, W. and Curry, T. H., J. Am. Chem. Soc., 1958, 80, 2252; CA 52, 15438i (1958).
25. Huntsman, W., Salomon, V. C. and Eros, D., J. Am. Chem. Soc., 1958, 80, 5455; CA 53, 9090c (1957).
26. Wilbur, Jr., J. M. and Marvel, C. S., J. Polym. Sci., A, 1964, 2, 4415; CA 62, 712e (1965).
27. Angelo, R. J., Belg. Pat. 632,632; Fr. Pat. 1,357,994, 1963 (Nov. 21); CA 61, 8437e (1964).
28. Dowbenko, R., Tetrahedron, 1964, 20, 1843; CA 61, 10604a (1964).
29. Modena, M., Bates, R. B. and Marvel, C. S., J. Polym. Sci. A, 1965, 3, 949; CA 62, 11921f (1965).

30. Montecatini Edison S.P.A, Ital. Pat. 808,080, 1968 (Mar. 15); CA 77, 47937r (1972).
31. Easterbrook, E., Brett, Jr., T. J. Loveless, F. C. and Matthews, D. N., IUPAC, Boston, 1971, 712.
32. Duck, E. W. and Cooper, W., IUPAC, Boston, MA, 1971, 722.
33. Aso, C. and O'Hara, O., Kobunshi kagaku, 1966, 23, 895; CA 66, 95453z (1967).
34. Mitchell, R., McLean, S. and Guillet, J. E., Macromolecules, 1968, 1, 417; CA 69, 107145c (1968).
35. Cesca, S., Priola, A. and Santi, G., J. Polym. Sci. B, 1970, 8, 573; CA 74, 42684j (1971).
36. Akhmedov, A. I., Sadykhov., Z. A., Levshina, A. M. and Vedeneeva, L. Y., Uch. Zap., Azerb. Univ., Ser. Khim. Nauk 1972, 75; CA 80, 83723p (1974).
37. Jones, J. F., J. Polym. Sci., 1958, 33, 513; CA 55, 6916h (1958).
38. Marvel, C. S., Kiener, P. E. and Vessel, E. D., J. Amer. Chem. Soc., 1958, 81, 4694; CA 54, 8589c (1958).
39. Marvel, C. S. and Kiener, P. E., J. Polym. Sci., 1962, 61, 311; CA 58, 1539e (1962).
40. Bardyshev, I., Gurichard, N. and Konshilov, Y., Zh. Prikl. Khim., 1968, 41, 399; CA 68, 96177r (1968).
41. Marvel, C. S. and Hwa, J. C. L., J. Polym. Sci., 1960, 45, 25; CA 55, 22896e (1961).
42. Julia, M. and Maumy, M., Bull. Soc. Chim. Fr., 1969, 2427; CA 71, 112112k (1969).
43. Walling, C. and Cioffari, A., J. Amer. Chem. Soc., 1972, 94, 6059; CA 77, 113411t (1972).
44. Bartlett, P. D., Bank, S., Crawford, R. J. and Schmid, G. H., J. Amer. Chem. Soc., 1965, 87, 1288; CA 62, 12998c (1965).
45. Bartlett, P. D. and Sargent, G. D., J. Amer. Chem. Soc., 1965, 87, 1297; CA 62, 12998f (1965).
46. Bartlett, P. D., Closson, W. D. and Cogdell, T. J., J. Amer. Chem. Soc., 1965, 87, 1308; CA 62, 12998g (1965).
47. Barlett, P. D., Trahanovsky, W. S., Bolon, D. A. and Schmid, G. H., J. Amer. Chem. Soc., 1965, 87, 1314; CA 62, 12999b (1965).
48. Julia, M. and Maumy, M., Bull. Soc. Chim. Fr., 1969, 2415; CA 71, 112111j (1969).
49. Butler, G. B. and Nash, Jr., J. L., J. Amer. Chem. Soc., 1951, 73, 2539; CA 46, 415h (1952).
50. Butler, G. B., J. Amer. Chem. Soc., 1955, 77, 482; CA 50, 858i (1956).
51. Pogorelko, V. Z., Karetnikove, N. A. and Ryabov, A. V., Tr. Khim. Khim. Tekhnol., 1972, 1, 65; CA 79, 5813h (1973).
52. Klesper, E., Strasilla, D. and Berg, M., Eur. Polym. J., 1979, 15, 587; CA 92, 23080q (1980).
53. Klesper, E., Strasilla, D. and Berg, M., Eur. Polym. J., 1979, 15, 593; CA 92, 23081r (1980).
54. Gaylord, N. G., Solomon, O., Stolka, M. and Patnaik, B. K., J. Macromol. Sci. (Chem.), 1972, 8, 983; CA 82, 98427k (1975).
55. Chu, S. C. and Butler, G. B., J. Polym. Sci. (Chem.), 1978, 16, 1375; CA 89, 110927c (1978).
56. Nishimura, J., Tanaka, N., Hayashi, N. and Yamashita, S., J. Polym. Sci., Polym. Chem. Ed., 1980, 18, 515; CA 92, 198854x (1980).
57. Stork, G. and Baine, N. H., J. Amer. Chem. Soc., 1982, 104, 2321; CA 96, 162176w (1982).
58. Tchen, T. and Block, K., J. Biol. Chem., 1957, 226, 921; CA 51, 15660i (1957).
59. Tchen, T. T. and Block, K., J. Biol. Chem., 1957, 226, 931; CA 51, 15661b (1957).
60. Miles, M. L., Dissertation Abstr., 1963, 24, 1835; CA 60, 12109a (1964).
61. Anghelina, M., Negoita, M., Corciovei, M. and Soloman, D., Mater. Plast., 1968, 5, 71; CA 69, 28218a (1968).

62. Baucom, K. B. and Butler, G. B., J. Org. Chem., 1972, 37, 1730; CA 77, 34014n (1972).
63. Pinazzi, C., Brosse, J. C. and Legeay, G., C. R. Acad. Sci., Ser. C, 1973, 277, 1327; CA 80, 121393r (1974).
64. Ohara, O., Aso, C. and Kunitake, T., J. Polym. Sci., Polym. Chem. Ed., 1973, 11, 1917; CA 80, 27516m (1974).
65. Ohara, O., Aso, C. and Kunitake, T., Polym. J., 1973, 5, 49; CA 80, 37523a (1974).
66. Takoda, Y. and Yokota, K., Jap. Pat. 73 02,936, 1973 (Jan. 27); CA 79, 126941g (1973).
67. Takada, Y. and Yokota, K., Jap. Pat. 73 02,935, 1973 (Jan. 27); CA 79, 126942h (1973).
68. Kawai, W., J. Polym. Sci. A1, 1966, 4, 1191; CA 64, 19802d (1966).
69. Trossarelli, L., Guaita, M. and Priola, A., J. Polym. Sci. B, 1967, 5, 129; CA 66, 95441u (1967).
70. Trossarelli, L., Guaita, M., Priola, A. and Sainel, G., Chimica Ind. Milan, 1964, 46, 1173; CA 62, 6562h (1965).
71. Trossarelli, L., Guaita, M. and Priola, A., Ricerca scient., 1965, 35, 429; CA 64, 5272f (1966).
72. Trossarelli, L., Guaita, M. and Priola, A., Makromol. Chem., 1967, 100, 147; CA 66, 46659n (1967).
73. Shcherbina, F. F. and Federova, I. P., Ukr. Khim. Zh., 1967, 33, 394; CA 67, 54694f (1967).
74. Shcherbina, F. F. and Fedorova, I. P., Vysokomolek Soedin., Ser. A, 1968, 10, 52; CA 68, 69376y (1968).
75. Trossarelli, L., Guaita, M., Priola, A. and Saini, G., Corsi Semin. Chim., 1968, (8), 115; CA 71, 124935u (1969).
76. Kaye, H., Macromolecules, 1971, 4(2), 147; CA 75, 21170u (1970).
77. Panzik, H. L., Dissertation, Univ. of Ariz., 1972; CA 77, 35001t (1972).
78. Pyriadi, T. M. and Harwood, H. J., ACS Polym. Preprints, Div. Polym. Chem., 1970, 11(1), 60; CA 76, 4188j (1972).
79. Kodaira, T., Ishikawa, M. and Murata, O., J. Polym. Sci., Polym. Chem. Ed., 1976, 14, 1107; CA 85, 33466q (1976).
80. Kodaira, T. and Murata, O., J. Polym. Sci., Polym. Chem., 1979, 17, 319; CA 91, 175780k (1979).
81. Panzik, H. L. and Mulvaney, J. E., J. Polym., Sci., Polym. Chem., 1972, 10, 3469; CA 78, 98050z (1973).
82. Yamada, B., Saya, T., Ohya, T. and Otsu, T., Makromol. Chem., 1982, 183, 963; CA 96, 218297f (1982).
83. Priola, A., Gatti, G. and Cesca, S., Makromol. Chem., 1979, 180, 1; CA 90, 122127a (1979).
84. Blount, D. H., U.S. Pat. US 4,215,303, 1975, (Oct. 15); CA 93, 187037c (1980).
85. Pyriadi, T. M. and Mutar, E. H., J. Polym. Sci., Chem., 1980, 18, 2535; CA 93, 205056q (1980).
86. Mathias, L. J., J. Polym. Sci., Polym. Lett. Ed., 1980, 18, 665; CA 94, 16163g (1981).
87. Fukuda, W., Takahashi, A., Takenaka, Y. and Kakiuchi, H., Polym. J. (Tokyo), 1988, 20, 337; CA 109, 23413v (1988).
88. Fukuda, W., Suzuki, Y. and Kakiuchi, H., J. Polym. Sci., Part C: Polymer Letters, 1988, 26, 305; CA 109, 129760w (1988).
89. Kodaira, T. and Sakaki, S., Makromol. Chem., 1988, 189, 1835; CA 109, 190940w (1988).
90. Behari, K., Raja, G. D. and Gupta, K. C., Polym. Commun., 1989, 30, 372; CA 112, 78024m (1990).
91. Roehm and Haas, Brit. Pat. 963,525; Fr. Pat. 1,325,266, 1964 (July 8); CA 61, 8462d (1964).

92. Moore, L. and Brown, R., U.S. Pat. 3,332,923, 1967 (July 25); CA 68, 49135r (1968).
93. Raetzsch, M. and Stephan, L., Plaste Kaut., 1971, 18, 572; CA 76, 46545d (1972).
94. Guaita, M., Camino, G. and Trossarelli, L., Makromolek. Chem., 1971, 146, 133; CA 75, 118630g (1971).
95. Fukuda, W., Nakao, M., Okumura, K. and Kakiuchi, H., J. Polym. Sci. A1, 1972, 10, 237; CA 76, 127509x (1972).
96. Yokota, K., Ogasawara, S. and Takada, Y., Kobunshi kagaku, 1970, 27(305), 611; CA 75, 6410d (1971).
97. Ichihashi, T. and Kawai, W., Kobunski kagaku, 1971, 28(311), 225; CA 75, 49641x (1971).
98. Gevorkyan, E. Ts., Sayadyan, A. G. and Matsoyan, S. G., Arm. khim. Zh., 1972, 25(2), 145; CA 77, 62367y (1972).
99. Sackmann, G. and Kolb, G., Makromolek. Chem., 1971, 149, 51; CA 76, 60121u (1972).
100. Yokota, K. and Takada, Y., Kobunshi kagaku, 1973, 30, 217; CA 79, 42921n (1973).
101. Yokota, K., Kaneko, N. and Takada, Y., Kobunshi kagaku, 1973, 30, 475; CA 80, 71157u (1974).
102. Yokota, K., Hirayama, N. and Takada, Y., Polym. J., 1975, 7, 629; CA 84, 31597p (1976).
103. Yokota, K., Kakuchi, T. and Takada, Y., Polym. J., 1976, 8, 495; CA 86, 90326p (1977).
104. Yokota, K., Kakuchi, T. and Takada, Y., Polym. J., 1978, 10, 19; CA 88, 137045e (1978).
105. Kakuchi, T., Yokota, K. and Takada, Y., Polym. J., 1979, 11, 7; CA 90, 204597t (1979).
106. Yokota, K., Kaneko, N., Iwata, J., Komuro, K. and Takada, Y., Polym. J., 1979, 11, 929; CA 92, 147289j (1980).
107. Arbuzova, I. A. and Plotkina, S. A., Vysokomol. Soedin., 1964, 6, 662; CA 61, 1946h (1964).
108. Arbuzova, I. A., Fedorova, E. F., Plotkina, S. A., and Minkova, R. M., Vysokomol. Soedin. Ser. A, 1967, 9, 189; CA 66, 86083b (1967).
109. Van Passchen, G., Janssen, R. and Hart, R., Makromol. Chem., 1960, 37, 46; CA 54, 16913f (1960).
110. Roovers, J. and Smets, G., Makromol. Chem., 1963, 60, 89; CA 58, 9237d (1963).
111. Trossarelli, L., Guaita, M. and Priola, A., Annali Chim., 1966, 56, 1065; CA 66, 29189s (1967).
112. Trossarelli, L., Guaita, M. and Priola, A., Ricerca scient., 1966, 36, 993; CA 66, 76328w (1967).
113. Trossarelli, L. and Guaita, M., Polymer, 1968, 9, 233; CA 69, 36480v (1968).
114. Rostovskii, E. N. and Barinova, A. N., Vysokomol. Soedin., 1959, 1, 1707; CA 54, 17952i (1960).
115. Romonov, L. M., Verkhoturova, A. P., Kissin, Yu. V. and Rakova, G. V., Vysokomol. Soedin., 1963, 5, 719; CA 60, 1844f (1964).
116. Overberger, C. G., Ringsdorf, H., and Avchen, B., J. Org. Chem., 1965, 30, 3088; CA 63, 11467h (1965).
117. Noma, K., Yosimiya, R., and Sakirada, I., Kobunshi kagaku, 1965, 22, 166; CA 65, 805h (1966).
118. Kawai, W., Makromol. Chem., 1966, 93, 255; CA 65, 3971b (1966).
119. Goethals, E. J., Bombeke, J., and De Witte, E., Makromol. Chem., 1967, 108, 312; CA 67, 117359g (1967).
120. Goethals, E. J., J. Polym. Sci. B, 1966, 4, 691; CA 65, 18689f (1966).
121. De Witte, E. and Goethals, E. J., Makromol. Chem., 1968, 115, 234; CA 69, 36494c (1968).
122. Kawai, W. and Ichihashi, T., Kogyo-kagaku zasshi, 1968, 70, 2004; CA 68, 115036n (1968).

123. Higgins, J. P. J. and Weale, K. E., J. Polym. Sci. A1, 1968, 6, 3007; CA 70, 4668s (1969).
124. Matsumoto, A., Tano, M., Ishido, H. and Oiwa, M., J. Polym. Sci., Polym. Chem. Ed., 1983, 21, 609; CA 98, 89994w (1983).
125. Fukuda, W., Yamano, Y., Tsuriya, M. and Kakiuchi, H., Polym. J. (Tokyo), 1982, 14, 127; CA 96, 163273n (1982).
126. Matsumoto, A., Ishido, H., Oiwa, M. and Urushido, K., J. Polym. Sci., Polym. Chem. Ed., 1982, 20, 3207; CA 97, 216791j (1982).
127. Fukuda, W., Sato, H., Imai, T. and Kakiuchi, H., Polym. J. (Tokyo), 1986, 18, 631; CA 105, 227394s (1986).
128. Yokota, K. and Takada, Y., Kobunshi kagaku, 1973, 30, 71; CA 79, 32339g (1973).
129. Urushido, K., Takizawa, Y., Matsumoto, A. and Oiwa, M., Makromol. Chem., Rapid Commun., 1983, 4(1), 21; CA 98, 107851d (1983).
130. Urushido, K., Iwasaki, F., Matsumoto, A. and Oiwa, M., Makromol. Chem., 1986, 187(4), 711; CA 105, 6877x (1986).
131. Gindin, L., Medvedev, S. and Fleshler, E., Zhur. Obshchei Khim., 1949, 19, 1694; CA 44, 1020a (1950).
132. Kinoshita, M., Kataoka, S. and Imoto, M., Kogyo -kagaku zasshi, 1969, 72, 969; CA 71, 61811c (1969).
133. Spektor, V. I. and Shur, A. M., Fiz.-Khim. Metody Anal. USSR, 1971, 77; CA 80, 48451b (1974).
134. Raetzsch, M., Plaste Kaut., 1972, 19, 169; CA 77, 35056q (1972).
135. Araki, M., Takeda, T. and Machida, S., Makromol. Chem., 1972, 162, 305; CA 78, 58868r (1972).
136. Stone, G. T., Brit. Pat. 1,336,869, 1973 (Nov. 14); CA 80, 96821u (1974).
137. Butler, G. B. and Baucom, K. B., J. Macromol. Sci. (Chem) A, 1974, 8, 1239; CA 82, 43843n (1975).
138. Baucom, K. B. and Butler, G. B., J. Macromol. Sci. (Chem), A, 1974, 8, 1205; CA 82, 43761j (1975).
139. Sackmann, G. and Funke, W., Makromol. Chem., 1969, 123, 4; CA 71, 3739z (1969).
140. Novichkova, L. M., Ospanova, K. M., zubko, N. V. and Rostovskii, E. E., Akad. Nauk Kaz. SSR, Ser. Khim., 1976, 26, 41; CA 86, 44082h (1977).
141. Dukhnenko, E. M., Komarov, N. V., Komarova, L. I. and Kovaleva, S. A., Izv. Sev. Kavk. Nauchn. Tsentra. Vyssh. Shk., Ser. Estesv. Nauk 1977, 5, 69; CA 88, 105852z (1978).
142. Urushido, K., Matsumoto, A. and Oiwa, M., J. Polym. Sci., Polym. Chem. Ed., 1978, 16, 1081; CA 89, 147278t (1978).
143. Otsu, T. and Horimoto, Y., Mem. Fac. Eng., Osaka City Univ., 1977, 18, 99; CA 89, 215844b (1978).
144. Krats, E. O., Novich'kova, L. M., Bondareva, N. S., Pokrovskii, E. I. and Rostovskii, E. N., Vysokomol. Soedin., Ser. B, 1974, 16, 105; CA 81, 37950h (1974).
145. Novichkova, L. M., Ospanova, K. M., Nazarova, O. V., Dmitrenko, A. V., Zubko, N. V, and Rostovskii, E. N. Izv. Akad. Nauk Kaz. SSR, Ser. Khim., 1978, 28, 36; CA 90, 55573c (1979).
146. Matsumoto, A., Ukon, J., Urushido, K. and Oiwa, M., Technol. Rep. Kansai. Univ., 1979, 20, 79; CA 91, 193717z (1979).
147. Matsumoto, A., Tamura, I, Yamawaki, M. and Oiwa, M., J. Polym. Sci., Polym. Chem. Ed., 1979, 17, 1419; CA 91, 92045y (1979).
148. Shimokawa, Y. and Miyama, H., Kobunshi Ronbunshu, 1979, 36, 415; CA 91, 108504k (1979).
149. Shimokawa, Y. and Miyama, H., Kobunshi Ronbunshu, 1979, 36, 407; CA 91, 108503j (1979).

150. Beckwith, A. L. J. and Wagner, R. D., J. Amer. Chem. Soc., 1979, 101, 7099; CA 92, 76365t (1980).
151. Sikkema, D. J. and Angad-Gaur, H., Makromol. Chem., 1980, 181, 2259; CA 93, 240059q (1980). 152. Urushido, K., Matsumoto, A. and Oiwa, M., J. Polym. Sci., Polym. Chem. Ed., 1980, 18, 1771; CA 93, 132875u (1980).
153. Shu, J. S., Lee, W., Varnell, G. I. and Bartelt, J. I. U.S. Pat. 4,318,976, 1982 (Mar. 9); CA 96, 182930r (1982).
154. Asanov, A., Mutalova, S. N., Tillyabaeva, Z. and Akhmedor, K. S. Dokl. Akad. Naak Uzb. SSR, 1982, 7, 37; CA 98, 126962g (1983).
155. Kawai, W. and Ichihashi, T., Kogyo-kwagaku zasshi, 1970, 73 (11), 2356; CA 74, 54254g (1971).
156. Hasegawa, A., Ger. Offen. DE 3,516,387, 1985 (Nov. 7); CA 104, 177776z (1986).
157. Butler, G. B. and Kimura, S., J. Macromol. Sci. (Chem), A, 1971, 5, 181; CA 73, 120932j (1970).
158. Shaaban, A.F., Arief, M. M. H., Mahmoud, A. A. and Messiha, N. N., Acta. Polym., 1987, 38, 461; CA 107, 237344f (1987).
159. Sasaki, I., Yamamoto, N. and Yanagase, A., Eur. Pat. Appl. EP 326,038, 1989 (Aug. 2); CA 112, 37386n (1990).
160. Bernstein, B. B., Orban G. and Odian, G. G., Polym. Prepr., 1964, 5, 887; CA 64, 12810d (1966).
161. Chiang, T., Pao, H., Sun, K. and Feng, H., Ko Fen Tzu T'ung Hsun, 1965, 7, 56; CA 63, 18270c (1965).
162. Kamogawa, H., Furuya, S. and Kato, M., J. Polym. Sci., C, 1966, 23, 655; CA 71, 13423f (1969).
163. Bywater, S., Block, P. and Wiles, D., Can. J. Chem., 1966, 44, 695; CA 64, 12807a (1966).
164. D'Alelio, G. F., U.S. Pat. US 3,364,282, 1968 (Jan. 16); CA 68, 50591t (1968).
165. Johnston, N. and Joesten, B., J. Polym. Sci. A1, 1972, 10, 1271; CA 77, 20176e (1972).
166. Oshiro, Y., Shirota, Y. and Mikawa, H., Polym. J., 1972, 3, 217; CA 77, 35117k (1972).
167. Kaye, H. and Chang, S.-H., J. Macromol. Sci. (Chem), 1973, 7, 1127; CA 79, 79311q (1973).
168. De Witte, E. and Goethals, E. J., J. Macromolec. Sci. (Chem.), 1971, 5(1), 1925; CA 73, 131381e (1970).
169. Goethals, E. J. and De Witte, E., J. Macromolec. Sci. (Chem.), 1971, 5, 1915; CA 73, 131382f (1970).
170. Yokota, K., Kakuchi, T., Taniguchi, Y. and Takada, Y., Makromol. Chem., Rapid Commun., 1985, 6, 155; CA 103, 142411d (1985).
171. Yokota, K., Kakuchi, T., Iiyarna, T. and Takada, Y., Polym. J. (Tokyo), 1984, 16, 145; CA 101, 38883g (1984).
172. Galeev, V. S., Skorobogatova, M. S. and Levin, Y. A., Mater. Nauch Konf., Inst. Org. Fiz., 1969 (Pub. 1970), 198; CA 78, 30612d (1973).
173. Sopar, B., Howard, R. and White, E., Polym. Preprints, 1972, 13, 422; CA 80, 83687e (1974).
174. Matsumoto, A., Inoue, I. and Oiwa, M., J. Polym. Sci., Polym. Chem. Ed., 1976, 14, 2383; CA 85, 193167w (1976).
175. Pinazzi, C. P. and Fernandez, A., ACS Symp. Ser., 1976, 25, 37; CA 85, 124407n (1976).
176. Fedorova, I. P., Shcherbina, F. F. and Paskal, L. P., Ukr. Khim. Zh. (Russ. Ed), 1976, 42, 1272; CA 86, 90338u (1977).
177. Tyuleneva, I. I., Boldyrev, A. G., Shefer, I. A., Rostovskii, E. N. and Kuvshinskii, E. v., Vysokomol. Soedin., Ser. B, 1977, 19, 789; CA 88, 23494v (1978).

178. Ito, N., Kobunshi Ronbunshu, 1977, 34, 625; CA 87, 202485u (1977).
179. Tanaka, A. and Yosomiya, R., Chiba Kog. Daig. Kenk. Hoko. Riko Hen, 1978, 23, 185; CA 90, 87927n (1979).
180. Lien, Q. S., Diss. Abstr. Int. B, 1978, 38, 5415; CA 89, 6614n (1978).
181. Yosomiya, R. and Tanaka, A., Chiba Kogyo Daigaku Kenk. Hok., Riko Hen 1978, 23, 185; CA 90, 87927n (1979).
182. Eginbaev, Zh. E., Ayapbergenov, K. A. and Muldakhmetov, Z. M., Deposited Doc., VINITI, 1977, 4371; CA 91, 124043q (1979).
183. Nishimura, J. and Yamashita, S., Polym. J., 1979, 11, 619; CA 91, 193705u (1979).
184. Dusek, K. and Spevacek, J., Polymer, 1980, 21, 750; CA 93, 240040b (1980).
185. Redpath, A. E. C. and Winnik, M. A., J. Am. Chem. Soc., 1980, 102, 6869; CA 93, 221277t (1980).
186. Schnarr, E. and Russell, K. E., J. Polym. Sci., Polym. Chem. Ed., 1980, 18, 913; CA 93, 8593m (1980).
187. Nishimura, J., Tanaka, N. and Yamashita, S., J. Polym. Sci.: Polym. Chem. Ed., 1980, 18, 1203; CA 93, 26866g (1980).
188. Yokota, K., Kakuchi, T. and Takada, Y., Hokka. Daig. Koga. Kenkyu Hokoku, 1980, 45; CA 94, 175657n (1981).
189. Stepanov, V. V., Barantseva, A. R., Skorokhodov, S. S. Lavrov, V. I. and Trofimov, B. A., Vysokomol. Soedin., Ser. B., 1984, 26, 741; CA 102, 25132g (1985).
190. Landin, D. T. and Macosko, C. W., Macromolecules, 1988, 21, 846; CA 108, 113009b (1988).
191. Shindo, Y., Sato,H., Sugimura, T., Horie, K. and Mita, I., Eur. Polym. J., 1989, 25, 1033; CA 112, 56921z (1990).
192. Kamogawa, H., J. Polym. Sci., A1, 1969, 7, 725; CA 71, 13433j (1969).
193. Berlin, A. A., Ivanov, V. V. and Enikolopyan, N. S., Vysokomol. Soedin.,Ser. B, 1967, 9, 61; CA 66, 95454a (1967).
194. Razumov, A. I.,Krivosheeva, I. A. and Chemodanova, L. A., Tr. Kazan Khim.-Tekhnol. Inst., 1967, 36, 468; CA 69, 107106r (1968).
195. Rasumov, A. I. and Krivosheeva, I. A., Tr. Kazan. Khim.-Tekhnol. Inst., 1967, 36, 475; CA 69, 107107s (1968).
196. Kato, M. and Kamogawa, H., J. Polym. Sci., A1, 1968, 6, 2993; CA 69, 107140x (1968).
197. Petropoulos, J. C. and Di Leone, R. R., U.S. Pat. 3,444,184, 1969 (May 13); CA 71 30131b (1969).
198. Krivosheeva, I., Borisova, I. A. and Razumov, A. I., Vysokomol. Soedin., Ser. A, 1969, 11, 259; CA 70, 97299k (1969).
199. Nowlin, G., Gannon, J. and Jungster, L. J. Appl. Polym. Sci., 1969, 13, 463; CA 71, 308742w (1969).
200. Ouchi, T., Imase, Y. and Oiwa, M., Bull. Chem. Soc. Jap., 1970, 43, 2863; CA 73, 110359p (1970).
201. Kodama, T. and Ide, F., Kobunshi kagaku, 1970, 27, 65; CA 72, 133499c (1970).
202. Matsumoto, A., Maeda, T. and Oiwa, M., Nippon Kagaku Zasshi, 1970, 91, 89; CA 72, 111887a (1970).
203. Matsumoto, A. and Oiwa, M., J. Polym. Sci., A1, 1970, 8, 751; CA 72, 111898e (1970).
204. Shaikhutdinov, E. M., Zhubanov, B. A., Kurmanaliev, O Sh. and Makarskaya, V. I., Izv. Akad. Nank. Kaz. SSSR, Ser. Khim, 1971, 21, 53; CA 75, 118644q (1971).
205. Ouchi, T., Yokai, K., Yoshimura, K. and Oiwa, M., Bull. Chem. Soc. Jap., 1971, 44, 1339; CA 75, 77308n (1971).
206. Kennedy, J. P. and Markowski, H. S., J. Polym. Sci. C, 1968, 22, 247; CA 69, 52521q (1968).
207. Bartlett, P. D. and Bank, S., J. Amer. Chem. Soc., 1961, 83, 2591; CA 55, 2459g (1961).

208. Malac, J., J. Polym. Sci., A1, 1971, 9, 3563; CA 76, 100579h (1972).
209. Beckwith, A. L. J. and Moad, G. J., J. Chem. Soc., Chem. Commun., 1974, 472; CA 82, 42688d (1975).
210. Beckwith, A. L. J., Blair, I. and Phillipou, G., J. Amer. Chem. Soc., 1974, 96, 1613; CA 80, 119868f (1974).
211. Beckwith, A. L. J. and Moad, G., J. Chem. Soc. Perkin 11, 1975, 1726; CA 84, 89182r (1976).
212. Beckwith, A. L. J. and Gara, W. B,, J. Chem. Soc., Perkin Trans., 1975, 2, 795: CA 83, 78106x, (1975).
213. Booth, H., Bostock, A. H., Franklin, N. C., Griffiths, D. V. and Little, J. H., J. Chem. Soc. Perkin, 1978, 11, 899; CA 90, 54120r (1979).
214. Pomerantseva, E. G., Perepletchikova, E. M. and Krats, E. O., Plast. Massy (USSR), 1979, 4, 39; CA 90, 205043w (1979).
215. Beckwith, A. L. J., Lawrence, T. and Serelis, A. K., J. Chem. Soc., Chem. Commun., 1980, 484; CA 93, 203922v (1980).
216. Beckwith, A. L. J. and Wagner, R. D., J. Chem. Soc., Chem. Commun., 1980, 485; CA 93, 204508b (1980).
217. Beckwith, A. L. J., Phillipou, G. and Serelis, A. K., Tetrahedron Lettr., 1981, 22, 2811; CA 96, 5880d (1982).
218. Beckwith, A. L. J. and Schiesser, C. H., Tetrahedron, 1985, 41, 3925; CA 105, 96710s (1986).
219. Beckwith, A. L. J. and Pigou, P. E., J. Chem. Soc., Chem. Commun., 1986, 85; CA 105, 42574g (1986).

4

Cyclopolymerization Leading to Polycyclic Systems

It has already been shown that polymers containing bicyclic and tricyclic units can be obtained by the polymerization of cis-1,3-divinylcyclohexane and *cis-1,3*-divinylcyclopentane,[2-12] o-divinylbenzene,[2-20 thru 2-24] 1,8-divinylnaphthalene[2-13] and other similar monomers. Several other types of monomers have been found to undergo cyclopolymerization with the formation of bicyclic or tricyclic units. Such cyclizations have been referred to as either a succession of ring closures or transannular polymerization.

SUCCESSION OF RING CLOSURES

The formation of bicyclic and tricyclic structures by the polymerization of acyclic, branched trienes, and tetraenes, the bicyclic and tricyclic structures having been formed by a succession of ring closures within a single monomer molecule before propagation to a neighboring molecule occurs was reported as early as 1961.[1-82] For example, triallylethyl- and tetraallylammonium bromides were converted to soluble polymers containing small amounts of residual unsaturation. A typical polymer of triallylethylammonium bromide (4–1) had an intrinsic viscosity of 0.14 dl/g and a molecular weight in

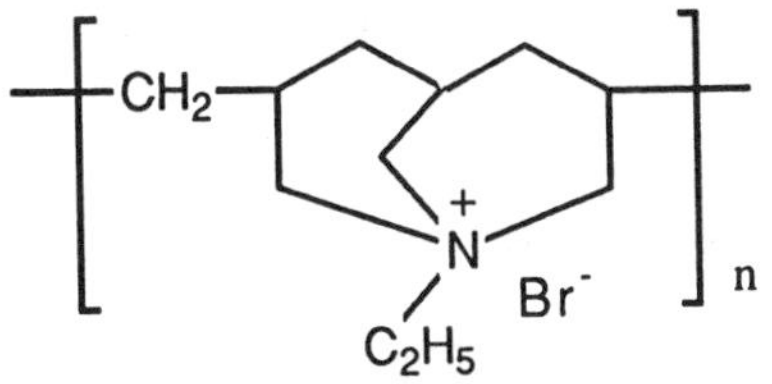

(4–1)

excess of 5000. A triolefinic monomer, 3-vinyl-1,5-hexadiene, was converted to a soluble polymer having little residual unsaturation. These results indicate an extensive linear chain structure containing 2,6-methylene-linked bicyclo[2.2.1]heptyl rings. Triallylmethylsilane was also polymerized to soluble, solid polymers showing little residual unsaturation, properties which are consistent with a large bicyclic [3.3.1] ring content. The radical polymerization of a number of branched triene compounds has also been studied.[1] Trivinyl orthoformate, trivinyl phosphate, triallylamine hydrobromide, triallylacetonitrile, triallylacetamide, and triallylcarbinol and its acetate, were polymerized in bulk and in solution by free radical initiation. It was established that polymerization of the first three compounds mentioned led to double intramolecular cyclization of the monomers with the formation of polymers containing bicyclic units. Polymerization of the other monomers gave both monocyclic and bicyclic chain growth in a ratio of 2:1 (4–2) where R = OH, CH_3 CO_2, CN or $CONH_2$). Soluble polymers were also

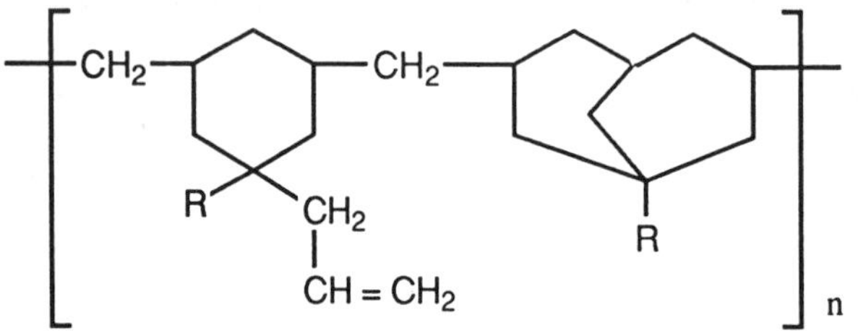

(4–2)

obtained[2] from triallylsilanes by γ-irradiation of the monomers in solution. The low residual unsaturation of the polymers was again interpreted in terms of the formation of monocyclic and bicyclic structural units.

Thus, it appears that monomers, functionally capable of polymerizing to yield cyclopolymers containing bicyclic or tricyclic units resulting from a double or triple intramolecular-intermolecular chain propagation, show a strong tendency to do so.

Trimethacrylamide[3] was synthesized and its cyclopolymerization characteristics studied.[4] Bulk polymerization by a free radical mechanism gave cyclic polytrimethacrylamide containing monocyclic units but no bicyclic structures. Hydrolysis of the polymer gave polydimethacrylamide, identical with the structure obtained by thermal cyclization of polymethacrylamide.

The earlier reports of the formation of cyclopolymers containing polycyclic structures by a succession of ring closures within a monomer[1-156] have been questioned.[1-20] It has been shown[2-177, 3-211] that the structure which was assigned to cyclopolymers of triallyl monomers may be incorrect, and structures resulting from 1,5–1,6 and 1,5–1,7 cyclizations, respectively, are more likely (4–3). Further, a much smaller rate constant was observed for the second than for the first cyclization, hence bicyclic units will only be favored at very low monomer concentrations.

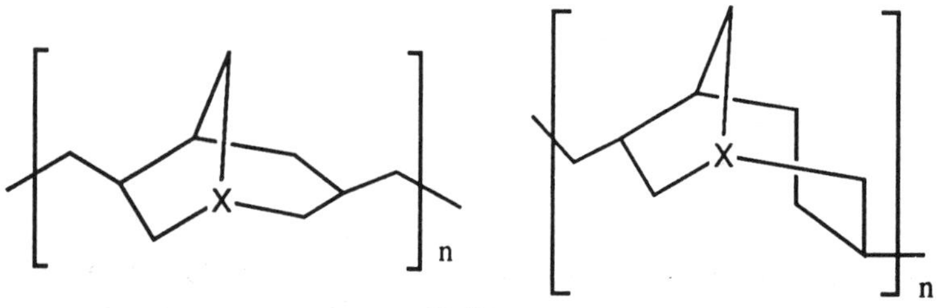

(4–3)

At higher concentrations, cross-linking will occur or residual unsaturation will remain. It was suggested[1-20] that earlier workers reported high extents of cyclization because of their inability to measure residual unsaturation accurately.

1,2,3-Triisocyanatopropane has been polymerized to give a linear polymer by a succession of ring closures for which a bicyclic recurring unit shown in Structure 4–4 was proposed.[1-45] *cis*–1,3-Diisocyanatocyclohexane has been cyclopolymerized under similar conditions. Evidence was presented for the bicyclic structure in the polymer.[5] Details have been discussed in an earlier review.[1-183]

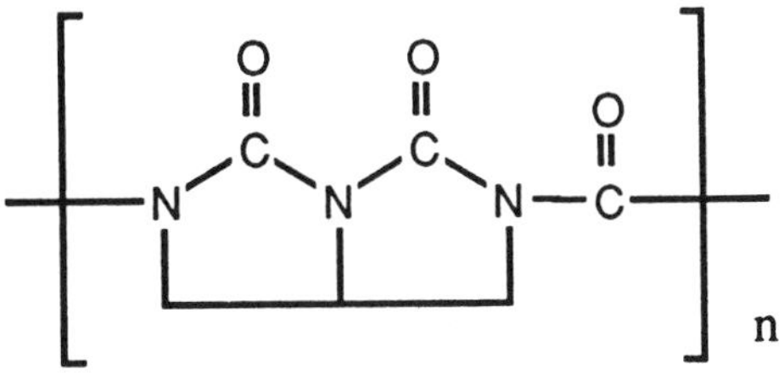

(4–4)

Copolymers of bicyclo[2.2.1]hepta-2,5-diene and ethylenically unsaturated monomers were prepared *via* tri-n-butylboron and oxygen.[6] The copolymers were shown to be made up of the nortricyclene group (4–5).

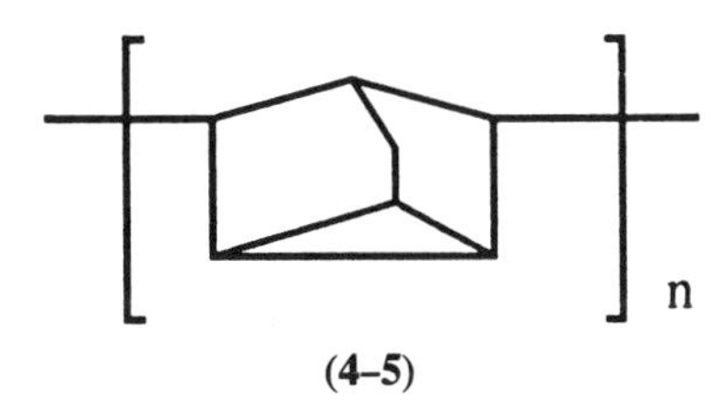

(4–5)

Bicyclic polymers were obtained by polymerization of 4-vinylcyclohexene in a solution containing acetyl chloride and aluminum chloride.[7] Spiropentane, an isomer of isoprene was shown to undergo cationic initiated ring opening, ring closure polymerization to give compounds of the cyclo-1,4-polyisoprene type, whereas in the presence of Ziegler-Natta catalysts higher molecular weight polymers of the cyclo-1,4-polyisoprene type shown (4–6) were obtained.[8]

The presence of internal 1,2-bis(long alkyl)cyclopentane units were shown to be present in tributyltin hydride reduced poly(vinyl chloride) and 1,3-butadiene-vinyl chloride copolymer.[9] These reactive polymers were postulated to have undergone a cyclopolymerization reaction which introduced cyclohexyl branches by a succession of ring closures. Terminally unsaturated polyalkenyl methanes, e.g. tetraallylmethane, were described along with their polymers and copolymers which were obtained by radical initiation.[10] No properties of the polymers were given. Tetraallylsilane and tetramethallylsilane were used as components of a catalyst system to polymerize ethylene or propylene.[11] A solid solution of titanium trichloride in aluminum chloride was mixed with tetraallylsilane in n-heptane at 80°C under an inert atmosphere; ethylene or propylene was then introduced at 20 atmospheres pressure, and the polymer isolated in the usual manner. The polyethylene had intrinsic viscosity of 4.1 dl/g.

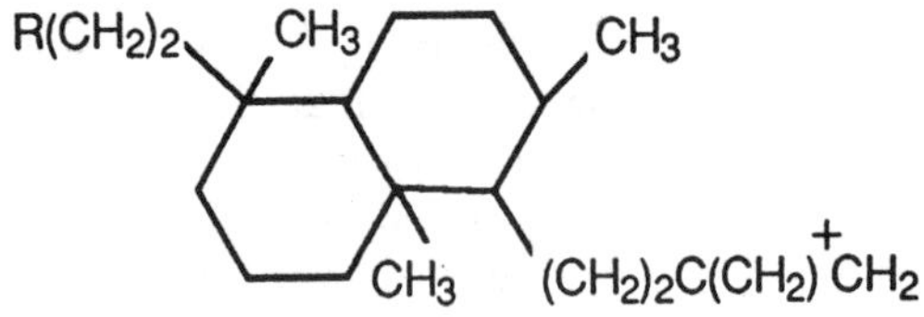

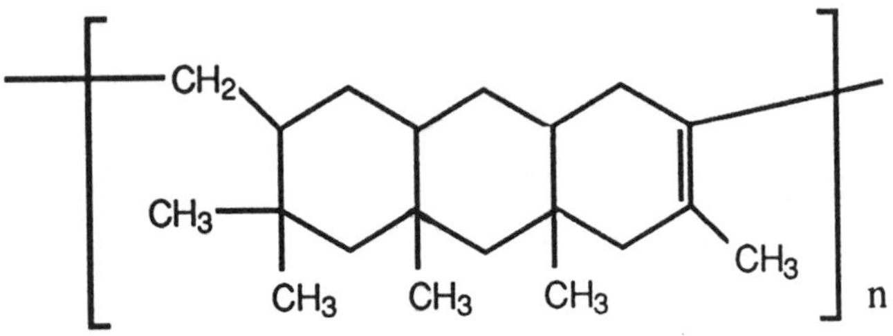

(4–6)

Cyclopolymerization of dimethylbis(vinylthio)silane and methyltri(vinylthio)silane occurred in radical initiated polymerization of the monomers.[12] The rate was independent of the number of vinylthio groups in the molecule.

The radically-initiated polymerization rates of the diallyl esters of fumaric, maleic, mesaconic and citraconic acids were determined.[13] The rate comparison led to the postulate that the monomers lactonized prior to polymerization. However, cyclopolymerization would also account for the results observed.

Diallyl fumarate and diallyl maleate were polymerized in presence of free radical initiators in a dilatometer.[14] Kinetic data were obtained to show that the fumarate was more reactive than the maleate. The molecular weights and intrinsic viscosities of the polymers were determined.

Copolymers of methyl methacrylate with methyl and ethyl α-hydroxymethylacrylate and with α-hydroxymethylstyrene were shown to eliminate alcohol by heating with the formation of six-membered lactone groups.[15] Attempts were made to incorporate similar lactone structures by cyclopolymerization of α-methacryloxymethylstyrene or ethyl α-methacryloxymethyl acrylate, but only crosslinked polymers or polymers with pendent unsaturation were formed.

It has been shown that the rate of cyclopolymerization (R_p) of triallyl citrate in benzene *via* radical initiation exhibited a linear relation between $R_p/[I]^{0.5}$ and $[I]^{0.5}$; [I] = initiator concentration.[16] A linear relation was also observed between monomer concentration and R_p/d.p., (d.p. = degree of polymerization). The ratio of the unimolecular cyclization reaction to the bimolecular propagation reaction of the uncyclized radical was estimated to be 4.5 mol/L.

A number of studies have been carried out on homopolymerization and copolymerization of allyl cyanurates. Copolymerization of triallyl cyanurate, triallyl isocyanurate, and diallylmelamine with methyl methacrylate, styrene and vinyl acetate have been studied.[17] Optimum conditions for incorporating the maximum amount of the triazine into the copolymers were established, and reactivity ratios of the several systems were reported. The extent of monomer cyclization in these systems was not reported.

Radiation-induced liquid- and solid-state polymerization of triallyl cyanurate was studied, and the two processes compared.[18] The rates were linear in each case but an acceleration was observed in the solid-state at temperatures near the melting point. The solid-state polymerization could be retarded by use of radical inhibitors, which was regarded as evidence for a radical mechanism. In both polymerizations, the polymer became insoluble after a definite conversion, but this point was characteristically different for the two systems and depended on the degree of conversion.

It was shown that the length and degree of branching of the alkyl radical influenced the reactivity of alkyl methacrylates in copolymerization with triallyl cyanurate; the reactivity ratios of ethyl methacrylate and n-butyl methacrylate were similar to but smaller than those of methyl methacrylate.[19] The reactivity ratios for the cyanurate were almost always the same and oscillated near zero.

The rate of benzoyl peroxide initiated polymerization of triallyl cyanurate was accelerated by the presence of $ZnCl_2$ with which the ester formed a complex; an increase in the $ZnCl_2$ ester ratio from 0 to 0.2 increased the polymerization rate from 5.33×10^{-5} to 7.69×10^{-5} mol/L.[20] The presence of $ZnCl_2$ in the reaction mixture did not affect the thermal stability of the resulting homopolymer. Incorporation of diallylamine units into the homopolymer increased the thermal stability of the resulting copolymer.

Triallylcyanurate or triallylisocyanurate based adhesive sealant systems have been described.[21] Single component, anaerobic adhesives showing high strength and water and heat resistances were based on copolymers of triallyl cyanurate or isocyanurate.

The kinetics of radical polymerization of the zinc chloride triallyl cyanurate complex have been studied.[22] Kinetic data for the radical polymerization of triallyl cyanurate in the presence and absence of $ZnCl_2$ showed that the polymerization rate was higher when the ester was complexed with $ZnCl_2$ than when $ZnCl_2$ was absent. A linear prepolymer which was soluble in organic solvents, and having 45% residual unsaturation was obtained *via* radical initiation of polymerization of triallyl acetyl citrate.[23] This prepolymer could be further polymerized and ultimately crosslinked to give thermosetting resins.

Radical polymerization of the zinc chloride-diallylcyanamide complex has been studied.[2-74] The presence of $ZnCl_2$ complexing agent led to increased polymerization rate and polymer yield. The polymerization occurred mainly with formation of cyanopyridine rings in the main chain.

Triallyl borate was homopolymerized and copolymerized with styrene in presence of acids or radical initiators.[24] The polymers were described as colorless and glasslike.

The polymerization of furfural divinyl acetal was investigated *via* free radical initiation.[25] Very low yields of polymer ($<$9%) were obtained.

The photopolymerization of triallylidene sorbitol in benzene initiated by the ester radical formed from the acetal radical by photolysis has been studied.[26] The rate of polymerization and the molecular weight of the polymer were small due to degradative chain transfer. The polymerization kinetics were studied in relation to degradative chain transfer by the allylidene group and the cyclization by three double bonds. The ratio of the rate constant of unimolecular cyclization to the total rate constant of bimolecular propagation and the chain transfer of uncyclized radical was 3.0 mol/dm^3. A small amount of cyclopolymerization took place.

Polytriallylamine is reported to undergo a succession of ring closures.[27] It has been reported to be superior to other weakly basic ion exchangers in the "Sirotherm" demineralization process. However, the earlier structures which had been reported were questioned in this work.

Tetraallylammonium chloride, often used as a crosslinking agent in other systems, was studied as a monomer for radical-initiated polymerization.[28] The rate of polymerization was only about one third that for diallyldimethylammonium chloride, and gelled only after about 20% conversion. The cyclopolymerizability was estimated kinetically as the ability of the initially formed five-membered monocyclic radical to form a bicyclic structure by a succession of ring closures giving a cyclization constant of 21 mol/L at [M] = 2 mol/L.

Cyclopolymerization studies of acetylenic compounds such as p-diethynylbenzene and 4,4′-diethynyldiphenylmethane have been carried out in the presence of catalytic systems [such as bromotetrakis(triethyl phosphite) cobalt].[29] Poly(p-diethynylbenzene) had trisubstituted C_6 H_6 rings. Polymerization in the presence of the catalyst containing electron-donor additives gave partially soluble polyphenylenes. The electron-donor additive accelerated the cyclopolymerization.

The possibility of observing cyclization reactions on addition of free radicals to squalene was investigated; however, it was concluded that no cyclization occurred.[30] Cationic initiated ring closures have been observed in squalene.

The polymerization of oligourethane acrylates has been studied. Gelation of polypropylene glycol-ethylene glycol monomethacrylate-2,4-tolylene diisocyanate copolymer occurred quite early during the copolymerization at low diisocyanate conversion.[31] This early gelling was reported to be due to the cyclization of the oligomeric chains and a decrease in their sequential mobilities. This configuration gives a higher probability of intramolecular cyclization than of intermolecular crosslinking.

A new crystalline polymer, poly(metacyclophane) (4–7) has been reported by polymerizing the dimer, obtained from spontaneous dimerization of 8,16-dihydroxy[2.2]metacyclophane in presence of oxygen and ultraviolet light, at 200–400°.[32]

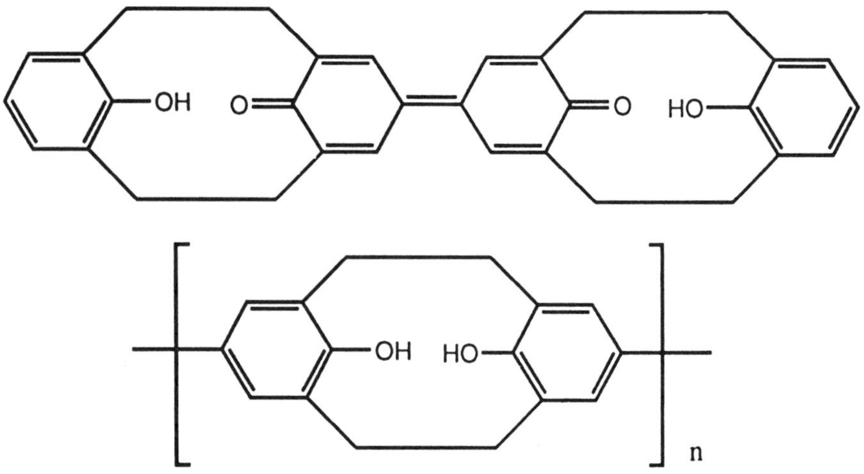

OH O O HO

OH HO n

(4–7)

TRANSANNULAR POLYMERIZATION

Cyclopolymerization involving a cyclic diene, the double bonds of which become involved in new cyclic structures or rings of different sizes has been referred to as transannular polymerization. Many novel polycyclic polymers have been synthesized by this mechanism. This section describes some of these monomers and their polymerizations.

It was found that 2-carbethoxy-bicyclo[2.2.1]-2,5-heptadiene (4–8) polymerized

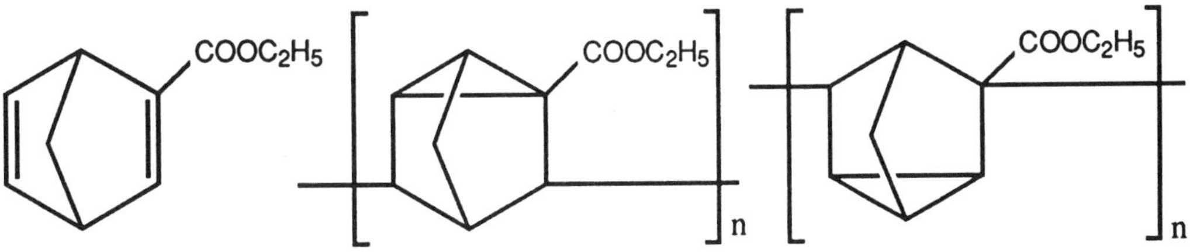

$COOC_2H_5$ $COOC_2H_5$ $COOC_2H_5$

n n

(4–8)

readily with free radical catalysts to give high molecular weight, soluble, essentially saturated polymers.[1-83] The infrared data on the polymers and the known chemistry of bicycloheptadienes suggested that the polymers consisted primarily of nortricyclene repeating units. In contrast, 2-carbethoxybicyclo[2.2.1]-2-heptene and 2-carbethoxybicyclo[2.2.1]-5-heptene did not undergo vinyl polymerization with free radical catalysts under the conditions used. Poly(bicyclo[2.2.1]-2,5-heptadiene) obtained by γ-radiation induced polymerization of the corresponding monomer was shown, by analysis, to contain the nortricyclene structure (4–5) as the principle unit in the polymer.[33] The cationic polymerization of this monomer in the temperature range of −123° to 40°C has been investigated[34]. The material obtained at the lowest temperature level was completely soluble, and analysis indicated a regular nortricyclene structure (4–5). Polymerization occurred readily even below −100°C. The high reactivity of this monomer was attributed to the highly strained ring structure. 2-Methylene-5-norbornene has been polymerized under a variety of cationic conditions[3-18] yielding soluble saturated polymers containing units of Structure 3–7. The relative reactivities and structural aspects in the cationic polymerization of mono- and diolefinic norbornanes were rationalized on a reasonable basis.[3-206]

It has been observed[1-84] that *cis,cis*-1,5-cyclooctadiene, in the presence of a Ziegler catalyst formed from triisobutylaluminum and titanium tetrachloride in a molar ratio of 0.5: 1, yielded a saturated, soluble polymer through a transannular polymerization. Spectral and other physical properties suggested it to be poly(2,6-bicyclo[3.3.0]octane). 1-Methyl-*cis,cis*-1,5-cyclooctadiene by use of the same catalyst also yielded a saturated, soluble polymer.[35] The structural units shown in Structure 4–9 were suggested for this polymer.

Polymers containing bicyclic units have also been obtained by the transannular polymerization of 1,4-dimethylenecyclohexane,[36] 1-methylene-4-vinylcyclohexane,[37] (4–10) 4-vinylcyclohexene,[38] and 1-allyl-2-methylenecyclohexane.[39] The polymerization of 1-methylene-4-vinylcyclohexane with a cationic initiator yielded a soluble polymer containing residual vinyl unsaturation. Nuclear magnetic resonance analysis indicated an *Sp2:Sp3* hydrogen ratio of 1:7.5, a ratio consistent with a copolymer structure in

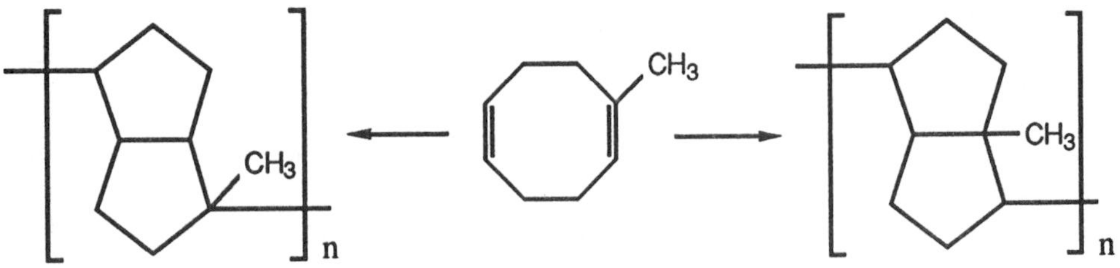

(4–9)

(4–10)

which about 40% of the monomer units have entered the polymer chain *via* the cyclopolymerization mechanism and the remaining 60% of the monomer units have entered the chain through the methylene group only, leaving the vinyl groups pendant. Ziegler catalyzed polymerization resulted in 88% cyclization. The remaining units were considered to result from propagation through the vinyl groups alone since Ziegler catalysts do not ordinarily initiate monomers of the isobutylene type. Similar determinations were carried out on the polymers of 4-vinylcyclohexene[38] and 1-allyl-2-methylenecyclohexane.[39]

Polymers with bicyclic units formed by transannular polymerizations have been produced from a number of other monomers. 4-Vinylcyclohexene[3-8, 3-9] has been polymerized using various initiators and the three different structural units quantified by infrared and NMR spectroscopy (See Structure 3–4).

The cyclopolymerization studies of 3-allylcyclopentene, 4-allylcyclopentene, 3-allylcyclohexene and 4-allylcyclohexene[3-22] have been carried out. It has also been shown that cationic and Ziegler-Natta catalyzed polymerizations of 1,2,6-cyclononatriene gave high yields of low molecular weight polymers containing 2,3,4,4a,5,7a-hexahydro-1H-indene-1,4-diyl groups produced by a transannular mechanism involving opening the carbon-carbon double bond and the allenic groups.[40]

Other hydrocarbon monomers which have been shown to lead to varying degrees of bicyclic structure in their polymers are 3-allylcyclopentene and 3-vinylcyclopentene.[1-183] A more recent study has shown that polybicyclic structures are obtained by polymerization of 4-vinylcyclohexene at $-15°$ to $-10°$ in a solution containing acetyl chloride and $AlCl_3$ with a CH_3COCl-monomer ratio = 1 and an $AlCl_3$-monomer ratio = (1.2–1.3):1.[7]

Another interesting cyclocopolymerization which involves transannular copolymerization has been reported.[41] Sulfur dioxide was copolymerized with *cis,cis*-1,5-cyclooctadiene using radical initiation in dilute solution. The copolymer shown in Structure 4–11 was obtained in film- and fiber-forming molecular weights, had no residual unsaturation, and contained two moles of sulfur dioxide per mole of cyclooctadiene. The copolymer decomposed at temperatures above 250° with gas evolution. A copolymer prepared at 10% weight concentration of monomers (SO_2:cyclooctadiene = 2.5:1) in tetramethylene sulfone in air at 25°C for 16 h, using methyl ethyl ketone peroxide as

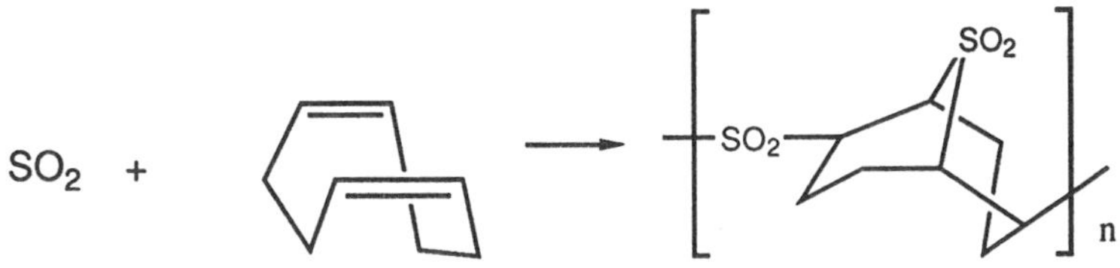

(4–11)

initiator was obtained in 98% yield and had an inherent viscosity of 2.2 at 0.5 g per 100 ml of dimethyl sulfoxide solution.

The thermal stability of this bicyclic copolymer has been studied.[42,1-91] Degradation followed first-order kinetics to a first approximation at 220°C or above. Random cleavage of the copolymer chain occurred during the pyrolysis. Pyrolysis of the copolymer up to 75% weight loss yielded the monomers as cracking products in the molar ratio of the copolymer composition. The activation energy E_a for the degradation process was found to be 41 kcal/mole with a frequency factor of $5.2 \times 10^{13}\ S^{-1}$. In comparison, E_a for the propylene-SO_2 copolymer is 32 kcal/mole. This copolymer decomposes approximately 100 times faster at 229° than the bicyclic copolymer does at 230°. It was concluded that the bicyclic sulfone structure is inherently more stable than the linear α,β-sulfone structure of the olefin-SO_2 copolymers.

Oxidative cyclization of 5,10-dimethyl-*cis-trans*-1,5,9-undecatriene was accomplished by use of mercuric acetate.[43] A mixture of products containing hydrindanes and cyclohexanols was obtained. A similar study was carried out on another similar monomer. Oxidative cyclization of 4-methyl-*trans*-1,5,9-decatriene was accomplished in presence of mercury salts to give mixtures containing perhydronaphthalenes and *cis*- and *trans*-butyl-4-methylcyclohexanone.[44] Homopolymers and copolymers of 1,5,9-cyclododecatriene were obtained in presence of organo-aluminum compound catalysts, *via* a transannular mechanism.[45] *Cis, trans, trans*-1,5,9-Cyclododecatriene was treated with ethylaluminum dichloride, in presence of trichloroacetic acid, to give bicyclo[6.4.0]dodec-4-ene, *via* a transannular ring closure reaction.[46] Another transannular cyclopolymerization study of 1,5,9-cyclododecatriene in presence of triethylaluminum titanium tetrachloride gave polymer consisting of the repeating units shown in Structure 4–12:[47]

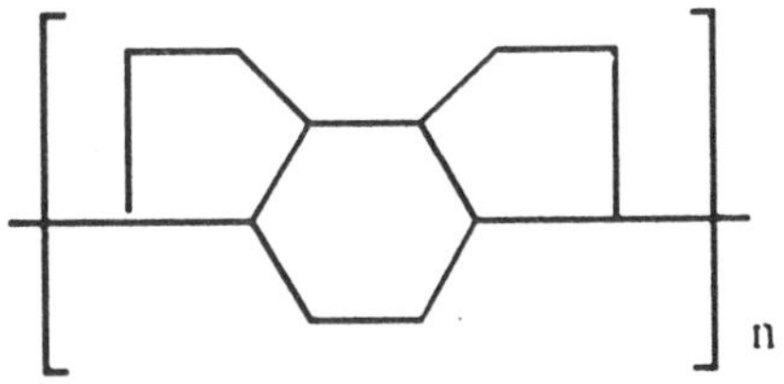

(4–12)

1,5,9-Cyclododecatriene was converted to 1,3,5-trivinylcyclohexane at high temperatures, a process structurally similar to cyclopolymerization but mechanistically quite different.[48] This mechanism most probably involves a series of Cope rearrangements. A copolymerization study of *trans, trans, cis*-1,5,9-cyclododecatriene with methyl methacrylate *via* radical initiation led to reactivity ratios of 4.84 for the triene of 4.84 for the triene and 0.20 for the ester.[49] No further structural information on the copolymer was given. Copolymerization of *trans, trans, cis*-1,5,9-cyclododecatriene with styrene *via* radical initiation was studied.[50] Reactivity ratios found were 2.54 for styrene and 0.22 for the triene. No further information on the polymer structure was given.

POLYMERIZATION OF 1,4-DIENES

The examples previously discussed of polymerizations involving monomers which undergo cyclopolymerization have involved ring closures in such a manner that the cyclic structures were derived from a single molecule of a monomer. The synthesis of N,N-divinylaniline was reported and its polymerization using a free radical initiator was studied.[1-89] It was found that the polymeric unit formed as the result of cyclization of two N,N-divinylaniline molecules and, depending on the method of polymerization, the reaction led to the formation of monocyclic or bicyclic units shown in Structure 4–13.

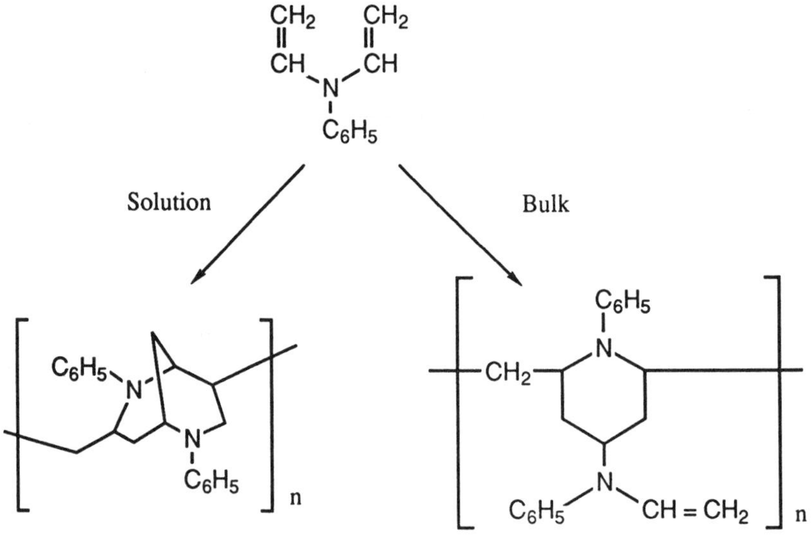

Solution Bulk

(4–13)

An infrared study of the CH stretching frequencies of the polymers in the 2900 cm^{-1} region ruled out the presence of the expected four-membered ring structure. The average of the asymmetric and symmetric CH_2 frequencies of the polymers lay in the range 2892–2916 cm^{-1} corresponding to a strainless straight chain or strainless ring structure. An average value in the range 2936–2969 cm^{-1} would be expected for a four-membered ring structure. A soluble polymer containing about 20% residual unsaturation was obtained[51] from the radical homopolymerization of divinyl ether. Assuming that radical initiation took place at the β-methylene group of the monomer, the structures shown (4–14) were postulated as the predominant units. About 1% succinnic acid was obtained by oxidation of the polymer with nitric acid. This indicated the presence of head-to-head bonds in the polymer. In addition, a bicyclic telomer composed of two moles of monomer and one mole of chloroform was obtained by the free radical telomerization of divinyl ether in chloroform. The infrared and nuclear magnetic resonance

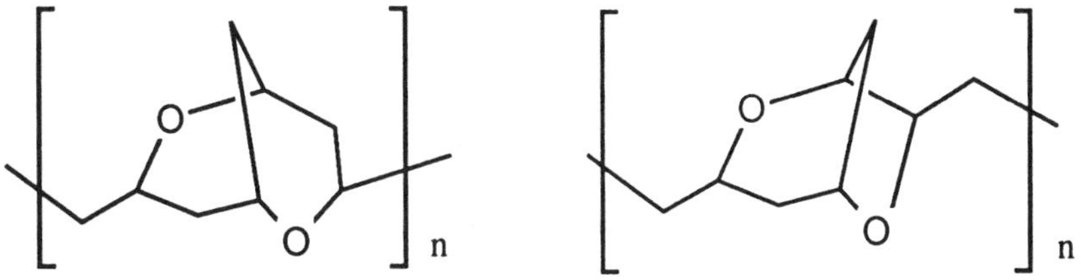

(4–14)

spectra indicated the presence of a methyl group in the telomer. From these results the bicyclic ring structure in the polymer was considered to have at least one five-membered ring.

The radical-initiated polymerization of substituted divinyl ketones led to the formation of solid, soluble polymers containing cyclopentane, or cyclohexane rings, depending on the structure of the monomers.[52–54] The polymer unit was formed by the cyclization of two monomer molecules. The polymerization of α-methylvinyl β-β-dimethylvinylketone and β-ethylvinyl vinyl ketone, for example, proceeded according to the proposed bimolecular cyclocopolymerization mechanism. A homopolymer of dibenzalacetone of molecular weight of 2500, soluble in benzene, acetone, dioxane, and chloroform, and containing no unsaturation, was obtained in 23% yield by prolonged heating of the monomer with di-tert-butyl peroxide.[55] The polymerization of divinyl ketone, however, yielded only insoluble polymers.[56]

1,4-Pentadiene has been polymerized by Ziegler catalysts to give soluble polymers containing a small amount of residual unsaturation.[57] Some main chain unsaturation as well as pendant group unsaturation was observed. The former was due to some isomerization of 1,4- to 1,3-pentadiene which then copolymerized with the 1,4-diene. The extent of isomerization varied with the catalyst system used. On the basis of an infrared examination of the CH_2 stretching frequencies in the 2900 cm^{-1} region, it was concluded that the polymer did not contain a methylene-linked four-membered ring structure. Analogous to the results obtained on N,N-divinylaniline (4–13), a mechanism, leading to bicyclo[3.3.1]-nonane rings was proposed.[1-89] The possibility of a five-membered ring structure was not considered.

Perfluoro-1,4-pentadiene has been polymerized under the influence of γ-radiation at temperatures of 100–170° and pressures of 8000–15,000 atm.[58,59] The compound was found to undergo double bond migration producing perfluoro-1,3-pentadiene. An average structure for the polymer was given, but no proof of the presence of a four-membered ring structure was provided.

Soluble polymers containing little residual unsaturation have recently been obtained from divinyl sulfone and several divinyl silanes.[56] Bicyclic cyclocopolymerization of other 1,4-dienes[60] has been investigated. The cyclocopolymerization was discussed mathematically on the basis of chain propagation reactions which occurred. The equations hold independently from the number of members constituting the rings in the cyclic radicals. A given cyclizable radical might undergo two different intramolecular reactions to give either five- or six-membered rings; thus two different cyclization ratios characterize the tendency of that radical toward cyclization. Divinyl ether as a representative 1,4-diene was considered.[61] The monomer was polymerized to give a product containing high fractions of monocyclic and bicyclic structural units (4–14). The residual unsaturation in the polymers belonged to the monocyclic units and were not directly pendant from the polymer backbone.

Another 1,4-diene monomer, divinylsulfide, was also studied.[62] Polymerization of this monomer in bulk and in benzene solution showed that the monomer proceeded by a bicyclic cyclocopolymerization process which accounted for the solubility of the resulting polymers. Evaluation of the cyclization ratios for the different cyclizable radicals involved in the chain propagation showed that the polymers consisted mainly of the bicyclic structural units and minor amounts of the monocyclic structural units.

A study of the temperature dependence of the cyclocopolymerization of divinyl ether has also been done.[63] The results showed that intramolecular cyclization is favored over intermolecular addition because the loss of internal rotational degrees of freedom is less effective in decreasing entropy than a loss of translational and rotational degrees of freedom. The cyclization was greater for the six-membered monocyclic radicals than for the five-membered monocyclic radicals and the activation energies for the cyclization of both radicals were greater than those for their addition onto monomer molecules. The dependence of the cyclization ratio on temperature was described by the Arrhenius equation.

Polymerization of 1,4-dienes can lead to bicyclic units which are formed from two molecules of monomer.[1-156, 1-19] Radical polymerization of divinyl ether yielded partially cyclized polymers, which, based on ^{13}C nuclear magnetic resonance spectroscopy, contained a five-membered ring monocyclic unit and a dioxobicyclo[3.3.0]octane unit in a 1:1 ratio.[64,65] The pendant vinyloxy group could be removed completely by dilute methanolic HCl. The ^{13}C NMR shifts expected for all the possible sterioisomers of these structural units were estimated, using a number of model compounds. A single sterioisomer was formed for both the monocyclic unit (trans ring closure) and the bicyclic unit (with the trans junction). The steric course of the cyclocopolymerization was compared with that of related systems.

The relationship between polymer composition and monomer concentration for the radical polymerization of divinyl ketone has been investigated.[66] The cyclization ratios were computed, and the polymer composition with linear, monocyclic and bicyclic structural units were calculated as a function of the monomer concentration at which the polymerization was carried out. Of the possible structural units, application of kinetic relationships provided evidence that six-membered monocyclic and bicyclic rings predominate.

The cyclopolymerization of α-α'-dimethoxycarbonyldivinylamine was reported and a five-membered ring bicyclic structure of Structure 4–15 was proposed from spectral data.[2-195]

An interesting and novel synthesis of polymetacyclophane of structure (4–7) has been described. 8,16-Dihydroxy[2.2]metacyclophane dimerized spontaneously in $CHCl_3$

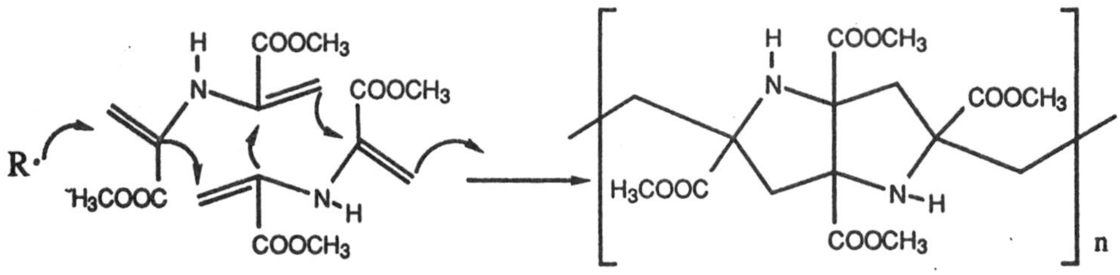

(4–15)

containing oxygen in the presence of sunlight to give the dimer which polymerized at 300–400°C to give the polymer.[40]

Cyclopolymerization of (E,E)-[6.2]paracyclophane-1,5-diene proceeded by an intra-intermolecular mechanism to give a polymer containing [3.2]paracyclophane (4–16) in the repeat unit. Thermal analysis of the polymer indicated a complex thermooxidative behavior in the presence of oxygen, while depolymerization occurred above approximately 400° in an inert atmosphere.[2-68]

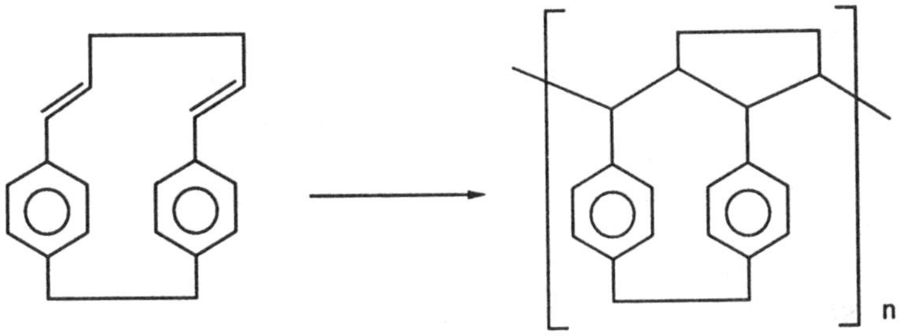

(4–16)

2-Ferrocenylbutadiene was prepared by dehydration of methylvinylferrocenylcarbinol and polymerized in the presence of $BF_3O(C_2H_5)_2$ or $TiCl_4$-iso-$(C_4H_9)_3Al$ catalysts.[67] The monomer was not polymerized by either anionic or radical mechanisms. The IR and NMR spectra of the polymer indicated absence of double bonds, suggesting cyclization. The polymer was thermally stable in an inert atmosphere at 270°C and in air to 300°C. Polymerization of ethyldivinylphosphinate proceeded according to the cyclopolymerization mechanism.[68] It also produced copolymers with styrene and methyl methacrylate. Poly(phenyldivinylphosphinate) and poly(N,N-diethyldivinylphosphinamide) were rubbery polymers which retained elasticity at 250°C. 1,3-Pentadiene-pentene copolymer was sulfurated with sulfuric acid to reduce its unsaturation and minimize its cyclization, which increased its chemical stability.[69] Cyclopolymerization of N,N-divinylaniline gave polymer of structure (See Structure 4–13).[70] Cyclocopolymerization with diethyl fumarate led to a 1:2 alternating cyclocopolymer. The copolymerization of N,N′-divinylureas was studied.[71] The copolymerizability of 1,3-divinylimidazolid-2-one and divinylhexahydropyrimid-2-one with ethyl acrylate was low as indicated by their reactivity ratios. This result was ascribed to resonance stabilization. The same argument was used to account for the inhibition of styrene polymerization caused by 1,3-diphenyl-1,3-divinylurea. Polymerization studies on N,N′-divinylureas and spectroscopic investigations of the polymer structure were carried out.[72] Homopolymerization of 1,3-divinyl-2-imidazolidone (R = vinyl) and of 1,3-divinylhexahydro-2-pyrimidinone of Structure 4–17 gave crosslinked, insoluble polymers. 1,3-Dimethyl-1,3-divinyl urea and 1,3-diphenyl-1,3-divinylurea did not polymerize.

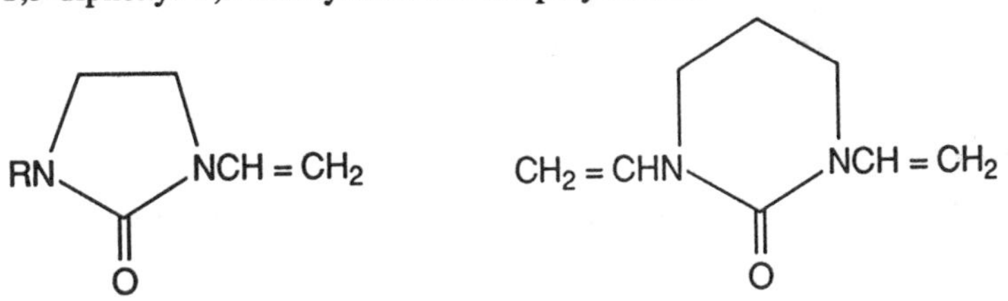

RN NCH = CH2 O   CH2 = CHN NCH = CH2 O

(4–17)

Spiro[2.4]hepta-4,6-diene is both a conjugated diene and a 1,3-diene.[73] When polymerized with triphenylmethyl salt initiators, the polymer contained only 1,2- and 1,4-addition structures. Formation of bicyclic structures from radical-initiated cyclopolymerization of divinyl ether as previously shown (See Structure 4–14) was confirmed.[74] Structural studies of poly(divinyl ether) have continued to appear. A carbon-13 NMR study of the structure of the homopolymer obtained from divinyl ether by radical polymerization showed that the cyclopolymer was composed of both a monocyclic unit and a bicyclic unit (See Structure 4–14).[65] The diene polymers, 1,2-poly(cis-1,4-hexadiene) and 1,2-poly(trans-1,4-hexadiene) having a predominately 1,8-diene structure, appeared to cyclize mainly by a [2+2] process of thermal cycloaddition of double bonds to yield bicycloheptane structures.[75] Bicyclo[2.2.1]heptadiene-2,5 and cycloheptatriene were polymerized *via* radical initiation under pressure to yield the corresponding polymers.[1-64] Higher pressures led to higher yields. In a study of vinyl compounds containing heterocyclic groups, thermal polymerization of 2-vinylfuran was investigated.[76] Kinetic data were obtained and the initiation mechanism was discussed in terms of the data and of the structure of the Diels-Alder type dimer formed thermally. 1,4-Pentadiene was polymerized *via* metal coordination catalysts to give amorphous polymers which were up to 64% soluble.[77] The polymerization proceeded *via* the cyclocopolymerization mechanism to introduce [3.3.1]bicyclo-linked chains. The polymers all had residual unsaturation. Cyclopolymerization of 1-vinyluracil has been proposed to yield polymers having the polycyclic repeating unit shown in Structure 4–18 *via* radical initiation.[78]

(4–18)

Interfacial esterification of poly(vinyl alcohol) with cinnamoyl chloride led to a poly(vinyl cinnamate) which could be cyclopolymerized through the cinnamoyl function to a ladder type polymer *via* photoinitiation.[79] Polymerization of bicyclo[2.2.1]hepta-2,5-diene with palladium pi-complexes has been reported.[80] The polymer was postulated to have been formed through one double bond of the monomer only, although other results reported earlier disagree with this postulate.

REFERENCES

1. Matsoyan, S. G., Pogosyan, G. M. and Eliasyan, M. A., Vysokomol. Soedin., 1963, 5, 777; CA 59, 7654f (1963).
2. Gusel'nikov, L. E., Nametkin, N. S. Polak, L. S. and Chernysheva, T. I., Izv. Akad. Nauk SSR Ser. Khim., 1964, 2072; CA 62, 9243f (1965).

3. Rudkovskaya, G. D. and Chizhenko, D. L., Izv. Akad. Nauk SSSR, Ser. Khim., 1970, (6), 1411; CA 73, 130596s (1970).
4. Rudkovskaya, G. D., Chizhenko, D. L. and Osipova, I. N., Vysokomolek Soedin.,Ser. B., 1971, 13, 423; CA 75, 118616g (1971).
5. Corfield, G. C. and Crawshaw, A., Chem. Commun., 1966, 85; CA 64, 12563d (1966).
6. Zutty, N. L., U.S. Pat. 3,287,327, 1966 (Nov. 22); CA 66, 29415n (1967).
7. Agaev, U. K., Guseinov, I. A., Tverdokhleb, I. P. and Rizaeva, S. Z., USSR Pat. SU 975,721, 1982 (Nov. 23); CA 98, 144002y (1983).
8. Pinazzi, C., Brosse, J. C. and Pluerdeau, A., Makromol. Chem., 1971, 142, 273; CA, 75 36774t (1971).
9. Starnes, W. H., Jr., Villacorta, G. M., Schilling, F. C., Park, G. S. and Sarema, A. H., Polym. Preprints, 1983, 24, 253; CA 100, 192315j (1984).
10. Mital, A. J. and Jones, J. F., U.S. Pat. 2,996,488; Brit. Pat. 877362, 1961 (Aug. 15); CA 56, 3649h (1962).
11. Montecatini Edison S.p.A., Ital. Pat. 792,246, 1967 (Nov. 15); CA 70, 20473e (1969).
12. Nozakura, S. Yamamoto, Y. and Murahashi, S., Polym. J., 1973, 5, 55; CA 80, 48432w (1974).
13. Sokova, F. M., Tr. Khim. Khim. Tekhnol., 1969, 197; CA 73, 99294r (1970).
14. Sokova, F. M., Tr. Khim. Khim. Tekhnol., 1968, 22; CA 71, 71022k (1969).
15. Powell, J., Whang, J., Owens, F. and Graham, R., J. Polym. Sci. A1, 1967, 5, 2655; CA 66, 105244y (1967).
16. Matsumoto, A., Ohata, T. and Oiwa, M., Bull. Chem. Soc. Japan, 1974, 47, 673; CA 82, 58212e (1975).
17. Roth, R. and Church, R., J. Polym. Sci., 1961, 55, 41; CA 56, 7498b (1962).
18. Hardy, G., Magy. Kem. Folyoirat, 1965, 71, 442; CA 64, 8320h (1966).
19. Kucharski, M. and Ryttel, A., J. Polym. Sci., Polym. Chem. Ed., 1978, 16, 3011; CA 90, 39329x (1979).
20. Danielyan, V. A., Oganesyan, Z. S. and Matsoyan, S. G., Arm. Khim. Zh., 1979, 32, 66; CA 91, 212048a (1979).
21. Brenner, W., U.S. Pat. US 4,216,134, 1980 (Aug. 5); CA 93, 221681v (1980).
22. Danielyan, V. A., Oganesyan, G. S. and Matsoyan, S. G., Arm. Khim. Zh., 1980, 33, 199; CA 93, 72375u (1980).
23. Bochert, P. J., Fr. 1,356,020; Br.101157L; US 3,242,143, 1964 (Mar. 20); CA 62, 5359f (1965).
24. Mikhant'ev, B. and Kretinin, S. A., Tr. Lab. Khim. Vyso. Soed. Voron. Univ., 1962, 37; CA 60, 10798f (1964).
25. Paniotov, I., Shapov, I. and Obreshkov, A., Izv. Inst. Org. Khim.; Bulg. Akad. Nauk, 1967, 3, 17; CA 68, 59929s (1968).
26. Ouchi, T., J. Macromol. Sci. (Chem.) A, 1981, 15, 417; CA 93, 221155b (1980).
27. Battaerd, H. A. J., U.S. Pat. US 3,619,394, (1971).
28. Matsumoto, A., Mano, H., Oiwa, M. and Butler, G. B., J. Polymer. Sci.: Part A: Polym. Chem., 1989, 27, 1811.
29. Sergeev, V. A., Korshak, V. V., Shitikov, V. K. and Vdovina, L. I., Dokl. Vses. Konf. Khim. Atsetilena, 4th. 1972, 3, 122; CA 81, 78284n (1974).
30. Breslow, R., Barrett, E. and Mohaesi, E., Tet. Letters, 1962, 1207; CA 59, 679a (1963).
31. Gudzera, S. S., Magdinets, V. V. and Spirin, Y. L., Dopov. Akad. Nauk Ukr. RSR, Ser. B, 1974, 36, 338; CA 81, 106412y (1974).
32. Mizogami, S. and Yoshimura, S., J. Chem. Soc., Chem. Commun., 1985, 21, 1736; CA 104, 207801y (1986).
33. Wiley, R. H., Rivera, W. H., Crawford, T. H. and Bray, N. F., J. Polym. Sci., 1962, 61, 538; CA 59, 4046g (1963).
34. Kennedy, J. P. and Hinlicky, J. A., Polymer, 1965, 6, 133; CA 63, 3047b (1965).

35. Valvassori, A. Sartori, G., Turba, V. and Lachi, M. P., J. Polym. Sci. C, 1967, 16, 23; CA 66, 66419n (1967).
36. Ball, L. E. and Harwood, H. J., Polym. Preprints, 1961, 2, 59; CA 56, 5846h (1961).
37. Butler, G. B., Miles, M. L. and Brey, W. S., Jr., J. Polym. Sci. A, 1965, 3, 723; CA 62, 13245g (1965).
38. Butler, G. B. and Miles, M. L., J. Polym. Sci. A, 1965, 3, 1609; CA 62, 16388g (1965).
39. Butler, G. B. and Miles, M. L., Polym. Engng. Sci., 1966, 6, 1; CA 64, 14284e (1966).
40. Pinazzi, C. P., Cattiaux, J. and Brosse, J. C., Eur. Polym. J., 1974, 10, 837; CA 83, 59547a (1975).
41. Frazer, A. H. and O'Neill, P., J. Am. Chem. Soc., 1963, 85, 2613; CA 59, 8877e (1963).
42. Frazer, A. H., J. Polym. Sci. A, 1964, 2, 4031; CA 61, 13443d (1964).
43. Julia, M., Fourneron, J. D. and Thal, C., An. Quim., 1974, 70, 888; CA 83, 131346d (1975).
44. Julia, M. and Fourneron, J. D., Bull. Soc. Chim. Fr., 1975, 3–4, Pt. 2 770; CA 83, 96805d (1975).
45. Itakura, J., Nakamura, Y. and Tanaka, H., Japan Pat. JP 75 03,485, 1975 (Jan. 14); CA 82, 171740p (1975).
46. Japan Synth. Rubber Co., Ltd., Japan Pat. JP 57 179,127, 1982 (Jul. 2); CA 98, 143976g (1983).
47. Yan, D. Y. and Hu, X., J. Polym. Sci., Part C: Polym. Lett., 1988, 26, 65; CA 109, 7017p (1988).
48. Wilke, G. and Rienacker, R., Ger. Pat. 1,065,413, 1959 (Sept. 17); CA 55, 8321g (1961).
49. Tokarzewski, L. and Wdowin, A., Polimery, 1969, 14, 533; CA 73, 4224f (1970).
50. Tokarzewski, L. and Wdowin, A., Polimery, 1969, 14, 25; CA 71, 30713t (1969).
51. Aso, C. and Ushio, S., Kogyo-kagaku zasshi, 1962, 65, 2085; CA 58, 12681f (1963).
52. Matsoyan, S. G. and Avetyan, M. G., Zh. Obsch. Khim., 1960, 30, 2431; CA 55, 8926e (1961).
53. Avetyan, M. G., Darbinyan, E. G., Saakyan, A. A., Kinoyan, F. S. and Matsoyan, S. G., Vysokomol. Soedin., 1964, 6, 3; CA 60, 14613h (1964).
54. Matsoyan, S. G., Avetyan, M. G. and Darbinyan, E. G., Izv. Akad. Nauk armyan SSR, Khim. Nauki 1964, 17, 412; CA 62, 1747g (1965).
55. Arbuzova, I. A. and Mosevich, I. K., Vysokomol. Soedin., 1964, 6, 13; CA 60, 15989f (1964).
56. Butler, G. B., Campus, A. F., Rogers, C. L., Sharpe, A. J. and Zweidinger, R., 1970; Results quoted in Ref. 1–156.
57. Trifan, D. S., Shelden, R. A. and Hoglen, J. J., J. Polym. Sci. A1, 1968, 6, 1605; CA 69, 3191y (1968).
58. Brown, D. W., Fearn, J. E. and Lowry, R. E., J. Polym. Sci. A, 1965, 3, 1641; CA 62, 16391f (1965).
59. Fearn, J. E., Brown, D. W. and Wall, L. A., J. Polym. Sci. A1, 1966, 4, 131; CA 64, 12812c (1966).
60. Kobayashi, H., Ohama, H. and Kitahama, R., Japan Pat. 69 03, 834, 1969 (Feb. 17); CA 70, 115737h (1969).
61. Guaita, M., Camino, G. and Trossarelli, L., Makromolek. Chem., 1969, 130, 252; CA 72, 44181a (1970).
62. Guaita, M., Camino, G. Chiantore, O. and Trossarelli, L., Makromolec. Chem., 1971, 143, 1; CA 75, 21181y (1971).
63. Guaita, M., Camino, G. and Trossarelli, L., Makromolec. Chem., 1971, 149, 75; CA 76, 60161g (1972).
64. Tsukino, M. and Kunitake, T., Macromolecules, 1979, 12, 387; CA 91, 21197p (1979).
65. Kunitake, T. and Tsukino, M., Makromol. Chem., 1976, 177, 303; CA 84, 74832d (1976).

66. Guaita, M., Camino, G., Chiantore, O., Revellino, M. and Trossarelli, L., Makromol. Chem., 1974, 175, 457; CA 81, 38193a (1974).
67. Korshak, V. V., Sosin, S. L., Alekseeva, V. P. and Afonina, R. I., Vysokomol. Soedin., Ser. B, 1975, 17, 779; CA 84, 44733e (1976).
68. Pyrkin, R. I., Levin, Y. A., Yagfarova, T. A., Kovalenko, V. I. and Usol'tseva, A. A., Mater. Nauch. Konf., Inst. Org. Fiz., 1970, 200; CA 78, 30613e (1973).
69. Klinov, I. Ya., Medvedeva, N. M., Ponomarenko, V. I. and Irkhin, B. L., Prom. Sin. Kauch. Nauch.-Tekh. Sb., 1972, 11, 16; CA 80, 60754k (1974).
70. Chang, E. Y. C. and Price, C. C., Macromol. Synth., 1974, 5, 55; CA 83, 79649p (1975).
71. Corfield, G. C., Monks, H. H. and Ellinger, L. P., Polymer, 1975, 16, 770; CA 84, 151046e (1976).
72. Corfield, G. C., Crawshaw, A. and Monks, H. H., J. Macromol. Sci. (Chem.), A, 1975, 9, 1085; CA 84, 17770h (1976).
73. Kunitake, T., Ochiai, T. and Ohara, O., J. Polym. Sci., Polym. Chem. Ed., 1975, 13, 2581; CA 84, 17796w (1976).
74. Guaita, M., Camino, G. and Trossarelli, L., IUPAC, Boston, 1971, 392.
75. Golub, M. A., J. Polym. Sci., Polym. Lett. Ed., 1978, 16, 253; CA 89, 24916s (1978).
76. Aso, C., Kunitake, T., Tanaka, Y. and Miyazaki, H., Kobunshi kagaku, 1967, 24, 187; CA 68, 22275z (1968).
77. Trifan, D. S., Shelden, R. and Hoglan, J., Polym. Prepr., Polym. Div., ACS, 1968, 9, 156; CA 69, 3191g (1968).
78. Kaye, H., Polym. Preprints, 1970, 11, 1027; CA 75, 21170u (1971).
79. Ryang, G. S. Li, J. S. and Choi, D. U., Hwahak Kwa Hwahak Kongop, 1971, 276; CA 77, 152796s (1972).
80. Hojabri, F., J. Appl. Chem. Biotechnol., 1973, 23, 601; CA 80, 133902u (1974).

5

Symmetrical Monomers Containing Other Multiple Bonds

The versatility of cyclopolymerization of nonconjugated dienes has been demonstrated. Compounds containing other polymerizable groupings have also been shown to undergo cyclopolymerization.

Among these are diynes, dialdehyes, diisocyanates, diepoxides, and dinitriles, as well as unsymmetrical monomers with respect to these functional groups. In addition, a variety of monomers are reported to lead to bicyclic structures.[1-156,1-191] Interest in cyclic structures derived from this group of monomers has increased in recent years because of the possibility that certain of these monomers can lead to conjugated polymers, of current interest in the semiconductor field.

CYCLOPOLYMERIZATION OF DIYNES

It was found as early as 1961 that 1,6-heptadiyne gave soluble polymers via Ziegler catalysis.[1] Linear cyclopolymers in which the backbone consisted of a continuous conjugated system of double bonds (5–1a) were proposed. However, it was later reported[2] that the results were more in agreement with the "polycyclotrimerized" structure (5–1b) which had been deduced[3] for the polymers obtained on nickel-carbonyl-phosphine complexes.

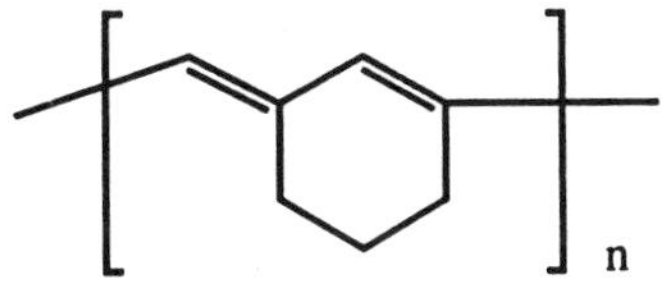

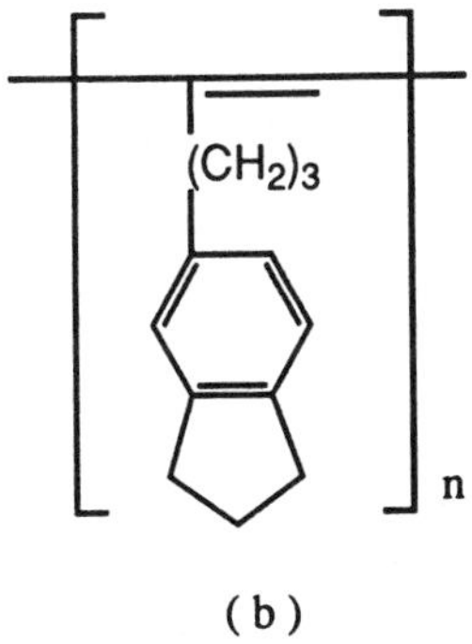

$(CH_2)_3$

n

n

(a)

(b)

(5–1)

The cyclopolymerization of propiolic anhydride in the presence of weak nucleophiles has been reported.[4] Using KCNS, KCN, KI, KBr, or KCl as catalyst, a black powder with a metallic gloss was obtained. The powder was soluble in dimethylformamide and the infrared spectrum showed a band corresponding to C=C at 1620 cm-[1], whereas the band at 2100–2300 cm-1 of C≡C was absent. The Structure 5–2 was proposed for the polymer.

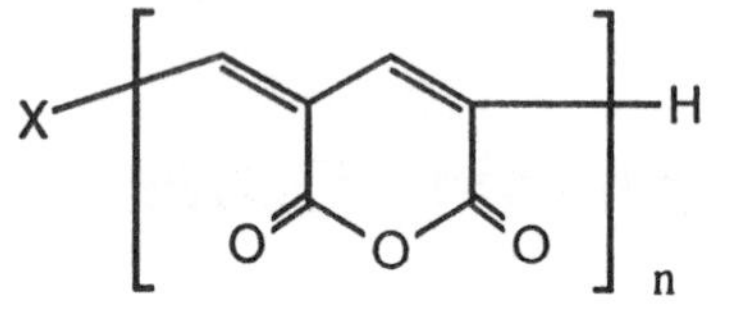

X–[ ]n–H

where x = CNS, CN, I, Br, or Cl

(5–2)

The polymerizations of the propargyl esters of several unsaturated acids, e.g. dipropargyl fumarate, propargyl crotonate, and propargyl methacrylate, have been reported.[5] The infrared spectra of the polyesters and their saponification products showed that the fumarate and the crotonate polymerized with the formation of lactone rings due to reaction between the double and triple bonds. The Structure 5–3 was given for the polymer from the fumarate. The polymer from the methacrylate contained no cyclic lactone structures in the chain.

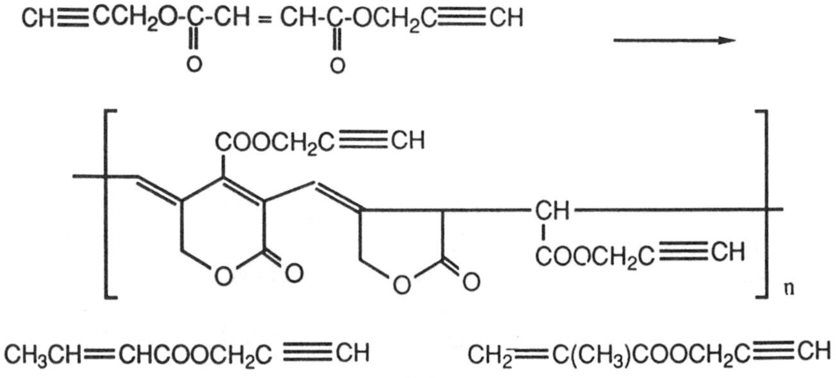

$CH{\equiv}CCH_2O\text{-}C(=O)\text{-}CH = CH\text{-}C(=O)\text{-}OCH_2C{\equiv}CH \longrightarrow$

$COOCH_2C{\equiv}CH$

CH

$COOCH_2C{\equiv}CH$

n

$CH_3CH{=}CHCOOCH_2C{\equiv}CH \qquad CH_2{=}C(CH_3)COOCH_2C{\equiv}CH$

(5–3)

Further studies on the cyclopolymerization of 1,6-heptadiyne on Ziegler Natta catalysts have been published.[6] Poly-1,6-heptadiyne and 1,6-heptadiyne-phenylacetylene copolymer were shown to contain 3-(5-indanyl)-propylvinylene groups; 1,6-heptadiyne-phenylacetylene copolymer also contained $C_6H_5C{=}CH$ groups. The absence of triphenylbenzene in the product of the copolymerization indicated that there was no homopolymerization of $C_6H_5C{\equiv}CH$.

The polymerization of o-diethynylbenzene by metal coordination and other catalysts was investigated; polymers were obtained readily with $AlR_nCl_{(3\text{-}n)}$- $TiCl_4$ and $AlR_nCl_{(3\text{-}n)}$ tris(acetylacetonato)titanium (n = 1-3, R = C_2H_5 or iso-C_4H_9).[7] The polymers obtained were pale brown to brown powders with molecular weights of 2000–3500, which did not melt. Poly(o-diethynylbenzene) obtained with coordination catalysts contained 30–50% unreacted ethynyl group. The methyleneindene unit was probably the predominant cyclic structure.

Anionic cyclopolymerization of nonconjugated dienes, bis-carbonyl compounds, diepoxy compounds, dinitriles, and diisocyanates was reviewed in 1972.[8]

The polymerization of a variety of diynes using palladium chloride as catalyst has been investigated.[9–12] Cyclopolymers with conjugation in the main chain have been obtained from monomers such as dipropargylacetic acid and dipropargyl ether. Electrical conductivities and other properties of the polymers were reported. Several soluble cyclic poly(dipropargyl ethers) and poly(propiolic anhydrides) were obtained in 66–89% yields by polymerization of the respective monomers in pyridine or DMF in the presence of $PdCl_2$.[9] All contained a conjugated, unsaturated backbone structure (See Structure 5–2).

Substituted dipropargyl ethers polymerized in high yield and formed soluble polymers in presence of palladium chloride.[10] Dipropargyl and other ethers with unsaturated end groups, which do not polymerize as fragments, were cyclopolymerized under the examined conditions.

Cyclopolymerization of substituted 1,6-heptadiynes has been reported. Acids, esters and ketoesters prepared from propargyl bromide and malonic ester were studied.[11] Soluble polymers containing substituted cyclohexene rings in the chain were obtained by $PdCl_2$-catalyzed polymerization.

Cyclopolymerization of 1,8-diacetylenic compounds has been studied in the presence of metal-complex catalysts.[12] Soluble conjugated polymers of Structure 5–4 (R = H, Br; R_1 = H, Br; R_2 = H, Pr, Ph) were obtained in the cyclopolymerization of the corresponding acetals, $RC{\equiv}CCH_2OCHR^2(OCH_2C{\equiv}CR_1)$, in the presence of $PdCl_2$ catalysts.

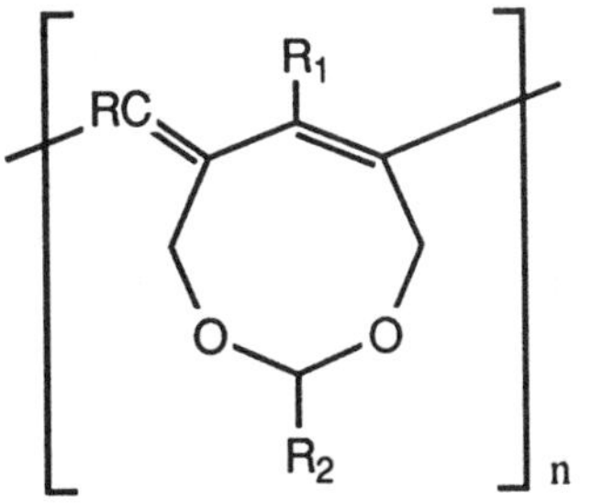

RC R1 O O R2 n

(5–4)

Although diacetylenes were studied as early as 1961[1] these structures have been studied more recently as the result of the present interest in semiconductors.[13] 1,6-Heptadiyne reportedly led to the polymer shown in Structure 5–1. The original investigation utilized coordination type initiators to yield soluble polymers in which the polymer backbone consisted of a continuous conjugated system of double bonds. A more recent investigation of this monomer was initiated as a result of the strong interest in electrically conducting films.[13,14] The results showed that 1,6-heptadiyne can be converted to a free-standing film which can be doped with iodine to yield an electrically conducting polymer. The conclusion was that the cyclopolymer possessed predominantly Structure 5–1, in accord with that proposed earlier.[1] The electrical conductivity was 10^{-12} $ohm^{-1}cm^{-1}$ but, when doped, the conductivity increased by eleven orders of magnitude.

The effect of free radicals on the kinetics of oxidation and doping of polyenes has been studied. Oxidation of polyenes initially doped them to form charge-transfer complexes which increased their conductivity; further oxidation gave degradation products that are not conductive. The decay was first-order, and poly(1,6-heptadiyne) was less stable to oxidative degradation than polyacetylene even though the absolute barrier controlling the rates is similar. Oxidation of the polyenes is related to radical formation, oxygen insertion at radicals being the primary mechanism.[15]

Polymerization of a variety of diynes using palladium chloride ($PdCl_2$) as initiator has been studied. Cyclopolymers with conjugation in the main chain were obtained from dipropargyl ether and 4-carboxy-1,6-heptadiyne. Electrical conductivities and other properties of the polymers were reported. The polymerization of N-substituted dipropargylamines in boiling pyridine or DMF containing $PdCl_2$ as the catalyst also gave soluble conjugated cyclic polymers[16] up to 91% yield. Better results were obtained if the amine salts were used.

The thermal polymerization of some propargyl derivatives has been studied.[17] m-Methylphenyl propargyl ether, p-tert-butylphenyl propargyl ether, and phenyl propargyl ether at 160–200° in air or under nitrogen gave mixtures of linear and cyclic products, with the ratio depending on the monomer and polymerization conditions.

It has been shown that 1,2,4-triphenylbenzene and 1,3,5-triphenylbenzene are formed during cyclopolymerization of phenylacetylene in the presence of $(C_2H_5)_3Al$-$TiCl_3$ system, depending on the $(C_2H_5)_3Al$ concentration.[18] A discussion of acetylene and its reactions, such as ethynylation, vinylation, carbonylation and cyclopolymerization, was published as early as 1958.[19] Cyclization trimerizations of alkynes by use of metal carbonyl compounds, as catalysts were studied.[20] Mono- and disubstituted acetylenes were converted to benzene derivatives. Unsymmetrical acetylenes were converted to benzene derivatives with identical substituents in the 1,2,5- and 3,5,6-positions, respectively.

The structures of the linear dienes and trimer products from the reaction of acetylenes with nickel carbonyl-phosphine complexes were determined.[21] Polymers from 1,5-hexadiyne decomposed at 300° and possessed corklike properties. 1,7-octadiyne or 1,8-nonadiyne gave rubbery high polymers, softening at about 250° with decomposition. Further study of polymerization of acetylenic compounds by nickel carbonyl-phosphine complexes and the scope of the reaction has been published.[22] Thirty-five different mono and disubstituted acetylenes were included in the study. Predominantly low molecular weight compounds were obtained. The kinetics of the polymerization of acetylenes via their reaction with nickel carbonyl-phosphine complexes were also studied and the mechanism discussed.[23] Mechanisms were proposed for the linear and cyclic polymerization of monoacetylenes.

A polymerization initiator for acetylenes, which consists of an hydridic reducing agent, such as sodium borohydride, plus a salt or complex of a Group VIII metal, such as nickel chloride, has been used.[24] Linear, *trans*, high-molecular-weight polymer was obtained from acetylene. Other polymers and copolymers of acetylenes were obtained in presence of a catalyst consisting of a metal hydride and a compound such as a phosphine-nickel bromide complex. Different catalyst combinations were used.[25]

Poly(diphenyldiacetylenes) were prepared by polymerization in presence of a titanium tetrachloride-tri-(isobutyl)aluminum catalyst at liquid nitrogen temperature.[26]

The polymer possessed semiconducting properties. A catalyst composition comprised of a mixture of a transition-metal complex with a metal hydride for initiating cyclic polymerization of diacetylenic compounds was described.[27] High yields of polymer were obtained.

Diphenyldiacetylene was polymerized thermally to yield a soluble polymer, postulated to possess the polycyclic structure shown in Structure 5–5.[28] The monomer was also polymerized in benzene with vanadyl acetylacetonate and triethylaluminum to yield polymer postulated to be of the structure shown in Structure 5–6. Thermal degradation of polymers of diphenylbutadiyne of partial ladderlike structure led to a black insoluble residue.[29] The products of the decomposition were identified.

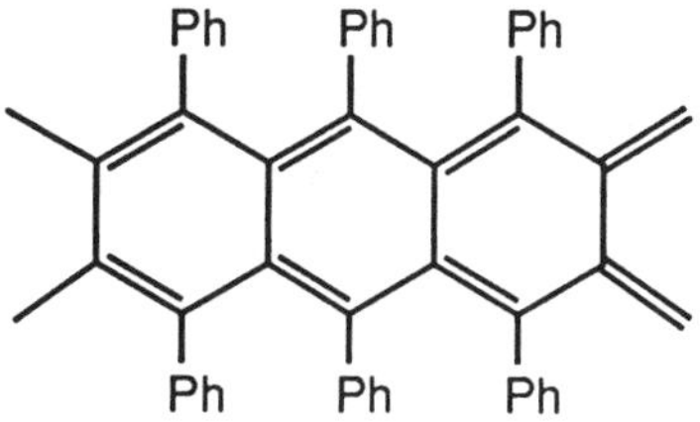

Ph Ph Ph

Ph Ph Ph

(5–5)

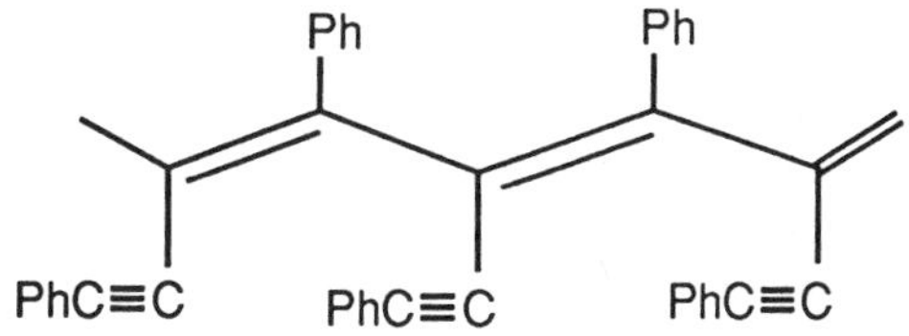

Ph Ph

PhC≡C PhC≡C PhC≡C

(5–6)

Catalytic polycyclotrimerzation of diethynylaromatic compounds gave branched polyphenylenes containing 1,3,5- or 1,3,4-trisubstituted benzene rings.[30] Copolymerization of diethynylaromatics with monoethynyl compounds gave similar, but less branched, polyphenylenes. Topochemical polymerization of monomers with conjugated triple bonds has been studied, and the properties of the polymers were discussed.[31] A structural similarity of these processes to cyclopolymerization exists however, mechanistically, the two processes are quite different. Poly(p-diethynylbenzene) and poly(p,p-diethynylbiphenyl), among others, were prepared by polycyclotrimerization of the corresponding monomers in the presence of transition-metal complexes which catalyze the trimerization of acetylene to benzene.[32]

Polycyclotrimerzation of aryl cyanates in presence of a zinc chloride catalyst gave highly crosslinked poly-s-triazines.[33] For example, bis(4-cyanatophenyl)methane gave a structure in which $R = R_1 = H$ (See Structure 5–7).

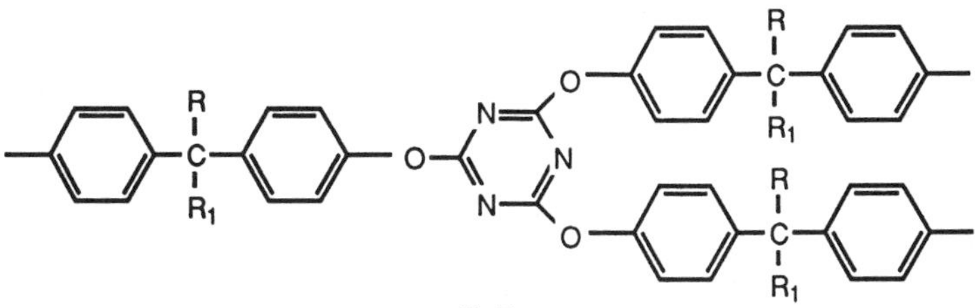

R C R_1 O N N N O O R C R_1 R C R_1

(5–7)

Poly(diphenylbutadiyne) and poly(dimethylbutadiyne) were polymerized both thermally and with organometallic coordination catalysts.[34] Polymers containing various fractions of the structural units shown in Structure 5–8 were obtained.

The reactivity ratios were calculated for copolymerization of acrylonitrile with propargyl acrylate, propargyl methacrylate and dipropargyl itaconate.[35] Propargyl methacrylate was more reactive than the itaconate, and all reactivity ratios indicated a

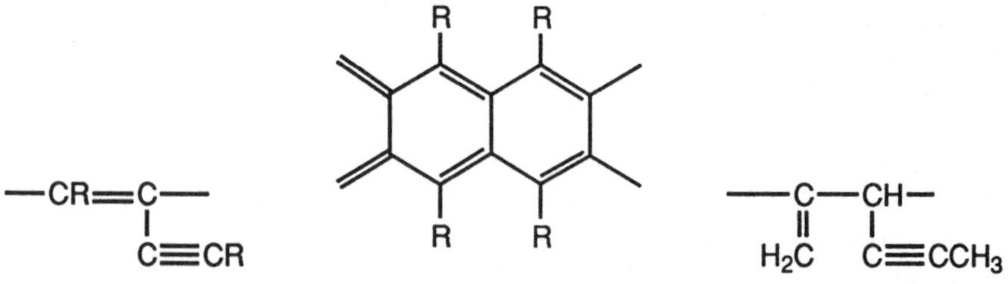

a b c

(5–8)

tendency toward alternation in copolymerization with acrylonitrile. Propargylic compounds, $CH{\equiv}CRR'X$ [where R and R′ are H or alkyl, and X is OH, Cl, Br, NR_2, CR_2OH, OR″, or O_2CR'' (R″ is alkyl or aryl)] were polymerized by $PdCl_2$ to give conjugated polymers.[36] In a study of Ziegler polymerization of alkynes, it was shown that the triple bond opening did not result in pure cis polymers due to cis-trans thermal isomerization, along with cyclization and chain scission.[37]

It was shown that a loss of crystallographic register between chains occurred during γ-ray polymerization of the cyclic tetradiyne monomer $[(CH_2)_2C{\equiv}C\text{-}C{\equiv}C(CH_2)_2]_4$ although crystallographic order in the chain-axis projection was retained.[38] The polymer had a butatriene backbone structure. A structure determination of the macromonomer, poly(1,11-dodecadiyne), and its cross-polymerized product was carried out by electron diffraction.[39] The unit cell of both macromonomer and cross-polymerized product was monoclinic, space group $P2_1/n$. Dipropargyldialkylsilanes were cyclopolymerized in presence of $MoCl_5$, which was superior to WCl_6, as a catalyst.[40] The diphenyl derivative gave more soluble polymers which were also more stable than the corresponding dialkyl structures.

Dipropargyl sulfide was cyclopolymerized in presence of tungsten hexachloride or molybdenum pentachloride to give a polymer of proposed structure (5–9).[41] Functionalized non-conjugated polyacetylenes were proposed from the copolymer of 1-(4-bromobutyldimethylsilyl)-1-propyne-1-trimethylsilylpropyne, which was prepared using $TaCl_5$-$(C_2H_5)_3Bi$ as catalyst.[42] Cyclopolymerization of dipropargylcarbinol was accomplished by use of the transition metal catalysts $MoCl_5$ or $Mo(OC_2H_5)_5$.[43] WCl_6 was not an effective catalyst and $PdCl_2$ showed only slight activity.

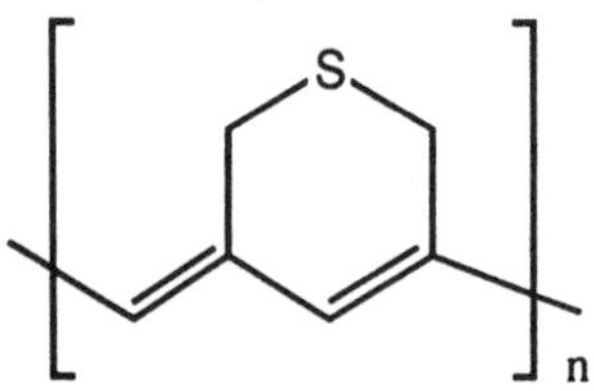

(5–9)

4,4-Diethynyldiphenylmethane was thermally polymerized by a free radical mechanism to a highly crosslinked structure of interest as a high temperature composite resin.[44] An initial rapid polymerization occurred, believed to involve formation of a cyclic trimer, followed by linear polymerization of the acetylenic end groups.

CYCLOPOLYMERIZATION OF DIALDEHYDES

As early as 1908, it had been reported that glutaraldehyde gave soluble solid products, on standing at room temperature.[45] However, the structures of the products were not elucidated. Both the spontaneous and cationic polymerizations of glutaraldehyde were reinvestigated and the structures of the products obtained were interpreted in accord with the cyclopolymerization mechanism.[1-49,46] Other workers independently reported similar results.[47,1-47,1-46] The polymers obtained were thermally unstable. On heating, depolymerization to monomer occurred. The polymers were soluble in such solvents as benzene and chloroform. The solubility and fusibility of the polyaldehydes were taken as proof for the linear nature of the products. A linear polymer of glutaraldehyde may be obtained by the cyclopolymerization mechanism leading to a polymer containing a tetrahydropyran ring in the structural unit shown in Structure 5–10 or, alternately, by polymerization through one of the aldehyde groups only (5–11). The infrared spectrum

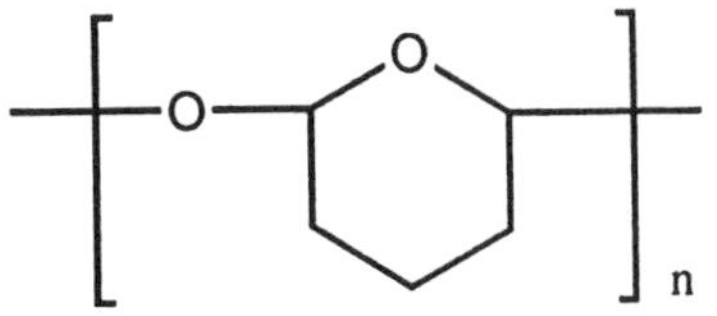

(5–10)

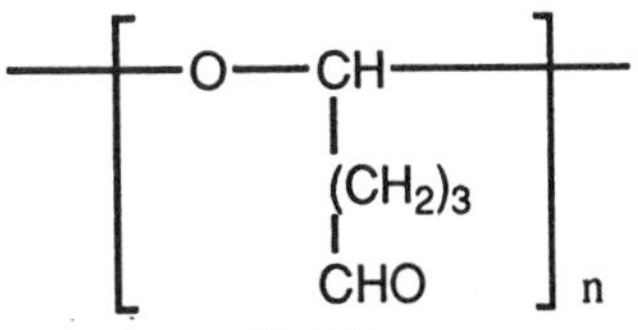

(5–11)

of the soluble fractions of the polymers showed a band at 1730 cm^{-1} due to the carbonyl group, indicating the presence of residual aldehyde and, in addition, there were strong bands in the acetal region between 1100 and 1200 cm^{-1}. The actual structure of the polymer was a copolymer of Structures 5–10 and 5–11. The ring structure was inferred to be six-membered, since the spectra of the polymers were similar to those of substituted tetrahydropyrans. The pendant aldehyde was determined which indicated that the degree of cyclization was >75%. Substituted glutaraldehydes have also been cyclopolymerized.[47,48]

The spontaneous and catalytic polymerization of succinnaldehyde proceeded predominantly via a cyclopolymerization mechanism to yield tetrahydrofuran ring units in the polymer. For the determination of the structure of the polymer the IR and NMR spectra were compared with the model compounds, 2,5-dimethoxy- and 2,5-diethoxytetrahydrofuran.[49,50]

Although adipaldehyde, $OCH(CH_2)_4CHO$, does not generally polymerize spontaneously, it has been polymerized using catalysts such as boron trifluoride diethyl etherate and triethylaluminum-water.[51] The polymer obtained, however, was only partially soluble in organic solvents, and the soluble fraction contained 60–70% residual aldehyde. It was concluded that cyclopolymerization took place but that the extent of cyclization was low since this required the formation of an energetically unfavorable seven-membered ring. Copolymerization with chloral resulted in a 1:1 copolymer having little residual aldehyde. The intramolecular reaction was considered to be promoted by the steric hindrance to intermolecular addition caused by the bulky trichloromethyl group.

The polymerization of suberaldehyde,[1-70] $OHC(CH_2)_6CHO$, using typical carbonyl polymerization catalysts [$Al(C_2H_5)_3$, $BF_3O(C_2H_5)_2$] has been reported. The rate of polymerization was extremely fast even when the monomer concentration was low.

Insoluble polymers were obtained showing carbonyl groups, ether groups, and a low concentration of hydroxyl groups in their spectra. An increase in the observed ratio of ether to carbonyl groups, at higher dilutions, provided evidence of some cyclopolymerization.

Polymerization of malealdehyde which, because of its multifunctional nature, may give rise to several structural units, (5–12a-e) has been investigated.[52–54] Cyclopolymers (5–12b) were obtained when alkali metal alkoxides, AlR_3, AlR_3-CH_3OH, AlR_3-H_2O, $Al(C_2H_5)_3$-$TiCl_4$, $BF_3O(C_2H_5)_2$, and pyridine were used as catalysts. The 1,2-addition

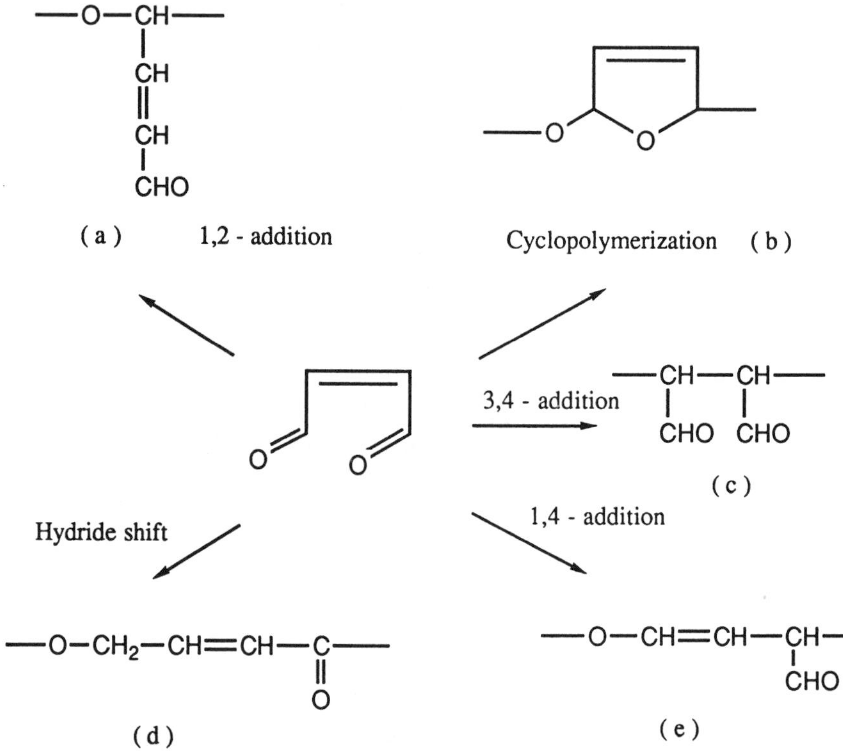

(a) 1,2 - addition

Cyclopolymerization (b)

3,4 - addition

(c)

Hydride shift

1,4 - addition

(d)

(e)

(5–12)

structure (5–12a) was also present to a small extent. The IR spectra of the polymers obtained with $Al(C_2H_5)_3$ and $Al(C_2H_5)_3$-$TiCl_4$ initiators were suggestive of the presence of five-membered lactone rings which may occur as a result of termination by hydride transfer (5–13). Spontaneous polymerization, which was probably induced by contaminating water, similarly produced cyclopolymers. Radical polymerization of malealdehyde with styrene or acrylonitrile yielded the 3,4-addition unit (5–12c) in the copolymer. Polymerization by 1,4-addition (5–12e) or by hydride transfer (5–12d) was not observed with any initiating system.

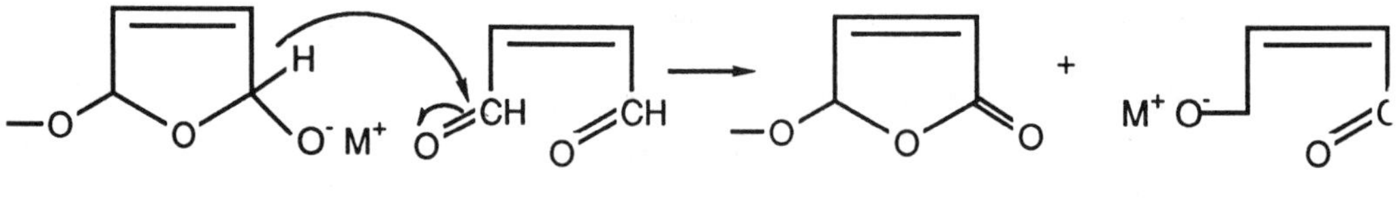

In general, the cyclization tendencies of the dialdehyde monomers corresponded fairly well to their ease of spontaneous polymerization (adipaldehyde < glutaraldehyde, succinnaldehyde < malealdehyde). This indicated that the cyclopolymerization process acted as a driving force in the spontaneous polymerization of dialdehydes.[53] The effect of dilution on the shift of the carbonyl IR stretching band was smaller for dialdehydes than for monoaldehydes. This difference was ascribed to an intramolecular dipole-dipole interaction in dialdehydes. Further, the inhibition of the spontaneous polymerization of malealdehyde by dilution led to the interpretation that the interaction (5–14) promoted cyclization.

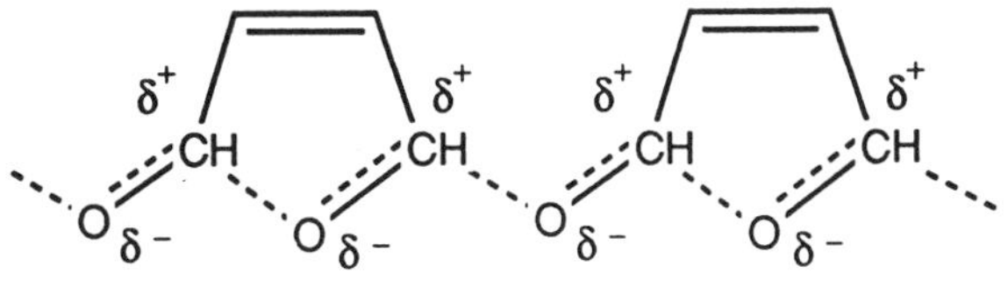

(5–14)

o-Phthalaldehyde homopolymerized readily to cyclopolymers using γ-ray irradiation, cationic catalysts, anionic catalysts and coordinated anionic catalysts (Table 5–1).[55,56] Terephthalaldehyde and isophthalaldehyde resisted polymerization. The ease of intramolecular cyclization of o-phthalaldehyde was demonstrated by the fact that cyclic 1,3-dialkoxyphthalans were obtained in good yield on reaction with various alcohols. It was also shown that the most stable conformation of o-phthalaldehyde anions is the meso form.[57] From the above considerations, it was proposed that dipole-dipole interactions between the two adjacent aldehyde groups promote the cyclopolymerization process through an intermediate or in a concerted manner, rather than the stepwise manner (5–15).

A second paper on cyclization of o-phthaldehyde, with the same type of initiation as those previously studied, was published in 1969. Various coordination catalysts ($Al(C_2H_5)_3$-transition metal compounds) and anionic compounds (tert-C_4H_9OLi, etc.) were employed. Molecular weights (number average) ranged from 1280 to 11,700. The mole % *cis*-content was determined using NMR techniques as described. The amount of *cis*-content decreased in the following order: γ-ray irradiation > cationic catalysts $\geqslant$ anionic catalysts > coordination catalysts. It was suggested that counterion effects may play an important role in configurations of the product.

Table 5–1 Polymerization of Phthalaldehyde

Initiator	Mole %	Monomer (g/ml)	Solvent	Time (h)	% Conversion	η	Softening point (°C)	% *cis*-content
γ-rays[a]		0.1	CH_2Cl_2	115	24	0.06	130-132	87
$Ph_3C^+BF_4^-$	0.5	0.1	CH_2Cl_2	24	87	0.11	125-127	77
t-BuOLi	1.6	0.1	THF	22	24	0.13	117-120	56
$AlEt_3$–$TiCl_4$(2/1)	Al 4.0	0.02	Toluene	70	28	—	152-154	37

[a] Dose: 9.27×10^6f.

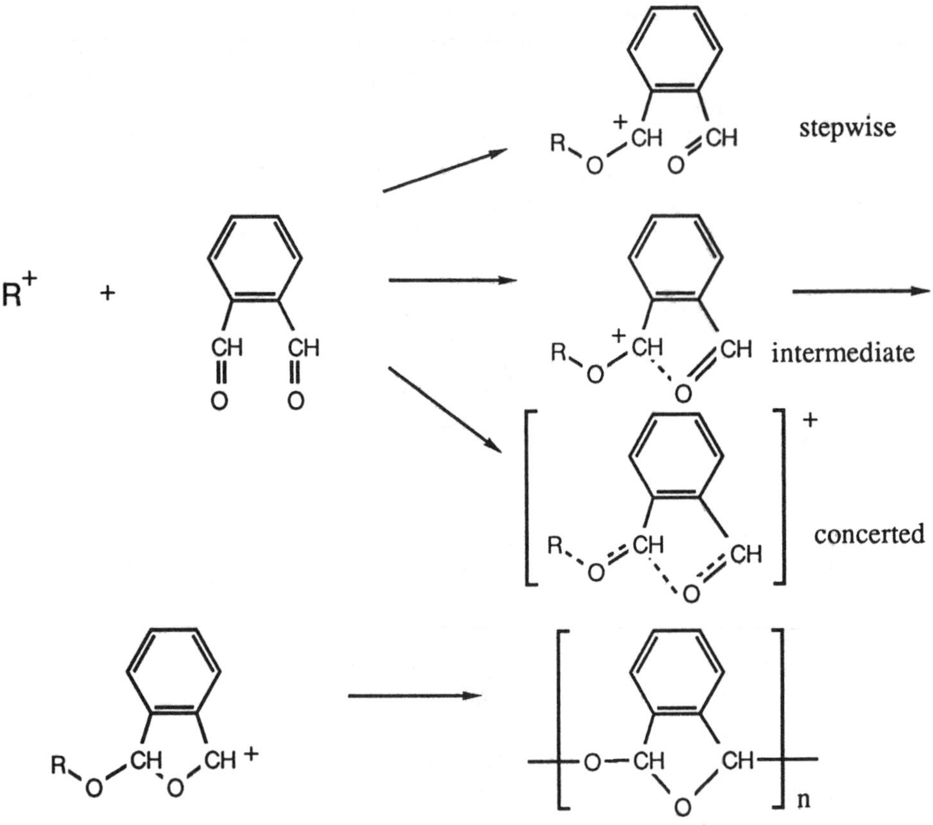

(5–15)

The cyclic unit formed in the cyclopolymerization of dialdehydes would possess two alkoxy substituents with either *cis*- or *trans*-configurations. The steric structure of the cyclic unit in polysuccinnaldehyde was determined from a comparison of the IR spectra of the polymer and model compounds, the *trans*-configuration being predominant with a $Zn(C_2H_5)_2$-CH_3OH catalyst and comparable amounts of the *cis*- and *trans*-configurations present with a $BF_3O(C_2H_5)_2$ catalyst.[49] A similar study was carried out for polymalealdehyde.[54] Determination of the configuration of the cyclic unit of poly(o-phthalaldehyde) was carried out, and it was found that the *trans*-configuration was predominant with coordinated anionic catalysts (See Table 5–1).[51]

The polymerization of *trans*-1,2-diformylcyclohexane[58] and *cis*-1,3-diformylcyclohexane,[59] using a cationic catalyst, led to soluble, thermally unstable polymers which, from their spectroscopic and chemical properties, consisted mainly of structural units containing the 8-oxabicyclo[4.3.O]nonane and 3-oxabicyclo[3.3.1]nonane ring, respectively, resulting from the cyclopolymerization mechanism.

The cyclopolymerization of *cis*-1,3-diformylcyclohexane, using a cationic initiator, has been reported.[2–32] The polymer, obtained in high yield was shown by spectroscopic data to contain 3-oxabicyclo[3.3.1]nonane rings formed through the cyclopolymerization mechanism. Soluble fractions of the polymer also contained some residual aldehyde groups.

A cyclopolymerization study of a variety of dialdehydes has been reported.[2-86] 1-Cyclohexene-4,5-dicarboxaldehyde and 1-methyl-l-cyclohexene-4,5-dicarboxaldehyde were polymerized using cationic, anionic, and coordination catalysts to yield cyclopolymers soluble in common organic solvents from all of the catalysts used.[2-286]

Some mechanistic aspects of cyclopolymerization of dialdehydes such as o-phthaldehyde and *cis*-cyclohexene-4,5-dicarboxaldehyde have been discussed.[2-27] A concerted process was proposed for the cyclization step in these cyclopolymerizations.

A review with fifty-six references dealing with the synthesis of polyethers by cyclopolymerization of dialdehydes was published in 1971.[60]

A study of cyclopolymerization of o-formylphenylacetaldehyde which gave a cyclic polymer or a cyclic trimer depending on the type of catalyst and the temperature was carried out.[61] Cyclopolymerization in the presence of $BF_3.O(C_2H_5)_2$ at $< -60°$ gave a benzopyran-type cyclopolymer as shown by IR and NMR spectra and elementary analyses. This polymer was converted to the cyclic trimer, a hexaoxadodecane, at O° in the presence of $BF_3.O(C_2H_5)_2$ in CH_2Cl_2 as shown by IR, NMR, and X-ray data. Cationic polymerization in the presence of $BF_3.O(C_2H_5)_2$ at $> -40°$ also gave the cyclic trimer. The cyclopolymer was also prepared at 0° and $-78°$ in the presence of LiOBu-tert. Anionic polymerization with $(C_2H_5)_3Al$ and $(C_2H_5)_3Al$-$C_6H_5NHOCCH_3$ catalysts gave mixed polymers containing cycloether and ester groups.

It has been reported that polymer containing both the cyclized tetrahydropyran and the open chain structures was obtained by cyclopolymerization of glutaraldehyde in the presence of an organometallic catalyst.[62] The polymer can be stabilized with hemiacetal "capping" reagents which have ester or ether end groups.

Other anionic polymerizations of monomers functionally capable of undergoing cyclic polymerization have been reported. At least five additional publications have appeared describing the cyclic polymerization of glutaraldehyde.[1-49,1-47,1-46,1-44,1-48] The properties of the polymers obtained were consistent with those predicted from an alternating intramolecular-intermolecular chain propagation, including always varying amounts of the open chain structure.

It has been shown that dialdehydes, such as glutaraldehyde, β-methyl-, and β-phenylglutaraldehyde, produced soluble, linear polymers containing tetrahydropyran units by spontaneous cyclopolymerization at low temperature.[1-48]

Research has been continued on the polymerization of aromatic aldehydes and it was observed that unsymmetrical and symmetrical bis-aldehydes shown, respectively, are partially cyclized by ionic initiators to polyacetals with seven-membered ring structures (5–16).[63]

It was shown that the cationic polymerization of 3-(o-formylphenyl)propionaldehyde gave polyacetals which contained 7-membered ring structures in mole fractions of 0.3–0.5 in addition to uncyclized oxymethylene units from polymerization of the aliphatic aldehyde group. Ziegler catalysts gave only uncyclized structures. Polymerization of o-phenylenediacetaldehyde (5–16) with BF_3-etherate, $LiOC(CH_3)_3$, or $(C_2H_5)_3Al$-$TiCl_4$ catalysts also gave polyacetals containing 7-membered cyclic structures (5–16) in mole fractions 0.5–0.8.

Other aliphatic dialdehydes which have been studied include succindialdehyde, malealdehyde, and *cis*- and *trans*-diformylcyclohexane.[1-183]

o-Phthalaldehyde has perhaps been studied more extensively than any other dialdehyde. This monomer homopolymerized readily to cyclopolymer using γ-ray irradiation, cationic catalysts, anionic catalysts, and coordinated anionic catalysts. Tere-

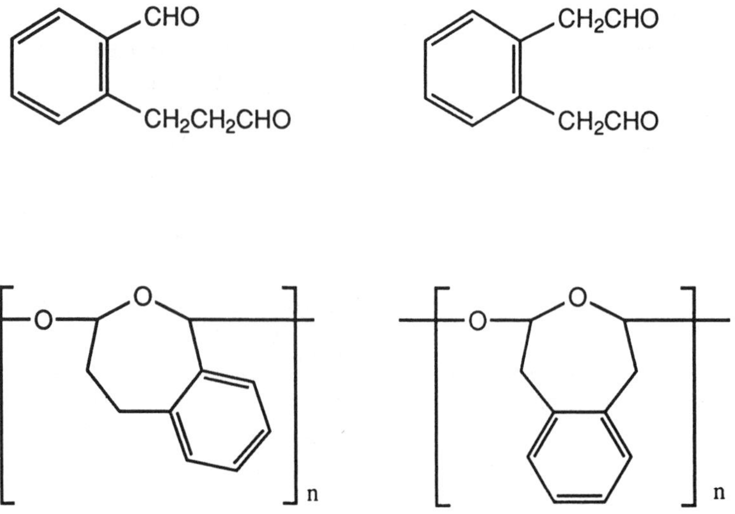

(5-16)

phthalaldehyde and isophthalaldehyde resisted polymerization. It was proposed that dipole-dipole interactions between the two adjacent aldehyde groups promoted the cyclopolymerization process through an intermediate or in a concerted manner, rather than the stepwise manner.[1-183]

The cationic initiated copolymerization of o-, m-, and p-phthaldehydes (r_2) with styrene (r_1) has been studied.[64] The monomer reactivity ratios were $r_1 = 0.77$, $r_2 = 0$ for the *meta*-isomer and $r_1 = 0.60$, $r_2 = 0$ for the *para*-isomer. Neither the *meta*- nor the *para*-isomer participated in the polymerization and acted simply as a compound having an electron-withdrawing group, thus reducing the cationic reactivity of these monomers.

Several investigations of anionic cyclopolymerizations of dialdehydes have been made. It was reported as early as 1908[45] that glutaraldehyde yielded a soluble solid substance on standing at room temperature. No structure was suggested for this substance. More recent polymerization of glutaraldehyde with several catalysts to yield a cyclic structure has been reported.[65] A head-to-tail intra-intermolecular mechanism was assumed in all catalyst systems studied—diethylzinc, aluminum triisobutyl, and boron trifluoride.

The IR data and solubility of the polymer in organic solvents supported the proposed structure of polyglutaraldehyde. The thermally initiated polymer slowly decomposed at room temperature to regenerate the monomer. The organometallic catalysts, on the other hand, yielded polymers stable to 160°C.

3-Substituted glutaraldehydes were studied.[48] 3-Methyl and 3-phenylglutaraldehyde were polymerized to give the predicted cyclic structures along with a certain amount of the open chain structures.

Tischenko condensation polymerization of glutaraldehyde has been investigated.[66] The resulting polymer molecular weight was low because of concurrence of Tischenko condensation polymerization and spontaneous cyclopolymerization initiated by traces of water. A structural study of the polymer of glyoxal, obtained by gas phase polymerization, led to the suggestion that the IR spectral data obtained could be attributed to condensed six-membered lactone rings of the polyether backbone and their equilibration in water with the corresponding hydroxy acid.[67] Dialdehyde monomers or their mixtures were polymerized in presence of water and basic catalysts having a pKa greater than 8.0.[68] The polymers were soluble in o-chlorophenol and benzyl alcohol.

2,2′-Biphenyldicarbaldehyde was copolymerized with styrene by cationic initiation to yield copolymers of molecular weight 6,000–7,000.[69] The monomer reactivities were reported. It was concluded, on the basis of pendant aldehyde groups, that the intramolecular cyclization was competitive with addition of comonomer during the propagation step. In a continuing study of bifunctional monomers, polymerization of adipaldehyde was studied in presence of boron trifluoride-etherate as initiator.[70] The polymer was partially soluble, but the pendant aldehyde content was rather high (60–70%). It was concluded that this aldehyde cyclopolymerized to a lesser extent than did glutaraldehyde or succinaldehyde. Copolymerization of adipaldehyde with chloral with the same initiator gave a copolymer of 1:1 composition that was almost completely cyclized.

Phthaldehyde was found to undergo cyclopolymerization with ease in presence of several cationic initiators as well as by X-ray irradiation.[71] The polymer was composed entirely of the dioxyphthalan unit. Enhanced polymerizability of phthaldehyde as compared with other aldehydes was observed, and was explained in terms of the intermediate type, or preferably, by a concerted propagation scheme. Kinetic parameters were reported, and the propagating chain end was considered to be "living". The ceiling temperature of the polymer was estimated to be −43°.

The nature of glutaraldehyde in aqueous solution has been studied.[72] It was proposed that polymers consisting of tetrahydrofuran rings linked by oxygen at the 2,2′-positions were formed spontaneously, apparently catalyzed by water. This structure was consistent with that proposed in earlier work.

Stereospecific polymerization of o-phthaldehyde was accomplished by use of a catalyst made up by addition of water to an aluminum complex catalyst.[73] This catalyst significantly increased the polymer yield and crystallinity.

Reduction of high-molecular-weight polymethacrolein of the predicted cyclic structure was accomplished by use of lithium aluminum hydride to poly(methallyl alcohols) still containing tetrahydropyran units, which were not reduced.[74] Poly(phthaldehyde) was obtained in 90% yield by polymerizing o-phthaldehyde with boron trifluoride diethyl etherate in methylene chloride at temperatures below −70°C, followed by quenching with pyridine.[75]

The formation of intramolecular and intermolecular crosslinks have been studied and examined for poly(vinyl alcohol) (PVA) of degree of polymerization of 325, 740, 1250, or 2595, crosslinked with terephthalaldehyde or glutaraldehyde in aqueous or DMSO solutions.[76] In solutions containing $\leqslant$0.1% (PVA), intramolecular crosslinks were favored. In solutions containing 0.6% PVA, intermolecular crosslinking was favored. The effects of the two types of crosslinking on viscosity and on gel chromatograms were discussed.

In a report on frontiers for photoresist chemistry, poly(phthaldehyde) was considered to have promise since it can be depolymerized and volatilized by the photochemical reaction.[77] The ceiling temperature for this polymer has been estimated to be −43°C.[71]

CYCLOPOLYMERIZATION OF DIISOCYANATES

Organic isocyanates are converted to their dimers (1,3-disubstituted uretidinediones) and their trimers (1,3,5-trisubstituted isocyanurates) in the presence of basic catalysts.[78,79] It has been shown that a basic catalyst such as sodium cyanide in dimethylformamide is also effective for the polymerization of both aliphatic and aromatic isocyanates to yield linear high molecular weight polymers which from a structural view point are sometimes referred to as "l-nylons."[80]

The cyclopolymerization of several aliphatic 1,2-diisocyanates using sodium cyanide in dimethylformamide as catalyst below $-30°C$ has been reported.[81] Soluble high molecular weight products were obtained which appeared to have the structure of poly(ketoethyleneureas) as shown by IR analysis (Table 5–2). 1,2,3-Triisocyanatopropane also polymerized to give a linear polymer for which a bicyclic recurring unit was assumed. An alternating intramolecular-intermolecular chain growth mechanism analogous to that postulated for 1,6-dienes was proposed. The 1,3-diisocyanates, 1,3-diisocyanato-2-oxapropane, and 1,3-diisocyanatopropane gave only minor amounts of cyclopolymerization, and the polymers were crosslinked. It was concluded that cyclopolymerization for diisocyanates strongly favors formation of five-membered rings, and this differs from that for diolefins where the preferred course appears to be the formation of six-membered rings. However, it had been reported earlier[1-44] that 1,3-diisocyanatopropane gave cyclopolymers which were soluble in sulfuric acid and cresol. The polymerization was initiated by use of anionic initiators such as sodium cyanide in dimethylformamide and butyllithium in heptane. The polymers were characterized by solubility determinations, differential thermal analysis, thermogravimetric analysis, infrared spectroscopy, x-ray diffraction, and viscosity measurements. The polymers were linear, crystalline, thermally stable materials having inherent viscosities up to 0.12 dl/g. The observed properties were consistent with that shown in Structure 5–17. This cyclopolymerization was reported to be the first for monomers containing heterodouble bonds.

In a similar type of cyclopolymerization, polymerization of 1,2-diisocyanates and 1,2,3-triisocyanates to cyclic polymers by an anionic mechanism have been described.[1-45] Sodium cyanide in dimethylformamide was used as initiator. As evidence for the cyclic nature of the polymer a split carbonyl absorption in the IR region at about 5.6 and 5.9μ was attributed to the presence in the polymer of two different types of carbonyl groups, one being intra-ring and the other being inter-ring. The structure obtained from propane-1,2,3-triisocyanate is shown in Structure 5–18. These polymers possessed inherent viscosities up to 1.04 dl/g.

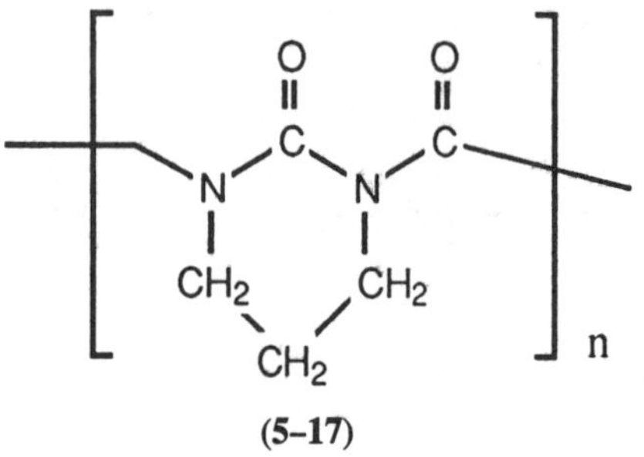

(5–17)

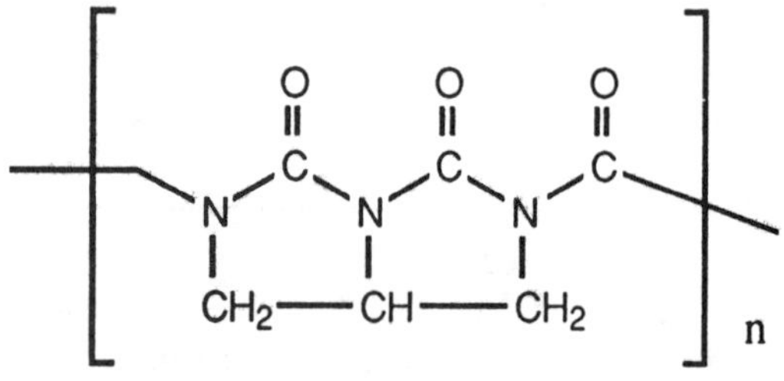

(5–18)

Table 5–2 Cyclopolymers of Aliphatic 1,2-Diisocyanates

Monomer[a]	Ring size	Melting point (°C)	η inh.[b]	Solvent
1,2-Diisocyanatoethane[c]	Five	365	0.85	Sulfuric acid, tetrachloroethanephenol (40/60 wt)
1,2-Diisocyanatopropane	Five	287	1.35 0.56[d] 0.72[e]	Formic acid, dimethyl sulfoxide nitromethane, dimethylformamide
1,2-Diisocyanatocyclohexane	[4.3.0]-Bicyclic	330	0.38	Pyridine, nitromethane, dimethyl sulfoxide
1,2,3-Triisocyanatopropane	[3.3.0]-Bicyclic	365-400	0.28[f]	Sulfuric acid

[a] Solvent, DMF; initiator: NaCN in DMF, −30° to −50°C.
[b] Inh. viscosity = (In η rel)/c; solvent, tetrachloroethanephenol (40/60 wt); conc'n, 0.5% w/v, 30°C.
[c] Polym. without catalyst.
[d] No. av. mol. wt. = 45,000 by osmotic pressure (nitromethane solution).
[e] Bulk polym. under N_2 (0°C).
[f] In sulfuric acid.

Cyclopolymerizations of α,ω-polymethylene diisocyanates [where n = no. of methylenes = 1, 2, 3, and 4] have been investigated.[82] All the polymers obtained in this work had essentially linear structures since they were soluble in some organic solvents (Table 5–3). Only polymer from monomer in which n = 1 gave an absorption band in the IR spectrum characteristic of the isocyanato group. Thus the degree of cyclization of polymers from monomers in which n = 2–4 appeared to be 100%. The mechanism of this cyclopolymerization was considered and it was indicated that the initiation step occurred by attack of the catalyst anion on the carbon atom of one of the isocyanato groups, followed by cyclization and intermolecular reaction of the hybrid anion with another monomer molecule (5–19). This was supported by the fact that the anionic polymerization of monoisocyanates gave "1-nylons" as pointed out earlier.[81,83] In the cyclic intermediate (5–19a), however, the negative charge is delocalized over the nitrogen and oxygen atoms. The succeeding intermolecular propagation reaction occurs by the opening of the carbon-nitrogen double bond of another diisocyanate molecule by the anion (5–19a). If the nitrogen in the hybrid anion has a greater negative charge than the oxygen, Structure 5–19b would be formed predominantly. In the reverse situation, the predominant formation of Structure 5–19c would be expected. The IR spectra of the polymers provided evidence which could correlate with either of these structures.

Quantitative aminolysis of the polymers using di-n-butylamine was investigated. The amine was consumed when the polymers were treated with an excess in dimethylformamide containing a catalytic amount of sodium cyanide. The amount of amine consumed was measured by titrating aliquots of the mixture with hydrochloric acid over a period of several days. A relatively fast reaction occurred over the first 10 hrs, followed by an extremely slow reaction continuing up to and beyond 50 hrs. By analogy with compounds containing similar groups, the rapid uptake of amine was believed to correspond with the reaction of the grouping -N=C-O-CO- with one mole of the amine, leading to fission of the oxygen carbonyl bond and the slow uptake with reaction of the grouping -CO-N-CO-N with fission of the nitrogen-carbonyl bond. The approximate amounts of each structure in the polymers are given in Table 5–3.

cis-1,3-Diisocyanatocyclohexane has been polymerized under similar conditions.[4–5] Polymers were obtained in high yield, as white powders with melting points varying

Table 5–3 Properties of Polymers of Polymethylene Diisocyanates[82]

	Decomposition temperature (°C)	Solubility	Structure (%)		
			5–19b	5–19c	Open chain
Polymer (n = 1)	350	Dimethyl sulfoxide	66	34	34
Polymer (n = 2)	270–280	Dimethyl sulfoxide	0	100	0
Polymer (n = 3)	350	*m*-Cresol	76	24	0
Polymer (n = 4)	350	Ethylene carbonate	40	60	0

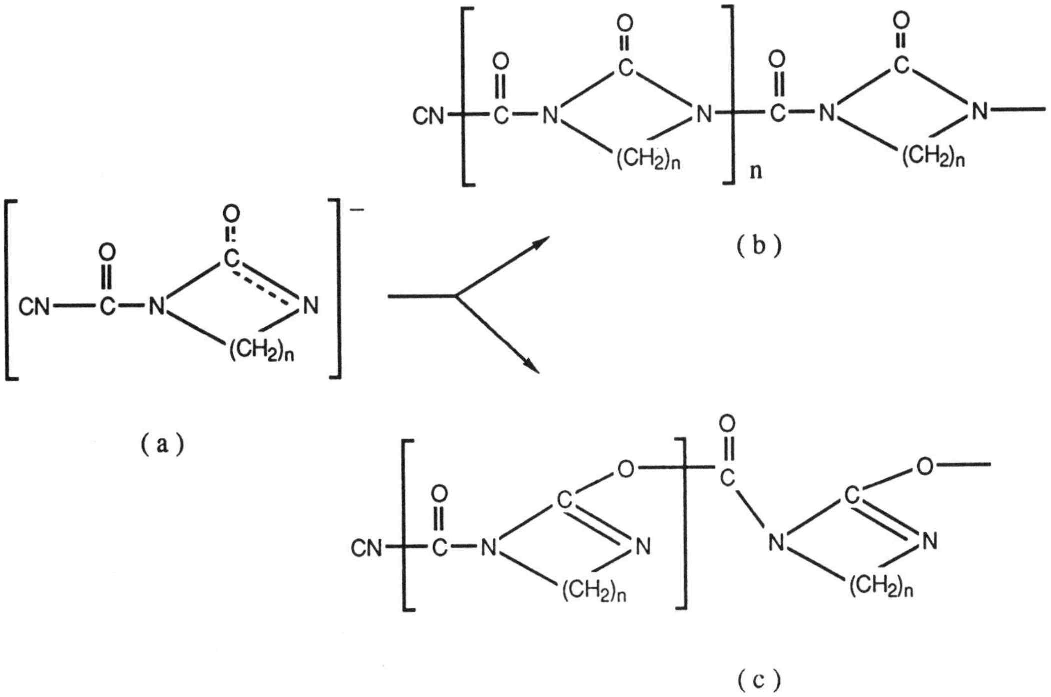

(a)

(b)

(c)

(5–19)

from 224 to 265°C, and [η] in formic acid (98–100%), of about 0.1 dl/g. The polymers contained little or no residual isocyanato groups, and were soluble in m-cresol and formic acid (98–100%), being recovered unchanged from such solutions. Thus the polymers have a linear structure consistent with the cyclopolymerization mechanism. Two different structural units (5–20) appear to be possible, depending upon whether polymerization occurred through the N=C or the C=O group in the intramolecular step. The proportions of these units were estimated by a quantitative aminolysis with di-n-butylamine, similar to the method developed earlier.[82] The results suggested that the structures (5–20) were present in the polymer in the ratio of 65:35, favoring reaction through the N=C bond.

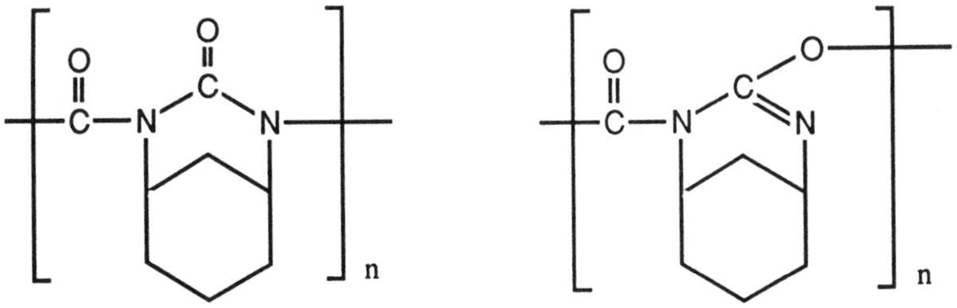

(5–20)

It has been observed that if the isocyanato grouping is attached to a secondary carbon atom, then polymerization does not occur using sodium cyanide in dimethylformamide as catalyst. For example, in the case of 2-isocyanatopropane,[80] isocyanatocyclohexane,[80] and *trans*-1,3-diisocyanatocyclohexane,[59] in which the axial-equatorial arrangement of the isocyanato groups does not allow the formation of a cyclopolymer. However, 1,2-diisocyanatocyclopropane,[81] 1,2-diisocyanatocyclohexane,[81] and cis-1,3-diisocyanatocyclohexane[4-5] readily polymerized. It appears therefore that formation of a five- or six-membered ring constitutes a powerful driving force to effect polymerization through an otherwise poorly polymerizable structure.

The cyanide initiated cyclopolymerization of *cis* and *trans*-1,3,5-triisocyanatocyclohexane was carried out to obtain polymers in both cases containing predominantly bicyclic structures, with one isocyanate group per original monomer molecule intact.[2-99]

cis-Diisocyanatocyclohexane was polymerized using NaCN as initiator to give soluble polymers which contained little or no residual isocyanato groups supporting that cyclopolymerization had occurred.[84] Two different structural units are possible for the cyclopolymer and the proportions of each unit were estimated by a quantitative aminolysis using di-(n-butyl)amine. The *trans* isomer did not polymerize under similar conditions.

The literature of the cyclopolymerization of diisocyanates has been reviewed in several publications.[1-156,1-135,85]

A pulsed NMR study of cyclopolymerization of 4,4′-diphenylmethane diisocyanate reported kinetics of cyclopolymerization of the monomer in the presence of $P[N(C_2H_5)_2]_3$ catalyst at 323–343°K. IR spectra of the polymer indicated that cyclotrimerization of the monomer was accompanied by cyclodimerization of the isocyanate groups[86] Cyclopolymerization of isocyanates in the presence of other organophosphorus compounds has been observed. The kinetics and mechanism of polymerization of 2,4-tolulene diisocyanate (2,4-TDI), 1,6-hexamethylene diisocyanate and 1,4-tetramethylene diisocyanate, in the presence of triethyl phosphite or triisopropyl phosphite catalysis at 20–80°, have been discussed.[87] The homo- and copolymerization of $OCN(CH_2)_4NCO$ and $OCN(CH_2)_6NCO$ in bulk in the presence of $(C_2H_5)_3N)_3P$ catalyst occurred with cyclization, giving heterochain polymers containing isocyanurate and uretidinedione rings.[2-118]

Cyclopolymerization of 2,4-TDI in the presence of phosphorus triamides has been studied. $(R_2N)_3P$ ($R = CH_3, C_2H_5, C_4H_9$) catalyzed both the cyclodimerization and cyclotrimerization of 2,4-TDI at 23–60° to give oligomers containing uretidinedione and isocyanurate units as soluble, white powders, stable to 280° and decomposing at 400–420°.[88]

The effects of intramolecular reactions on network properties have been studied. The correlations between the modulus of polyurethane networks formed at complete reaction and the extent of intramolecular reactions at gelation were interpreted via use of mathematical equations. The occurrence of post-gel intramolecular reaction prevented attainment of the perfect network.[89]

A synthetic approach to "ladder"-type polymers via a two-stage polymerization of vinyl isocyanate (5–21) has been reported.[90] Vinyl isocyanate (5–21a) was polymerized through the isocyanate group by using sodium cyanide as initiator in a solution of dimethylacetamide at a temperature of −55°C. The N-vinyl-1-nylon (5–21b) was isolated, characterized, and redissolved in dimethylacetamide. It was then polymerized with azobisisobutyronitrile and UV light to the structure 5–21d having 90% ladder structure. The alternate pathway was followed by spontaneous polymerization to 5–21c at 25°C and then irradiation with ^{60}Co at 30° for 48 hrs to a polymer with 85% ladder structure.

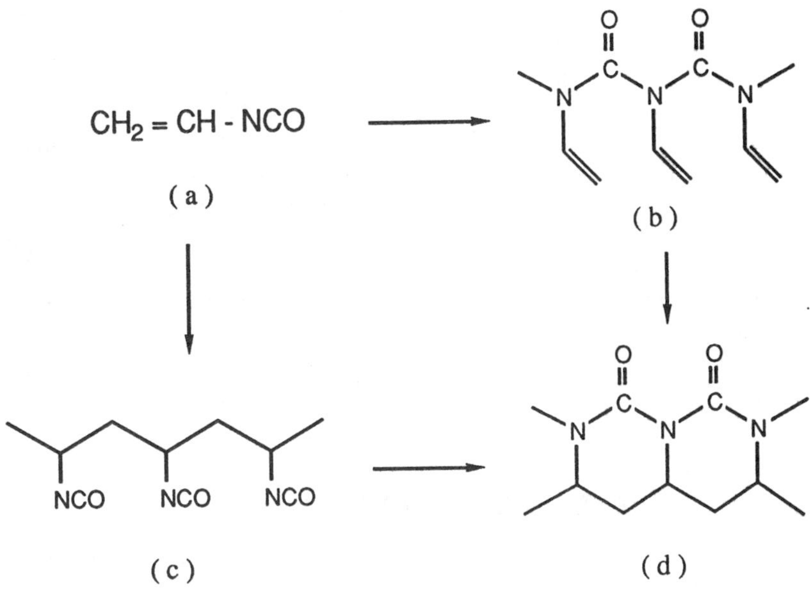

$CH_2 = CH - NCO$

(a)

(b)

NCO NCO NCO

(c)

(d)

(5–21)

A study of the cyclopolymerization of 2,4-TDI in the presence of triethyl- and triisopropyl phosphites has been conducted.[91] Cyclopolymerization in the presence of $(C_2H_5O)_3P$ or (iso-$C_3H_7O)_3P$ catalyst, formation of both a cyclic dimer and phosphorylated polyisocyanate oligomers (containing isocyanurate rings with a hetero P atom) occurred. The reaction could be shifted in one direction or the other by varying reaction conditions. Dimerization rate constants and properties for the phosphorylated oligomers were reported.

Isocyanurate polymers have been synthesized from aliphatic diisocyanates.[92] For example, hexamethylene diisocyanate was polymerized, in presence of sodium or potassium salts of fatty acids, to produce the polymers. 2,3-Diisocyanato-1-propene was synthesized and its polymerization characteristics studied.[93] Polymerization occurred at room temperature in absence of light. 1,2-Diisocyanatoalkanes, suitable as cyclopolymerizable monomers, were synthesized.[94] For example, 1,2-diisocyanatoethane was obtained via thermal decomposition of 2-imidazolidinone-N-carbonyl chloride.

In an organic chemical study of the reactions of ethylene diisocyanate, it was observed that with active hydrogen compounds such as primary and secondary amines, alcohols, and mercaptans, novel cyclic 1:1 adducts, 1-substituted-2-imidazolidinones, were obtained in high yields.[95] Linear polymers were prepared by polymerizing trimethylene diisocyanate via an anionic initiator.[96] The high-melting polymer was soluble in sulfuric acid with slight decomposition and was postulated to have the structure described earlier (5–17). Aromatic o-diisocyanates, a new class of compounds, were synthesized and described.[97] No polymerization reactions were reported.

1,3,5-Cyclohexane triisocyanate was synthesized as a crosslinking agent for use in the manufacture of polyurethane foams.[98] 1,3,5-Cyclohexanetricarbonyl chloride[99] was synthesized as an intermediate for synthesis of the 1,3,5-cyclohexane triisocyanate, reported earlier.[98]

Cyclopolymerization of 2,4-TDI with 1,6-hexamethylene diisocyanate, in presence of sodium diethyl phosphite, has been reported.[100] Copolycyclotrimerization of 2,4-TDI with hexamethylene diisocyanate in presence of glycidyl phenyl ether-triethylamine catalysts did not contain linear polymers and products of homocyclotrimerization, but isocyanurate rings were shown to be present.[101] The reactivity ratios of the monomers were determined. The rate of cyclopolymerization of toluene diisocyanate in the presence of potassium acetate accelerated with temperature increase up to 100°C.[102] Cyclopolymerization was more controllable in presence of lithium acetate. A possible mechanism was discussed.

A polarographic method for monitoring the cyclopolymerization of isocyanates has been developed.[103] The reaction was studied using the model system phenyl isocyanate-lithium acetate or potassium acetate triphenylisocyanurate. It appears that the terms "cyclopolymerization" and "cyclotrimerization polymerization" were used interchangeably in this publication.

CYCLOPOLYMERIZATION OF DIEPOXIDES

1,2,5,6-Diepoxyhexane was polymerized as early as 1964 under the influence of a variety of catalysts to afford soluble polymers for which tetrahydropyran recurring units were proposed. Aluminum isopropoxide, phosphorus pentafluoride-water, diethylzinc and triisobutyl aluminum were employed as catalysts. A diethylzinc-water catalyst system gave the highest molecular weight polymer with limited solubility {η = 0.45 dl/g] for 1,2,5,6-diepoxyhexane. The IR spectrum was consistent with a cyclic ether. An anionic mechanism for polymerization of 1,2,5,6-diepoxyhexane (5–22) was proposed.

$$CH_2\overset{O}{—}CH—(CH_2)_2—CH\overset{O}{—}CH_2 \longrightarrow \left[CH_2—\text{(tetrahydropyran ring)}—O \right]_n$$

(5–22)

A rigid structure was proposed to explain the limited solubility of the polymer. Polymerization of 1,2,5,6-diepoxyhexane and o-(diepoxyethyl)benzene were also investigated.[104] The properties of the polymers suggested that cyclopolymerization had occurred. Results of a subsequent study confirmed the proposed cyclopolymerization of 1,2,5,6-diepoxyhexane. Strong evidence was given in support of a linear polymer structure composed of symmetrically substituted tetrahydrofuran rings.[105] It has also been shown that methenolysis of 1,2,5,6-diepoxyhexane yielded tetrahydrofuran-2,5-dimethanol.[106]

Cyclopolymerization of N,N-diglycidylaniline has also been carried out, and an analogous structure consisting of symmetrically substituted enchained morpholine units was proposed. Molecular weights up to 81,800 were obtained. The *meso-* and *dl* diastereoisomers of o-di(epoxyethyl)benzene were polymerized by a variety of catalysts.[107] A cyclopolymer was obtained which consisted mainly of isochroman recurring units.

It has been reported that polymerization of 4-vinylcyclohexene diepoxide in presence of trimethylamine and water as catalysts yielded polymers soluble in organic solvents.[108] The polymer contained 4.5 milliequivalents of epoxide per gram.

N,N-Bis(2,3-epoxypropyl)aniline (N,N-diglycidylaniline) was reinvestigated and shown to undergo ring opening and cyclopolymerization to give linear polymers containing 2,4,6-substituted morpholindiyl units (5–23), using potassium *tert*-butoxide as

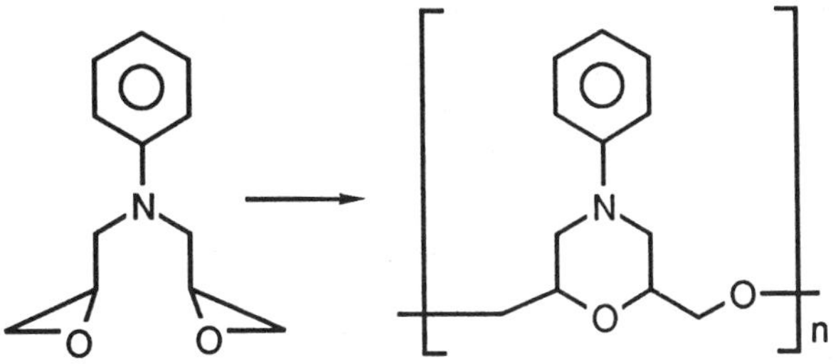

(5–23)

initiator,[109,110] in tris(dimethylamino)phosphoramide as solvent. The polymers were soluble in benzene, chloroform, tetrahydrofuran and dimethylformamide but were insoluble in acetone and methanol. These and other studies on a variety of diepoxides and their participation in cyclopolymerization have been reported in earlier reviews.[1-183,1-178]

During three-dimensional polymerization of diglycidyl phthalate with aromatic diamines, side reactions involving cyclization and aminolysis can result in network defects. The presence of such reactions and their relative contribution to the overall reaction was established by gel chromatographic analysis of model systems of diglycidyl phthalate with aniline or o-anisidine.[111]

The effects of cyclization reactions on topological structure and properties of crosslinked polymers prepared from diglycidyl ethers and aromatic amines have been studied. Ring formation in the crosslinking of epoxy resins was studied using a statistical model. Systems based on ortho diglycidyl ethers or esters exhibited a marked tendency to form cyclic defects in the linear network.[112]

The synthesis of polymers with thiacrown ether units via cyclopolymerization of 1,2-bis(2,3-epithiopropoxy)benzene and 1,2-bis[2-(2,3-epithiopropoxy)ethoxy]benzene has been reported.[113] The higher metal-binding capacity of the 1,2-bis[2-(2,3-epithiopropoxy)ethoxy]benzene homopolymer compared to the 1,2-bis(2,3-epithiopropoxy)benzene homopolymer was related to a larger size of the thiacrown ether rings of the former polymer.

Cyclization in the reaction between diglycidylaniline and amines has been studied. The reaction of diglycidylaniline with N-methylaniline or aniline was investigated by HPLC, and the products were identified by mass spectrometry. An eight-membered ring formed by the intramolecular reaction of the secondary amino group and the epoxy group in the diglycidylaniline-aniline monoadduct was the main cyclic product.[114]

The kinetics of cyclopolymerization of N,N-bis(2,3-epoxypropyl) aniline were studied.[115] High initial rates of the polymerization, with oxazine ring formation, initiated by potassium t-butoxide, were observed. After a 20% conversion, the reaction followed first order kinetics in monomer. The order in catalyst was 1.8 indicating termolecular reactions. Another report on cyclopolymerization of diepoxy amines, has been published.[116] Soluble polymers were obtained. This study of cyclopolymerization of diepoxy amines was continued with determination of the cyclization constant.[117]

The concept of cyclopolymerization was applied to the polymerization of o-bis(epoxyethyl)benzene and 1,5-hexadiene oxide.[118] By use of basic or Lewis acid catalysts, both monomers led to linear soluble polymers with only a few epoxy groups remaining, suggesting that a high degree of cyclopolymerization had occurred. The polymer formed from the benzene derivative was postulated to consist of isochroman ring units connected with each other through the 1- and 4- positions.

Glycidyl ethers of 3,3′-diallydiene[2,2-bis(4-hydroxy-3-allylphenyl)propane] were prepared by reaction with epichlorohydrine in presence of sodium hydroxide.[119] Use of this glycidyl ether produced a high degree of crosslinking when mixed with hexahydro-1,3,5-triacryloyltriazine m-phenylenediamine and benzoyl peroxide and heated. The glycidyl ether of 2-allylphenol could be substituted for the triazine.

Transannular reactions of cis,cis-1,5-cyclooctadiene occurred upon reaction with perbenzoic acid, followed by lithium aluminum hydride reduction.[120] In addition to the normal products, the transannular products, 1,4-epoxycyclooctan-8-ols and 1,5-epoxycyclooctan-8-ols were identified.

A study of cyclization of bis(2,3-epoxypropyl)ether in the reaction with butanol catalyzed with sodium butoxide gave the *cis*- and *trans*-isomers of the dioxane and dioxepane structures (5–24).[121] The ratios of the two structures varied with solvent and temperature. Bis(glycidylether) of polyethylene glycols in which the degree of polymerization was four to 20 were used in finishing of polyamide fiber materials.[122] No discussion of cyclization was offered.

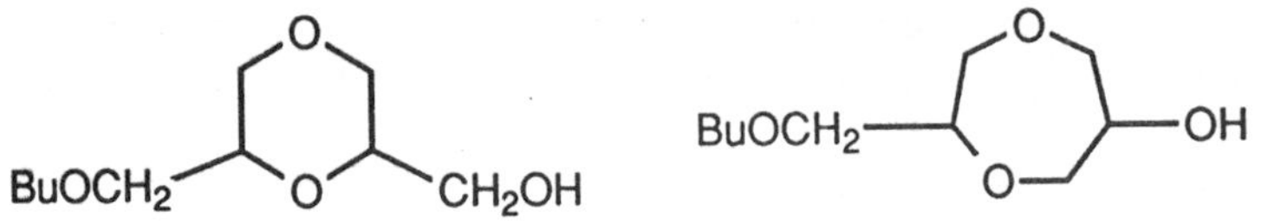

$BuOCH_2$ O O CH_2OH $BuOCH_2$ O O OH

(5–24)

Bis-glycidyl esters were synthesized by treating the diallyl ether of tetrachloroxylenes with perbenzoic acids.[2-426] Polymers containing crown ether units were synthesized by cyclopolymerization of diepoxide monomers such as 1,2-bis(2,3-epoxypropyl)benzene producing polymers with 13- and 9-membered rings. The polymers were evaluated for their cation-binding ability.[2-483]

CYCLOPOLYMERIZATION OF DINITRILES

It has been shown that succinonitrile can be polymerized by basic catalysts such as sodium methoxide in methanol at atmospheric pressure.[123] At room temperature, the reaction was very slow, requiring several weeks, and yielded a black solid. The structure (5–25) was postulated where n was from six to 65 according to the conditions. The

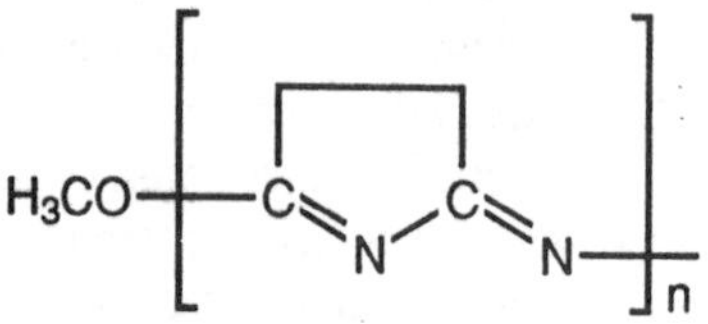

H_3CO C N C N n

(5–25)

extreme intractibility of the polymer prevented proof of structure. Reaction occurred with explosive violence when malononitrile, succinonitrile, glutaronitrile, and adiponitrile were heated under high pressure. The polymerization of perfluoroglutaronitrile, treated with 0.3 mole % diethylamine as catalyst, was investigated at pressures from 10,000 to 20,000 kg/cm^2. The polymer may have either a ring or chain form, the former being a network of triazine rings connected by -$(CF_2)_3$- while the latter was considered to have the cyclic structure analogous to 5–25. At pressures above 20,000 kg/cm^2 the pure polytriazine structure was produced. The product was a transparent yellow or light amber solid. At lower pressures the amount of the cyclic structure increased. At 7000 kg/cm^2 the product was a black solid.

Succinonitrile has been polymerized, in bulk or in solution, at elevated temperatures in the presence of basic catalysts, to form a brown-black polymer.[124] The reaction was slow at 70°C, requiring several days, whereas at 135°C in dimethyl sulfoxide (1:1) the mixture became solid after a few hours. The reaction was catalyzed by traces of oxygen, but polymerization occurred even after exhaustive exclusion of oxygen. The reaction was characterized by a long induction period followed by a rapid autoacceleration to almost complete reaction, forming a black solid, which may explain the explosive nature of the reaction at high pressure, observed earlier.[123] The polymer was partly soluble in methanol, 6 N hydrochloric acid, acetic acid, formic acid, phenol, dimethyl sulfoxide, and succinonitrile. The NMR spectrum of the polymer was consistent with structure 5–25. The polymer reacted rapidly with atmospheric oxygen to form the nitrone, where the ratio of imine and nitrone bonds was dependent on the conditions. Glutaronitrile turned black after prolonged heating at 145°C in the presence of base. Under the same conditions, malononitrile darkened and thickened while standing at room temperature for extended periods of time; adiponitrile did not darken under these conditions.

Homopolymers[125,126] and copolymers of fumaronitrile , maleonitrile, succinonitrile, diphenylmaleonitrile, *meso*-, and *dl*-diphenylsuccinonitrile in the presence of free radical initiators of medium-high decomposition temperature (Table 5–4) were obtained. Black, infusible but soluble polymers, were obtained. All fumaronitrile and maleonitrile polymers were soluble in different solvents depending upon their inherent viscosity: η inh. up to about 0.20, soluble in dimethylformamide, dimethylsulfoxide, hexamethylphosphoramide; η inh. 0.20–0.40, soluble in sulfuric, phosphoric, formic, and methanesulfonic acids; η inh. above 0.40 soluble in fuming sulfuric acid. Low conversion but high molecular weight (e.g. η inh. = 1.3) succinonitrile homopolymers were easily soluble in acidic solvents. The other polymers were only partly soluble in dimethylformamide and dimethyl sulfoxide but completely soluble in sulfuric, phosphoric, and methanesulfonic acids. None of the materials had a melting point below 530°C or seemed to melt in an open flame. On the basis of the physical appearance of the polymers, their solubility, and analytical data, as well as chemical and spectroscopic evidence, the completely conjugated structures (5–26) were proposed. The polymers possessed good thermal stability.

The structure of cyclopolymers of dinitriles has been investigated.[127] By use of pulsed and wide-line NMR, some properties of the homopolymer of fumaronitrile were determined. The method was applicable to determination of number-average molecular weight of polymers that may be insoluble or infusible but not crosslinked, e.g. 1,2-dinitrile cyclopolymers derived from 1,2-dinitriles, such as fumaronitrile. The method involved studies on the solid polymers.

Table 5–4 Polymers from 1,2-Dinitriles[a]

Monomers, (wt.%)							Temp.	Time	Yield	
(1)	(2)	(3)	(4)	(5)	(6)	(7)	(C°)	(hrs)	(%)	η inh.[b]
100	—	—	—	—	—	—	160	43	72	0.35
—	100	—	—	—	—	—	150	48	17	0.31
—	—	100	—	—	—	—	150	24	15	1.28
—	—	—	100[c]	—	—	—	330	2	85	0.07
—	—	—	—	100[d]	—	—	285	7	43	0.06
—	—	—	—	—	100[d]	—	285	7	43	0.06
—	—	—	—	—	—	100	160	29	30[e]	—
50	—	—	—	—	—	—	160	47	45	0.18
50	—	—	50	—	—	—	160	47	30	0.15
50	—	—	—	50	—	—	160	47	43	0.28
50	—	—	—	—	—	50	160	27	90	0.12
—	50	50	—	—	—	—	160	23	90	0.67
—	—	—	50	—	—	50	160	74	20	—
—	—	—	—	50	—	50	160	23	27	—

[a] 4% di-t-butyl peroxide as initiator unless indicated.
[b] Approximately 0.4 g per 100 ml methanesulfonic acid at 30°C.
[c] 8% t-butyl hydroperoxide as initiator.
[d] 4% t-butyl hydroperoxide as initiator.
[e] Product; tetramer, phthalocyanine.
Note: (1) = Fumaronitrile; (2) = Maleonitrile; (3) = Succinonitrile; (4) = Diphenylmaleonitrile, (5) meso-Diphenylsuccinonitrile; (6) = dl-Diphenylsuccinonitrile; (7) = Phthalonitrile.

A number of nitriles, including succinonitrile, malonitrile, fumaronitrile, tetracyanoethylene, benzonitrile, benzyl cyanide, o-phthalonitrile, and 1,2,4,5-tetracyanobenzene were polymerized by cationic, anionic, radical, and Ziegler-Natta catalysts, giving polymers which contained in-chain C=N units.[128] Succinonitrile polymer was reported to possess a cyclic structure.

Succinonitrile has also been polymerized with cationic initiators to polymers containing cyclic units analogous to the repeat unit of 5–25.[129] The polymers were also semiconductors. Cationic initiators used were Lewis acids such as zinc chloride. The polymerization proceeded by a second order reaction. This monomer was also polymerized coordinatively by Ziegler-Natta catalysts in a reaction that was approximately first order. Polysuccinonitrile prepared by the former method was branched and contained C:N linkages in the chain, while that prepared by the latter method contained a large proportion of units having the structure shown. The polymers were semiconduc-

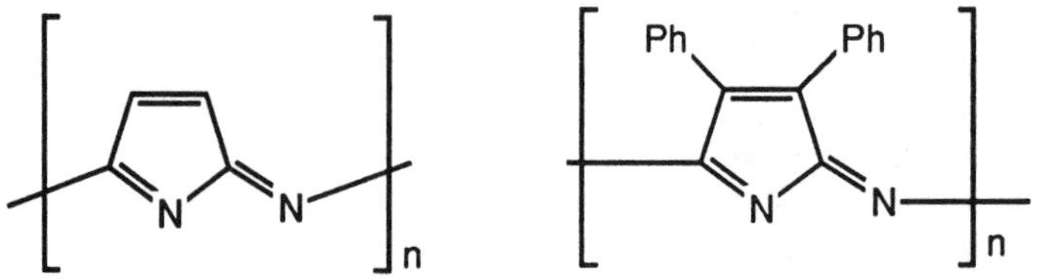

(5–26)

tors, with conductivities $1.5 \times 10^{-5} - 1.8 \times 10^{-10}$ $ohm^{-1}cm^{-1}$ at 300°K, and had high thermal stability, with weight loss ~20% in nitrogen at 800°C for the most stable sample.

Cyclopolymerization of aromatic nitriles to polymers with ring-chain structures has been observed. A review, with thirteen references, of the preparation, properties, and uses of heat-resistant polymers obtained by the triaryl-s-triazine ring-forming reaction of terminal nitrile groups on polyimides, polybenzimidazoles and polyimidazopyrrolones has summarized this work.[130]

A non- or partially crosslinked triaryltriazine ring-containing polymer or copolymer was prepared by cyclopolymerizing an aromatic nitrile-modified oligomer precursor in the presence of an excess of aromatic nitrile (chain extender) at 100° and >200 psi.[131]

It has been shown that dinitriles such as 3-hydroxyglutaronitrile can be cyclized with anhydrous hydrogen bromide to 2-amino-5-bromopyridine.[132] This facile cyclization is pertinent to the cyclopolymerization of dinitriles.

Adiponitrile was thermally polymerized in presence of potassium hydroxide to yield a dimer of known structure and a polymer of proposed structure shown (5–27):[133]

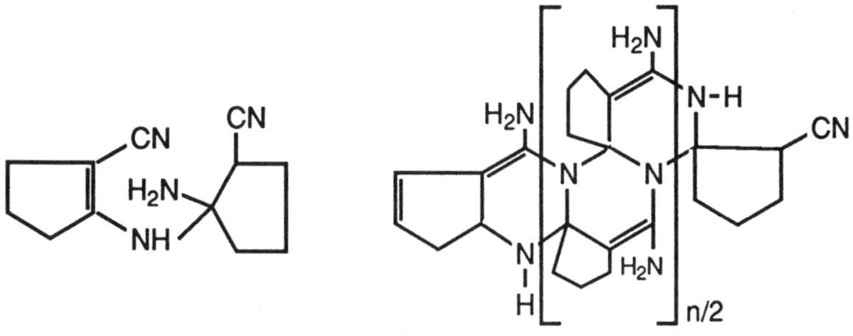

CN CN H_2N NH H_2N N-H H_2N CN N N N H H_2N n/2

(5–27)

Homopolymers and copolymers of fumaronitrile, maleonitrile, and succinonitrile were prepared by using medium-high-temperature free-radical initiators.[134] Black, nonfusible but soluble polymers were obtained. The evidence obtained indicated a structure containing α-pyrrolenine rings and free nitrile groups in the fumaro- and maleo- derivatives and 1-pyrroline rings and free nitrile groups in the succino derivative.

An aromatic, diether-linked resin, prepared from 4,4′-bis(3,4-dicyanophenoxy)biphenyl of Structure 5–28[135] has been found to exhibit excellent thermo-

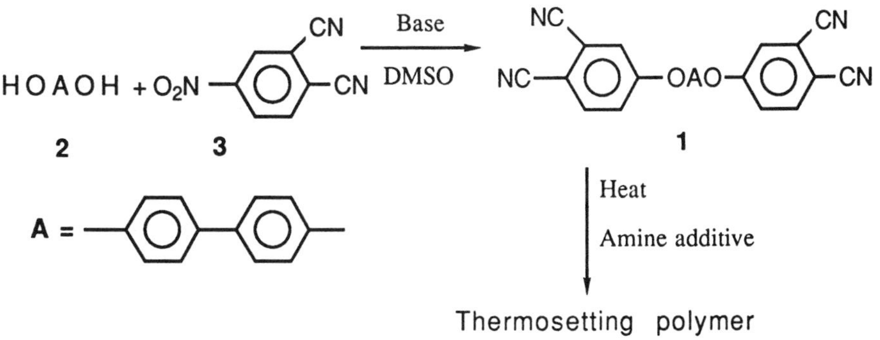

HOAOH + O_2N–(C₆H₃)(CN)–CN —Base, DMSO→ NC(NC)–C₆H₃–OAO–C₆H₃–(CN)CN

2 3 1

A = –C₆H₄–C₆H₄–

Heat

Amine additive

Thermosetting polymer

(5–28)

oxidative properties. Polymerization occurs by a cyclic addition reaction without the formation of volatile by-products. The structural relationship of this monomer to phthalonitrile and maleonitrile suggests that similar structural units were formed as in those cases.

REFERENCES

1. Stille, J. K. and Frey, D. A., J. Am. Chem. Soc., 1961, 83, 1697; CA 55, 21648a (1961).
2. Hubert, A. J. and Dale, J., J. Chem. Soc., 1965, 3160; CA 63, 5572c (1965).
3. Colthup, E. C. and Meriwether, L. S., J. Org. Chem., 1961, 26, 5169; CA 57, 1049f (1962).
4. Yakhimovich, R. I., Shilov, E. A. and Dvorko, G. F., Dokl. Akad. Nauk SSSR, 1966, 166, 388; CA 64, 12800b (1966).
5. Medvedeva, L. I., Fedorova, E. F. and Arbuzova, I. A., Vysokomol. Soedin., Ser. A, 1967, 9, 2042; CA 67, 117383b (1967).
6. Berlin, A.A., Ermakova, V. D. and Cherkashin, M.I., Vysokomolek. Soedin. Ser. B, 1972, 14, 305; CA 77, 48944c (1972).
7. Aso, C., Kunitake, T. and Saiki, K., Makromolek. Chem., 1972, 151, 265; CA 76, 113688t (1972).
8. McCormick, C. L. and Butler, G. B., J. Macromol. Sci., Rev. Macromol. Chem., 1972, 8(2), 201; CA 78, 4528m (1973).
9. Akopyan, L. A., Ambartsumyan, G. V., Ovakimyan, E. V. and Matsoyan, S. G., Vysokomol. Soedin., Ser. A, 1977, 19, 271; CA 86, 140508x (1977).
10. Akopyan, L. A., Ambartsumyan, G. V., Grigoryan, S. G. and Matsoyan, S. G., Vysokomol. Soedin., Ser. A, 1977, 19, 1068; CA 87, 53648v (1977).
11. Akopyan, L. A., Ambartsumyan, G. V., Matsoyan, M. S., Ovakimyan, E. V. and Matsoyan, S. G., Arm. Khim. Zh., 1977, 30, 771; CA 88, 137025y (1978).
12. Akopyan, L. A., Ambartsumyan, G. V., Ovakimyan, E. V., Gevorkyan, S. B. and Matsoyan, S. G., Arm. Khim Zh., 1978, 31, 510; CA 89, 180410c (1978).
13. Gibson, H. W., Bailey, F. C., Epstein, A. J., Rommermann, H. and Pochan, J. M., J. Chem. Soc., Chem. Commun., 1980, 426; CA 93, 115142k (1980).
14. Gibson, H. W., Bailey, F. C., Pochan, J. M. and Harbour, J., Polym. Preprs., ACS, Div. Polym. Chem., 1981, 22, 35; CA 93, 115142k (1980).
15. Pochan, J. M. and Gibson, H. W., Org. Coat. Plast. Chem., 1980, 43, 872; CA 96, 181778k (1982).
16. Ambartsumyan, G. V., Gevorkayn, S. B., Kharatyan, V. G., Gavalyan, V. B., Saakyan, A. A., Grigoryan, S. G. and Akopyan, L. A., Arm. Khim. Zh., 1984, 37, 188; CA 101, 73144p (1984).
17. Nishanbaeva, S., Yusupbekov, A. K. and Abdurashidov, T. R., Tr. Tashk. Politekh. Inst., 1972, 90, 101; CA 83, 131988q (1975).
18. Bantsyrev, G. I. and Cherkashin, M. I., Dokl. Vses. Konf. Khim. Atsetilena, 4th. 1972, 3, 115; CA 81, 78285p (1974).
19. Cheng, S.-S., Hua Hsueh T'ung Pao, 1958, 15; CA 56, 1328g (1962).
20. Hubel W. and Hoogzand, C., Chem. Ber., 1960, 93, 103; CA 54, 9839f (1960).
21. Meriwether, L. S., Colthup, E. C. and Kennerly, G. W., J. Org. Chem., 1961, 26, 5163; CA 57, 1048i (1961).
22. Meriwether, L. S., Colthup, E. C., Kennerly, G. W. and Reusch, R. N., J. Org. Chem., 1961, 26, 5155; CA 57, 1048e (1961).
23. Meriwether, L. S., Leto, M. F., Colthup, E. C. and Kennerly, G. W., J. Org. Chem., 1962, 27, 3930; CA 58, 429b (1962).

24. Luttinger, L. B., J. Org. Chem., 1962, 27, 1591; CA 57, 2045i (1962).
25. Luttinger, L., U.S. Pat. 3,098,843; Brit. Pat. 889,730 1962 (Feb. 21); CA 58, 5513d (1963).
26. Teyssie, P. and Korn-Girard, A., J. Polym. Sci., A2, 1964, 2, 2849; CA 61, 5777f (1964).
27. Luttinger, L., U.S. Pat. 3,131,155, 1964 (Apr. 28); CA 61, 2060h (1964).
28. Davydov, B. E., Demidova, G. N., Pintskhalava, R. N. and Rozenshtein, L. D., Elektrokhimiya, 1965, 1, 876; CA 63, 14988f (1965).
29. Chauser, M. G., Cherkashin, M. I., Kushnerev, M. Ya., Protsuk, T. I. and Berlin, A. A., Vysokomol. Soedin., Ser. A, 1968, 10, 916; CA 69, 19726n (1968).
30. Korshak, V. V., Sergeev, V. A., Shitikov, V. K., Vol'pin, M. E. and Kolomnikov, I. S., Vysokomol. Soedin., Ser. B, 1971, 13, 873; CA 76, 113588K (1972).
31. Wegner, G., IUPAC, Boston, 1971, 908.
32. Korshak, V. V., Sergeev, V. A., Shitikov, V. K. and Kolomnikov, I. S., Dokl. Akad. Nauk SSSR, 1971, 201, 112; CA 76, 86186u (1972).
33. Korshak, V. V., Vinogradova, S. V., Pankratov, V. A. and Puchin, A. G., Dokl. Akad. Nauk SSSR, 1972, 202, 347; CA 76, 154525f (1972).
34. Chauser, M. G., Cherkashin, M. and Berlin, A. A., Khim. Atsetilena, Tr. Vses. Konf., 3rd. 1972, 320; CA 79, 53822g (1973).
35. Usmanova, M. M. and Rashidova, S. Sh., Uzb. Khim. Zh., 1974, 18, 51; CA 82, 98505j (1975).
36. Akopyan, L. A., Grigoryan, S. G. and Matsoyan, S. G., USSR Pat. SU 554,269, 1977 (Apr. 15); CA 87, 6659q (1977).
37. Simionescu, C., Dumitrescu, S. and Percec, V., J. Polym. Sci., Polym. Symp., 1978, 64, 209; CA 90, 138282h (1979).
38. Banerjie, A., Lando, J. B., Yee, K. C. and Baughman, R. H., J. Polym. Sci., Polym. Phys. Ed., 1979, 17, 655; CA 90, 204610s (1979).
39. Lando, J. and Thakur, M., Macromolecules, 1983, 16, 143; CA 98, 35223a (1983).
40. Kim, Y-H, Gal. Y-S, Kim, U-Y. and Choi, S-K., Macromolecules, 1988, 21, 1991; CA 109, 38273y (1988).
41. Gal, Y. S. and Choi, S. K., J. Polym. Sci., Part C: Polym. Lett., 1988, 26, 115; CA 109, 7019r (1988).
42. Kunzler, J. and Percec, V., Polym. Prepr. (A.C.S. Div. Polym. Chem.) 1988, 29, 221; CA 109, 38373f (1988).
43. Kim, Y-H., Choi, K-Y. and Choi, S-W, J. Polym. Sci., Part C: Polym. Lett., 1989, 27, 443; CA 112, 7988t (1990).
44. Nguyen, H. X. and Ishida, H., J. Polym. Sci.: Part B: Polym. Physics, 1989, 27, 1611.
45. Harries, C. and Tank, L., Bericht, 1908, 41, 1701; CA 2, 2375 (1908).
46. Aso, C. and Aito, Y., Makromol. Chem., 1962, 58, 195; CA 58, 9238c (1963).
47. Meyersen, K., Schulz, R. C. and Kern, W., Makromol. Chem., 1962, 58, 204; CA 58, 5792b (1963).
48. Aso, C. and Aito, Y., Bull. Chem. Soc. Japan, 1964, 37, 456; CA 61, 12097h (1964).
49. Aso, C., Furuta, A. and Aito, Y., Makromol. Chem., 1965, 84, 126; CA 63, 5750b (1965).
50. Aito, Y., Matsuo, T. and Aso, C., Bull. Chem. Soc. Japan, 1967, 40, 130; CA 66, 95540a (1967).
51. Aso, C. and Aito, Y., Kobunshi kagaku, 1966, 23, 564; CA 67, 73904r (1967).
52. Aso, C. and Miura, M., J. Polym. Sci. B, 1966, 4, 171; CA 64, 14291g, (1966).
53. Aso, C. and Miura, M., Kobunshi kagaku, 1967, 24, 178; CA 68, 3211j (1968).
54. Aso, C. Kunitake, T., Miura, M. and Koyama, K., Makromol. Chem., 1968, 117, 153; CA 70, 4673q (1969).
55. Aso, C. and Tagami, S., J. Polym. Sci. B, 1967, 5, 217; CA 66, 115902e, (1967).

56. Aso, C. and Tagami, S., Polym. Prepr., 1967, 8, 906; CA 70, 97283a (1969).
57. Stone, E. W. and Maki, A. H., J. Chem. Phys., 1963, 38, 1999; CA 58, 10886h (1963).
58. Overberger, C. G. and Ishida, S., Polym. Preprints, 1964, 5, 210; CA 64, 2170g (1966).
59. Corfield, G. C., Ph.D. Thesis, Univ. of London, 1967.
60. Aso, C., Kobunski, 1971, 20, 94; CA 75, 6379a (1971).
61. Tagami, S., Kagiyama, T. and Aso, C., Polym. J., 1971, 2, 101; CA 75, 21154s (1971).
62. Moyer, W. W., Jr., U.S. Pat. 3,395,125, 1968 (July 30); CA 69, 59801q (1968).
63. Tagami, M. and Kunitake, T., Makromol. Chem., 1974, 175, 3367; CA 82, 58207g (1975).
64. Aso, C., Tagami, S. and Kunitake, T., J. Polym. Sci. A1, 1970, 8, 1323; CA 73, 15299k (1970).
65. Overberger. C. G., Ishida, I. and Ringsdorf, H., J. Polym. Sci., 1962, 62, 173; CA 58, 10306f (1963).
66. Yokota, K., Ito, Y. and Ishii, Y., Kogyo-kwagaku zasshi, 1963, 66, 1112; CA 60, 9362b (1964).
67. Hay, J. and Kerr, C. M. L., J. Polym. Sci. B, 1965, 3, 19; CA 62, 6569c (1965).
68. Koral, J. N., U.S. Pat. 3,214,410, 1965 (Oct. 26); CA 63, 18300g (1965).
69. Kunitake, T., Tsugawa, S. and Aso, C., Makromol. Chem. 1973, 169, 95; CA 80, 3863w (1974).
70. Aso, C. and Aito, Y., Makromol. Chem., 1967, 100, 100; CA 67, 73904r (1967).
71. Aso, C., Tagami, S. and Kunitake, T., J. Polym. Sci. A1, 1969, 7, 497; CA 71, 13439r (1969).
72. Hardy, P., Nicholls, A. C. and Rydon, H. N., Chem. Commun., 1969, 565; CA 71, 29894w (1969).
73. Yasuda, H. and Tani, H., Macromolecules, 1973, 6, 303; CA 79, 42872x (1973).
74. Andreeva, I. V., Koton, M.M. and Madorskaya, L. Ya., Vysokomol. Soedin., Ser.A, 1967, 9, 2496; CA 68, 30340g (1968).
75. Aso, C. and Tagami, S., Macromol. Synth., 1972, 4, 117; CA 84, 44729h (1976).
76. Braun, D. and Walter, E., Colloid Polym. Sci., 1980, 258, 795; CA 93, 187241q (1980).
77. Stinson, S. C., Chem. and Eng. News, 1983 (Sept. 26), 61, 23.
78. Saunders, J. H. and Slocombe, R. J., Chem. Rev., 1948, 43, 203; CA 43, 1014e (1949).
79. Arnold, R. G., Nelson, J. A. and Verbanc, J. J., Chem. Rev., 1957, 57, 47; CA 51, 9511f (1957).
80. Shashoua, V. E., Sweeney, W. and Tietz,R. F., J. Am. Chem. Soc., 1960, 82, 866; CA 55, 16416h (1961).
81. King, C., J. Am. Chem. Soc., 1964, 86, 437; CA 60, 8136a (1964).
82. Iwakura, Y., Uno, K. and Ichikawa, K., J. Polym. Sci. A, 1964, 2, 3387; CA 61, 12097c (1964).
83. Iwakura, Y, Uno, K. and Kobayashi, N., J. Polym. Sci. A, 1968, 6, 1087; CA 69, 28014f (1968).
84. Corfield, G.C. and Crawshaw, A., J. Macromolec. Sci. (Chem.), 1971, 5, 1855; CA 73, 131386k (1970).
85. Bur, A. J. and Fetters, L. J., Chem. Rev., 1974, 76, 727; CA 80, 27600j (1974).
86. Prokop'ev, V. P., Bakhitov, M. I., Davlethaev, I. G., Kupreishvili, T. I. and Kuznetsov, E. V., Vysokomol. Soedin., Ser. B, 1981, 23, 688; CA 96, 35765n (1982).
87. Bakhitov, M. I., Zainutdinova, L. S., Kuznetsov, E, V. and Kligman, F. L., Sint. Poliuretanov, Eds.Omel'chenko etc. 1981, 114; CA 97, 72867p (1982).
88. Emelyanova, N. K., Bakhitov, M. I. and Kuznetsov, E. V., Vysokomol. Soedin., Ser. A, 1985, 27, 1434; CA 103, 196442h (1985).
89. Stepto, R. F. T., Polym. Prepr., Polym. Div., ACS, 1985, 26, 46; CA 104, 89127h (1986).
90. Overberger, C. G., Ozaki, S. and Mukamal, H., J. Polym. Sci., Pt. B, 1964, 2, 627; CA 61, 5776c (1964).

91. Bakhitov, M. I., Kuznetsov, E. V., Zainutdinova, L. S. and Batalina, M. V., Vysokomol. Soedin., Ser. B, 1980, 22, 540; CA 93, 205092y (1980).
92. Nippon Polyurethane Industry Co., Ltd., Jpn. Kokal Tokkyo Koho JP 82 47,319, 1982 (Mar. 18); CA 97, 56387b (1982).
93. Ahaski, H., Kogyo-kwagaku zasshi, 1960, 63, 368; CA 56, 2327f (1962).
94. Sayigh, A., Tilley, J. and Ulrich, H., J. Org. Chem., 1964, 29, 3344; CA 62, 546c (1965).
95. Tilley, J. N. and Sayigh, A. A. R., J. Org. Chem., 1964, 29, 3347; CA 62, 546d (1965).
96. Black, W. B. and Miller, W. L., Fr. Pat. 1,360,460; U.S. Pat. 3,163,624, 1964 (May 8); CA 62, 7777a (1965).
97. Schnabel, W. J. and Kober, E., J. Org. Chem., 1969, 34, 1162; CA 71, 13055n (1969).
98. Weyland, H. and Hamel, E., U.S. Pat. 3,551,469, 1970 (Dec. 29); CA 74, 64954z (1971).
99. Weyland, H. and Hamel, E., U.S. Pat. 3,649,687, 1972 (Mar. 14); CA 76, 140053p (1972).
100. Zainutdinova, L. Sh., Bakhitov, M. I. and Kuznetsov, E. V., [Uch. Zap.] Gor'kov. Unst., 1976, 1, 16; CA 87, 202403r (1977).
101. Sorokin, M. F., Shode, L. G., Onosova, L. A., Chan Thanh Chon. and Emel'yanova, L. K., Deposited Doc., VINITI, 1976, 793; CA 88, 62786m (1978).
102. Mirkind, L.A., Postnikova, V. A., Blagonravova, A. A., Soirin, Yu. L. and Sporykhina, V. S., FATIPEC Cong., 1976, 13, 514; CA 85, 160673e (1976).
103. Sporykhina, V. S., Mirkind, L. A. and Postnikova, V. A., Metody Anal. Kont. Kach. Prod. Khim., 1978, 47; CA 89, 148225k (1978).
104. Aso, C. and Aito, Y., Makromol. Chem., 1964, 73, 141; CA 61, 3210c (1964).
105. Bauer. R. S., J. Polym. Sci. A-1, 1967, 5, 2192; CA 67, 82456y (1967).
106. Wiggins, L. F. and Woods, D. J. C., J. Chem. Soc., 1950, 1950, 1567; CA 45, 1023g (1951).
107. Stille, J. K. and Hillman, J. J., J. Polym. Sci. A-1, 1967, 5, 2067; CA 67, 82415j (1967).
108. Statton, G.L. and Sauer, R.W., Ger. Offen. 2,135,486, 1972 (April 6); CA 77, 49127a (1972).
109. Onosov, G. V., Sorokin, M. F., Shode, L. G. and Dobrovinskii, L. A., Tr. Mosk. Khim. Tekhnol. Inst., 1975, 86, 106; CA 85, 109028w (1976).
110. Sorokin, M. F., Shode, L. G. and Onosov, G. V., Deposited Doc., VINITI, 1976, 2197; CA 89, 110581k (1978).
111. Chepel, L. M., Zelenetskii, A. N., Salamatina, O. B., Zaitseva, N. P., Trofimova, G. M. and Novikov, D. D., Vysokomol. Soedin., Ser. A, 1983, 25, 410; CA 98, 143882y (1983).
112. Chepel, L. M., Topolkaraev, V. A., Zelenetskii, A. N., Prut, E. V., Trofimova, G. M., Novikow, D. D. and Berlin, A. A., Vysokomol. Soedin., Ser. A, 1982, 24, 1646; CA 98, 126763t (1983).
113. Yokota, K., Hashimoto, H., Kakuchi, T. and Takada, Y., Makromol. Chem., Rapid Commun., 1984, 5, 767; CA 102, 79401n (1985).
114. Matejka, L., Tkaczyk, M., Pokoray, S. and Dusek, K., Polym. Bull. (Berlin), 1986, 15, 389; CA 105, 115430u (1986).
115. Onosov, G. V., Sorokin, M. F. and Shode, L. G., Tr. Mosk. Khim.-Tekhnol Inst., 1975, 86, 109; CA 85, 94760c (1976).
116. Sorokin, M. F., Shode, L. G. and Onosov, G. V., Khim. Tek. Sint. Iss. Plenk. Vest. Pig., 1976, 24; CA 87, 118191n (1977).
117. Sorokin, M. F., Shode, L. G. and Onosov, G. V.. Kmim. Khim. Tek. Sint. I. Pl. Vest. Pig. 1976, 29; CA 87, 118192p (1977).
118. Aso, C., Kamao, H. and Aito, Y., Kogyo-kwagaku zasshi, 1964, 67, 974; CA 62, 10517h (1964).
119. Korshak, V. V., Kamenskii, I. V., Solov'eva, L. K. and Utkina, S. I., Tr. Mosk. Khim.-Tekhnol. Inst., 1967, 52, 199; CA 68, 115332f (1968).

120. Cope, A., Fisher, B., Funke, W., McIntosh, J. and McKervey, M., J. Org. Chem., 1969, 34, 2231; CA 71, 38836f (1969).
121. Sorokin, M. F., Shode, L. G. and Dobrovinskii, L. A., Zh. Org. Khim., 1974, 10, 1603; CA 81, 136122w (1974).
122. Nikka Chemical Industry Co., Ltd., Jap. Pat. 59 199,870 [84 199,870], 1984 (Nov. 13); CA 102, 115112e (1985).
123. Johns, I. B., Polym. Preprints, 1964, 5, 239; CA 64, 2175e (1966).
124. Peebles, Jr., L. H. and Brandrup, J., Makromol. Chem., 1966, 98, 189; CA 66, 2872m (1967).
125. Liepins, R. Campell, D. and Walker, C., J. Polym. Sci. A-1, 1968, 6, 3059; CA 69, 107108t (1968).
126. Liepins, R., Makromol. Chem., 1968, 118, 36; CA 70, 29406t (1969).
127. Liepins, R., Buckley, Jr., C. and Olf, H. G., J. Polym. Sci., A1, 1970, 8, 2049; CA 73, 77737k (1970).
128. Woehrle, D. and Manecke, G., Makromolek. Chem., 1970, 138, 283; CA 74, 3907q (1971).
129. Woehrle, D., Makromol. Chem., 1972, 161, 121.
130. Hsu, L. C. and Philipp, W. H., Proc. IUPAC, Macromol. Spmp., 28th., 1982, 170; CA 99, 88589b (1983).
131. Hsu, L. C., U.S. Pat. US 4,555,565, 1985, (Nov. 26); CA 104, 149671t (1986).
132. Johnson, F., Panella, J. P., Carlson, A. A. and Hunneman, D. H., J. Org. Chem., 1962, 27, 2473; CA 57, 8536i (1962).
133. Zil'bernon, E., Feller, K. and Sergeeva, M., Vysokomol. Soedin., Ser. B, 1968, 10, 44; CA 68, 69411f (1968).
134. Liepins, R., Campbell, D. and Walker, C., Polym. Preprints, 1968, 9, 765; CA 69, 107108t (1968).
135. Keller, T. M., J. Polym. Sci., Part A. Polym. Chem., 1988, 26, 3199.

6

Monomers Containing Different Multiple Bonds (Functionally Unsymmetrical Monomers)

ALKENE-ALDEHYDES

A variety of monomers having two kinds of functional groups have been studied in cyclopolymerization. For example, o-vinylbenzaldehyde has been cyclopolymerized *via* cationic initiation to yield polymers containing both of the predicted cyclized structures shown (6–1).[1] However, the latter cyclic structure predominates since the aldehyde

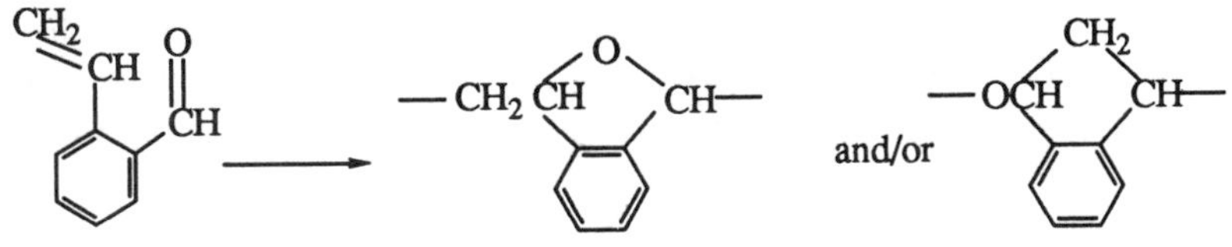

and/or

(6–1)

group is more reactive toward the cationic initiator than the vinyl group. The cyclopolymer also contained 20% of pendant vinyl group and 5% of pendant aldehyde group.

The remainder of the monomer unit was consumed in the cyclized structure. On the basis of these structural data, a cationic propagation scheme of o-vinylbenzaldehyde was proposed in which the neighboring aldehyde group reacted exclusively with the growing cation to form a cyclic unit and the intramolecular cyclization of the neighboring vinyl group with the cation was competitive with the intermolecular propagation. This scheme was consistent with that expected from the cyclopolymerization schemes of o-divinylbenzene and o-phthalaldehyde.

Polymerization of acrolein dimer with an initiator system such as $Al(C_2H_5)_3$-H_2O or $Al(C_2H_5)_2Cl$ gave a polymer consisting in part of 6,8-dioxabicyclo[3.2.1]octane units (6–2, R = H).[2] Methacrolein dimer was cyclopolymerized, giving a soluble polymer having only 1,4-dimethyl-6,8-dioxabicyclo[3.2.1]octane (6–2, R = CH_3) recurring units using $BF_3O(C_2H_5)_2$ as initiator.[3]

It has been reported that the polymerization of acrolein dimer with a catalyst system such as $AlEt_2Cl$ gave a polymer consisting, in part, of 6,8-dioxabicyclo[3.2.1]octane units [(See 6–2), R=H].[2-283] Methacrolein dimer was also cyclopolymerized giving a

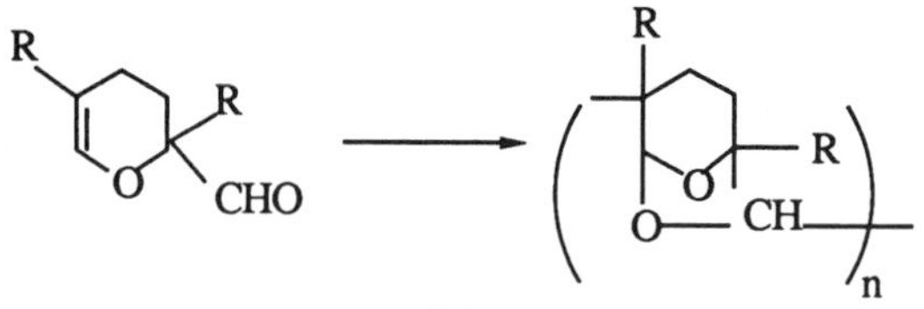

(6–2)

soluble polymer having only 1,4-dimethyl-6,8-dioxabicyclo [3.2.1]octane [See 6–2, $R=CH_3$] recurring units, using $BF_3O(C_2H_5)2$ as initiator.[4] 5,6-Dihydro-2,6-dimethyl-3-vinyl-2H-pyran from 5,6-dihydro-2,6-dimethyl-2H-pyran-3-carboxaldehyde was prepared and its polymerization and copolymerization studied.[5] 5,6-Dihydro-2,6-dimethyl-3-vinyl-2H-pyran was homopolymerized in the presence of either ionic or radical initiators. Copolymerizations of 5,6-dihydro-2,6-dimethyl-3-vinyl-2H-pyran (M_1) with methyl methacrylate or styrene proceeded smoothly using azobisisobutyronitrile as the initiator; the monomer reactivity ratios were $r_1 = 0.53$ and $r_2 = 1.03$ for the methyl methacrylate copolymerization and $r_1 = 0.0$ and $r_2 = 0.9$ for the styrene copolymerization.

Polymers of 3,4-dihydro-2H-pyran-2-carboxaldehyde were synthesized by use of an organoaluminum halide in combination with a metal halide.[6] The polymer is soluble in benzene, toluene, chloroform, and pyridine and the specific viscosity at 30° in toluene was 0.9. The absorption spectra indicated that it contained 7% free aldehydes and 42% vinyl ether radicals. The structure was believed to be linear and the proposed unit structure as shown (6–2,R=H)

The polymerization of 2,5-dimethyl-3,4-dihydro-2H-pyran-2-carboxaldehyde (methacrolein dimer) in the presence of cationic initiators to give the cyclic structure (See Structure 6–2, $R=CH_3$), which was soluble in organic solvents, has also been described.[7]

Polymerization of 3,4-dihydropyran-2-carboxaldehyde by use of $Cr(C_6H_6)_2$ gave a crystalline polymer with the vinyl ether group intact.[8] The polymer contained no aldehyde group.

By use of the reaction products of organic aluminum compounds with water, and acid anhydrides or acyl halides as initiators, a polymer which contains 60% residual vinyl ether groups was obtained.[9] This same monomer had been polymerized by coordination catalysts to give a polymer that contained an acetone-soluble fraction.[10]

A novel type of anionic cyclopolymerization has been reported.[1-150] It was found that tertiary phosphines, having a pK_a value greater than 8.0, were active catalysts for the cyclic polymerization of crotonaldehyde. A light yellow polymer was obtained in 66% yield and a number-average molecular weight of 2240. In a later study, a structure for the polymer and a mechanism for its formation were proposed.[11] A systematic investigation of the effects of temperature, solvent, catalyst, and monomer concentration on rate of polymerization and molecular weight of the polymers was conducted. The structure proposed was consistent with that obtained through an anionic-initiated vinyl polymerization followed by cyclization of the resulting pendant aldehyde groups.

The following facts were cited as evidence supporting the proposed mechanism:

1. Only certain tertiary phosphines were effective catalysts.
2. Tributyl and tripropyl phosphine gave the same rates of polymerization.

3. Larger catalyst concentrations resulted in increased rates.
4. Catalyst concentration had little effect on molecular weights.
5. Polymerization occurred from −50 to 25°C.
6. Increased temperature resulted in faster rates.
7. The molecular weight of the polymers increased slightly with lower temperature.
8. Higher dielectric constant solvents gave higher rates of polymerization.
9. Higher monomer concentrations, likewise, gave higher rates.
10. Monomer concentration had little effect on molecular weights of the polymers.
11. The molecular weight of the polymer increased with conversion and leveled off after 30% reaction.

The initiation step is represented as involving attack of the tertiary phosphine on the vinyl group of crotonaldehyde to form a zwitterion which then may attack another molecule of monomer, etc. (6–3). The proposed propagation step involves a continuing process of charge separation—not energetically favorable.

The proposal, however, is supported by evidence from two sources. In 1955, it had been proposed that formation of macrozwitterions from acrylonitrile and nonionic Lewis bases such as triethylphosphine occurred.[12] Confirmation of this theory was claimed.[13, 14] However, recent studies have cast doubt upon zwitterionic species in certain systems.[15]

It has been shown that, in polymerization of acrylonitrile with phosphines, the zwitterionic chains did not grow beyond a degree of polymerization of 1, even in polar solvents.[15] The polymers were studied by fractional precipitation and gel chromatography.

The large amount of energy required for charge separation (See Structure 6–3b) allowed step 6–4d to occur (6–4). This chain transfer to monomer then resulted in 6–4d being the species responsible for anionic initiation. This possibility must be considered for the similar crotonaldehyde monomer. Detection of low-molecular-weight phosphines would be considered strong evidence supporting the transfer mechanism.

Polymerization of acrolein has also been studied on numerous occasions and by a large number of investigators.[16] Radical initiated polymerization of this monomer led to gel formation at conversions greater than 16%. However, methacrolein polymers were soluble up to high conversions. The following equation relating viscosity number, Z(in L/g) and molecular weight, M, of polymethacrolein in the range of 5,000–20,000: $Z = 2.8 \times 10^{-6} M^{0.97}$ was derived. The currently accepted structure of polyacrolein, depending upon method of synthesis, history, and other conditions is shown (6–5). These structures, including the accepted mechanisms for polymerization of acrolein under different conditions of initiation, have been discussed in greater detail in a later section.

Acrolein has also been polymerized *via* a radical redox system, and the polymers were reduced with sodium borohydride to give poly(allyl alcohols) with a residue of four mole % aldehyde groups, and still containing four mole % of double bonds.[17] Redox polymers of acrolein were subjected to a Knoevenagel reaction with active methylene compounds such as malononitrile, ethyl acetoacetate, etc. using piperidine as catalyst.[18] Soluble polyacrolein derivatives were obtained.

Copolymerization of acrolein and methacrolein in solution by radical initiation was investigated.[19] Copolymerization parameters ($r_1 r_2$) of the following systems were detera- Initiation

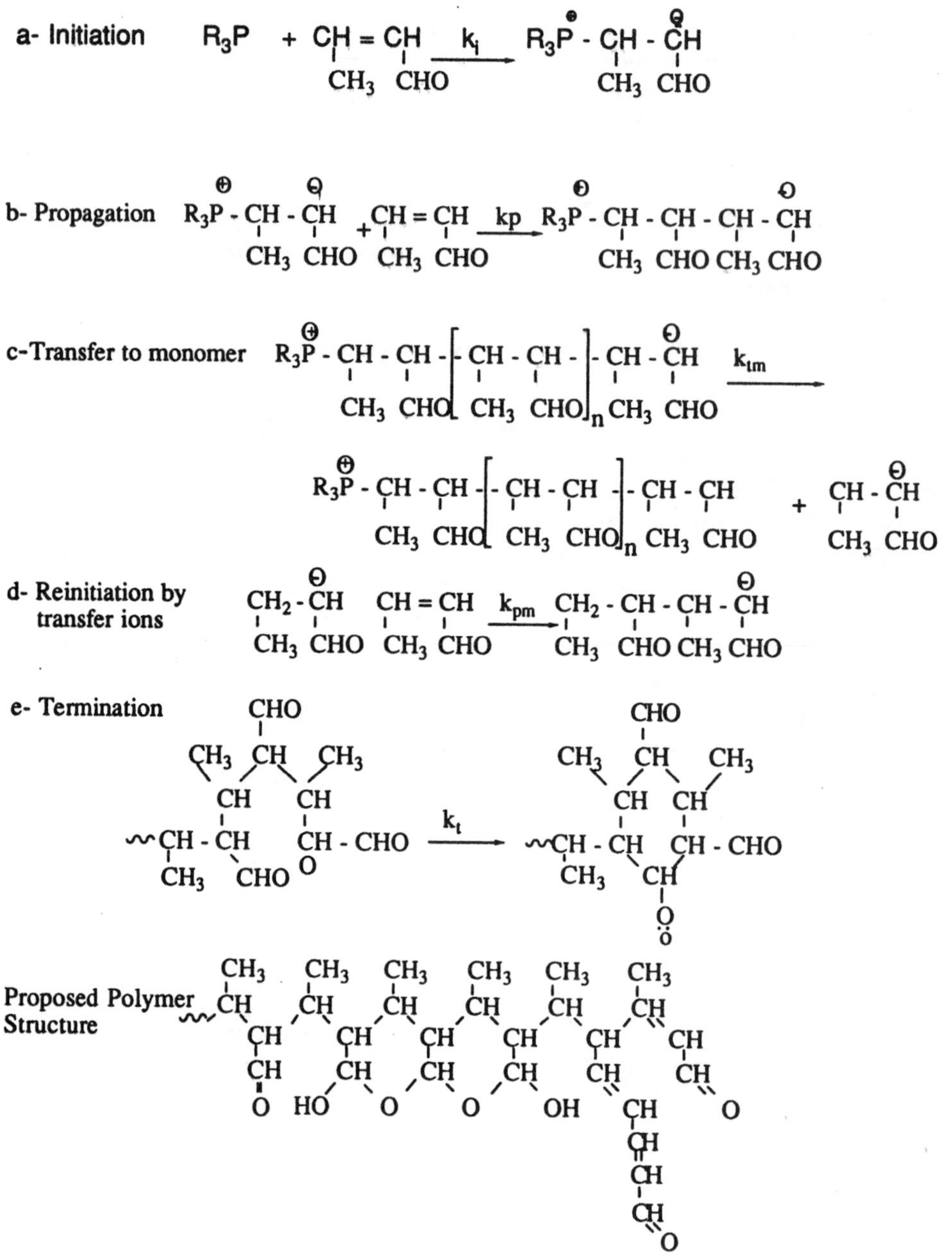

$$R_3P + \underset{CH_3}{CH}=\underset{CHO}{CH} \xrightarrow{k_i} R_3\overset{\oplus}{P}-\underset{CH_3}{CH}-\underset{CHO}{\overset{\ominus}{CH}}$$

b- Propagation

$$R_3\overset{\oplus}{P}-\underset{CH_3}{CH}-\underset{CHO}{\overset{\ominus}{CH}} + \underset{CH_3}{CH}=\underset{CHO}{CH} \xrightarrow{kp} R_3\overset{\oplus}{P}-\underset{CH_3}{CH}-\underset{CHO}{CH}-\underset{CH_3}{CH}-\underset{CHO}{\overset{\ominus}{CH}}$$

c-Transfer to monomer

$$R_3\overset{\oplus}{P}-\underset{CH_3}{CH}-\underset{CHO}{CH}-\left[\underset{CH_3}{CH}-\underset{CHO}{CH}-\right]_n-\underset{CH_3}{CH}-\underset{CHO}{\overset{\ominus}{CH}} \xrightarrow{k_{tm}}$$

$$R_3\overset{\oplus}{P}-\underset{CH_3}{CH}-\underset{CHO}{CH}-\left[\underset{CH_3}{CH}-\underset{CHO}{CH}-\right]_n-\underset{CH_3}{CH}-\underset{CHO}{CH} + \underset{CH_3}{CH}-\underset{CHO}{\overset{\ominus}{CH}}$$

d- Reinitiation by transfer ions

$$\underset{CH_3}{CH_2}-\underset{CHO}{\overset{\ominus}{CH}} \quad \underset{CH_3}{CH}=\underset{CHO}{CH} \xrightarrow{k_{pm}} \underset{CH_3}{CH_2}-\underset{CHO}{CH}-\underset{CH_3}{CH}-\underset{CHO}{\overset{\ominus}{CH}}$$

e- Termination

CHO / CH₃ CH CH₃ / CH CH / ~CH - CH CH - CHO / CH₃ CHO O $\xrightarrow{k_t}$ CHO / CH₃ CH CH₃ / CH CH / ~CH - CH CH - CHO / CH₃ CH / O:

Proposed Polymer Structure

CH₃ CH₃ CH₃ CH₃ CH₃ CH₃
~CH CH CH CH CH CH
CH CH CH CH CH CH
CH CH CH CH CH CH
O HO O O OH CH O
CH
CH
CH
O

(6–3)

mined: acrolein-acrylonitrile, 1.60, 0.52; acrolein-acrylamide, 1.69, 0.21; acrolein-methacrylonitrile, 0.72, 1.20; acrolein-2-vinylpyridine, ≈4, ≈0; and methacrolein-methacrylonitrile, 1.78, 0.40. Q and e values were also reported.

In a continuing study of the structure of polyacrolein, a number of low-molecular-weight model compounds were used.[20] Among these were valeraldehyde, glutaraldehyde, 1-hydroxytetrahydrofuran, 1-(2-tetrahydropyranyloxy)tetrahydropyran, and

$$1)\ R_3P + CH_2{=}\underset{\underset{CN}{|}}{CH} \longrightarrow R_3\overset{\oplus}{P}-CH_2-\underset{\underset{CN}{|}}{\overset{\ominus}{C}H} \xrightarrow{AN} R_3\overset{\oplus}{P}-\left[-CH_2-\underset{\underset{CN}{|}}{CH}-\right]-CH_2-\underset{\underset{CN}{|}}{\overset{\ominus}{C}H}$$

a b

$$2)\ a + \underset{\underset{CN}{|}}{\overset{H}{C}}{=}CH_2 \longrightarrow R_3\overset{\oplus}{P}-CH_2-\underset{\underset{CN}{|}}{CH} + \underset{NC}{\overset{\ominus}{C}}{=}CH_2$$

c d

(6–4)

$$-CH_2\underset{\underset{CHO}{|}}{\overset{R}{C}}-CH_2 \quad \text{R R OH O O OH}_n \qquad \left(CH_2\underset{\underset{CH_2OH}{|}}{\overset{R}{C}}\right)_n \left(CH_2 \text{ R R O O}\right)_n$$

R = H, CH_3

(6–5)

2,6-dialkoxytetrahydropyrans. Spectral comparisons were made with polyacrolein. It was concluded that polyacrolein reactions led either to tetrahydropyran derivatives or to acidic aldehyde derivatives.

Various cyclopolymerizable monomers were synthesized by reaction of acrolein as a diene with several vinyl compounds as dienophiles.[21] Thus, a series of N-(2,3-dihydro-2-pyranyl)carbamic acid esters and ureas were prepared by addition of various N-vinylcarbamic acid esters and N-vinylureas to α,β-unsaturated carbonyl derivatives, and hydrolyzed to $CH_2(CH_2CHO)_2$ and $CH_3CO(CH_2)_3CHO$. Hydrogenation of these monomers led to the cyclized tetrahydropyran structures.

Radical-initiated polymerization of acrolein led to polymers containing aldehyde groups which produced polyhydroxy-polycarboxylic acids by the Cannizzaro reaction with each other.[22] The reactions took place on the surface of solid polyacroleins. Polyacrolein was converted to a polyol by reaction of the polymer with formaldehyde in presence of basic catalysts.[23]

A variety of model compounds, including isovaleraldehyde, glutaraldehyde, and various tetrahydropyran derivatives, were allowed to react with active hydrogen compounds, e.g. malononitrile, in presence of piperidine, in a continuing study of the structure and reactivity of polyacrolein.[24] Copolymers of methacrylonitrile with acrolein or methacrolein were studied *via* the UV and IR absorption spectra of the carbonyl groups of the copolymers, properties of which depend on the sequence length distribution of aldehyde groups in the copolymers.[25]

The structure of copolymers of acrolein and methacrolein with methacrylonitrile was studied and the results were interpreted in terms of the statistical theory of sequence length distribution in polymers.[26] At sequence lengths of two or three, the aldehyde groups reacted to form cyclic aldehyde hydrate ethers (tetrahydropyran structures). A mixture of styrene and acrolein was prepared and the acrolein polymerized by sodium methoxide.[27] A radical initiator was then added to copolymerize the styrene with the pendant vinyl groups of the polyacrolein. A crosslinked copolymer was reported.

3-Methylenebicyclo[3.3.1]nonane-7-one was postulated to undergo cationic initiated cyclopolymerizations to produce the adamantane polymer structure shown in Structure 6–6.[28] 3,7-Dimethylenebicyclo[3.3.1]nonane was also postulated to undergo a similar cyclopolymerization to produce the corresponding adamantane polymer.

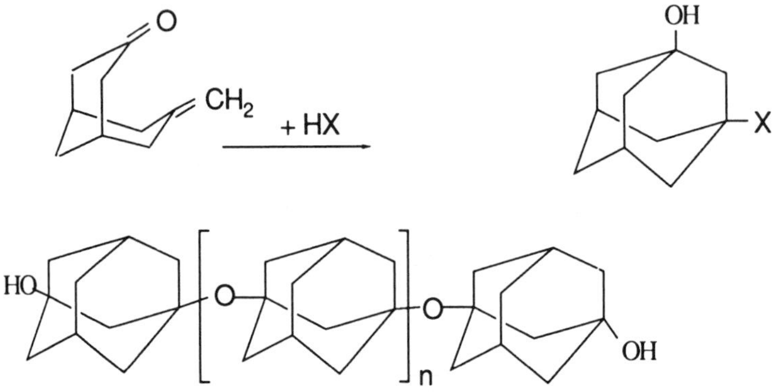

O CH2 + HX OH X

HO O O n OH

(6–6)

Poly-3,4-acrolein was degraded with sulfuric acid by a mechanism suggested to be a statistical chain scission of the polymer without complete depolymerization.[29] The activation energy for the process was determined. Polymethacrolein was reduced with lithium aluminum hydride to the corresponding poly(methallyl alcohol).[5-74] The molecular weight of the reduced polymer was 18,000–20,000 and structural evidence obtained supported the following structure (6–7).

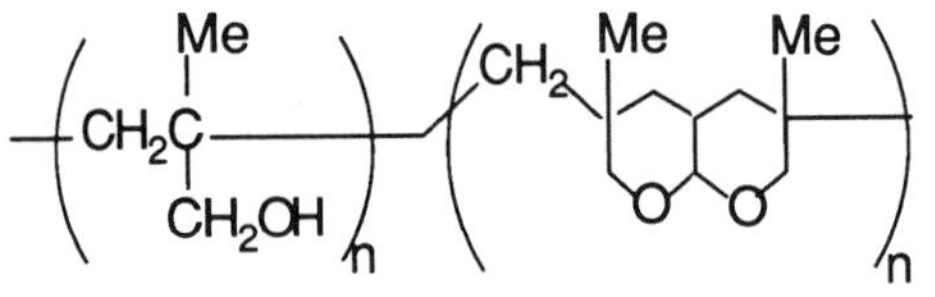

Me CH2C CH2OH n CH2 Me Me O O n

(6–7)

The importance of cyclopolymerization in producing polymers unobtainable from other sources was discussed.[30] For example, reactions on polyacroleins produced many new polymers. Free radical polymerization of allyl acrylate led to a cyclopolymer which could be hydrolyzed to yield an alternating copolymer of acrylic acid and allyl alcohol. Polymerization of acrolein induced by the pyridine-water system was studied.[31] One polymer was analyzed to show that it consisted of about 26 units, only four of which had free aldehyde groups.

By use of model compounds, the cyclized product obtained by depolymerizing poly(α-cyanostyrene) with sodium methoxide under oxygen-free, anhydrous conditions was identified.[3-139]

Copolymers of 3,4-dihydro-2H-pyran-2-carboxaldehyde (acrolein dimer) with phenylisocyanate were obtained under varying conditions.[32] An alternating copolymer was obtained *via* anionic initiation, independent of the monomer ratio. The isocyanate content of the copolymer obtained *via* diethylaluminum chloride catalysis increased with increasing polymerization temperature. The latter catalyst also introduced structural units into the chain formed *via* cyclopolymerization (6–8).

(6–8)

Of interest in a discussion of cyclopolymerization is the result of intramolecular photocycloaddition of 3-(allyloxy)- and 3-(allylamino)-2-cyclohexenones to the corresponding oxa- and azabicyclo[2.2.1]hexanes.[33] 2-Methyleneglutaronitrile was synthesized *via* a dimerization reaction of acrylonitrile in the presence of a tertiary phosphine catalyst.[34] This monomer may undergo a cyclopolymerization *via* intramolecular reaction between the C-C double bond and the C-N triple bond; however, no such reactions have been reported.

Some peculiarities in redox polymerization of methacrolein in aqueous media with a metabisulfite ion-containing redox system have been observed.[35] A decrease in the pH of the reaction medium or an increase in the metabisulfite ion concentration caused a decrease in the percent of free aldehyde groups in the polymer as well as an increase in the intrinsic viscosity due to an increase in the rigidity of the polymer chain. The polymerization occurred by a radical mechanism with a simultaneous heterolytic process at the carbonyl group leading to the formation of a cyclized structure of aldehyde groups of the semiacetal ester type.

Furfural has been cationically copolymerized with p-tolyl vinyl ether, dihydropyran or divinyl ether over a wide range of monomer feeds; however, the mole fraction of furfural in the comonomer was always <0.50.[36] At $-78°$, furfural copolymerized with vinyl ethers selectively through the aldehyde group, while at $\approx 0°$ deeply-colored copolymers of complex structures were formed. The monomer reactivity ratios for furfural (M_1) and p-tolyl vinyl ether were $r_1 = 0.15 \pm 0.15$ and $r_{11} = 0.25 \pm 0.05$, respectively. More than half of the second vinyl groups were consumed in divinyl ether-furfural copolymers indicating formation of 1,3-dioxane rings.

It has been shown that polyacrolein and polymethacrolein obtained in the presence of dialkylbis(dipyridyl)iron, $FeR_2(dipy)_2$ ($R = CH_3$, C_2H_5, C_3H_7) contained fractions with C-C backbone and pendant aldehyde groups, of polyether type, and of polycyclic structure.[37] An unusual exothermal behavior of poly(allyloxyacetaldehyde), when subjected to differential scanning calorimetry in air, was observed in the 110–40°C temperature range.[38] This behavior was attributed to a chain reaction of neighboring allyl side groups in the polymer chain. The extent of cyclization in the polymerization was not quantified.

Polymerization of methylvinyl ketone by use of sodium or sodium alkoxides led to partly cyclized polymer.[39] The mechanism of cyclization was discussed. Statistical copolymers of styrene and o-vinylbenzaldehyde were prepared by a radical mechanism.[40] These copolymers were subjected to subsequent grafting reactions on the pendant aldehyde function assumed to be present, using photolytic excitation in presence of methyl methacrylate. Grafting to the copolymer backbone was quite low. No explanation for the low value was given. Possibly the aldehyde function was consumed in a cyclopolymerization process involving the o-vinyl function.

ALKENE-EPOXIDES

Alkene-epoxides lend themselves to cyclopolymerization in appropriately substituted molecules, since both structures are subject to cationic initiation. Glycidyl acrylate and methacrylate have been reported to cyclopolymerize, using $BF_3O(C_2H_5)_2$ as initiator. Although some disagreement still exists as to the exact mechanism, there appears to be general agreement that both functional groups of the monomers participate in a cyclopolymerization mechanism.[1-71]

It was reported that both the epoxy ring and the double bond of the monomer participated with the formation of cyclic tetrahydropyran units in the main chain.[1-17a] On the other hand, the free radical polymerization of glycidyl methacrylate and glycidyl acrylate has been reported to yield linear polymers with reaction taking place without participation of the glycidyl group.[1-72] It was later reported, however, that in the cationic polymerization of glycidyl methacrylate, catalyzed by $BF_3O(C_2H_5)_2$, only the epoxy group participated in the polymerization.[41] The cyclopolymerization described earlier apparently did not occur.[1-71] It has also been reported that the cationic polymerization of glycidyl crotonate yielded cyclopolymers, as the result of reaction of the crotonic bond with the epoxide ring (6–9).[42]

where R = H, Me

(6–9)

Allyl glycidyl ether was converted to homopolymer when it was used as a sensitive, difunctional trapping agency in a reaction which involved the noncatalyzed addition of highly fluorinated perhalo ketones to substituted olefins containing allylic hydrogen atoms, a reaction which led to 3,4-unsubstituted alcohols.[43] The reaction mechanism was discussed.

Further studies on polymerization of molecules containing epoxy and alkene functionalities are obviously necessary before a final conclusion can be drawn on this subject.

ALKENE-ALKYNES

Polymerization of both conjugated and nonconjugated alkene-alkynes has been studied quite extensively. The nonconjugated structures have been investigated with emphasis on their cyclopolymerization capabilities.

Polymerization of vinylethynyl carbinols and their derivatives have been fully investigated.[44–58] Soluble, high molecular weight polymers were obtained which were fully stable in air even when heated.[57] Earlier reports had indicated that only moderate molecular weights (1000) of polymers containing cyclobutene rings were obtained.[59] It was shown by chemical and spectroscopic methods that the polymer unit was formed by the cyclization of two monomer molecules and that it contained a nonconjugated double bond (in a five-membered ring) and a triple bond.[57] From these data, the mechanism (6–10) was proposed for the polymerization. Cyclopolymerization, involving both the

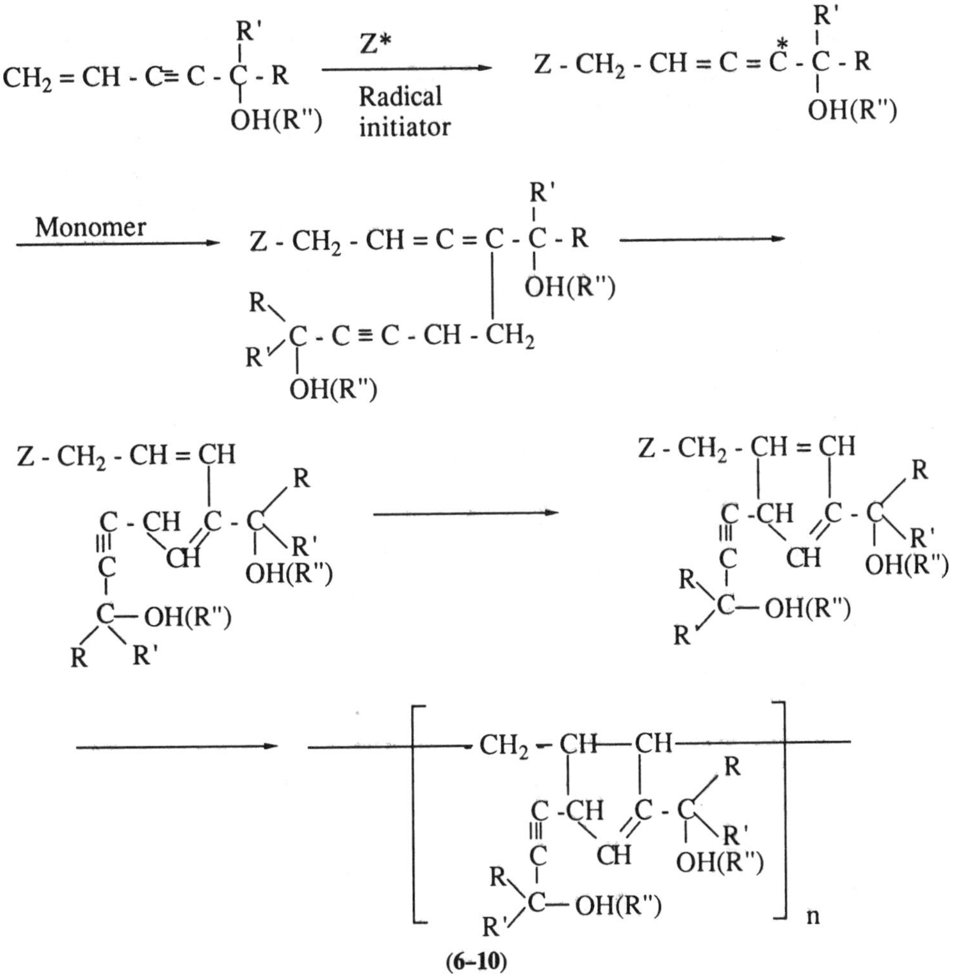

(6–10)

double and triple bonds, has been reported in radical-initiated polymerization of dipropargyl maleate and fumarate, and propargyl crotonate and methacrylate.

Radical polymerizations of glycidyl ethers of (vinylethynyl)carbinols were studied.[60] The polymers which were soluble in a variety of solvents, were proposed to have the structure shown (6–11).

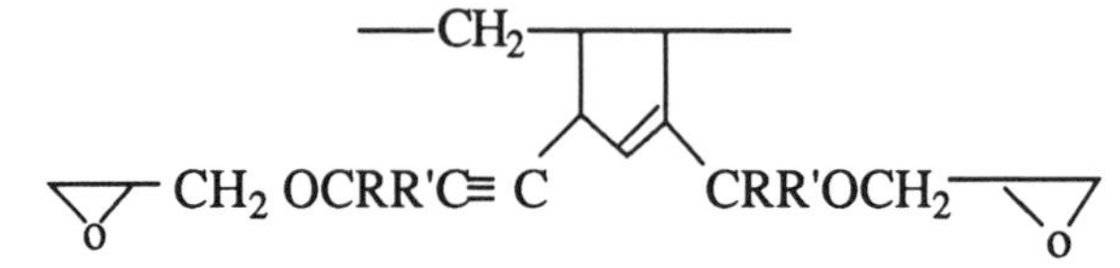

—CH_2—

CH_2 OCRR'C≡ C CRR'OCH_2

O O

(6–11)

There are still controversies existing with regard to the copolymerizability of double and triple carbon-carbon bonds by free radical initiation. Dimethyl(vinylethynyl)carbinol was converted to a polymer *via* free-radical initiation, and the structure of the polymers studied by IR and Raman spectroscopy.[61] The polymers were found to contain acetylenic bonds in each repeat unit, and the polymerization was postulated to occur through the vinyl group with no cyclopolymerization. In polymerization studies of glycols and carbinols of the isopropenylacetylene series *via* radical initiation, it was concluded that the polymerization proceeded through the olefinic double bond only and no cyclization occurred.[62]

The structures of polymers obtained from dimethyl(vinylethynyl)carbinol and its methyl and trimethylsilyl ethers, using benzoyl peroxide as initiator were determined; it was shown that the carbon-to-carbon triple bond did not take part in polymer formation,[63] inconsistent with the hypothesis of cyclopolymerization of vinylacetylene monomers. However, it had been shown much earlier that triple carbon-carbon bonds did not participate in radical initiated cyclopolymerization under the conditions used in that investigation.

A soluble polymer with a high concentration of acetylenic groups was obtained by treating vinylethynylcarbinol with alkali in an appropriate solvent.[64] Radical initiated polymerization of methacrolein was shown to lead to a cyclic polymer terminated by hydroxyl groups.[65] Anionic polymerization led to a linear structure, postulated to be formed *via* a 1,4-addition polymerization mechanism.

Oxygen, sulfur and nitrogen-containing heterocyclic vinylethynylcarbinols were polymerized in bulk or in solution in presence of free radical initiators, and the properties of the polymers presented; their structures were also discussed.[66]

Polymerization of 2-methyl-1,5-hexadiene-3-yne *via* cationic and radical initiation was studied.[67] The polymer was crosslinked, and spectral data supported the cyclic ladder structure shown (6–12).

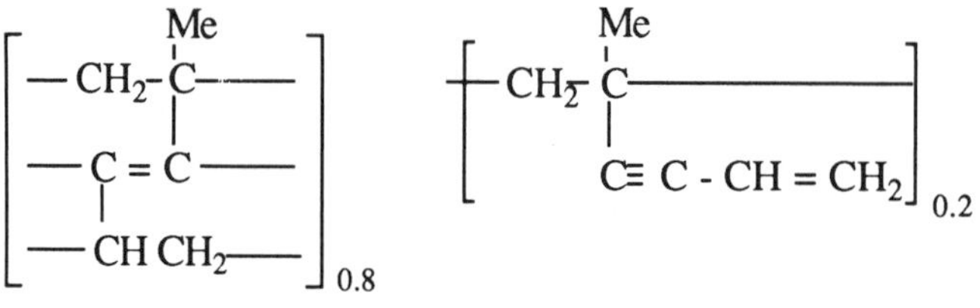

Me Me

—CH_2–C— —CH_2– C—

—C = C— C≡ C - CH = CH_2 0.2

—CH CH_2— 0.8

(6–12)

In a structural study of polymers of dimethyl(vinylethynyl)carbinol and its derivatives, and, in an attempt to resolve the conflicting literature reports on the polymer structure, it was concluded that the triple bond of the monomer does not participate in polymer formation.[68]

The polymers derived from the methyl ether of dimethyl(vinylethynyl)carbinol *via* various cationic and metal-coordination initiators have been investigated. Polymers with conjugated double bonds were stable up to 250° without melting.[69] However, copolymers of the monomer with a comonomer derived by elimination of methanol from the monomer were soluble and had molecular weights of about 2,000. Radical initiated polymerization of a silicon-containing vinylethynylcarbinol was studied.[70] The polymer was thermoplastic and stable up to 200°C. It was concluded that polymerization took place at the vinyl group while the triple bond remained intact. Block polymerization of butyl dimethyl(vinylethynyl)carbinyl acetal in presence of radical initiators yielded a transparent elastic polymer.[71] After hydrolysis to remove the acetal groups, it was compared with the polymer of dimethyl(vinylethynyl)carbinol and found to be identical. It was concluded that the polymerization mechanism proceeds without involving the acetylenic bonds, although the residual unsaturation was not reported.

The monomers, propargyl acrylate, propargyl methacrylate, and propargyl crotonate were polymerized and copolymerized with styrene and methyl methacrylate *via* free radical initiation.[72] The monomer reactivity ratios were reported. The extent of cyclopolymerization, if any, was not disclosed. Heating of $H_2C{:}CHC{\equiv}CC(CH_3)_2OH$ in dichloroethane containing boron fluoride etherate or stannic chloride gave first $H_2C{:}CHC{\equiv}CCH_3{:}CH_2$ and 1,1,3,3-tetramethyl-4-vinylisobenzofuran, which copolymerized to give vinylisopropenylacetylene-1,1,3,3-tetramethyl-4-vinylisobenzofuran copolymer.[73] This ladder copolymer consisted partially of the structure shown (6–13).

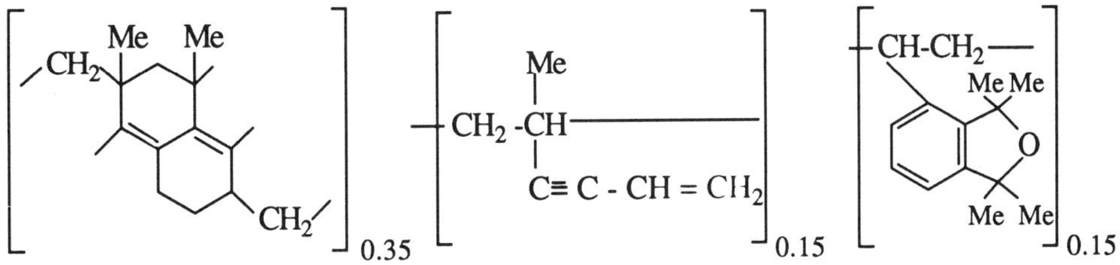

(6–13)

Vinylacetylene and related compounds were polymerized *via* cationic initiation to give ladder polymers, having a system of conjugated double bonds.[74] Propargyl acrylate and propargyl methacrylate can be bulk polymerized in the presence of benzoyl peroxide and azoisobutyronitrile to give insoluble polymers.[75] Conducting the radical polymerization of the acrylate in dioxane solution also gave an insoluble polymer; the methacrylate, under these conditions, gave a soluble polymer which underwent a transition to an insoluble polymer after 35–40% yield. IR data indicated that the soluble polymer formed initially was of a linear-cyclic structure which underwent crosslinking *via* $C{\equiv}C$ during intensification of polymerization or increased temperature.

Intramolecular cyclization of allyl ethers of vinylacetylenic alcohols underwent cyclization in xylene containing 4-(t-butyl)catechol.[76] The compound, $CH_2{=}CHCH_2OCRR'{-}C{\equiv}CC(CH_3){=}CH_2$, gave the dihydrophthalan structure shown

(6–14). The compound, $CH_2{=}CHCH_2OC(CH_3)_2C{\equiv}CCH{=}CH_2$, gave the oxatricyclododecene shown in Structure 6–14. Vinylacetylenic ketones undergo acid catalyzed cyclization to give γ-hydroxypyrones.[77] For example, the α-(β,β-dichlorovinyl)-α′-

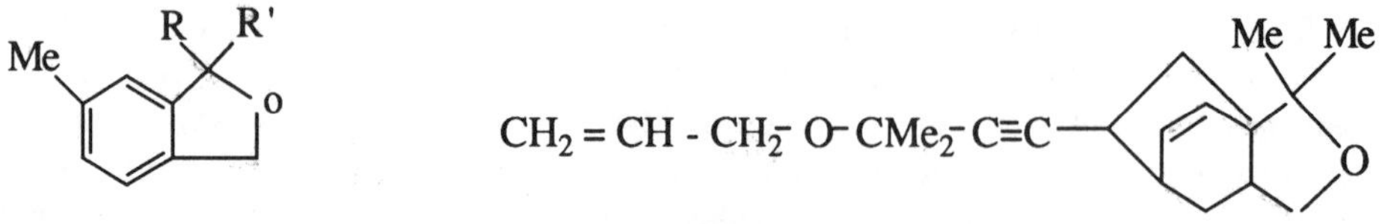

Me R R' O

$CH_2 = CH - CH_2 - O - CMe_2 - C{\equiv}C$ Me Me O

(6–14)

acetylenic ketone, 3,4-$R^1R^2C_6H_3C \equiv C\text{-}COCR{=}CCl_2$ (R = H, CH_3, R^1R^2 = H_2, OCH_2O) gave the pyrone ether structure shown (R^3 = H). This compound, when methylated, gave a mixture of the two structures shown (R^3 = CH_3) (6–15). Polymeri-

R^1 O O R^2 R OR^3 R^1 O OMe R^2 R O

(6–15)

zation of (9-anthryl)methyl methacrylate or (9-anthryl)methyl propiolate at room temperature, in the presence of $TiCl_4$, gave polymers (72 and 64% yield, respectively) with spectral properties indicating the presence of units of structure (6–16), formed by Diels-Alder polymerization.[2–65]

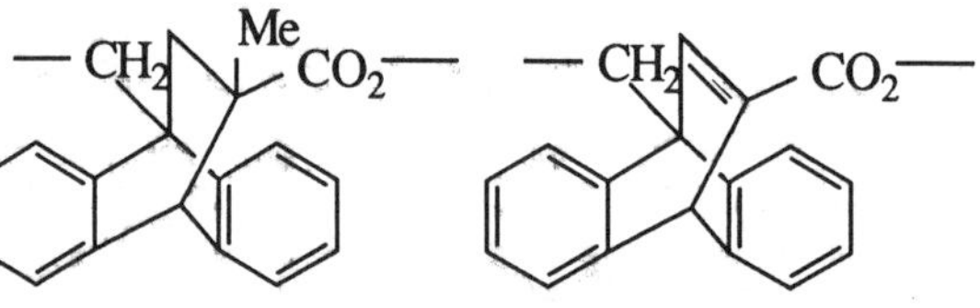

— CH_2 Me CO_2— — CH_2 CO_2—

(6–16)

It is apparent from the uncertainties enumerated in this presentation of cooperative participation of alkene and alkyne bonds that further investigations are necessary before fully reliable conclusions can be drawn about this topic.

OTHER FUNCTIONALLY UNSYMMETRICAL MONOMERS

A number of other cyclopolymerizable monomers containing functionally unsymmetrical bonds have been studied. These examples are summarized in this section.

Cyclic oligomers, which were essentially pure trimers, were isolated from the alkoxide-catalyzed hydrogen-transfer polymerization of methacrylamide.[78]

2-Vinylfuran was polymerized by use of iodine in methylene chloride, as solvent, and the polymer was reported to be of the structure shown (6–17).[79] Poly(4-cyano-1-butene) was obtained by polymerizing 4-cyano-1-butene in liquid ammonia in presence of a metal amide at −80 to 33°C.[80] The structure of the polymer was not revealed;

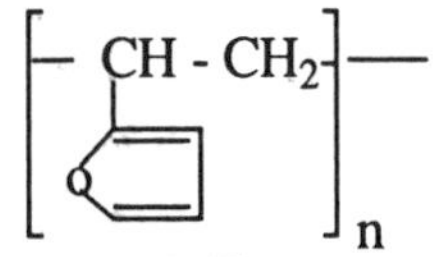

$$\left[-CH-CH_2- \right]_n$$

(6–17)

however, this monomer is structurally capable of cyclopolymerizing if the two independent multiple bonds should interact. Hydrolysis of an organolithium aminosilicon compound yielded cyclic polysiloxanes of the structure shown (6–18).[81]

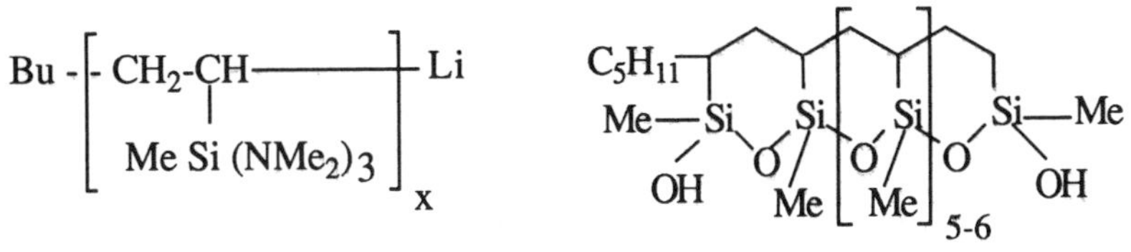

$$Bu-\left[CH_2-\underset{MeSi(NMe_2)_3}{CH} \right]_x-Li$$

C_5H_{11} Me—Si Si Si Si—Me OH O Me O Me O OH 5-6

(6–18)

Radical initiated copolymerization of maleic anhydride with 1,1-dichloro-2-vinylcyclopropane was studied.[82] Based on the composition and structure of the copolymer (6–19), rearrangement of the radical arising from the cyclopropane monomer and cyclization of the growing chain was suggested. In a continuing study of polymerization of vinylcyclopropanes, radical initiation of copolymerization of 1,1-dichloro-2-vinylcyclopropane with monosubstituted ethylenes was carried out.[83] A special copolymer composition equation, applicable to this system, was developed. The theory was applied to methyl acrylate, methyl methacrylate, and styrene. Reactivity ratios were determined. n-Butyllithium initiated polymerization of methyl isopropenyl ketone has been shown to lead to a polymer of limited molecular weight; the polymer of higher molecular weight contained the intramolecular cyclized units shown (6–20).[84] The process of cyclization eliminates water, which retards the polymerization and limits the molecular weight. A bimodal molecular weight distribution was observed. Triethylaluminium initiated polymerization of methyl isopropenyl ketone was also studied.[85] It was proposed that a catalyst-monomer complex was formed which rearranged to the initiating species. A coordination mechanism for the polymerization was proposed.

Vinyl isocyanate was copolymerized with styrene or methyl methacrylate.[86] Linear, soluble copolymers were obtained when the comonomer ratio of the isocyanate monomer was $>9:1$; however, when this ratio was $<2:1$, the copolymers were insoluble. The soluble copolymers retained the active isocyanate function.

An in-depth review of the literature on potential chemical systems for intramolecular cycloaddition cures has been conducted.[87] The literature, from 1957 to 1978, of intramolecular cycloaddition (IMC) reactions includes synthesis of hydrocarbon and heterohydrocarbon systems and their evaluation for the potential use in IMC-curing matrix resins.

The kinetics of bulk, thermal polymerization of (3-phenoxyphenyl)acetylene at 400–600° K have been determined by differential scanning calorimetry.[88] Gel chromatography of the products showed that molecular weight was independent of temperature. This fact was discussed in terms of a biradical mechanism in which the kinetic and molecular chain lengths were controlled by a first-order termination step involving cycl-

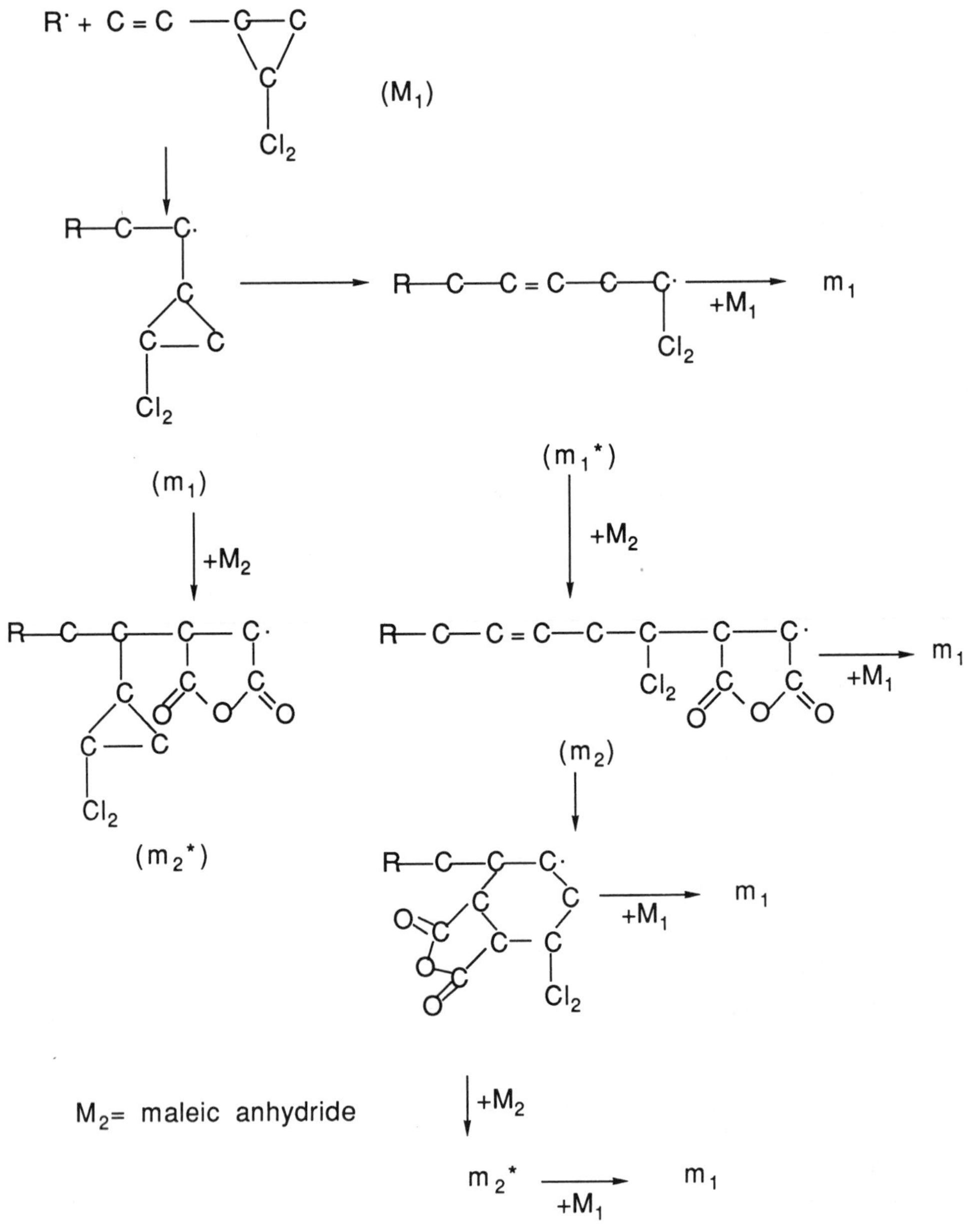

$$
\begin{aligned}
&R^{\cdot} + C{=}C{-}\underset{\underset{\text{Cl}_2}{|}}{\text{(cyclopropyl)}} \quad (M_1) \\
&\downarrow \\
&R{-}C{-}\dot{C}{-}\text{(cyclopropyl-Cl}_2) \longrightarrow R{-}C{-}C{=}C{-}C{-}\underset{\underset{\text{Cl}_2}{|}}{\dot{C}} \xrightarrow{+M_1} m_1 \\
&(m_1) \xrightarrow{+M_2} (m_2^{*}) \qquad (m_1^{*}) \xrightarrow{+M_2} (m_2) \xrightarrow{+M_1} m_1 \\
&(m_2) \longrightarrow \text{cyclized radical} \xrightarrow{+M_1} m_1 \\
&\downarrow +M_2 \\
&m_2^{*} \xrightarrow{+M_1} m_1
\end{aligned}
$$

M_2= maleic anhydride

(6–19)

ization of the growing chain. The molecular weight of the polymer was controlled predominantly by steric and thermochemical factors rather than by energetics.

The spontaneous polymerization of epichlorohydrin with allylamine in dioxane, CCl_4, $CHCl_3$, C_2H_5OH, or H_2O at -2 to $+30°C$ was shown to be a first order reaction, having an activation energy of 12.2 kcal/mol.[89] The interaction at -2 to $0°C$ gave monomeric salts, whereas at 25–30°C the spontaneous polymerization gave H_2O-soluble cationic allylamine-epichlorohydrin copolymer polyelectrolyte. The reduced viscos-

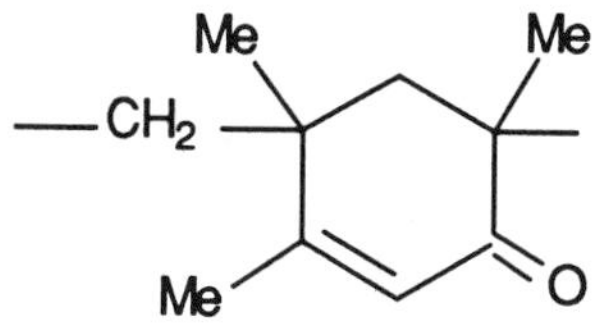

(6–20)

ity of the polymer solutions decreased with increasing polymerization temperature. The polymerization rate increased with increasing dielectric constant of the solvent. A tentative mechanism for the polymerization was proposed.

Copolymers of maleic anhydride and dicyclopentadiene with molar ratios of 1:1 to 2:1 were prepared by free radical initiation in a polar solvent.[90] The copolymer was saturated when the molar ratio was 2:1 and was soluble in acetone. A review with 45 references of current studies of the cyclopolymerization of carbon suboxide (C_3O_2), in which a mechanism involving a zwitterionic intermediate with competing propagation or termination steps was proposed to explain the details of the mechanistic pathway. (See Structure 2–41).[2-294] 4,7-Dihydro-1,3-dioxepins, as electron donors, undergo free-radical initiated copolymerization with maleic anhydride, as an electron-acceptor to give 1:1 alternating copolymers.[91] The dioxepins, however, undergo copolymerization with maleimides in a nonequimolar fashion to give soluble copolymers. Total hydrolysis of the copolymers produced head-to-head poly(4-hydroxycrotonic acid) which readily cyclized to give head-to-head poly(γ-crotonolactone). In the presence of strong acids, the copolymer gave the poly(γ-crotonolactone) directly (6–21).

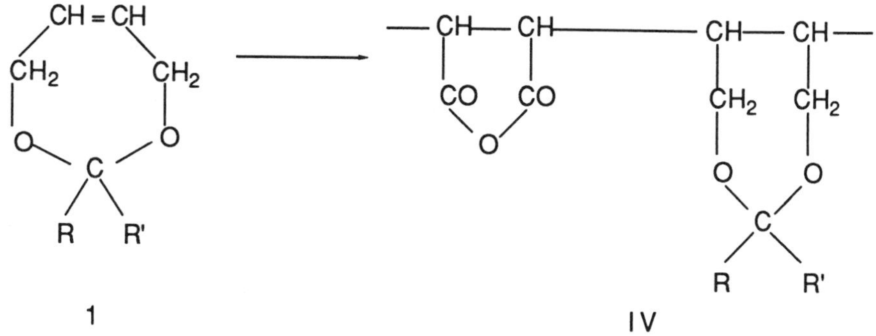

(6–21)

REFERENCES

1. Aso, C., Tagami, S. and Kunitake, T., Polym. J., 1970, 1, 395; CA 73, 110212k (1970).
2. Kitahama, Y., Ohama, H. and Kobayashi, H., Prepr. IUPAC Symp. Macro. Chem., Tokyo, 1966, 1-204; CA 71, 13440j (1969).
3. Kitahama, Y., J. Polym. Sci. A-1, 1968, 6, 2309; CA 69, 67795f (1968).
4. Kitahama, Y., Ohama, H. and Kobayashi, H., J. Polym. Sci. B, 1967, 5, 1019; CA 67, 117379e (1967).
5. Takemoto, K., Wada, H. and Imoto, M., Kobunshi kagaku, 1968, 25, 614; CA 70, 88319q (1969).
6. Kobayashi, H., Ohama, H. and Kitahama, R., Jap. Pat. 69 00, 865, 1969 (Jan. 16); CA 70, 115735f (1969).

7. Kobayashi, H., Ohama, H. and Kitahama, R., Jap. Pat. 70 25, 313, 1970 (Aug. 21); CA 74, 4010d (1971).
8. Guaita, M., Camino, G. and Trossarelli, L., Makromolek. Chem., 1969, 130, 243; CA 72, 44182b (1970).
9. Kobayashi, H., Ohama, H. and Kitahama, R., Japan Pat. 70 25,312, 1970 (Aug. 21); CA 73, 121188h (1970).
10. Barabas, E. S. and Manson, J., U.S. Pat. US 3,408,330, 1968 (Oct. 29); CA 70, 12140d (1969).
11. Koral, J. N., Makromolek. Chem., 1963, 62, 148; CA 59, 768h (1963).
12. Horner, L. H., Jurgeleit, H. and Klupfel, K., Justus Liebigs Ann. Chem., 1955, 591, 153; CA 50, 2494f (1956).
13. Kochetov, E. V., Markevitch, M. A. and Enikolopyan, N. S., Dokl. Akad. Nauk SSR, 1968, 180, 143.
14. Kochetov, E. V., Berlin, A. A., Masalakaya, E. M. and Enikolopyan, N. S., Vysokomol. Soedin., A, 1970, 12, 1118.
15. Franzmann, G., Eisenbach, C. D. and Jaacks, V., Angew. Chem., Int. Ed., 1971, 10, 349.
16. Schulz, R. C., Suzuki, S., Cherdron, H. and Kern, W., Makromol. Chem., 1962, 53, 145; CA 57, 13974e (1962).
17. Schulz, R. C., Kovacs, J. and Kern, W., Makromol. Chem., 1962, 54, 146; CA 57, 13974e (1962).
18. Schulz, R. C., Meyersen, K. and Kern, W., Makromol. Chem., 1962, 53, 58; CA 55, 8308c (1962).
19. Schulz, R. C., Kaiser, E. and Kern, W., Makromol. Chem., 1962, 58, 160; CA 57, 13974f (1963).
20. Schulz, R. C., Meyersen, K. and Kern, W., Makromol. Chem., 1962, 54, 156; CA 57, 13974f (1962).
21. Schulz, R. C. and Hartmann, H., Ber., 1962, 95, 2735; CA 58, 6778h (1963).
22. Schulz, R. C., Kolloid-Z., 1962, 182, 99; CA 57, 10023e (1962).
23. Schulz, R. C., Kovacs, J. and Kern, W., Makromol. Chem., 1963, 67, 187; CA 60, 1856b (1964).
24. Schulz, R. C., Meyersen, K. and Kern, W., Makromol. Chem., 1963, 59, 123; CA 57, 13974e (1963).
25. Schulz, R.C., Kaiser, E. and Kern, W., Makromol. Chem., 1964, 76, 99; CA 61, 10788g (1964).
26. Schulz, R. C., Kaiser, E. and Kern, W., Makromol. Chem., 1964, 76, 99; CA 60, 16055e (1964).
27. Schulz, R. C. and Passman, W., Makromol. Chem., 1964, 72, 198; CA 60, 16055e (1964).
28. Stetter, H., Gartner, J. and Tacke, P., Angew. Chem. Int. Ed. Engl., 1965, 4, 153; CA 62, 10349g (1965).
29. Schulz, R. C. and Wegner, G., Makromol. Chem., 1967, 104, 185; CA 67, 64776g (1967).
30. Schulz, R. C., Pure Appl. Chem., 1968, 16, 433; CA 69, 87648p (1968).
31. Yamashita, N., Sumitomo, H. and Maeshima, T., Kogyo-kwagaku zasshi, 1968, 71, 1723; CA 70, 42943v (1969).
32. Kitahama, Y., Ohama, H. and Kobayashi, H., J. Polym. Sci., A1, 1969, 7, 935; CA 71, 39503g (1969).
33. Tamura, Y., Kita, Y., Ishibashi, H. and Ikeda, M., J. Chem. Soc. D., 1971, 1167; CA 75, 151631f (1971).
34. McClure, J. D., U.S. Pat. 3,562,311, 1971 (Feb. 9); CA 74, 88354e (1971).

35. Andreeva, I. V., Madorskaya, L. Y., Sidorovich, A. V. and Koton, M. M., J. Polym. Sci., A-1, 1972, 10, 1467; CA 77, 48905r (1972).
36. Kunitake, T., Yamaguchi, K. and Aso, C., Makromol. Chem., 1973, 172, 85; CA 80, 71187d (1974).
37. Yamamoto, T., Noda, S. and Yamamoto, A., Kobunshi kagaku, 1973, 30, 94; CA 78, 136744p (1973).
38. Sumitomo, H., Hashimoto, K. and Kitao, O., J. Polym. Sci., Polym. Chem. Ed., 1975, 13, 327; CA 82, 171493k (1975).
39. Nasrallah, E. and Baylouzian, S., Polymer, 1977, 18, 1173; CA 88, 191560g (1978).
40. Van Ballegooie, P. and Rudin, A., J. Polym. Sci., Part A: Polym. Chem., 1988, 26, 2449; CA 109, 150175u (1988).
41. Otsu, T., Goto, K. and Imoto, M., Kobunshi Kagaku, 1964, 21, 723; CA 62, 18425f (1965).
42. Rostovskii, E. N., Lis, A. L. and Arbuzova, I. A., Vysokomol. Soedin., 1965, 7, 1792; CA 64, 5215e (1966).
43. Adelman, R., J. Org. Chem., 1968, 33, 1400; CA 68, 104519w (1968).
44. Matsoyan, S. G. and Morlyan, N. M., Izv. Akad. Nauk armyan. SSR, Khim. Nauki 1963, 16, 571; CA 61, 726c (1964).
45. Matsoyan, S. G. and Morlyan, N. M., Izv. Akad. Nauk armyan. SSR, Khim. Nauki 1964, 17, 319; CA 61, 13430b (1964).
46. Matsoyan, S. G., Morlyan, N.M. and Saakyan, A. A., Izv. Akad. Nauk armyan. SSR, Khim. Nauki 1962, 15, 405; CA 58, 6930d (1963).
47. Matsoyan, S. G. and Morlyan, N. M., Izv. Akad. Nauk armyan. SSR, Khim. Nauki 1963, 16, 347; CA 60, 10796c (1964).
48. Matsoyan, S. G. and Norlyan, N. M., Vysokomol. Soedin., 1964, 6, 945; CA 61, 7104b (1964).
49. Matsoyan, S. G., Norlyan, N. M. and Kinoyan, F. S., Vysokomol. Soedin., 1965, 7, 1159; CA 63, 14982e (1965).
50. Matsoyan, S. G. and Saakyan, A. A., Izv. Akad. Nauk armyan. SSR, Khim. Nauki 1965, 18, 60; CA 63, 8493h (1965).
51. Matsoyan, S. G. and Saakyan, A. A., Izv. Akad. Nauk armyan. SSR, Khim. Nauki 1963, 16, 159; CA 59, 8879f (1963).
52. Matsoyan, S. G. and Saakyan, A. A., Izv. Akad. Nauk armyan. SSR, Khim. Nauki 1964, 17, 676; CA 63, 7114b (1965).
53. Matsoyan, S. G. and Saakyan, A. A., Zh. Obsch. Khim., 1963, 33, 3795; CA 60, 9268d (1964).
54. Matsoyan, S. G., Morlyan, N. M. and Saakyan, A. A., Izv. Akad. Nauk armyan. SSR, Khim. Nauki 1965, 18, 68; CA 63, 8494c (1965).
55. Kocharyan, N. M., Matsoyan, S. G., Barsamyan, S. T., Pikalova, V. N., Tolapchyan, L. S. and Morlyan, N. M., Dokl. Akad. Nauk armyan. SSR, 1963, 37, 7; CA 59, 12269a (1963).
56. Matsoyan, S. G. and Morlyan, N. M., Izv. Akad. Nauk armyan. SSR, Khim. Nauki 1964, 17, 522; CA 62, 14832h (1965).
57. Matsoyan, S. G. and Morlyan, N. M., Izv. Akad. Nauk armyan. SSR, Khim. Nauki 1964, 17, 329; CA 61, 13430e (1964).
58. Akopyan, L. A., Avetyan, M. G. and Matsoyan, S. G., Izv. Akad. Nauk armyan. SSR, Khim. Nauki 1964, 17, 703; CA 63, 4230d (1965).
59. Nazarov, I. N. and Terekhova, L. N., Izv. Akad. Nauk SSSR, Otdel. Khim. Nauk 1950, 66; CA 44, 8904i (1950).
60. Matsoyan, S. G. and Akopyan, L. A., Arm. Khim. Zh., 1967, 20, 719; CA 68, 50141w (1968).

61. Brodskaya, E., Koyazhev, Y., Shergina, N. and Okladnikova, Z., Vysokomol. Soedin. Ser. B, 1968, 10, 895; CA 70, 38246t (1969).
62. Glazunova, E., Narnitskaya, M. A., Grigina, I. N., Yasenkova, L. S. and Nikitin, V. I., Vysokomol. Soedin. , Ser. A, 1968, 10, 1235; CA 69, 36535s (1968).
63. Kryazhev, Yu. G., Okladnikova, Z. A., Rzhepka, A. V., Brodskaya, E. I. and Shostakovskii, M. F., Vysokomol. Soedin., Ser. A, 1968, 10, 2366; CA 70, 29396q (1969).
64. Shostakovskii, M., Kryazhev, Y., Okladnikova, Z. and Rzhepka, A., USSR Pat. 215,509, 1968 (Sept. 28), CA 70, 97410q (1969).
65. Brodskaya, E., Koyazhev, Y., Frolov, Y., Kalikhman, I. and Yushmanova, T., Vysokomol. Soedin. , Ser. A, 1969, 11, 655; CA 70, 115656f (1969).
66. Matsoyan, S. G. and Saakyan, A., Arm. Khim. Zh., 1969, 22, 161; CA 71, 39530p (1969).
67. Shostakovskii, M. F., Kryazhev, Y. G., Yushmavova, T. I., Brodskaya, E. I. and Kalikhman, I. D., Vysokomol. Soedin. , Ser. A, 1969, 11, 1558; CA 71, 92182c (1969).
68. Kalikhman, I. D., Kryazhev, Yu. G. and Rzhepka, A. V., Vysokomol. Soedin., Ser. B, 1969, 11, 234; CA 70, 115664g (1969).
69. Kryazhev, Y.G., Cherkashin, M.I., Yushmanova, T.I., Kalikhman, I.D., Baiborodina, E.N. and Shostakovskii, M.F., Vysokomol. Soedin., Ser. A, 1969, 11, 700; CA 71, 22379y (1969).
70. Shostakovskii, M. F., Kryazhev, Y. G., Okladnikova, Z. A., Rzhepka, A. V. and Komarov, N. V., Vysokomol. Soedin., Ser. B, 1969, 11, 174; CA 71, 3773f (1969).
71. Shostakovskii, M., Kryazhev, Y., Rzhepka, A. and Okladnikova, Z., Izv. Sib. Otd. Akad. Nauk. SSSR, Ser. Khim. Nauk 1969, 134; CA 72, 90907q (1970).
72. Moriya, M., Mano, S. and Yamashita, T., Kobunshi kagaku, 1971, 28, 143; CA 75, 36751h (1971).
73. Yushmanova, T. I., Kryazhev, Y. G., Kalikhman, I. D. and Brodskaya, E. I., Vysokomol. Soedin., 1971, 13, 2064; CA 76, 14985f (1972).
74. Shostakovskii, M.F., Yushmanova, T.I., Kryazhev, Y.G., Kalikhman, I.D. and Brodskaya. E.I., Khim. Atsetilena, Tr. Vses. Konf. 1972, 3rd, 323; CA 79, 53823h (1973).
75. Askarov, M., Ilkhamov, M. and Syltanov, K., Khim. Atsetilena, Tr. Vses. Konf., 1972, 37D, 339; CA 79, 53825k (1973).
76. Akopyan, L. A., Gezalyan, D. I. and Matsoyan, S. G., Arm. Khim. Zh., 1974, 27, 768; CA 82, 125204u (1975).
77. Julia, M. and Binet du Jassonneix, C., Bull. Soc. Chim. Fr., Pt. 2, 1975, 3, 751; CA 83, 96932t (1975).
78. Trossarelli, L, Guaita, M. and Priola, A., Makromol. Chem., 1967, 109, 253; CA 68, 13454e (1968).
79. Stoicescu, C. and Dimonie, M., Rev. Roum. Chim., 1968, 13, 109; CA 69, 27901f (1968).
80. Farbenfabirken Bayer A-G, Brit. Pat. 1,132,428, 1969 (Oct. 30); CA 70, 12125c (1969).
81. Speier, J. L., U.S. Pat. 3,445,425, 1969 (May 20); CA 72, 3940d (1970).
82. Takahashi, T., J. Polym. Sci. A1, 1970, 8, 617; CA 72, 111863q (1970).
83. Takahashi, T., J. Polym. Sci. A1, 1970, 8, 739; CA 72, 111840e (1970).
84. Catterall, E. and Lyons, A., Eur. Polym. J., 1971, 7, 839; CA 75, 152166v (1971).
85. Catterall, E. and Lyons, A., Eur. Polym. J., 1971, 7, 849; CA 75, 152167w (1971).
86. Butler, G. B. and Monroe, S., J. Macromol. Sci. (Chem), 1971, 5, 1063; CA 75, 141438j (1971).
87. Lau, K. S. Y. and Arnold, F. E., Rept. AFML-TR-79-4065, 1979, CA 92, 147565w (1980).
88. Pickard, J. M. and Jones, E. G., Macromolecules, 1979, 12, 895; CA 91, 158180x (1979).

89. Zainutdinov, S. S., Dzhalilov, S. T. and Rakhmatullaev, K., Uzb. Khim. Zh., 1979, 2, 44; CA 91, 39971w (1979).
90. Gaylord, N. G., U.S. Pat. US 4,168,359, 1979 (Sept. 18); CA 92, 7372f (1980).
91. Culbertson, B. M. and Aulabaugh, A. E., ACS Symp. Ser., 1982, 195, 371; CA 97, 182964a (1982).

7

Copolymerization of Nonconjugated Dienes with Conventional Vinyl Monomers

Numerous examples of the copolymerization of 1,6-, or other nonconjugated dienes, with vinyl monomers, or other co-monomers such as sulphur dioxide, are reported in the literature. In most cases, the diene cyclopolymerized and soluble polymers were obtained. For example, soluble copolymers of diallylcyanamide with vinyl chloride and vinyl acetate[1,2,3] have been produced in which the diallyl monomers were postulated to cyclize to *N*-cyanopiperidine units. Further cyclic copolymers of diallyl compounds and sulphur dioxide[4,5] have been produced. Copolymerization of 1,6-dienes with conventional vinyl monomers can proceed by two distinctly different pathways: (a) the nonconjugated diene alone can form the cyclic structure followed by reaction of the resulting cyclic radical (or other reactive intermediate) with the comonomer, or (b) the comonomer can contribute to formation of the cyclic structure along with the nonconjugated diene. The latter type is generally defined as "cyclocopolymerization" and is discussed in detail in Chapter 8.

Various copolymerizations of 1,6-dienes with conventional vinyl monomers have been studied. For example, copolymerization of methacrylic anhydride with a variety of vinyl monomers was studied[1-66] as early as 1961. Methyl methacrylate, styrene, ethyl acrylate, vinyl acetate, 2-chloroethyl vinyl ether, and allyl chloroacetate were among the comonomers studied. It was concluded that, in general, higher conversion to soluble copolymers was realized (a) the less reactive the comonomer in radical-initiated copolymerizations, (b) the greater the dilution, (c) the greater the difference in the molar concentration of the two monomers, and (d) the lower the conversion. According to reactivity ratio determinations, based upon a copolymer composition equation derived in this study for the special divinyl system and a comparison with published data, it was concluded that the anhydride ring radical behaved like a methyl methacrylate radical.

In a study of copolymerization of symmetrical nonconjugated dienes with vinyl monomers, a copolymer composition equation was developed and reactivity ratios determined. By applying the composition equation, values of reactivity ratios for 12 sets of monomer pairs were obtained and found to be the best values with a permissible range of errors.[6]

The alternating equimolar copolymer of isoprene and maleic anhydride has been synthesized by trapping the maleic anhydride generated by retrograde dissociation of the furan-maleic anhydride Diels-Alder adduct at 60°C in the presence of isoprene and a radical initiator.[7]

COPOLYMERS OF SYMMETRICAL 1,5- OR 1,6-DIENES WITH VINYL MONOMERS

Of the numerous reports which have appeared in the literature on copolymerization of nonconjugated dienes with other monomers, one of the earliest studies[1-66] established that cyclopolymerization can proceed by two distinctly different pathways: (a) the nonconjugated diene alone can form the cyclic structure followed by reaction of the resulting cyclic radical (or other reactive intermediate) with the comonomer, or (b) the comonomer can contribute to formation of the cyclic structure along with the nonconjugated diene. This chapter is limited to the former type. Chapter 8 is devoted to the latter type of "cyclocopolymerization." Because of the large number of copolymerization studies between nonconjugated dienes and conventional vinyl monomers, this section has been divided into four subsections which deal with those monomers which form: (1) carbocyclic rings, (2) rings containing nitrogen, (3) rings containing oxygen and (4) rings containing other elements.

Copolymers of Dienes which Form Carbocyclic Rings

A copolymerization study of 2,6-disubstituted 1,6-heptadienes of the following structure (7–1) with acrylonitrile, vinylidene chloride, styrene, acrylamide, and methyl acrylate

$$CH_2=\underset{\displaystyle R}{\underset{|}{C}}CH_2CH_2CH_2\underset{\displaystyle R}{\underset{|}{C}}=CH_2$$

where R is

$$-\overset{\displaystyle O}{\overset{\|}{C}}\text{-}OCH_2CH_3,\quad -\overset{\displaystyle O}{\overset{\|}{C}}\text{-}OH,\quad -\overset{\displaystyle O}{\overset{\|}{C}}\text{-}NH_2,\ \text{and}\ -CN$$

(7–1)

was undertaken.[2-2] When the total concentration of monomers was about 10%, either in emulsion or solution, soluble copolymers were obtained. The absence of unsaturation in the copolymers and their excellent solubility indicated that the disubstituted heptadienes underwent cyclopolymerization, even in presence of the reactive monomers studied.

It has been shown that substituted 1,6-heptadiene monomers undergo copolymerization with acrylonitrile to produce copolymers having fiber-forming properties superior to those of polyacrylonitrile.[1-26] For example, a copolymer of 15% 2,6-dicyano-1,6-heptadiene with 85% acrylonitrile was spun and drawn fourfold to strong filaments having a wet initial modulus of 10 g/den at 90°C and a recovery from 3% elongation in 50°C water of 56%. Typical values for fiber prepared in a similar manner from acrylonitrile homopolymer are 3g/den and 40%, respectively. The preparation of other copolymers of these dienes is tabulated in Table 7–1.

Table 7–1 Copolymers[a] of 2,6-Disubstituted 1,6-Heptadienes with Monofunctional Monomers

X in heptadiene	Y in comonomer	Solvent	Initiator	Temp. (°C)	Time (h)	Conversion (%)	Intrinsic viscosity[b]	% Heptadiene
$-COOCH_3$ (0.5 g)	-CN (11.5 g)	Emulsion (100 ml)	Peroxydisulfate (0.03 g)-metabisulfite (0.25 g)-β-Mercaptoethanol (0.05 g)-peroxydisulfate (0.01 g)	55	3.0	67.3	1.28	5.2
$-COOCH_3$ (0.5 g)	-CN (09.5 g)	Emulsion	Peroxydisulfate	55		95	4.28	4.5
-CN (0.3 g)	$-C_6H_5$ (0.9 g)	Tetramethylene sulfone (6 ml)	UV, benzoin (0.002 g)	20	22	58.5	1.26	31.2
-CN (0.4 g)	$-COOCH_3$ (0.4 g)	Tetramethylene sulfone (5 ml)	UV, benzoin (0.001 g)	20	8	62.5	1.30	34.5

[a] Copolymer is $CH_2 = \underset{X}{C}\text{-}(CH_2)_3\text{-}\underset{X}{C} = CH_2$ with $CH_2 = CHY$.

[b] dl/g in dimethylformamide

Crystalline copolymers of ethylene and 1,5-hexadiene have been reported.[8, 1-60] These copolymers, prepared through use of a modified Ziegler catalyst derived from titanium chloride, aluminum chloride and triethylaluminum, were crystalline over the compositional range of 15–93 mole % of ethylene. Also reported was an unusual mutual synergistic effect on polymerization rate in the system. The addition of ethylene to the 1,5-hexadiene monomer resulted in an increase of 1,5-hexadiene conversion up to about 10 wt. % of ethylene. An increased rate of ethylene polymerization by addition of 1,5-hexadiene was also observed. The reasons for these phenomena were not determined as X-ray analysis of the copolymers showed them to be essentially block copolymers. More random copolymers might have been expected, as suggested by the above observations, if a 1,5-hexadiene chain end preferred to react with ethylene while an ethylene chain end preferred to react with 1,5-hexadiene.

X-ray diffraction patterns showed that the copolymers crystallized in blocks of poly-1,5-hexadiene and polyethylene units. The minimum block of polyethylene units consisted of about 20 monomers long while that for poly-1,5-hexadiene along the chain consisted of about 10 - 11 monomers. The poly- 1, 5-hexadiene unit consisted of alternating methylene and cis-1,3-cyclopentane groups formed by cyclopolymerization of the 1,5-diene. The same conclusion was reached from the melting point behavior of the copolymers. The melting point curve resembled that of a eutectic composition. Other effects on physical properties such as inherent viscosity, density, solubility and mechanical properties were also determined.

The copolymerization of o-divinylbenzene (M_1) and 2-methyl-5-vinylpyridine has also been studied.[9] The copolymerization was conducted in benzene solution at 60°C in presence of benzoyl peroxide as initiator. The copolymer was soluble in benzene. Its composition was determined by nitrogen analysis, and the residual double bond content, determined by infrared spectroscopy, was found to be 70–80% of the double bonds calculated from the divinylbenzene content of the copolymer. Intramolecular cyclization was thus shown to have occurred. The reactivity ratios estimated by using the Roovers-Smets copolymer composition equation were: $r_1 = 0.62 \pm 0.05$, $r_2 = 0.78 \pm 0.15$, and r_3 (cyclized M_1) $= 0.2 \pm .3$. These were compared with reactivity ratios, $r_1 = 0.45 \pm 0.15$ and $r_2 = 0.90 \pm 0.10$ which were calculated by use of the Fineman-Ross equation. The cyclization constant was 2.6 moles/l.

The monomer reactivity ratios for the styrene-o-divinylbenzene system[1-67] have also been determined; cyclopolymerization was observed to some extent. The copolymers were soluble and the copolymerization showed no signs of crosslinking, unlike those from the m- and p-isomers which swell but do not dissolve. Other evidence showed that there was less than one vinyl group per aromatic ring which indicated that cyclopolymerization had occurred.

Vulcanizable copolymers of olefins, e.g. ethylene and propylene, and polyenes were synthesized by use of an appropriate amount of a polyene containing at least two double bonds, one of which is in a terminal position, and the other of which is non-reactive in the copolymerizing system, but remains intact as a vulcanization site.[10] A metal-coordination catalyst system was employed. Use of trans-1,2-divinylcyclobutane for this purpose was described. A hydrocarbon vulcanizable elastomer consisting of a terpolymer of ethylene, propylene and a nonconjugated diene was described and its properties reported.[11] The terpolymer retained a fraction of the unsaturation of the diene, thus providing a vulcanization site. Other reports have shown that nonconjugated dienes such as 1,4-hexadiene undergo some cyclization, thus reducing the residual unsaturation on a molar basis.

Mixtures of a 1-alkene and a triolefin, such as *trans, trans, cis*-1,5,9-cyclododecatriene, were copolymerized in presence of a metal-coordination catalyst to give products which can be vulcanized.[12] Neither the nature of residual double bond content nor its ratio to the total was given. Other elastomeric copolymers of α-olefins and nonconjugated cyclic polyenes were prepared by a metal-coordination catalyst system; 1,5,9-Cyclododecatriene was used to provide residual double bond reactivity in the copolymer as a vulcanization site.[13] Copolymers of ethylene, propylene and controlled amounts of 1,5-cyclooctadiene were prepared in presence of a metal-coordination catalyst system as vulcanizable elastomers.[14] The copolymers were amorphous. Maleic anhydride and similar compounds were thermally copolymerized with 1,5-cyclooctadiene to give 1:1 alternating cyclocopolymers.[15]

A variety of copolymers and terpolymers of 1,6-diolefins and monovinyl compounds was prepared via radical initiation.[16] As an example, 2,2-diallyldiethylmalonate and ethyl acrylate were copolymerized to a soluble material, proposed to possess a cyclic structure formed via cyclopolymerization of the diallyl monomers. Copolymers of 1,5-cyclooctadiene and trioxane have been described.[17] A 1:2 cyclopentadiene-maleic anhydride copolymer was synthesized by copolymerizing the monomers at 80–205° in the presence of tert-butyl peroxide or peracetate.[18] The copolymer had molecular weight 530–1000 and softening temperature 260–80°C, independent of temperature and catalyst concentration. Cyclocopolymerization of d-limonene with maleic anhydride gave 1:2 alternating cyclocopolymers which were optically active.[19] 1,5-Hexadiene and 2,5-dimethyl-1,5-hexadiene were cyclocopolymerized with maleic anhydride and the structures of the copolymers were compared with those derived from 1,4-dienes.[20] The polymers contained residual unsaturation but approximated the 2:1 molar ratio of maleic anhydride predicted by a bimolecular alternating intra-intermolecular chain mechanism.

Copolymers of Dienes which Form Nitrogen-Containing Rings

Because of strong interest in the diallyl quaternary ammonium polymers, an exceptionally large number of copolymerization studies have been conducted on these structures.

Copolymerization of several diallyl monomers with acrylamide and acrylonitrile were studied rather early.[1-39] Diallyldimethylammonium chloride, diallylmethylamine hydrobromide, diallylamine, diallylmethylamine, diallylmethylamine oxide, diallylmethylsulfonium methyl sulfate, diallylsulfide, diallylmelamine, and the zwitterion obtained by quaternizing diallylmethylamine with α-haloacetic acid were among those compounds studied. In general, soluble copolymers of high molecular weight were obtained; however, diallyl sulfide and diallylmelamine occasionally gave gelled copolymers, depending upon the conditions. Molecular weights of these copolymers were in the hundred thousand to million range. The soluble copolymers with acrylamide were effective strengthening agents for paper, and one of the copolymers with acrylonitrile could be spun into fibers. The acrylonitrile copolymers were receptive to acid dyes. Copolymers with triallylamine, triallylmethylammonium bromide, and tetraallylammonium bromide were crosslinked.

1-(Diallylamino)-2,3-epoxypropane was prepared and its co-polymerization with styrene was studied.[21] Copolymerization of this monomer at 48.2 mole % with styrene in benzene solution at 60°C, using AIBN as initiator, resulted in a copolymer containing 3.90 mole % of the amine. Copolymerization at 10.2 mole % of the amine with styrene under similar conditions resulted in isolation of essentially pure polystyrene. It was concluded that this monomer, being a diallylic system, entered into copolymerization reluctantly and behaved as a degradative chain transfer agent.

Monomer reactivity ratios were determined for the copolymerization of lauryl methacrylate (M_1) with N-allylurea, N,N-diallylurea, and N,N′-diallylurea at 100°C with benzoyl peroxide.[22] The following values were reported: N-allylurea, $r_1 = 29.9$, $r_2 = 0.22$; N,N′-diallylurea, $r_1 = 27.4$, $r_2 = 0.14$; N,N-diallylurea, $r_1 = 14.5$, $r_2 = 0.12$. The monomer reactivity ratios were considered as evidence for cyclic propagation reactions in the copolymerization involving N,N′-diallylurea.

Some of the experimental procedures for making copolymers of dimethyldiallylammonium chloride, as well as some of the experimental variables and some procedures for evaluating the copolymers in flocculation and sludge dewatering,[2-91] have been summarized and discussed.

The preparation of acrylamide-dimethyldiallylammonium chloride-diethyldiallylammonium chloride terpolymer has been reported by terpolymerizing the corresponding monomers in the presence of ethylenediamine tetraacetic acid (EDTA), Na salicylate, $(NH_4)_2S_2O_4$, $NaHSO_3$, and $CuSO_4$.[23] The partially hydrolyzed polymer was useful as a retention aid in paper making.

The preparation of polysalts of structure (7–2) by adding poly(dimethyldiallylammonium chloride) to the sodium salt of poly(2-hydroxy-3-methacryloyloxypropane-l-sulfonic acid) has been described.[24]

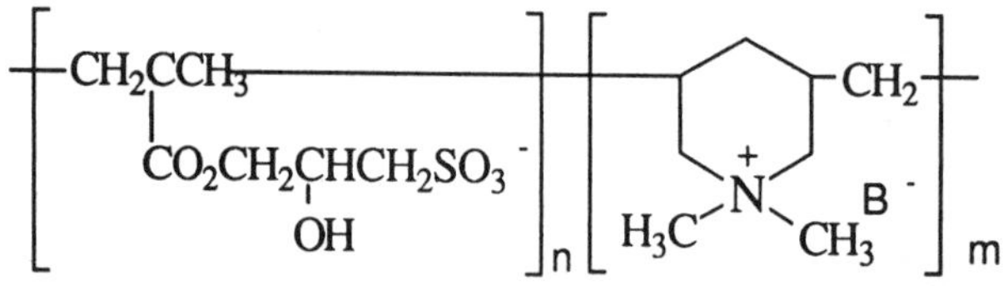

$$\left[-CH_2C(CH_3)(CO_2CH_2CH(OH)CH_2SO_3^-)- \right]_n \left[-\text{(piperidinium ring, } N^+(CH_3)_2\text{)}-CH_2- \;\; B^- \right]_m$$

(7–2)

Water-soluble copolymers of dimethyldiallylammonium chloride (DADMAC) and methylstearyldiallylammonium chloride have been reported by radical initiated copolymerization.[25] The copolymers are useful as antistatic agents for textiles, electrical conductive coatings for paper, and softening agents for paper and textiles.

The copolymerization of diallylcyanamide with vinyl acetate by radical initiation yielded copolymers of Structure 7–3.[26]

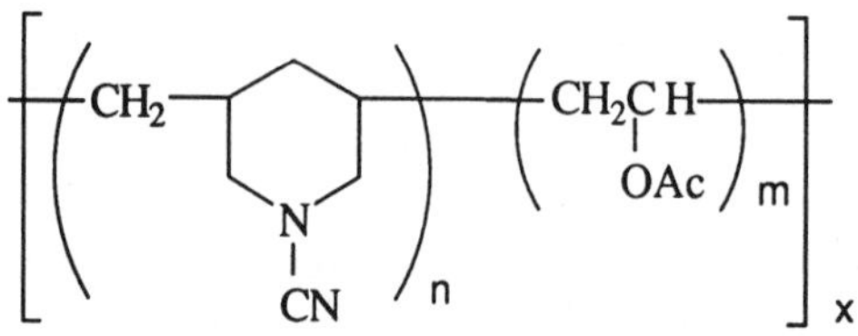

$$\left[\left(-CH_2-\text{(piperidine ring, } N-CN\text{)}- \right)_n \left(-CH_2CH(OAc)- \right)_m \right]_x$$

(7–3)

The copolymerization of N,N-diallylglycinonitrile hydrochloride with acrylic acid has been shown to yield copolymers of Structure 7–4.[27] The copolymers were shown to be useful as scale formation inhibiting agents.

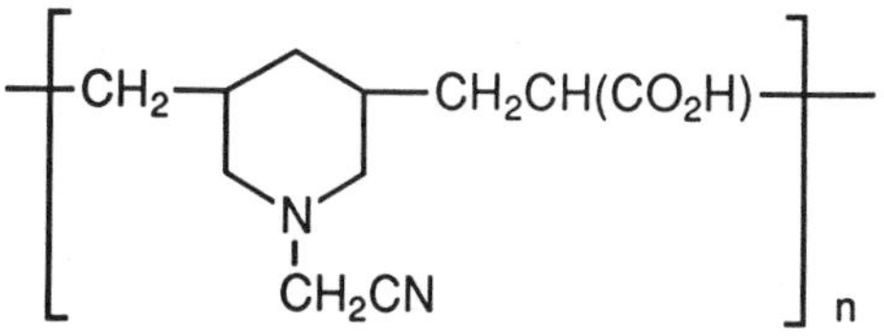

CH_2 — $CH_2CH(CO_2H)$ — N — CH_2CN — n

(7–4)

Copolymers of bis-ethylenically unsaturated amines such as diallylamine, dimethylallylamine, and diallylalkylamines with olefins such as styrene, alkyl acrylates, acrylamide, acrylonitrile, methacrylonitrile, and vinyl acetate have been prepared.[1–57] Copolymerization of these amines with acrylamide produced copolymers of extremely high molecular weight. For example, a linear copolymer prepared from a mixture of 98.5 mole % of acrylamide and 1.5 mole % of diallylbenzylamine had a viscosity of over 100,000 cP at a concentration of 11.1% in water. Terpolymers of acrylic acid, acrylamide, and diallylethylamine, or acrylonitrile, vinyl acetate, and diallylamine, were also prepared. In all cases, the copolymers and terpolymers were low in residual unsaturation. Although the structures of the copolymers were not determined, in view of the properties observed and the large number of examples of cyclopolymerization of diene monomers now on record, cyclic structures for the diene monomer units in these copolymers are reasonably certain. It was stated that: "The reason for this monofunctionality of a bis-ethylenically unsaturated compound of the kind used in practicing this invention is not fully understood."

Substantially linear copolymers of diallylbarbituric acid with unsaturated compounds such as maleic anhydride, vinyl acetate, vinylene carbonate, methyl acrylate, acrylonitrile, styrene, and vinylidene chloride in the presence of conventional free-radical initiators have been prepared.[1–62] The copolymers contained no residual double bond, indicating that the diallyl monomer underwent cyclopolymerization.

Copolymerization of N,N-dimethylaminoethyl methacrylate with DADMAC gave perfectly soluble copolymers and no residual vinyl groups were observed by physical methods.[2–131]

Reactivity ratios (r) have been determined for copolymerization of DADMAC (M_1) and acrylamide (M_2). Both r_1 and r_2 depended on DADMAC concentration in the initial comonomer solution, and increased with decreasing DADMAC concentration. An increase in total monomer concentration increased r_1, but r_2 remained nearly constant. The results were explained by association of the cationic monomer and electrostatic interactions.[28]

A stable microbiocidal aqueous dispersion comprising methylene bisthiocyanate and a self-inverting emulsion of a copolyampholyte containing DADMAC was suitable for inhibiting the growth of slime-forming bacteria, fungi, and algae.[29] Emulsions containing a saponified DADMAC-vinyl acetate copolymer or allyltrimethylammonium chloride-vinyl acetate copolymer were prepared and mixed with thickeners, freeze-thaw stabilizers, rustproofing agents, preservatives, defoamers, emulsifiers or dispersants, and water.[30] Polyampholytes and their salts were prepared from anionic, cationic, and some nonionic monomers and used as scale and corrosion inhibitors, in drilling fluids, and in waterflood oil recovery. A typical example of such a polyampholyte is an acrylic acid-DADMAC-N,N-dimethylacrylamide terpolymer (70:20:10).[31]

Cationic polyelectrolytes from acrylamide and DADMAC were prepared in aqueous solutions and their flocculating capacities were tested in aqueous suspensions of TiO_2. Acrylamide homopolymer of molecular weight $<10^6$ had virtually no flocculating capacity. DADMAC homopolymer accelerated the sedimentation of TiO_2, and this acceleration increased with increasing molecular weight. The optimum dose of the homopolymer decreased with increasing molecular weight. The flocculating capacity of the copolymers increased and the optimum dose decreased with increasing molecular weight, but larger doses than those of the homopolymer were required for producing the same sedimentation rates.[32]

A copolymer of DADMAC with acrylic or methacrylic acid has been used as a bactericide for water-oil base cutting fluids.[33] The effect of polyelectrolyte flocculants on microorganisms in receiving streams has been studied. Tests of polyelectrolytic toxicity to aquatic microorganisms demonstrated that poly(DADMAC) did not inhibit the cumulative oxygen uptake of a culture of mixed aquatic microorganisms, as measured with an electrolyte respirometer. Although oxygen uptake curves showed that the polymer could not support microbial activity by the mixed aquatic microbe, a Pseudomonas species was isolated from the St. Lawrence River which could biodegrade and assimilate the polyelectrolyte. This bacterial species was also able to utilize the monomer as a source of both carbon and nitrogen.[34]

A graft copolymer of cellulose and DADMAC was prepared by impregnating cellulose with aqueous monomer containing $(NH_4)_2S_2O_8$, removing the excess and heating the impregnated cellulose. The extent of grafting of the cationic monomer could be increased to 17.5% and monomer conversion was as high as 25%.[35]

Terpolymers containing N,N-diallylacetamide for use as high temperature fluid loss additives and rheology stabilizers for high pressure, high temperature oil well drilling fluids have been synthesized.[36]

Certain copolymers of diallylamine have been used in developing dye fastness improvements. Copolymers having structural units (7–5) ($R_{1\text{-}4}$ = H, CH_3) in 95–50:5–50 ratio were prepared and used to improve dye fastness of textiles.[37]

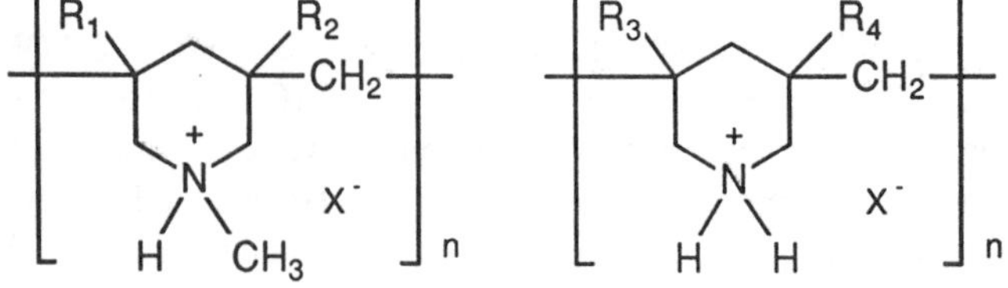

(7–5)

Linear copolymers of diallylamine oxides with vinyl monomers via radical initiation were described.[1-34] Critical and controlled molar proportions of the tetrafunctional monomers in the comonomer mixtures were deemed necessary, and it was stated that, "The reason for this monofunctionality of the bis-ethylenically unsaturated compound of the kind used in practicing this invention is not fully understood."!

Copolymers of vinyl chloride with poly-N-alkyl substituted aminotriazine monomers of small but critical quantities were obtained by free radical initiation.[38] Monomers used were N,N-diallylmelamine, 2,4-di(methallylamino)-6-chloro-1,3,5-triazine and N,N',N''-triallylmelamine, among others. The copolymers were soluble, indicating that cyclopolymerization had occurred. Diallylamine, diallylaniline, N,N'-diallylpiperazine, dimethallylamine, and N-methyldiallylamine, when copolymerized with vinyl chloride, gave soluble copolymers, free of gels having higher intrinsic viscosities than the corresponding vinyl chloride homopolymer.[39]

Cross-linked, water-swellable carboxylic acid copolymers were prepared by copolymerization of acrylic acid in presence of polyallyl tertiary amines.[40] Examples of the amines used were triallylamine, trimethallylamine, methallyldiallyl amine, N,N,N',N'-tetraallylethylenediamine, N,N,N',N'-tetramethallylethylenediamine, and N,N-diallyl-N-octylamine. Solution viscosities of the polymers were reported, indicating that cyclopolymerization was the major route of the "cross-linking" comonomers, and that little or no cross-linking occurred.

Diallylamine derivatives readily polymerized and copolymerized with certain other monomers, presumably via cyclopolymerization. On the other hand, monoallyl derivatives were reluctant to polymerize, possibly due to degradative chain transfer.[41] An attempt was made to synthesize poly(allylammonium hydrochloride) and a copolymer of allylammonium hydrochloride with diallylammonium hydrochloride. A polymer of low molecular weight was obtained in the former case, whereas the copolymerization attempt was inconclusive although strong indications were present that some copolymerization had occurred. Copolymers containing 0.1–0.5% nitrogen were obtained from ethylene, acrylonitrile and diallylamine with metal-coordination catalysts.[42] Existence of the comonomers in the copolymers was shown by infrared (IR) spectra.

Diallyl cyanurate or dimethallyl cyanurate were used as crosslinking agents for unsaturated polyesters in molding compositions.[43] The extent of cyclization in these comonomers was not discussed. Cyclocopolymerization of vinyl acetate with diallylcyanamide in aqueous emulsion in presence of free-radical initiators was studied.[44] The kinetics were studied as well, and the polymers examined. The absence of unsaturation in the soluble copolymer suggested that the diallyl compound had undergone intramolecular cyclization. Further studies on the cyclocopolymerization of diallyl cyanamide with vinyl acetate in emulsion led to the development of an improved synthesis of diallyl cyanamide.[45] N,N-Diallylcyanamide-styrene copolymer, prepared via radical initiation was treated with hexamethylenediamine to give a heat-stable resin.[46] Diallylcyanamide was copolymerized with vinyl acetate to give linear copolymers containing up to 12% cyano-substituted ring structures.[47]

Copolymers of N,N-diallylurea with vinyl acetate were synthesized as intermediates for further reaction with formaldehyde to produce heat-stable resins.[48] Polymers or copolymers derived via cyclopolymerization from acrylimide or methacrylimide were treated with various derivatizing agents to displace the imide hydrogen atom.[49] Useful products were obtained. Cationic copolymers, soluble in cold water, were prepared from a variety of diallyl quaternary ammonium monomers containing the 3-acetyl-4-hydroxybenzyl group in the side chain with sulfur dioxide.[50]

Amphoteric ion-exchange resins of improved capacity were prepared by free-radical copolymerization of triallylamine hydrochloride, 1,4-bis-(N,N-diallylamino-ethyl) benzene, methyldiallylamine hydrochloride, and 1,6-bis(N,N-diallylamino) hexane dihydrochloride with methacrylic derivatives.[51] A thermally regeneratable ion-exchange resin was obtained in the triallylamine case. N,N-Diallylacetamide cationic derivatives when treated with SO_2 gave polysulfones [(7–6), R = N^+Me_3, N^+Et_3, N-

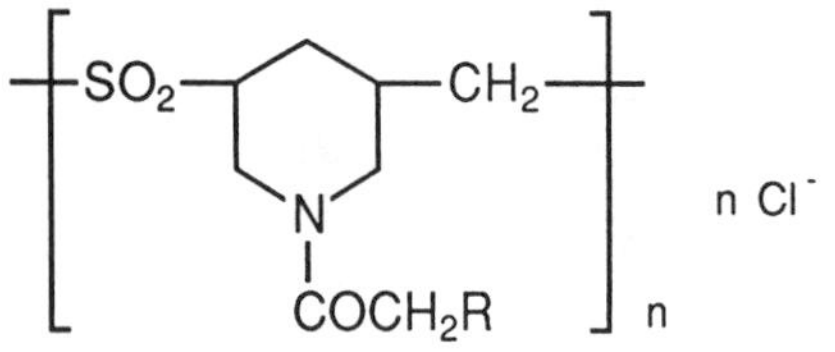

$$\left[SO_2 - \text{(piperidine ring, N-COCH}_2\text{R)} - CH_2 \right]_n \quad n\ Cl^-$$

(7–6)

methyl morpholinium, pyridinium, S^+Me_2, or $S^+(CH_2CH_3OH)_2$] of structure shown, which were useful as paints, adhesives, sizes, dye additives, antistatic agents, and coagulants.[52]

Polysulfones, useful as paper sizes, textile finishes, polymer additives, coagulents and thickeners have been prepared by copolymerizing 2-chlorodiallylacetamide, sulfur dioxide and DADMAC to give terpolymers of the termonomer compositions dependent upon termonomer feeds and other variables (7–7).[53] Vinyl alcohol copolymers containing cyclic units in the backbone were prepared by acetic acid or alkali saponification of diallylcyanamide-vinyl acetate copolymers.[54]

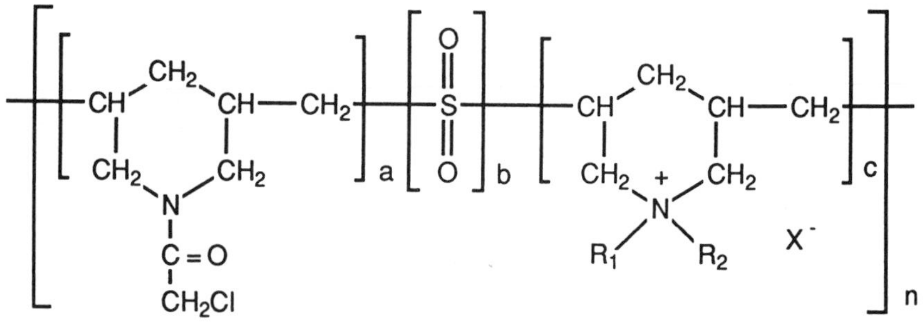

$$\left[\left[\text{CH}(CH_2)\text{CH}-CH_2;\ CH_2-N(C{=}O,\ CH_2Cl)-CH_2 \right]_a \left[SO_2 \right]_b \left[\text{CH}(CH_2)\text{CH}-CH_2;\ CH_2-N^+(R_1)(R_2)-CH_2 \right]_c \ X^- \right]_n$$

(7–7)

Polymers and copolymers from diallylamine derivatives have been prepared. Salts of diallylamine derivatives were photocyclohomopolymerized, in the presence of photosensitizers, or photocyclocopolymerized with unsaturated compounds in the presence of photosensitizers to give water soluble polymers.[2-170] The photosensitizers were SO_2 or mixtures, in alcohol solvents, of benzoyl derivatives and hydroquinone, hydroquinone monomethyl ether, dioxane, morpholine, or tertiary amines. A polyethylene bag, containing an aqueous dispersion of diallyldimethylammonium chloride was irradiated by a fluorescent lamp to give 89.4% water soluble poly(DADMAC).

Quaternary ammonium monomers have been prepared and copolymerized with acrylic monomers.[55] Allyldimethylamine was treated with allyl chloride to give a yellow, transparent solution of DADMAC. Acrylonitrile was copolymerized with the quaternary ammonium monomer to give 89.4% yield of copolymer having $[\eta] = 2.91$ dl/g. Cyclic polymers and copolymers with low viscosity have been studied.[56] Poly(DADMAC) and diacetoneacrylamide-DADMAC copolymers with low intrinsic viscosity and monomer content and containing ring structures were manufactured using redox catalysts and glycerol chain transfer agent, for use as flocculating agents for waste water and as coatings in the manufacture of paper insulators.

A variety of allylamines containing three or more allyl groups were converted to crosslinked polymers by radical initiation.[2-112] In addition to triallylamine, the bis(diallylaminoalkanes) in which the alkane bridge contained from 2–10 carbon atoms were polymerized. All of these polymers had been prepared and described earlier. Cyclopolymerization as the major mode of polymer formation was confirmed.

An attempt to prepare amphoteric copolymers for use in desalination by copolymerization of N-substituted diallylamines and methacrylic acid was reported to be unsuccessful, although such copolymers have been reliably reported elsewhere.[57] Copolymerization of diallylcyanamide with vinyl chloride has been studied.[1] Radical copolymerization gave the corresponding copolymer containing N-cyanopiperidine units. The absence of crosslinks and unsaturation indicated that the diallyl monomer polymerized exclusively via cyclization. The reactivity ratios were $r_1 = 0.68 \pm 0.05$ and $r_2 = 0.44 \pm 0.05$, respectively. Attempts to graft triallylamine hydrochloride to trapped macroradicals from acrylamide and to an autoxidized polyisopropylstyrene sample containing hydroperoxide groups were unsuccessful.[58]

Heat regenerable, amphoteric ion exchange resins, having good salt uptake and exchange capacity, were prepared by terpolymerizing an appropriate ratio of divinylbenzene, ethyl acrylate and triallylamine, followed by hydrolysis.[59] The ion exchange acid capacity was 3.58 meq./g, base capacity 2.64 meq/g. and salt uptake 1.32 meq./g. Graft copolymers of polysaccharides with mixtures of acrylamide, acrylic acid, and DADMAC were prepared by radical polymerization of aqueous suspensions in water-immiscible organic solvents containing surfactants.[60] A dilute solution study of DADMAC-diethylacrylamide copolymers by use of light scattering has been reported.[61]

It has been shown that in preparation of thermally regenerable ion-exchange resins, due to the difficulties experienced in attempting to obtain resins in bead form by emulsion polymerizations of mixtures of acrylic esters and allylamines, the approach of incorporating resin plums in a matrix may be the best.[2-94] The best beads were obtained from the methyl acrylate-triallylamine and diallylamine systems partially polymerized in paraffin oil containing talc, but these had the poorest thermally regenerable capacities. Polymeric dispersants gave more stable emulsions than did conventional surfactants, but monomer migration into the third phase was too rapid to permit the preparation of satisfactory beads. Although partial prepolymerization of the monomers reduces their migration, it is impracticable on a large scale.

Graft copolymers of a water-soluble monomer and polysaccharide have been prepared using a two-phase reaction system under controlled conditions.[62] The graft copolymers were formed at high grafting efficiency and high conversion. An acrylamide-acrylic acid-amylopectin-DADMAC graft copolymer was prepared. Vinylbenzyl and bis-(vinylbenzyl)amine derivatives were prepared and polymerized to give ion exchange resins in one step from the polymerization.[63] A mixture of N-methylbis-(vinylbenzyl)amine and DADMAC were copolymerized to give copolymer beads with 52.6% total solids, 5.80 mequiv./g anion exchange capacity, 0.06 mequiv./g strong base capacity and 270% swelling (acid-base cycling).

The reactions of hydrolyzed vinyl acetate-diallylcyanamide copolymer with aldehydes have been studied in an aqueous medium.[2] For a modified poly(vinyl alcohol) containing piperidine units in the main chain and OH and CH_3CO_2 side groups, reaction with HCHO, CH_3CHO, or C_3H_7CHO in aqueous phase occurred at both the OH and piperidine reactive groups to form modified poly(vinyl acetals) of three-dimensional structure. Poly(vinyl alcohol) containing piperidine units in the main chain and OH and

CH_3CO_2 side groups derived via hydrolysis of the vinyl acetate-diallylcyanamide cyclocopolymer have been synthesized.[3] The hydrolyzed cyclocopolymer was acetylated with butyraldehyde in methyl alcohol at 60° for ~12 hours to give a piperidine- and acetate-containing poly(vinyl butyral). Films prepared from the product had tensile strength 250–420 kg/cm^2, relative elongation 180–260%, adhesion 80–100 kg/cm^2, and dielectric strength 83.6–87.0 kV/mm^2.

An UV spectrophotometric method for analyzing copolymers of vinyl acetate-maleic anhydride and poly(diallylammonium chloride) has been reported.[64] Maleic anhydride-vinyl acetate copolymer and poly(diallylammonium chloride), useful as coagulating agents for purification of feed waters for steam boilers, could be determined quantitatively by measuring the extinction at their wavelength maxima in UV light. Manufacture of nontoxic flocculants for water purification by polymerization of DADMAC with N,N-dialkyl(meth)acrylamides has been described.[65] The copolymers were obtained as a 25% aqueous solution.

Water soluble cationic block copolymers of DADMAC were prepared which contained a block having predominantly quaternized repeat units and a block having predominantly primary or secondary amine groups.[66] A study of the polymerization process for monoallyl and diallylamine derivatives has been reported.[2-136]

Copolymers of (diallylamino)pyridine with dimethylacrylamide and DADMAC were prepared as candidates for supernucleophilic catalysts.[67] The reactivity ratios of the monomers were determined. Poly(DADMAC) and acrylamide-DADMAC copolymers were studied by using a low-angle laser light scattering instrument as a gel chromatographic detector coupled with the usual refractive index detector.[68] Molecular weights and molecular weight distributions were obtained. Poly(DADMAC) with high molecular weight was prepared by the inverse emulsion technique.[69] Conversion greater than 90% and polymer of intrinsic viscosity of 1.7 were obtained. Graft copolymers of poly(DADMAC) with acrylamide were synthesized using ceric salt/nitric acid initiation system.[70] Graft copolymers of molecular weight up to 1.7×10^6 were obtained. A copolymerization study of DADMAC with acrylamide and quaternized dimethylaminoethyl methacrylate was carried out in solution and reverse microsuspension.[71] The copolymer reactivity ratios were determined.

Copolymers of low molecular weight but in high conversion were synthesized from DADMAC and maleic anhydride.[72] Copolymerization of DADMAC with acrylamide led to crosslinking at all feed ratios at total monomer concentration of 4.0 mole/l, but to gelation only when acrylamide ratio reached 80 mole %.[73] The gel point occurred at 51% conversion, and swelling ratios were as high as 1400. Addition of isopropyl alcohol as chain-transfer agent prevented gelation. A study of certain factors involved in the cyclopolymerization of DADMAC and acrylic acid has been reported.[74]

Alternating ampholytic copolymers of maleic acid with allylamines, for example, allylamine, diallylamine, methyldiallylamine, and DADMAC were prepared by radical polymerization and were characterized by viscometry, potentiometry, and turbidimetry with respect to pH and ionic strength of the aqueous solutions.[75] The effect of the chemical structure of cationic units on solution properties of the polyampholyte was discussed predominantly in terms of zwitterion formation. The isoelectric points of the four prepared polymers differed only slightly, but the values depended strongly on the ionic strength of the system and shifted to lower pH with increasing ionic strength of the system. The analytical composition and the pKa values obtained confirmed the alternating structures.

The electrochemical properties of poly(DADMAC) membranes in nonaqueous media depended on the polymer and medium nature.[76] The increase of donor properties of solvent decreased the ion exchangeable sorption of salts and led to the increase of the membrane selectivity. The electrical conductivity and transference numbers were measured in H_2O-DMSO mixtures. The addition of aprotic solvent allowed membrane formation with high swelling ability and electrical conductivity. Maleic anhydride copolymers of N-methyldiallylamine and DADMAC were studied at the isoelectric points of the amphoteric copolymers.[77] The copolymers were completely soluble and existed in zwitterion form at their isoelectric points. The diallylamine copolymer had a broad pH range of insolubility.

Many copolymers utilizing nonconjugated dienes as comonomers have been found to be useful. For example, poly(DADMAC) and/or its copolymer with acrylamide has been used as a cationic coagulant in a variety of aqueous systems.[78] The unique applications of homopolymers and copolymers of this monomer are discussed at length in Chapter 12. Graft copolymerization has also included cyclopolymerization as an integral part of the process. For example, acrylic acid-DADMAC copolymer was esterified with 1-chloromethyl-3,4-cyclohexanediol to give a prepolymer which was treated with acrylamide to give a graft copolymer.[79]

The zwitterionic structures of maleic acid alternating copolymers with allylamine, diallylamine, methyldiallylamine, and dimethyldiallylammonium hydroxide were determined by IR, UV, Raman and ^{13}C NMR spectroscopy, and the pH dependence of the structures was studied.[80] Significant differences in the structures were observed at the isoelectric point in dependence of the basicity of the amine function. Coulombic interaction increased with basicity of the cation units, while in the same order, the tendency of H bond formation between cationic and anionic units decreased. Symplex formation between maleic acid-methyldiallylamine alternating copolymer and poly(DADMAC) was studied using IR and ^{13}C NMR.[81] A zwitterionic bond, originally existing in the polyampholyte at the isoelectric point, was cleaved in symplex formation and was subsequently replaced by an electrostatic bond between carboxylate groups and ammonium groups.

Alternating copolymers of maleic acid and allylamine derivatives formed symplexes with cationic polyelectrolytes, but not with anionic polyelectrolytes.[82] Varying the basicity of the cationic units in the polyampholyte chain as well as the pH and the ionic strength of the aqueous medium affected the symplex formation. The symplex formation exhibited significant differences between the polyampholyte with tertiary amine groups and that with quaternary ammonium groups. The results are interpreted on the basis of a zwitterionic interaction between one of the carboxyl groups and the basic N-function of the polyampholyte. A copolymer of DADMAC and acrylamide of molecular weight 500,000 was prepared and found to be a useful component of a creamy skin-cleansing composition.[83]

Copolymers of Dienes which Form Oxygen-Containing Rings

A large number of copolymerization studies have been conducted on non-conjugated dienes such as esters, ethers, acetals, and anhydrides. This section includes discussions on a number of these systems.

Copolymerizations of methacrylic anhydride and a variety of common types of vinyl monomers were carried out rather early.[1-66] Methyl methacrylate, n-hexyl methacrylate, styrene, lauryl methacrylate, ethyl acrylate, vinyl acetate, diisobutylene, benzyl vinyl sulfide, 2-chloroethyl vinyl ether, allyl urea, and allyl chloroacetate were among the comonomers studied. Both bulk and solution polymerizations at 60–80°C, with benzoyl peroxide as initiator, were used. Both soluble and insoluble copolymers were obtained, depending on the comonomer used and experimental conditions under which the copolymerization was carried out. In general, soluble copolymers were formed under the following conditions: (a) the less reactive the comonomer in free radical copolymerizations; (b) the greater the dilution; (c) the greater the difference in the moles of the two components in the charge; and (d) the lower the conversion. According to reactivity ratio determinations, based upon a copolymer composition equation derived for the special divinyl-vinyl system and a comparison with published data, the anhydride ring radical behaved like a methyl methacrylate radical.

Copolymerizations of acrylic (AA) and methacrylic (MAA) anhydrides (M_1) with styrene (St) and methacrylonitrile (MAN) and AA with allyl chloride has been studied.[84] In these systems, it was shown that the reactivity ratio of an open unit radical (r_1) was equal to that of a cyclized radical (r_3) and the usual Alfrey-Price equation was valid. The ratios of the cross-propagation rate constants to the cyclization rate constants (k_{12}/k_c) were also evaluated. The following values for r_1, r_2 and k_{12}/k_c, respectively, were found: AA-ST, 0.1, 0.17, 4; AAMAN, 0.9, 0.04, 0.14; AA-allyl chloride, 11.5, 0.01, 0; MAA-ST, 0.26, 0.12, 0.2; and AA-MAN, 1.6, 0.27, ~0. These data led to the conclusion that the cyclization reaction was always more pronounced for MAA than for AA.

Cyclopolymerization of aliphatic divinyl acetals with vinyl acetate by free radical initiation has been described.[1-58,85] It was concluded that the copolymerization followed a cyclic mechanism almost exclusively as the polymers contained practically no residual double bonds. The divinyl acetal units of the copolymers form 1,3-dioxane rings. Of all the systems studied, only divinyl formal gave a crosslinked copolymer when copolymerized with vinyl acetate beyond a degree of conversion of 30%. This specific behavior of divinylformal was explained on the basis of a *trans*-(hindered) conformation of the monomer. The *trans*-conformation was indicated by a strong doublet in the Raman spectrum of the monomer, whereas the other monomers showed only a single frequency. Copolymers of the other acetals containing up to 90 mole % of the acetals could be carried to high conversions without crosslinking. An increase in the molar ratio of the divinyl acetal in the initial monomer mixture lowered the rate of copolymerization. The cyclic copolymerization of these acetals with vinyl acetate in any molar ratio and at any degree of conversion produced copolymers of practically the same composition as the initial monomer mixture. The reactivity ratios of the monomers were determined and it was shown that the divinyl acetals and vinyl acetals have the same relative reactivities, forming an azeotropic copolymer at any monomer ratio ($r_1 \cong r_2 \cong 1$).

Copolymerization of divinylformal with vinyl acetate was also studied rather early.[2-264] The copolymer was hydrolyzed to yield polyvinyl alcohol which possessed a higher 1,2-glycol content than the usual polyvinyl alcohol prepared at the same temperature. It may be concluded on the basis of these results that some tail-to-tail polymerization occurs during cyclopolymerization to yield some five-membered rings.

Divinyl acetals were synthesized and by use of radical initiation, the monomers were converted to polymers through cyclopolymerization.[2-255, 1-59] It was also reported that copolymerization studies of these monomers with vinyl monomers produced copolymers. The structures of the copolymers obtained with styrene and divinyl butyral (DVB) are shown in Table 7–2.

Copolymerization of mixtures of divinyl acetals was also studied. For example, copolymers of divinyl formal with divinyl acetal and divinyl acetal with divinyl butyral were reported. The structure of the polymers was demonstrated by hydrolysis with a water-alcohol solution of hydroxylamine hydrochloride. It was concluded that the predominant ring size was six-membered since the β-glycol structure was shown by its oxidation with 20% nitric acid to oxalic acid. These results are in support of the postulate that the two vinyl groups react in a head-to-tail manner.

The copolymerization of divinyl formal, divinyl acetal, and divinyl butyral with 2-methyl-5-vinylpyridine, and divinyl butyral with dimethyl crotonyl carbinol and dimethyl crotonyl carbinyl acetate have been studied.[86] An increase in the divinyl acetal content of the initial mixture of monomers significantly reduced the rate of copolymerization with 2-methyl-5-vinylpyridine, and an increase in the number of divinyl acetal groups in the copolymer reduced the melting point and the molecular weight of the polymer. In the copolymerizations with dimethyl crotonyl carbinol and dimethyl crotonyl carbinyl acetate with divinylbutyral, increasing the amount of divinylbutyral up to 50–60% decreased the rate of copolymerization after which it remained constant. Cyclopolymerization of the divinyl acetate was reported to occur in all cases.

Radical telomerization of diallyl ether in carbon tetrachloride or chloroform using benzoyl peroxide or azobisisobutyronitrile as initiators has been studied.[87] A cyclized product obtained in chloroform was identified as 3-(2,2,2-trichloroethyl)-4-methyltetrahydrofuran instead of the tetrahydropyran derivative expected from earlier results using bromotrichloromethane.[88] Diallyl ether (M_1) was copolymerized with 2-methyl-5-vinylpyridine and acrylonitrile. The following reactivities were determined: 2-methyl-5-vinylpyridine, $r_1 = 0.01$, $r_2 = 80$ at 70°C; acrylonitrile, $r_1 = 0.01$, $r_2 = 4.9$ at 80°C. Diallyl ether acted as a chain transfer reagent with acrylonitrile giving an apparent transfer constant of 2.14×10^{-3}.

Divinyl carbonate and ethyl vinyl carbonate were copolymerized with *p*-chlorostyrene and vinyl acetate.[89] The Alfrey-Price Q and e values were found to be 0.035 and 0.23, respectively, for divinyl carbonate, and 0.0025 and −0.26, respectively for ethyl vinyl carbonate. These values indicated that copolymerization reactivities for these monomers are similar to those for vinyl acetate. The cyclopolymerization reaction

Table 7–2 Copolymerization of DVB with Styrene

Molar ratio of DVB/styrene in initial mixture	Molar ratio of DVB/styrene in copolymer	Copolymer M.P. (°C)
30/70	12.32/87.68	115–120
50/50	14.49/85.51	107–116
60/40	15.99/81.01	105–115
70/30	24.90/75.10	104–106
80/20	35.70/61.30	95–98
90/10	57.18/42.72	56–58

of divinyl carbonate had little effect on the copolymerization parameters. The extent of cyclization of divinyl carbonate was found to be lower in copolymerization with *p*-chlorostyrene than with vinyl acetate.

Copolymers of divinylbutyral with epoxy olefins have been claimed.[90] Epoxy olefins employed were glycidyl methacrylate, allyl glycidyl ether, and vinyl glycidyl ether. The copolymer with glycidyl methacrylate was postulated to have the following structure (7–8): Free radical initiated cyclopolymerization of divinyl acetal with other 1,6-dienes, e.g. diallyl phenyl phosphonate, has been reported to yield cyclic copolymers of the two monomers.[91]

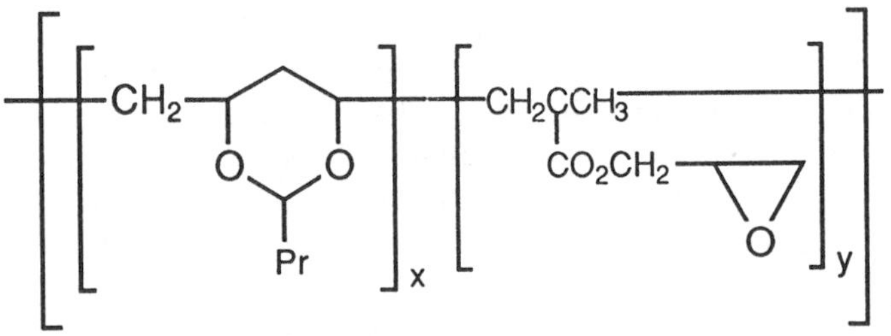

(7–8)

Copolymerization of methacrylic anhydride with styrene, methyl methacrylate, and methyl acrylate, has been studied and their relative reactivities determined.[92] Copolymerization of ethylene with a variety of dienes via radical initiation was described as early as 1946.[93] Divinyl formal, divinylbutyral and methacrylic anhydride were described as comonomers with ethylene in a process which lead to a "thermoplastic polymer" in the case of divinyl formal, a "polymer which melted fairly sharply at about 100°C" in the case of divinyl butyral, and a polymer "insoluble" in a variety of organic solvents, but "soluble on warming with N-sodium hydroxide," in the case of methacrylic anhydride. No explanation was offered for the total absence of crosslinking.

Alternating 1:1 copolymers of acrylic and methacrylic acids were obtained by radical initiated cyclopolymerizing the mixed anhydride of acrylic and methacrylic acids and hydrolyzing the product.[94] Allyl crotonate and allyl cinnamate were converted to soluble copolymers free of gels, with vinyl chloride.[1-80] The copolymers possessed higher intrinsic viscosities than the corresponding vinyl chloride homopolymer prepared under identical conditions and were shown to be useful for insulation of underground cables.

Radically initiated cyclopolymerization of the mixed anhydride of acrylic and methacrylic acids gave 1:1 alternating copolymers of acrylic and methacrylic acids after hydrolysis of the polyanhydride.[95] The molecular weight of a typical polymer was 850,000. In a study of copolymers of maleic anhydride, radical copolymerization of styrene and diallyl maleate was investigated.[96] The main structure of the copolymer as shown in Structure 7–9 indicates that cyclopolymerization occurred to produce the lactone repeat unit. A sequence distribution study of methyl methacrylate-styrene copolymers derived from methacrylic anhydride-styrene copolymers has been published.[97]

In a study of copolymers of maleic anhydride or its derivatives, radical copolymerization of diallyl fumarate with styrene at 80° in benzene was investigated.[98] The degree of cyclization in the copolymer increased from 0% in copolymer prepared at high monomer concentrations (3.00 mole/L) to 100% for copolymer prepared at low monomer concentration (0.30 mole/L).

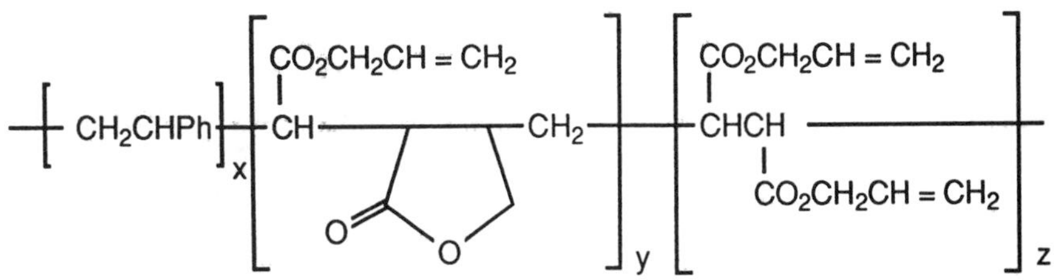

(7-9)

The cyclocopolymerization of diallyl ether with maleic anhydride has been studied.[99] Based on copolymer composition and solubility studies, the mechanism for the cyclocopolymerization consisted of the formation of cyclized 1:1 and 1:2 units, due to the proximity of the pendant double bond in the growing radical. Cyclization was not an exclusive reaction, and the reaction to form cyclized 1:2 units could dominate when the concentration of the comonomer was large. The observed limiting conversion with respect to the feed composition agreed with the theoretical limit of 1:1 copolymer when there was more diallyl ether in the feed, and a theoretical limit of 1:2 copolymer when maleic anhydride was richer in the feed.

A kinetic equation for copolymer composition for products of the cyclocopolymerization of a 1,6-diene with a monoolefin that does not homopolymerize has been derived and was applied to the free radical copolymerization of diallyl ether with maleic anhydride and fumaronitrile.[100] With both systems, no unreacted C:C bonds remained in the copolymer. The copolymers contained 1:1–1:2 diene-monoolefin ratio over a wide range of monomer feed composition. The reactivity ratios obtained from the equation indicated that in the maleic anhydride system, intramolecular cyclization of uncyclized diallyl ether was 2.9 times greater than addition of maleic anhydride monomer to the uncyclized radical. In the fumaronitrile system, intramolecular cyclization of the radical was 6.7 times greater than comonomer addition.

Various mono- and diallyl esters of maleic anhydride copolymers, (comonomers: styrene, n-butyl vinyl ether, iso-butyl vinyl ether, or vinyl acetate) have been prepared by reaction of allyl alcohol with the copolymers, and their crosslinking by electron beams or UV light was evaluated.[3-149] The sensitivity of the allyl-esterified copolymers to electron beams depended on the weight-average molecular weight (M_w) of the copolymer and on the allyl group concentration. Maleic anhydride-styrene copolymer allyl ester showed the highest sensitivity at the same intrinsic viscosity and the allyl group concentration. By adding photosensitizers such as 4,4′-bis(dimethylamino)pentaphenone, the allyl esters could be photocrosslinked. The photosensitivity was proportional to M_w of the allyl-esterified copolymer and to the square of the allyl group concentration.

Copolymerization of 2-[allyl-(n-propyl)oxycarbonyloxy]acrylic acid esters with methyl methacrylate and styrene has been studied.[101] Reactivity ratios and Q-e values in the copolymerization of $CH_2{:}C(OCO_2CH_2CH{:}CH_2)CO_2CH_2CH{:}CH_2$ and $CH_2{:}C(OCO_2C_3H_7)CO_2CH_2CH_2CH_3$ with styrene and methyl methacrylate showed that the allyl ester had higher relative reactivity toward the polymer radicals than did the propyl ester. Both monomers formed an alternating copolymer with styrene and a random copolymer with methyl methacrylate. Both esters were electron-acceptor monomers with practically equivalent Q-e values. The reactivity ratios and Q-e values for both esters were similar to those for ethyl 2-acetoxyacrylate in similar copolymerizations, indicating the absence of a substituent effect by the α- substituent.

Cyclopolymerization of diallyl maleate has been studied. Radical polymerization in C_6H_6 at 60° in the presence of AIBN was kinetically analyzed in terms of cyclopolymerization.[102] Kinetic equations involving bicyclointramolecular cyclization were derived by assuming steady-state conditions for the different types of radicals; various parameters involved in the equations were estimated from an extension of model experimental results. The validity of the treatments was confirmed by comparison with experimental data, including the relation between the content of the unreacted allylic and maleic double bonds and the monomer concentrations. The bicyclic ring content was estimated to be 5.45% for the bulk polymerization of the monomer; it reached 47.2% at 0.55 mol/h, corresponding to a 10% dilution of the pure monomer.

The kinetics of the cyclopolymerization of diallyl fumarate has been analyzed and the sequence distribution of noncyclic, monocyclic, and bicyclic units was estimated.[103]

Styrene-methyl methacrylate copolymers derived from a styrene-methacrylic anhydride copolymer have been studied.[104] Reactivity ratios and cyclization constants for the copolymerization were determined from the structures of the methacrylate copolymers.

Cyclopolymerization of diallyl ether with sulfur dioxide gave soluble copolymers containing 1:2 molar diallyl ether:SO_2.[105] The formation of rings from diallyl ether and SO_2 and of 3,4-disubstituted tetrahydrofuran rings by cyclized diallyl ether units on the main chain was proposed, based on evidence obtained. Vinyl cinnamate was cyclocopolymerized with vinyl acetate, and the monomer distributions in the copolymer were determined. Reactivity ratios were calculated.[106] Participation by the cinnamate double bond was demonstrated. In a copolymerization study of allyl cinnamate with methyl methacrylate, styrene and maleic anhydride, it was observed that the nature of the comonomer influenced the extent of formation of cyclic structure by the allyl cinnamate.[107] Methacrylates did not change the extent of cyclization but no cyclization occurred in the styrene copolymerization.

Copolymers of Dienes which Form Rings Containing Other Elements

Only a limited amount of attention has been given to copolymerization of nonconjugated dienes containing other elements than carbon, nitrogen, or oxygen. This section illustrates some of these systems.

Copolymerization studies of diallylphenylphosphine oxide with lauryl methacrylate have reported and monomer reactivity ratios for the copolymerization were measured.[108] The reactivity ratios at 140°C varied with the composition of the starting mixture and the value of r_2 was significantly negative. Diallylphenylphosphine oxide had previously been shown to undergo cyclopolymerization.[1-33]

The reactivity ratios for the copolymerization of lauryl methacrylate with diallyl phenylphosphonate, diallyl butylphosphonate, and diethyl allylphosphonate, respectively, have been determined.[1-68] Diallyl phenylphosphonate and diallyl butylphosphonate have similar activities, which are greater than that of diethyl allylphosphonate. Intrinsic viscosities and molecular weights were measured for some diallyl phenylphosphonate-lauryl methacrylate copolymers and these values were found to be dependent upon the phosphorus content of the copolymers.

Soluble copolymers were obtained by copolymerizing vinyl monomers, e.g. acrylic acid with minor amounts of tetraallylsilane and tetraallylgermane.[109] Unsaturated copolymers were obtained from one mole of a mixture of olefins with 0.2 mole or less of polyunsaturated silane, e.g. dimethyldivinylsilane, dimethyldiallylsilane, tetravinylsilane, tetraallylsilane, diethyldiallylsilane, diphenyldiallylsilane, etc., in presence of metal-coordination catalysts.[110] The copolymers were useful as vulcanizable elastomers.

Heat-resistant acrylic fibers have been prepared by polymerizing acrylonitrile with a 1,6-diene compound, $CH_2{:}CRCH_2ZCH_2CR{:}CH_2$, (Z = O, S, SO_2, NR^1, $N^+R^2R^3X^-$, NCN; R, R^1, R^2, R^3 = H, alkyl; X = Cl, Br, SO_4, NO_3) and wet spinning the polymer.[111] For example, acrylonitrile was polymerized with diallylamine to give the copolymer.

COPOLYMERS OF 1,6-DIENES WITH SULFUR DIOXIDE

The copolymerization of diallylamine hydrochloride with sulfur dioxide at temperatures varying from O° to 50°C in solvents such as dimethylsulfoxide, dimethylformamide, and methanol have been studied.[112] Both ammonium persulfate and azobisisobutyronitrile were used as initiators. The composition of the copolymer was always 1:1 regardless of the feed ratio. Conversions ranged from 14.3 to 90.1% when calculated on the basis of the total monomers charged. However, when calculated on the basis of the 1:1 diallylamine hydrochloride-sulfur dioxide complex, the conversions ranged up to 100%. Intrinsic viscosities [η], were as high as 1.58 dl/g when measured in 0.1 *N* sodium chloride at 30°C. It was postulated that the following structure (7–10) adequately describes the copolymer:

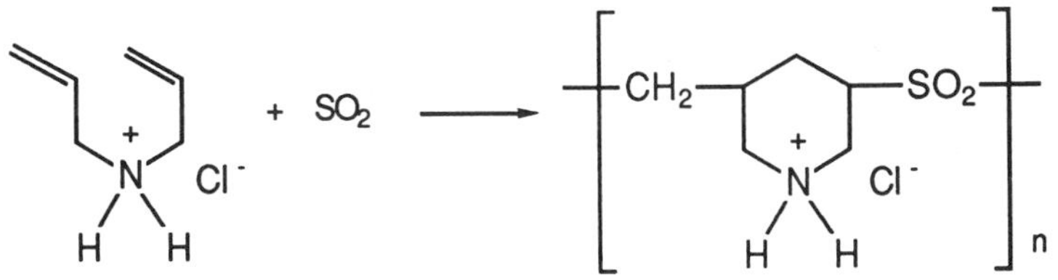

(7–10)

The copolymerization of DADMAC with sulfur dioxide has also been studied.[113] Solvents were dimethylsulfoxide, methanol, or acetone, and radical initiators such as ammonium persulfate, azobisisobutyronitrile, and dilauroyl peroxide were used. Temperature ranges were 30–50°C, and conversions ranged up to 100% based upon the 1:1 DADMAC-sulfur dioxide complex. The composition of the copolymer was always 1:1 molar in the two monomers regardless of the composition of the feed. The inherent viscosity (ln $\eta_{rel/c}$ for 0.5 g of copolymer per 100 ml of 0.1 *N* aqueous NaCl solution) of a copolymer prepared in methanol as solvent using ammonium persulfate as initiator at 30°C for 72 hrs was 1.98. Two possible structures were proposed for the copolymer (7–11); however, the evidence could not distinguish between these structures.

This study of copolymerization of diallyl compounds with sulfur dioxide has been extended to include a variety of additional copolymer structures.[114] In general, the inherent viscosities of these copolymers were lower than those previously described.[112,113] Except for the copolymers of diallylmorpholinium chloride and diallylpiperidinium chloride with sulfur dioxide, the copolymers were 1:1 molar in composition of the monomers. However, in these cases, the composition approached 2:1 molar

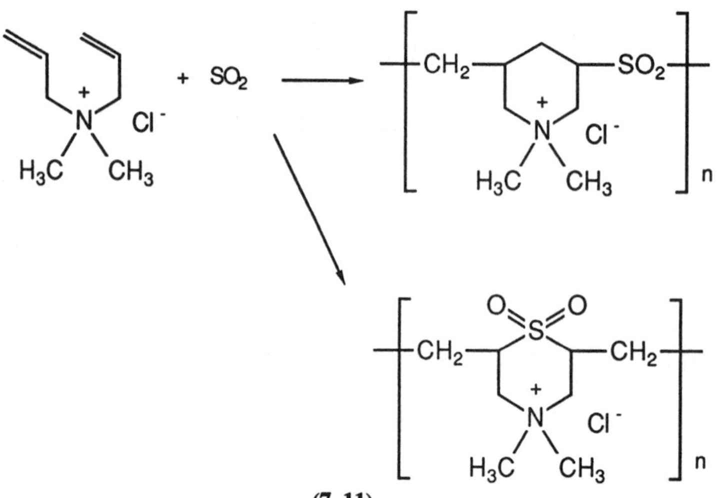

(7-11)

in sulfur dioxide. The following structures were proposed to account for this unusual result (7-12).

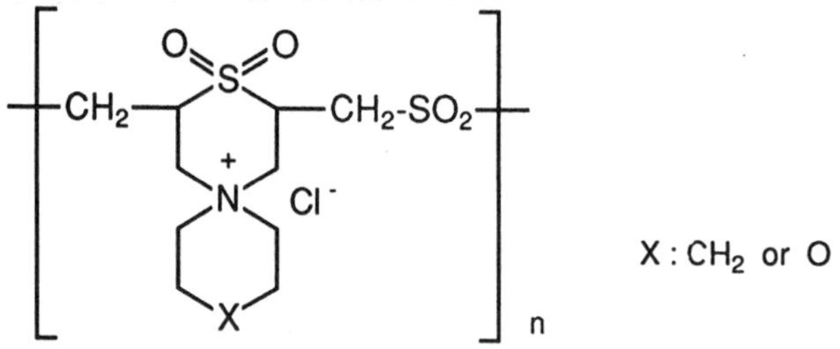

X : CH_2 or O

(7-12)

The cationic polysulfones described in the previous papers[112-114] were shown to be good flocculants[115] for aqueous suspensions of kaolinite. However, their effectiveness varied with the pH of the solution depending upon the structure of the copolymer. The molecular weights of the copolymers studied ranged from 14,000 to 330,000.

It was also shown that the copolymer of DADMAC and sulfur dioxide was absorbed on bentonite, a hydrophilic and negatively charged clay.[116] From these studies, the authors concluded that the absorption of the polycation on bentonite occurred by an ion exchange process. The polymer was not adsorbed from the complex by treatment with sodium chloride or calcium chloride.

Alternating copolymers of dialkyldiallylammonium chlorides and sulfur dioxide of Structure 7-11 have also been synthesized by photoinitiation.[117,118] The copolymers were useful for the coagulation and precipitation of materials suspended in water and as paper sizes, antistatic agents, surfactants, and thickeners. Also described was the synthesis of copolymers of Structure 7-13 by radical copolymerization of diallylsulfamic acid or its salts with sulfur dioxide.[119] In addition to the above uses, these copolymers were also reported to be useful as fireproofing agents for textiles, as acidic catalysts, and as herbicides.

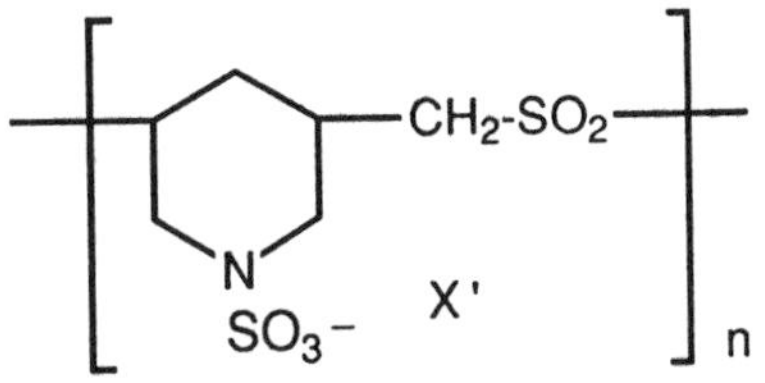

CH_2-SO_2

N

SO_3^- X'

n

(7–13)

Copolymers of a wide variety of 1,6-dienes with sulfur dioxide have been prepared.[1-61] Among these were diallyl sulfide, diallyl diethyl malonate, diallyl ethylcyanoacetate, diallylmalononitrile, a variety of N,N-diallyl alkanamides, diallyl cyanamide, diallylaminoacetonitrile, a variety of N,N-diallylalkyl- or arylsulfonamides, and N,N-diallyl perfluorooctanesulfonamide. These copolymers were found to have inherent viscosities up to 1.67 dl/g (1.11 g/100 ml, dimethylformamide), and degrees of polymerization up to 1000. The molar ratio of diene to sulfur dioxide was found to be 1:1, and, since the copolymers were soluble and essentially saturated, it was assumed that the diene had entered the polymer chain as a cyclohexane ring, formed by the cyclopolymerization mechanism. Phosphoric acid amides containing ⩾ one N-allyl or N-methallyl group have been radically copolymerized with SO_2 to give polysulfones (7–14).[2-173] Thus, 3.29 g di-phenyl N,N-diallylphosphoramidate in 10 ml dimethylsulfoxide (DMSO) was mixed with 0.32 g SO_2 and 0.0456 g $(NH_4)_2S_2O_8$ and polymerized 48 hours at 39°, to give a white, powdery polymer containing units of Structure 7–14 and having intrinsic viscosity 1.193 (0.5 g/dl, DMSO, 30°).

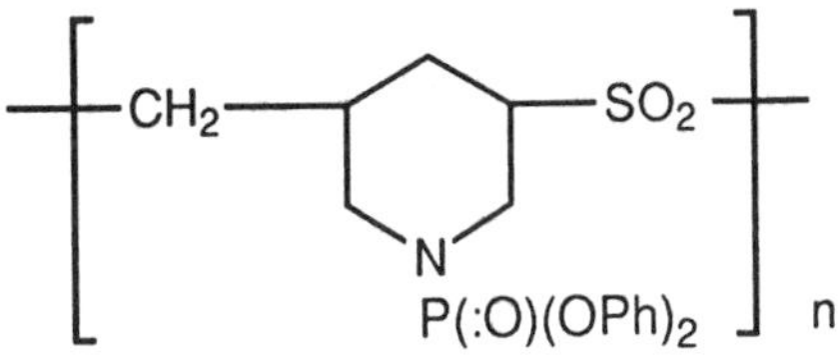

CH_2 SO_2

N

P(:O)$(OPh)_2$ n

(7–14)

The cyclocopolymerization of 1,6-heptadiene and sulfur dioxide has been reported.[120] Copolymers of 1,6-heptadiene and SO_2 of various mole ratios were prepared, and the structures of the copolymers were studied by elemental analyses, IR and ^1H NMR spectroscopy. Structures containing a cyclopentane ring and a thiane ring in the main chain were proposed. The copolymer had no melting or softening point, but decomposed at 250°. It was easily soluble in DMSO or H_2SO_4.

Cyclocopolymerization of diallyl compounds[4] and sulfur dioxide has been studied. Copolymers of 4-(4-pyridyl)-1,6-heptadiene hydrochloride and 4-(1-buten-4-yl)pyridine hydrochloride with SO_2 were prepared, and the effects of polymerization conditions on their properties were studied by elemental, IR, and NMR analysis.

In a study of differential dyeing of textiles, it was shown that a cellulosic, vinal, wool, silk, polyamide, and/or acrylic textile blended or not blended with polyester fibers when treated partially or totally with a copolymer of (3-chloro-2-hydroxypropyl)diallylamine hydrochloride, SO_2 and DADMAC could be dyed with anionic, cationic, or fiber reactive dyes (7–15).[121]

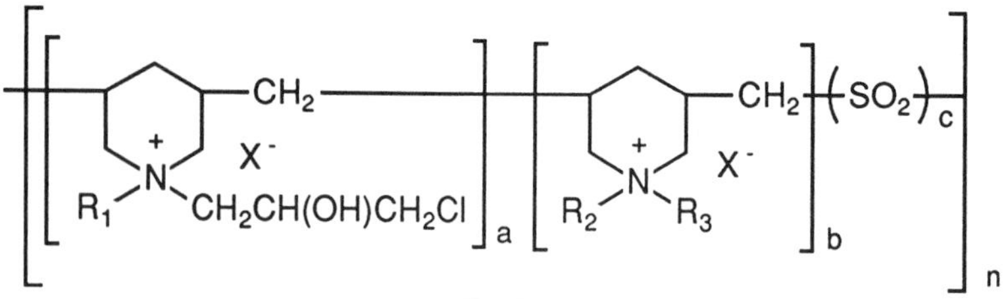

(7–15)

The cyclocopolymerization of some 4-substituted 1,6-heptadiene[5] derivatives with sulfur dioxide has been studied. Copolymers of SO_2 with N-substituted 4-(1,6-heptadien-4-yl)pyridinium chlorides and bromides and N-substituted 4-(3-butenyl)pyridinium chlorides and bromides, and some other 1,6-heptadiene derivatives substituted in the 4-position were prepared. The effects of copolymerization conditions on the conversions and viscosities of the copolymers were studied and their structures were characterized by elemental analysis, IR and ^{1}H NMR spectroscopy. The thermal stabilities of the copolymers were also studied.

Poly(1,5-cyclooctadiene sulfone) was obtained by a radical-initiated transannular copolymerization of 1,5-cyclooctadiene and sulfur dioxide.[122] The copolymer was more stable thermally than single stranded alkene-sulfur dioxide copolymers.

Terpolymers of α-olefins and *cis,cis*-1,5-cyclooctadiene with sulfur dioxide and carbon monoxide were synthesized and characterized.[123] The terpolymerizations could be carried out above the ceiling temperature for the sulfur dioxide (or carbon monoxide) α-olefin copolymer, which is indicative of greater thermal stability in polymers containing the bicyclic repeat unit produced by the transannular mechanism.

Copolymerization of DADMAC with sulfur dioxide led to oligomers of inherent viscosity of 0.2–0.24 which had high catalytic activity in the reaction of sulfur dioxide with hydrogen sulfide.[124] The oligomers of structure shown earlier (See Structure 7–11) formed from cyclocopolymerization of DADMAC with sulfur dioxide were shown to form a complex with copper sulfate, which could be used as a heterogeneous catalyst.[125]

Polysulfones of norbornadiene were obtained without catalyst, and by using nickel chloride and hydrogen peroxide, respectively, as catalysts.[126] Only the peroxide catalyst gave copolymers having a high proportion of nortricyclene units in the chain in support of the cyclopolymerization or transannular mechanism for insertion of the norbornadiene comonomer into the propagating chain. Linear poly[spiro[3′-sulfonylmethylcyclohexyl-1′,5-(barbituric acid)]] was synthesized from diallylbarbituric acid and sulfur dioxide *via* free radical initiation.[127] The number of repeating units varied up to 300.

1-5-Dienes such as 1,5-hexadiene, 1,2-divinylcyclohexane and isopropenyl allyl ether were copolymerized with sulfur dioxide to yield alternating cyclocopolymers in the molar ratio of 1:2.[128] Radical initiators were used. 1,5-Cyclooctadiene was converted to a 1:1 alternating copolymer of Structure 7–16 by mixing the diene with sul-

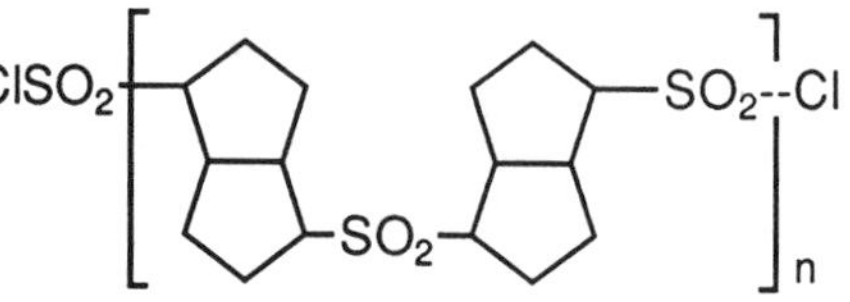

(7–16)

furyl chloride and irradiating with an ultraviolet lamp.[129] The cyclocopolymer structure was confirmed by chemical and spectroscopic analysis. 1,5-Cyclooctadiene has also been copolymerized with sulfur dioxide via radical initiation to yield a copolymer soluble in DMSO (See Structure 7–16).[130] Other copolymers were synthesized from the cyclic dienes, dicyclopentadiene and 1,4-dimethylenecyclohexane. These copolymers were also soluble in suitable solvents, indicating the absence of excessive crosslinking, and cyclopolymerization of the respective dienes.

Copolymerization of dicyclopentadiene with sulfur dioxide via radical initiation gave a copolymer which had no residual double bonds and contained two moles of sulfur dioxide per mole of diene.[131] The overall activation energy for the copolymerization in methanol was 7.4 kcal/mole. The copolymer structure was not discussed further.

COPOLYMERS OF UNSYMMETRICAL 1,5- OR 1,6-DIENES

The kinetics of polymerization and copolymerization of vinyl-*trans*-cinnamate (VC), a monomer which contains two double bonds of very different chemical nature, have been studied.[3-110] By determination of the residual unsaturation of the polymer at different concentrations, it was possible to show that the ratio of the cyclization rate constant to the vinyl propagation rate constant was equal to 13. In the range of low monomer concentrations, the polymerization of VC was first order with respect to monomer concentration, and one half order with respect to initiator concentration. At monomer concentrations higher than 3 molar, the reaction became independent of monomer concentration, and first order with respect to the initiator. Modified copolymer composition equations were derived taking into account the different reactivity of the monomer radical before and after cyclization. The reactivity ratios r_1, r_2, and r_3, were defined and the experimental methods for their evaluation described.

On the basis of these equations, the following reactivity ratios were determined: VC (M_1)-vinyl acetate, $r_3 = 1.2 \pm 0.1$, $r_2 = 0.04$; M_1-vinyl pyrrolidine, $r_3 = 1.15 - 1.30$, $r_1 = 0.01$; M_1-methacrylonitrile, $r_3 = 0.15$, $r_2 = 4$; and M_1-styrene, $r_1 = 0.25 \pm 0.1$, $r_2 = 1.25 \pm 0.1$. It was suggested that VC formed a pseudo-cyclic benzyl radical stabilized by resonance before addition of the second vinyl monomer.

Copolymerization of vinyl acrylate with ethyl acrylate, methyl methacrylate, and styrene in emulsion has been studied.[132] The molecular weights of the copolymers were much lower than the molecular weights of the homopolymers of these monomers prepared under the same conditions. Since there was no apparent decrease in rate in the copolymerizations, and conversions to polymer were greater than 90%, it was postulated that vinyl acrylate was functioning as a chain transfer agent. The mechanism for the chain transfer action was postulated to involve formation of a cyclic radical and disproportionation to a $\Delta^{\alpha,\gamma}$-butenolide structure as shown (7–17):

$$n\,M\cdot + CH_2{=}CHC({=}O)OCH{=}CH_2 \longrightarrow n\,M{-}CH_2{-}\overset{\cdot}{C}H \;\; [\text{ring: } C(=O){-}O{-}CH{=}CH_2] \longrightarrow$$

$$n\,M{-}CH_2{-}CH{-}CH_2 \;[\text{ring: } C(=O){-}O{-}\overset{\cdot}{C}H] + M \longrightarrow n\,M{-}CH_2{-}CH{-}CH_2 \;[\text{ring: } C(=O){-}O{-}CH{=}] + HM\cdot$$

(7–17)

Support for this mechanism was provided by the IR spectrum of poly(styrene-vinyl acrylate).

The cyclocopolymerization of 4-vinylcyclohexene with p-chlorostyrene by use of coordination catalysts has been described.[133] By use of the Roovers-Smets equation, reactivity ratio parameters, K_e, K'_e, r_2 and r_3 were determined.

Cyclic copolymers have been prepared by cyclopolymerization of unsaturated cyclic anhydrides or imides with allyl alcohol or allylamine derivatives.[134] Thus polymerization of 1:1 maleic anhydride:methallyl alcohol with azobisisobutyronitrile gave 100% maleic anhydride-methallyl alcohol polymer, intrinsic viscosity 0.228 dl/g, believed to have Structure 7–18.

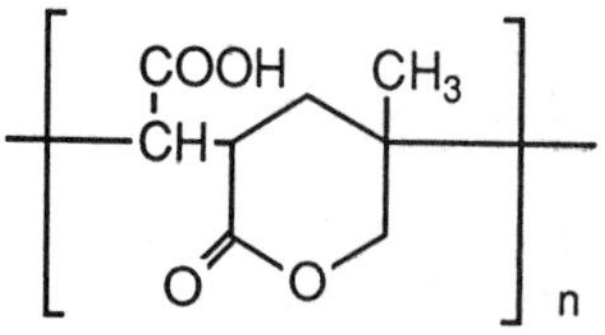

COOH CH3 CH O O n

(7–18)

Maleic anhydride-4-vinylcyclohexene copolymers of Structure 7–19 by radical initiated copolymerization of the monomers in dioxane or without solvent have been reported.[135]

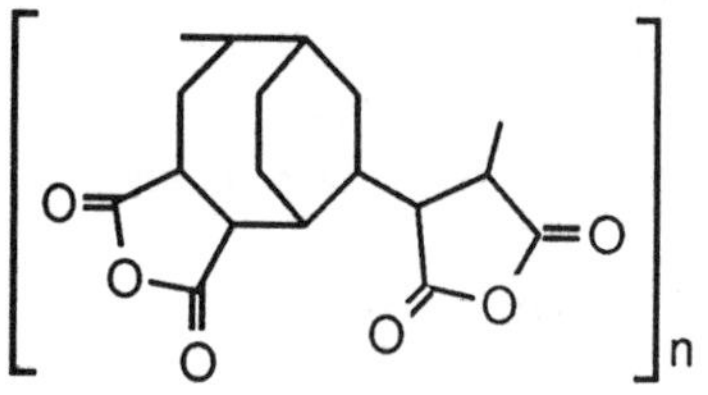

O O O O O O n

(7–19)

It was shown that the addition of tetracyanoethylene did not change the composition of allyl methacrylate-Me methacrylate copolymer, but did affect the tacticity, mean sequence length, and sequence length distribution.[136] Radical homopolymerization of 6-allyloxy-3(2H)-pyridazinone does not occur, but copolymerization with styrene gave styrene-6-allyloxy-3(2H)-pyridazinone copolymer having a cyclic structure shown (7–20) through intramolecular propagation.[137] Copolymerization with acrylonitrile gave acrylonitrile-6-allyloxy-3(2H)-pyridazinone copolymer.

O O N N H

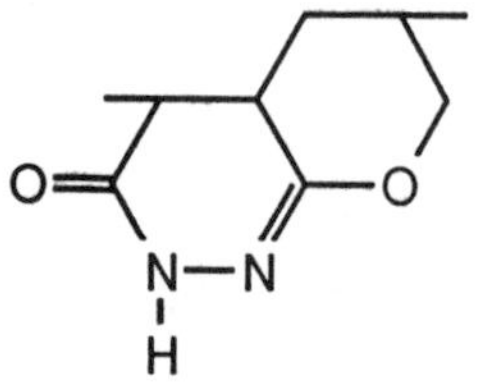

(7–20)

Block copolymers of divinyl butyral and vinyl acetate were obtained by copolymerizing the monomers in presence of radical initiators.[138] The copolymers could be converted by basic hydrolysis to products containing hydroxyl groups along the chain.

1,4-Hexadiene was used as a termonomer with ethylene and propylene to produce elastomers with residual unsaturation for vulcanization sites.[139] More recent studies have shown that a fraction of the diene enters the chain via cyclopolymerization to reduce its effectiveness.

The copolymerization of allyl acrylate with vinyl acetate via radical initiation in solution was studied.[140] The copolymer was shown to be crosslinked. IR spectroscopy showed that the copolymer contained unreacted allyl groups. The difference in the relative reactivities of the two double bonds of allyl acrylate was considered too great to allow a high degree of cyclocopolymerization. Bicyclocopolymers of Structure 7–21

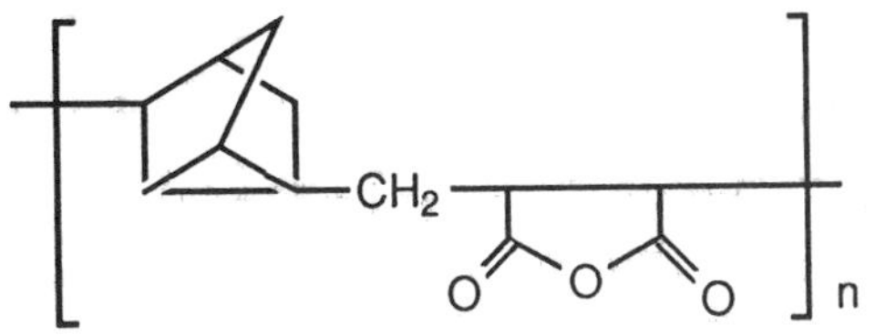

(7–21)

were obtained by free radical initiation from 5-methylenebicyclo[2.2.1]-2-heptene and maleic anhydride.[141] The copolymers were of constant composition with an alternating arrangement of comonomers but varied in solubility, degree of cross-linking and molecular weights.

Diallyl maleate-p-acetoxystyrene copolymers were hydrolyzed to produce copolymers of p-vinylphenol and allyl esters of ethylenically unsaturated acids.[142] The copolymer of allylmethacrylate and sodium acrylate was used as a component of waterless presensitized lithographic plates.[143]

COPOLYMERS OF SYMMETRICAL 1,X-DIENES WHERE X > 6

Copolymers of styrene with various dimethacrylates were studied by measuring their moduli of elasticity.[144] Network densities, calculated from these moduli, permitted determination of the degree of cyclization of the divinyl monomers. This value was 63% for methacrylic anhydride, 40% for butanediol dimethacrylate, 22% for hydroquinone dimethacrylate, and 11% for 4,4′-dihydroxydiphenyl dimethacrylate.

Copolymerization of diallyl phthalate (M_1) with styrene was studied using benzoyl peroxide as initiator at 60°, 70°, 80°, and 90°C, and with tert-butyl peroxide as initiator at 100°, 110°, 120°, and 130°C.[145] Monomer reactivity ratios calculated by means of the Mayo-Lewis equation were: $r_1 = 0.057$, $r_2 = 32.8$ (60°); $r_1 = 0.105$, $r_2 = 23.8$ (80°); $r_1 = 0.053$, $r_2 = 21.4$ (100°); $r_1 = 0.105$, $r_2 = 16.5$ (120°). Cyclopolymerization of diallyl phthalate occurred as well as some homopolymerization; however, the amount of cyclopolymerization decreased as the mole fraction of styrene increased

The reactivity ratios for the copolymerization of lauryl methacrylate (M_1) with diallyl phenylphosphonate, diallyl butylphosphonate, and diethyl allylphosphonate were determined at 80°C.[1-68] The values obtained were: diallyl phenyl-phosphonate, $r_1 = 19.5$, $r_2 = 0.072$; diallyl butylphosphonate, $r_1 = 18.7$, $r_2 = 0.091$; diethyl allylphosphonate, $r_1 = 52.5$, $r_2 = 0.066$. There was no evidence for double bonds in the copolymers from the diallyl monomers, and it appeared that both double bonds of the monomers were involved in the copolymerization. These results are consistent with the cyclopolymerization mechanism in which the diallyl phosphonate monomers propagate by an intramolecular cyclization. A copolymer containing 1.24% phosphorus had an intrinsic viscosity of 0.23 dl/g in decalin at 37.8°C and a number average molecular weight of 83,000.

The reactivity ratios for the copolymerization of diallyl benzenephosphonate (M_2) and methyl methacrylate and styrene at 70°C were determined. For styrene, r_1 = 28.97, r_2 = 0.027; for methyl methacrylate, r_1 = 22.96, r_2 = 0.135.[146] Viscosities of the copolymers in benzene were dependent upon the phosphonate content of the monomer mixtures. Evidence was presented that polymerization of diallyl benzenephosphonate occurred according to the cyclopolymerization mechanism. Ring sizes were not postulated nor determined.

Radical initiated copolymerization of triallylidene sorbitol and tricrotylidene sorbitol with styrene and acrylonitrile were studied and the monomer reactivity ratios and *Q* and *e* values determined.[2-379]

Copolymerization of styrene with dimethylsilyl dimethacrylate, $(CH_3)_2Si[O_2CC(CH_3): CH_2]_2$, and with diethylsilyl dimethacrylate in dioxane and Bz_2O_2 has been described.[147] The copolymers prepared were colorless solutions with variable viscosity, which depended on the starting concentrations and compositions. Infrared spectroscopic data suggested that the copolymers possessed a linear structure and contained cyclic units of dimethylsilyl dimethacrylate or diethylsilyl dimethacrylate.

It has been reported that copolymers, which were soluble and free of gels, were obtained from allyl crotonate and allyl cinnamate, with vinyl chloride.[1-80] In another example of copolymers of vinyl chloride with unsymmetrical dienes, soluble copolymers free of gel were obtained with allyl β-allyloxypropionate. Other unsymmetrical monomers used in forming similar copolymers were allyl allyloxyacetate, allyl o-allyloxybenzoate, allyl p-allyloxybenzoate, and allyl γ-allyloxybutanoate.

It was shown via a limiting viscosity study that intramolecular cyclization reactions occurred to some extent in copolymerization of styrene with divinyl benzene.[1-14] Since the divinyl benzene was not defined with respect to o, m, or p-content, it must be assumed that the commercial product was used. More recent work has shown that o-divinyl benzene cyclopolymerized almost exclusive of other modes of propagation. Diethylene glycol bis(allylmaleate), neopentyl glycol bis(allyl maleate) and ethylene glycol allyl maleate or fumarate were copolymerized with vinyl chloride to give copolymers of improved properties.[148] Allyl and methallyl esters of polycarboxylic acids were used as components in a system designed to produce thermosetting copolymers of styrene, acrylate esters, etc.[149] The nature of the polycarboxylic acid was not described.

The copolymerization of nonconjugated diolefins with vinyl monomers involved departures from conventional vinyl copolymerization kinetics and the standard forms of the binary or ternary composition equations were inadequate.[150] These departures include intramolecular cyclization reactions and generally the formation of some pendant double bonds. A general treatment of the copolymerization of symmetrical nonconjugated diolefins with vinyl monomers gave closed form relationships which took into account such reactions and their consequences. The general relationships may be approximated by a new series of composition equations which take into account the formation of both cyclized and noncyclized units in the chain and are simple enough for experimental use. The latter expressions are:

$dm_1: dm_3: dm_4::$

$$[M_1]([M_1] + \alpha_c + \alpha_4[M_4]): - [M_1([M_1] + a_4[M_4]):$$

$$[M_4]([M_1] + [M_4]/\delta_1)\ \{\alpha_4 + \alpha_c/([M_1]/\beta_4 + [M_4])\}$$

where $\alpha_c = k_c/k_{11}$, $\alpha_4 = k_{14}/k_{11}$, $\delta_1 = k_{41}/k_{44}$, and $\beta_4 = k_{24}/k_{21}$, and $[M_1]$, $[M_3]$, and $[M_4]$ refer to the concentrations of diolefin double bonds, pendant double bonds, and comonomer, respectively.

A study has been made of copolymerization of monovinyl monomers with divinyl benzene.[151] The composition relations were derived both for the general case, including crosslinking reactions, and for the approximate case where the concentration of pendant unsaturation may be assumed to be unaffected by crosslinking reactions. The divinyl monomer must be incapable of undergoing short-range cyclization, or alternating intra-intermolecular propagation. The general case composition equation has a similarity to that for ternary polymerization of three vinyl monomers. The approximate case closely resembles the usual binary copolymerization relationship.

Copolymerization of diallyl methylphosphonate with ethylene in presence of AIBN, at 400 atm. pressure and 70–80°C was studied and the properties of the copolymers discussed.[152] The structure of the polymer (residual double bond) was reported to depend on the content of the phosphorus monomer in the copolymer, an observation consistent with cyclopolymerization.

Diallyl esters of aliphatic dicarboxylic acids were copolymerized with styrene via free radical initiation.[153] Reactivity ratios were given for bulk polymerization of diallyl esters of oxalic, malonic, succinic and sebacic acids.

Copolymerization of divinyl adipate with methyl acrylate was studied by polarography.[154] The results of the method were used to construct kinetic curves and to determine reactivity ratios for the monomers. The properties of the copolymers were not described.

Monomer reactivity ratios for the copolymerization of 1,2,4-trivinylbenzene with styrene were determined.[155] The copolymerization kinetics of these two monomers clearly do not resemble those of o-, m-, or p-divinylbenzene, structured elements of which are present in the trivinyl monomer.

Ethylene was copolymerized with the diallyl ester of methylphosphonic acid via radical initiation.[156] Concentrations of the ester $>5\%$ increased the reaction rate but decreased the molecular weight. The presence of the comonomer in the copolymers decreased their crystallinity, lowered the melting point, and reduced the elasticity. Although not postulated, cyclopolymerization may be assumed since crosslinking did not occur.

Polymerization and copolymerization of 1-(3-vinylphenyl)-4,5-dichloro-6(1H)-pyridazinone) was carried out in presence of free radical initiators to produce crosslinked polymers.[157] Copolymerization of allyl and dimethallyl esters of isophthalic and terephthalic acids with styrene in presence of radical initiators was studied.[158] Monomer reactivities for the four systems were reported. The degree of cyclization of the esters decreased with an increase in styrene concentration.

The divinyl ether of N-acetyldiethanolamine was copolymerized with glycols in presence of sulfuric acid of low concentration.[159] The dilute acid solution may catalyze the reaction to form the polyether more effectively than it does the vinyl ether polymerization which should have occurred via cyclopolymerization to some extent. Dimethallidenepentaerythritol and dicrotylidenepentaerythritol were copolymerized with maleic anhydride by free radical initiation.[160] From these results, it was proposed that the copolymerization proceeded via a pseudo-terpolymerization of the charge transfer complex between the anhydride and the allidene monomers. The reactivity ratios were calculated for the two systems.

Anionic initiated copolymerization of styrene with p-divinylbenzene was studied.[161] The reaction to gel exhibits simpler kinetics than the corresponding radical reaction. The gel point was delayed considerably. From the data obtained, it was possible to calculate the theoretical gel point which differs considerably from the observed gel point due to an abnormal occurrence of intramolecular crosslinking.

Copolymerization of diethylene glycol divinyl ether with styrene, via radical initiation, gave soluble copolymer.[162] The reactivity ratios were 0 for the vinyl ether and 16.8 ± 0.5 for styrene. Diallyl phosphonates were copolymerized with vinyl chloride via a radical mechanism to give fusible copolymers.[163]

Dimethallyl phthalate was copolymerized with vinyl acetate by radical initiation.[164] The rate and degree of copolymerization increased with an increase in the mole fraction of vinyl acetate. The monomer reactivity ratios were determined on the basis of the copolymer equation in which the intramolecular cyclization was considered, and were reported.

Copolymerization of diallyl phthalate (M_1) with vinyl acetate or methyl acrylate (M_2) in the presence of benzoyl peroxide has been studied.[2-436] The copolymerization rate and the intrinsic viscosity of the copolymers decreased with increasing mole fraction of M_1 in the feed. The copolymerization rate depended on the ease of abstraction of an allylic hydrogen atom by the growing polymer chain radicals and of reinitiation by the allylic radical. The copolymerization parameters were determined (comonomer, temperature, r_1, r_2, Q_1, e_1; where r_1 and r_2 are reactivity ratios, Q_1 and e_1 are Alfrey-Price values, given): vinyl acetate, 60°, 1.20, 0.72, 0.033, 0.16; vinyl acetate, 80°, 0.93, 0.83, 0.028, 0.29; Me acrylate, 60°, 0.058, 12.9, 0.024, 0.06; Me acrylate, 80°, 0.050, 11.5, 0.023, −1.14.

Styrene was polymerized via radical initiation in presence of diallyl acetals and diallyl esters, giving chain transfer constant values for diallyl esters, which were small compared to those for the diallyl acetals.[165] The mechanism of polymerization of styrene in presence of diallyl compounds was discussed. Cyclocopolymerization of diallyl phthalate with vinyl chloride was studied with free radical initiation.[166] The copolymerization rate and degree of polymerization decreased with increasing mole fraction of the ester in the feed. The monomer reactivity ratios were determined on the basis of the cyclopolymerization mechanism and were reported. Diallyl isophthalate or diallyl terephthalate was bulk copolymerized with allyl benzoate via free radical initiation.[167] The copolymerization rate decreased with increasing mole fraction of allyl benzoate. The monomer reactivities were estimated on the basis of the cyclopolymerization mechanism.

Intramolecular cyclization of styrene-p-divinylbenzene copolymers has been studied in the copolymer formation in toluene and cyclohexane at low conversion.[168] The first polymer chains formed were extensively cyclized with the formation of a relatively large number of small rings. The molecular weight of polystyrene formed in the absence of p-divinylbenzene was controlled by chain transfer. At very low conversions, the number average molecular weight (M_n) and the weight average molecular weight (M_w) were 2.2×10^5 and 4.4×10^5, respectively, and were unaffected by substituting some of the styrene by p-divinyl benzene; but on continued reaction, M_n increased slowly and M_w increased rapidly. Even at the lowest conversions, intrinsic viscosity and radius of gyration were decreased by introduction of p-divinylbenzene.

It has been shown that in the copolymerization of diallyl phthalate with methylallyl benzoate at 60°, the residual unsaturation of the copolymer decreased with increasing the benzoate mole fraction in the feed; for diallyl isophthalate copolymerization, the unsaturation was nearly constant, and with diallyl terephthalate it increased with increasing mole fraction of the benzoate.[2-442] Intramolecular cyclization and steric hindrance differences in the diallyl esters accounted for these results. Monomer reactivity ratios were calculated for cyclized and uncyclized diallyl ester radicals.

It has been shown that the free radical copolymerizability decreased in the order diallyl phthalate > vinyl acetate > allyl benzoate with essentially no difference between diallyl tetrachlorophthalate and diallyl tetrabromophthalate.[2-445] The cyclization during polymerization decreased in the order of the tetrabromophthalate > the tetrachlorophthalate > the phthalate. Copolymerization parameters were given for various combinations of the monomers.

It was also shown that in the cyclocopolymerizations of diallyl phthalate with monovinyl monomers, e.g. allyl benzoate, vinyl acetate, methyl acrylate, and vinyl chloride, the reactivity ratio of cyclized radical (r_c) was smaller than that of the uncyclized radical (r_1), owing to steric hindrance in the addition reaction of the cyclized radical with diallyl phthalate.[2-444] The validity of the values of r_c and r_1 was discussed on the basis of copolymerization of diallyl phthalate in dilute solution corresponding to r_c, and that of allyl propyl phthalate, to r_1. The copolymerizations of methyl allyl phthalate and allyl octyl phthalate with vinyl acetate were also presented. Both the rate of copolymerization and the copolymerizability of vinyl acetate systems decreased in the order, methyl allyl > allyl propyl > allyl octylphthalate and was ascribed to the steric effect of the alkyl group.

In a continuing study of allyl ester polymerization, it was shown that the monomer reactivity ratio of diallyl oxalate in copolymerization with allyl benzoate at 60° was 0.89.[2-315] The monomer reactivity ratios of diallyl succinate and diallyl adipate during copolymerization with allyl benzoate were both 0.92 and those of the benzoate were 1.08 and 0.99, respectively. The monomer reactivity ratios for cyclization (r_c) of the succinate and the adipate were 1.38 and 1.39, respectively.

Copolymerization of allyl phthalate and styrene has been studied by other investigators. The copolymer was obtained in bulk polymerization initiated by benzoyl peroxide at 353°K and at variable concentration of the monomers.[2-447] The monomer reactivity ratios of diallyl phthalate and styrene were $r_1 = 0.145 \pm 0.099$ and $r_2 = 25.5 \pm 2.6$, respectively. The ratio of the rate constant of unimolecular cyclization to the rate constant of bimolecular chain propagation of the uncyclized radicals derived from diallyl phthalate was 0.27 mole/dm^3. Continuous addition of styrene during the copolymerization gave copolymer of constant composition.

Crosslinking and cyclization in copolymerization of vinyl acetate and divinyl adipate has been studied.[169] The reactivity ratios for the copolymerization were close to unity. The adjacent pendant double bonds, resulting from incorporation of the divinyl monomer, more readily underwent intramolecular cyclization than intermolecular reaction. This cyclization occurred with 10% of the divinyl adipate units when very concentrated (95% monomers, 5% acetone) solutions were polymerized and increased to 23% when a 50% solution of monomers in $CHCl_3$ was used.

Radical polymerization of allyl octyl oxalate in the presence of benzoyl peroxide proceeded via a cyclic mechanism and the chain transfer constant of the growing radical to monomer was estimated to be 0.063.[170] In the copolymerization of allyl octyl acetate with diallyl oxalate, the rate and degree of polymerization decreased with an increase in the mole fraction of the monoallyl ester in the feed, and a similar tendency was also observed with respect to residual unsaturation in the copolymer. The monomer reactivity ratios were r_1 (monoallyl ester) = 0.088 and r_2 = 0.91 based on the general copolymer composition equation, since the cyclopolymerization mechanism did not seem to apply to this copolymerization system.

Kinetic and reactivity data for the radical copolymerization of diethylene glycol divinyl ether with CH_2:$CHCO_2R$ ($R = CH_3$, C_2H_5, C_3H_7, C_4H_9, C_5H_{11}, C_6H_{13}) showed that the predominant influence on the rate of polymerization was steric and not electronic factors.[3-182] Steric hindrance in the copolymerization increased with increasing length of the alkyl chain in the acrylate. The effective number of active centers in the divinyl ether was calculated to be 1.6 when steric screening factors were taken into consideration. Kinetic equations were derived to describe the rate of copolymerization.

Copolymerization of the divinyl ether of diethylene glycol and maleic anhydride led to diphilic ion exchangers.[171] Copolymerization study of diallyl terephthalate with dialkyl maleates produced polymers with alternating structures.[172] Transparent plastics with various refractive indices and dispersions were obtained by varying the comonomer compositions. The best mechanistic and thermal properties were obtained when the terephthalate-maleate monomer molar ratio was 55:45.

The kinetics of the bulk polymerization of monoallyl phthalate with allyl benzoate or diallyl phthalate in the presence of benzoyl peroxide at 70° has been studied.[173] The copolymer composition was interpreted in terms of the kinetics, including the partial elimination of phthalic anhydride from the monoallyl phthalate growing chain end in its propagation reaction with another monomer. The gel point of the monoallyl-diallyl phthalate copolymer was determined and its ionic crosslinking with zinc acetate was discussed.

Copolymerization of diallyl tartrate with several vinyl monomers has been studied.[2-459] The radical polymerizations of diallyl tartrate (M_1) with diallyl succinate, diallyl phthalate, allyl benzoate, vinyl acetate or styrene were studied to determine the characteristic hydroxyl group effect observed in the homopolymerization of diallyl tartrate. In polymerization of the tartrate with the succinate or the phthalate, the dependence of the rate of polymerization on monomer composition was different for different copolymerization systems and unusual values larger than unity for the product of the monomer reactivity ratios were obtained. The r_1 and r_2 values were 1.50 and 0.64 for the tartrate-benzoate system and 0.76 and 2.34 for the tartrate-acetate system, respectively. The product r_1r_2 for the latter polymerization system was larger than unity. In the copolymerization with styrene, the largest effect due to the high polarity of the tartrate was observed. Solvent effects were examined to improve the copolymerizability of the diallyl tartrate. Results were discussed in terms of the hydrogen bonding ability of the tartrate.

Some kinetic data on the three-dimensional copolymerization of hydroquinone divinyl ether with maleic anhydride have been determined.[174] The rate of AIBN-initiated copolymerization of maleic anhydride with hydroquinone divinyl ether in acetic anhydride reached maximum value in 40–50 minutes and then rapidly decreased; an increase in the concentration of the divinyl ether increased the maximum and the overall copolymerization rates and shifted the gel point to shorter reaction time. The yield of the gel depended little on the ratio of the monomers. Even at relatively low conversions, only insoluble copolymer was formed, suggesting alternating copolymerization.

Cyclization in vinyl-divinyl copolymerization has been studied.[3-184] Chemical analysis and NMR studies of low-conversion soluble ethylene dimethacrylate-styrene copolymer containing 5–60% ethylene dimethacrylate showed that cyclization predominated in the crosslinking copolymerization. Compact structures were formed with the character of microgel particles, especially at higher styrene contents; whereas at higher dimethacrylate contents, this component was incorporated into the copolymer only as a unit with a pendant double bond, and the number of crosslinks did not further increase with increasing dimethacrylate. The pendant vinyls in the particle cores were so immobilized or sterically hindered that they were unable to react even with small monomer molecules. Procedures for determining copolymerization parameters for pairs of mono- and divinyl monomers, which usually neglect cyclization, were revised. An analysis of the theoretical model of crosslinking copolymerization with weak cyclization, and experimental results with this system show that the copolymerization parameters need not be too distorted if they are determined from the mono- and divinyl component contents in the copolymer.

Ethylene glycol dimethacrylate was used as a tetrafunctional monomer in radiation crosslinking of poly(vinyl chloride).[175] A mechanism for the results obtained was proposed and discussed. Diallylphthalate was copolymerized with lower α-olefins via radical initiation.[176] The reaction proceeded mainly by cyclopolymerization. Monomer reactivity ratios were determined and it was shown that the reactivity ratio of the cyclized radical was much larger than that of the uncyclized radical.

A study was carried out to determine the effect of dilution and the structure of dimethacrylate systems on the gel point in copolymerization of methyl methacrylate with polymethylene dimethacrylates.[177] The relations between crosslinking and cyclization in the systems were discussed. o-o′-Diallylbisphenol was copolymerized with bismaleimides to produce useful coatings and films.[178] Thermal copolymerization of N,N′-diallyl-1,6-diazaspiro[4,4]-nonane-2,7-dione with N,N′-(methylenedi-p-phenylene)bis-maleimide gave a product with glass temperature of 237°.[179]

COPOLYMERS OF BICYCLIC DIENES WITH VINYL MONOMERS

Some properties of the vinyl chloride-bicyclo[2.2.1]-hepta-2,5-diene copolymers have been reported.[180] The copolymers were prepared in acetone solution with acetyl peroxide as initiator at 45°C and contained about 88 wt% of vinyl chloride. The bicycloheptadiene monomer entered the copolymer essentially as the substituted nortricyclene. The monomer-polymer composition curve indicated that the copolymer was alternating in character. The glass transition temperature of the copolymer was raised by about 1° for each wt% of the bicycloheptadiene monomer incorporated into the copolymer. For each wt % of vinyl chloride copolymerized, the melting point of the copolymer was lowered by about 3°.

Unsaturated polyanhydrides, useful as crosslinking agents for epoxy resins were prepared by the mercaptan-initiated free radical addition polymerization of 5-vinyl-2-norbornene with maleic anhydride.[181] A mixture of this polyanhydride when heated with a novolac epoxy resin gave a cured polymer with good thermal stability. Bicyclo[2.2.1]hepta-2,5-diene and maleic anhydride were copolymerized in presence of a radical initiator.[182] A sample of the copolymer, after hydrolysis was shown to contain a ratio of diene to maleic anhydride equal to 1.51:1.

Copolymers of ethylene with several bicyclic dienes of structures shown (7–22)

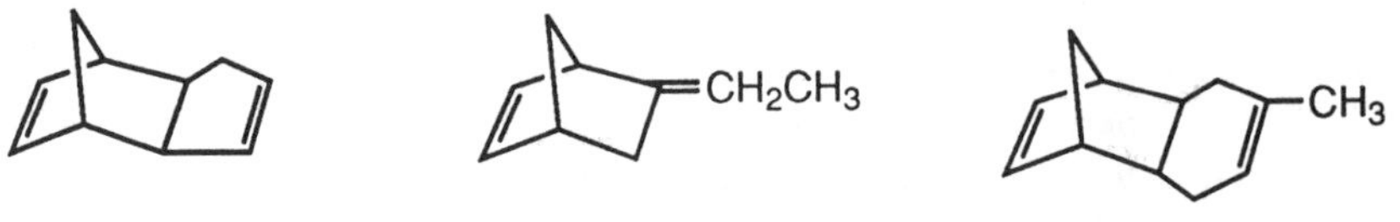

$=CH_2CH_3$ $-CH_3$

(7–22)

were synthesized by use of metal-coordination catalysts.[183] The catalyst efficiency, copolymer composition, diene content and viscosity were a function of the diene monomer concentration and type. Copolymers of maleic anhydride and dicyclopentadiene with a molar ratio of 1:1 to 2:1 prepared by free radical initiation in a polar solvent have been described.[6–90] The copolymer was saturated when the molar ratio was 2:1.

OTHER COPOLYMERS WITH NONCONJUGATED DIENES

A number of nonconjugated dienes have been copolymerized with monomers containing other functional groups, for example, carbonyl compounds, alkynes, and α,β-disubstituted alkenes. This section describes some of these copolymers.

The cationic initiated cyclocopolymerization of benzaldehyde and divinyl ether has been reported to contain units of the 1:1 and 1:2 structures shown (7–23).[184] Polymers prepared at low benzaldehyde ratios (7–23) also contained units with pendant double bonds.

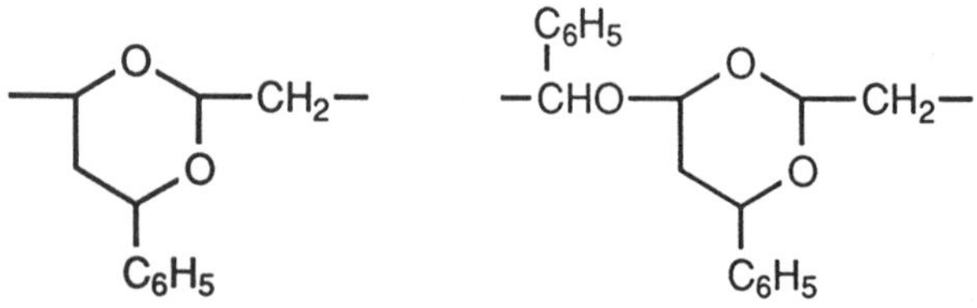

O $-CH_2-$ O C_6H_5 C_6H_5 $-CHO-$ O $-CH_2-$ O C_6H_5

(7–23)

Dialkylsilyl dimethacrylates $R_2Si[O_2CC(CH_3){:}CH_2]_2$ ($R = CH_3$, C_2H_5, C_3H_7, C_4H_9) have been polymerized to give soluble cyclic polymers, which were hydrolyzed to give poly(methacrylic acid) with higher pKa than atactic poly(methacrylic acid).[185] Copolymerization of the dialkylsilyl dimethacrylates with styrene under similar conditions gave copolymers enriched in diester units in increased yield with decreasing styrene mole fraction. Bulk copolymerization of diethylsilyl dimethacrylate with styrene and with methyl methacrylate, followed by hydrolysis, gave copolymers with randomly situated pairs of carboxyl side groups.

It has been shown that isoprene-monomethylmaleate copolymer, obtained by radical polymerization, had a mixture of alternating, block polyisoprene, and cyclized polyisoprene structures.[186] The isoprene units in the copolymer were of 1,4-configuration.

It has been shown that in the radical copolymerization of 1-vinylindole with maleic anhydride, a charge-transfer complex was formed which may be involved in the initiation and propagation steps of the copolymerization.[187] The molecular structure of the copolymer was shown by ^{13}C NMR to be complicated by the formation of indoline rings along with chains by an attack of the propagating maleic anhydride radical at the 2-position of the indole ring. An excess of maleic anhydride units was, therefore, usually found in the copolymer. The mechanism of the cyclization reaction was clarified by copolymerizing the anhydride with 2-methyl-1-vinylindole, 3-methyl-1-vinylindole, or 2,3-dimethyl-1-vinylindole; an alternating copolymer was obtained only with the last monomer. Radical polymerization of 1-vinylindole and 2-methyl-1-vinylindole, 3-methyl-1-vinylindole or 2,3-dimethyl-1-vinylindole gave linear polymers from the 2-methyl and 2,3-dimethyl derivatives and cyclic indoline structures from the 1-vinylindole and the 3-methyl derivative.[3-83] It was postulated that cyclization involved the 2-position of the indole ring.

Copolymers of allyl methacrylate and propargyl methacrylate have a very high speed as electron-beam and X-ray resists.[188] The polymers have a structured complexity not often encountered in addition polymerization.

The reaction of syndiotactic poly(methacrylic acid) with dicyclohexylcarbodiimide gave cyclic methacrylic anhydride-methacrylic acid copolymers.[189] The statistical distribution of cyclized and noncyclized groups was determined by esterification with diazomethane, which did not react with the anhydride groups. Hydrolysis of the anhydride groups gave a methyl methacrylate-methacrylic acid copolymer, the sequence distribution of which was determined by NMR.

The mode of ring opening during methanolysis of syndiotactic methacrylic anhydride-methacrylic acid copolymers was studied by statistical calculation, Monte Carlo simulation and ^{1}H NMR.[190] Significant deviations from random ring opening were observed, indicating that the nature of the monomer unit adjacent to the anhydride ring may exert an influence.

REFERENCES

1. Sayadyan, A. G., Dzhanikyan, O. A. and Mirzoyan, V. A., Arm. Khim. Zh., 1976, 29, 708; CA 86, 30141q (1977).
2. Safaryan, E. B. and Sayadyan, A. G., Arm. Khim. Zh., 1979, 32, 401; CA 91, 193961z (1979).
3. Safaryan, E. B. and Sayadyan, A. G., Arm. Khim. Zh., 1979,2, 405; CA 91, 193962a (1979).
4. Amemiya, Y., Katayama, M. and Harada, S., Makromol. Chem., 1975, 176, 1289; CA 83, 114973p (1975).
5. Amemiya, Y., Katayama, M. and Harada, S., Makromol. Chem., 1977, 178, 289; CA 86, 90300a (1977).
6. Chang, H-C., and Feng, H-T., Ko Fe T 'ung Hsun, 1965, 7, 448; CA 64, 12810f (1965).
7. Gaylord, N. G. and Maiti, S., J. Macromol. Sci. (Chem) 1972, 6, 1481; CA 78, 124964x (1973).
8. Makowski, H. S., Shim, B. K. C., and Wilchinsky, Z. W., J. Polym. Sci. A, 1964, 2, 4973; CA 62, 5346c (1965).
9. Aso, C. and Nawata, T., Kogyo-kagaku zasshi, 1965, 68, 549; CA 63, 8485b (1965).

10. Natta, G., Mazzanti, G., Valassori, A., Sartori, G. and Turba, V., Belg. Pat. 631,165; Fr. 1,359,982; 1963 (Aug. 16); CA 60, 16087d (1964).
11. Verbanc, J. J., Fawcett, M. S. and Goldberg, E. J., I & EC Prod. Res. Dev.; U.S. 2,933,490, 1962, 1, 70; CA 57, 3600h (1962).
12. Stedefeder, J. and Weber, H., Belg. Pat. 636,945, 1963 (Dec. 31); CA 62, 9258c (1965).
13. Natta, G., Mazzanti, G., Valvassori, A., Dall'Asta, G., Sartori, G. and Cameli, N., Belg. Pat. 623,741, 1963 (Feb. 14); CA 59, 1821h (1963).
14. Natta, G., Mazzanti, G., Valvassori, A., Sartori, G. and Cameli, N., Belg. Pat. 623,698, 1963 (Feb. 14); CA 59, 1821f (1963).
15. Dowbenko, R., U.S. Pat. US 3,261,815, 1966 (July 9); CA 65, 15539b (1966).
16. Wright, C. D., U.S. Pat. 3,247,170, 1966 (April 19); CA 64, 19922h (1966).
17. Ethyl Corp., Brit. Pat. 1,012,552, 1968 (Nov. 21); CA 68, 115190h (1968).
18. Gaylord, N. G., Deshpande, A. B. and Martan, M., J. Polym. Sci., Polym. Lett. Ed., 1976, 14, 679; CA 85, 193171t (1976).
19. Doiuchi, T., Yamaguchi, H. and Minoura, Y., Eur. Polym. J., 1981, 17, 961; CA 96, 52712k (1982). 20. Kodaira, T. and Nishioka, K., Makromol. Chem., Part 1, 1987, 281; CA 196, 156915v (1987).
21. Michelotti, F. W., J. Polym. Sci., 1962, 59, 51; CA 57, 1054a (1962).
22. Beynon, K. I. and Hayward, E. J., J. Polym. Sci. A, 1965, 3, 1793; CA 63, 1876d (1965).
23. Varveri, F. H., Jula, R. J. and Hoover, M. F., U.S. Pat. 3,639,208, 1972 (Feb. 1); CA 77, 7588w (1972).
24. Schaper, R. J. and Hoover, M. F., U.S. Pat. 3,579,613, 1971 (May 18); CA 75, 64872s (1971).
25. Hoover, M. F., Ger. Offen. 1,814,597, 1969 (July 24); CA 71, 82113s (1969).
26. Sayadyan, A. G. and Simonyan, D. A., Arm. khim. Zh., 1968, 21, 1041; CA 71, 30720t (1969).
27. Woodward, D. G., U.S. Pat. 3,574,175, 1971 (April 6); CA 75, 6657q (1971).
28. Wandrey, C. and Jaeger, W., Acta Polym., 1984, 36, 100; CA 102, 79381f (1985).
29. Weber, G. A., Pettebone, R. H. and D'Errico, M. J., Eur. Pat. Appl. EP 69,573, 1983 (Jan. 12); CA 98, 174866a (1983).
30. Nippon Synthetic Chemical Industry Co., Ltd., Sanso, K. K., Ube Ind., Ltd., Japan Pat. 58 15,546, 1983 (Jan. 28); CA 99, 54544t (1983).
31. Costello, C. A., Matz, G. F., Sherwood, N. S., Boffardi, B. P., Yorke, M. A. and Amjad, Z., Eur. Pat. Appl. EP 82,657, 1983 (Jun. 29); CA 99, 125373z (1983).
32. Wyroba, A., Przem. Chem., 1983, 62, 681; CA 100, 86513n (1984).
33. Ginberg, A. M., Dremyatakaya, L. D., Ivanova, T. A.,Vladiminski, R. A., Klyachko, Y. A., Schnaider, M. A., Kamenskaya, E. V., Konstantinov, V. A., Gorodilin, V. V. and Khokhlov, B. A., USSR Pat. SU 1,071,630, 1984 (Feb. 7); CA 100, 177625s (1984).
34. Mourato, D. and Gehr, R., Sci. Tech. Eau, 1983, 16, 323; 327; CA 100, 108802k (1984).
35. Shkurnikova, I. S., Penenshik, M. A., Virnk, A. D. and Topchiev, D. A., Izv. Akad. Nauk SSSR, Ser. Khim., 1984, 4, 928; CA 101, 55619t (1984).
36. Giddings, D. M., Ries, D. G. and Syrinek, A. R., U.S. Pat. US 4,502,966, 1985 (Mar. 5); CA 102, 187828r (1985).
37. Iwata, M., Saka, T., Ikada, T. & Kondo, N., Jap. Pat. JP 61 133,213 [86 133,213], 1986 (Jun. 20); CA 105, 210367n (1986).
38. Martin, Jr., R. H., U.S. Pat. 3,012,019, 1961 (Dec. 5); CA 56, 6183g (1962).
39. Martin, Jr., R. M., U.S. Pat. 3,043,816, 1962 (July 10); CA 57, 11342b (1962).
40. Spaulding, D. E. and Horne, Jr., S. E., U.S. Pat. 3,032,538, 1962 (May 1); CA 57, 3638i (1962).

41. Butler, G. B. and Brooks, T. W., J. Macromol. Chem., 1960, 1, 231; CA 65, 5537f (1966).
42. Matkovskii P. E., Leonov, I. D., Kissin, Yu. V., Chirkov, N. M., Pomogailo, A. D. and Beikhol'd, G. A., Izv. Akad. Nauk SSSR, Ser. Khim., 1968, 930; CA 69, 19566k (1968).
43. Wende, A., Priebe, H. and Deutsch, K., Ger. Pat. 1,266,495, 1968 (Apr. 18); CA 69, 11109t (1968).
44. Sayadyan, A. G. and Simonyan, D. A., Arm. Khim. Zh., 1969, 22, 528; CA 71, 102273z (1969).
45. Sayadyan, A. G., Simonyan, D. A. and Safaryan, E., Arm. Khim. Zh., 1970, 23, 757; CA 75, 77301e (1971).
46. Iwakura, Y., Uno, K. and Tsuruoka, K., Jap. Pat. JP 71 03,175, 1971 (Jan. 26); CA 74, 112782r (1971).
47. Simonyan, D. and Sayadyan, A., Plast. Massy., 1971, 10; CA 76, 14987h (1972).
48. Iwakura, Y., Uno, K. and Tsuruoka, K., Jap. Pat. JP 71 03,174, 1971 (Jan. 26); CA 74, 126415k (1971).
49. Lonza Ltd., Fr. Pat. 2,098,394, 1972 (Apr. 14); CA 77, 165327q (1972).
50. Haruta, M., Kageno, K. and Harada S., Makromol. Chem., 1973, 170, 11; CA 80, 15245j (1974).
51. Battaerd, H., Botto, B. and Samadoni, P., Ger. Offen. 2,327,433, 1974 (Dec. 13); CA 80, 121728d (1974).
52. Harada, T. and Kato, T., Jap. Pat. JP 73 103,700, 1973 (Dec. 26); CA 81, 121391j (1974).
53. Harada, S. and Kato, T., Jap. Pat. JP 73 102,900, 1973 (Dec. 24); CA 80, 109091h (1974).
54. Sayadyan, A. G., Safaryan, E. B. and Oganesyan, G. P., USSR 425,924, 1974 (April 30); CA 81, 170251x (1974).
55. Fujii, Y. and Maekawa, H., Jap. Pat. JP 75 23,001, 1975 (Aug. 5); CA 84, 18074w (1976).
56. Sharpe, A. J., Jr., Sinkovitz, G. and Noren, G. K., Ger. Offen. 2,544,840, 1976 (April 22); CA 85, 48581x (1976).
57. Jeffery, E. A., Hodgkin, J. H. and Solomon, D. H., J. Macromol. Sci. (Chem.), A, 1976, 10, 943; CA 85, 109022q (1976).
58. Bolto, B. A. and Jackson, M. B., J. Macromol. Sci. (Chem) A, 1978, 12, 745.
59. Wade, K. O. and Brown, J. H., Australian Pat. 502,873, 1979 (Aug. 9); CA 91, 212327r (1979).
60. Iovine, C. P. and Ray Chaudhuri, D. K., Fr. Demande 2,411,848, 1979 (Jul. 13); CA 91, 212002f (1979).
61. Ioan, S., Mocanu, G. and Maxim, S., Eur. Polym. J., 1979, 15, 667; CA 92, 77020p (1980).
62. Iovine, C. P. and Ray Chaudhuri, D. K., U.S. Pat. US 131,576, 1978 (Dec. 26); CA 90, 122289e (1979).
63. Clemens, D. H. and Glavis, F. J., U.S. Pat. US 4,140,659, 1979 (Feb. 20); CA 90, 187907z (1979).
64. Croitoru, V. and Vladescu, L., Bul. Inst. Politeh Ser. Chim.-Metal, 1979, 41, 25; CA 93, 8661g (1980).
65. Mocanu, G. C., Maxim, S., Carpov, A., Maftei, M. and Zamfir, A., Rom. Pat. 68,697, 1978 (Dec. 15); CA 93, 133320w (1980).
66. Boevink, J. E., Derrick, A. P. and Moodie, J. A., U.S. Pat. US 4,347,339, 1982 (Aug. 31); CA 97, 183424m (1982).

67. Mathias, L. J. and Cei, G., Polym. Prepr. (ACS Div. Polym. Chem.) 1987, 28, 124; CA 106, 214434a (1987).
68. Lin, F. C. and Getman, G. D., Int. GPC Symp. '87, 1987 225; CA 109, 150402r (1988).
69. Bhattacharyya, B. R. and Dalsin, P. D., U.S. Pat. US 4,713,431, 1987 (Dec. 15); CA 108, 151170e (1988).
70. Lin, Y.-Q., Pledger,Jr., H. and Butler. G. B., J. Macromol. Sci. (Chem.) A, 1988, 25, 999; CA 109, 171036q (1988).
71. Hunkeler, D., Hamielec, A. E. and Baade, W., Adv. Chem. Ser. (Polym. Aqueous Media), 1989, 223, 175; CA 112, 139875p (1990).
72. Chen, S. R. T. and Vaughan, C. W., Eur. Pat. Appl. EP 344,841, 1989 (Dec. 6); CA 112, 199356e (1990).
73. Matsumoto,A., Wakabayashi, S., Oiwa, M. and Butler, G. B., J. Macromol. Sci. (Chem.) A, 1989, 26, 1475; CA 112, 21386b (1990).
74. Parkhamovich, E. S., Boyarkina, N. M., Kryuchkov, V. V. and Topchiev, D. A., Vopr. Tekhnol. Fenol. Smol, Fenopl. Ion. 1989, 69; CA 113, 59916h (1990).
75. Hahn, M., Kotz, J., Linow, K. J. and Philipp, B., Acta. Polym., 1989, 40, 36; CA 110, 115429d (1989).
76. Kolosnitsyn, V. S. and Yakovleva, A. A., Elektrokhimiya, 1989, 25, 544; CA 110, 223403z (1989).
77. Kudaibergenov, S. E., Sigitov, V. B., Bekturov, E. A., Ketts, I. and Filipp, B., Vysokomol. Soedin., Ser. B, 1989, 31, 132; CA 110, 213749c (1989).
78. Machida, M. and Sato, T., Jap. Pat. 63,252,600 [88,252,600], 1988 (Oct.19); CA 111, 12097k (1989).
79. Lin, Y. Q. and Butler, G. B., J. Macromol. Sci. (Chem.) A, 1989, 26, 681; CA 111, 78738b (1989).
80. Hahn, M., Koetz, J., Ebert, A., Schmolke, R., Phillip, B. Kudaibergenov, S., Shigitov, V. and Bekturov, E. A., Acta Polym., 1989, 40, 331; CA 111, 58641m (1989).
81. Koetz, J., Hahn, M., Philipp, B., Kudaibergenov, S., Sigitov, V. and Bekturov, E. A., Acta Polym., 1989, 40, 405; CA 111, 97910t (1989).
82. Koetz, J., Hahn, M. and V. and Philipp, B., Acta Polym., 1989, 40, 401; CA 111, 97909z (1989).
83. Nakamura, S., Yui, S. and Muramatau, K., Jap. Pat. 01 157,910 [89 157,910], 1989 (June 21); CA 112, 11801u (1990).
84. Smets, G., Deval, N. and Hous, P., J. Polym. Sci. A, 1964, 2, 4825; CA 62, 6557g (1965).
85. Matsoyan, S. G., Voskanyan, M. G. and Cholakyan, M. G., Vysokomol. Soed., 1963, 5, 1035; CA 59 11668d (1963).
86. Matsoyan, S. G., Voskanyan, M. G., Geverkyan, E. T. and Choklakyan, A. A., Izv. Akad. Nauk armyan. SSR, Khim. Nauki 1964, 17, 420; CA 62, 1748b (1965).
87. Aso, C. and Sogabe, M., Kogyo-kagaku zasshi, 1965, 68, 1970; CA 64, 11324e (1966).
88. Friedlander, W. S., Chem. & Eng. News, 3/10/58 and 4/21/58, 1958, 36, 44.
89. Kikukawa, K., Nozakura, S. and Murahashi, S., Kobunshi Kagaku, 1967, 24, 801; CA 68, 115007d (1968).
90. Raymond, M. A., U.S. Pat. 3,640,968, 1972 (Feb. 8); CA 76, 141771h (1972).
91. Sultanov, K., Ilkhamov, M. Kh., Askarov, M. A. and Tadzhibaev, A., USSR Pat. 203,901, 1967 (Oct. 9); CA 69, 19801h (1968).
92. Baines, F. C. and Bevington, J. C., Polymer, 1970, 11, 647; CA 74, 31983e (1971).
93. Hanford, W., U.S. Pat. US 2,396,785, 1946 (Mar. 19); CA 40, 34594 (1946).
94. Hwa, J. C. H., Ger. Pat. 1,124,695, 1962 (Mar. 1); CA 56, 14484c (1962).
95. Hwa, J. C. H., U.S. Pat. US 3,239,493, 1966 (Mar. 8); CA 64, 17802e (1966).
96. Noma, K. and Niwa, M.. Dosh. Daig. Riko. Kenk. Hokoku, 1969, 10, 1; CA 71, 102269c (1969).

97. Neumann, D. L., Akron, Univ. of, Thesis, 1971, 192; CA 76, 127560g, (1972).
98. Noma, K. and Niwa, M., Dosh. Daig. Riko. Kenk. Hokoku, 1972, 12, 204; CA 77, 88887t (1972).
99. Fujimori, K. and Khoo, M., Makromol. Chem., 1978, 179, 859; CA 88, 137056j (1978).
100. Fujimori, K., Makromol. Chem., 1978, 179, 2867; CA 90, 23837a (1979).
101. Kholodenko, G. E., Likhterov, V. R., Etlis, V. S., Pomerantseva, E. G. and Chernov, A. N., Vysokomol. Soedin., Ser. B, 1979, 21, 851: CA 92, 94741s (1980).
102. Urushido, K., Matsumoto, A. and Oiwa, M., J. Polym. Sci., Polym. Chem. Ed., 1979, 17, 4089; CA 92, 198841r (1980).
103. Urushido, K., Matsumoto, A. and Oiwa, M., J. Polym. Sci., Polym. Chem. Ed., 1980, 18, 1131; CA 92, 215856u (1980).
104. Neumann, D. L. and Harwood, H. J., ACS Symp. Ser., 1982, 195, 43; CA 97, 182959c (1982).
105. Fujimori, K., J. Polym. Sci., Chem. Ed., 1983, 21, 25; CA 98, 54562b (1983).
106. Maudgal, S., J. Macromol. Sci. (Chem.), A., 1984, 21, 631; CA 100, 210515f (1984).
107. Novichkova, L M., Ospanova, K. M. and Ishmukhanbetova, N. K., Izu. Akad. Nauk. Kaz. SSR, Ser. Khim., 1987, 68; CA 108, 95018x (1988).
108. Beynon, K. I., J. Polym. Sci. A, 1963, 1, 3357; CA 59, 15387c (1963).
109. Jones, J. F. and Mital, A. J., U.S. Pat. US 2,985,631, 1961; CA 55, 22924b (1961).
110. Natta, G., Valvassori, A. and Sartori, G., Fr.1,359,995; Brit. 1,001,838; 1964 (Apr. 30); CA 62, 4195h (1965).
111. Yamamoto, K., Takenaka, Y. and Iwasa, T., Japan Pat. 78 126,322, 1978 (Nov. 4); CA 90, 122948u (1979).
112. Harada, S. and Katayama, M., Makromol. Chem., 1966, 90, 177; CA 64, 9824d (1966).
113. Harada, S. and Arai, K., Makromol. Chem.. 1967, 107, 64; CA 68, 50098n (1968).
114. Harada, S. and Arai, K., Makromol. Chem., 1967, 107, 78; CA 68, 50099p (1968).
115. Ueda, T. and Harada, S., J. Appl, Polym. Sci., 1968, 12, 2395; CA 70, 29796v (1969).
116. Ueda, T. and Harada, S., J. Appl. Polym. Sci., 1968, 12, 2383; CA 70, 20569r (1969).
117. Harada, S. and Arai, K., Ger. Offen. 1,957,756, 1970 (May 21); CA 73, 26214e (1970).
118. Harada, T. and Arai, K., Japan Pat. JP 70 37,033, 1970 (Nov. 25); CA 74, 64690k (1971).
119. Harada, T. and Arai, K., Japan Pat. JP 70 35,792, 1970 (Nov. 14); CA 74, 64688r (1971).
120. Amemiya, Y., Katayama, M. and Harada, S., Makromol. Chem., 1977, 178, 2499; CA 88, 23504y (1978).
121. Harada, S., Suzuhi, K. and Yanagita, T., Japan Pat. JP 74 80,379, 1974 (Aug. 2); CA 83, 116779d (1975).
122. Frazer, A. H., U.S. Pat. US 3,133,903, 1964 (May 19); CA 61, 19168h (1964).
123. Frazer, A. H., J. Polym. Sci., A, 1965, 3, 3699; CA 64, 15990f (1966).
124. Leplyanin, G. V., Tolstikov, G. A., Vorob'eva, A. L., Shurupov, E. V., Abdrashitov, Y. M., Bikbaeva, G. G., Sysoeva, L. B., Sataeva, F. A. and Kozlov, V. G., USSR SU 1,530,631, 1989 (Dec. 23); CA 113, 7014t (1990).
125. Shaul'skii, Yu M., Leplyanin, G. V., Vorob'eva, A. I., Sysoeva, L. B., Fakhretdinov, R. N., Marvanov, R. M., Dzhemilev, U. M. and Tolstikov, G. A., USSR SU 1,495,339, 1989 (Jul. 23); CA 112, 55589k (1990).
126. Alexander, R. J. and Doyle, J. R., J. Polym. Sci. B, 1963, 1, 625; CA 60, 4262b, (1964).
127. Mattson, J. R., U.S. Pat. 3,134,756, 1964 (May 26); CA 61, 8436e (1964).
128. Peninsular ChemResearch, Inc., Brit. Pat. 1,013,230, 1965 (Dec. 15); CA, 64 6787c (1966).
129. Dowbenko, R., U.S. Pat. US 3,317,490, 1967 (May 2); CA 67, 21509w (1967).
130. Spamhour, J. D., U.S. Pat. 3,331,819, 1967 (July 18); CA 67, 74012k (1967).

131. Yamaguchi, T., Ono, T. and Kudo, E., Kobunshi kagaku, 1969, 26, 714; CA 72, 67327s (1970).
132. Delmonte, D. W. and Hays, J. T., Abstr. ACS Meet., Atlantic City, N.J., 1959, 1.
133. Kawai, W. and Katsuta, S., J. Polym. Sci. A1, 1970, 8(9), 2421; CA 73, 99262d (1970).
134. Sackmann, G. and Kolb, G., Ger. Offen. 2,058,879, 1972 (May, 31); CA 77, 75794k (1972).
135. Ellinger, L. P., Brit. Pat. 1,255,838, 1971 (Dec. 1); CA 76, 86405q (1972).
136. Heublein, G., Boerner, R. and Schuetz, H., Faserforsch Textiltech, 1978, 29, 616; CA 90, 23818v (1979).
137. Matsubara, Y., Yoshihara, M. and Maeshima, T., Kogyo-kwagaku zasshi, 1971, 74, 1909; CA 76, 25643u (1972).
138. Matsoyan, S. G., Avetyan, M. G. and Voskanyan, M. G., USSR Pat. 134,866, 1961 (Jan. 10); CA 58, 12942b (1961).
139. Tamey, R. and Verbanc, J., Ger. Offen. 1,953,101, 1970 (May 14); CA 73, 26418z (1970).
140. Sayadyan, A. G., Boyakhchyan, M. and Kinoyan, F., Arm. Khim. Zh., 1971, 24, 732; CA 76, 46553e (1972).
141. Beck, W., Spell, H. and Pledger, H.,Jr., J. Macromol. Sci. (Chem), 1971, 5, 491; CA 74, 64454m (1971).
142. Gupta, B., U.S. Pat. US 4,877,843, 1989 (Oct. 31); CA 112, 180180w (1990).
143. Yoshida, S., Takahashi, H. and Kita, N., Japan Pat. JP 01 118,843 [89 118,843], 1989 (May 11); CA 112, 45727q (1990).
144. Wesslau, H., Makromol. Chim., 1966, 93, 55; CA 65, 865f (1966).
145. Matsumoto, A. and Oiwa, M., Kogyo-kagaku zasshi, 1967, 70, 360; CA 68, 13435z (1968).
146. Hashimoto, S. and Furukawa, I., Kobunshi kagaku, 1964, 21, 647; CA 62, 7874c (1965).
147. Andreev, D. N., Sokolova, N. P., Alekseeva, D. N. and Pavlovskaya, V. E., Vysokomolek. Soedin, Series A, 1969, 11, 492; CA 70, 115595k (1969).
148. Martin, Jr., R. H., U.S. Pat. 3,012,012, 1959 (July 31); CA 56, 8939a (1962).
149. Hauschild, R., Fr. Pat. 1,343,540, 1963 (Nov. 22); CA 60, 13413h (1964).
150. Gibbs, W. E. and McHenry, R. J., J. Polym. Sci. A, 1964, 2, 5277; CA 62, 10518g (1964).
151. Gibbs, W. E., J. Polym. Sci. A, 1964, 2, 1809; CA 62, 1748d (1964).
152. Atakozova, M., Gusev, V., Kuznetsov, E., Monostyrskii,V. and Terteryan, R., Tr. Kazan Khim. Tekhnol. Inst., 1967, 36, 381; CA 69, 107115t (1968).
153. Matsumoto, A. and Oiwa, M., Kogyo-kwagaku zasshi, 1968, 71, 2063; CA 71, 91953t (1969).
154. Shur, A.M., Filimonova, M. M. and Filimonova, B. F., Vysokomol. Soedin., Ser. A., 1967, 9, 2193; CA 68, 3217r (1968).
155. Wiley, R. H. and Ahn, T., J. Polym. Sci., A1, 1968, 6, 1293; CA 69, 3156r (1968).
156. Terteryan, R. A., Vysokomol. Soedin., Ser. A, 1969, 11, 1850; CA 71, 125009g (1969).
157. Manecke, G. and Wehr, G., Makromol. Chem., 1970, 138, 289; CA 74, 4145b (1971).
158. Matsumoto, A. and Oiwa, M., Kogyo-kwagaku zasshi, 1970, 73, 228; CA 72, 121965p (1970).
159. Chekulaeva, I. A., Talypine, G. V. and Ponomarenko, V. A., Izv. Akad. Nauk SSSR. Ser Khim., 1969, 11, 2577; CA 72, 67322m (1970).
160. Ouchi, T. and Oiwa, M., Bull. Chem. Soc. Jap., 1970, 43, 2858; CA 73, 110440h (1970).
161. Worsfold, D., Macromolecules, 1970, 3, 574; CA 73, 121159z (1970).
162. Zhubanov, B., Kurmanaliev, O. and Shaikutdinov, E., Izv. Akad. Nauk Kaz. SSSR Ser. Khim., 1971, 21, 74; CA 75, 118758e (1971).
163. Koyanagi, S., Kitamura, H. and Shimizu, T., Japan Pat. 71 21,454, 1971 (June 17); CA 75, 130296k (1971).

164. Matsumoto, A. and Oiwa, M., J. Polym. Sci., A1, 1971, 9, 3607; CA 76, 113644a (1972).
165. Matsumoto, A. and Oiwa, M., J. Polym. Sci., A1, 1972, 10, 103; CA 76, 127547h (1972).
166. Matsumoto, A., Ise, T. and Oiwa, M., Nippon Kagaku Kaishi, 1972, 209; CA 76, 154181r (1972).
167. Matsumoto, A. and Oiwa, M., Nippon Kagaku Kaishi, 1972, 166; CA 76, 154179w (1972).
168. Soper, B., Haward, R. N. and White, E. F. T., J. Polym. Sci., Pt A1, 1972, 10, 2545; CA 78, 30254p (1973).
169. Breitenbach, J. W. and Gleixner, G., Monatsh. Chem., 1976, 107, 1315; CA 86, 73195c (1977).
170. Matsumoto, A., Nishimura, M. and Oiwa, M., Technol. Rep. Kansai Univ. and 1977, 18, 53; CA 87, 202186x (1977).
171. Musabekov, K.B., Birimzhanova, Z.S., Pasechnik, V.A., Kurmanaliev, O.S. and Dumpis, J., Reaktskii v Zhidk. Faze, Alma-Ata 1979, 127; CA 93, 115306s (1980).
172. Pfeiffer, K. and Lorkowski, H. J., Plaste Kautsch (E. Ger.), 1979, 26, 67; CA 90, 204928b, (1979).
173. Imai, I., Matsumoto, A. and Oiwa, M., J. Polym. Sci., Polym. Chem. Ed., 1979, 17, 3435; CA 92, 22905a (1980).
174. Tsarik, L. Ya., Antsiferova, L. I. and Kalabina, A. V., Vysokomol. Soedin., Ser. B., 1980, 22, 416; CA 93, 150941s (1980).
175. Palma, G., Carenza, M. and Pollini, C., Eur. Polym. J., 1980, 16, 333; CA 93, 95843j (1980).
176. Matsumoto, A., Sako, S. and Oiwa, M., Technol. Rep. Kansai Univ., 1980, 21, 99; CA 93, 240032a (1980).
177. Rabadeux, J. C., Durand, D. and Bruneau, C., Makromol. Chem., Rapid Commun., 1984. 5, 191; CA 101, 7898m (1984).
178. Repecka, L. N., Eur. Pat. Appl. EP 339,489, 1989 (Nov. 2); CA 112, 180641x (1990).
179. Wang, P. C., U.S. Pat. 4,885,351, 1989 (Dec. 5); CA 112, 199311m (1990).
180. Zutty, N. L. and Whitworth, C. J., SPE Trans., 1964, 4, 22; CA 60, 10873g (1964).
181. McCartney, R. L., U.S. Pat. 3,703,501, 1972 (Nov. 21); CA 78, 30844f (1973).
182. Pledger, H., Jr., U.S. Pat. 3,240,763, 1966 (Mar. 15); CA 64, 17802b (1966).
183. Schnecko, H., Caspary, R. and Degler, G., Angew. Makromol. Chem., 1971, 20, 141; CA 76, 86221b (1972).
184. Aso, C., Tagami, S. and Butler, G. B., J. Polym. Sci. A1, 1972, 10, 1851; CA 77, 88891q (1972).
185. Asinovskaya, D. N., Andreev, D. N. and Davidyuk, L. N., Vysokomol. Soedin., Ser. A, 1977, 19, 1325; CA 87, 53634n (1977).
186. Andrei, C. and Barboiu, V., Rev. Roum. Chim., 1979, 24, 11; CA 90, 204630y (1979).
187. Priola, A., Gatti, G., Santi, G and Cesca, S., Makromol. Chem., 1979, 180, 13; CA 90, 122128b (1979).
188. Daly, R. C., Hanrahan, M. J. and Blevins, R. W., Advances in Resist Tech. and Proc. II, 1985, 539, 138.
189. Berg-Feld, M. C., Mano, E. B. and Klesper, E., Polym. Bull. (Berlin), 1982, 6, 493; CA 96, 200320y (1982).
190. Berg-Feld, M. C., Mano, E. B. and Klesper, E., Polym. Bull., 1986, 16, 487.

8
Cyclopolymerization of Dienes with Selected Vinyl Monomers

A type of copolymerization distinctly different from that described in earlier chapters now constitutes a significant portion of the cyclopolymerization literature.[1, 1-16, 1-17, 1-87, 1-88] This process, referred to as cyclocopolymerization, incorporates both comonomers into the developing cyclic structure. Perhaps the most extensively studied example of this unusual type of copolymerization is the cyclocopolymerization of divinyl ether and maleic anhydride (Equation 8–1). This copolymer has been extensively studied for its biological properties; it exhibits a broad spectrum of biological activities, including antitumor, antiviral, antibacterial, anticoagulant, and antiarthritic properties, as well as being capable of generating interferon, inhibiting inverse transcriptase, activating macrophages, and eliminating plutonium.[2]

CYCLOCOPOLYMERS IN WHICH BOTH COMONOMERS CONTRIBUTE TO THE CYCLIC STRUCTURE

The first example of cyclocopolymerization in which both monomers were involved in the cyclic structure was the copolymer obtained by copolymerization of the electron donor diene divinyl ether and the electron acceptor olefin, maleic anhydride.[1-19, 1-156, 1-171] This copolymerization has been illustrated. Although this copolymerization had been reported to yield noncrosslinked copolymer before the cyclopolymerization mechanism had been postulated,[1-8, 1-23, 3] it was not until the latter mechanism had become well established that a plausible explanation for this most unusual copolymerization of 1,4-dienes with alkenes was forthcoming. The postulated mechanism accounted for (a) the failure of the system to crosslink in accord with the widely accepted (at the time) theory of Staudinger;[2-1] (b) the essential absence of carbon-carbon double bond content in the copolymer; (c) copolymer composition equivalent to a diene-olefin molar ratio of 1:2; and (d) essentially quantitative conversion of monomers to copolymer. Evidence in support of the postulated mechanism included the following: (a) the elemental analysis for the copolymer over a wide range of conversions was consistent with a 1:2 diene-olefin ratio; (b) the infrared spectrum was essentially initiation (1)

(2) selective intermolecular propagation

(3) selective intramolecular propagation

and

(4) selective intermolecular propagation

and

(8–1)

alternate repetition of (1-4)

and

devoid of residual double bond absorption and contained characteristic absorption bands for cyclic anhydride and six-membered cyclic ether structures; (c) the copolymer composition was consistent with the known reactivity ratios of such monomers; and (d) the presence of cyclic ether groups was confirmed by chemical evidence which involved cleavage by hydriodic acid and incorporation of iodine into the copolymer.

The original copolymer softened at 350°C with decomposition. It had an intrinsic viscosity of 0.176 dl/g in dimethylformamide (DMF) and the molecular weight was estimated to be about 15,000 through use of the Staudinger constants reported for polyacrylic acid.[1-30]

A wide variety of cyclopolymerizable pairs of monomers were investigated during the early work on this topic (Table 8–1). Further investigations have been carried out on a number of the above systems, the results of which are reported here.

Saturated, linear high molecular weight cyclopolymers could be obtained by radical initiation of the mixture of monomers consisting of a 1,4-diene and an alkene.[1] Both monomers contributed to the cyclic structure, which was originally proposed to be a six-membered ring. However, the ring size may also be five-membered or a mixture of the two. The system works most effectively if the monomer pair consists of an electron donor and an electron acceptor. Either the 1,4-diene or the alkene may be donor or acceptor.

Table 8–1 Copolymers of 1,4-Diolefins and Olefins

1,4-Diene	Olefin	[η] dl/g DMF	% Yield	Solvent
Divinyl ether	Fumaro-nitrile	—	69.2	Acetone
Divinyldimethylsilane[a]	Maleic anhydride	0.121	—	DMF
Divinyl sulfone[b]	Maleic anhydride	0.064	—	Acetone
Divinylcyclopenta-methylenesilane	Maleic anhydride	0.106	24.3	Benzene
Divinyl ether	Dimethyl fumarate	0.123	—	Benzene
Divinyl ether	Diethyl maleate	0.235	62.1	Benzene
Divinyldimethylsilane	Vinyl acetate	0.063	—	
Divinyl ether	Acrylo-nitrile	—	71.0	—
Divinyl sulfone	Dimethyl fumarate	—	84.7	Benzene
1,4-Pentadiene	Maleic anhydride	—	100	Acetone

[a] [η] = 0.082 dl/g (2N NaOH).
[b] [η] = 0.052 dl/g (2N NaOH).

Cyclopolymerization of divinyl sulfone, and 1,4-pentadiene with maleic anhydride, dimethyl maleate, acrylonitrile and vinyl acetate was studied in greater detail and under a greater variety of conditions.[4] High molecular weight copolymers of divinyl sulfone with maleic anhydride were not prepared. The highest inherent viscosity that was obtained with this system was 0.13. It should be noted, however, that both monomers in this case are electron acceptors. The soluble copolymers obtained were shown by C, H, and S analyses to be 2:1 maleic anhydride-divinyl sulfone copolymers. Aliphatic unsaturation was present only in low concentration.

Other copolymers of divinyl sulfone were prepared using acrylonitrile, vinyl acetate, and dimethyl maleate as the comonomers. At low conversions, soluble copolymers could be obtained when acrylonitrile or vinyl acetate was used. At 13% conversion, a 2:1 vinyl acetate-divinyl sulfone copolymer was obtained which had an inherent viscosity of 0.21 in DMF. The copolymer analyzed as 2.2:1 vinyl acetate to divinyl sulfone and showed little aliphatic unsaturation. This material was soluble in DMF and DMSO and melted at 290° to 300°C with decomposition. Similarly, 7% conversion to soluble copolymer was obtained using acrylonitrile as the comonomer. The copolymer analyzed as a 3.0:1 acrylonitrile-divinyl sulfone copolymer, showed only little aliphatic unsaturation, was soluble in DMF, and had an inherent viscosity of 0.29. Upon heating to 230 to 250°C the copolymer began to darken. Copolymers of divinyl sulfone with dimethyl maleate were insoluble in all the solvents tested.

Copolymers of 1,4-pentadiene and MA were prepared in 50% conversion and with inherent viscosities as high as 0.29 in DMF. Analyses indicated that this composition was 2:1 maleic anhydride-1,4-pentadiene and the IR spectra indicated little aliphatic unsaturation. Films could be obtained from DMF solutions of this copolymer. The copolymer began to soften at 118°C, decomposition without darkening began at 250°C, and noticeable discoloration started at 310°C.

Further copolymerization studies of 1,4-hexadiene with maleic anhydride in the presence of benzoyl peroxide gave polymers containing 66.7 mole % of the diene, and only cyclic structures.[5] In polymerization of the diene with 5-vinyl-2-norbornene, dicyclopentadiene, and 5-ethylidene-2-norbornene, the double bond in the bicycloheptene portion was particularly reactive. The copolymers had 1,4-hexadiene contents of 51–60 mole %, owing to branching reactions of this double bond, with the exception of the 5-ethylidene-2-norbornene copolymer, which contained ≈50 mole % of the hexadiene, regardless of monomer concentration. The latter was the only polymer in which extensive unimolecular cyclization was probable.

Copolymers of divinyl ether with dimethyl maleate, vinyl acetate, and acrylonitrile were also prepared. All of the copolymers prepared using acrylonitrile were insoluble, but only low yields were obtained using vinyl acetate. The yields of copolymer obtained with each of these comonomers were low, varying from 3 to 8% for vinyl acetate to 8 to 18% for acrylonitrile. Both solution and emulsion polymerization systems were utilized.

Copolymers of DVE and dimethyl maleate were also prepared in both solution and emulsion systems. The largest conversions were obtained in an emulsion system using a persulfate-metabisulfite initiator system. At conversions of 60% or less the copolymer was soluble in hexamethyl phosphoramide. Inherent viscosities as high as 0.80 were obtained. IR spectra of these copolymers showed somewhat more aliphatic unsaturation than the DVE-MA copolymers.

Additional work on cyclocopolymerization included the monomer pair, dimethyldivinylsilane and maleic anhydride.[6] In a parallel study, copolymerization of trimethylvinylsilane and maleic anhydride was also investigated. Consistent with the theory that donor-acceptor complexes are involved in certain copolymerizations, it was shown in this work that both of the vinyl silanes studied formed complexes with maleic anhydride. The stiochiometric composition of these complexes was shown to be 1:1 molar. The equilibrium constants for complexation of trimethylvinylsilane and dimethyldivinylsilane with maleic anhydride were determined to be 0.061 and 0.107 l/mole respectively. The data for the copolymerization study of trimethylvinylsilane and maleic anhydride confirmed that alternating copolymers were formed, and that the maximum rate of polymerization occurred at monomer mole fraction of each monomer = 0.5; this was also the concentration at which the concentration of charge-transfer complex was maximum. Intrinsic viscosities for representative samples were in the range of 0.08 dl/g (acetone, 30°).

Data for the copolymerization of dimethyldivinylsilane and maleic anhydride confirmed that cyclopolymers were formed in the molar ratio of 1:2 with respect to silane and maleic anhydride, and that this copolymerization was 5 to 15 times as fast as the corresponding trimethylvinylsilane copolymerization. Intrinsic viscosities for representative samples were in the range of 0.07 to 0.13 dl/g (acetone 30°).

In an extension of the work on divinyl ether copolymerizations, the copolymerization possibilities of furan and maleic anhydride were investigated.[7] It appears that prior to this effort, others had assumed that the driving force toward the Diels Alder reaction between furan, an electron rich diene, and maleic anhydride, an excellent dienophile, was so great that copolymerization of the pair could not be accomplished. The electronic and structural similarities between furan and divinyl ether prompted this investigation of the copolymerizability of this comonomer pair. It was shown that copolymerization occurred readily in presence of azobisisobutyronitrile, that the rate of copolymerization was maximum at monomer mole fractions of 0.5, and that 1:1 alternating copolymers were obtained. These results are consistent with participation of a donor-acceptor complex which was shown to exist, with a maximum absorption of 291 nm, and an equilibrium complex of formation of 6.9×10^{-2}. The structure of the copolymer involved 1,4-addition by furan and 1,2-addition by maleic anhydride.[8] Further experiments also showed that the Diels Alder adduct could be converted to copolymer of the same structure under the same conditions that the monomer pair could be copolymerized. These results suggested that a common intermediate may be formed in both the Diels Alder adduct formation and the copolymerization.

Further modification of the structure of the divinyl ether-maleic anhydride and similar copolymers by post reaction of the preformed copolymers has been carried out. Conversion of the divinyl ether-maleic anhydride copolymer to a variety of imides could be readily accomplished with certain substituted amines.[9] Thus, it was possible to modify the polymer structure, either through partial or total conversion of the anhydride units to selected imide units, which may also possess other functional groups capable of enhancing the effectiveness of the material as an antitumor agent. Other physiological characteristics of the material could be modified in a similar manner by careful selection of coreactant. Certain substituted amines, however, did not lend themselves to conversion to the cyclic imide. For example, 2-aminothiazole appeared to be readily converted to the half-amide; however, all efforts to close the cyclic imide ring were unsuccessful.

Copolymers of divinyl ether with maleimides of certain α-amino acids have also been prepared.[10] Copolymers prepared and studied were those of divinyl ether with N-carbethoxymaleimide, maleoylglycine, maleoyl-DL-alanine, maleoyl-DL-phenylalanine, maleoyl-L-phenylalanine, maleoyl-DL-methionine and maleoyl-DL-leucine. These copolymers were characterized by elemental analyses to establish the comonomer ratios, by viscosity measurements, and by determination of molecular weights by membrane osmometry after esterification with diazomethane. Polymer conversions ranged from 75 to 97%. Inherent viscosities of the carboxyl-containing copolymers ranged from 0.06 to 0.33 as sodium salt in 1M NaCl solution. The copolymer of divinyl ether and maleoyl-DL-alanine with an inherent viscosity of 0.27 had a number average molecular weight of 26,000 after methylation with diazomethane. The copolymer with N-carbethoxymaleimide with an inherent viscosity of 0.19 in acetone had a number average molecular weight of 170,000. These comonomer pairs also exhibited donor-acceptor character.

A general copolymer equation was developed for the cyclocopolymerization of 1,4-dienes and monoolefins.[11] The kinetic scheme considered was as follows (Equation 8-2):

$$
\begin{array}{ll}
m_1^{\cdot} + M_1 \xrightarrow{k_{11}} m_1^{\cdot} & m_3^{\cdot} + M_2 \xrightarrow{k_{32}} m_2^{\cdot} \\
m_1^{\cdot} + M_2 \xrightarrow{k_{12}} m_3^{\cdot} & m_c^{\cdot} + M_1 \xrightarrow{k_{c1}} m_1^{\cdot} \\
m_3^{\cdot} \xrightarrow{k_c} m_c^{\cdot} & m_c^{\cdot} + M_2 \xrightarrow{k_{c2}} m_2^{\cdot} \\
m_3^{\cdot} + M_1 \xrightarrow{k_{31}} m_1^{\cdot} & m_2^{\cdot} + M_1 \xrightarrow{k_{21}} m_1^{\cdot} \\
 & m_2^{\cdot} + M_2 \xrightarrow{k_{22}} m_2^{\cdot}
\end{array}
\qquad (8\text{–}2)
$$

The following differential copolymer composition equation, which may be applied by putting $n \approx [m_1]/[m_2]$, the fractional ratio of monomers combined in the copolymer at low conversions, was obtained (Equation 8-3).

$$
n = \frac{(1 + r_1 x)\{1/[M_2] + (1/a)(1 + x/r_3)\}}{(1/a)\{(x/r_3) + (r_2/x) + 2\} + (1/[M_2])\{1 + (1 + r_2/x)(1 + r_c x)^{-1}\}} \qquad (8\text{–}3)
$$

A similar equation was derived which related the relative rate of addition of diene and the rate of cyclization, as follows

$$
\begin{aligned}
d[M_1]/d[M_c] &= [(K_{11} + K_{12})/K_{12}k_c](K_{31} + K_{32} + k_c) \\
&= (r_1 x + 1)\{([M_1]/r_3 a) + ([M_2]/a) + 1\}
\end{aligned} \qquad (8\text{–}4)
$$

Equation 8-4 applies at low conversions, where $d[m_1]/d[m_2] \approx [m_1]/[m_2]$, the ratio of the total fraction of diene (unsaturated and cyclic) to the fraction of diene in cyclized units, in the copolymer.

If precise analytical methods are available for determining both the total fraction of diene in the copolymer and the fraction of either cyclic units or pendant vinyl groups, then, by making a series of such measurements for different initial monomer feed compositions, values for r_1, r_2 and a could be obtained from Equation 8-4. Then the remaining two parameters r_2, and r_c, could be obtained from Equation 8-3.

In certain special cases Equation 8–3 may be approximated to simpler forms, as in the following examples.

(a) If $k_c >> k_{32}$ so that a is very large and cyclization is the predominant reaction of the radicals m_3, then Equation 8–3 gives

$$n = (1 + r_1x)(1 + r_cx)/[r_cx + (r_2/x) + 2] \tag{8–4a}$$

This is equivalent to considering the addition of monoolefin to diene radicals to be a concerted bimolecular step proceeding through a cyclic transition state and producing the cyclic repeating unit.

(b) If in addition there is a strong alternating tendency so that $(r_1,r_2,r_c) \rightarrow 0$, then Equation 8–4a reduces in the limit to (8–4b).

$$n = 1/2 \tag{8–4b}$$

This predicts an alternating copolymer composition of 2:1 m in contrast to 1:1 for the similar limiting case of the classical binary copolymer composition equation.

(c) If the diene has a negligible tendency to add to its own radicals and $r_1 \approx r_c, \approx 0$, and there is also predominant cyclization, then Equation 8–4a gives

$$n = 1/[r_2/x) + 2] \tag{8–4c}$$

A plot of 1/n against 1/x should be linear with a slope r_2 and an intercept 2.0.

The application of each of the special cases of the theory was shown for experimental systems. Divinyl ether with both maleic anhydride and N-phenylmaleimide and 1,4-pentadiene with maleic anhydride were found to correspond to special case (b). The divinyl ether-acrylonitrile system was found to correspond to special case (a), with $r_1 = 0.024$, $r_2 = 0.938$, and $r_c = 0.017$. The 1,4-pentadiene-acrylonitrile system was found to correspond to special case (c), with $r_1 \approx r_c \approx 0$ and $r_2 = 1.13$.

In order to further support the relevance of the general copolymer composition equation for cyclocopolymerization, copolymerizations of (a) 3,3-dimethyl-1,4-pentadiene (M_1) with acrylonitrile (M_2) and (b) divinyl sulfone (M_1) with acrylonitrile (M_2) were studied.[12] The reactivity ratio parameters for system (a) were: $r_1 \approx r_c \approx 0$, $r_2 = 3.31$. For system (b), $r_1 = 0.364$, $r_2 = 1.94$, and $r_c = 0.067$. The copolymers of system (a) were always richer in M_2 than the feed. The intrinsic viscosity $[\eta]$ of a copolymer of system (a) having a mole fraction of $M_2 = 0.97$ was 0.71 dl/g. A monomer feed having mole fraction of $M_2 = 0.67$ led to copolymer having M_2 mole fraction $= 0.88$. System (a) was found to fit the conditions of special case (c) above [See Equation 8–4c], thus confirming that the diene has a negligible tendency to add to its own radicals, but that there is predominant cyclization.

For system (b), the divinyl sulfone (M_1) appeared to exhibit chain transfer properties as the copolymers having high M_1 content were of very low intrinsic viscosity. For example, when M_2 in copolymer $= 0.35$, $[\eta] = 0.08$ dl/g; however, when M_2 in copolymer $= 0.93$, $[\eta] = 0.64$ dl/g. System (b) was found to fit the requirement for special case (a) (See Equation 8–4a) which confirms that cyclization predominates over competitive reactions such as branching or crosslinking; this is equivalent to considering the addition of monoolefin to diene radical to be a concerted bimolecular step proceeding through a cyclic transition state and producing the cyclic repeating unit.

Further studies on the divinyl ether (M_1)-fumaronitrile (M_2) [system (a′)] system have now been reported.[13] Also studied was the copolymerization of divinyl ether (M_1)

with tetracyanoethylene (M_2) [system (b′)] and with 4-vinylpyridine (M_2) [system (c′)]. Over a wide range of initial monomer composition, the mole fraction of M_2 in the copolymer of system (a′) was in the range of 0.55–0.63, and the copolymers contained only 2–3 % unsaturation, indicating a high degree of cyclization. The composition of the copolymer of system (c′) was in the range of 0.85 and 0.99 mole fraction of M_2. The copolymers of system (a′) were found to correspond to special case (b) of the general copolymer equation. The copolymers of system (b′) were found to correspond to the Alfrey-Price equation, since apparently little cyclization occurred. Reactivity ratios for this system were: $r_1 = 0.23$, $r_2 = 0.12$. System (c′) in this study was found to most closely correspond with special case (c) of the general copolymer equation with $r_1 \approx r_c \approx 0$, and $r_2 = 32.0$.

The differences in polarity, Δe between the diene and the monoolefin have been correlated with the extent of cyclization in copolymerization of 1,6-dienes with olefins.[13] As shown in Table 8–2, the ratio of the cyclization rate constant of the 1,6-diene to the rate constant of the addition of the 1,6-diene to the monoolefin decreased as the value of Δe decreased.

These results suggested that the relative polarities of the monomers and radicals involved in cyclocopolymerization have an important bearing on the structure of the copolymer and the rate of copolymerization. For example, it was pointed out that as the values of e, the measure of polarity of the monomer increased, both the copolymerization rate and the extent of incorporation of divinyl ether into the copolymer increased. p-Chlorostyrene, 4-vinylpyridine, acrylonitrile, fumaronitrile, and tetracyanoethylene were compared. The ease with which divinyl ether added to the monoolefins increased from 4-vinylpyridine to tetracyanoethylene. p-Chlorostyrene led only to the homopolymer. These results are shown in Table 8–3.

The same order of reactivity was observed when the copolymerization rates were compared as shown in Table 8–4.

In the same way, the percentage of cyclization was related to the polarity difference of the two monomers. It was observed that the percentage of cyclization decreased when the difference between the two monomers decreased as shown in Table 8–5.

However, the failure of the highly sterically hindered tetracyanoethylene radical to cyclize may be the result of the steric factor rather than strictly a polarity factor.

Radical initiated copolymerization studies of N,N-divinylaniline with styrene, p-methylstyrene, methyl methacrylate, acrylonitrile, and vinyl acetate have been reported.[1-89] With the exception of the vinyl acetate case, the mole fraction of divinylaniline in the copolymer was less than that of the monomer feed. The Q and e values for this monomer were found to be 0.15 and -1.6 respectively. For comparison, copolymerization of divinyl sulfone and divinyl ether with methyl methacrylate and styrene was studied. The Q and e values for divinyl sulfone were found to be 0.14 and 1.4 respectively, and those for divinyl ether were 0.04 and -1.3. The higher Q values of the divinyl monomers compared with those of the corresponding monovinyl monomers were postulated to be the result of direct interaction of neighboring double bonds. N,N-Divinylaniline showed a chain transfer constant of $C = 0.034$ in methyl methacrylate copolymerization and $C = 0.013$ in styrene copolymerization. A copolymer of divinylaniline with styrene containing 17% divinylaniline had an inherent viscosity ($\ln \eta_{rel/c}$) of 0.36. A copolymer of divinylaniline with methyl methacrylate containing 50 % divinylaniline had an inherent viscosity of 0.15 and a molecular weight of 10,000–22,000.

Table 8–2 Effect of Polarity Difference of Comonomers on Cyclization of 1,6-Diene

Comonomer	e	Δe	k_c/k_{12}[a]
Acrylic anhydride	1.2		
		0.2	7.1
Methyl methacrylate	1.0		
Styrene	−0.8		
		2.0	0.25
Acrylic anhydride	1.2		

[a] Ratio of the cyclization rate constant of the 1,6-diene to the rate constant of the addition of the 1,6-diene to the monoolefin.

Table 8–3 Relative Composition of the Copolymers ($[M_1] = [M_2] = 0.5$)

Monoolefin	M_1[a]	e[b]
p-Chlorostyrene	0.0000	−0.3
4-Vinylpyridine	0.025	−0.19
Acrylonitrile	0.251	+1.11
Fumaronitrile	0.393	1.2
Tetracyanoethylene	0.523	1.2

[a] m_1 is the molar fraction of divinyl ether in the copolymer.

[b] Value of e with respect to styrene.

Table 8–4 Relative Copolymerization Rates

Monoolefin[a]	Temperature (°C)	Reaction time (min)	Conversion (%)
4-Vinylpyridine	60	80	3.68
Fumaronitrile	60	90	8.56
Tetracyanoethylene	50	60	11.56

[a] Initial feed monomer molar fraction: $[M_1] = [M_2] = 0.5$.

Table 8–5 Effect of Polarity Difference of Comonomers on Cyclization in 1,4-Diene

Monoolefin[a]	Cyclization (%)
Acrylonitrile	100
Fumaronitrile	96.3
Tetracyanoethylene	18.4

[a] Initial monomer concentration: $[M_1] = [M_2] = 0.5$.

More recently, high molecular weight copolymers of divinyl ether and maleic anhydride have been reported.[1-90] A copolymer prepared at 1.32 moles/l total monomer concentration in benzene, using 0.25% (based on monomer) of dichlorobenzoyl peroxide as initiator, at 53 % conversion, had an inherent viscosity of 1.69 at 0.4 g per 100 ml in DMF. The mole fraction of maleic anhydride in the copolymer was found to be 0.67. Additional results on divinyl ether copolymers with dimethyl maleate, acrylonitrile and vinyl acetate, divinyl sulfone copolymers with maleic anhydride, dimethyl maleate, acrylonitrile and vinyl acetate, and 1,4-pentadiene copolymers with maleic anhydride and vinyl acetate were included. Only in the case of the divinyl ether-maleic anhydride copolymer were satisfactory inherent viscosities reported.

Copolymerization studies of divinyl ether with N-substituted maleimides having the following substituents on nitrogen: (a) hydrogen, (b) methyl, (c) ethyl, (d) n-propyl, (e) isopropyl, (f) n-butyl, (g) isobutyl, (h) benzyl, (i) α-naphthyl, (j) phenyl, and (k) p-methoxyphenyl have been reported.[14] A copolymer of divinyl ether and isobutyl maleimide, prepared and taken to 80% conversion in benzene as solvent at 1.5 molar total monomer concentration (1.0 molar in maleimide, 0.5 molar in divinyl ether) at 50.0°C, using a concentration of 2.4×10^{-2} moles/l of azobisisobutyronitrile as initiator had a molecular weight of 12,600 as determined by vapor pressure osmometry, an intrinsic viscosity of 0.60 dl/g (benzene), and a melting point of 200–205°.

The cyclocopolymerization of divinylphenylphosphine with acrylonitrile has been reported and evidence presented that this 1,4-diene undergoes cyclocopolymerization in accordance with the proposed mechanism.[2-341]

Cyclocopolymerization of benzaldehyde with divinyl ether by cationic initiation has been described.[7-184] The cyclocopolymer contained both structural units at monomer mole ratio 1:1 and at 2:1. Polymers prepared at low benzaldehyde contents also contained units with pendant double bonds.

It has been shown that divinyl ether and divinyl sulfone undergo spontaneous copolymerization at room temperature to yield a copolymer having the bicyclic structure (8–1) and to have $[\eta] = 0.192$ (DMF).[15] Some solvent and steric effects have been

(8–1)

described in a study of cyclocopolymerization of 1,4-dienes with a variety of olefins.[16] Tricyclo-copolymers have been obtained via radical cyclopolymerization of maleic anhydride with 5-ethylidene-, 5-methylene-, and 5-vinylnorbornene.[17]

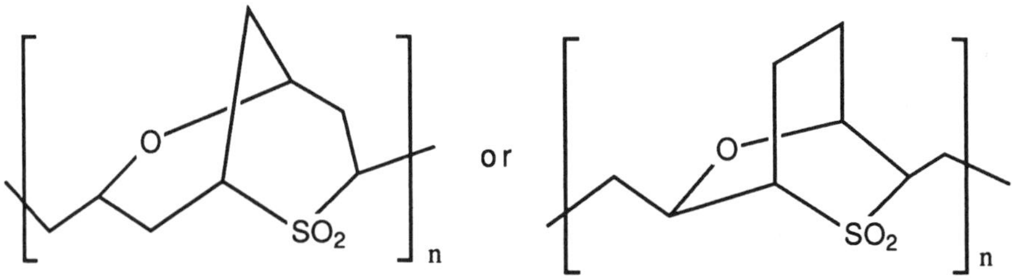

Further evidence for the participation of a molecular complex in the cyclocopolymerization of 1,4-dienes with suitable olefins has been presented.[18] For example, a cyclic 2:1 maleic anhydride-divinyl ether copolymer was obtained in 8.5% yield by heating the monomers at 60° for 18 hrs. in the absence of a radical initiator, indicating that the molecular complex initiated the reaction. A 1:1 cyclic citraconic anhydride-divinyl sulfone copolymer was also prepared in 11% yield by treating the anhydride with the sulfone and azobisisobutyronitrile in dioxane at 45° for 26 hrs.

The radical initiated cyclocopolymerization of divinyl ether with chloromaleic anhydride was studied in an effort to determine the stereochemistry of the anhydride unit in the copolymer.[19] The ease of loss of hydrogen chloride from the copolymer indicated incorporation of the anhydride *via* a *trans* addition.

Some physical properties of the cyclopolymer of divinyl ether and maleic anhydride have been published.[20] Intrinsic viscosities were determined in two solvents. The weight average molecular weight was determined by light-scattering measurements and the number average molecular weight was determined by osmometry. Sedimentation velocity indicated that the chain length distribution was much broader than random, which was in agreement with the ratio of the two molecular weight determinations. The polymer was fractionated by precipitation from acetone with hexane. Addition of sodium tetraphenylboron was necessary to obtain separation.

A description of the original experiments which led to postulation of the cyclocopolymerization mechanism as a means of accounting for the failure of 1,4-dienes and olefins to yield crosslinked polymers has been published.[1-16]

Cyclopolymers of divinyl ether and tetrahydronaphthoquinone have been prepared, in which copolymer structures were (8–2) obtained.[21] The cyclocopolymerization of

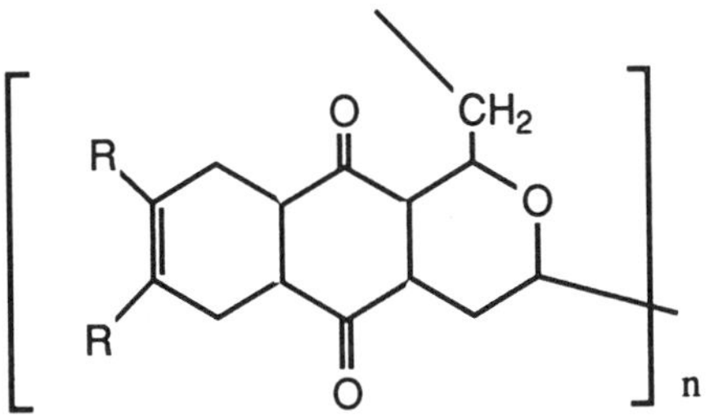

R = H, CH_3

(8–2)

divinyl ether and several substituted maleic anhydrides has also been studied.[22] Regular cyclocopolymers having 1:1 or 1:2 (divinyl ether: anhydride) compositions were obtained. Acrylonitrile spontaneously copolymerized with divinyl ether, in the presence of zinc chloride, with a strong tendency to form a 1:2 alternating cyclocopolymer (8–3).[23] Other Lewis acids as complexing agents, e.g., triethylaluminum were also studied.

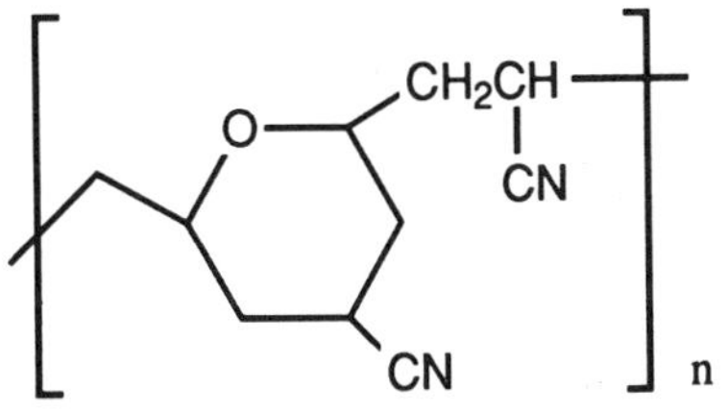

(8–3)

When acrylonitrile, methacrylonitrile, 2-vinylpyridine and 4-vinylpyridine were copolymerized with divinyl ether in presence of one of the complexing agents, both the rate of copolymerization and alternating tendency of the cyclopolymer increased with the amount of complexing agent.[23] Identification of donor-acceptor complexation between divinyl ether and bis(acrylonitrile)-zinc chloride complex was made. Cyclocopolymers have been obtained also by copolymerizations of divinyl sulphone with ethyl vinyl ether or dihydropyran,[24] divinyl ether with furfural,[5-69] and 1,6-heptadiene with sulphur dioxide.[7-120]

The use of synthetic polymers as biologically active materials is increasing. Among those polymers receiving most attention by investigators has been the alternating cyclopolymer of divinyl ether and maleic anhydride, incorrectly and inadvertently known as "pyran copolymer" or preferably, DIVEMA. This copolymer has been investigated extensively for a variety of biological effects since the initial observation by scientists at the National Cancer Institute that the hydrolyzed and neutralized copolymer had considerable antitumor activity along with a much reduced toxicity in comparison with other polyanions investigated. DIVEMA has also been shown to be an interferon inducer, to possess antiviral, antibacterial, antifungal, anticoagulant and antiarthritic activity, to aid in removing polymeric plutonium from the liver, to inhibit viral RNA-dependent DNA polymerase (reverse transcriptase), and to activate macrophages in affecting the immune response of test animals.[2] The interferon-inducing capability was initially observed during a clinical investigation of DIVEMA as an antitumor drug. The importance of molecular weight (MW) and molecular weight distribution (MWD) has been emphasized. A biphasic response of the reticuloendothelial system to the copolymer drug had been observed which led to the postulate that the copolymer may consist of a toxic molecular weight fraction, and another fraction of lower toxicity. Other evidence for molecular weight dependence was based upon the observation that antiviral activity of the copolymer required a higher molecular weight than for the antitumor activity. The latter was optimum with low molecular weight samples of narrow molecular weight distribution. These results led to development of a method for synthesizing a copolymer of narrow MWD by photochemically initiating solution polymerization in acetone, with tetrahydrofuran as a chain-transfer agent. This method gave samples having MWD (M_w/M_n) ranging from $1 \cdot 6$ to $2 \cdot 6$, in contrast to $3 \cdot 7$ to $7 \cdot 7$ for earlier prepared samples.

Recent progress in the biological applications of DIVEMA has been reviewed.[25] The biological activity of polymers structurally related to DIVEMA has also been reviewed.[1-171] A modified procedure has been described for synthesis of the 2:1 copolymer of maleic anhydride and divinyl ether for study as a tumor inhibitor.[26]

Linear polymers having cyclic repeat units in the chain were synthesized by copolymerizing 1,4-dienes with suitable alkenes in the presence of radical initiators.[27] Monomer pairs were made up of symmetrical electron-donor dienes and electron-acceptor alkenes, or symmetrical electron-acceptor dienes and electron-donor alkenes. Suitable unsymmetrical 1,4-dienes were also used.

In a continuing study of the mechanism of cyclocopolymerization, solvent effects on the equilibrium constant for donor:acceptor complex formation between maleic anhydride and divinyl ether were studied.[28] The constant increased with polarity of solvent as did the rate of polymerization and MW. Solid, cyclic 1:2 divinyl ether-maleic anhydride copolymer of number average molecular weight (M_n) 5,000–30,000 and weight average (M_w)/number average molecular weight ratio = 1.5–2.5, prepared by benzoyl peroxide initiated copolymerization of the monomers in benzene-carbon tetrachloride mixed solvent, after fractionation on a sand column, was effective in treating tumors and viruses in mice, but did not have the undesirable side effects of the same copolymer having MW and MWD properties outside the range referred to above.[29]

Investigation of the molecular structure and biological activities of DIVEMA and modification of its structure have continued.[30] Incorporation of 5-fluorouracil into the copolymer led to a polymer which exhibited antileukemia activity. Results of these investigations are reported later.

A modified form of DIVEMA of controlled MW and MWD (^{14}C-MVE-2) was studied from the standpoint of toxicity and tissue distribution when administered intraperitoneally as an alternative to the usual intravenous route of administration.[31] The results indicated the alternative route of intrapertioneal administration to be superior in its overall positive effects.

The cyclocopolymer of divinyl ether and citraconic anhydride was resynthesized and fractionated for the purpose of optimizing its antitumor activity with respect to MW and MWD.[32] An optimum fraction resulted in an increase in survival times of treated vs. control animals of 267% against P-388 lymphocytic leukemia.

The cyclocopolymerization of divinyl ether and maleic anhydride was investigated in terms of mechanism and control of the ratio of five- and six-membered ring structures.[33] The content of six-membered ring increased with increasing copolymerization temperature and with decreasing concentration of monomers, indicating a reversible cyclization process controlled thermodynamically at higher temperatures and kinetically at lower temperatures. The six-membered ring content could be controlled at 46–79%.

An extensive study of cyclocopolymerization has been conducted. An ^{1}H and ^{13}C NMR study of the structure of DIVEMA led to the proposal that the polymer consists of a six-membered ring in chair-like conformation with predominantly trans geometry in the anhydride ring (See Equation 8–1).[34]

Propenyl isopropenyl ketone and vinyl isobutenyl ketone were copolymerized with acrylonitrile, vinylidene chloride and 2-methyl-5-vinylpyridine in bulk in presence of a radical initiator.[35] The copolymers were soluble in a variety of organic solvents. Cyclization took place with formation of cyclopentane rings. A variety of β-aryl substituted divinyl ketones were polymerized via radical initiation.[36] All polymers were soluble in organic solvents, and evidence was reported to support a cyclization mechanism for the polymerization. A homopolymer of dibenzalacetone of molecular weight 25,000, soluble in benzene, acetone, chloroform, and containing no unsaturation was obtained in 23% yield by prolonged heating of the monomer with di-tert-butyl peroxide.[4-55] The polymer structure was consistent with the cyclocopolymerization mechanism.

The optimal conditions for the cyclocopolymerization of butadiene and styrene were 210° at 60 atmospheres, 2.7:1 mole ratio of butadiene: styrene, and contact time of six hours.[37] Also obtained were the Diels-Alder adducts, phenylcyclohexene and vinylcyclohexene. The structures of the cyclocopolymers were not included.

In a continuing study of radical polymerization of tetravinylsilane, copolymerization of this monomer with methyl acrylate was described.[38] The total rate of the copolymerization was indirectly related to the concentration of the silane component. The reactivity ratios of the comonomer pair were reported.

Participation of Electron-Donor Dienes and Electron-Acceptor Alkenes in Cyclocopolymerization

Examples, previously discussed, of cyclopolymerizations and copolymerizations involving monomers which undergo cyclopolymerization have involved ring closures in such a manner that all members of the cyclic structure were derived from a single monomer.

In 1958, an example of a bimolecular alternating intramolecular-intermolecular copolymerization was reported.[1-16, 1-87]

It was shown that divinyl ether copolymerized with maleic anhydride in such a manner that a six-membered cyclic structure was formed in which the divinyl ether contributed four members and the maleic anhydride contributed the remaining two. (See Equation 8–1) Because of the reluctance of either monomer to undergo free-radical homopolymerization, and the marked tendency of both to undergo copolymerization with each other, the cyclic copolymer appears to be the preferred product. Divinyl ether is known to be an electron-donor diene and maleic anhydride is an electron-acceptor alkene. This section deals with cyclocopolymerization involving pairs of monomers which fit this pattern.

Perhaps the most extensively studied example of this unusual type of copolymerization is the cyclocopolymer of divinyl ether and maleic anhydride. This copolymer (DIVEMA) has been extensively studied for its biological properties; it exhibits a broad spectrum of biological activities.[1-171]

Cyclocopolymerization of a wide variety of comonomers has been studied, and it has been concluded that cyclocopolymerization is most likely to occur when one monomer possesses strong electron-donor properties and the other strong electron-acceptor properties.[1-171, 1-178] Thus, it appears reasonable to suspect that an intermediate donor-acceptor complex plays a significant role in controlling the regularly alternating cyclocopolymerization.

The participation of donor-acceptor complexes in the radical polymerization of maleic anhydride with divinyl ether has been evaluated by a method based on polymerization rate constants experimentally determined at two different total monomer concentrations and two different monomer concentration ratios. The participation of the complex in the copolymerization was only 1–17% despite a significantly higher complex reactivity compared to monomer reactivity.[39]

Cyclopolymerization of d-limonene with maleic anhydride in tetrahydrofuran has been reported recently. The composition of the copolymer showed that d-limonene readily underwent cyclocopolymerization giving 1:2 alternating copolymers which were optically active.[7-19]

Evidence for the participation of a donor-acceptor complex in the cyclocopolymerization of 1,4-dienes with alkenes was presented.[2-341] The existence of such complexes in the following comonomer pairs: divinyl ether-maleic anhydride, divinyl ether-maleimide, divinyl ether-fumaronitrile, and dihydropyran-maleic anhydride was substantiated. The complex was shown by the appearance in each case of a new absorption band in the ultraviolet spectrum of a mixture of the two compounds. The equilibrium constants for complex formation were determined and were reported as follows (in the same order as the above presentation): 0.014, 0.037, 0.015, and 0.02. Reasonable structures for the intermediate complexes were postulated based upon the structure of the product in accord with general practice in investigations dealing with such complexes.[40–42]

Additional studies of cyclocopolymerization of dimethyldivinylsilane with maleic anhydride have been published.[6] This system has been related to the theory that a complex may be involved. The equilibrium constant for complexation of dimethyldivinylsilane with maleic anhydride was determined by NMR to be 0.107 l/mole. Consistent with previous cyclocopolymerizations postulated to proceed through complexes, the complex between dimethyldivinylsilane and maleic anhydride was shown to be a 1:1 molar complex; however, the copolymer contains maleic anhydride in a 2:1 molar ratio. The postulate that the complex undergoes a 1:1 alternating copolymerization with maleic anhydride can thus account for the structure of the product. The maleic anhydride content of the copolymer over a wide range of comonomer feed ratios was found to vary between molar ratios of 0.660 and 0.672. The copolymer had an intrinsic viscosity of 0.13 dl/g at 30°C.

Also studied was the copolymerization of trimethylvinylsilane with maleic anhydride. The monomer pair formed a 1:1 molar complex. The equilibrium constant for complexation in this case was found by NMR to be 0.061 l/mole as compared to 0.107 for the complex of maleic anhydride with dimethyldivinylsilane. A comparison of these constants with those of the ethyl vinyl ether-maleic anhydride complexation constant, 0.041 l/mole, and the divinyl ether-maleic anhydride complexation constant, 0.036 l/mole, suggested that the dπ-pπ bonding in the silicon compounds may lead to a decrease of the electron density of the double bonds of the vinylsilanes. The λ_{max} values for these four complexes are shown in Table 8–6.

It was concluded that the ionization process requires less energy in the case of the vinyl ethers in comparison with the silanes since for a series of complexes of different donors with the same acceptor, λ_{max} is linearly related to the ionization potential of the donor. This can be understood by the fact that the double bonds of the vinyl ethers are more electron-rich due to delocalization of the nonbonding pairs of the oxygen atom towards the vinyl group. The opposite effect would be expected from the dπ-pπ bonding in the vinylsilanes.

The composition of the trimethylvinylsilane-maleic anhydride copolymer was shown to be 0.5 mole fraction of each monomer, and to have an intrinsic viscosity of 0.08 dl/g at 30°C. The rate of copolymerization of this monomer pair was found to be maximum at a monomer feed ratio of 0.5 mole fraction of each monomer, consistent with the theory that complex participation had occurred. In contrast, the maximum rate of copolymerization of the dimethyldivinylsilane-maleic anhydride system occurred at a monomer feed ratio of 0.67 mole fraction of maleic anhydride; however, the maximum rate in the cyclocopolymerization was more than four times the maximum rate for the simple copolymerization system. These observations are consistent with previous comparisons of relative rates of cyclopolymerizations in general and the noncyclization counterpart.

Table 8–6 Comparison Between Vinylsilanes and Vinyl Ethers Complexes with Maleic Anhydride

Complex	λ_{max} (mμ)	K (l/mole)
Me_3VinSi-MA	< 250	0.061
Me_2Vin_2Si-MA	< 250	0.107
EVE-MA	277.5	0.041
DVE-MA	278	0.036

A decrease in both the rate of copolymerization and the nitrogen content of the copolymer of divinyl ether with fumaronitrile upon dilution has been observed.[43] The dilution effect was interpreted to be the result of the decreased concentration of the complex. The role of complex participation in cyclocopolymerization is presented in detail below.

The copolymers of divinyl ether with maleic anhydride obtained in a study using dichlorobenzoyl peroxide as initiator were found to have much higher inherent viscosities than those copolymers previously reported.[4] The copolymers were soluble in concentrated H_2SO_4 and DMF and dissolved with reaction in dilute aqueous alkali, methanol, and ethanol. Inherent viscosities of 1.7 were obtained for DMF solutions of these copolymers. Concentrated solutions prepared in DMF were extruded into water-DMF solutions from a hypodermic syringe to give fibers and cast on glass plates to give films.

Solutions could be prepared in 97% H_2SO_4, which showed no viscosity change on standing at room temperature for five days. The copolymer could be recovered from this solvent apparently unchanged. These copolymers do not melt but softened at about 230°C and began to decompose at 290 to 300°C.

Several derivatives of this copolymer were prepared and characterized. Treatment of a copolymer, (inherent viscosity of 1.31 in 97% H_2SO_4) with hot water gave a polyacid which was soluble in water and ethanol. The polyacid had an inherent viscosity of 1.51 in ethanol and 1.76 in water. This derivative did not melt but began to decompose at 290 to 300°C.

Half-ester copolymers were prepared by reaction of the copolyanhydride with ethanol or methanol. Both of these derivatives melted with decomposition between 285° and 300°C and gave clear films and fibers from concentrated solutions. Elemental analyses and IR spectra indicated that these derivatives contained two ester and two acid groups per repeat unit.

Treatment of the copolyanhydride with t-butylamine gave a product which analyzed for one amide and three acid groups per repeat unit. This amide-acid was soluble in DMF, DMSO, and formic acid. The derivative, on heating to 135 to 150°C, changed from white to a deep red color. Continued heating to 210°C resulted in decomposition as evidenced by gas evolution and a loss of color.

In an effort to further characterize certain high molecular weight samples of the copolymer, use of DMF as solvent for intrinsic viscosity and osmotic pressure determinations gave unexpected results.[44] The intrinsic viscosity increased sharply at higher dilution, suggestive of polyelectrolyte behavior. During a more recent investigation to utilize gel permeation chromatography to fractionate this copolymer, it was shown that DMF does indeed react with the anhydride units of the copolymer chain to give a polyelectrolyte.[45]

The IR spectrum of the reaction product of the copolymer with dimethyl formamide showed the presence of strong carboxylate anion absorption at 1400–1600 cm^{-1}. Maleic anhydride was shown to undergo reaction with DMF in a similar manner.

As indicated in the reaction scheme, both Structures 8–4 and 8–5 could be present in DMF. When the copolymer concentration was decreased, the charges on the polymer may be separated to some extent, resulting in mutual repulsion of the same charge in the main chain, and hence its rapid expansion.[46] Because of this chain expansion at high dilution, the intrinsic viscosity of the copolymer would increase immensely, as was observed.[44]

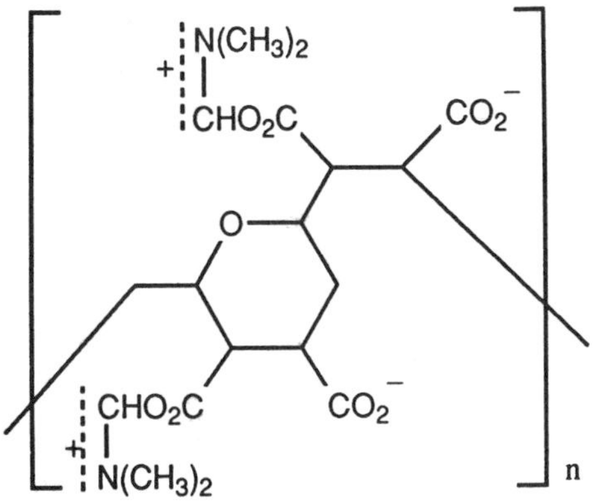

(8–4) 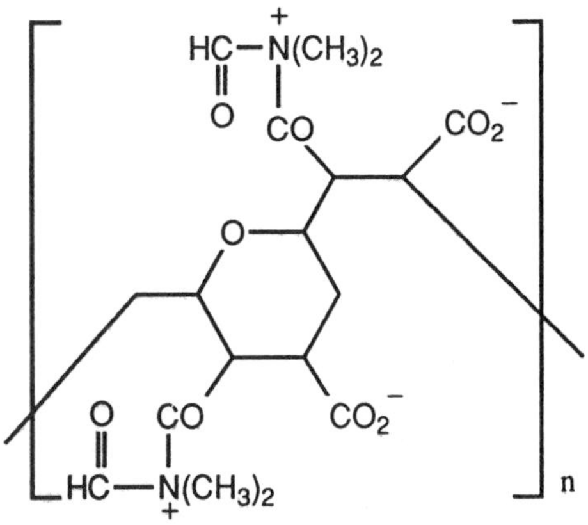 (8–5)

Attempts to fractionate the anhydride form of the copolymer made it necessary to resort to a complete methylation of the copolymer.[45] It was shown that the copolymer of divinyl ether and methyl fumarate could be fractionated on Styragel columns using a mixed solvent system of 80% acetone and 20% tetrahydrofuran. The anhydride copolymer could be converted completely to the methyl ester by the two step procedure which involved synthesis of the half-ester by reaction with anhydrous methanol followed by esterification of the remaining carboxyl groups with diazomethane. Since the molecular weight of the anhydride copolymer increased by 19.22% upon complete esterification, its original molecular weight was readily available.

Absolutely anhydrous solvents were used with the anhydride in order to prevent hydrolysis.[45] With the ester, the polymer chain tended to expand because of steric crowding of adjacent carbomethoxy groups, resulting in unusually high calculated MWs.

The MWD of DIVEMA has been studied.[44] M_w was determined by light-scattering measurements and M_n by osmometry of several samples of DIVEMA. Also, results from an ultracentrifuge sedimentation velocity pattern were reported. These results indicated an M_w/M_n ratio of four to six for most samples. The polymer was fractionated by precipitation from acetone using hexane as precipitant, with the addition of a small amount of sodium tetraphenyl boron. Nine samples were obtained having $[\eta]$ values (dl/g) varying from 0.02 to 0.49 from an original sample of $[\eta] = 0.28$ dl/g and $M_w = 300{,}000$. A sample of $[\eta] = 0.17$ dl/g had $M_n = 47{,}500$ by membrane osmometry and $M_w = 250{,}000$ for $M_w/M_n = 5.2$. Correlation of intrinsic viscosity and M_w was not obtained.

The importance of MW and MWD in biological samples of DIVEMA was emphasized by a study which showed a biphasic response of the reticuloendothelial system to DIVEMA.[47] These results led to the postulate that the copolymer might consist of two materials, one toxic, and the other less toxic.[48] It was assumed and confirmed that the toxic fraction was material of higher MW. Earlier work[49] had shown that most of a polymer prepared with ^{14}C-labeled DIVEMA was excreted by mice fairly rapidly, but a smaller portion was excreted very slowly. This obvious importance of MW had also been demonstrated in the antiviral activity of DIVEMA. The antiviral activity toward encephalomyocarditis in mice required a higher MW than for antitumor activity. The latter was optimum with low MW samples of narrow MWD.

With these results in mind, studies were initiated to develop experimental methods to fractionate suitable samples of DIVEMA and to synthesize samples of narrow MWD. Because of the ease of hydrolysis of the anhydride form of DIVEMA, problems were encountered in GPC fractionation of this structural form. Ultimately, gel permeation chromatographic (GPC) studies on the methylated form of the copolymer were resorted to and satisfactory results were obtained. A similar copolymer had been reported earlier via cyclocopolymerization of divinyl ether and dimethyl fumarate.[1-88] The results of the study led to the conclusion that the sample of DIVEMA prepared by the original slurry method[1-87] yielded a polymer with broad MWD when taken to high conversions. Thus, a method was developed for synthesizing a copolymer of narrow MWD by photochemically initiated solution polymerization in acetone as solvent and tetrahydrofuran as a chain transfer agent.[48] Nine samples of molecular weight ranging from 25,000 to 45,000 and having M_w/M_n ranging from 1.6 to 2.6 were obtained. A sample of MW = 32,500 prepared by the slurry method gave a value of 7.7 for M_w/M_n while the sample designated as NSC 46015 of MW = 22,500, and the sample approved for clinical evaluation, gave a value of 3.7 using the GPC method on the methyl ester.

In a quantitative evaluation of the structure and properties of DIVEMA the completely methylated ester (DME) of DIVEMA was synthesized[50] as a molecular probe, using procedures similar to those reported earlier.[45] By use of solution light scattering, GPC and intrinsic viscosity measurements, the conclusion was drawn that tetrahydrofuran at 30°C is a theta solvent for DME, and that DME has a random coil conformation, with possible long-chain branching at higher molecular weights. The different biological behavior of higher MW fractions than the lower material was tentatively attributed to this structural variation. Based upon the characteristic ratio, C^{∞}, for DME, which gives a measure of chain flexibility, and molecular model studies, the proposal was made that DIVEMA consisted of the five-membered tetrahydrofuran structure rather than the originally proposed six-membered tetrahydropyran structure. However, as was pointed out above, this proposal is not in accord with earlier results which showed that one of the carboxyl groups in DIVEMA is different from the others, consistent with the tetrahydropyran structure, but inconsistent with the tetrahydrofuran structure.[2]

A systematic kinetic study of the cyclocopolymerization of vinyl ethers with maleic anhydride in different solvents has been carried out.[51] Although the kinetics of formation of 1:2-divinylether-maleic anhydride and 1:1-ethylvinyl ether-maleic anhydride copolymers could be explained without intervention of donor-acceptor complexes, the complexes may participate in competing reactions. In the propagation of the divinyl ether copolymer, addition of ether is slower than addition of the first anhydride molecule, but addition of the second anhydride molecule is slower than the first.

An extensive physicochemical characterization of DIVEMA has been conducted.[52] Solution properties of the 1:2 divinyl ether-maleic anhydride copolymer were determined by IR spectroscopy, light scattering, and by determining refractive index increments and viscosity. The IR spectra indicated the hydrolysis was nearly complete in the sample as received. Light scattering data showed the weight-average molecular weight to be 146,000 and showed a tendency for the copolymer to interact with the solvent water. The polymer expanded on dilution. The root-mean-square end-to-end distance was 241 . Under physiologic conditions (pH 6.8) the intrinsic viscosity was 0.53 dl/g, which gave a root-mean-square end-to-end distance of 333°A.

Proton and ^{13}C NMR spectra of DIVEMA led to the proposal of a chair-form, six-membered ring structure with predominately *trans* geometry in the anhydride ring.[53] There was no significant change in structure with polymerization temperature. An intramolecular cyclization mechanism of cyclocopolymerization was proposed on the basis of HOMO-LUMO interaction of the comonomers and intramolecular radical addition on the preoriented double bond. This mechanism suggested the formation of the six-membered ring structure. Other evidence showed the presence of the five-membered ring structure as well.

A five-membered ring structure was assigned to DIVEMA obtained in $CHCl_3$, based on ^{13}C NMR analysis and several model compounds.[54] The copolymer obtained in a mixed solvent of acetone and carbon disulfide was predominantly of five-membered ring, but also contained other ($\approx 10\%$) repeating units. Plausible mechanisms for the cyclocopolymerization were discussed. The structure generally accepted includes both five- and six-membered rings, the ratio depending on the conditions used for synthesis of the copolymer.

Photoinitiated cyclocopolymerization studies of divinyl ether and fumaronitrile were continued.[55] Comparison of the initiation mechanism in photopolymerization with azobisisobutyronitrile initiated polymerization of the same system indicated that the rate of photopolymerization was first order in the number of polymerization-initiating excited species formed per second. In the excited state, the charge-transfer complex species, either alone or together with non-complex species, initiated polymerization, but the non-complex species alone did not initiate polymerization. The structure of a polymer obtained from maleic anhydride in presence of organic bases was studied.[56] From triethylamine, a polymer of the following suggested structure (8–6) was obtained: If this *cis*-poly(vinylene ketoanhydride) structure is correct it constitutes a divinyl ketone repeating unit, and would simulate divinyl ketone in cyclocopolymerization.

$$\left[\overset{O}{\overset{\|}{C}}CH{=}CH\overset{O}{\overset{\|}{C}}CH{=}CH\overset{O}{\overset{\|}{C}}O \right]$$

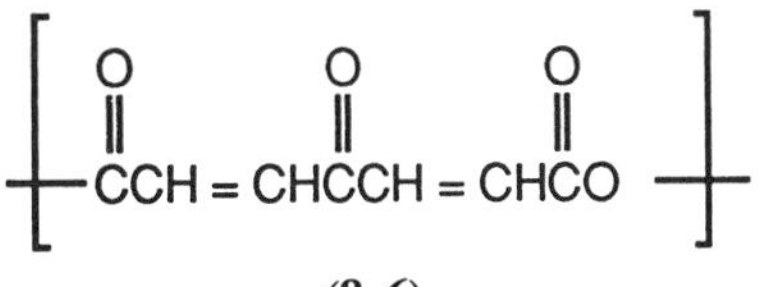

(8–6)

A soluble cyclocopolymer of 1:2 molar composition was obtained by UV irradiation of divinyl ether and fumaronitrile.[57] Calculations, based on the use of different UV filters, showed that both the donor-acceptor complex and the noncomplexed species are capable of initiating polymerization. The mechanism is similar to that of free-radical initiation. Further evidence for participation of donor-acceptor complexes in the cyclocopolymerization of the divinyl ether-fumaronitrile system was presented.[58] The fumaronitrile content of the copolymer decreased in a linear manner when the total monomer concentration was progressively lowered from 3 mole/l to 0.5 mole/l while keeping the monomer feed ratio of 1:2 constant, consistent with donor:acceptor complex participation. The role of molecular complexes in cyclopolymerization has been studied in detail as reported in a doctoral dissertation.[59]

^{13}C NMR studies of the structure of DIVEMA have led to the conclusion that the copolymer contains five-membered tetrahydrofuran rings and six-membered tetrahydropyran rings in a ratio of 0.8:1.[60]

Cationic-initiated copolymerizations of 2-thiophenealdehyde with dihydropyran and divinyl ether have been studied.[61] By use of boron trifluoride-etherate as initiator, it was shown that the cationic reactivity of this aldehyde is intermediate between benzaldehyde and furfural.

A detailed synthetic procedure for DIVEMA, {poly[2,4-(7,9-dioxo-3,8-dioxabicyclo[4.3.0]nonanediyl)-(3,4-ioxooxolanediyl) methylene]} has been published in *Macromolecular Syntheses*, along with methods of characterization and spectral evidence in support of the polymer structure.[62] The intrinsic viscosity of the sample reported was 0.75 dl/g in acetone at 30°C.

A low-molecular-weight polyelectrolyte with antitumor properties and low toxicity has been reported.[63] Carboxyimamidate ("Carbethimer") (8–7) is a derivative of an

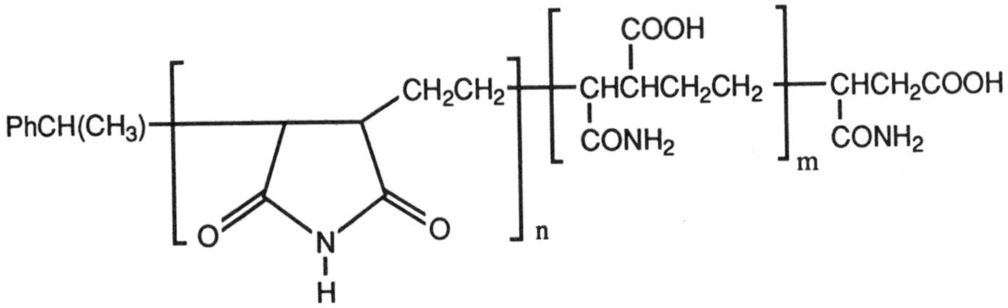

(8–7)

alternating 1:1 ethylene-maleic anhydride copolymer related in its structure to DIVEMA. The copolymer inhibited growth of Lewis lung carcinoma, Madison 109 lung carcinoma, M5076 ovarian tumor, colon carcinoma 26, B16 melanoma, and P815 mastocytoma in vivo. The antitumor activity was dose related between 30 and 2,000 mg/kg intraperitoneally, nontoxic dose levels.

A polyanion bicyclic ether polymer (8–8) was synthesized as a potential antineoplastic agent.[64] Toxicity studies were performed for various equivalent weight copolymer fractions (8–8). This polyanion was found to be less toxic than DIVEMA. The

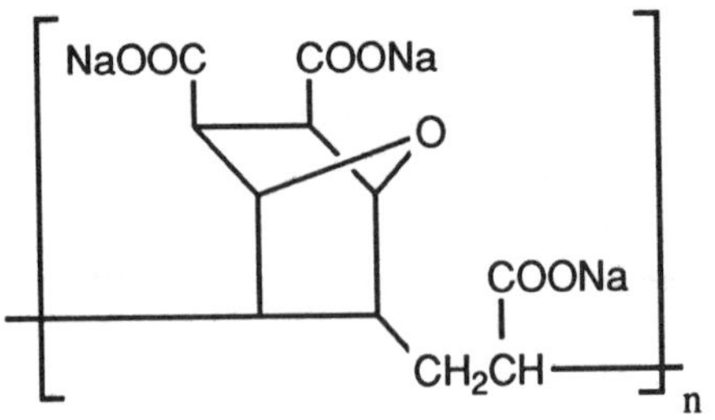

(8–8)

structure was derived by converting a 1:1 alternating copolymer of the Diels-Alder adduct of furan and maleic anhydride with acrylic acid to the corresponding trisodium salt.

Copolymerization of divinyl sulfoxide with dimercaptans by a base catalyzed mechanism gave oligomers of molecular weight 1100–1300 in 86–98% yield.[65] Oligomers with vinyl, mercapto, or vinyl and mercapto end groups could be obtained by control of the ratio of the monomers. Divinyl sulfoxide is analogous in structure to divinyl ether and divinyl sulfone; both of which have been studied in cyclocopolymerization reactions. It does not appear that this monomer has been studied as a comonomer in cyclocopolymerization.

The synthesis and results of biological testing of copolymers of divinyl ether with citraconic anhydride and itaconic anhydride were reported.[66] Also reported was the synthesis and biological evaluation of the 1:1 alternating copolymer of furan and maleic anhydride, its half amide, and the half amides of the previously mentioned cyclocopolymers with divinyl ether. The effectiveness of the copolymers was reported as the ratio of survival times of test animals vs. that of control animals in percent for the optimum dose. These values ranged from 130–150.

Other evidence for contribution of donor-acceptor complexes to the mechanism of cyclocopolymerization has been obtained. The monomer pairs DVE-(maleimides of α-amino acids) also exhibited donor-acceptor character.[10] In the study of copolymerization of 1,4-dienes with alkenes, both copolymerization of symmetrical electron-donor dienes with electron-acceptor alkenes and symmetrical electron-acceptor dienes with electron-donor alkenes were studied.[27] In a copolymerization study of tetravinylsilane with methyl acrylate, the silane was regarded as a weak electron-donor while the acrylate as a weak electron- acceptor.[38]

Electron-Acceptor 1,4-Dienes and Electron-Donor Vinyl Monomers

Only a relatively few monomer pairs in which the 1,4-diene is an electron acceptor and the alkene is an electron donor have been investigated. Donor-acceptor complexes have been identified between divinyl sulphone and vinyl ethers and it was suggested that they are involved in the cyclocopolymerization mechanism.[24]

Radical cyclopolymerization of divinyl sulfone (I) with $C_2H_5OCH{:}CH_2$ (II) or dihydropyran (III) which gave soluble copolymers richer in vinyl ether over a wide range of monomer compositions was investigated although the ethers do not homopolymerize under these conditions.[20] Because the polymers contain almost no double bonds, the polymerization was considered to be cyclopolymerization in which the diene propagates and the vinyl ether does not. The reactivity ratios for polymerization of I with II are R_1 0.02 and R_c 3.0, and for I with III, 0.02 and 5.0, respectively. Formation of 1:1 charge transfer complexes was detected, with maximum absorption in $CHCl_3$ at 23°, 39,350 and 39,050 cm^{-1} for I-II and I-III, respectively. Great deviations from the Q-e equation for binary copolymerization suggested participation of monomer complexation. However, the equation predicts the composition very well, in random cyclopolymerization of I with acrylonitrile, in which no charge transfer complex was detected.

Other electron-acceptor dienes which have been reported to participate in cyclocopolymerization reactions are the divinyl ketones referred to earlier.[35,36,4-55]

Although a variety of electron-acceptor dienes have been synthesized and studied under other conditions, 1,4-dienes such as the divinyl ketones, divinyl quaternary ammonium salts, divinyl phosphonium salts, divinyl amines, divinyl sulfoxides, and 1,4-pentadiene have not been studied extensively as comonomers with vinyl ethers, vinyl halides, styrenes, vinyldialkylamines, vinyl esters or other similar electron-donor alkenes. (See Equation 8-5) Only one system other than the divinyl sulfone-vinyl ether case that has been considered seriously appears to be the divinyl sulfone-vinyl acetate system.[1-171]

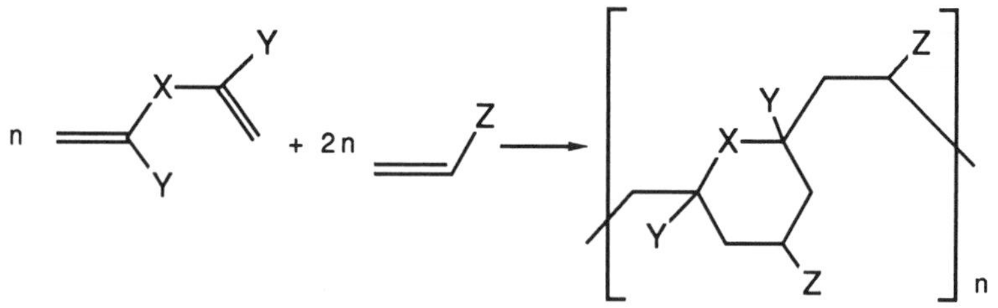

$$n\ \text{(X-linked diene bearing Y groups)} + 2n\ \text{CH}_2{=}\text{CH–Z} \longrightarrow [\text{cyclocopolymer}]_n \qquad (8\text{–}5)$$

where: x = monoatomic connector or electron acceptor functions
y = electron acceptor group, eg. CN
z = electron donor group, eg. OCH_3

Propenyl isopropenyl ketone and vinyl isobutenyl ketone were copolymerized with vinyldiene chloride and 2-methyl-5-vinylpyridine by radical initiation.[35] The copolymers were soluble and cyclization was reported to occur with formation of cyclopentane rings. These comonomer pairs represent acceptor 1,4-dienes and relatively strong electron acceptor alkenes. Copolymerization constants r_1 and r_2, polarity, and specific activity were calculated. In some cases, the MW of the copolymers decreased with decrease in the diene content. The vinylpyridine monomer was more reactive than vinylidene chloride with vinyl isobutenyl ketone.

CYCLOCOPOLYMERS OF 1,5-DIENES AND SULFUR DIOXIDE

Cyclocopolymerization of 1,4-dienes with alkenes leads to cyclization in which the cyclic structure is formed through the addition of four atoms from the diene to two atoms of the alkene. It is conceivable that the six-membered cyclic structures could be formed by the addition of other combinations which total six. It has been reported that 1,5-hexadiene, a diene capable of inserting five members into the cyclic structure, would undergo cyclocopolymerization with sulfur dioxide, a monomer capable of inserting one member into the cyclic structure (Equation 8–6).[1-17]

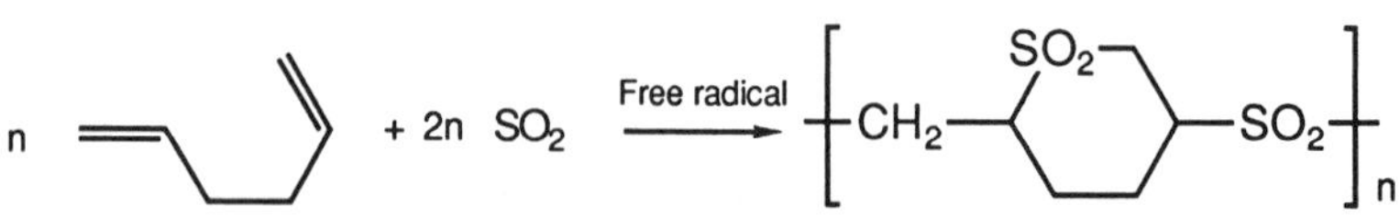

$$n\ \text{CH}_2{=}\text{CH–CH}_2\text{–CH}_2\text{–CH}{=}\text{CH}_2 + 2n\ SO_2 \xrightarrow{\text{Free radical}} \left[CH_2\text{–(cyclic }SO_2\text{ ring)–}SO_2 \right]_n \qquad (8\text{–}6)$$

This system has been studied in greater detail.[67] One sample of the copolymer had an inherent viscosity (0.25 g in 100 ml methyl ethyl ketone, 25°C) of 1.532; another sample which had inherent viscosity of 0.692 (in acetophenone) was found to have an average MW of 125,000. Softening points were in the range of 307–357°. IR analysis indicated the cyclic structure to contain six members.

An interesting transannular cyclocopolymerization has been reported.[1-91,4-41] Sulfur dioxide was copolymerized with *cis,cis*-1,5-cyclooctadiene using radical initiation in dilute solution. The copolymer was obtained in film- and fiber-forming molecular weights, had no residual unsaturation, and contained two moles of sulfur dioxide per mole of cyclooctadiene.

The copolymer decomposed at temperatures above 250° with gas evolution. A copolymer prepared at 10% weight concentration of monomers (SO_2: cyclooctadiene = 2.5:1) in tetramethylene sulfone in air at 25°C for 16 hrs., using methyl ethyl ketone peroxide as initiator in 98% yield had an inherent viscosity (ln $\eta_{rel/c}$) of 2.2 at 0.5 g per 100 ml of dimethyl sulfoxide solution.

The thermal stability of this bicyclic copolymer has been determined.[4-42] Degradation followed first-order kinetics to a first approximation at 220°C or above. Random cleavage of the copolymer chain occurred during the pyrolysis. Pyrolysis of the copolymer up to 75% loss of weight yielded the monomers as cracking products in the molar ratio of the copolymer composition. The activation energy E_a, for the degradation process was found to be 41 kcal/mole with a frequency factor of $5.2 \times 10^{13}\ S^{-1}$. In comparison, E_a, for the propylene-SO_2 copolymer is 32 kcal/mole. This copolymer decomposes approximately 100 times faster at 229° than the bicyclic copolymer does at 230°. It was concluded that the bicyclic sulfone structure is inherently more stable than the linear α,β-sulfone structure of the olefin-SO_2 copolymers.

Cyclocopolymerization of bicyclopentene and other bicyclic dienes with sulfur dioxide to yield fused ring systems has been disclosed.[68] The four dienes synthesized and studied were bicyclopentene, bicyclohexene, dicyclopentenyl ether, and dicyclohexenyl ether. One tetraene, quatercyclopentene was also synthesized and studied.

Completely soluble copolymers were reported from bicyclopentene in high conversions down to −39°C; the copolymer structure is represented below (Equation 8–7):

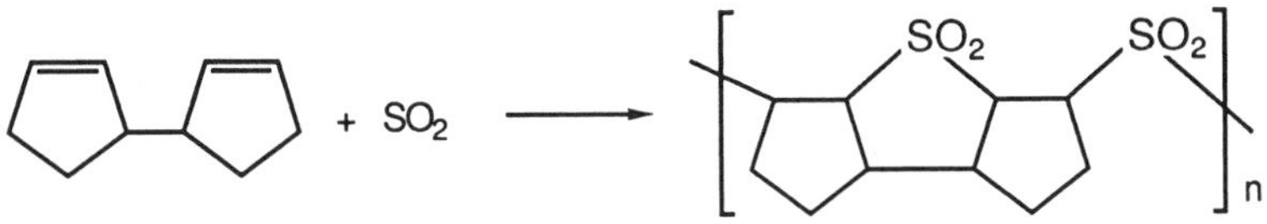

+ SO_2 ⟶ [SO_2 SO_2]$_n$ (8–7)

A copolymer prepared in diethyl ether under argon at O°C using methyl ethyl ketone peroxide as initiator for a period of 24 hrs., gave 87.6% yield of a 1:1 copolymer which had an inherent viscosity (0.5 g per 100 ml DMSO) of 1.72.

Bicyclohexene also was reported to give soluble copolymers but in a much slower reaction and in low conversion. Quatercyclopentene was reported to give only insoluble polymers. The four dienes and the tetraene discussed earlier as comonomers with sulfur dioxide, were also studied as comonomers with maleic anhydride.[69] Soluble, low molecular weight copolymers were reported from all five compounds. Their compositions approached the 2:1 comonomer ratio suggested by the fused ring structures which were proposed as the main repeating units. The structures predicted by the originally postulated cyclocopolymerization mechanism and supported by evidence obtained are shown (8–9 to 8–13).

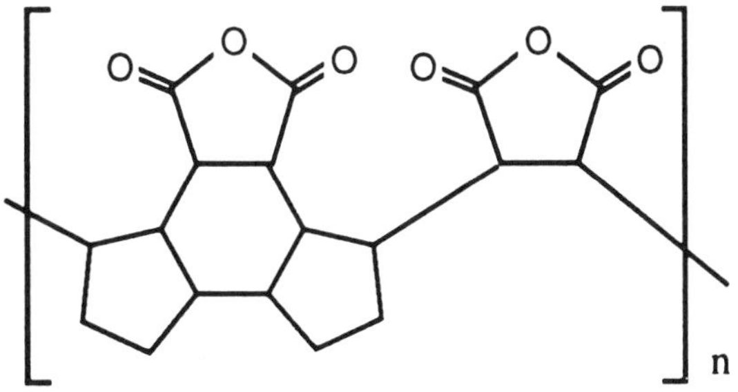

O O O O O O]$_n$

(8–9)

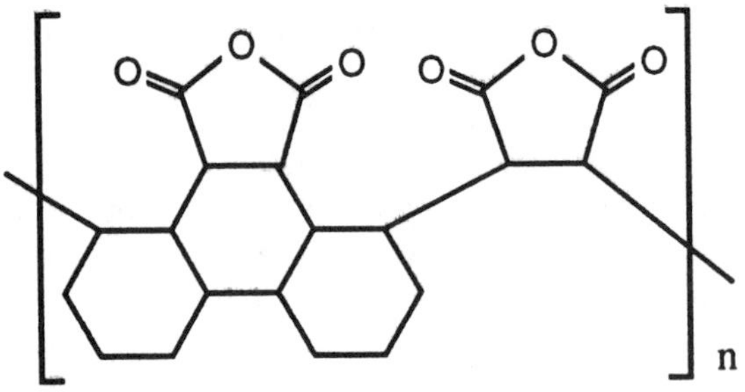

(8–10)

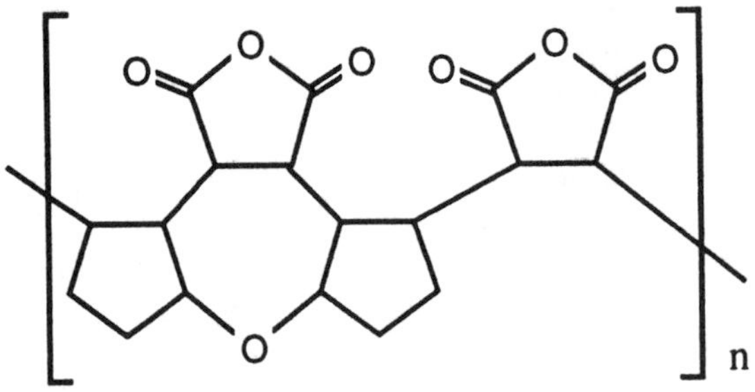

(8–11)

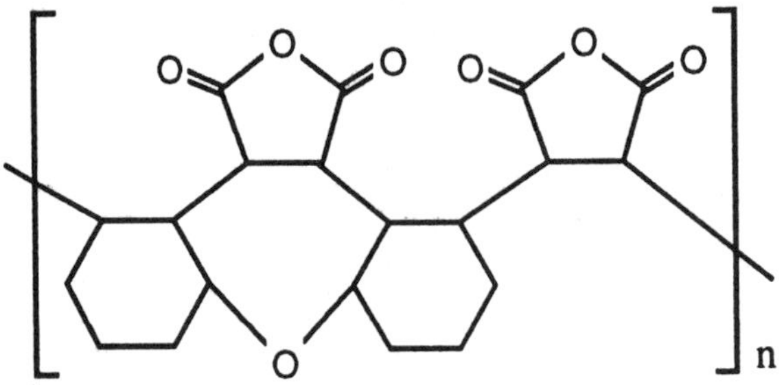

(8–12)

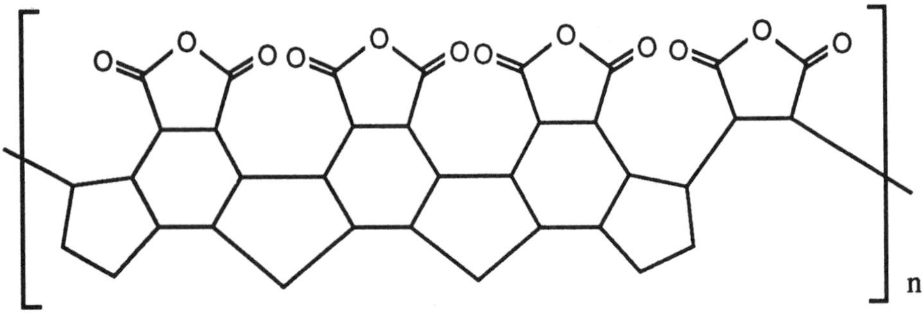

(8–13)

Apparently, MWs for these copolymers were low as inherent viscosities (0.5 g per 100 ml in tetrahydrofuran) were in the range of 0.029–0.039. Softening points for the copolymers were in the range of 190–220°C.

Copolymerization of dicyclopentadiene with sulfur dioxide has been reported to occur.[70] Excess liquid sulfur dioxide was used as solvent and the copolymerization was initiated with UV light or with radical initiators. The preferred temperature for highest MW appeared to be $-30°C$. In an experiment at $0°C$ for eight hrs. at a monomer mole ratio of 2.9 sulfur dioxide to dicyclopentadiene, the yield of copolymer, having a reduced viscosity of 1.87 (0.5 g per 100 ml phenol containing 2% α-pinene, $60°C$) was 35.6%. The elemental analysis and IR of the copolymer indicated a composition ratio of dicyclopentadiene to sulfur dioxide of 1:2, and essentially no residual double bonds. The structure shown was assumed for the copolymer (8–14).

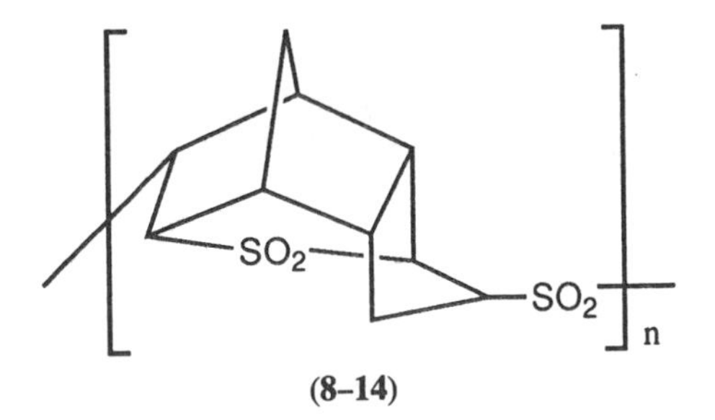

(8–14)

Copolymerization of bicyclo[2.2.1]-2,5-heptadiene (BCHP) with sulfur dioxide has been reported.[71] With no added initiator in spectrograde cyclohexanone or dioxane as solvent, a flash copolymerization occurred, leading to 1:1 copolymer. However, if the solvents were distilled under nitrogen just before use, no flash copolymerization occurred. Cyclohexanone peroxide or oxygen added to the freshly distilled cyclohexanone solvent led to extremely rapid flash copolymerizations, even at low temperatures. No flash copolymerization occurred with toluene, n-heptane, or acetonitrile as diluent.

It was also observed that a strong yellow-colored solution resulted upon mixing pure BCHP with liquid sulfur dioxide in the absence of an initiator indicating the formation of a 1:1 donor- acceptor complex, represented as follows (8–15):

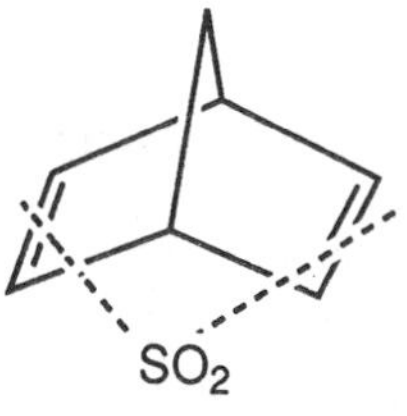

SO_2

(8–15)

The 1:1 copolymer composition was in good agreement with the postulated mechanism which included initiation by formation of the 1:1 complex and addition of these 1:1 complexes to each other in the propagation step.

In order to check the role of oxygen in the initiation step, t-butylhydroperoxide was studied as initiator. In these experiments, an exothermic reaction occurred leading to flash copolymerization and 1:1 copolymer. Use of benzoyl peroxide in similar experiments gave no flash copolymerization. An oxidation reduction system, in which formation of free radicals were formed, was proposed.

Spectral evidence to support these postulates was obtained. In support of a radical mechanism, transfer reactions with halomethane, reactions in presence of diphenylpicrylhydrazyl radical and presence of benzoquinone, and copolymerization of the complex with maleic anhydride, were carried out. It was concluded that the strongly electronegative $\cdot(SO_3H)$ radical was the initiating species in the t-butylhydroperoxide experiments, and that in the cyclohexanone or dioxane experiments, solvent hydroperoxides or peroxides were responsible for initiation in a similar manner.

Based on IR, NMR, and copolymer composition evidence, the following structure for the copolymer was proposed (8–16):

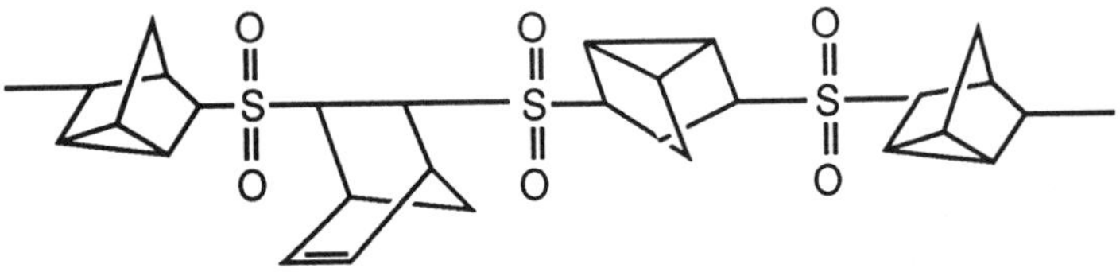

(8–16)

The absence of signals at $\delta = 0.95$ for the C-H absorption of the cyclopropane ring was attributed to a possible shift due to the SO_2 unit.

Cyclocopolymerization of *cis,trans*-1,5-cyclodecadiene with sulfur dioxide has been reported to yield copolymer of 1:1 M comonomer composition with transannular bond formation in the carbocyclic ring (8–17).[72] This is in contrast to the cyclocopolymeriza-

$+ SO_2 \longrightarrow$

(8–17)

tion of *cis,cis*-1,5-cyclooctadiene with sulfur dioxide which yielded a copolymer of 1:2 M composition in diene: SO_2. A copolymer of *cis,trans*-1,5-cyclododecadiene with sulfur dioxide, prepared at 8.3% diene concentration in chloroform (diene: SO_2 = 1:2), at 25°C, using aqueous ammonium nitrate as initiator, for 70 hrs., gave 96% yield of copolymer having an inherent viscosity (0.4 g per 100 ml chloroform) of 1.03 and M_n of 60,000. Although the polymer has four saturated carbon atoms between the sulfone groups, the polymer degraded thermally above 150°C with random chain cleavage; however, the copolymer was found to possess much improved alkali stability over the usual olefin-sulfur dioxide copolymer. This copolymer remained unaffected by 5% alcoholic KOH after several days, while the butadiene-sulfur dioxide copolymer dissolves in 1/2 hrs.

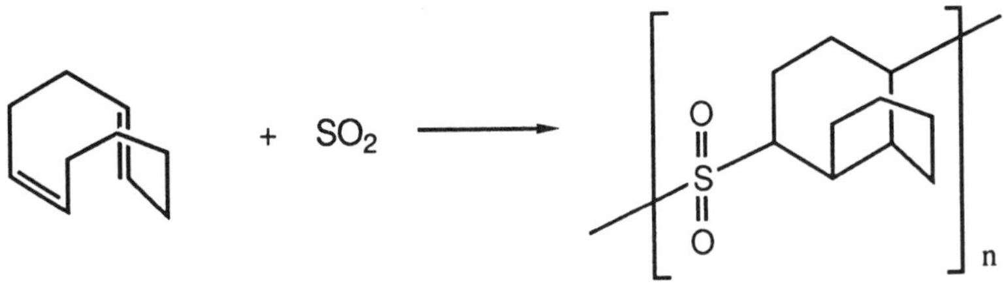

A cyclic copolymer of maleic anhydride and *cis,trans,trans*-cyclododecatriene has also been reported.[73] The propagation steps proposed are shown in the following series of reactions in (8–18). In contrast to the results reported earlier in which *cis,cis*-1,5-cyclooctadiene was shown to undergo cyclocopolymerization with sulfur dioxide to yield a copolymer having a diene-SO_2 molar ratio of 1:2[74], copolymers of this diene

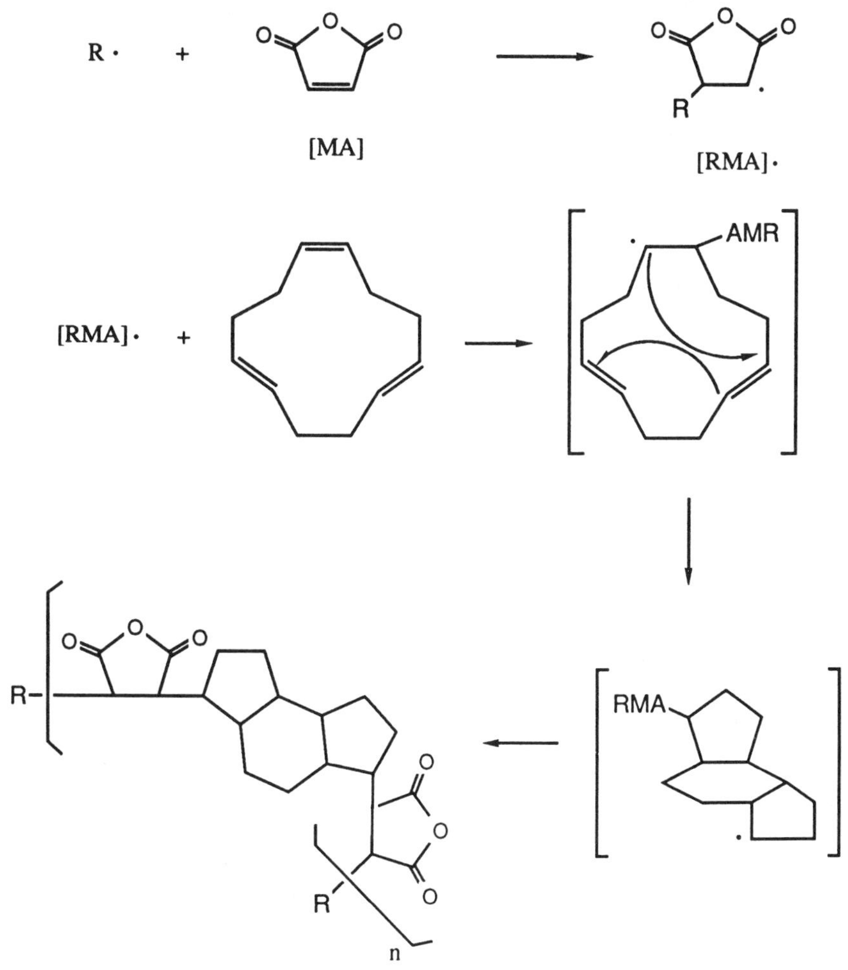

R· + [MA] → [RMA]·

[RMA]· + → [·AMR] → [RMA] → R–[ ]n

(8–18)

with maleic anhydride in which the transannular reaction does not incorporate a molecule of the comonomer have been reported. The composition of the copolymer was reported to be 1:1 M and a copolymer prepared in xylene had a melting point of approximately 220°C. The MW was reported to be 3550, corresponding to about eleven maleic anhydride-1,5-cyclooctadiene repeating units. There does not appear to be any published data on the biological evaluation of these copolymers.

Cyclocopolymerizations have been reported in which the ring closure or transannular step does not insert a molecule of the comonomer.[1-63] Such a copolymerization is analogous to copolymerization of 1,5- or 1,6-dienes with olefins; however, in this case, the transannular reaction results in the closure of a three-membered ring. Bicyclo [2.2.1]-hepta-2,5-diene was shown to undergo copolymerization with monomers such as vinyl chloride, vinylidene chloride, acrylonitrile, ethyl acrylate, and methyl methacrylate to copolymers in which only the transannular product, 3,5-disubstituted nortricyclene, resulted from the bicycloheptadiene monomer.

These results are particularly significant in view of the fact that bicycloheptadiene had been shown to undergo homopolymerization by a radical initiated mechanism to yield a polymer which was believed to be an approximately alternating copolymer containing both the 3,5-disubstituted nortricyclene and 5,6-disubstituted bicyclo[2.2.1]-hept-2-ene units.

This homopolymerization was first reported, however, without structural data in 1960.[75] Small yields of low molecular weight, soluble polymers were formed when the monomer was heated with radical sources.

These structural data, along with the copolymerization data obtained, led to the postulate that formation of the nortricyclene structure is energetically more favorable than reaction through only one double bond but that steric requirements make it improbable for two nortricyclene units to enter the propagating chain adjacent to each other. This postulate that the cyclization step is energetically more favorable than a competitive, but noncyclization step is consistent with the postulate advanced earlier.[1-17]

Reactivity ratios for the copolymerizations were for M_2 = bicycloheptadiene, M_1 = comonomer were reported as follows: M_1 = vinyl chloride, $r_1 = 0.74$, $r_2 = 0.35$; M_1 = vinylidene chloride, $r_1 = 1.41$, $r_2 = 0.08$; M_1 = acrylonitrile, $r_1 = 0.67$, r_2 0.08; M_1 = ethyl acrylate, $r_1 = 3.05$, $r_2 = 0.01$; M_1 = methyl methacrylate, r_1 10.0, $r_2 = 0$. From the data, the average Q and e values for bicyclo [2.2.1]heptadiene were reported as 0.09 and -1.04 respectively. The copolymer reduced viscosities were reported to be greater than 0.3 when measured at 0.2 wt.% concentration in an appropriate solvent at 30°C.

Cyclopolymerization of diallyl ether with sulfur dioxide has also been shown to occur. Radical copolymerization in acetone gave soluble copolymers containing 1–2 mol diallyl ether/mol SO_2. The formation of rings from both monomers and of 3,4-disubstituted tetrahydrofuran rings by cyclized diallyl ether units on the main chain was proposed.[7-105]

Cyclopolymerization of 1,X-Dienes which Lead to Larger Rings

Some large ring-containing cyclocopolymers have recently been reported. Cyclocopolymerization of the divinyl monomer, 1,2-bis-(2-ethenyloxyethoxy)benzene (a 1,13-diene) with maleic anhydride via radical initiation has been shown to produce the 1:2 alternating cyclocopolymer (Equation 8-8).[2-464] As postulated, a similar divinyl ether has also been prepared and studied[2-383] in an earlier investigation;[3-157] considerable steric and

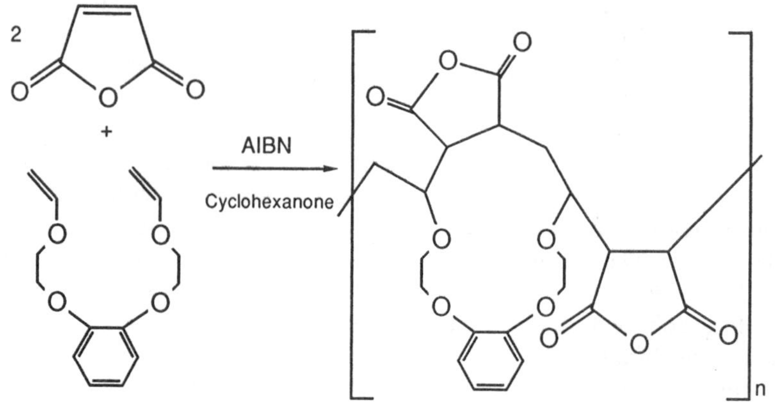

AIBN

Cyclohexanone

(8–8)

conformational control is possible in this case as well. Also, the monomer pair was shown by UV spectroscopy to form a donor-acceptor complex having an equilibrium constant of 0.280, suggesting participation of the complex in the copolymerization. Also consistent with this proposal was the observation that the maximum rate of copolymerization occurred at the monomer molar ratio of 1:2.

The polymers would be expected to simulate the properties of crown ethers.[2-385, 2-464] A 1,7-diene, 1,2-bis(ethenyloxy)-benzene, was also synthesized. Homopolymerization of the 1,7-diene by both radical and cationic initiation led to cyclopolymers of different ring sizes. Homopolymerization of the 1,13-diene led to cyclopolymer only by cationic initiation. Both monomers were copolymerized with maleic anhydride to yield predominantly alternating copolymer, having macrocyclic ether-containing rings in the polymer backbone.

New phase transfer catalysts containing oxyethylene oligomers have been synthesized by cyclopolymerization of the divinyl ether of tetraethylene glycol (8–19) a 1,16-diene.[2-466] The polymer has a phase transfer activity equal to that of 18-crown-6 as a catalyst for the cyanation of 1-bromohexane by sodium cyanide.

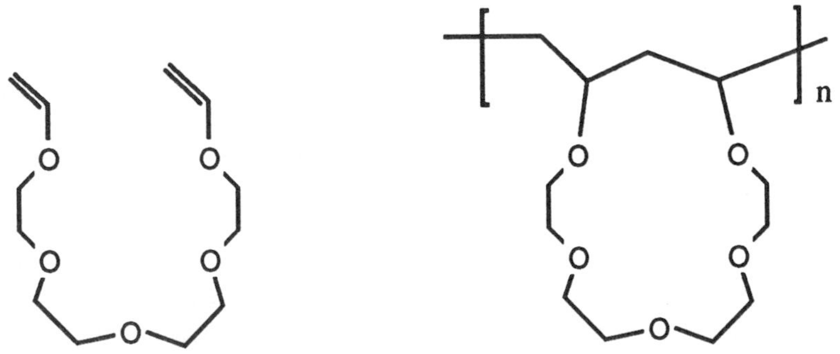

(8–19)

A study of the polymerization kinetics of divinyl glutarate, divinyl adipate, divinyl suberate, divinyl azelate and divinyl sebacate, initiated by free radicals was undertaken.[76] An equation was obtained which gives the reaction rate in terms of initiator concentration, monomer concentration, the ratio of linear chain propagation rate constants to the chain termination constants, the ratio of cyclic radical rate constants to the termination constants and the peroxy radical formation rate constants.

THE ROLE OF DONOR-ACCEPTOR COMPLEXES IN CYCLOCOPOLYMERIZATION

The unusual and apparent general nature of this cyclocopolymerization in which bimolecular ring closure with the monoolefin even under otherwise unfavorable conditions, is highly favored prompted the proposal that the comonomer pair which leads to ring closure may form a donor-acceptor complex prior to initiation, and that the propagating species is the complex rather than the individual comonomer molecules.[14] This proposal was supported by the establishment of the existence of complexes in several comonomer pairs by use of UV spectrometry, and subsequent calculation of the equilibrium constants of formation of these complexes.

The alternating copolymerization of dihydropyran (DHP) and maleic anhydride (MA) had been reported.[77] The similarity of this system to the divinyl ether (DVE)-MA system prompted an investigation of this monomer pair.[14] It was observed that a yellow color was formed when MA was added to DHP. Since this color diminished as the polymerization reached completion and since DVE represented a structure similar in electronic value to DHP, research was begun to evaluate the nature of this phenomenon. Vinyl ethers possess electron donor double bonds, whereas MA and similar structures possess electron-acceptor double bonds. Double bond electron-density calculations substantiated the π-electron acid-base character of this system, which could well give rise to a donor-acceptor complex.[78]

The characterization of complexes is a well known procedure. Early work was done on iodine-aromatic hydrocarbon complexes.[79] Since that time, a great deal of theoretical and experimental interest has been generated on these and other nonionic complexes.

The complexes are characterized by an intense electronic absorption in the visible or near ultraviolet spectrum that is attributable to neither component of the complex alone, but to a new molecular species, the complex itself.[80] Such complexes have been considered to arise from a Lewis acid-base type of interaction, the bond between the components of the complex arising from partial transfer of the π-electron from the π-base (aromatic molecule) to orbitals of the π-acids (I_2, MA, etc.) The theory of donor-acceptor complexes has recently been discussed.[42]

Reliable equilibrium constants were obtained experimentally in accord with theory and published criteria.[81,82] The values of the complexation constants, K, obtained in this study were: DVE:MA, 0.0137; DVE:MI 0.0374; DHP:MA, 0.02; and DVE:FN, 0.0151 (MI = maleimide; and FN = fumaronitrile). Because of the nature of copolymerization products resulting from mixtures of the above mentioned monomers that show donor-acceptor characteristics, one is led to consider the possibility that the complex is involved as a reactive intermediate in the polymerization mechanism. DHP and MA give a perfect 1:1 copolymer as shown.[14,77] Alternation, in this case, could well be explained by reaction of the donor- acceptor complex with the propagating free radical.

In a similar manner, polymerization of a 1:1 complex (8–20) of divinyl ether and maleic anhydride could explain the novel formation of cyclic copolymer.[1-156]

Since the initial derivation of the cyclocopolymer composition equation,[11] kinetic data for numerous combinations of monomers have been determined. In all cases very high values for α have resulted (α is the ratio of the relative rates of cyclizations and addition to another monoolefin). This high value is remarkable in the case of monomers such as acrylonitrile and indicates that almost all the radicals in position to form a ring prefer to do so. The results of this research suggested a reaction involving a donor-acceptor complex. The geometry of this complex is unknown, but a 1:1 stoichiometry limits the number of possible structures.

One possible complex in the system of divinyl ether and maleic anhydride has only one double bond in the ether complexed (8–20). The intermediate radical represents a situation analogous to the ring-closing propagation step in the 1,6-diene intra-intermolecular mechanism, previously described. The geometry of organic donor-acceptor complexes has been determined simply by producing the structure that gives the maximum π-orbital overlap for the donor and acceptor.[80] A complex formed from one double bond of divinyl ether and maleic anhydride is geometrically sound. When coupled with the kinetic data that show extremely rapid ring closure another reaction scheme, involving reaction of the donor-acceptor complex which would include both double bonds of the ether must be considered (8–20).

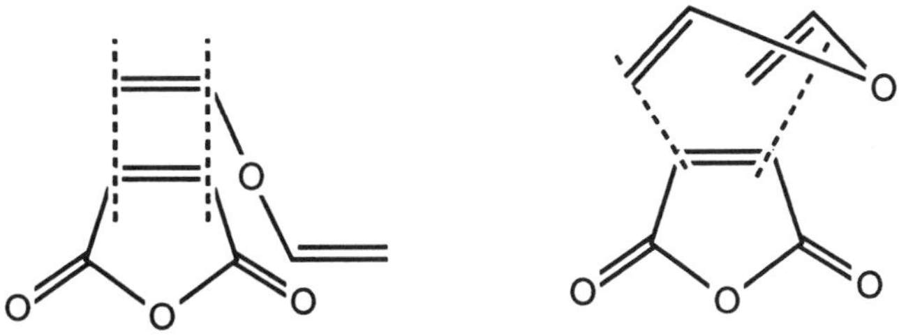

(8–20)

Orientation of monomers, through a donor-acceptor complex prior to free-radical reaction, explains this unusual cyclic structure and also accommodates the kinetic data. A random reaction of monomers in this polymerization is totally eliminated on the basis of all analytical data that has been presented. The probability, that these monomers in the absence of complex participation would attain the correct orientation in solution to give the product, can be assumed to be very low.

The reaction of such a complex with a propagating free radical to produce a linear polymer with cyclic units, offers the best explanation of the formation of this polymer. Identification of a donor-acceptor complex, however, in a reaction system does not necessarily mean that it is on the reaction coordinate, although many proposals of this kind have been made.

In an attempt to explain the alternating copolymer of styrene and maleic anhydride, a donor-acceptor complex was identified which was proposed to be the reactive intermediate.[83] This work was later supported by additional evidence.[84] Complexes of a series of substituted styrenes and maleic anhydride and their corresponding copolymers were studied later.[85] In this work it was concluded that the complex was determining the nature of the polymeric product of the free-radical initiation.

In other fields, analogous work lends support to this theory. Photochemical reactions have been studied in which appropriate filters were used to exclude all reactions except those of the complexes.[86] The systems studied are not unlike those of interest here—cyclohexene, the maleate and fumarate esters, and maleic anhydride. These yield cyclo-addition products when photoexcited. In all cases, when a donor-acceptor complex has been identified, the stereochemistry of those reactions is much more specific than otherwise.

Investigations in the field of polymer chemistry independently led to a conclusion that was similar to the conclusion arrived at in these earlier studies.[87] It was concluded that in the system of p-dioxene and maleic anhydride the limiting factor was the reaction of the donor-acceptor complex. This complex was identified spectrophotometrically. Evidence of the inclusion of the complex in the reaction comes from both the alternating copolymerization of p-dioxene and maleic anhydride and the unique idea of investigation of terpolymerization of acrylonitrile, p-dioxene and maleic anhydride. The terpolymerization still yielded a 1:1 ratio of p-dioxene and maleic anhydride; normal terpolymerization kinetics would not yield this 1:1 ratio.

Later work led to the conclusion that the alternating copolymerization could be reduced to a homopolymerization of the donor-acceptor complex formed between the comonomers. It was also concluded that the terpolymerization of these alternating copolymerizable monomers with a third monomer, which has little or no interaction with either monomer of the pair, could be reduced to a copolymerization of the charge-transfer complex between the alternating copolymerizable monomers and the third monomer.[88]

To provide further evidence for the formation of a complex from the comonomers of a cyclocopolymerizable pair and its participation in the copolymerization, both a copolymerization study of a suitable comonomer pair and a terpolymerization study of this comonomer pair with a third monomer was undertaken.[43]

The 1,4-diene used was divinyl ether (DVE) and the monoolefins were maleic anhydride (MA) and fumaronitrile (FN). Acrylonitrile (AN) was used as the third monomer in the terpolymerization experiments. When a solution of maleic anhydride in chloroform (or DMF) was added to a solution of a 1,4-diene (DVE) in the same solvent, there appeared in the UV spectrum of the mixture a new and broad band. The appearance and position of this band was attributed to the formation of electron-donor-acceptor complexes. The interpretation of the spectra of these complexes permitted the determination of their stoichiometric composition using the continuous variation method.[89,90]

The donor-acceptor bond maximum was at 278 nm for the DVE:MA complex. The maximum absorbance was found to be for a molar fraction of 0.5 in acceptor; hence it could be concluded that the stoichiometry of the complex DVE-MA (in $CHCl_3$ and DMF), was 1:1. The geometry of the complex is unknown as well as whether one or both double bonds of the diene are involved in the complex; however, maximum absorption occurred at mole fraction of acceptor = 0.5.

Some observations have been reported that could be of interest in understanding the structure of the complex. It has been reported that alkyl vinyl ethers as well as other electron-rich alkenes react with tetracyanoethylene (TCNE) to give a 1:1 addition product having a 1,1,2,2,-tetracyanocyclobutane structure.[91] While the reaction of ethyl vinyl ether (EVE) with TCNE was very fast, the reaction of DVE-TCNE was much slower and only the 1:1 addition product could be isolated. The structure of the DVE-TCNE adduct was determined by NMR to be 1,1,2,2-tetracyano-3-vinyloxycyclobutane.[43]

This structure is analogous to the structure of the adduct of EVE-TCNE, without the second olefinic bond becoming involved; the remaining double bond of the DVE-TCNE adduct did not react with an additional amount of TCNE. Moreover, the e value given for EVE($e = -16$) was greater than that observed in the case of DVE($e = -13$).[92] Such observations led to the conclusion that the electron density of the double bond is greater for EVE than for DVE, which can be attributed to the effect of the ethoxy group.

On the other hand, in the charge-transfer complex theory it is well known[93] that when different donors interact with the same acceptor, the charge-transfer energy varies linearly with the ionization potential of the donors and, since the ionization potential for alkene derivatives is directly related to the electron density of the double bond, it is reasonable to expect that the complex of DVE:MA would not absorb at a significantly longer wave length then the complex of EVE:MA. Thus, τ_{max} for the EVE:MA and

DVE:MA complexes appeared near 278 nm, while that for the furan;MA complex appeared at 291 nm.

The fact that the maximum absorption of the DVE-MA complex compared with the EVE-MA complex was not found at a significantly longer wave length suggests that the DVE-MA complex may have some other configuration than that of the EVE-MA complex and that the presence of both double bonds could have some effect on the donor-acceptor interaction. Since the cyclic structure is known to appear in the copolymer, it seems that the donor (DVE), in the complex, has a pseudocyclic structure analogous to the furan molecule. As shown above, the furan-MA complex has its maximum at 291 nm, a longer wave length than the noncyclic ethers. These observations do not establish the absolute configuration of the complex, but since all of them substantiate each other, it is postulated that both double bonds of DVE participate in complex formation with maleic anhydride in a more or less pseudocyclic form similar to the furan structure. Methods to determine the equilibrium constant of donor-acceptor complexes both by UV spectroscopy using the Benesi Hildebrand equation [79] or by NMR[94,95] have been published. The NMR method, where it can be applied, is more accurate and easier to use in the case of weak complexes where the UV absorption maximum is near that of one of the components.[94]

The results of the NMR method gave the values: $K(lm^{-1})(DVE{:}MA) = 0.036$; $K(lm^{-1})(DVE{:}FN) = 0.008$. For comparison, the value reported for the complex between benzene and trinitrobenzene in CCl_4 at 35.5° is 0.31 lm^{-1}.[95]

There has been some disagreement in the case of alternating free-radical copolymerization, concerning participation of the molecular complex in the mechanism of the reaction, because of the absence of a dilution effect on the monomer reactivities.[85] Since complex formation is a bimolecular reaction and dissociation of the complex unimolecular, dilution would reduce complex concentrations. If the molecular complex is responsible to some extent for the regularity of the 1:2 structure in the cyclopolymer, it is quite reasonable to expect a decrease of this regularity by dilution.

The copolymerization of divinyl ether with fumaronitrile had been reported earlier.[13] In alternating copolymerization, a maximum in the rate of polymerization for an initial monomer feed of 1:1 (medium in which the concentration of complex is highest) has been observed.[87] However, in this case a polymerization rate maximum for a monomer feed of DVE-FN equal to 1:2, similar to the composition of the polymer obtained, was observed; however, the stoichiometry of the complex was 1:1. The two steps of cyclocopolymerization, the formation of the cyclic radical and the addition of the electron poor monomer to this cyclic radical, seem to be dependent upon the electronegativity difference between the two monomers responsible for complex formation. Using the optimal monomer feed for DVE:FN of 1:2, copolymerizations were carried out for different total concentrations in the same solvent (DMF), in order to determine whether there was a dilution effect in this system. The results are given in Figure 8–1 and in Table 8–7.

On diluting the medium of copolymerization, not only a decrease of the rate of polymerization as expected was observed, but also a decrease in the nitrogen content of the copolymer, and consequently in the fumaronitrile content was observed. The upper limit of FN content in the copolymer corresponds thus to the 1:2 Structure 8–21a previously described.[13]

Since the nitrogen content of the polymer decreased on dilution, it seemed reasonable to assume that the 1:2 regularity of the copolymer may also decrease. NMR and IR

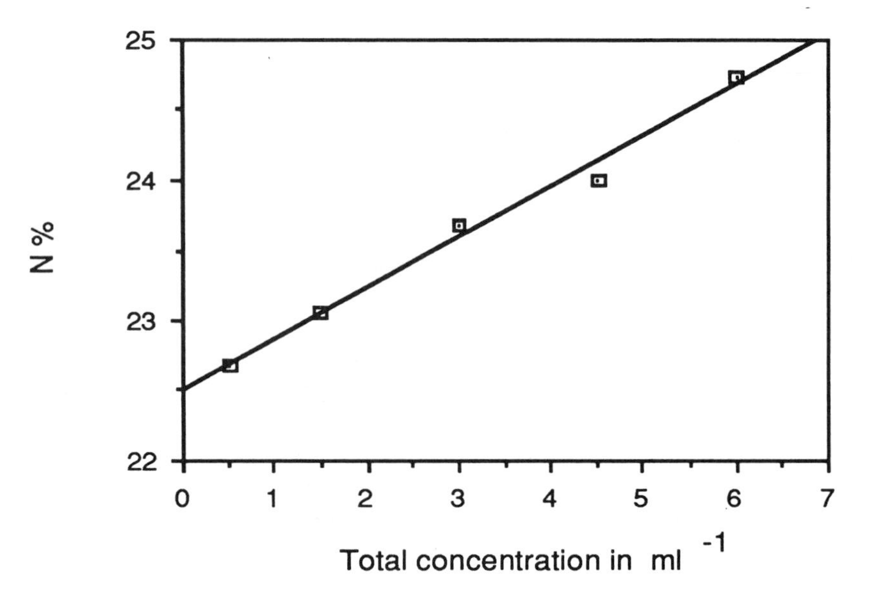

Figure 8–1 Dilution effect on the copolymerization of divinyl ether (M_1) with fumaronitrile (M_2).

studies, however, did not show any detectable change in the spectra of the copolymers and particularly in the unsaturation content. Two remaining possibilities which could explain the decrease of FN in the copolymer are (a) the decrease of selectivity of addition to the cyclic radical, and (b) homopolymerization of a limited number of divinyl ether units. Both require addition of vinyl ether radicals to vinyl ether double bonds. This has been shown to occur in a homopolymerization study of DVE;[4–51] however, the rate of this homopolymerization is very low compared to its copolymerization with FN, but in the absence of sufficient FN this alternative course would be favored. The first possibility would lead to a limited sequence of Structure 8–21b units: The second possibility would lead to a limited sequence of Structure 8–21c units:[4–51]

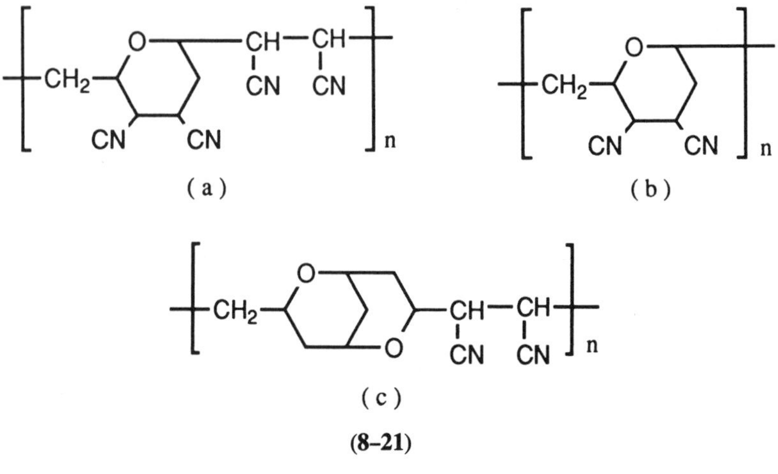

(a)

(b)

(c)

(8–21)

Table 8–7 Dilution Effect on the Copolymerization of Divinyl Ether (M_1) with Fumaronitrile (M_2)[a]

Experiment no.	Total concentration (molel/l)	N (%)[b]	M_2	[η] (dl/g)[c]
1	6	24.74	0.660	
2	4.5	24.00	0.642	0.13
3	3	23.68	0.630	0.11
4	1.5	23.05	0.615	
5	0.5	22.66	0.600	

[a] Solvent, DMF; 60°; DVE/FN = 1:2; initiator, AIBN (1 mole%).

[b] Nitrogen percentage was corrected on the basis of 95% found for polyacrylonitrile.

[c] In acetone, 30°C.

The structure of the copolymer was thus represented as $[8\text{–}21a]_x$ – $[8\text{–}21b]_y$ – $[8\text{–}21c]_z$ in which x >>> y, z. Consequently, it was concluded that addition to the cyclic radical was less selective than the cyclization step. An analogous dilution effect was also observed in the case of terpolymerization as reported in the discussion which follows. (See Table 8–8).

The copolymerization of DVE:acrylonitrile had been reported earlier.[96] While the copolymer DVE:MA had a true 1:2 structure, the copolymer DVE:AN was generally richer in acrylonitrile than the 1:2 composition expected; however, in this case also, the divinyl ether could be incorporated into the copolymer in a six-member ring formed by copolymerization with AN. The high concentration in AN in the copolymer was explained by the fact that AN readily homopolymerizes by a free radical mechanism while MA does not.

Another explanation may be that the double bond of acrylonitrile is less electron deficient (e = 2.25) than fumaronitrile (e = 1.96); the molecular association between AN and DVE would then be much weaker than the association of DVE:MA and therefore the effect of this association on the polymerization mechanism, if any, would be greater in the case of MA and FN than in the case of AN. It was not possible to detect any obvious molecular association between DVE and AN, while it was shown earlier that DVE and MA form a stable donor-acceptor complex. On the other hand, acryloni-

Table 8–8 Dilution Effect on the Terpolymerization of Divinyl Ether-Maleic Anhydride-Acrylonitrile[a]

Total initial monomer concentration (mole/l)	Conversion (%)	Polymer composition (%)			
		AN	MA	DVE	n
6	7.5	27.2	43.6	29.2	49.3
1.2	8.8	31.3	39.7	29.0	36.9

[a] Solvent, DMF; 60°C; initiator, AIBN (1 mole%); ratio DVE:MA:AN = 1:1:1

trile (M_1) copolymerizes only to a small extent, if at all, with maleic anhydride (M_2). The values of the reactivity ratios given are $r_1 = 6$, $r_2 = 0$ (T = 60°C).[97]

In the first series of experiments involving terpolymerization of DVE, MA, and AN,[43] the donor-acceptor ratio was kept constant at the optimal value of 1:2, the donor being divinyl ether, and the acceptor being the sum of maleic anhydride and acrylonitrile; in every experiment the relative molar concentration in MA and AN was varied. Their sum was always twice the molar concentration of DVE. Table 8–8 shows the results of this series of experiments. In the second series of experiments, different ratios of DVE:MA:AN were used, and the total concentration was the same in all the experiments, equal to six moles per liter.[43] The results are given in Figure 8–2. A dilution effect is also illustrated in Table 8–8.

From these experiments, it was observed that when the ratio donor-acceptor was kept constant (1:2), while varying the relative concentration in MA and AN, a terpolymer was obtained. Although AN can homopolymerize readily by a free radical mechanism, the terpolymer was always richer in MA and always poorer in AN than the monomer feed. NMR and IR studies showed no detectable remaining double bond in the terpolymer. In all the experiments, regardless of the composition of the monomer feed used, the ratio DVE:MA in the terpolymer was always less than 1:1 and had the 1:2

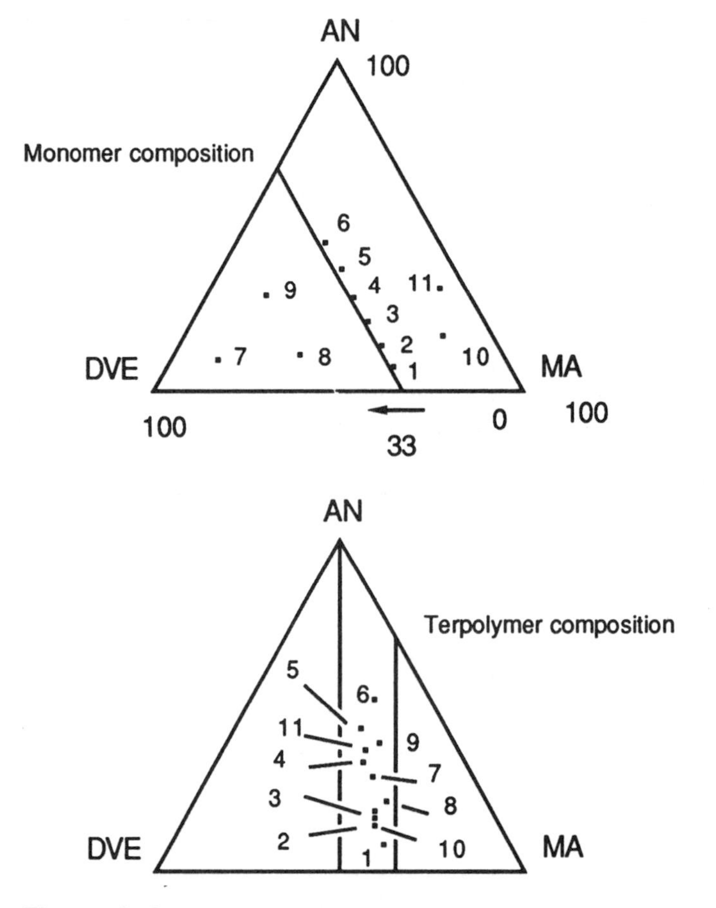

Figure 8–2 Radical terpolymerization of divinyl ether (DVE)-maleic anhydride (MA)-acrylonitrile (AN).

structure as an upper limit. Since the ratio DVE:MA in the terpolymer varied from 1:1 to 1:2, and MA does not homopolymerize nor copolymerize with AN under the conditions used, the conclusion was drawn that every molecule of DVE in the terpolymer chain was cyclized by a MA molecule to form the 1:1 DVE:MA cyclic unit.

If the cyclic radical reacts with another molecule of MA in order to give the 1:2 DVE:MA structure, this step, because of the non-reactivity of the MA radical with another molecule of MA or with AN, must be followed by an addition of a new DVE:MA couple. If the cyclic radical reacts with AN a growing chain terminated by an AN radical would be formed; this AN radical could react either with a DVE:MA couple or with another molecule of AN. The homopolymerization of the AN radical would result in a sequence of AN units within the chain of the terpolymer. This would increase the nitrogen content of the terpolymer and decrease the 1:2 regularity of the monomer pair DVE:MA, similar to the copolymerization of DVE:AN discussed earlier. The structure of the terpolymer can thus be illustrated (8–22) where n represents the 1:2 DVE:MA unit in the cycloterpolymer, and m the unit containing acrylonitrile. In this

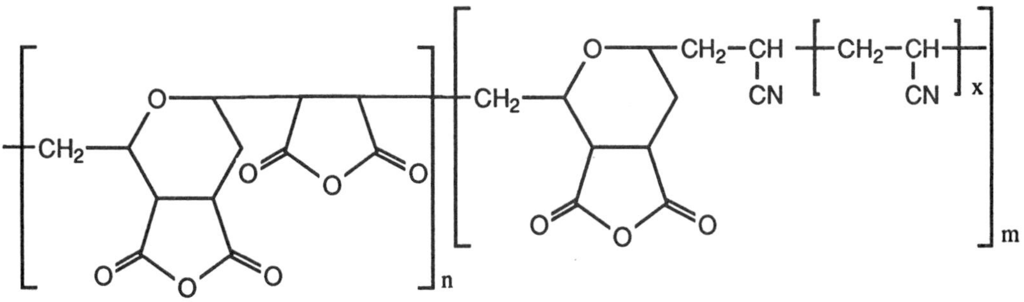

(8–22)

series of experiments in which the donor-acceptor ratio was kept constant to the 1:2 value, it was observed that the value of n decreased with the decrease of the MA:DVE ratio in the monomer feed. The results are given in Figure 8–3. Here again in the

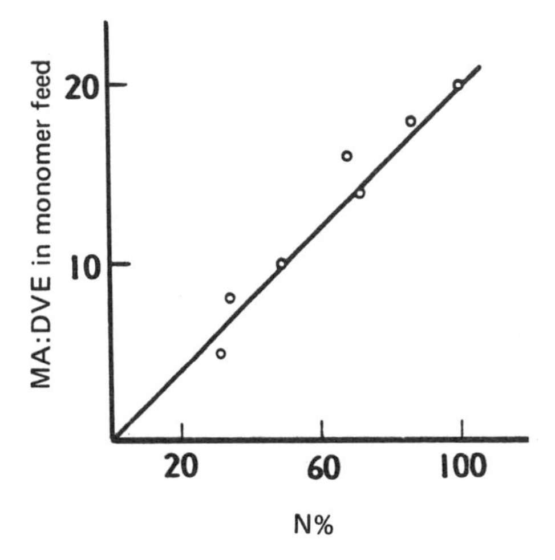

Figure 8–3 Variation of the DVE-MA content of the terpolymer (η) vs. the DVE-MA ratio in the monomer feed.

terpolymerization system, it was observed that by diluting the polymerization solution, a decrease in the selectivity of the monomer addition was observed. For a monomer feed of DVE:MA:AN equal to 1:1:1, the value of the 1:2 DVE:MA content in the terpolymer (n) decreased with dilution, while the AN content increased for the same content of DVE, (Table 8–9).

Thus, it was shown in these cycloterpolymerization experiments that divinyl ether reacted preferably with maleic anhydride to give a cyclopolymer. This verifies the statement made earlier, that the molecular association between maleic anhydride and divinyl ether, being greater (e value smaller) than the association between divinyl ether and acrylonitrile, has a greater effect in controlling the composition of the terpolymer. Therefore, this molecular association, which has donor-acceptor complex character, has a remarkable effect on the cyclopolymerization mechanism. It was also observed that while the cyclization step is very selective (DVE:MA), the next step, the addition of the monoalkene to the cyclic radical is only slightly so, which allows the inclusion of AN units in the terpolymer chain.

Solvent effects in the cyclopolymerization of MA with DVE have been systematically studied.[28] Evidence was presented that supports a donor-acceptor (DA) complex formed between the comonomers as the active species in the cyclocopolymerization. The equilibrium constants (K) of DA complexation with MA was measured in n-heptane by UV spectrophotometry for tetrahydrofuran (THF), ethyl vinyl ether, DVE, furan, and dihydropyran. K-values increased in the above order. K of the MA-DVE pair was measured by UV and NMR in polyhaloalkanes as solvents. K-values decreased with increase of the dielectric constant of the solvent. Values reported were: THF, K(l/m) = 8.6×10^{-2}; ethyl vinyl ether, 1.5×10^{-1}; DVE, 1.8×10^{-1}; furan, 4.7×10^{-1}; and diydropyran, 1.1×10^{-1}. The rate of copolymerization and number-average-molecular weights decreased in more polar solvents. The initial rate was about 100 times faster in $CHCl_3$ than in DMF. In dilute solution, $[M_1 + M_2] = 0.57$ mole/liter, the initial overall rate of copolymerization was maximum in DVE-rich feed in $CHCl_3$, CH_2Cl_2, and DMF. Kinetic derivation failed to explain the rate profile. Strong solvation of MA by the more polar solvent has been proposed as an explanation. The overall rate of copolymerization was proportional to one-half order of AIBN concentration in DMF. The overall energy of activation was 27 kcal/mole in $CHCl_3$ and in DMF. Thermal autopolymerization, photopolymerization and τ-ray polymerization of the MA-DVE pair in bulk gave the same 2:1 (=[MA]:[DVE]) copolymer. Retardation of rate by

Table 8–9 Variation of the 1:2 DVE-MA Content in the Terpolymer (η) with the MA:DVE Ratio in the Monomer Feed

Experiment no.	MA-DVE in monomer feed	n (%)
1	1.8	84.6
2	1.6	65.4
3	1.4	66.9
4	1.0	43.2
5	0.8	28.5
6	0.5	22.8

hydroquinone was observed in thermal autopolymerization. The dative state of the DA complex was considered to be the initiating species.

The energy of the DA transition, which can be calculated from τ_{max}, is a function of, among other minor variables, the electron affinity of the acceptor (E_A) and the ionization potential (I_D) of the donor.[98] Empirically, the DA transition energy is proportional to the ionization potentials of the donors when complexing with the same acceptor:

$$h\nu_{CT} = mI_D + n \tag{8–9}$$

where m and n are constants.

These results show that I_D is smaller when there is more orbital overlap between the orbitals of nonbonding electrons of the oxygen and π-electrons of the vinyl group of ether. The reversed order of furan and dihydropyran may be due to the aromaticity of furan, which lowers the energy level of the highest occupied molecular orbital of furan. It is interesting to note that there is a somewhat parallel relation between τ_{max}, hence DA transition energy, and the equilibrium constant of complexation.

Since DA complexation follows polar association of the components, the more polar solvent would reduce the magnitude of K. The K-values of MA-DVE were measured by UV and by NMR to determine the change of the K-values by the solvent polarities. The following results were reported: (solvent, dielectric constant, K), n-heptane, 1.924, 1.08×10^{-1}; CCl_4, 2.238, 9.8×10^{-2}; $CHCl_3$, 4.806, 3.6×10^{-2}; CH_2Cl_2, 9.08, 1.4×10^{-2}; and $(CH_2Cl)_2$, 10.36, 1.3×10^{-2}. The equilibrium constant decreased with increase of polarity of solvents. The same effect has been observed by other authors for various DA complexes.[99,100]

If the DA complex is the propagating species in the cyclocopolymerization of the MA-DVE pair, the rate of polymerization should be greater with higher K-values, thus producing higher molecular weight polymer. This point was tested by carrying out the copolymerization using solvents of widely different dielectric constants. The following results were reported: (solvent, dielectric constant, $M_n \times 10^{-3}$)dioxane, 2.209, 10.0; chloroform, 4.806, 10.5; ethyl acetate, 6.02, 6.0; acetophenone, 17.4, 3.7; methyl ethyl ketone, 18.51, 5.3; nitromethane, 28.06, 2.8; and DMF, 37.0, 1.9.

The effect of the polarity of the polymerization medium is evident, indicating the importance of the complex as an intermediate in this copolymerization. The more polar solvent results in a lower K-value, and also a lower rate of polymerization. Relative yields of copolymer under identical conditions varied from 36 in dioxane to one in DMF.

The initial rate of copolymerization was determined at [MA]/[DVE] = 1 in $CHCl_3$ (dielectric constant = 4.906) and in DMF (dielectric constant = 37) at 50, 60 and 70°C. The results reported were: [solvent, T, °C, R_p(M/l sec)] $CHCl_3$, 50, 3.13×10^{-5}; 60, 1.09×10^{-4}; 70, 3.54×10^{-4}; DMF, 50,4.07×10^{-7}; 60, 1.62×10^{-6}; 70, 4.49×10^{-6}. The difference in the initial rate was as much as 100-fold in $CHCl_3$ and in DMF. This was explained by the great difference of dielectric constant and the powerful solvating ability of DMF which solvates the free monomer(s), preventing them from DA complexation.

The overall energy of activation, E_{act}, was calculated using the initial rate data by use of Arrhenius plots. The values were: $CHCl_3$, 26.9 kcal/mole; and DMF, 26.7 kcal/mole. Despite the great difference in solvating property of the solvent, the overall energy of activation obtained by this method was nearly the same in $CHCl_3$ and DMF. Since, in copolymerization the expression of the overall rate of polymerization includes several steps, it is usually not possible to interpret and assign an activation energy to each elemental reaction. It appears, however, that the rate constants of initiation, propagation, and termination of this radical copolymerization may not be affected by the change in solvent; this is in agreement with the fact that the effect of solvent on the rate constant of radical reactions is generally very little.[101]

The agreement of ΔE_{act} values in these solvents suggests that the equilibrium constant of complexation does not change independently in each solvent; that is, ΔH of CT complexation is small and approximately the same in both solvents.[100] This consideration of the overall energy of activation led to a study of spontaneous copolymerization of this system in $CHCl_3$, and CH_2Cl_2, which is discussed below.

The number-average molecular weights of copolymer samples obtained in $CHCl_3$, reported earlier, were measured and plotted as a function of conversion. The molecular weight decreased with conversion, since less monomer was available in the system at higher conversion. The molecular weight was lower at a higher temperature due to the increased rate of the thermal decomposition of AIBN. The results showed that there was no living propagating species in the copolymerization.

The initial rate of copolymerization was measured in $CHCl_3$ (dielectric constant = 4.806), CH_2Cl_2 (9.08) and DMF (37) as a function of feed composition. $CHCl_3$ and CH_2Cl_2 were chosen to compare the initial rate, because the measured equilibrium constant of complexation differed by a factor of two in these solvents, while the mode of solvation may not be drastically different because they are structurally similar. The results showed a decrease in relative initial rate from 4.5 to 1.5 for $CHCl_3$, and from 3.3 to 1.0 for CH_2Cl_2 over the feed MA mole fraction from 0.175 to 0.825.

Although the observed rate was not exactly proportional to the measured K-values reported, the rate was much faster in $CHCl_3$ at all feed compositions, which is consistent with the higher K-value in $CHCl_3$. This difference in initial rate in these solvents can be attributed to the different concentrations of the complex, which appears to be the polymerizing species.

The initial rate was also determined in DMF using a larger concentration of total monomers and AIBN. It was significant that under the conditions used for polymerization, the rate was faster in DVE-rich feed in all three solvents. These results suggest that competitive mechanisms involving both the complex and free monomers may be occurring. This concept has subsequently been thoroughly investigated and published in two separate papers.[102, 103]

The kinetic derivation of the overall rate of copolymerization, assuming the simplest propagation reactions[11], gives a rather complicated expression (Equation 8–10):

$$R_p = \frac{2K_{12}K_{21}[M_1][M_2]R_i^{1/2}}{\{t_{11}(K_{21}[M_1])^2 + t_{22}(k_{12}[M_2])^2 + 2K_{12}K_{21}t_{12}[M_1][M_2]\}^{1/2}} \tag{8–10}$$

where $R_1 = k_d f[I_2]$ is rate of initiation, t_{11} and t_{22} are rate constants of termination of the same types of radicals, t_{12} is the rate constant of cross- termination, $[M_1]$ and $[M_2]$ are concentrations of DA complex and MA, respectively, and k_{12} and k_{21} are rate constants of propagation.

If cross-termination were the only termination reaction, a much simpler rate expression would be obtained for this scheme (Equations 8–11 and 8–12):

$$R_p = \left(\frac{2K_{12}K_{21}k_d f\,[I_2]}{t_{12}} \right)^{1/2} \left([M_1][M_2] \right)^{1/2} \tag{8–11}$$

where $[M_1] = K[MA][DVE]$. Therefore,

$$R_p = (\text{constants})^{1/2} \left(K \frac{[DVE]}{2} \right)^{1/2} [MA] \tag{8–12}$$

Under these assumptions, the overall energy of activation reported earlier could be interpreted as a sum of enthalpies (Equation 8-13):

$$\Delta E_{act} = \frac{1}{2} \left(\Delta H_{CT} + \Delta H_{k21} + \Delta H_{k12} + \Delta H_{kd} - \Delta H_{t21} \right) \tag{8–13}$$

Equation 8–12 predicts rate maximum at feed composition of 2:1 (=[MA]:[DVE]) and slower rate in DVE-rich feed.

If only the complex is assumed to be involved in the polymerization producing 2:1 (MA:DVE) copolymer composition through the proposed "escape reaction", in which only one of the components of the CT complex reacts and the other "escapes" from the reaction[104] Equations 8–14 and 8–15, the overall rate of copolymerization would be proportional to the concentration of the CT complex and the rate would be maximum at 1:1 feed composition (Equation 8–16): where [C] = K[MA][DVE].

CT complex with escape reaction k_{11} → $m_2\cdot$ + free DVE (8–14)

CT complex with cyclization k_{2C} → $m_1\cdot$ (8–15)

$$R_p = 2K_{2c}K_{1c}[C] \left(\frac{R_i}{t_{12}K_{1c}K_{2c} + t_{11}k_{2c}^{2} + t_{22}K_{1c}^{2}} \right)^{1/2} \tag{8–16}$$

If only cross-termination were the termination reaction, the overall energy of activation could be interpreted (Equation 8–17), under these assumptions, as:

$$\Delta E_{act} = \Delta H_{CT} + \frac{1}{2}(\Delta H_{k2c} + \Delta H_{k1c} + \Delta H_{kd} - \Delta H_{t12}) \quad 8\text{–}17$$

The observed rate profile did not totally conform to either reaction scheme. This reflects the argument made earlier that competitive mechanisms may be operative in which the free monomers compete in an alternating copolymerization with homopolymerization of the DA complex.

Number-average molecular weights and nitrogen contents, due mainly to AIBN fragments, were measured for the copolymers obtained in DMF. The average molecular weight as measured by vapor pressure osmometry (VPO) was higher when the rate of polymerization was faster. The nitrogen content confirmed this trend. Since the rate of decomposition of AIBN was not altered by the feed composition,[101] this showed that the kinetic chain length was larger in DVE-rich feed. Since the termination reaction did not change with feed composition, and the 2:1 copolymer composition was constant, it appears that the propagating radical was always the same, therefore, the longer kinetic chain length in DVE-rich feed means that there was a greater chance of propagation in DVE-rich feed.

The measured number-average molecular weight (M_n) by vapor pressure osmometry (VPO) did not agree with the calculated number- average molecular weight as determined by nitrogen content, assuming combination as the only termination reaction. M_n by VPO was much less. The NMR spectrum of copolymer prepared in DMF showed a very small amount of DMF tightly solvated to the copolymer even after five repeated purifications by dissolving the copolymer in acetone and precipitating into a large amount of boiling diethyl ether. Since M_n measured by VPO will be very low when even a trace of low molecular weight impurity is present in the sample, the disagreement between the measured M_n by VPO and calculated M_n by nitrogen content could be due to the very small amount of tightly solvated DMF in the copolymer, which seemed impossible to separate.[6] The NMR spectrum also confirmed the consumption of all the vinyl groups of the comonomers, since only a negligible amount of vinyl proton absorptions could be seen.

The strong solvation of the copolymer suggested a pronounced effect of solvation of MA on the rate of polymerization, since MA is an extremely polar monomer. In DVE-rich feed, MA may be solvated by DVE molecules, eventually forming DA complex. The radical of MA (m_2) could easily penetrate the solvent cage of the DVE molecules to polymerize the DA complex. But in MA-rich feed, the excess free MA may be tightly solvated by solvent, reducing the effective concentration of MA for DA complexation with DVE. The cyclized radical of DVE (m_1) may not be able to easily penetrate the solvent cage around the MA molecule because of polar repulsions between the approaching M_1 radical and polar groups of the solvating solvent, and between the MA unit of m_1 radical and the MA molecule in the cage. This solvation effect of the solvent could be an explanation for the faster rate of propagation in DVE-rich feed and the slower rate in MA-rich feed. As a consequence, the position of maximum rate would have shifted toward DVE-rich feed in dilute systems.

The limiting conversion in different solvents may support this effect of solvation to MA. The conversion was measured after a long period of polymerization. If the copolymerization proceeds with consumption of MA and DVE in a 2:1 ratio, the limiting conversion will be the point where one of the comonomers has been completely consumed. When the initial comonomer ratio of [MA]/[DVE] in the feed = 1, the limiting conversion would be 79.2 wt%. In relatively poorly solvating $CHCl_3$ and CH_2Cl_2, the observed values were 75.7 and 79.6%, respectively after 24 hours, in good agreement with the theoretical value, supporting the alternating structure of 2:1 composition of the copolymer. But in DMF, only 26.7%, less than half the theoretical conversion, was obtained during the much longer period of 74 hours. These results suggest that DMF powerfully solvates the very polar MA molecule, preventing DA complexation with DVE. The MA may have remained unattacked by the growing radical, because it was strongly protected in the solvent cage. Dead copolymer exerted no significant effect on the yield in DMF.

If the termination reaction is bimolecular, the overall rate of copolymerization should be proportional to one-half order of initiator concentration (Equation 8–10 or 8–16). The rate of copolymerization was measured as a function of concentration of AIBN at 60°C in DMF. The rate was closely proportional to one-half order of AIBN concentration, verifying the bimolecular termination reaction.

An energy diagram for the DA complex has been proposed.[42] The DA complex is predominantly in a no-bond state in the ground state of the complex. When energy is supplied, the complex will be excited to predominantly the dative state.[41,80] The thermal transition from the no-bond state to the dative state occurs at E_{Th} from the excited vibrational levels of the no-bond state and λ_{max} shows the transition from the most populated vibrational level of the no-bond state of the complex. If enough energy is supplied, the dative state would be dissociated (Figure 8–4). Since the dative state is charged, solvation and polarity of solvent stabilize the dative state more than the no-bond state and the energy of transition from the no-bond state to the dative state would be smaller (Figure 8–5).

Spontaneous polymerization of donor-acceptor pairs has been known to be initiated either cationically or radically, or both.[105,106] The dative state of the complex, which is formed by the thermal excitation of the vibrationally excited no-bond state, has been thought of as the initiating species. Therefore, when photoenergy is supplied to the system to bring the no-bond state to the dative state, initiation would be accomplished[51] and the maximum rate of initiation could be expected at the wave length of maximum DA transition (τ_{max}). Moreover, if higher energy is supplied to dissociate the dative state, such as τ- ray of ^{60}Co, initiation should be much easier.

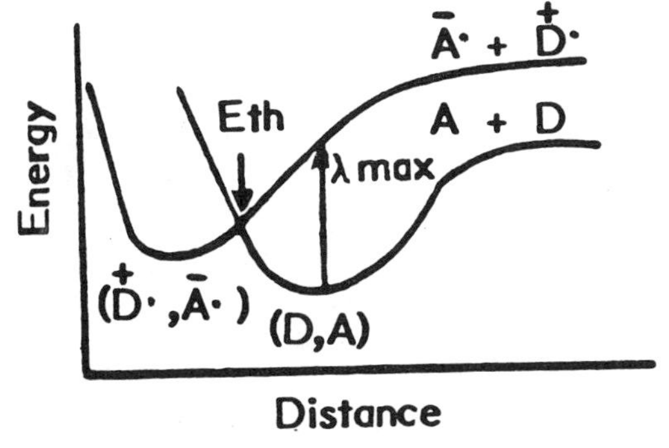

Figure 8–4 Relationship between the no-bond state and the dative state of the CT-complex.

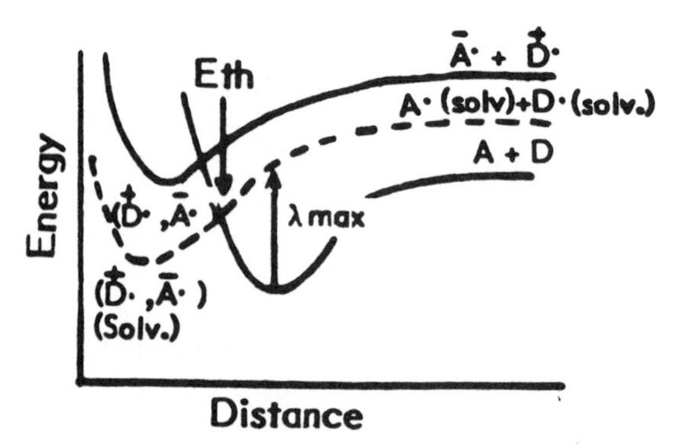

Figure 8–5 Relationship between the no-bond state and the solvated dative state of the CT-complex.

Thermal autopolymerization was carried out in bulk at 80°C in darkness. Photopolymerization was conducted in sealed Pyrex glass tubes using a Hanovia high-pressure UV lamp at room temperature. Although the Pyrex glass filters out UV radiation of shorter wavelength than 290 nm, the UV spectrum of the DA complex of the DVE-MA pair would exhibit singificant absorbance at wavelengths longer than 290 mn. Therefore, excitation of a portion of the no-bond state is still possible. High-energy polymerization was carried out in sealed Pyrex tubes continuously irradiated by τ-rays of a ^{60}Co source at room temperature. The energy supplied was the same in each run.

The results of thermal autopolymerization (Th), photopolymerization (Ph), and τ-ray polymerization (τ-R) were as follows: [Type of polymer, T°C, % yield/time] Th, 80°, 0.5/40 hrs.; Ph, 24°, 3/7 hrs,; τ-R, 25°, 35/1.5 hrs. Copolymerization occurred at a constant rate in each. In thermal autopolymerization, retardation was observed with hydroquinone. In addition, the structure and composition of the copolymers obtained in these three polymerizations were exactly the same as those of copolymers prepared with AIBN. Since the monomers could be thermally excited to initiate polymerization, these observations suggested a radical mechanism for the copolymerization, at least in thermal autopolymerization, since the dative state of the DA complex was not investigated, for initiation by photopolymerization or τ-ray polymerization.

Spontaneous polymerization of this system was examined because the two factors involved could be separately observed, that is, the overall amount of DA complex which does not change significantly with temperature,[100] and the fraction of the dative state of the complex which depends on the population of vibrationally excited state of the no-bond state of the DA complex, which was found to be dependent on temperature (Figures 8-4 and 8-5).

In the more polar solvent, K is smaller and the overall amount of DA complex would also be smaller; therefore, the rate of propagation would be less. In more polar solvent, however, the fraction of DA in the dative state, which is a charged state, would be larger, since it is better stabilized by the polar solvent, and therefore, a larger rate of initiation would be expected. Of course, in the ground state the amount of the dative state is very small compared with the amount of the DA complex itself.

Chloroform (dielectric constant = 4.806, $K^{24°C} = 3.6 \times 10^{-2}$) and methylene chloride (dielectric constant = 9.08, $K^{38°C} = 1.4 \times 10^{-2}$) were chosen here as the solvents to compare the thermal autopolymerization because the polyhalomethanes may not be greatly different in solvation behavior, but have a twofold difference in dielectric constant. The results showed that the initial rate was about three times higher in CH_2Cl_2 at 50°C than in $CHCl_3$, about the same in both solvents at 60°C, and about two times higher in $CHCl_3$ at 70°C than in CH_2Cl_2. This explanation was offered: At 50°C the amount of the dative state was very small in both solvents but much more in polar CH_2Cl_2 than in $CHCl_3$. Initiation seemed to be more important than propagation. Therefore, the rate reflected the difference in the amount of initiating species and was faster in CH_2Cl_2. But at 70°C, vibrational excitation from the no-bond state of the complex to the dative state occurred much more frequently and therefore there was relatively more DA in the dative state in both solvents than at 50°C. The amount of initiation species was not so greatly different and was no longer the controlling factor, but it was the amount of overall DA complex that played a role in the propagation that made the difference in rate. Thus the rate was faster in the less polar $CHCl_3$ in which there was more DA complex for the propagation reaction. Hence the cross-over in the rate-temperature plot of the spontaneous copolymerization that was observed may be explained on this basis.

In an attempt to substitute the isoelectronic molecule, 4-phenyl-1,2,4-triazoline-3,5-dione, to N-phenylmaleimide, for maleic anhydride in the divinyl ether-maleic anhydride copolymerization, it was found that sponteneous copolymerization of the triazolinedione with divinyl ether occurred very rapidly to give a 1:1 alternating copolymer having one residual double bond. In a further study of the spontaneous copolymerization, the intermediate was identified as a 1,4-dipole.[107, 108] This 1,4-dipolar species was found to undergo a cyclization reaction with weakly dipolarophilic alkyl ketones to form a new 1,3,4-tetrahydrooxadiazine ring structure.

This powerful dienophile, 4-phenyl-1,2,4-triazoline-3,5-dione, was found to undergo spontaneous copolymerization with tert-butyl N-vinylcarbamate, *via* a 1,4-dipolar mechanism.[109] Other examples were also described. In a continuing study designed to substitute the powerful dienophile 4-phenyl-1,2,4-triazoline-3,5-dione, for maleic anhydride in a structural study of DIVEMA, a novel intramolecular rearrangement of the 1,4-dipole, derived from the diene and vinyl acetate, was observed, and the structure of the product reported.[110]

Solvent effects in earlier copolymerization studies of electron-donor dienes and electron-acceptor alkenes were in accord with participation of the donor-acceptor complex in the reaction.[28] Evidence for participation of both the donor-acceptor complex and the free monomers was obtained in a photo-initiated copolymerization study of divinyl ether and fumaronitrile by use of different UV filters.[57] Large-ring-containing copolymers formed from 1,13-bis(vinyl ethers) and maleic anhydride were postulated to proceed without cross-linking as a result of the electron-donor character of the diene and the electron-acceptor character of the anhydride.[2-383]

CYCLOCOPOLYMERS AS BIOLOGICALLY ACTIVE AGENTS

A large number of naturally occurring polyanionic materials are known and each performs its designated role in natural processes. In recent years, a wide variety of synthetic polyanions have become available, and investigators have compared their physiological properties with the naturally occurring polyanions. Perhaps the most widely investigated synthetic polyanion is the 1:2 regularly alternating cyclocopolymer

(DIVEMA) of divinyl ether (DVE) and maleic anhydride (MA),[1-87] first discovered in 1951 (See Equation 8–1). This material has been investigated extensively, both from the standpoint of its chemical structure[1-156] and its biological activity.[96, 111, 112] Although certain aspects of its structure remain undetermined,[53] it has been shown to possess a wide spectrum of biological activity.[2, 1-164] It possesses antitumor as well as other activities and is an interferon inducer.[111] Many structural modifications of DIVEMA have been synthesized most of which possess antitumor activity and interferon-inducing capability.[66, 10, 113]

The results of biological investigation on other synthetic polyanions[114-116] prompted submission of DIVEMA of Mn = 15–20,000 to the Cancer Chemotherapy National Service Center, National Institutes of Health for evaluation as a potential antitumor agent. The results of this study[117] in one example showed that the weight of the tumor developed by the test animals was only 11% of that of the control animals.

The level of interest in DIVEMA[2, 1-164] has remained high even though the concept of cyclocopolymerization has long been well established.[1, 1-16, 1-17, 1-87, 1-88] This continued high level of interest has been the result of: (a) continuing studies on the fine structure of the cyclocopolymer and mechanistic details of the cyclopolymerization process,[53] (b) continued interest in the physiological properties of this cyclocopolymer, particularly its antitumor properties[2, 1-164]; (c) its ability to induce the generation of interferon.[111, 112]

Interferon, first discovered in 1957,[118] is a substance of protein-like structure that is produced in the cells of vertebrates in response to viral infection, and possesses antiviral action. Today, it is generally accepted that interferon plays an essential role in the formation of a host's nonspecific resistance to super-infection with a second virus.[119] Thus, the role of interferon in combating viral infections may be similar to that of antibodies toward bacterial infections. It is immediately apparent that a successful and readily available interferon inducer, such as DIVEMA, could play an extremely significant role not only in aiding recovery from a viral infection, but also in its use in a prior generation of interferon to prevent a viral attack.

DIVEMA has often been designated by the misnomer "Pyran copolymer," and unfortunately much of the literature dealing with its physiological properties uses this misnomer. The phenomenal breadth of the physiological properties of this material was recently discussed[48] in a paper that emphasized not only its antitumor, antiviral, and interferon-generating action, but also the fact that it possesses antibacterial and antifungal properties as well. In a more recent review of the biological activity of synthetic polymers,[2, 1-164] it was disclosed that DIVEMA also possesses anticoagulant, and antiarthritic properties, and inhibits inverse transcripture, activates macrophages, and is capable of effecting elimination of plutonium.

The antitumor activity of DIVEMA has been demonstrated by its effectiveness against adenocarcinoma 755, Lewis lung carcinoma, Friend leukemia virus, and Dunning ascites leukemia.[2, 96] The respective minimum effective doses (mg/kg) for the above were 10, 4, 10, and 7.5 and the respective therapeutic indices (maximum tolerated dose/minimum effective dose) were 4, 8, 8, and 8. A comparison of the activity of DIVEMA against Lewis lung carcinoma, a slow-growing solid tumor that is difficult to control with certain other antitumor agents[2] showed that DIVEMA is comparable to cyclophosphamide, a widely used alkylating agent, and far superior to 6-mercaptopurine, an antimetabolite that showed essentially no activity. It has also been shown to be effective against a number of other solid tumors.

The cost of efforts by the scientific and medical community of the world to solve the cancer problem is phenomenal and may never be known. However, in spite of the cost of these efforts, the problem is far from being solved. Much progress has been made, although results have been somewhat discouraging, and support for many areas of related research has been considerably reduced or discontinued. New approaches are being taken, but implications occasionally appear that many consider that problem to be so diverse that a single solution is highly improbable. Consequently, much effort and funding has been diverted to other more immediate health problems.

Although radical surgery was the major therapeutic approach to most cancers during the later nineteenth and early twentieth centuries, the cure rate for many types of cancer by surgical techniques remains the same today as at the turn of the century.[1-182] Modern radiation therapy provided the first realistic alternative to surgery and has been widely used to complement the surgical approach. However, radiation has its limitations since many regions of the body cannot realistically tolerate high doses of radiation therapy due to the accompanying toxicity. For these and other reasons, prevention, early detection, and improved systemic therapy appear to offer the greatest hope for future advances in the control of cancer. The development of chemotherapy has been the most important step in the control of systemic cancer. Immunotherapy as a developing field is also becoming important. Since the introduction of chemotherapy and its extensive use during the last twenty years, the number of cures among newly diagnosed cancer patients has risen from a figure of about 33% as a result of combined surgery and radiation to approximately 55% by 1980. DIVEMA has played a significant role in the chemotherapeutic investigations directed against cancer. Some aspects of these studies are discussed in this section.

DIVEMA was first reported in 1958, although it was discovered in 1951. Structure 1–5 was assigned to this copolymer on the basis of the well-accepted theory that in radical propagation reactions, the more stable radical will predominate as the major propagating species.[1-17, 1-156] However, recent studies using ^{13}C NMR spectroscopy have shown DIVEMA to contain a mixture of five- and six-membered rings with the six-membered ring predominating in the ratio of 56:44.[60]

Several acronyms have been used for the copolymer. The misnomer, "pyran copolymer," was used during the early studies. More recently DIVEMA has been employed. In recent years, a wide variety of synthetic polyanions have become available, and investigators have compared their physiological properties with the naturally occurring polyanions.[1-17, 1-87] DIVEMA, in the form of its sodium salt, may be the most extensively investigated polyanion. More than 300 references deal with its biological properties. The material has been investigated extensively, both from the standpoint of its chemical structure[1-156] and its biological activity.[111, 112] It has been shown to possess a wide spectrum of biological activity.[2]

The broad spectrum of antiviral activity of DIVEMA is further demonstrated by its activity against more than 20 viruses including herpes simplex[120], Rauscher leukemia,[121] Maloney sarcoma[96], vesicular stomatitis,[122] Mengo,[123] MM,[124] and foot and mouth disease.[125] *In vitro* activity against HIV has been recently demonstrated.

The antibacterial activity of DIVEMA is demonstrated against both gram-positive[126] and gram-negative[127] bacteria. It was found that mice treated with DIVEMA become resistant to *Listeria monocytogenes*, an intracellar bacterium, after four days and retained their resistance up to two months.[126, 128] Intraperitoneal (IP) injection of DIVEMA markedly enhanced the survival of mice challenged with *Klebsiella pneu-*

monia.[127] The copolymer also provides mice with enhanced resistance to *Diplococcus pneumonia*[128, 130] and *Pasteurella tularemius*.[128]

DIVEMA also inhibits adjuvant disease,[130] a disease believed to be the result of a hypersensitivity reaction to mycobacterial antigens, and has similarities to rheumatoid arthritis.

The extensive biological investigation of DIVEMA has also shown it to be antifungal[127, 129] and to be an anticoagulant.[131] It provides protection to mice against infection with the fungus, *Cryptococcus neoformans*.[129, 131] Enhanced survival of mice challenged with *Trypanosoma duttoni*[129] or the sporozoite stage of *Plasmodium berghei* was observed following treatment with DIVEMA.[132]

The interferon inducing properties of DIVEMA depend on MW,[133] as do anti-viral properties.[120] The polymer shows appreciable toxicity for MW's in excess of 18,000 to 20,000. It was shown that low MW copolymers with narrow MWD were not only low in toxicity but retained the anti-tumor activity shown by higher MW samples against both Ehrlich adenocarcinoma and Lewis lung carcinoma.[134, 135]

A series of investigations to study the physicochemical interactions of DIVEMA with divalent cations has been carried out.[52, 136, 137] This interaction followed an order of affinity for site binding: $Mn^{2+} > Ca^{2+} > Mg^{2+}$. The formation of a soluble chelate complex was indicated and specific conditions in terms of pH and degree of neutralization were found under which complex formation was preceded by precipitation of microspheres. DIVEMA appears to precipitate under conditions that are close to physiologic. A mechanism was suggested for microsphere formation that was consistent with a hypothesized mechanism of pharmacologic action of DIVEMA-divalent cation complexes.

An unusual example of duplicative effort occurred as part of the background scenario for DIVEMA. It became apparent in 1965 that "quite an active material supplied from an industrial company" and designated as NSC-46015, "Pyran-2-succinic anhydride, 4,5-carboxytetrahydro-6-methylene dianhydride"[138] was the same material as DIVEMA, the structure of which had been published much earlier.[1-87] Further inquiry revealed[139] that the above sample had been submitted as a "'commercial discreet material' and was so classified until recently when the submitters allowed us to declassify their material." Although the structure of DIVEMA had been supplied to NIH,[117] it was further stated[139] "we did not have sufficient information on the polymer supplied... to assign a rigorous structure, thus it was carried in our files with no correlation to NSC-46015."

The results of one specific test with NSC-46015, the conditions of which were comparable to those reported earlier, for the original sample of DIVEMA submitted from the original source,[1-87] showed that the tumor weight developed by the test animals was 14% of that of the control animals, in excellent agreement with 11% reported earlier.[117] However, based on these and additional tests, DIVEMA designated NSC-46051 from the industrial source, was approved for clinical evaluation. It was during the clinical study that DIVEMA was found to possess the ability to induce the generation of interferon.[111, 112]

This broad spectrum of activity quite naturally led to considerable effort by biologists and physicians to explain the polymer's mode of action. One of the first discoveries showed that, under certain conditions, it greatly accelerated the rate of phagocytosis.[140] The extensive biological investigation of DIVEMA has also shown it to stimulate immune response.[141-144] Quite surprisingly, it showed the same activity in immunosuppressed mice as in normal mice.[143] It was shown to inhibit reverse transcriptase in birds, reptiles, and mammals.[145] In a study of a specific avian virus, it was also found to be a potent inhibitor of purified DNA polymerase (deoxynucleosidetriphosphate:DNA deoxynucleotidyltransferase; EC2.7.7.7) isolated from avian myeloblastosis virus (AMV). DIVEMA showed unique features of inhibition unlike other known inhibitors, by interacting with the polymerase at a region other than the template site. The inhibitory effect was somewhat molecular weight dependent, the higher molecular weight samples, $[\eta] = 0.425$ dl/g, being most effective.

Furthermore, in a recent study on therapy of peritoneal murine cancer (Renca) with biological response modifiers (BRM), both MVE-2 (DIVEMA at controlled MW and MWD) and interleukin-2 were used.[146] While both significantly increased the survival time of tumor-bearing mice, only MVE-2 led to definite cures. A single injection of MVE-2 cured 20% of the tumor-bearing mice, while repeated administration of the drug at 12 day intervals cured 70% of the mice. Combined therapy with doxorubicin hydrochloride and a single dose of MVE-2 cured 90% of the tumor-bearing animals.

The superior therapeutic efficiency of MVE-2 compared to that of interleukin-2 may be due to its ability, after inoculation, to generate and maintain high levels of cytotoxic effector cell activity for an elevated period of time within the peritoneal cell population. Additionally, MVE-2 augments effector cell activity in the liver, lungs, spleen, and blood and may therefore more efficiently interfere with metastasis formation in those compartments. The additive effects of MVE-2 and the chemotherapeutic agent suggest that more effective therapy may be achieved by the combination of immunotherapy with BRMs and chemotherapeutic drugs.

A more recent study dealing with interleukin-2 has produced significant and interesting results.[147] The adoptive transfer of tumor-infiltrating lymphocytes (TIL) expanded in interleukin-2 (IL-2) to mice bearing micrometastases from various types of tumors showed that TIL are 50 to 100 times more effective in their therapeutic potency than are lymphokine-activated killer (LAK) cells. Therefore the use of TIL was explored for the treatment of mice with large pulmonary and hepatic metastatic tumors that do not respond to LAK cell therapy. Although treatment of animals with TIL alone or cyclophosphamide alone had little impact, these two modalities together mediated the elimination of large metastatic cancer deposits in the liver and lung. The combination of TIL and cyclophosphamide was further potentiated by the simultaneous administration of IL-2. With the combination of cyclophosphamide, TIL, and IL-2, 100% of mice ($n = 12$) bearing the MC-38 colon adenocarcinoma were cured of advance hepatic metastases, and up to 50% of mice were cured of advance pulmonary metastases. Techniques have been developed to isolate TIL from human tumors. These experiments provide a rationale for the use of TIL in the treatment of humans with advance cancer.

In view of recent reports[145] in which MVE-2 was shown to be superior to IL-2, it follows that MVE-2 should be studied with tumor-infiltrating lymphocytes plus cyclophosphamide, as reported above.[147] A considerable amount of work has shown that DIVEMA activates macrophages.[148, 149] It has also been shown to stimulate T- and B-cells,[150-152] NK cells, and the colony stimulating factors, (CSF).[153, 154] In general, these activities were found to be molecular weight dependent, higher molecular weights leading to greater activity.

The initial polymer investigated was prepared using benzoyl peroxide in a nonsolvent for the polymer, and the polymerization was carried to a very high conversion. As a result, the polymer had a very broad molecular weight distribution (MWD). This polymer underwent Phase 1 clinical evaluation as a chemotherapeutic agent against cancer. Unfortunately, it showed a number of undesirable side effects,[155] and efforts were made to decrease the toxicity without losing the anti-tumor activity. Degradation of the polymer showed that decreasing the molecular weight did indeed decrease the toxicity of the polymer.[48] The polymer was characterized by converting it to its methyl ester and fractionating by size exclusion chromatography (SEC). These fractions, which had reasonably narrow molecular weight distributions, were then compared with polystyrene standards and shown to fit the universal calibration curve. Sufficient quantities of polymer for both characterization and biological evaluation were obtained by a solution polymerization procedure with limited conversion; a comparison of the two polymers by SEC showed that the new polymer designated as MVE-2 had a much narrower MWD than the original polymer. Biological testing showed that this new polymer did indeed have low toxicity and maintained its anti-tumor activity.[48] Meanwhile, however, research done at the National Cancer Institute, as well as at other laboratories, demonstrated that the most desirable way of using this material was as an immunoadjuvant in conjunction with other treatments.[156, 157] The goal was to reduce the tumor burden by some means-surgery, radiation, chemotherapy—and then to use the polymer to stimulate the body to get rid of the last few remaining cancer cells. An IND (Investigative New Drug) clearance was obtained from the FDA and the polymer was investigated clinically in a Phase 1 study in several hospitals. The results of these studies are available.

Recent investigations have shown that MVE-2 prolongs the lifetime of mice inoculated with P-388 lymphocytic leukemia tumor cells by 230% at dosage of 50 mg/kg.[158]

A study of the interaction of DIVEMA with a natural polycation (histone HI) as a model for its *in vivo* function has been carried out.[159] The mechanism of the anti-tumor effect of this anionic polymeric drug, at both the cellular and molecular level is unknown. In this study, a nephelometric titration procedure was used to evaluate the titration of histone HI by DIVEMA. At a constant concentration of histone (10 ug/ml), maximum turbidity was obtained at a DIVEMA concentration of 2 ug/ml. This ratio of DIVEMA to histone, 1:5, was used throughout as representing the most favorable concentrations for complex formation.

By varying the complex concentration, while maintaining the 1:5 ratio, three distinct phases could be demonstrated. At low concentrations, a soluble phase existed. At the intermediate concentration, the components formed a coacervate phase, and at high complex concentration, an aggregate phase formed. The effect of ionic strength on the stability of preformed DIVEMA-histone complex was seen to be minimal over the physiologic range of sodium as judged by particle size determined from small angle light scattering. The titration of histone by DIVEMA was found to be much more sensitive to ionic strength. Concentrations of sodium ion higher than 200 Mm blocked the apparent interaction of the components. Magnesium ions were also effective in blocking the DIVEMA interaction, and formed complexes with DIVEMA at concentrations above 3 Mm.

An effort was made to identify the stoichiometry of the DIVEMA-histone complex using a dye-titration method with meso-tetra(4-N-methylpyridyl)porphyrin. This cationic dye binds to DIVEMA by an electrostatic interaction. When the DIVEMA-histone complex was present at the 1:5 ratio, the binding characteristics indicated that one porphyrin molecule was bound every 6.43 DIVEMA monomer units, compared to one porphyrin every 1.93 DIVEMA monomer units in the control.

The mode of administration used in the clinical studies of DIVEMA and MVE-2 was by intravenous (IV) injection. An alternate route of administration [intraperitoneal (IP) injection] of MVE-2 (mean mol. wt. of 11,000) was studied[31] which may be less toxic and more efficacious than the present IV route of administration. Renal toxicity and proteinuria were observed after IV administration of MVE-2 (25 mg/kg) to dogs twice a week for up to one month. In contrast, when MVE-2 was given IP, there was no evidence of renal toxicity. Furthermore, macrophase accumulation in the mesenteric lymph nodes, mediastinal lymph nodes and the peritoneal cavity were increased. Tissue concentration of ^{14}C in mesenteric lymph nodes and thymus were much higher after an IP than after an IV dose of [^{14}C]MVE-2. Remarkably high concentrations of ^{14}C were also observed in the mediastinal lymph nodes following an IP dose. These distribution studies indicated that a portion of the drug administered IP was absorbed directly into the lymphatic system draining the peritoneal cavity. Furthermore, IP administration may provide direct stimulation of peritoneal macrophages and also modulate immune function through direct effects on lymph nodes trapping the drug.

Many attempts to modify DIVEMA to enhance its effectiveness as well as reduce its toxicity have been made. Among these is a recent patent which describes a procedure for administration of DIVEMA in the form of its partial calcium salt.[160] Solid DIVEMA Ca^{2+} salts (mol. wt. 2000–100,000, degree of neutralization 5–70%) were as effective as the corresponding alkali metal salts in treating tumors or viruses, and were much less toxic. Thus, a solution of 10 parts of DIVEMA (mol. wt. 10,000 polydispensity 2.0) in 100 parts H_2O was neutralized with 2N NaOH to pH 7.2. Adding 20 parts 10% aq. $CaCl_2$ slowly to 50 parts of this solution with stirring and adding 30 parts physiological saline solution gave a solution of polymer with 20% CO_2H groups neutralized by Ca^{2+} salt. The toxicity (mortality to mice in 14 days) of this solution was 0/10 at 750 mg/kg body weight, compared with 9/10 at 90 mg/kg for the Na salt. The average increase in survival time for mice with Colon-26 tumors treated with 12.5 mg/kg 40% neutralized Ca salt was 210%, compared with 218 with the Na salt.

In addition to DIVEMA, a wide variety of related copolymers have been synthesized and evaluated.[32, 161] Many of these have shown potential to be as effective as anti-tumor agents as DIVEMA. For example, the copolymer of furan and maleic anhydride has been found to be quite effective, as well as the copolymer of divinyl ether and citraconic anhydride. Also, 5-fluorouracil, a proven and effective anti-tumor agent, has been immobilized on DIVEMA by covalent bonding, as well as certain related copolymers, and all have been shown to be quite effective.

A number of copolymers prepared have been shown to possess specific anti-tumor activity; however, these materials were evaluated initially without regard to molecular weight or molecular weight distribution. Recent investigations have dramatically demonstrated the importance of these properties in controlling both activity and toxicity of such materials.[2] Of 58 copolymers synthesized in this earlier study, 15 were shown to possess significant anti-tumor properties (See Table 8–10). In view of these studies, it was considered highly desirable that these materials be reevaluated as anti-viral agents, with due consideration being given to these important variables.

The samples originally submitted to NIH were those obtained in the initial preparative experiment in most cases; and although their fundamental properties were determined for further identification purposes, they were submitted without regard to MW or MWD. In view of the recent observations of the importance of both MW and MWD of

DIVEMA on its biological properties such as toxicity and anti-tumor effectiveness, a program of reevaluation of the 15 above-identified copolymers was initiated with particular emphasis being placed on control of MW and MWD. The resulting samples were then evaluated against P-388 lymphocytic leukemia in CD_2F_1 mice.

Following resynthesis and fractionation of these 15 copolymers,[32, 161] considerably reduced toxicity and increased effectiveness were observed in certain of the samples, consistent with the postulate that higher molecular weights are toxic. Typical results of these studies are summarized in Table 8–11.

The most effective copolymer was that derived from divinyl ether and citraconic anhydride (NSC 133788). The most probable structure for this material is shown (8–23):

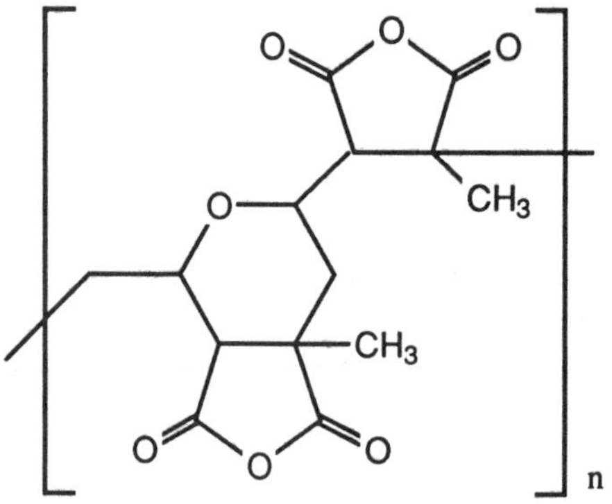

1 : 2 Copolymer of Divinyl Ether and Citraconic Anhydride

(8–23)

The 1:1 copolymer of furan and maleic anhydride, designated as NSC 119166, and its half-amide, half-ammonium salt, designated as NSC 119167, were evaluated[162] for their antitumor activity. The test results showed a T/C value of 130 at the optimum dose of 400 mg/kg for NSC 119166. The result for NSC 119167 was 125 at optimum dose of 400 mg/kg. As shown in Table 8–11, the optimum sample after fractionation gave T/C = 147 at optimum dose of 50 mg/kg for both NSC 119166 and NSC 119167.

In addition, the copolymer of furan and itaconic anhydride, designated as NSC 199165, and its half-amide, half-ammonium salt, designated as NSC 119168, were synthesized,[7] and the copolymer derivative was evaluated.[162] The test result showed T/C = 150 at optimum dose of 600 mg/kg. This polymer has not been fractionated for further evaluation.

The ratios of survival times for test animals to control animals in all cases were greater than one, and in the latter case this ratio reached approximately 1.5 at the optimum dosage level of 600 mg/kg of body weight. The results of these studies have been summarized.[163]

The antitumor activity of a number of structures closely related to DIVEMA were recently discussed.[66] The copolymers were prepared in a manner similar to that reported earlier for synthesis of DIVEMA.[163]

Antitumor activity of varying degree appears to be a general property of copolymers related in structure to DIVEMA. The corresponding copolymer of divinyl ether and citraconic anhydride, designated as NSC 133788, (8–23) has been evaluated by the

Table 8–10 Identification of Copolymers Prepared That Showed Significant Antitumor Properties

NSC No.	Copolymer name	Passed NCI intital test against
59195	Divinylcyclopentamethlyenesilane-MA[b]	1
59198D	Dimethyldivinylsilane-MA	1
84650	1,4-Pentadiene-MA, $[\eta] = 0.26$dl/g	5
84653	DVE-N-ethylmaleimide	4
99425	1,4-Pentadiene-MA-$BrCCl_3$ telomer	2
104304	4-Vinylcyclohexene-MA	6
119166	Furan-maleic anhydride	2
119167	Furan-maleic anydride-half amide	2
119168	Furan-itaconic anhydride-half amide	2,3
133788	DVE-citraconic anhydride	2
133789	β-Chloroethylvinylether-citraconic anhydride	2
255081	Styrene-CMAFU	3
255082	β-Chloroethylvinylether-CMAFU	3
255083	DVE-CMAFU	3
266066	DVE-maleimide of d,l-leucine	3

[a] Tumor identification:
1. Adenocarcinoma 755
2. Ll210 Lymphoid Leukemia
3. P388 Lymphocytic Leukemia
4. Sarcoma 180
5. Lewis Lung Carcinoma
6. Walker Carcinosarcoma

[b] MA = maleic anhydride

[c] CMAFU = 2-carboethoxyacryloyl-1-(5-fluroouracil)

Program Analysis Branch for Drug Research and Development, National Cancer Institute.[162] In evaluating these copolymers, lymphoid leukemia cells were injected into test animals by the intraperitoneal route on day zero. Dosages of the test drug were calculated on a mg/kg of body weight basis, dissolved in saline, and injected by the intraperitoneal route on day one. The mean survival time in days of the test group and the control group was determined and a ratio of test animals to control animals was calculated. In all tests, the animals were evaluated at five days, for survival, as a measure of drug toxicity. All data presented represented six of six survivors in each test group. A T/C value of 130 at optimum dose of 400 mg/kg was obtained for this material. After fractionation, a TC value of 267 at optimum dose of 6 mg/kg was obtained for this NSC No. 133788 (See Table 8–11). These results emphasize the importance of removing the higher molecular weight toxic fraction, and using a fraction of narrow MWD.

DVE, styrene and β-chloroethyl vinyl ether were copolymerized with 1-(2-carbomethoxyacryloyl)-5-fluouracil (CMAFU)[113] to yield copolymers of molecular weights in the range of 5000 to 7500. The styrene (St) and chloroethyl vinyl ether (CEVE) copolymers were shown by analysis to be 1:1 alternating copolymers, but the

Table 8–11 Typical Results of Anti-Tumor Evaluation of Copolymer Fractions

NSC No.	Sample No.	M_n^a	M_w^b	MWD[c]	Min. Effect. Dose mg/kg	T/D[d] %
59198	CHP-011-3	0.069[e] (2N NaOH)			25	123
84650	83-29G	8,386	10,231	1.22	50	121
84653	T9024B-2	18,400	30,100	1.64	50	121
119166	83-38A	6,500	12,500	1.92	50	147
119167	83-38B	6,100	11,300	1.85	50	147
	HFAS					
119168	11-NMT-13	4,400	8,200	1.86	100	147
133788	DHMT-6A	—	—	—	6	267
	DHMT-72B	22,400	28,900	1.29	50	207
133789	9-7-2	6,420	14,600	2.28	25	197
255081	CHP-007-5	0.028[e](THF)			50	152
255082	CHP-006-3	0.029[e](THF)			100	180
255083	CHP-005-1	0.175[e] (2N NaOH)			25	126

[a] M_n = Number average molecular weiqht.

[b] M_w = Weight average molecular weight.

[c] MWD = M_w/M_n = molecular weight distribution.

[d] T/C(%) = Increase in mean survival time of treated over control animals. Criterion for pass T/C $\geqslant$ 120%.

[e] $[\eta]$ = Intrinsic viscosity, dl/g; ( ) = solvent.

Table 8–12 Results of Evaluation of Polymers by Development Therapeutics Program, Division of Cancer Treatment, National Cancer Institute[194]

Dose[a] (mg/kg)	T/C[b] (%) St/CMAFU (NSC 255081)	CEVE/CMAFU (NSC 255082)	DVE/CMAFU (NSC 255083)
400	94	129	112
200	192	165	189
	144	141	172
100	171	165	137
	127	127	140
50	133	123	125
25	115	117	105

[a] Dose in mg/kg of animal weight *via* intraperitoneal injection in CD_2F_1 male mice, 6 of 6 survivors after fifth day against P 388 lymphocytic leukemia.

[b] Ratio of test animal (T) to control (C) survival times.

DVE copolymer was 2:1 molar in CMAFU:DVE content. The results of the evaluation of these copolymers by the Development Therapeutics Program, Division of Cancer Treatment, National Cancer Institute[162] are shown in Table 8–12.

5-Fluorouracil was released from these copolymers hydrolytically, and their respective rates of release were determined from dispersions in 0.5 M NaCl. Under these conditions, the monomer (CMAFU) released 5-FU almost instantly and completely. On the other hand, the St:CMAFU and CEVE:CMAFU copolymers were resistant to hydrolysis and released 5-FU gradually, DVE:CMAFU was hydrolyzed faster than the other two copolymers, but much slower than the monomer. The hydrophobic character of the St:CMAFU copolymer appears to be important for slow release of 5-FU.

Polymers structurally related to DIVEMA have been synthesized and their biological activities evaluated by the National Institute of Health[162] program. These polymers are presented in this section along with a typical test result, based upon six of six survivors in each case. To select a result meeting the criteria of six of six survivors, unfortunately, it will not be possible to make meaningful comparisons among the wide variety of polymer structures presented. However, it should be kept in mind that extensive test results are available on each polymer evaluated. These results can be made available upon request.

Polymers containing carboxyl or carboxylic anhydride groups along with typical anti-tumor activities[162] based upon six of six survivors are included in Table 8–13.

A variety of imide derivatives were also prepared by copolymerization of DVE with the appropriate maleimides. The polymers may be considered as derivatives of DIVEMA. Although in a few isolated cases, the method used for synthesis may have been via reaction of DIVEMA with the appropriate primary amine, in general, the method of synthesis was via copolymerization of DVE with the appropriate maleimide derivative. Thus, it can be assumed that the polymers were structurally homogeneous rather than possessing the variety of functional groups that would be expected to be present via the derivatization method. The antitumor evaluations were carried out by the National Cancer Institute, National Institutes of Health.[162] The results of this aspect of the study were not particularly encouraging and are not shown.

Several copolymers of divinylsulfone were evaluated against L-1210 Lymphoid leukemia or Sarcoma 180; however, the results were not encouraging, and no further tests were carried out. None of these copolymers has been fractionated for reevaluation of an optimum sample.

DIVEMA[164] of molecular weight 5,000–30,000 was useful in the treatment of Ehrlich adenocarcinoma and encephalomycarditis virus infection in mice. The molecular weight distribution was 1.5–2.5. It was also utilized as a major component of an aqueous dispersion,[165] and as an inhibitor of Friend leukemia virus infection in mice.[116] In the latter application, it was postulated to involve interferon induction. It has also been shown to be effective is suppressing foot-and-mouth disease[166] when administered at 0.5–300 mg/kg daily in physiologically tolerated salt solutions. It not only suppressed growth of the virus in infected animals but also increased resistance to virus infection in healthy animals[167] by stimulating the production of interferon.

A large number of additional studies on the biological properties of DIVEMA and its modified forms have been reported. These studies are summarized in Section 11.3 of this volume.

A summary of the biological activity of DIVEMA including its anti-tumor properties has been published.[28] Improved performance and lower toxicity were exhibited by a

Table 8–13 Selected Antitumor Activity Results on Carboxyl-Containing Polymers of Copolymers

NSC No.	Polymer or Copolymer	Tumor	Dose mg/kg	T/C %
D59195	Divinylcyclopentamethylenesilane	CA	100	117
D59196	DVE-MA	LE	45.0	122
D59197	Divinyl sulfone-MA	LE	100	110
D59198	Dimethyldivinysilane-MA	SA	12.5	81
D59199	1,4-Pentadiene-MA	SA	30	49
84645	$DVSO_2$-Methacrylic acid	LE	400	101
84649	1,4-Pentadiene-MA $[\eta]$ = 0.18 dl/g	LE	50.0	95
84650	1,4-Pentadiene-MA $[\eta]$ = 0.26 dl/g	LL	10.0	43
99425	1,4-Pentadiene-MA-$BrCCl_3$ (telomer)	LE	100	146
99427	DVE-MA-$BrCCl_3$ (telomer)	LE	200	107
104,304	4-Vinylcylcohexane-MA	WM	100	92
119168	Furan-Itaconic anhydride (half amide)	LE	177	137
133,788	DVE=Citraconic anhydride	LE	132	103
148132	2-Vinylnorbornene-MA	LE	80.5	106
148134	DVE-p-carboxyphenylmaleimide	PS	3.12	110
255083	DVE-CMAFU	PS	200	189
			50	125
266062	DVE-Maleimide of d,1-alanine	PS	20.0	117
266066	DVE-Maleimide of d,1-leucine	PS	7.5	125

MA = Maleic anhydride DVE = Divinyl ether CMAFU = 2-Carboethoxyacryloyl-1-(5-fluorouracil) LL = Lewis lung carcinoma LE = 1–1210 Lymphoid leukemia CA = Adenocarcinoma 755 SA = Sarcoma 180 WM = Walker carcinosarcoma 256 PS = P388 Lymphocytic leukemia

modified form of DIVEMA, designated as MVE-2, when both MW and MWD of the copolymer were controlled by the synthetic method.[31] The initial observation that polycarboxylates possess anti-tumor properties and continual studies has led to a derivative of an ethylene-maleic anhydride copolymer known as "Carbethimer" of improved anti-tumor properties and lowered toxicity.[63]

Polyanions, in general, appear to possess anti-tumor properties, depending on molecular weight, molecular weight distribution and toxicity. A polyanion having lowered toxicity relative to DIVEMA has been synthesized.[64] During an early study of the anti- tumor properties of cyclocopolymers related to DIVEMA it was shown that the analogous copolymer of divinyl ether with citraconic anhydride produced potentially useful anti-tumor agents as well.[66] DIVEMA was found to be a potent inhibitor of purified DNA topoisomerase I from human spleen.[168] The polymer completely inhibited the relaxing activity of the topoisomerase at a concentration of 0.09 μg/ml. The inhibition was reversible.

REFERENCES

1. Butler, G. B., U.S. Pat. US 3,320,216, (Reissue 26,407), 1967 (May 16); CA 67, 22329z (1967).
2. Breslow, D. S., Pure & Appl. Chem., 1976, 46, 103; CA 86, 91163v (1977).
3. Butler, G. B., Gropp, A. H. Angelo, R. J. Husa, W. J., and Jorolan, E. P., 5th Quart. Rep. AEC Contr. AT(40-1)1353, 1953, (Sept. 15).

4. Stackman, R. W., J. Macromol. Sci. (Chem.), 1971, A5, 251.
5. Raetzch, M. and Schmieder, H., Plaste Kaut., 1974, 21, 182; CA 81, 92002a (1974).
6. Butler, G. B. and Campus, A. F., J. Polym. Sci. A1, 1970, 8, 523; CA 72, 111892y (1970).
7. Butler, G. B., Badgett, T. and Sharabash, M., J. Macromol. Sci. (Chem), 1970, 4, 51; CA 72, 21963t (1970).
8. Ragab, Y. A. and Butler, G. B., Polym. Lett., 1976, 14, 273.
9. Umrigar, P. P., M. S. Thesis, Univ. of Fla., 1972.
10. Butler, G. B. and Zampini, A., J. Macromol. Sci. (Chem.), A, 1977, 11, 491; CA 86, 121818y (1977).
11. Barton, J. M., Butler, G. B., and Chapin, E. C., J. Polym. Sci. A, 1965, 3, 501; CA 62, 14830g (1965).
12. Butler, G. B. and Kasat, R. B., J. Polym. Sci. A, 1965, 3, 4205; Ca 64, 11321h (1966).
13a. Butler, G. B., VanHaeren, G. and Ramadier, M. F., J. Polym. Sci. A1, 1967, 5, 1265; CA 67, 32975w (1967).
13b. Ham, G. E., Copolymerization, Intersci. Publishers, 1964, 81.
14. Butler, G. B. and Joyce, K. C., J. Polym. Sci. C, 1968, 22, 45; CA 69, 52488j (1968).
15. Butler, G. B. and Sharpe, Jr., A. J., J. Polym. Sci. B, 1971, 9, 125; CA 74, 112433u (1971).
16. Fujimori, K., Ph.D Dissertation, Univ. of Florida, 1971; CA 77, 34999n (1972)
17. Pledger, H. Jr. and Butler, G. B., J. Macromolec. Sci. (Chem.), 1971, 5 (8), 1351; CA 76, 60145e (1972).
18. Butler, G. B. and Sharpe, A. J. Jr., ACS Polym. Preprints, Div. Polym. Chem., 1970, 11 (1), 42; CA 76, 4202i (1972).
19a. Guilbault, L. J. and Butler, G. B., J. Macromolec. Sci. (Chem.), 1971, 5, 1219; CA 76, 4194h (1972).
19b. Hine, J., Physical Organic Chemistry, 2nd.Ed., 1962, Chapter 8.
20. Allen, V. R. and Turner, S. R., J. Macromolec. Sci. (Chem.), 1971, 5, 227; CA 73, 121002t (1970).
21. Fujimori, K. and Butler, G. B., J. Macromol. Sci. (Chem), A, 1972, 6, 1609; CA 78, 44062m (1973).
22. Fujimori, K. and Butler, G. B., J. Macromol. Sci. (Chem) 1973, 7, 415; CA 78, 84916z (1973).
23. Fujimori, K. and Butler, G. B., J. Macromol. Sci. (Chem) 1973, 7, 387; CA 78, 84915y (1973).
24. Fujimori, K., J. Macromol. Sci. (Chem), A, 1976, 10, 999; CA 85, 160602f (1976).
25. Breslow, D. S., Polym. Prepr., 1981, 22, 24.
26. Espy, H. H., U.S. Pat. US 3,244,943, 1965 (Dec. 21); CA 64, 14038f (1966).
27. Peninsular ChemResearch, Inc., Brit. Pat. 921,462, 1963 (Mar. 20); CA 59, 1776c (1963).
28. Butler, G. B. and Fujimori, K., J. Macromol. Sci. (Chem), 1972, 6, 1533; CA 78, 44061k (1973).
29. Breslow, D, S., U.S. Pat. US 3,794,622, 1972 (Feb. 26); CA 81, 106369q (1974).
30. Butler, G. B., J. Macromol. Sci. (Chem.A), 1979, 13, 351; CA 91, 96530a (1979).
31. Baldwin, J. R., Carrano, R. A., Imondi, A. R., Iuliucci, J.D., Hagerman, J. R., Polym. Sci. Technol. (Plenum), 1985, 32, 139; CA 105, 90887b (1986).
32. Butler, G. B., Tollefson, N. M., Gifford, G. E. and Flick, D. A., J. Bioactive and Compatible Polymers, 1987, 2, 206; CA 108, 118811s (1988).
33. Kodaira, T. and Akiguchi, T., Kenk. Hok-Asa. Gara. Kog. Gijut. Shore., 1987, 51, 77; CA 109, 150148n (1988).
34. Chu, Y. C., Diss. Abstr. Int. B, 1978, 38, 3218; CA 88, 90116m (1978).
35. Avetyan, M. G., Darbinyan, E. G. and Matsoyan, S. G., Izv. Akad. Nauk armyan. SSR, Khim. Nauki 1963, 16, 247; CA 60, 666c (1964).

36. Darbinyan, E. G., Avetyan, M. G. and Matsoyan, S. G., Arm. Khim, Zh., 1966, 19, 527; CA 66, 37578h (1967).
37. Paushkin, Y., Terekhova, G., Ponomarenko, V. and Shoronov, V., Neftepererab, Neftekhim (Moscow), 1971, 30; CA 76, 24799n (1972).
38. Simek, I. and Komova, L., Chem. Zvesti., 1963, 17, 757; CA 60, 9362g (1964).
39. Georgiev, G. S., God. Sofii. Univ., Khim. Fak., 1977 (Pub. 1981), 72 Pt.1, 131; CA 96, 123404z (1982).
40. Foster, R., Organic Charge Transfer Complexes,Ac.P., 1969, CA 72, 30757f (1970)
41. Mulliken, R. S. and Person, W. B., Molecular Complexes, Wiley-Int., 1969; CA 73, 10303x (1970).
42. Kosower, E. M., Progress in Physical Organic Chemistry, 1965, (Wiley), 3, 81; CA 65, 5348c (1966).
43. Butler, G. B. and Campus, A. F., J. Polym. Sci. A1, 1970, 8, 545; CA 72, 111891x (1970).
44. Allen, V. R. and Turner, S. R., J. Macromol. Sci. (Chem.), 1971, A5, 227.
45. Butler, G. B. and Wu, C., Polym. Sci. and Tech.(Water Sol. Polym), 1973, 2, 369; CA 81, 136638u (1974).
46. Morawetz, H., Macromolecules in Solution, Wiley, 1965, Ch. VII.
47. Munson, A. E., Regelson, W., Lawrence, Jr., W. and Wooles, W. R., J. Reticuloendothel. Soc., 1970, 7, 375.
48. Breslow, D. S., Edwards, E. I. and Newburg, N. R., Nature, 1974, 246, 160; CA 80, 115950d (1974).
49. Regelson, W., Miller, G., Breslow, D. S. and Engle,III, E. J., Reticuloendothel. Soc., Abstr. 6 An. M., 1969, Paper 13.
50. Samuels, R. J., Polymer, 1977, 18, 452; CA 87, 185064m (1977).
51. Zeegers, B. and Butler, G. B., J. Macromol. Sci. (Chem), 1972, 6, 1569; CA 78, 124965y (1973).
52. Levine, H. I., Mark, E. H. and Fiel, R. J., Polym. Prepr., ACS, Div. Polym. Chem., 1978, 19, 570; CA 93, 168750p (1980).
53. Butler, G. B. and Chu, Y. C., J. Polym. Sci., Polym. Chem. Edn. 1979, 17, 859; CA 90, 187494n (1979).
54. Kunitake, T. and Tsukino, M., J. Polym. Sci., Polym. Chem. Ed., 1979, 17, 877; CA 90, 204609y (1979).
55. Butler, G. B. and Zeegers, B., Polym. Prepr., ACS, Div. Polym. Chem., 1971, 12, 420; CA 78, 30286a (1973).
56. Shopov, I., Izv. Otd. Khim. Nauki, Bulg. Akad. Nauk. 1970, 3, 483; CA 74, 126123p (1971).
57. Zeegers, B. and Butler, G. B., J. Macromol. Sci. (Chem) 1973, 7, 349; CA 78, 84914x (1973).
58. Butler, G. B. and Campus, A. F., Polymer Preprints, 1968, 9 (2), 1266; CA 72, 111891x (1970).
59. Sharpe, A. J., Jr., Ph.D. Thesis, U. of Florida, 1970, CA 75, 130144j (1971).
60. Freeman, W. J. and Breslow, D. S., ACS Symposium Ser., 1982, 186, 243; CA 97, 56440p (1982).
61. Kunitake, T. and Yamaguchi, K., J. Polym. Sci., Chem. Ed., 1973, 11, 2077; CA 80, 27491z (1974).
62. Butler, G. B. and Wu, C. C., Macromol. Synth., 1982, 8, 89; CA 98, 35058a (1983).
63. Fields, J. E., Asculai, S. S., Johnson, J. H. and Johnson, R. K., J. Med. Chem., 1982, 25, 1060; CA 97, 66070g (1982).
64. Ottenbrite, R. M., Enright, N., Munson, A. and Kaplan, A., Polym. Preprints, 1979, 20, 600; CA 94, 167605d (1981).

65. Trofimov, B. A., Gusarova, N. K., Nikol'skaya, A. N., Amosova, S. V. and Baranskaya, N. A., Zh. Prikt. Khim. (Leningrad), 1984, 57, 1574; CA 101, 152426k (1984).
66. Butler, G. B., J. Polym. Sci., C; IUPAC Symp., Madrid, 1975, 50, 163; CA, 44685r (1976).
67. Stille, J. K. and Thompson, D. W., J. Polym. Sci., 1962, 62, S118; CA 60, 671d (1964).
68. Meyersen, K. and Wang, J. Y. C., J. Polym. Sci. A1, 1967, 5, 1827; CA 67, 82419p (1967).
69. Meyersen, K. and Wang, J.Y.C., J. Polym. Sci. A1, 1967, 5, 1845; CA 68, 50100g (1968).
70. Yamaguchi, T. and Ono, T. Chemy. Ind., 1968, 769; CA 69, 27825a (1968).
71. VanHaeren, G. and Butler, G. B., Polym. Preprints, 1965, 6, 709; CA 66, 46634a (1967).
72. Ramp, F. L., J. Macromol. Sci. (Chem.) A, 1967, 1, 603; CA 66, 29137y (1967).
73. Duck, E. W., Locke, J. M., and Thomas, M. E., Polymer, 1968, 9, 60; CA 68, 59956y (1968).
74. Dowbenko, R., Brit. Pat. 1,015,215, 1966, CA 64, 8399d (1966).
75. Schmerling, L., U.S. Pat. 2,930,781, 1960, (Mar. 29), CA 54, 16927a, (1960).
76. Freidlin, G. N. and Solop, K. A., Vysokomol. Soedin. , Ser. A, 1973, 15, 575; CA 79, 19213z (1973).
77. Kimbrough, R. D., J. Polym. Sci.B., 1964, 2(B), 85.
78. Fukui, K., Hayashi, K., Yonezawa, T., Nagata, C. and Okamura, S., J. Polym. Sci., 1956, 20, 537.
79. Benesi, H. A. and Hildebrand, J., J. Am. Chem. Soc., 1949, 71, 2703.
80. Mulliken, R. S., J. Am. Chem. Soc., 1952, 74, 811.
81. Scott, R. L., Rec. Trav. Chim., 1956, 75, 787.
82. Person, W. B., J. Amer. Chem. Soc., 1965, 87, 167.
83. Bartlett, P. and Nozaki, K., J. Amer. Chem. Soc., 1946, 68, 1495.
84. Barb, W., Trans. Faraday Soc., 1953, 49, 143.
85. Walling, C., Briggs, E., Wolfstirn, K. and Mayo, F. R., J. Amer. Chem. Soc., 1948, 70, 1537.
85. Robson, R., Grubb, P. W. and Baltrop, J. A., J. Amer. Chem. Soc., 1964, 2153.
87. Iwatsuki, S. and Yamashita, Y., Makromol. Chem., 1965, 89, 205.
88. Iwatsuki, S. and Yamashita, Y., Makromol. Chem., 1967, 102, 232.
89a. Job, P., Compt. Rend., 1925, 180, 928.
89b. Job, P., Ann. Chim. Phys., 1926, 9, 113.
90. Vosburgh, W. C. and Cooper, G. R., J. Amer. Chem. Soc., 1941, 65, 437.
91. Williams, J. K., Wiley, D. W. and McKusick, B. C., J. Amer. Chem. Soc., 1962, 84, 2210.
92. Price, C. C. and Schwan, T. C., J. Polym. Sci., 1955, 16, 577.
93. Briegleb, G., Springer Verlag, Publishers, 1961.
94. Hanna, M. W. and Ashbaugh, A. L., J. Phys. Chem., 1964, 68, 811.
95. Foster, R. and Fyfe, C. A., Trans. Faraday Soc., 1965, 61, 1626.
96. Chirigos, M. A., Comparative Leukemia Research 1969, 1970, 36, 278.
97. Mayo, F. R., Lewis, F. M. and Walling,, C., J. Amer. Chem. Soc., 1948, 70, 1529.
98. McConnell, H., Ham, J. S. and Platt, J. R., J. Chem. Phys., 1953, 21, 66.
99. Huong, H. V., Platzers, N. and Josein, M. L., J. Amer. Chem. Soc., 1969, 91, 3669.
100. Kokubo, T., Iwatsuki, S. and Yamashita, Y., Macromolecules, 1968, 1, 482.
101. Pryor, W. A., Free Radicals, McGraw-Hill, NY, 1966, 13.
102. Yoshimura, M., Mikawa, H. and Shirota, Y., Macromolecules, 1978, 11, 1085.
103. Georgiev, G. S. and Zubov, V. P., European Poly. J., 1978, 14, 93.
104. Fujimori, K., Master's Thesis, Univ. of Florida, 1968.
105a. Tazuke, S. and Okamura, S., J. Polym. Sci., Part B, 1965, 3, 923.

105b. Tazuke, S. and Okamura, S, J. Polym. Sci., Part B, 1967, 5, 453.
106a. Ellinger, L. P., Polymer, 1964, 5, 559.
106b. Ellinger, L. P., Polymer, 1965, 6, 549.
107. Butler, G. B., Turner, S. R. and Guilbault, L., J. Org. Chem., 1971, 36, 2838; CA 75, 129777m (1971).
108. Butler, G. B., Guilbault, L. and Turner, S. R., J. Polym. Sci.,B, 1971, 9, 115; CA 74, 112476u (1971).
109. Guilbault, L. J., Turner, S. R. and Butler, G. B., J. Polym. Sci. B, 1972, 10, 1; CA 76, 113610m (1972).
110. Wagener, K., Turner, S. R. and Butler, G. B., J. Org. Chem., 1972, 37, 1454; CA 77, 75176h (1972).
111. Merigan, T. C., Nature, 1967, 214, 416.
112. Merigan, T. C., New Eng. J. Med., 1967, 1283.
113. Umrigar, P. P., Ohashi, S. and Butler, G. B., J. Polym. Sci., Chem., 1979, 17, 351.
114. Breslow, D. S. and Hulse, G. E., J. Am. Chem. Soc., 1954, 76, 6399.
115. Regelson, W. and Holland, J. F., Nature (Lond.), 1958, 181, 46.
116. Regelson, W., Advan. Exp. Med. Biol., 1967, 31, 315, CA 69, 75392n (1968).
117. Butler, G. B., National Institutes of Health, 1959 (Oct. 12).
118. Isaacs, A. and Lindenmann, J., Proc. Roy. Soc. (London) Ser. B, 1957, 147, 258.
119. Murthy. Y. K. S. and Anders, H. P., Angew. Chem. Int. Ed., 1970, 9, 480.
120. Morahan, P. S., Cline, P. F., Breining, M. C. and Murray, B. K., Antimicrob. Agents Chemother., 1979, 15, 547.
121. Chirigos, M. A., Turner, W., Pearson, J. and Griffin, W., Int. J. Cancer, 1969, 4, 267.
122. DeClercq, E. and Merigan, T. C., J. Gen. Virol., 1969, 5, 359.
123. Merigan, T. C. and Finkelstein, M. S., Virology, 1968, 35, 363.
124. Schmidt, J. P., Pindak, F. F., Giron, D. J. and Ibarra, R. R., Texas Rep. Biol. Med., 1971, 29, 133.
125. Richmond, J. Y., Infect. Immun., 1971, 3, 249.
126. Remington, J. S. and Merigan, T. C., Nature (London),1970, 226, 361.
127. Pindak, F. F., Infect. Immun., 1970, 1, 271.
128. Giron, D. J., Schmidt, J. P., Ball, R. J. and Pindak, F. F., Antimicrob. Agents Chemother., 1972, 1, 80.
129. Regelson, W., Munson, A. and Wooles, W., Symp. Series Immunobiolog. Standards, 1970, 14, 227.
130. Kapusta, M. A. and Mendelson, J., Arthr. Rheum., 1969, 12, 463.
131. Shamash, Y. and Alexander, B., Biochem. Ciophys. Acta., 1969, 194, 449.
132. Van Dijck, P. I., Clausen, M. and DeSomer P., Ann. Trop. Med. Parasitol, 1970, 65, 5.
133. National Institute of Health., Cancer Chemo. Nat. Ser. Center Reports, 1961 (July).
134. Karrer, K., Humphreys, S. R. and Goldin, A., Int. J. Cancer., 1967, 2, 213.
135. Hellmann, K. and Burrage, K., Nature, 1969, 224, 273.
136. Levine, H. I., Mark, E. H. and Fiel, R. J., Arch. Biochem. and Biophysics., 1977, 184, 156.
137. Scheintaub, H. M. and Fiel, R. J., Arch. Biochem. and Biophysics., 1973, 158, 171.
138. Nelson, E. R., Letter from Drug. Dev. B. Can. Chemo.NIH 1965 (Aug. 25).
139. Nelson, E. R.,. Drug Dev. Branch, Cancer Chem. NIH, Let. 1965.
140. Munson, A. E., Regelson, W., Lawrence, W. and Wooles, W. R., J. Reticuloendoethl. Soc., 1970, 134, 309.
141. Regelson, W. and Munson, A. E., Ann. N. Y. Acad. Sci., 1970, 173, 831.
142. Braun, W., Regelson, W., Yajima, Y. and Ishizuko., Proc. Soc. Exp. Biol. Med., 1970, 133, 181.
143. Hirsch, N. S., Black, P. H., Wood, M. L. and Monaco, A. P., Proc. Soc. Exp. Med., 1970, 134, 309.

144. Hirsch, M. S., Black, P. H., Wood, M. L. and Monaco, A. P., J. Immun., 1972, 108, 1312.
145. Papas, T. S., Pry, T. W. and Chirigos, M. A., Proc. Nat. Acad. Sci. USA, 1972, 71, 367.
146. Salup, R. R., Heberman, R. B., Chirigos, M. A., Back, T. and Wiltrout, J., Immunol., 1985, 7, 417; CA 104, 86827a (1986).
147. Rosenberg, S. A., Spiess, P. and Lafreniere, P., Science, 1986, 233, 1318.
148. Braun, W., Regelson, W., Yajima, Y. and Ishizuka, M., Proc. Soc. Exp. Biol. Med., 1970, 133, 171.
149. Schultz, R. M., Papamatheakis, J. D., Stulos, W. A. and Chirigos, M. A., Cell. Immunol., 1976, 25, 309.
150. Stylos, W. A., Chirigos, M. A., Lengel, C. R. and Lvng, P. I., Cancer Immunol. Immunoter., 1978, 5, 165.
151. Stylos, W. A., Chirigos, M. A., Lengel, C. R. and Weiss, J. F., Coll. Immunol., 1978, 40, 437.
152. Weu, J. Y., Heinbough, J. A., Holden, H. T. and Herbermah, R. B., J. Immunol., 1979, 122, 175.
153. Chirigos, M. A., Schlick, E., Piccoli, M., Read, E., Hertwig, K. and Bartocai, A., Adv. Immunopharmacol.2, Proc. Int. Cont. 1983, 669.
154. Zander, A. R., Templeton, J., Gray, K. N., Spitzer, D. S., Verma, D. S. and Dicke, K. A., Biomed. Pharmacother., 1984, 38, 107; CA 101, 103788w (1984).
155. Regelson, W., Shnider, B. I., Colsky, J., Olson, K. B., Holland, J. F., Johnson, C. L. Dennis, L. H., Imm. Mod. and Contr. Neopl. Adj. Therapy 1978, 469.
156. Mohr, S. J., Chirigos, M. A., Fuhrman, F. S. and Pryor, J. W., Cancer Res., 1975, 35, 3750.
157. Mohr, S. J., Chirigos, M. A., Smith, G. T. and Fuhrman, F. S., Cancer Res., 1976, 36, 2035.
158. Gifford, G. E., Letter to Dr. G. B. Butler, 1984 (May 11).
159. Reczek, P. R., Ph.D Dissert., State U. of N.Y. at Buff. 1979 (Sept.).
160. Wolgemuth, R. L., U.S. Pat. 4,223,109, 1980 (Sept. 16); CA 98, 35183n (1983)
161. Butler, G. B., Xing, Y., Gifford, G. E. and Flick, D. A., Ann. N. Y. Acad. Sci., 1985, 446, 149; CA 104, 341c (1986).
162. NIH, Drug Research and Development, Prog. Analysis Branch, Report of Antitumor Evaluation.
163. Rogers, C. L., Pharmacol. Rep., Univ. Fla. Med. Colleg. 1973 (March).
164. Breslow, D. S., U.S. Pat. 3,794,622; Ger. Off. 2,262,449 1973 (July 5); CA 80, 22867s (1974).
165. Floyd, J. D., U.S. Pat. US 3,085,077, 1963 (Apr. 9); CA 59, 4125f (1963).
166. Regelson, W., U.S. Pat. 3,624,218, 1971 (Nov. 30), CA 76, 114404j (1972).
167. Regelson, W., U.S. Pat. 3,749,771, 1973 (July 31), CA 79, 129096n (1973).
168. Tang, B-S., Wang, L-K. and Xu, X-L., Private Communication, 1989.

9

Cyclopolymerization Leading to Ladder Polymers

Many conjugated dienes and α,β-unsaturated compounds have been shown to undergo an unusual type of cyclopolymerization. Among these are the 1,3-dienes, α,β-unsaturated aldehydes, α,β-unsaturated nitriles, and α,β-unsaturated isocyanates. In a generalized way, the reaction can be represented as follows (9–1).

1,2 - Addition Polym. → Multiple Cyclopolymerization →

$$-X-Y = -CH=CH_2, \quad -C=O, \quad -C\equiv N \quad \text{or} \quad -N=C=O$$

(9–1)

1,2-Addition polymerization through the vinyl group is usually accomplished first, followed by a second stage, "zipping-up", or multiple-stage cyclopolymerization of the structure in which each multiple-bonded side chain bears the same relationship to each of its neighbors that the double bonds of a 1,6-diene do to each other. In the case of vinyl isocyanate in which -X-Y is -N=C=O, the initial step has been accomplished through both the vinyl group and the -N=C group followed by the second stage cyclopolymerization on the remaining multiple bond.

A large amount of work in the area of this special type of cyclopolymerization has led to some interesting and unique structures. Several investigators [104–110] have reported evidence for ladder or double-strand segments in polymers obtained by treating either various polyisoprenes or the monomer itself[1-107-1-109] with cationic initiators. It was reported that cationic cyclization of the polymer led to crosslinking unless dilute solutions were used. The soluble polymers were obtained by using boron trifluoride or phosphorus oxychloride in polymer concentrations of about 0.25% in benzene. The resulting material consisted of segments of fused structure, although chain scission also resulted, reducing the molecular weight to about one-fourth of the polyisoprene, or from about 160,000 to about 40,000.[1-111, 1-112]

Copolymers of ethylene and isoprene[1-105] were synthesized in presence of metal-coordination catalysts and their structures studied. The polymer, which was soluble in hot tetrachloroethylene showed the absence of attached vinyl and isopropenyl groups, had a sharp melting point of 146°, and an inherent viscosity of 5.8. A cyclized structure was proposed.

A study of cyclopolymerization of isoprene[1-107] in presence of ethylaluminum dichloride has shown that the rate of polymerization is linearly dependent on the monomer concentration, and the initial rate is proportional.

In a study of polymerization of conjugated dienes, e.g. butadiene and isoprene,[1-108] with Ziegler-Natta type catalysts under conditions favoring cyclic polymers, it was concluded that the poly(butadiene) obtained, contained at least 85% of the cyclic ladder structure, and the poly(isoprene) obtained, contained at least 70% of an analogous structure.

Cationic-initiated cyclization of 3,4-polyisoprene[1-111] to yield a ladder or double chain linear hydrocarbon polymer *via* a cyclopolymerization mechanism was publicized. The ladder or double chain polymers[1-112] were discussed in greater detail and a mechanism proposed.

Isoprene, 1,3-butadiene, and 2-chloro-1,3-butadiene[1-110] were polymerized with complex catalysts consisting of alkyl- or aryl-magnesium bromide or triethylaluminum and excess titanium tetrachloride to obtain powdery, insoluble, probably crosslinked polymers with high density and high heat resistance. However, the cyclic structure content of these polymers ranged from 60–90%. Evidence was presented that these polymer chains consisted of fused six-membered saturated rings in the form of a linear ladder or a spiral ladder structure. Structural evidence indicates that these polymers contain residual linear segments with 1,4 units which can be isomerized to cyclic form by the action of sulfuric acid. It was proposed that the cyclic structures were formed during the polymerization itself from intermediate 1,2-polymeric structures rather than as the result of initiator on first-formed linear chains. Two alternative mechanisms were proposed to account for this single-step cyclopolymerization of conjugated diene monomers to ladder polymers. Since the metal halide was present in considerable excess, the polymerization was believed to be cationic. The first of the proposed mechanisms involved a reversal of the direction of polymerization, resulting in an intramolecular cyclopolymerization to yield a sequence of fused six-membered saturated rings, as shown in (9–2). The second proposed mechanism involved a cyclization reaction initiated by attack of an active center of a monomer unit or of a growing chain on the pendant double bonds of a polymer molecule, as indicated in Structure 9–3. It was shown that the catalyst components individually or together failed to cyclize either 3,4-polyisoprene or *cis*-1,4-polyisoprene. Both proposed mechanisms for the extensive degree of cyclization are consistent with these observations.

In an extension of this work, cyclizations of *cis*-1,4, *trans*-1,4-, and 3,4-polyisoprene[1-113] with sulfuric acid were carried out and the infrared spectra of these cyclized polymers compared with those of the cyclopolymers of isoprene, butadiene, and chloroprene prepared from monomers by use of Ziegler-Natta catalysts. The similarity of all the spectra indicated that both types of polymer have the same polycyclic structure. However, it was concluded that even though these similarities in the IR spec-

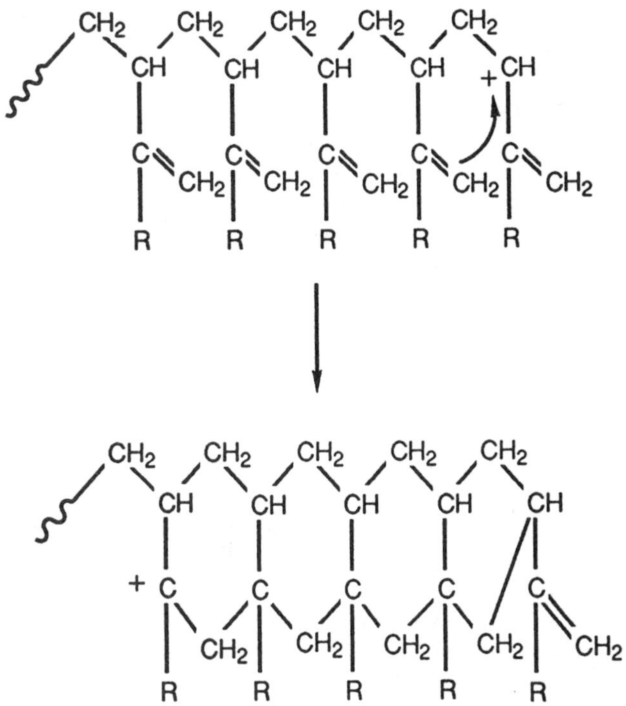

(9–2)

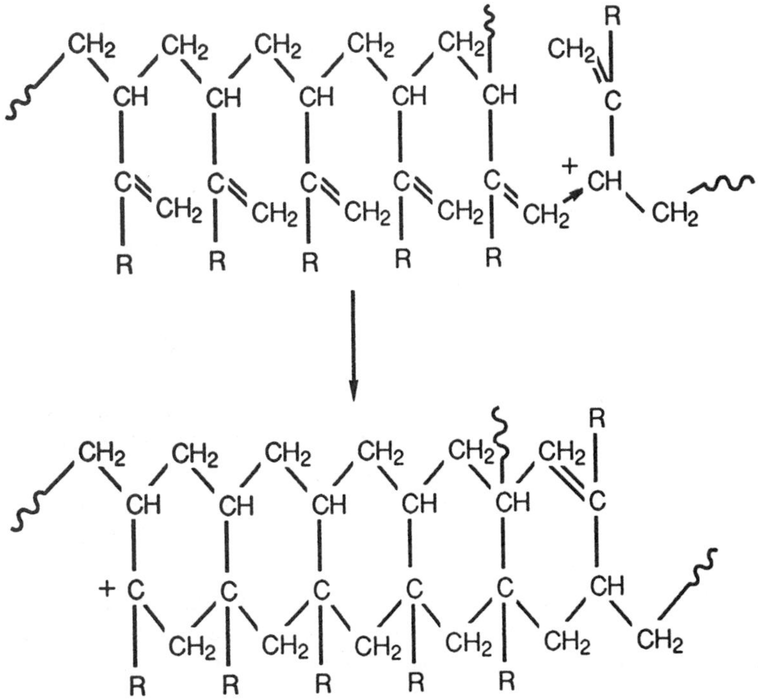

(9–3)

tra exist, further evidence would be required to show that the independently derived polymers possess identical structures.

Since both isotactic and syndiotactic 1,2-polybutadiene[1-101,1-102] have been reported, second-stage cyclization experiments on each of these stereoregular polymers should lead to interesting results.[1-98,1-135] As the structures shown in Chapter 1 (See Structures 1-27, 1-28, 1-29, 1-30) indicated, there should be large energetic factors to favor one of the two stereoregular cyclized structures in each case over the other possibility. For example, from isotactic, 1,2-polybutadiene, the *cis*-1,3 *trans*-1,2 conformation can readily assume the conformationally favored chair form of the derived cyclohexane rings while the other stereoregular form, the *cis*-1,3 *cis*-1,2 conformation, can only assume the higher energy boat form of the derived cyclohexane rings. The difference in strain energy between the chair and boat forms of a cyclohexane rings is usually considered to be 6 kcal/mole.

From syndiotactic, 1,2-polybutadiene, the *trans*-1,3 *cis*-1,2 conformation can readily assume the conformationally favored chair form of the derived cyclohexane rings whereas the other stereoregular form, the *trans*-1,3 *trans*-1,2 conformation, can only assume the higher energy boat form of the cyclohexane rings. Thus, from each stereoregular form of 1,2-polybutadiene, one of the two stereoregular forms derivable from each should be energetically favored over the other by approximately $6n$ kcal/mole of polymer of average degree of polymerization, n.

Formation of a double-strand polymer obtained through a two-stage polymerization of vinylisocyanate[1-114,1-115] was reported in 1963. It was shown that polymers containing a large portion of ladder polymer structure could be obtained by either of the following routes: (a) polymerization of vinylisocyanate through the vinyl double bond by allowing the monomer to stand at 25°C for two days, followed by cobalt-60 irradiation at 30° for 48 hours. The resulting polymer was found to contain 85% ladder structure although it was insoluble. (b) polymerization of vinylisocyanate to *N*-vinyl-1-nylon, initiated by sodium cyanide in dimethylformamide at −55°C, followed by treatment of the vinyl nylon with azobisisobutyronitrile at 80° for 4 hours. The resulting polymer (See Structure 1-31) was found to contain 90% ladder structure and was soluble in several solvents.

Natural rubber[1] has received considerable attention with respect to cyclization. The kinetics of cyclization of this material initiated by sulfuric acid or phenol were studied by measuring residual unsaturation. The relations between the experimentally determined and calculated values were discussed. The best fitting value of the average length of reaction sequence was found to be 0.4. These results indicated that the cyclized rubber was tricyclic in nature.

Glass transitions for a number of polymers and copolymers, including cyclopolybutadiene[2] were studied by use of calorimetry, thermogravimetry and differential thermal analysis. The theory of glass transition was discussed.

The general process of statistical degradation of double stranded polymers[3] was considered from the probability standpoint and treated theoretically. Useful predictions can be made based on the theory.

Thermal stability studies of these unusual structures have been prevalent. Thermal degradation of partial ladderlike polymers of diphenylbutadiene[5-29] led to a black insoluble residue. The products of the decomposition were identified.

In a study of double-stranded molecules, a [6]beltene derivative[4] (I) and the corresponding ladder polymer (II) were prepared by the Diels-Alder reaction from III (R - hexamethylene) by heating at 110°C. I and II were soluble and the crystal structure of I was determined (9–4).

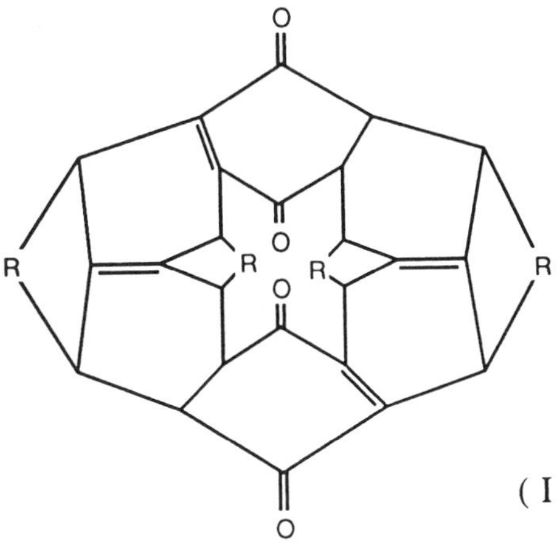

(I)

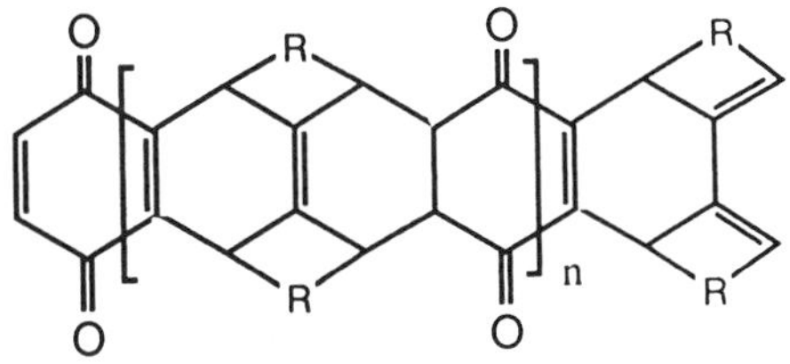

(II 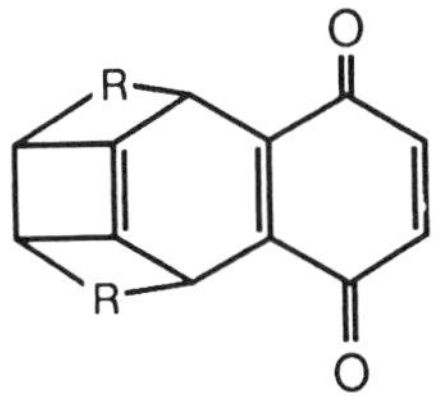) (III)

(9–4)

CYCLOPOLYMERIZATION OF CONJUGATED DIENES AND DIENE POLYMERS

The first published reports of ladder polymer synthesis from conjugated dienes appear to be those which appeared in 1963.[1-104, 1-106, 1-109] It was shown that copolymers of ethylene and isoprene having a monomer mole ratio of 25:1, and in which the isoprene units were incorporated into the chain through the 3,4-double bond, when treated with phosphorus oxychloride in tetrachloroethylene, yielded soluble copolymer containing linearly fused cyclohexane units (See Structure 9–3).

Through use of a Grignard reagent-titanium tetrachloride catalyst system at a Mg/Ti ratio of less than two to polymerize butadiene, isoprene, and chloroprene, insoluble, powdery polymers which contained infrared absorption peaks indicative of cyclic structure were obtained.[1-109] The cyclic structure content of these copolymers ranged from 60 to 90%. Since this catalyst system did not cyclize 3,4-polyisoprene or natural cis-1,4-polyisoprene under the polymerization conditions, it was postulated that a cationic mechanism in which cyclization resulted from either reversal of the direction of the polymerization reaction (a), or copolymerization of additional monomer on a growing chain with pendant unsaturated groups (b) had occurred. The structures and the polymerization mechanism proposed are shown (See Structures 9–2 and 9–3).

The earlier reports were followed shortly by other publications[1–104] in which it was shown that cationic initiated cyclization ($POCl_3$ or BF_3) of poly-3,4-isoprene in dilute solution led to a completely soluble, noncrosslinked polymer whose structure consisted largely of cyclohexane rings fused in a linear fashion. The polymer was aromatized to yield structures containing several types of linearly fused aromatic compounds, thus providing further substantiating evidence for a fused cyclic structure.

In the presence of vanadium oxychloride and $(CH_3)_3Al_2Br_3$ in a V/Al ratio of 1:8, it was shown[5] that isoprene and its copolymer with ethylene could be converted in a single-stage reaction to polymers and copolymers having the cyclohexane recurring unit (9–5).

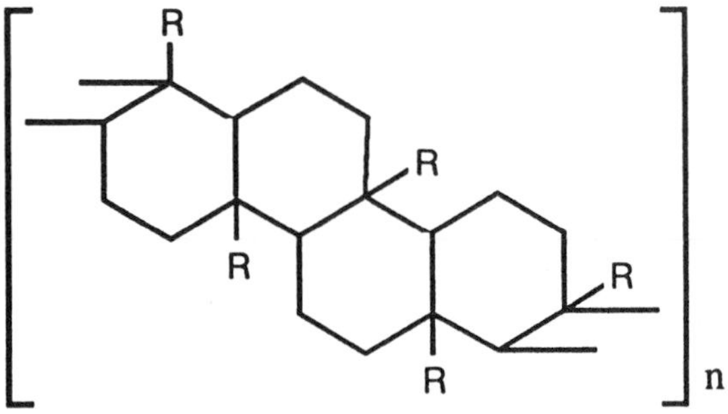

(9–5)

This original work has been followed by a sequence of papers[6] in which additional evidence for the cyclopolymerization mechanism was presented, along with effects of varying catalyst ratios, polymerization conditions, and additional structural evidence.

The use of titanium tetrachloride as initiator for cyclopolymerization of 3,4-polyisoprene has been studied.[7] Although this work essentially substantiated the results of previous workers, minor structural differences were apparent indicating that the microstructures of polymers prepared via different initiators are not identical.

The efforts to obtain ladder structures either by a second-stage cyclopolymerization of poly-1,2-butadienes (or poly-3,4-isoprenes), or by a single-stage cyclopolymerization of conjugated dienes have been very closely associated with earlier work on cyclization of natural *cis*-1,4-polyisoprene and related structures. Evidence has been presented that the structure of cyclized polybutadiene,[8] prepared by treatment with sulfuric acid under rather drastic conditions, may be considered to include polycyclic structures joined by uncyclized segments of the chain together with isolated six-membered rings.

Significant contributions to the structure of cyclized 1,4-polyisoprenes[9, 10] have been made. More recently, it has been shown through a NMR study of cyclized cis-1,4-polybutadiene and cis-1,4-polyisoprene[11] that the polymer consisted of linear and cyclic portions. The cyclic polymer was reported to have a structure similar to that obtained in the single-stage cyclopolymerization of the conjugated dienes.[5]

A novel cyclic structure (9–6) has been observed in polymerization of butadiene[12] by the catalyst system consisting of triethylaluminum, Lewis acid, and metal salt.

Cyclization Isomerization CH_3 n CH_3

(9–6)

It was suggested that this structure arose from 1,2-polybutadiene by cyclization and isomerization reactions. The cyclization was postulated to result from attack of the acidic catalyst on a vinyl group of 1,2-polybutadiene as had been suggested.[13]

Chloranil has been used to aromatize cyclized 1,2-polybutadiene.[2-30] The aromatized structure was found to possess excellent thermal stability.

In a study of the microstructure of the polymers of *cis*- and *trans*-1,3-pentadiene prepared in the presence of alkali metals, it was shown that a significant amount of the monomer appeared in the polymer as α-methylcyclohexene groups.[2-28, 2-29]

The literature on cyclopolymerization of 1,3-dienes and cyclization of polydienes to ladder polymers has become quite voluminous during the intervening years. Consequently, this section is presented in subsections as follows: (1) cyclopolymerization of monomeric isoprene; (2) cyclization studies of polyisoprene; (3) cyclopolymerization of 1,3-butadiene; (4) cyclization studies of polybutadiene; (5) cyclopolymerization of other 1,3-dienes; and (6) cyclopolymerization of vinylacetylenes and closely related structures to ladder polymers.

Cyclopolymerization of Isoprene

Cyclopolymerization of isoprene[14] was readily accomplished with the stable carbocation salts, $C_7H_7^+$ $SbCl_6^-$ and $(C_6H_5)_3C^+SbCl_6^-$ in nitrobenzene. Relatively non-polar solvents were not effective. The proposed initiation mechanism involves the formation of a catalyst-nitrobenzene complex which undergoes a one-electron transfer with isoprene to yield a monomer cation-radical.

Cyclopolymerization of isoprene[15] was studied in the presence of aluminum chloride, ethylaluminum dichloride, titanium tetrachloride, antimony pentachloride, or phosphorus pentafluoride, using n-heptane, benzene, acetonitrile, nitromethane or nitrobenzene. Nitrobenzene had a tremendous activating effect on the polymerization. All polymers had a predominantly cyclic structure with a 30% *trans*-1,4-content. Only traces of polymer were obtained with solvents other than nitrobenzene.

Cyclopolymerization of butadiene or isoprene[16] with Ziegler type catalysts has been shown to lead to saturated polymers containing fused cyclohexane rings. The amount of cyclic structure and non-terminal linear unsaturation depended on the electron accepting ability of the catalyst which was influenced by the nature of the solvent.

In a study of the ethylaluminum dichloride catalyzed polymerization of isoprene,[17] the maximum conversion to cyclized polyisoprene and the polymerization rate depended on the square of initial monomer concentration in heptane and on the first power of initial monomer concentration in benzene. The latter was attributed to both isoprene and benzene in the primary complex.

The cyclopolymerization of isoprene[18] in aromatic solvents proceeded via a complex of the monomer with the catalyst, chain growth occurring via a monomer cation-radical.

The proposed mechanism of cyclopolymerization of isoprene[19] with catalysts containing ethylaluminum halides and the nature of the active species were discussed. The polymerization to solid cyclopolyisoprene was accelerated by addition of $TiCl_4$. A proposed mechanism involved monomer activation by conversion to a cation radical by a one-electron transfer to catalyst cation, followed by additional steps which led to classification of the system as a "living polymerization."

A study of cationic polymerization of isoprene[20] with ethylaluminum dichloride showed that cyclopolyisoprene was obtained in low conversion in equilibrium with unreacted monomer and catalyst. The addition of monomer resulted in renewed polymerization and addition of styrene gave block copolymers.

Isoprene[21] has been cyclopolymerized in presence of ethyl aluminum dichloride in benzene, n-heptane, or in bulk. The rate of polymerization was found to be linearly dependent on the monomer concentration. The initial rate was proportional to the catalyst concentration. The rate constant of propagation at 20°C in benzene was 1.28 l $mole^{-1}$ $min.^{-1}$ and its activation energy was found to be 3.7 kcal/mole.

Cationic copolymerization of isoprene[22] with styrene in n-heptane or benzene medium formed copolymers in which isoprene units were linked to form cyclic segments. The kinetic course was found to depend on the solvent. Relative monomer reactivity ratios determined in both solvents showed that in n-heptane, styrene was the more reactive, with the reverse being the case in benzene. Block copolymers could be formed.

In a continuing study of cyclo- and cyclized diene polymers, the nature of the active species and the mechanism of cyclopolymerization of isoprene[23] with catalysts containing ethylaluminum halides were further investigated. Addition of $TiCl_4$ to the polymerization medium markedly accelerated the polymerization to solid cyclopolyisoprene.

Styrene-isoprene block and graft cyclopolymers[24] were synthesized by cationic initiated polymerization in heptane at 20°C with ethylaluminium dichloride. Isoprene alone gave cyclopolyisoprene in low conversion which formed an equilibrium with unreacted monomer and catalyst. Further addition of isoprene produced renewed polymerization. Introduction of styrene to the dormant system gave block copolymers and reinitiated isoprene polymerization. Graft cyclopolymers were formed by polymerizing isoprene in presence of polystyrene. The isoprene units in the graft copolymers were present as cyclic segments.

Cyclopolyisoprenes[25] were formed through use of Ziegler-type catalysts in heptane. The reaction rate and polymer microstructure were strongly dependent on the Al/Ti ratio. Cyclopolyisoprene was the main product at Al/Ti ratios 1.5 and maximum cyclic polymer was obtained at a ratio of 0.75.

The cyclopolymerization of isoprene[26] has been studied with vanadium tetrachloride-diethylaluminum bromide catalyst system at 30° in C_6H_{14}. The polymerization rate was a linear function of isoprene and catalyst concentrations. The overall activation energy was 8.96 kcal/mole. The presence of cyclic structures suggested a cationic polymerization mechanism.

Organoaluminum halides and alkyl phosphate zirconium salts have been used as polymerization catalysts for isoprene[27] to give cyclized isoprene rubber.

High-purity cyclized polyisoprene was prepared by living polymerization of isoprene[28] in the presence of an organometallic catalyst to give cyclized rubber, and bringing the polymer into contact with a catalyst composed of BF_3-ether complex and a carboxylic acid under mild conditions.

Cyclopolymerization of Polyisoprenes

This section deals with postcyclization of polyisoprenes, postulated to be of the nature of a "zipping up" process, also considered to be analogous to a multiple ring closure via cyclopolymerization.

A phenol solution of natural rubber[29] was treated with phosphorus pentoxide at elevated temperatures to yield cyclized rubber in which the original degree of unsaturation had been decreased to 25%.

Polymerization of butadiene and isoprene[1-103] with certain selected organometallic coordination catalysts led to 1,2-polybutadiene and 3,4-polyisoprene, respectively. Fractionation of the polybutadiene yielded a crystalline polymer with a syndiotactic structure. The polyisoprenes were completely amorphous products.

Polymerization of isoprene[30] in n-heptane with Lewis acids was shown to be a process leading to a quasiequilibrium which was characterized at very low conversion. In aromatic solvents, polymerization rates were much higher, and higher still in halogenated solvents. Cyclopolyisoprene of molecular weights up to 100,000 were obtained. Residual unsaturation in the polymers did not exceed 30%.

Photocyclization of 1,2-polybutadiene and 3,4-polyisoprene[31] was studied. A variety of cyclized structures were observed; however, the process illustrated in the Structure 9–7 was considered very unlikely:

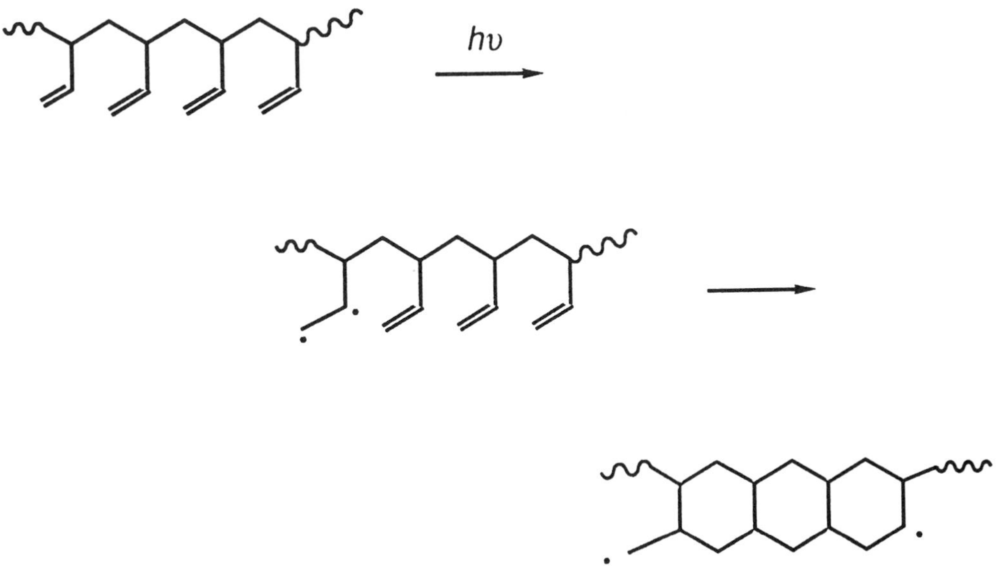

hυ

(9–7)

Modified precipitation chromatography making use of periodical temperature changes has been shown to efficiently fractionate relatively low molecular weight materials such as the pyrolysis of a variety of hydrocarbon polymer products, including cyclopolyisoprene.[32]

Cyclized *cis*- and *trans*-1,4-polyisoprene and cyclopolyisoprene[33] were studied. The IR spectra of cyclized 1,4-polyisoprenes indicated that they possess individual rings at least as large as six carbons, in which the rings contain sequences of at least two methylene groups. Corresponding spectra of cyclopolyisoprene indicated that such rings are few or absent.

NMR studies of cyclopolyisoprene[34] prepared by polymerization of isoprene in methylene chloride in presence of Lewis and Broensted acids, showed the presence of cyclic structures in the polymer.

The character of cationically prepared microgels of cyclopolyisoprene[35] was studied. The compactness and the quasi-spherical shape of the microgels were due to intramolecular crosslinking.

In a pyrolysis gas chromatographic study of cyclopolyisoprene,[36] it was shown that the pyrolytic products were essentially the same as those from linear polyisoprene.

Cyclization of synthetic *cis*-1,4-polyisoprene[37] was carried out by use of titanium tetrachloride and p-toluenesulfonic acid catalyst. The average cyclicity and structure of the polymers were studied, and methods of distinguishing among mono-, bi-, tri, or polycyclic polymer were discussed.

Cyclized polyisoprene,[38] cyclized to the extent of 98%, and melting at 170–80°C, was produced by treating a liquid polyisoprene in toluene with free-radical-generating compounds at high temperatures.

The homogeneity of cyclized rubber[39] was improved by removing the yellow fraction from the natural rubber before cyclization with p-toluenesulfonic acid.

An infrared study of out-of-plane deformation vibrations of linear and cyclized 1,2- and 3,4-polyisoprenes[40] have been conducted. The results indicated that the influence of configurational changes on the intensity of absorption bands for out-of-plane vibrations of double bonds is not a general phenomenon.

It has been shown that the molecular weight of *cis*-1,4-polyisoprene[41] in benzene when treated with iodine, decreased at a low mole ratio of iodine/double bonds. At high polymer concentrations iodine catalyzed the cyclization of both this polymer and natural rubber.

Linear 1,2- and 3,4-polyisoprenes[42, 43] were cyclized at 20° in dilute solutions of aromatic and chlorinated aliphatic solvents in presence of $TiCl_4$ and $POCl_3$ catalysts. Light scattering and viscosimetric analysis of the cyclized polymer showed an increase in the steric hindrance factor with progressive cyclization, although at high cyclization degrees, the characteristic dimensions decreased due to internal crosslinking.

All prominent resonances in the ^{13}C NMR spectra of epoxidized *cis*- and *trans*-1,4-polyisoprenes[44] and *cis*- and *trans*-1,4-polybutadienes have been assigned.

Both pyrolytic and differential thermal gravimetric methods (TGA) of distinguishing between natural and synthetic *cis*-1,4-polyisoprene rubbers[45] depend on the presence of nonrubbers. The markedly different TGA traces for the natural rubber and the synthetics are due to cyclization of the polymers to different extents.

A method of cyclized rubber production has been developed in which polyisoprene was treated with p-toluenesulfonic acid in an aromatic hydrocarbon.[46]

Many of the research programs which led to development of cyclopolymers of conjugated dienes were directed toward certain polymer applications. A limited number of applications which were introduced along with the synthetic methods or structural investigations have been summarized.

Cyclized natural rubber[47] has been shown to be a useful component of UV-curable printing inks for polyolefins.

Compositions of natural or synthetic rubber[48] containing up to 30% cyclized diene rubber were found to be useful as tread materials having good resistance to impact cutting.

Cyclized isoprene rubber[49] has been found to be superior to other materials in priming semiconductor devices followed by sealing with epoxy resins.

The uses of cyclized rubber,[50] as well as those of chlorinated rubber, chlorinated polypropylene, and poly(vinyl chloride) in coatings technology was reviewed.

Cyclized rubber[51] has been used in the development of correction fluid compositions. These fluids are applicable to water- and oil-based printing and writing inks and also contain TiO_2, rubber binders, and hydrocarbon solvents.

A developer for cyclized rubber[52] photoresists has been described. The method provides high-resolution negative type photoresists and permits carrying out simultaneous developing and rinsing processes.

Protective coatings based on cyclized rubber have been developed. Formation, physical properties, and coating applications of cyclized rubber[53] are discussed with an emphasis on its use in antifouling and marine paints. Some formation aspects are explored, including the choice of solvents, pigments, and plasticizers.

Cyclopolymerization of 1,3-Butadiene

The early work on cyclopolymerization of isoprene has, quite naturally, led to similar investigations on the conjugated butadiene. Many studies have concentrated on both monomers simultaneously, as was the case in the report which follows.

Cyclopolymerization of isoprene and butadiene[54] with an ethylaluminum dichloride-titanium tetrachloride catalyst system yields ladder polymers containing fused cyclic structures. Polymers prepared in n-heptane are generally insoluble powders, while those prepared in aromatic solvents are soluble even when the molecular weight exceeds 1×10^6. The cyclopolymers contain residual unsaturation, 10–40% of which is cycloalkenyl, and methyl groups.

1,2-Polybutadiene[55] was synthesized via complex catalysts under pressure of liquid ethylene to yield this polymer without formation of polyethylene.

An EPR study of the photoinitiated polymerization of butadiene[56] has been conducted in the presence of vanadium tetrachloride. Cyclic polybutadiene was the main product of the polymerization.

A cyclized copolymer from butadiene[57] and an alpha olefin has been obtained in the presence of a catalyst consisting of equimolar amounts of BuNa and tetramethylethylenediamine. In the absence of the diamine, low monomer conversion and little cyclization were observed.

Cyclized polybutadiene[58] was prepared from the monomer in toluene solution by a combination of catalysts, isolated and molded at 170°C to prepare an electric insulator having high electrical resistivity.

Cyclopolymerization during the synthesis of poly(vinyl chloride-co-1,3-butadiene)[4–9] has been observed. The presence of other important structural defects in butadiene-vinyl chloride copolymers, including a cyclohexyl branch arrangement that results from a cyclopolymerization reaction, was also observed.

Cyclopolymerization of Polybutadienes

Postcyclization studies have been carried out on essentially all of the polybutadienes, including both *cis*- and *trans*-poly-1,4-butadiene, and poly-1,2-butadiene. Apparently, no attempts have been made to evaluate separately the tendencies for the isotactic and syndiotactic poly-1,2-butadiene to cyclize (See Structures 1–27 through 1–30).

A study of the preparation of bistranded or ladder polymers from polybutadienes[59] and their aromatization has been carried out. Specifically, 1,2-polybutadiene was studied and by use of metal-coordination catalysts, an intramolecular cyclization to an essentially saturated, thermally stable polymer was accomplished.

This paper tends to refute the assignment of NMR peaks to cyclic structures in polymers from butadiene[60] and isoprene in an earlier publication.

Oxidation of cyclopolydienes[61] at elevated temperatures was shown to differ from the oxidation of linear, single-chain polydienes. The former is a radical-induced oxidation in which the reactions of peroxide radicals, leading to rupture of the chain, dominate over those which lead to formation of hydroperoxides and branching.

Cyclization of *cis*-1,4-polybutadiene[62] by titanium tetrachloride was studied. The degree of cyclization increased with increasing amount of catalyst and with increasing temperature. The residual unsaturation could be reduced to 15%. Spectral studies showed that the cyclization was accompanied by *cis-trans* isomerization.

Cyclization of *cis*-1,4- and *trans*-1,4-polybutadiene[63] and of poly(2,3-dimethyl-1,3-butadiene) was carried out with sulfuric acid. The decrease of total unsaturation and the kinetics of cyclization were measured. The length of cyclic segments was indicated to be 25–45 cycles, except in the case of the 2,3-dimethyl derivative, only monocyclics were formed.

Thermosetting cyclized polyurethanes were obtained from α,ω-dihydroxy- or α,ω-dicarboxy-1,2-polybutadiene[64] or from the corresponding isoprene derivatives, a chain extender, (e.g. tolylene diisocyanate), 1,5-hexadiene, and a free radical initiator.

Several samples of cyclized polybutadiene,[65] cyclized to a high degree, were studied by light scattering, viscometry and liquid phase partition chromatography. Remarkable changes in the properties of the polymer, relative to the parent structure, were observed. The catalyst, aluminum bromide in o-xylene, was postulated to cause some degradation of the chain as well as cause cyclization.

The addition of α,β-unsaturated carboxylic acids to polybutadienes[66] and polypentenamer was investigated in presence of acid catalysts to obtain photosensitive elastomers. The structure of the products was identified as cyclized polydiene rubbers having pendant α,β-unsaturated carboxylate groups in amounts up to 20 mol%.

Polybutadiene[67] was cyclized in presence of a catalyst consisting of ethylaluminum dichloride and benzyl chloride in toluene. After recovery of the polymer it was found to be 36% cyclized with viscosity of 1.36.

Selectively cyclized block copolymer elastomers were prepared by cyclizing a block copolymer of butadiene[68] and isoprene in presence of a Lewis acid and trichloroacetic acid. A living polymer of butadiene was converted to the block copolymer by adding isoprene. Improved performance characteristics were obtained in the cyclized polymer.

In a study of thermal decomposition of polybutadiene[69] by thermogravimetric analysis, it was shown that the isothermal decomposition of butadiene rubber was significantly different from that in the heating mode and involved an exothermic reaction occurring at 35–75°C. This decomposition was shown to be rapid and temperature specific and appeared to be related to a cyclization reaction.

A vinylcyclopentane structure was discovered by use of NMR spectroscopy in anionically polymerized polybutadiene[70] with lithium counterions in presence of N, N, N′,N′-tetramethylethylenediamine. The structure apparently was formed from intramolecular cyclization during propagation.

A discussion of cyclic polybutadiene oligomers[71] was included in a review, with 39 references on functional chemicals via homogeneous catalysis.

The reactions responsible for structural changes of polybutadiene[72] at <250°C were reviewed and discussed. Thermal degradation of samples having different initial contents of *cis*-1,4-, *trans*-1,4- and 1,2-units were investigated at <250°C where cyclization, crosslinking, and *cis-trans* isomerization occurred without depolymerization. The rates of double bond loss and isomerization were determined.

Conjugated diene polymers[73, 3-21] having an unsaturated bond in the side chain have been cyclized in the presence of catalysts containing a dialkylaluminum halide and a halogenated acetic acid, to give heat resistant polymers.

Butadiene polymers[74] have been cyclized in the presence of cationic polymerizable monomers and aluminum trihalides or their complexes to give a cyclic product in 56% yield.

It has been shown that crosslinking of *cis*-1,4-poly-butadiene[75] by dicumyl peroxide occurred by a cyclizing polymerization of the chain double bonds.

Proton signal assignments and derived relations between signal intensity and degree of cyclization have been made based on the ^{1}H NMR spectra of *cis*-1,4-polybutadiene[76] before and after cyclization.

The cyclization and crosslinking of polybutadiene[77] in solution have been studied by electron beam irradiation. With irradiation, crosslinking and loss of double bond by cyclization occurred at the same time.

A cationic cyclization study of *cis*-1,4-polybutadiene[78] and a technological evaluation of the polymer have been conducted. Controlled cyclization in xylene by aluminum chloroalkyl-organic halide catalysts at >100°C gave gel-free elastomers when 35–40% of the initial unsaturation had disappeared.

Further studies on cationic cyclization of *cis*-1,4-polybutadiene[79] and a physicochemical characterization of the polymer have been conducted. Fractionation of the cyclized rubber showed an equal distribution of unsaturation in all fractions. No aromatic residues or conjugated double bonds were found at degrees of cyclization <40%. The average length of cyclized sequences was 3–4 cyclohexane rings at degree of cyclization 15–60%.

During the course of many investigations on cyclopolymers of butadiene, the program also included an evaluation of the products in specific applications. A number of these studies are included.

The physical and chemical properties and processing characteristics of laminates and molding compounds prepared from elastomeric filled and unfilled polybutadiene diol resins extended by diisocyanates were summarized.[80] Polybutadiene is reported to contain a large fraction of poly-1,2-butadiene, and the curing process to involve the multiple cyclopolymerization of the pendant vinyl groups to yield a partial ladder structure.

The synthesis of thermoplastic cyclized 1,2-polybutadienes,[81] useful as coatings, by cyclization of polybutadienes containing $\geqslant 90\%$ 1,2-isomer in the presence of BF_3 etherate, has been reported.

Partially cyclized polybutadiene,[82] having 95% 1,2-linkage before cyclization, cyclization content up to 60%, and molecular weight = 100,000, was found to be useful in fabricating flotation materials.

Cyclopolymerization of Other Conjugated Dienes

In attempts to extend the cyclization studies on isoprene and butadiene, a limited number of other conjugated dienes have been investigated. This section deals with these publications.

It has been reported that bromination of polymers of diphenylbutadiyne[83] of Structure 9–8a promoted intramolecular cyclization with the formation of Structure 9–8b containing bromine end groups. However, the cyclization was not complete even at 400°C, when equivalent amounts of bromine were used.

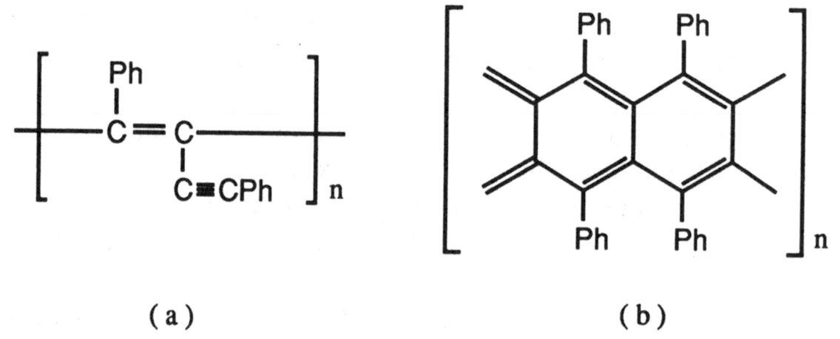

(a) (b)

(9–8)

2,3-Dimethyl-1,3-butadiene[84] was converted to linear polymers with cyclic structure in the chain, and copolymers with the dimer of the monomer were obtained. Cationic and Ziegler type catalysts as well as free radical initiators were effective.

Polymers containing 20% *trans*-1,2-isomers, 80% *trans*-1,4-isomers and approximately 3% cyclic structures were obtained when 1,3-pentadiene[85] was polymerized with rhodium tris(acetylacetonate) and a diethylaluminum catalyst and in emulsion using tetraallyldirhodium chloride catalyst.

Polymerization of 2,3-dimethyl-1,3-butadiene[86] with Ziegler-Natta type catalysts was shown to yield cyclic poly(dimethylbutadiene) with reduced unsaturation and some *trans*-1,4-units in the chain when the Al/Ti ratio was one or less.

The structure of anionic poly(2-phenylbutadiene)[87] has been studied. Polar modifiers, e.g. THF, triglyme, or hexamethylphosphoramide, increased the vinyl repeat unit content of anionic solution polymerized poly(2-phenyl-1,3-butadiene). Intramolecular cyclization reactions involving 3,4-structural units occurred during polymerization in polar media.

Bicyclopropyl[88] in the presence of Lewis acids or Ziegler-Natta catalysts gave polymers containing monocyclohexene groups, like those obtained from 1,4-polyisoprene cyclization at low temperature. At higher temperatures, the rate of conversion decreased and the polymerization proceeded by ring opening of two cyclopropanes followed by cyclization between two monomers.

Cationic polymerization of *cis*-1,3-pentadiene[89] has been studied and it was shown that polymers obtained by polymerization of this monomer and its *trans* isomer consisted of *trans*-1,4-, 1,2-units and cyclic structures independent of the nature of the original 1,3-isomer.

It has been postulated that the thermal anaerobic cyclization of poly(*trans*-1,3-pentadiene)[90] at 260° occurred by a sigmatropic rearrangement with H shift, an intramolecular ene reaction. The thermally induced loss of *trans*-1,2-unsaturation followed second-order kinetics, and the entropy of activation was −16 and 17 eu at 260 and 422°C, respectively.

Cyclopolymerization of Vinylacetylenes to Ladder Polymers

Although vinylacetylene and its derivatives have been extensively studied, the advent of ladder or double chain polymers has generated new interest in these monomers. The possibility of synthesis of conjugated polymers, of interest to the electroconductive polymer area, has also sparked renewed interest in these monomers. This section deals briefly with studies along these lines.

The synthesis of ladder or double-chain polymers by cationic polymerization of conjugated vinylacetylenes[91] in the presence of trichloroacetic acid or boron trifluoride:etherate, followed by heat treatment of the polymers, has been described. The products were reported to be stable in air at temperatures of 400°C and were readily graphitized at higher temperatures.

Poly(α-methoxyisopropylvinylacetylene), poly(phenyl-vinylacetylene) or poly(isopropenylvinylacetylene),[92] when annealed in vacuum at 100–400°C were gradually converted to the ladder polymers consisting of the polycyclic aromatic polymers, which become nuclei of graphite-like structures.

Thermal degradation of poly(vinylacetylenes)[93] gave products the nature of which depended on the structure of the starting ladder polymer and the conditions of the degradation. The products were analyzed by mass spectroscopy. A path of thermal rearrangement based on other data was suggested.

Poly[dimethyl(vinylethynyl)carbinol][94] containing an acetylenic bond in each monomer unit when heated to 360–420°C yielded a ladder-type polyene by polycyclization through the acetylenic bonds which exhibited paramagnetism. Dehydration to give aromatic rings began at 360° and increased with increasing temperature.

Random copolymerization of acetylenic monomers[95] and 1,3-dienes has been studied by using nickel naphthenate-diethylaluminum chloride catalysts. Propyne and dienes tended to form cyclized copolymers.

LADDER POLYMERS OF α,β-UNSATURATED ALDEHYDES

Acrolein, as an α,β-unsaturated aldehyde contains a conjugated double-bond system, and can be considered as a 1,3-diene. In contrast to symmetrical 1,3-dienes, the double bonds in acrolein possess unequal reactivities. Polymerizations initiated with water soluble redox systems produced polymers almost exclusively through the vinyl group.[96] Ionic catalysts are effective agents for initiating polymerization of acrolein through the carbonyl double bond; however, these polymers also contain free carbonyl groups which show that polymerization also occurs to some extent through the vinyl double bond. The content of double bonds and carbonyl groups depends on the initiator and polymerization conditions, and the structure can be represented as a copolymer of the following structure (9–9).[96] When sodium naphthalene was used in nonaqueous media, more than 70 mole% of the copolymer consisted of bond formation through the carbonyl group.[1-130]

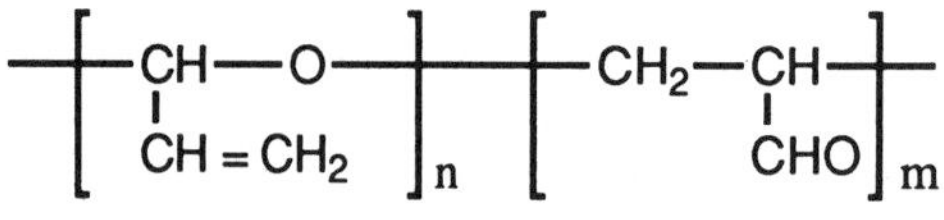

$$\left[-\underset{\displaystyle CH=CH_2}{\underset{|}{CH}}-O- \right]_n \left[-CH_2-\underset{\displaystyle CHO}{\underset{|}{CH}}- \right]_m$$

(9–9)

It has been shown[97] that initiation of polymerization of acrolein with sodium cyanide at temperatures less than −10°C in tetrahydrofuran or toluene have a uniform structure, containing only the repeating unit (9–10):

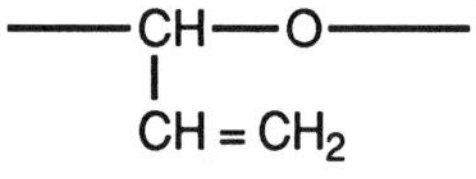

$$-\underset{\displaystyle CH=CH_2}{\underset{|}{CH}}-O-$$

(9–10)

However, at temperatures higher than −10°C, polymerization of the vinyl group was also observed and copolymers were obtained.[98] Acrolein does not appear to undergo 1,4-addition polymerization.[96] However, methacrolein and α-ethylacrolein have been reported[99] to contain units indicative of 1,4-addition. The concentration of tetrahydropyran structure was 57–60% in these polymers.

The free-radical initiated polymerization, particularly when accomplished with redox systems in aqueous solution, yields an insoluble white powdery polymer.[100, 101, 1-51] The structure of acrolein polymers can best be described as macromolecules containing the following structural units (9–11):

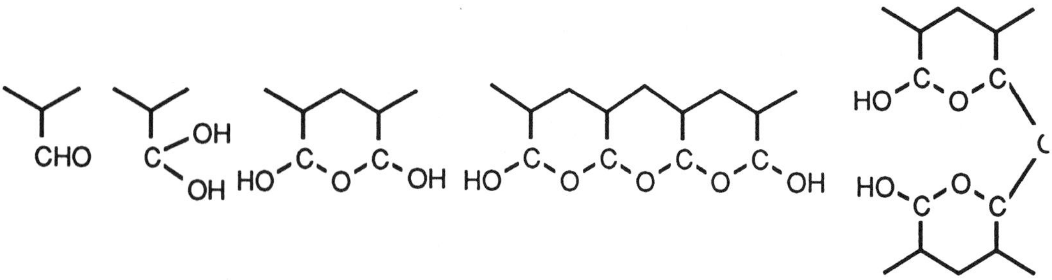

(9–11)

While it has not been completely established exactly what course the polymerization takes, the following structural relationships indicate possible routes to the observed Structure 9–12.

Regardless of which course the polymerization takes, the relationship to cyclopolymerization is apparent, as the former can be considered as a cyclopolymerization of a 1,6-diene, while the latter can be considered a cyclopolymerization in which the glutaraldehyde[5-47] units undergo cyclopolymerization to the polyacetal structure.[102, 103]

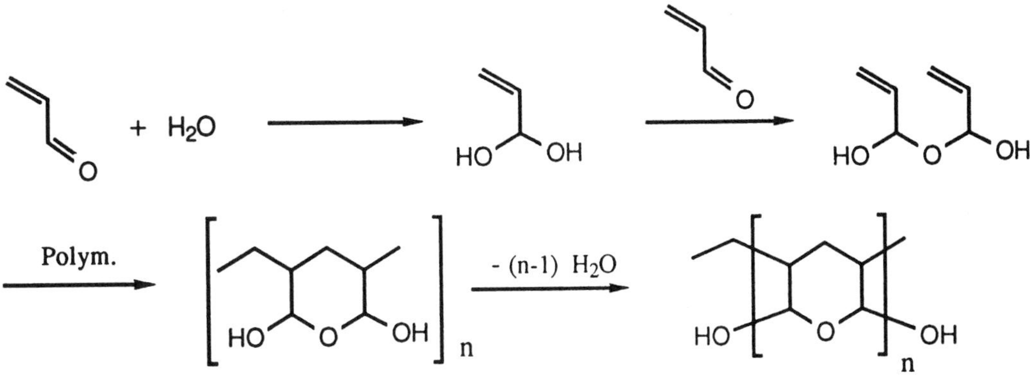

(9–12)

The reactions of high molecular weight polyacrolein[104] prepared by redox[102] or by x-ray[105] initiation have been studied. These polymers had viscosity average molecular weights greater than 200,000 and intrinsic viscosities greater than 1.0 dl/g when measured at 25°C in aqueous sulfur dioxide containing sodium sulfate. Among other reactions, the polyacrolein gave 2,6-disubstituted tetrahydropyran structures (9–13).

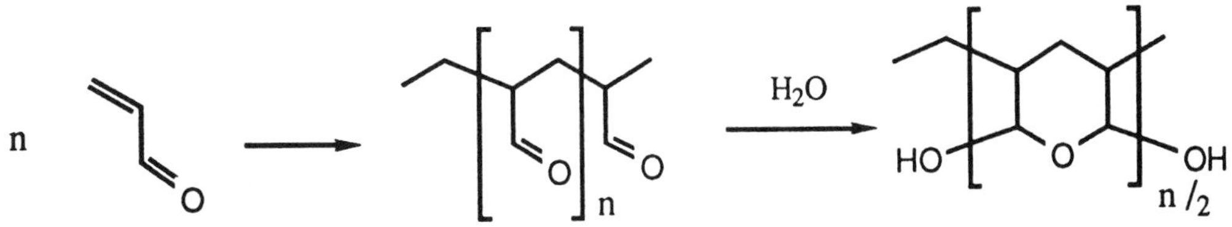

(9–13)

Crotonaldehyde, also an α,β-unsaturated aldehyde, has been reported to undergo anionic polymerization.[1-150, 6-11] Initiation was by tertiary phosphines having a pKa greater than 8.0, and conversion to polymer was 66%. The polymer had a number average molecular weight of 2240 and an intrinsic viscosity of 0.048 dl/g at 30°C in methyl ethyl ketone. The infrared spectral and other analytical data indicated the polymer to have the 4-methyltetrahydropyran ladder structure. This structure is postulated to be the result of a vinyl type polymerization of crotonaldehyde with a simultaneous cyclization of some vicinal, pendant aldehyde groups. The details of this proposed mechanism for synthesis of ladder polymers from this unsaturated aldehyde were presented in Chapter 6 [References 6–13 through 6–17 and Structure 6–3)].

Synthesis of a polymer of methyl-substituted tetrahydropyranyl structure terminated by hydroxyl groups, by radical polymerization of methacrolein,[106] has been described. Anionic polymerization initiated by naphthalene sodium and triphenylmethyl sodium led to a linear structure $[\text{-}CH_2CMe\text{:}CHO\text{-}]_n$ which was confirmed by hydrolysis and ozonolysis.

Bulk and solution polymerization of acrolein[107] were carried out in vacuo at room temperature by γ irradiation. In the case of bulk polymerization, a white, insoluble, solid polymer was obtained at 96% conversion. Solution polymerization to various conversions and in various solvents led to soluble polymers the properties of which led to the postulate that a pyran structure containing hydroxyl and residual aldehyde groups could well represent the polyacrolein obtained.

The polymerization of acrolein and α-methylacrolein[108] in the presence of gaseous boron trifluoride was investigated. The reaction was carried out both in bulk, and in solution and both in complete absence of moisture, and with water added. Bulk polymerization gave soluble polymers at 10–15% conversion. At the higher degrees of conversion, the polymers were insoluble and did not contain free aldehyde groups. Low temperature polymerization gave completely soluble polymers.

Cationic initiated copolymerization of acrolein[109] with 1,4-butynediol gave 88% of a 1:1 alternating copolymer with the aldehyde function intact. However, fractionation of this copolymer gave a small fraction of insoluble copolymer, postulated to be a polyacetal of the structure shown (9–14).

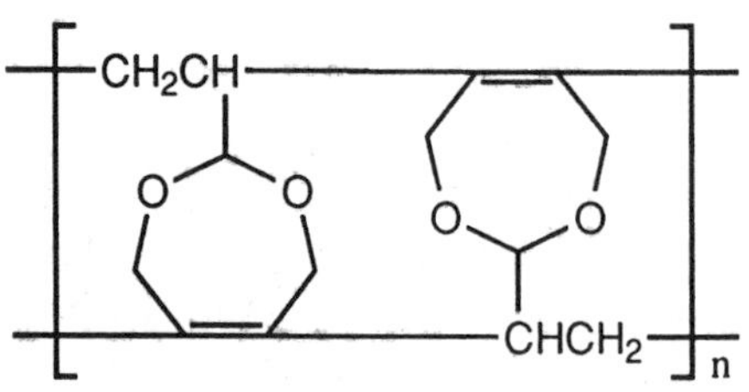

(9–14)

In a study of ladder polymers from poly(isopropenyl methyl ketone)[110] an NMR investigation of the polymer prepared from poly(isopropenyl methyl ketone) at 300°C, and containing up to 80% 3-methyl-1-buten-1,4:2,3-tetrayl units, indicated that the condensation reaction was random and that higher conjugated ring formation was a consecutive reaction.

Polyacrolein[6–23] was converted to products which contained mainly primary alcoholic groups by reaction with formaldehyde in presence of basic catalysts. It was concluded that the polymers consisted of essentially methylene 1,3-propanediol units.

A mixture of styrene and acrolein[6–27] was subjected to sodium methoxide which polymerized the acrolein. Subsequent copolymerization of the pendent vinyl groups of the polyacrolein and styrene was accomplished by free radical initiation to yield a crosslinked polymer.

Copolymers of methacrylonitrile and acrolein or methacrolein[6–25] were investigated by use of absorption spectra of the carbonyl group. The main contribution to the spectra arose from aldehyde functions in isolated positions. However, at sequence lengths of two or three consecutive aldehyde groups, cyclic aldehyde ether units were formed by the secondary cyclopolymerization to effect ring closure.

In a review of the reactions of polymers with 69 references, the reactions of polyacrolein[6–30] as a means of producing new polymers having partial ladder structures were emphasized.

It has been shown that formation of complexes between the propagating polymer end and the monomer molecule in the anion-radical polymerization of methacrylaldehyde[111] was energetically advantageous and led to the formation of a polymer with a high content of tetrahydropyran rings. The content of free aldehyde groups increased with polymerization temperature and decreased C:C bond content.

LADDER POLYMERS OF ACRYLONITRILE AND RELATED STRUCTURES

Although acrylonitrile is a conjugated structure, polymers from this monomer are formed predominantly through the vinyl double bond. However, when polyacrylonitrile is subjected to heat or to treatment with base, postpolymerization occurs and the polymer becomes colored. Much speculation and effort has been devoted to the nature of this reaction. The first suggestion of structural changes in the polymer to account for this phenomenon was made in 1950. It was suggested that the black color was an indication that a secondary reaction between the nitrile groups had occurred. The following structure for the heat-treated polymer was proposed (9–15).[112] This structure is believed

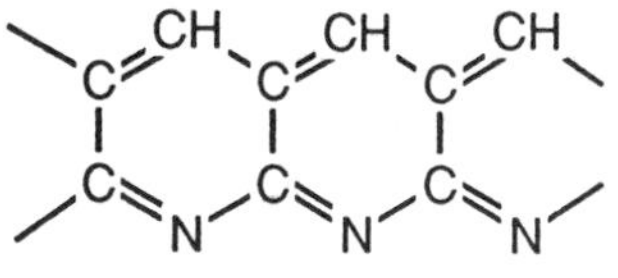

(9–15)

to be the predominant one in "Black Orlon." The following structure for the base-treated polymer (9–16) was proposed.[113] These postulates were extended in several subsequent papers[114–117]

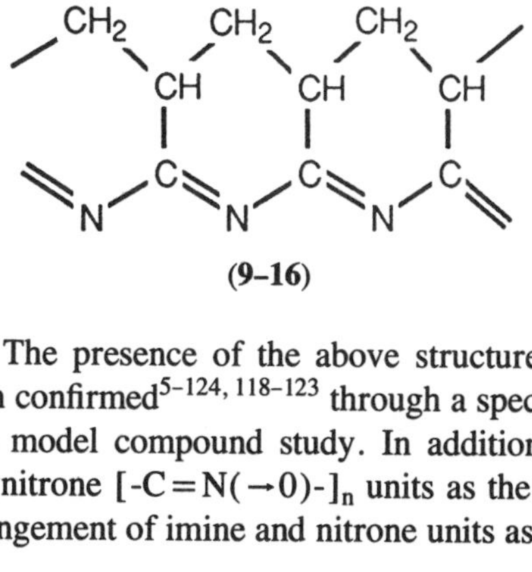

(9–16)

The presence of the above structure in heat or base-treated polyacrylonitrile has been confirmed[5-124, 118-123] through a spectral investigation in conjunction with an extensive model compound study. In addition, evidence was obtained for the presence of polynitrone $[-C=N(\rightarrow 0)-]_n$ units as the result of the absorption of oxygen. A random arrangement of imine and nitrone units as shown in structure (9–17) was postulated:

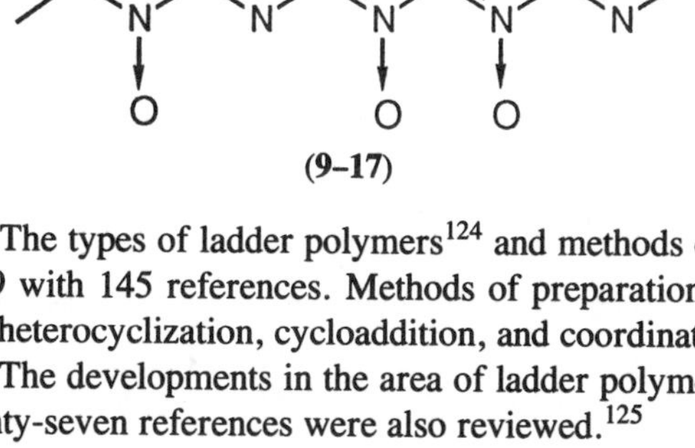

(9–17)

The types of ladder polymers[124] and methods of their preparation were reviewed in 1969 with 145 references. Methods of preparation include equilibration, "zipping-up," polyheterocyclization, cycloaddition, and coordination polymerization.

The developments in the area of ladder polymers for the period 1965–70, including twenty-seven references were also reviewed.[125]

It is not the purpose of this volume to review ladder polymers beyond the purpose of showing the relationship of formation of certain polymers of this group to cyclopolymerization. It can be seen that formation of such polymers could be visualized as two successive polymerizations in which one or both steps may be initiated under anionic conditions. 1,2-Addition polymerization is usually carried out first, followed by a succession of cyclopolymerization steps.

The effect of the photosensitizing action of eosin yellow in the absence of a reducing agent and in presence of defined quantities of oxygen on the polymerization of aqueous solutions of acrylonitrile[126] was investigated. The effects of polymer yield, reaction rate and polymer-molecular-weight with changing temperatures and duration of irradiation were established. The structure of the polymer, whether cyclized, was not discussed.

In a study of the mechanism of thermal conversion of polyacrylonitrile,[127] ultraviolet absorption was monitored as the temperature passed through the 200–300° temperature range. Attempts were made to correlate structural changes in the polymer with each significant change in the ultraviolet absorption.

Concentrated solutions of silver nitrate were reported to polymerize acrylonitrile[128] explosively. The report of this study included 22 references.

Polymers of the structure shown (9–18)[129] were reported to be obtained via an

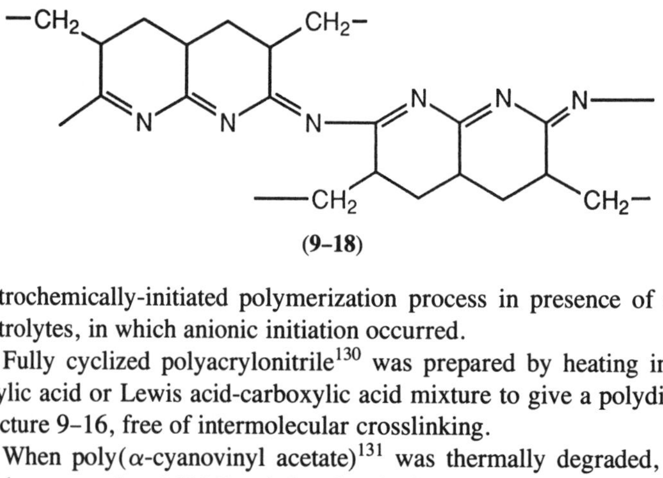

(9–18)

electrochemically-initiated polymerization process in presence of alkali persulfates as electrolytes, in which anionic initiation occurred.

Fully cyclized polyacrylonitrile[130] was prepared by heating in presence of a carboxylic acid or Lewis acid-carboxylic acid mixture to give a polydihydro-pyridacene of Structure 9–16, free of intermolecular crosslinking.

When poly(α-cyanovinyl acetate)[131] was thermally degraded, a rapid decrease in weight occurred at 250°C and the electrical conductivity of the product increased. The results of the study supported the following course of reactions for formation of the final product (9–19).

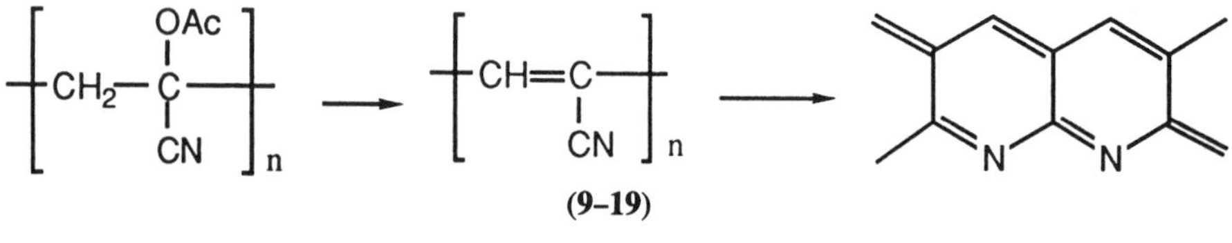

$$\left[CH_2-\underset{CN}{\overset{OAc}{C}} \right]_n \longrightarrow \left[CH{=}\underset{CN}{C} \right]_n \longrightarrow$$

(9–19)

Poly (α-chloroacrylonitrile)[132] was prepared and dehydrohalogenated to the conjugated polymer (analogous in structure to that predicted by vinyl polymerization of vinylacetylene). This polymer was then cyclized to produce ladder structures analogous to a 4-chlorosubstituted derivative of Structure 9–19. The ladder structure was shown to be a semiconductor, and was stable up to 600°.

A study of the structure of the products formed during the low-temperature (−196°) polymerization of γ-irradiated crystalline acrylonitrile has been reported.[133] The less soluble fraction of the isolated polymer (molecular weight ~1500) was shown to have the Structure 9–16.

The changes of the structure of polyacrylonitrile[134] during pyrolysis were studied. The pyrolysis proceeds in two stages: at low temperatures only C=NH bridges are formed between chains. At 150°-200°C further crosslinking and aromatization result in formation of three-dimensional polyconjugated systems of partial ladder-partial open chain structure.

Poly(α-cyanovinyl acetate)[135] was completely hydrolyzed and 60% of the freed carboxyl and hydroxyl groups took part in lactonization. Sodium hydroxide caused hydrolysis and elimination of acetic acid resulting in formation of double bonds conjugated with the nitrile triple bonds. The nitrile groups not hydrolyzed formed conjugated rings of Structure 9–15.

In a study of the thermal and alkaline degradations of polyacrylonitrile,[136] it was proposed that the main process involved in the thermal decomposition was the one-step formation of condensed 1,8-naphthyridine rings, which occurred in vacuo as a fast exothermal autocatalytic process with rapid heating.

The observed shrinkage of polyacrylonitrile fiber during heat treatment at $\geqslant$200°C was attributed to formation of imperfect ladder polymers as a first step in carbon fiber fabrication.[137] A model cyclization reaction leading to angled step ladder polymers with sequences of $\leqslant$ four rings was proposed.

Low temperature degradation of polyacrylonitrile[138] by use of sodium cyanide gave a poly(1,4-dihydropyridine) intermediate (9–20).

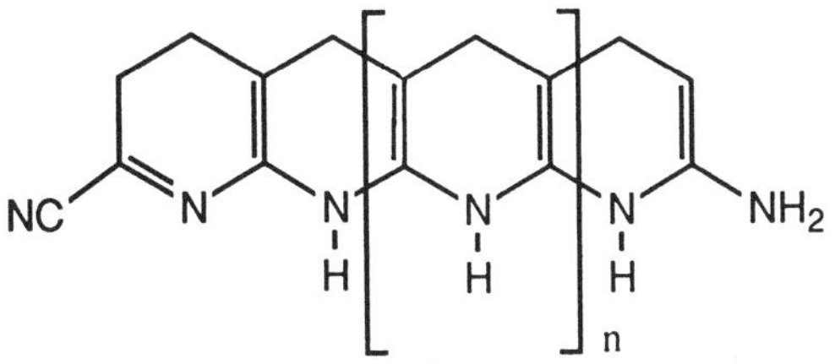

(9–20)

Cyclopolymerization of acrylonitrile in the glassy state (−196°) under the action of γ- and UV-radiation was observed.[139] A 50% solid solution of acrylonitrile in the diethyl ether of diethylene glycol was irradiated to give polymer of Structure 9–16.

Cyclization of polyacrylonitrile[140] was investigated in presence of a variety of organic acids and alcohols. The concentration of cyclic structures in the polymers increased with decreasing pK of the acids, and the logarithm of the relative concentration of the cyclic structures was linearly related to their Taft constants. In strong acids, cyclization followed a mechanism of acid catalysis, but weak acids and alcohols initiated cyclization by a nucleophilic adduct on the CN group.

The thermal behavior of poly(methacrylonitrile)[141] has been studied. The effect of polymerization catalyst on intramolecular cyclization and thermal degradation (rate of formation of C:N bonds) at 200° was in the order azobisisobutyronitrile > benzoyl peroxide > butyl lithium > diethyl magnesium. Formation of C:N was faster in methacrylic acid-methacrylonitrile copolymer than in the homopolymer.

The vitrification, color development, thermal degradation and carbonization of polyacrylonitrile[142] was reviewed with 77 references.

Acrylic polymers such as polyacrylonitrile[143] and acrylonitrile-styrene copolymers were cyclized in presence of 2-pyrrolidinone to give heat resistant dihydropyridacene polymers free of intermolecular crosslinking.

Cyclization in polyacrylonitrile[144] and polymethacrylonitrile has been studied. Insertion polymethacrylonitrile and polyacrylonitrile prepared by adsorption of the monomers on montmorillonite and free radical polymerization by exposure to γ-rays, were extensively cyclized. Polymers prepared in bulk and in solution by γ-ray initiation showed no evidence of cyclization.

The influence of the comonomer on the intramolecular cyclization reaction in acrylonitrile[145] copolymerizations has been studied. The extent of the reaction between acrylonitrile and vinyl acetate, vinyl chloride, styrene, butadiene, vinylidene chloride, methyl acrylate, and methyl methacrylate was governed by the reactivity of the comonomer-unit ended radical, but the cyclization reaction could not explain all the kinetic deviations observed. No correlation was observed between the copolymer absorption at 290 nm and the composition deviations calculated from kinetic results, except for the vinyl chloride copolymers.

The radical concentrations, reactivity ratios and homopropagation rate constants were determined for the copolymerizations of styrene with methyl methacrylate, vinyl chloride (I) with vinyl acetate (II), and acrylonitrile[146] with I and with II. The number of acrylonitrile radicals calculated from these data was larger than the theoretical for I but smaller for II, possibly due to intramolecular cyclization reaction.

Block acrylonitrile-p-dichlorobenzene polymers[147] in acrylonitrile-$PhCl_2$-Li catalyst systems in THF solution have been prepared with an increase in polyphenylene block observed on the complete polymerization of acrylonitrile (I). A cyclization of CN groups to a conjugated system was observed in polymerization of I followed by dehydrogenation of polyvinyl chains. An equimolar ratio of $PhCl_2$ and I showed a predominant I polymerization in the initial 17 hours of polymerization, followed by a sharp rise in polyphenylene fractions.

In a spectroscopic examination of the alkaline degradation of polyacrylonitrile[148] and its oligomers in dimethylsulfoxide under nitrogen, it was shown that the model compounds, $RCH(CN)[CH_2CH(CN)]_nR'$, ($R, R' = CH_3$, $n = 1$–4) when treated with KOH proceeded with formation of an anion conjugated with a series of conjugated $C=N$ bonds, resulting in intramolecular cyclization of the CN groups.

An infrared investigation of the structure of copolymers of acrylonitrile[149] with itaconic acid methyl esters has been conducted. The structure of copolymers with monomethyl itaconate and with dimethyl itaconate, obtained in the presence and in the absence of $ZnCl_2$, was studied by IR spectroscopy. The observed decrease in the $-C\equiv N$ absorption band and the formation of the :C:N-band was more clearly pronounced in the copolymers obtained in the presence of $ZnCl_2$. The results are explained by assuming the formation of a six-membered ring as a result of the transfer of one α-methylene H atom from the itaconates to the $-C\equiv N$ bond.

It has been shown that the main product from thermal decomposition of polyacrylonitrile[150] at $>250°$ in air was HCN. The amount of HCN evolved was temperature dependent, decreasing from 17.7 mol% (based on nitrile groups) at decomposition temperature 200° to 10 mol% at 300°. The activation energy for HCN formation was 15 kcal/mol at 250–300°. Nitrile group cyclization competed with HCN evolution, and had activation energy 29 kcal/mol.

The behavior of systems based on acrylonitrile[151] radio-frequency plasmas have shown that the important reactions of monomers, binary and ternary polymers of acrylonitrile with vinyl acetate and α-methylstyrene were nitrile-group cyclization, chain scission, crosslinking and oxidation.

In a study of oxidation processes of polyacrylonitrile[152] and acrylonitrile-itaconic acid copolymer fibers, it was concluded that the copolymer fiber mainly cyclized by a free radical process but could also be induced by an ion transfer process in the case of partial hydrolysis.

A stoppered, vacuum tube for blood handling was prepared from an acrylonitrile-type polymer[153] having the inner surface coated with selective additives including DADMAC-SO_2 copolymer to facilitate the separation of blood serum from the blood clot.

The process of thermal stabilization of polyacrylonitrile[154] fibers pretreated with cuprous chloride was studied in air and under nitrogen. The rate of cyclization proceeded furiously at 240°-300°C and was much faster than the rate of dehydrogenation. The incorporation of the salt enhanced the rate of cyclization by ~1.4 times in air at 260° and by ~43% under nitrogen at 340°.

LADDER POLYMERS OF VINYL ISOCYANATE

Vinyl isocyanate and its polymers have been studied. This monomer was polymerized through the isocyanate group by using sodium cyanide as initiator at −55°C to yield N-vinyl-l-nylon.[1-114, 1-115, 155, 5-90] When the polymer, in dimethylacetamide solution, was treated with AIBN and ultraviolet light at room temperature, the polymer was soluble in solvents of high dielectric strength, and was shown to consist of 90% ladder polymer structure (9–21). When the monomer was permitted to stand for 2 days at 25°C, in-

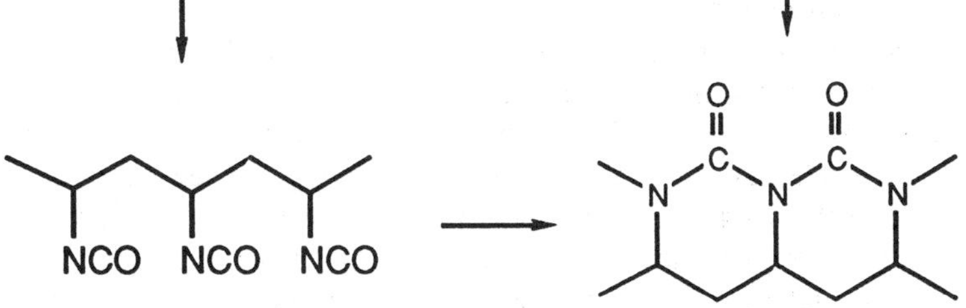

$CH_2{=}CH{-}NCO$

(9–21)

soluble polyvinyl isocyanate was obtained, which on treatment with Co^{60} irradiation at 30°C for 48 h, gave an insoluble polymer. This polymer was shown to contain about 85% of the ladder structure. Further studies on polymers and copolymers of vinyl isocyanate have been reported.[156, 1-116, 6-86] Attempts to synthesize triple strand polymers by polymerization of *cis*- and *trans*-1,3,5-triisocyanatocyclohexane[157] have been reported. However, the obtained polymers contained predominantly bicyclic structures with one pendant isocyanate group per cyclohexane unit.

A variety of N-vinyl compounds were obtained through use of the cyclopolymerizable monomer, vinyl isocyanate.[158] Most of the compounds were polymerizable via radical initiation.

The bisulfite adduct of vinyl isocyanate[159] was prepared and polymerized via radical initiation to avoid the insoluble polymers obtained by direct polymerization of vinyl isocyanate due to reactions of the isocyanate group.

Optically active N-vinyl compounds were prepared by reaction of vinyl isocyanate[160] with L-amino acids to give N-vinyl urea derivatives and by a similar reaction with L-menthol to give L-menthyl N-vinylcarbamate.

Copolymers from ethylene and vinyl isocyanate or isopropenyl isocyanate[161] with acrylic esters, vinyl acetate or acrylamide via radical initiation have been described. A typical product softened at 83°C, and was soluble in decahydronaphthalene.

Vapor-phase photosensitized polymerization of vinyl isocyanate[162] and acrylic acid on polyethylene films has been studied. Orientation of the grafted layers was studied by analysis of the IR spectrum. Curves showing the dichroism and IR bands were given for the polymers and showed that the photosensitized graft polymers gave oriented structures.

MACROCYCLIC LADDER POLYMERS VIA CYCLOPOLYMERIZATION

Although a variety of macrocyclic ladder polymers have been reported via condensation polymerization techniques, e.g. the polysiloxanes, this section is limited to those structures believed to be formed via cyclopolymerization reactions.

The ladder polymers discussed previously are postulated to consist of predominantly six-membered fused rings. Multiple cyclopolymerizations of unsaturated side chains or pendant groups in which the ring size is greater than six have been demonstrated. It has been shown[163] that when 4-vinylpyridine in excess was added to an aqueous solution of polystyrene sulfonic acid, a precipitate formed which consisted of stoichiometric amounts of polystyrenesulfonic acid and of poly-4-vinylpyridine. The authors postulated that this interesting cyclopolymerization (9–22) occurred as the 4-vinylpyridine molecules became quaternized through proton, initiated by a strong polarization of the vinyl double bond.

An interesting type of cyclopolymerization was proposed and demonstrated.[164–167] It has been referred to as "intramolecular cyclooligomerization". The process is illustrated in the Structure 9–23. Polynuclear compounds also containing four and five p-cresol units per molecule have been used. Alkaline hydrolysis of the ladder oligomer containing four p-cresol units gave a molecularly homogeneous hexaacrylic acid.

Diallyl- and divinylpolysilmethylenes[168] were found not to undergo polymerization with ionic or radical initiators; however, Ziegler catalysts yielded a non-crosslinked polymer which contained unsaturation in the side chains from 1,3-diallyldisilmethylene.

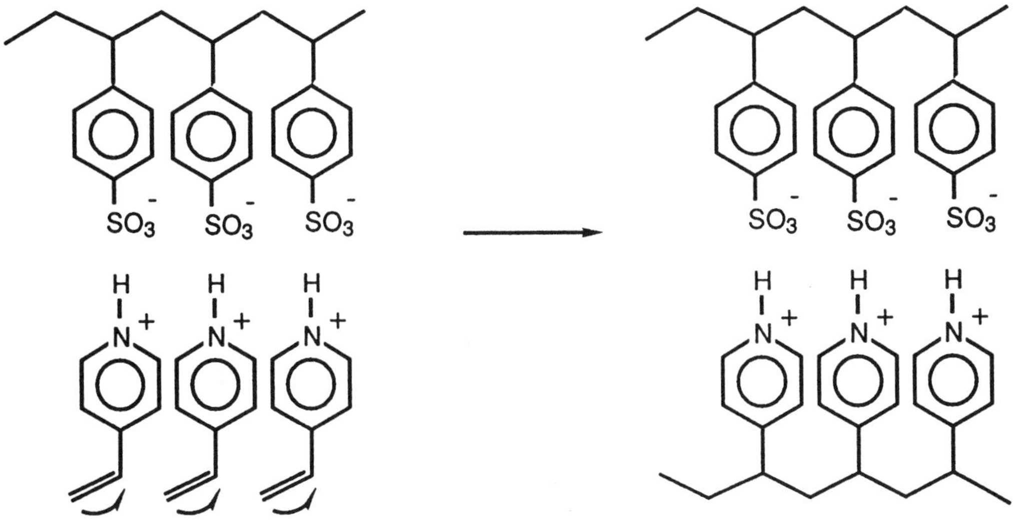

(9–22)

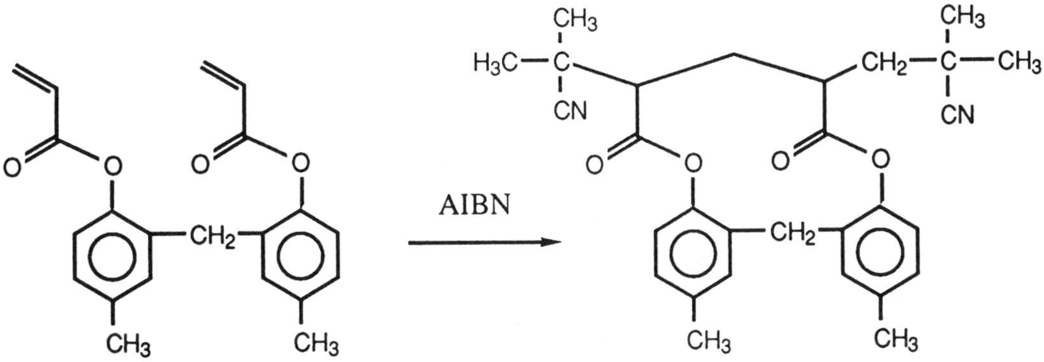

(9–23)

The kinetics of polymerization of 4-vinylpyridine[169] on macromolecule "templates" were studied and the mechanism studied. Macrocyclic structures were reported.

Preparation of molecularly homogeneous tetramethacrylic acids[170] with different matrices were accomplished in a matrix reaction model program. Alkaline hydrolysis of the structure shown (9–24) depicts such a model compound.

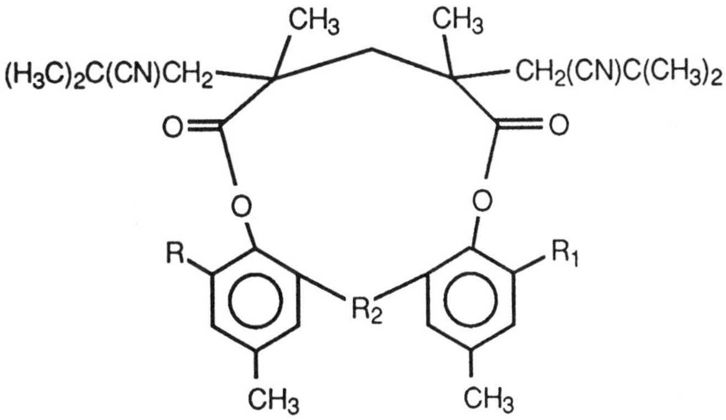

(9–24)

Molecularly homogeneous tri-, tetra-, and penta-methacrylic acids[171] were described in a continued study of matrix reaction models. The models were synthesized via alkaline hydrolysis of structures shown (See Structure 9–23).

A thermosetting cyclic polybutadiene hydrocarbon[172] was obtained by mixing a polyfunctional diene prepolymer with a polyfunctional hydrocarbon prepolymer, a chain extender, e.g. toluene diisocyanate, and a radical initiator followed by heating in a mold. A number of examples were given.

Cyclized polydienediols, e.g., 1,2-polybutadienediol,[173] were used in conjunction with a polyetherdiol, or -triol in presence of a chain extender, e.g. toluenediisocyanate, and an organic peroxide to produce opaque rubbery resins with good properties. A variety of processes were presented.

Semiladder block copolymers[174] were obtained by esterifying p-cresol-formaldehyde trimer or octamer with acryloyl chloride or methacryloyl chloride, followed by copolymerization with styrene.

Cationic polymerization of vinyl p-(vinyloxy)benzoate[3–132,175] led to a linear polymer having the vinyl ester group intact. Radical post-polymerization was reported to lead to a macrocyclic ladder polymer of molecular weight of 2000. Hydrolysis of the prepolymer gave p-(vinyloxy)benzoic acid.

In a study of models for matrix reactions, the reaction of a mixture of acrylic esters with polynuclear phenolic compounds[176] was investigated. The phenols were treated with acryloyl or methacryloyl chloride, and polymerized to yield the ester, a trimeric structure analogous to Structure 9–23. Decomposition with lithium borohydride regenerated the phenol and gave oligoalcohols having 2–5 repeat units.

An equation was derived for calculating the cyclization probability of crosslinked polymers.[177] The validity of the equation was confirmed by crosslinking poly(vinyl butyral) with hexamethylene diisocyanate in pyridine.

A theoretical approach to simulation of degradation processes[3–208] examined some relations of random crosslinking without cyclization. The problem was considered very complex and no conclusions were drawn.

o-Isopropenylstyrene[178] could be selectively polymerized through the vinyl groups with butyllithium and the linear polymer polymerized further with aluminum bromide. In this second stage all of the double bond was used, but the molecular weight of the polymer did not change, and the polymers were soluble.

In a two-stage polymerization study of o-allylstyrene and o-propenylstyrene[179] the vinyl groups were shown to be selectively polymerized with butyllithium to give the respective unsaturated polymers. The linear polymers were polymerized further by aluminum bromide to consume all of the unsaturation. Poly(o-allylstyrene) gave a soluble polymer but the poly(o-propenylstyrene) gave mostly crosslinked polymer.

The synthesis of poly(carbosiloxanes) from dichloromethylvinyl silane[180] was carried out in a two-step process. First, the vinyl group was polymerized in the presence of a radical initiator. The second step subjected the oligomer obtained in the first step to hydrolytic polycondensation. The structure shown (9–25) was confirmed by IR spectroscopy.

Intramolecular reactions to produce macrocyclic structures occurred when terephthaldehyde was used as a crosslinking agent for poly(vinylalcohol)[181] for solutions containing concentrations of the poly(vinyl alcohol) $\leqslant 0.1\%$. However, intermolecular crosslinking occurred in solutions containing $> 0.6\%$ of the polymer.

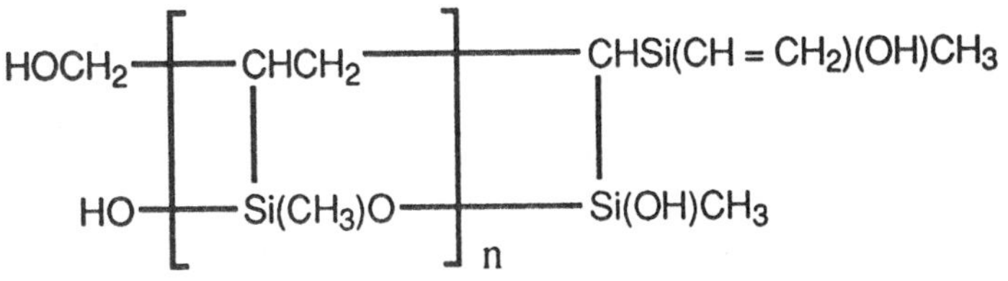

$HOCH_2-[CHCH_2]-CHSi(CH=CH_2)(OH)CH_3$

$HO-[Si(CH_3)O]_n-Si(OH)CH_3$

(9–25)

In a study of side reactions in radical cyclization of 2,2′-methylene-bis(4-methyl-1,2-phenylene) dimethacrylate,[182] it was shown that 55% of the cyclic ester (See Structure 9-23) polymerized by head-to-tail addition when the bis-dimethacrylate was refluxed with an excess of azobisisobutyronitrile. Four percent of the isomeric derivative was obtained by head-to-head addition.

A loss of crystallagraphic register between chains occurred during γ-ray polymerization of the cyclic tetradiyne monomer $[(CH_2)_2C\equiv C\text{-}C\equiv C(CH_2)_2]_4$[5-38] although crystallographic order in the chain-axis projection was retained. The polymer has a macrocyclic butatriene backbone structure.

Poly(7,16-dihydroheptacenes),[183] a new type of obligate ladder polymer conformationally restricted to two dimensions was obtained by polymerization of the heptacene derivative shown in Structure 9–26 by sunlight or artificial light. The process was

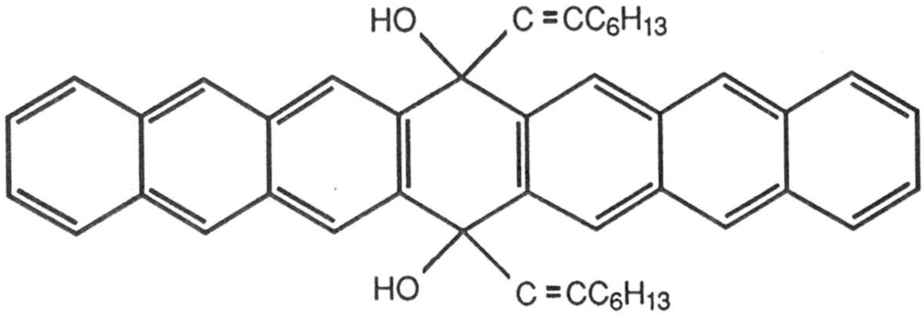

HO, $C\equiv CC_6H_{13}$

HO, $C\equiv CC_6H_{13}$

(9–26)

accompanied by the loss of the anthracene-like UV absorption of the monomer. The molecular weight was 5600, and the properties indicated the polymer to be a semirigid ladder confined locally to two dimensions by the rigidity of the dianthracene units.

Macrocyclic ladder polymers were postulated to arise from the cross-polymerized product of 1,11-dodecadiyne.[5-39] Electron diffraction patterns were obtained from the macromonomer and cross-polymerized crystals of the polymer. The unit cell of both macromonomer and cross-polymerized material was monoclinic, space group $p2_1/n$ (9–27).

The predicted enhancement of thermal stability of double strand polymers in comparison to their single strand counterparts generated considerable interest among polymer scientists during the period of emphasis on thermal stability by the airline industries and the armed forces laboratories. During the early publication of successful efforts to synthesize ladder polymers from isoprene or 1,3-butadiene, several news stories appeared on this work.[184, 185] A Belgian Patent has also been issued on cyclization polymerization.[186]

As early as 1966, the presence of cyclic structures in polybutadienes and polyisoprenes was observed in the NMR spectra. The structures were identified by compari-

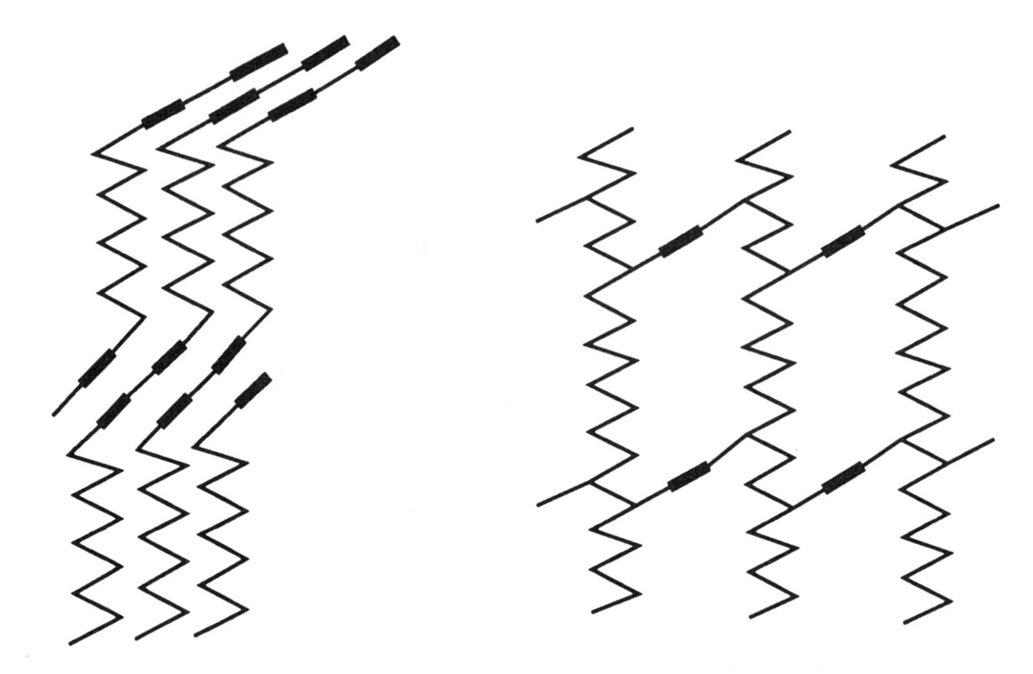

Model of cross-polymerization reaction

(9–27)

son with spectra of ladder polymers obtained from 3,4-polyisoprene and 1,2-polybutadiene.[187] The amount of cyclic structures in *cis*-1,4-polybutadiene was also determined from the total amounts found by microstructure determinations. Additional IR evidence was provided by changes in the polymer molar absorptivities. In the long-wavelength region, IR spectra also appeared to be promising for these investigations. The results of this paper have been challenged.[60]

REFERENCES

1. Banerjee, B., Lab. Dev., 1967, 5, 284; CA 68, 13822y (1968).
2. Goinetti, E., Vinciguerra, N. and Cerbo, D., Russ. Chem., 1969, 21, 274; CA 72, 67383g (1970).
3. Schweitz, H., J. Chim. Phys. Physicochem. Biol., 1969, 66, 657; CA 71, 39407d (1969).
4. Godt, A., Enkelmann, V. and Schlueter, A. D., Angew. Chem., 1989, 101, 1704; CA 112, 78090e (1990).
5. Bell, V. L., Jr., Belg.Pat. 623,940; 623,941, 1963, CA 60, 12133f (1964).
6. Gaylord, N. G., Kossler, I., and Stolka, M., J. Macromol. Sci. (Chem.) A, 1968, 2, 1105; CA 70, 5063c (1969).
7. Golub, M. A. and Heller, J., J. Polym. Sci. B, 1966, 4, 469; CA 65, 5540e (1966).
8. Shelton, J. R. and Lee, L. H., Rubber Chem. Technol., 1958, 31, 415; CA 52, 19217c (1958),
9. Golub, M. A. and Heller, J., Can. J. Chem., 1963, 41, 937; CA 58, 10379g (1963).
10. Golub, M. A. and Heller, J., Tetrahedron Lett., 1963, 2137; CA 60, 8212b (1964).
11. Koltsov, A. I., Vysokomolek. Soedin., Ser. B, 1967, 9,(2), 97; CA 66, 95544e (1967).

12. Taniguchi, M., Kawabata, N. and Furukawa, J., J. Polym, Sci. B, 1967, 5, 1025; CA 67, 117856h (1967).
13. Carbonaro, A. and Greco, A., Chimica Ind. Milan, 1966, 48, 363; CA 65, 5537a (1966).
14. Gaylord, N. G. and Svestka, M., J. Macromol. Sci. (Chem), 1969, 3, 897; CA 71, 30728b (1969).
15. Gaylord, N. G. and Svestka, M., J. Polym. Sci., B, 1969, 7, 55; CA 70, 97267y (1969).
16. Gaylord, N., Pure Appl. Chem., 1970, 23, 305; CA 74, 142452a (1971).
17. Matyska, B., Dolezal, I. and Koessler, I., Collect. Czech. Chem. Commun., 1970, 36, 2924; CA 76, 25661y (1972).
18. Matyska, B., Antropiusova, H., Svestka. M. and Gaylord, N. G., J. Macromol. Sci. (Chem), 1970, 4, 1529; CA 73, 78269w (1970).
19. Gaylord, N. G., Kossler, I., Matyska, B. and Mach, k., J. Polym. Sci., A1, 1968, 6, 125.
20. Gaylord, N. G., Matyska, B., Kossler, I. and Krauserova, J. Polym. Sci., C, 1968, 24, 277; CA 67, 3497m (1967).
21. Kossler, I., Stolka, M. and Mach, K., J. Polym. Sci., C, 1963, 4, 977; CA 60, 7005h (1964).
22. Krauserova, H., Kossler, I., Matyska, B. and Gaylord, N. G., J. Polym. Sci., Part C, 1968, 23, 327; CA 69, 87513r (1968).
23. Gaylord, N. G., Kossler, I., Matyska, B. and Mach, K., Polym. Preprints, 1967, 8(1), 174; CA 67, 3496k (1967).
24. Gaylord, N. G., Matyska, B., Kossler, I. and Krauserova, H., Polym. Preprints, 1967, 8(1), 830; CA 67, 3497m (1967).
25. Krauserova, H., Mach, K., Matyska, B. and Kossler, I., J. Polym. Sci., C, 1967, 16, 469; CA 67, 33593g (1967).
26. Gandhi, V. G., Deshpande, A. B. and Kapur, S. L., J. Polym. Sci., Chem. Ed., 1974, 12, 1257; CA 81, 12223w (1974).
27. Shikata, K. and Mizumoto, Y., Jap. Pat. 79 52,193, 1979 (Apr. 24); CA 91, 108879m (1979).
28. Todoko, M. and Watanabe, H., Eur. Pat. Appl., EP 110,356, 1984 (Jun. 13); CA 101, 74112p (1984).
29. Tutorskii, I. A., Morkov, V. V., Fomina, L. P., Belyanin, V. B. and Dogadkin, B. A., Vysokomol. Soedin., 1963, 5, 593; CA 59, 2996f (1963).
30. Gaylord, N. G., Matyska, B., Mach, K. and Vodehnal, J., J. Polym. Sci., A1, 1966, 4, 2493; CA 65, 17171g (1966).
31. Golub, M. A., Macromolecules, 1969, 2, 550; CA 72, 4155p (1970).
32. Dankelka, J., Polacek, J. and Koessler, I., Collect. Czech. Chem. Commun., 1969, 34, 2833; CA 71, 113341w (1969).
33. Clark, J., Lal, J. and Henderson, J., J. Polym. Sci. B, 1971, 9, 49; CA 74, 77210d (1971).
34. Drahoradova, E., Doskocilova, D. and Matyska, B., Collect. Czech. Chem. Commun., 1971, 36, 1301; CA 75, 22135y (1971).
35. Vohlidal, J., Bohackova, V. and Matyska, B., J. Chim.Phys. Physicochim. Biol., 1972, 69, 556; CA 77, 76352b (1972).
36. Van Stratum, P. G. M., J. Chromatography, 1972, 71, 9; CA 77, 153462s (1972).
37. Agnihotri, R., Falcon, D. and Fredericks, D., J. Polym. Sci. A1, 1972, 10, 1839; CA 77, 89659g (1972).
38. Halasa, A. F., U.S. Pat. US 4,248,988, 1981 (Feb. 3), CA 94, 140458w (1981).
39. Nadarajah, M., Harangoda, H., Blasingham, C.G., Kirubakaran, J.K. and Jayasinghe, P.P., Rubber Res. Inst. Ceylon, Q. J., 1973, 50(Pt.3-4) 157; CA 83, 180712d (1975).
40. Vodehnal, J. and Drahoradova, E., J. Polym. Sci., A-2, 1972, 10, 2033; CA 78, 5178j (1973).

41. Tutorskii, I. A., Sokolova, L. V., Izyumnikov, A. L., Kharlamov, V. M. and Dogadkin, B. A., Vysokomol. Soedin., Ser. A, 1973, 15, 1587; CA 80, 16073p (1974).
42. Vohlidal, J., Bohackova, V. and Matyska, B., J. Polym. Sci., Polym. Symp., 1973, 42, 901; CA 82, 86706n (1975).
43. Kossler, I. and Vodehnal, J., Anal. Chem., 1968, 40, 825; CA 68, 87976y (1968).
44. Gemmer, R. V. and Golub, M. A., J. Polym. Sci., Polym. Chem. Ed., 1978, 16, 2985; CA 90, 39376k (1979).
45. Gelling, I. R., Loadman, M. J. and Sidek, B. D., J. Polym. Sci., Polym. Chem. Ed., 1979, 17, 1383; CA 90, 205504x (1979).
46. Vainer, A. Y. and Dryakina, T. A., USSR 730,696, 1980 (Apr. 30); CA 93, 115706x (1980).
47. Shinohara, S. and Ozuru, I., Jap. Pat. 80 38,814, 1980 (Mar. 18); CA 93, 48748x (1980).
48. Sumitomo Chemical Co., Ltd., Jap. Pat. 80,152,733, 1980 (Nov. 28); CA 92, 122921n (1981).
49. Toyo Electronics Industry Corp., Jap. Pat. 80,150,258, 1980 (Nov. 22); CA 94, 123211t (1981).
50. Van Oeteren, K. A., Seifen, Oele, Fette, Wachse, 1977, 103, 342; CA 87, 119364h (1977),
51. Carbon Paper Co., Ltd., Jap. Pat. JP 59 24,765 [84 24,765], 1984 (Feb. 8); CA 101, 40014t (1984).
52. Fujitsu, Ltd., Jap. Pat. JP 59 02,043 [84 02,043], 1984 (Jan. 7}; CA 101, 31146k (1984).
53. Van der Klooster, J., J. Oil Colour Chem. Assoc., 1984, 67(4), 105; CA 101, 39884v (1984).
54. Gaylord, N. G., Koessler, I. and Stolka, M., J. Macromol. Sci. (Chem), 1968, 2, 421; CA 68, 96445b (1968).
55. Iwamoto, M. and Yuguchi, S., Jap. Pat. JP 69 01,217, 1969 (Mar. 5); CA 71, 3927j (1969).
56. Pilar, J., Toman, L. and Marek, M., J. Polym. Sci., Polym. Chem. Ed., 1976, 14, 2399; CA 85, 178069x (1976).
57. Halasa, A. F. and Cheng, T. C., U.S. Pat. US 3,931,127, 1976 (Jan. 6); CA 84, 122580u (1976).
58. Iketani, T, Watanabe, T. and Harita, Y., Jap. Pat. JP 78,124,557, 1978 (Oct. 31); CA 90, 73025e (1979).
59. Biesenberger, J. A., Ph.D. Thesis, Princeton Univ., 1963; CA 60, 12204e (1964).
60. Chen, H. Y., J. Polym. Sci., Polym. Ltrs., 1966, 4, 1007; CA 66, 46598s (1967).
61. Dvorak, J., Matyska, B., Vodehnal, J. and Kossler, I., Collect. Czech. Chem. Commun., 1967, 32, 1561; CA 67, 3313y (1967).
62. Shagov, V. S. and Yakubchik, A. I., Vestn. Leningrad Univ., Fiz. Khim., 1967, 22, 157; CA 68, 30817t (1968).
63. Koessler, I., Vodehnal, J., Stolka, M., Kalal, J. and Hartlova, E., J. Polym. Sci. , C, 1967, 10, 1311; CA 68, 40714w (1968).
64. Kendrick, W. P., Lubowitz, H. R. and Burns, E. A., Fr. Pat. 1,540,983, 1968 (Oct. 4); CA 71, 22586p (1969).
65. Boheckovoy, V., Figerova, E., Gallot-Grubasic, Z., Polacek, J. and Stolka, M., J. Chim. Phys. Physicochem. Biol., 1970, 67, 777; CA 73, 66974j (1970).
66. Azuma, C., Sanui, K. and Ogata, N. J. Appl. Polym. Sci., 1980, 25, 1273; CA 93, 115644a (1980).
67. Ichikawa, M., Harita, Y. and Ito, H., Jap. Pat. JP 73 66,684, 1973 (Sep. 12); CA 80, 146822t (1974).
68. Lal, J., U.S. Pat. 4,242,471, 1980 (Dec. 30); CA 94, 104685j (1981).
69. McCreedy, K. and Keskkula, H., J. Appl. Polym. Sci., 1978, 22, 999; CA 89, 7351m (1978).

70. Quack, G. and Fetters, L. J., Macromolecules, 1978, 11, 369; CA 88, 191715m (1978).
71. Parshall, G. W. and Nugent, W. A., Chemtech, 1988, 18, 314; CA 109, 8329d (1988).
72. Chiantore, O., Luda di Cortemiglia, M. P., Guaita, M. and Rendina, G., Makromol. Chem., 1989, 190, 3143; CA 112, 56934f (1990).
73. Ichikawa, M., Harita, Y. and Tashiro, M., Japan Pat. JP 73 29,879, 1973 (Sept. 13); CA 80, 134553t (1974).
74. Harita, Y., Nagaoka, I., Kurokawa, M. and Komatsu, K., Jap. Pat. JP 74 112,997, 1974 (Oct. 28); CA 82, 140981p (1975).
75. Ast, W., Bosch, H. and Kerber, R., Angew. Makromol. Chem., 1979, 76/77, 67; CA 90, 205499z (1979).
76. Tanaka, Y., Sato, H. and Gonzalez, I. G., J. Polym. Sci., Polym. Chem. Ed., 1979, 17, 3027; CA 91, 176036j (1979).
77. Hayashi, K., Tachibana, M. and Okamura, S., J. Polym. Sci., Polym. Chem. Ed., 1980, 18, 2785; CA 93, 205889p (1980).
78. Priola, A., Bruzzone, M., Mistrali, F. and Cesca, S., Angew. Makromol. Chem., 1980, 88, 1; CA 93, 151411f (1980).
79. Priola, A., Passerini, N., Bruzzone, M. and Cesca, S., Angew. Makromol. Chem., 1980, 88, 21; CA 93, 151412g (1980).
80. Vaughan, R. W., Jones, J. W. and Lubowitz, H. R., Proc. Anniv. Conf. SPI Reinf. Pl. D. 25th 1970; CA 73, 15669f (1970).
81. Schaffhauser, R. J. and Parris, C. L., Ger. Offen. 2,026,531, 1970 (Dec. 17); CA 74, 65079e (1971).
82. Institut Francais du Petrole, Jap. Pat. JP 80 54,343, 1980 (Apr. 21); CA 93, 73300c (1980).
83. Chauser, M. G., Kalikhman, I. D., Cherkashin, M. I. and Berlin, A. A., Vysokomolek. Soedin. Ser. A, 1970, 12 (5), 1022; CA 73, 67014h (1970).
84. Gaylord, N. G. and Koessler, I., J. Polym. Sci., C, 1965, (Pub. 1968), 16, 3097; CA 70, 4658p (1969).
85. Zachoval, J., Krepelka, J. and Klimova, M., Collect. Czech. Chem. Commun., 1972, 37, 3271; CA 80, 3895h (1974).
86. Gaylord, N. G., Stolka, M., Stepan, V. and Kossler, I., J. Polym. Sci., C, 1968, 23, 317; CA 69, 87541y (1968).
87. Ambrose, R. J. and Hergenrother, W. L., Macromolecules, 1972, 5, 275; CA 77, 75626g (1972).
88. Pinazzi, C. P., Legeay, G. and Brosse, J. C., Makromol. Chem., 1973, 164, 135; CA 78, 136787e (1973).
89. Denisova, T. T., Livshits, I. A. and Gershtein, E. R., Vysokomol. Soedin., Ser A, 1974, 16, 880; CA 81, 50086a (1974).
90. Golub, M. A., J. Polym. Sci., Polym. Lett. Ed., 1977, 15, 369; CA 87, 23909p (1977).
91. Kryazhev, Yu. G., Yushmanova, T. I. and Borodin, L. I., Vysokomolek. Soedin., Ser. B, 1970, 12 (7), 487; CA 73, 88223r (1970).
92. Vakul'skaya, T. I. and Kryazhev, Y. G., Vysokomol. Soedin., Ser. A., 1973, 15, 1783; CA 80 48523b (1974).
93. Vitkouskii, V. Yu., Salaurov, V. N., Doshlov, O. L. and Myachin, Yu. A., Khim. Vysokomol. Soedin. Neflekhim., 1973, 88; CA 81, 64252p (1974).
94. Kryazhev, Yu. G., Borodin, L. I., Kalikhman, I. D. and Keiko, V. V., Vysokomol. Soedin., Ser. A, 1974, 16, 119; CA 81, 64232g (1974).
95. Furukawa, J., Kobayashi, E. and Kawagoe, T., J. Polym. Sci., Polym. Chem. Ed., 1978, 16, 1609; CA 89, 147288w (1978).
96. Schulz, R. C., Angew. Chem.,Int. Edn., 1964, 3, 416; CA 61, 1950a (1964).
97. Schulz, R. C., Wegner, G. and Kern, W., J. Polym. Sci. C16/2, 1967, 98, 9; CA 67, 64775x (1967).

98. Schulz, R. C., Wegner, G. and Kern, W., Makromol. Chem., 1967, 100, 208; CA 66, 55817q (1967).
99. Koton, M.M., Andreev, I.V., Getmanchuk, Y.P., Madorskaya, L.Y., Pakrovskii, E.I., Kol'tsov, A. I. and Filatova, V.A., Vysokomolek. Soedin., 1965, 7, 2039; CA 64, 11321d (1966).
100. Hunter, L. and Forbes, J. W., J. Polym. Sci. A, 1965, 3, 3471; CA 64, 2165c (1966).
101. Schulz, R. C. and Kern, W., Makromolek. Chem., 1956, 18-19, 4; CA 51, 1090d (1957).
102. Ryder,Jr., E. E. and Pezzaglia, P., J. Polym. Sci. A, 1965, 3, 3459; CA 64, 2166b (1966).
103. Fischer, R. F. and Stewart, Jr., A. T., J. Polym. Sci. A, 1965, 3, 3495; CA 64, 870a (1966).
104. Bergman, E., Tsatsos, W. T. and Fischer, R. F., J. Polym. Sci. A, 1965, 3, 3485; CA 64, 867h (1966).
105. Bell, E. R., Campanil, V. A.and Bergman, E., U.S. Pat. US 3,105,801 to Shell Oil Co., 1963; CA 59, 14133a (1963).
106. Andreeva, I.V., Koton, M.M., Getmanchuk, Y.P., Madorskaya, L.Y., Pikrovskii, E.I. and Kol'tsov, A.I., Khim. Atsetilena, 1968, 386; CA 70, 115658h (1969.
107. Toi, Y. and Hachihama, Y., Kogyo-kwagaku Zasshi, 1959, 62, 1924; CA 57, 12707f (1962).
108. Andreeva, I.V., Koton, M. M. and Kovaleva, K., Vysokomol. Soedin., 1962, 4, 528; CA 57, 12707i (1962).
109. Shostakovskii, M., Kubretsova, T. and Annenkova, V., Vysokomol. Soedin, Ser B, 1970, 12, 848; CA 74, 32046p (1971).
110. Kador, U. and Mehnert, P., Makromol. Chem., 1971, 144, 37; CA 75, 36794z (1971).
111. Rashkov, I., Spasov, S. and Panaiotov, I., Makromol. Chem., 1973, 170, 39; CA 80, 15275u (1974).
112. Houtz, R. C., Textile Res. J., 1950, 20, 786; CA 45, 898b (1951).
113. McCartney, J. R., Mod. Plastics, 1953, 30(11), 118; CA 48, 5076i (1954).
114. Burlant, W. J. and Parsons, J. L., J. Polym. Sci., 1956, 22, 249; CA 51, 3173c (1957).
115. Grassie, N.and McNeill, I. C., J. Polym. Sci., 1958, 27, 207; CA 53, 8694b (1959).
116. Silins, E. A., Motorykina, V. P.,Shmit, I.K., Geiderikh, M. A., Davydov, G. E. and Krentsel, B. A., Elektrokhimiya, 1966, 2, 117; CA 64, 14347e (1966).
117. Silins, E. A., Motorykina, V. P., Geiderikh, M. A., Davydov, G. E. and Krentsel, B. A., Zh. fiz. Khim., 1967, 41, 309; CA 66, 116104d (1967).
118. Peebles, Jr., L. H., J. Polym. Sci., A1, 1967, 5, 2637; CA 67, 117667x (1967).
119. Kirby, J. R., Brandrup, J. and Peebles, L. H., Jr., Macromolecules, 1968, 1, 53; CA 68, 115141c (1968).
120. Brandrup, J., Kirby, J. R. and Peebles, L. H., Jr., Macromolecules, 1968, 1, 59; CA 68, 115142u (1968).
121. Brandrup, J. and Peebles, L. H., Macromolecules, 1968, 1, 64; CA 68, 115143v (1968).
122. Brandrup, J., Macromolecules, 1968, 1, 72; CA 68, 115144w (1968).
123. Friedlander, H. N., Peebles, L. H., Brandrup, J. and Kirby, J. R., Macromolecules, 1968, 1, 79; CA 68, 115145x (1968).
124. Overberger, C. G. and Moore, J. A., Fortschr. Hochpolym. Forsch, 1969, 7 (1), 113; CA 72, 44160t (1970).
125. De Winter, W., Ind. Chim. Belge, 1970, 35 (12), 1097; CA 74, 42645x (1971).
126. Simionescu, C. and Ungureanu, C., Rev. Roumaine Chim., 1964, 9, 627; CA 62, 13238g (1965).
127. Drabkin, I. A., Rozenshtein, L., Geiderikh, M. and Davydov, B., Dold. Akad. Nauk. SSR, 1964, 154, 197; CA 60, 13389d (1964).
128. Schnecko, H., Chimia, 1965, 19, 113; CA 62, 14828c (1965).

129. Ungureanu, C., Ungureanu, D. A. and Simionescu, C., Rev. Roum. Chim., 1968, 13, 913; CA 69, 97238u (1968).
130. Gump, K. H. and Stuetz, D. E., U.S. Pat. US 3,736,309, 1973 (May 29); CA 79, 79762n (1973).
131. Ota, T., Masuda, S. and Kobayashi, M., Kogyo-kwagaku zasshi, 1970, 73, 1866; CA 74, 32117n (1971).
132. Chukhadzhyan, G., Kalaidzhyan, A. and Petrosian, V., Vysokomol. Soedin., Ser. A, 1970, 12, 171; CA 72, 91100h (1970).
133. Kotin, E., Gerasimov, G. N. and Abkin, A. D., Vysokomol. Soedin., Ser. B, 1970, 12, 860; CA 74, 64537r (1971).
134. Gochkovskii, V., Vysokomol. Soedin., Ser. A, 1971, 13, 2207; CA 76, 46642h (1972).
135. Ota, T., Masuda, S., Isomichi, M. and Ebisudani, M., Kogyo-kwagaku zasshi, 1971, 74, 1470; CA 75 118770c (1971).
136. Danner, B. and Meybeck, J., Plast. Polym. Conf. Suppl., 1971, 5, 36; CA 80, 15407p (1974).
137. Mueller, D. J., Fitzer, E. and Fiedler, A. K., Plast. Polym. Conf. Suppl., 1971, 5, 10; CA 80, 16188e (1974).
138. Potter, W. D. and Scott, G., Nature, Phys. Sci., 1972, 236, 30; CA 77, 5983x (1972).
139. Kotin, E., Gerasimov, G. N. and Abkin, A. D., Vysokomol. Soedin., Ser. B, 1972, 14, 282; CA 77, 48842t (1972).
140. Pogorelko, V. Z., Bazhenova, N. N. and Ryabov, A. V., Tr. Khim. Khim. Tekhnol., 1972, 44; CA 79, 19341q (1973).
141. Nakamura, S., Otake, T. and Matsuzaki, K., J. Appl. Polym. Sci., 1972, 16, 1817; CA 77, 102541j (1972).
142. Vasile, C., Stud. Cercet. Chim., 1973, 21, 81; CA 79, 42850p (1973).
143. Gump, K. H. and Stuetz, D. E., U.S. Pat. US 3,736,310, 1973 (May 29); CA 79, 79763p (1973).
144. Blumstein, R., Blumstein, A. and Parikh, K. K., Appl. Polym. Symp., 1974, 25, 81; CA 82, 17181a (1975).
145. Guyot, A., Dumont, M., Graillat, C., Guillot, J. and Pichot, C., J. Macromol. Sci. (Chem.), A, 1975, 9, 483; CA 83, 132021f (1975).
146. Guyot, A., Guillot, J. and Pichot, C., J. Macromol. Sci. (Chem.), A, 1975, 9, 469; CA 83, 132020e (1975).
147. Kryazhev, Y.G., Ermakova, T.G., Tatarova, L.A., Brodskaya, E.I., Vakul'skaya, T.L. and Salaurov, V.N., Vysokomol. Soedin., Ser. A, 1975, 17, 2181; CA 84, 44759t (1976).
148. Balard, H. and Meybeck, J., Bull. Soc. Chim. Fr., Pt. 2, 1977, 11-12, 1147; CA 88, 191737v (1978).
149. Budevska, K. and Plachkova, S., Angew. Makromol. Chem., 1978, 70, 1; CA 89, 130056w (1978).
150. Braun, D. and Disselhoff, R., Angew. Makromol. Chem., 1978, 74, 225; CA 90, 55477z (1979).
151. Vasile, C., Macoveanu, M. and Odachian, L., Mater. Plast.(Bucharest), 1980, 17, 226, 249; CA 94, 104278d (1981).
152. Liu, Q., Hou, G., Xu, Z. and Wu, R., Gaofenzi Tongxun, 1986, 214; CA 105, 228370t (1986).
153. Anraku, H. and Shoji, Y., Jap. Pat. JP 61,195,357 [86,195,357], 1986 (Aug. 29); CA 106, 64030z (1987).
154. Zhao, G. and Chen, B., Gaofenzi Xuebao, (3), 1989, 363; CA 112, 120465h (1990).
155. Overberger, C. G., Ozaki, S. and Mukamal, H., Polym. Lett., 1964, 2, 627; CA 60, 10796a (1964).
156. Monroe, S. B., Ph.D. Dissertation, Univ. of Florida, 1962; CA 58, 2366e (1963).

157. Butler, G. B. and Corfield, G. C., U.S. Clear H. Fed. Sci. Tech. Inf. AD-694123 1969; CA 72, 79712b (1970).
158. Schulz, R. C. and Hartmann, H., Monatsh., 1961, 92, 303; CA 55, 27161a (1961).
159. Schulz, R. C. and Hartmann, H., Makromol. Chem., 1962, 55, 227; CA 57, 13973g (1962).
160. Schulz, R. C. and Hartmann, H., Makromol. Chem., 1963, 65, 106; CA 59, 7658i (1963).
161. Naarmann, H. and Kastning, E. G., Brit. Pat. 947,472, 1964 (Jan. 22); CA 60, 12134c (1964).
162. Egorov, Y. P., Kachan, A. A., Kostyleva, Z. A., Lebo., Y. G., Khranovskii, V. A. and Shrubovich, V. A., Teor. Eksp. Khim., 1967, 3, 843; CA 68, 59931m (1968).
163. Kabanov, V. A., Aliev, K. V., Kargina, O. V., Petrikoeva, T. J. and Kargin, V. A., IUPAC, Symp. Macro. Chem. Prague, 1965, 129; CA 67, 64766v (1967).
164. Kern, W. and Kammerer, H., Pure Appl. Chem., 1967, 15, 421; CA 69, 19551b (1968).
165. Kammerer, H. and Jung, A., Makromolek. Chem., 1967, 101, 284; CA 66, 95452y (1967).
166. Kammerer, H.,Skukla, J.,Oender, N. and Schewermann, Y., J. Polym. Sci.C, 1968, 22, 213; CA 69, 52689a (1968).
167. Kammerer,H. and Oender, N., Makromolek. Chem., 1968, 111, 67; CA 68, 59925n (1968).
168. Greber, G. and Degler, G., Makromol. Chem., 1962, 52, 174; CA 57, 12526e (1962).
169. Kabanov, V. A., Petrovskaya, V. A. and Kargin, V. A., Vysokomol. Soedin., Ser. A, 1968, 10,; 925; CA 69, 19561e (1968).
170. Kaemmerer, H., Shukla, J. S. and Scheuermann, G., Makromol. Chem., 1968, 116, 72; CA 69, 67773x (1968).
171. Kaemmerer, H. and Shukla, J. S., Makromol. Chem., 1968, 116, 62; CA 69, 77775p (1968).
172. Lubowitz, H. R. and Burns, E. A., Fr. Pat. 1,531,023, 1968 (June 28); CA 71, 13685t (1969).
173. Lubowitz, H. and Burns, E. A., Fr. Pat. 1,535,848, 1968 (Aug. 9); CA 71, 4153r (1969).
174. Polowinski, S., Int. Symp. Macromol. Chem., Preprs., 1969, 3, 171; CA 75, 64686j (1971).
175. Kinoshita, M., Kataoka, S. and Imoto, M., Kogyo-kagaku zasshi, 1969, 72, 972; CA 71, 61809h (1969).
176. Kaemmerer, H. and Hegemann, G., Makromol. Chem., 1970, 139, 17; CA 74, 42654z (1971).
177. Irzhak, V. I., Kuzub, L. I. and Enikolopyan, N. S., Dokl. Akad. Nauk SSSR, 1971, 201, 1382; CA 76, 127969x (1972).
178. Yokota, K., Ogawasara, S. and Takada, Y., Kobunshi kagaku, 1972, 29, 587; CA 77, 165149h (1972).
179. Yokota, K., Ogasawara, S. and Takada, Y., Kobunshi kagaku, 1972, 29, 593; CA 77, 165150b (1972).
180. Andrianov, K. A.,Delazari, N. and Emel'kina, N., Izv. Akad. Nauk armyan. SSR, Khim. Nauki 1973, 2295; CA 80, 37498w (1974).
181. Braun, D., and Walter, E., Colloid Polym. Sci., 1976, 254, 396; CA 85, 47476e (1976).
182. Kaemmerer, H. Makromol. Chem., 1977, 178, 1659; CA 87, 118144z (1977).
183. Sastri, V. R., Schulman, R. and Roberts, D. C., Macromolecules, 1982, 15, 939; CA 97, 72874p (1982).
184. Anon., Chem. Eng. News, 1963, 41, 50.
185. Anon., Chem. Eng. News, 1963, 41, 37.
186. Angelo, R. J., Belg. Pat. 633,279, 1963.
187. Binder, J. L., J. Polym. Sci., Pt. B, 1966, 4, 19; CA 65, 7416d (1966).

10

Mechanism of Cyclopolymerization

Cyclopolymerization of 1,5- and 1,6-dienes readily occurs to yield linear, noncrosslinked polymers in contrast to earlier well-accepted theories of crosslinking.[2–1] Mechanistic studies dealing with various aspects of the cyclopolymerization process are extensive and cannot be presented in detail here. However, important aspects of the process are embodied in a kinetic study carried out on methacrylic anhydride.[2–317] The kinetic relationship between intramolecular and intermolecular propagation was derived from the reaction schemes shown in Structure 10–1. This study showed that the energy

M $\xrightarrow[\text{intermolecular}]{K_{11}}$

$\xrightarrow[\text{intramolecular}]{K_{C}}$ $\xrightarrow[\text{intermolecular}]{K_{C1}}$ Polymer

(10–1)

of activation for the intramolecular cyclization step is higher than that for the intermolecular step by about 2.6 kcal/mol. The rate of cyclization, however, was found to be considerably faster than the intermolecular propagation step, in support of a very high steric factor favoring cyclization. The value k_c/k_{11} was found to be 2.4 mol/L, and the Arrhenius frequency factor ratio was found to be 256 mol/L in favor of the cyclization step. Also, it was shown that increasing steric factors, higher temperature, lower monomer concentration, poorer solvent, and higher conversion favor cyclization.

Numerous other kinetic studies have reported k_c/k_{11} values to vary from 2.8 for radical-initiated polymerization of o-divinylbenzene in benzene at 50°C to 200 for radical-initiated polymerization of divinylformal in benzene at 50°C. Arrhenius frequency factor ratios have been reported to vary from 50 mol/L for radical-initiated polymerization of o-divinylbenzene in benzene at 50°C up to 2.2×10^4 for cationic-initiated polymerization of o-divinylbenzene in toluene. The values for $E_c - E_{11}$ have been reported to vary from approximately zero for radical-initiated polymerization of vinyl *trans*-cinnamate at 70°C in benzene to 5.3 kcal/mol for cationic-initiated polymerization of o-divinylbenzene in toluene.[1-155]

Dinitriles, diynes, dicarbonyls, diisocyanates, diepoxides, and other similar compounds have been shown to cyclopolymerize.

Many investigations have been made, especially in cationic and radical systems, to elucidate the mechanism of cyclopolymerization. At the present time, however, no single mechanistic theory has been postulated which can explain all of the aspects of the perfectly alternating intra-intermolecular pathway "cyclopolymerization" follows. The most important factors seem to be the relative stabilities of the intermediate radicals and the relative rates of the intra- and intermolecular steps. A comparison of the results obtained in some of the important papers in the anionic cyclopolymerization literature with studies in radical and cationic systems has given valuable insight into certain aspects of the mechanism of cyclopolymerization. Obviously, many early experiments were limited by techniques then available for kinetic study and product analysis. Later, many experimental methods were made available, which encouraged reinvestigation of several of these systems.[1]

In order to satisfactorily evaluate the mechanistic studies on the remarkably diverse aspects of cyclopolymerization, a number of basic principles of science have been relied upon and resorted to in support of these investigations. Among these are elucidation of the details of the most probable mechanism of free-radical initiated polymerization of olefins;[3-1] addition of hydrogen bromide to alkenes in presence of oxygen or peroxides proceeds inversely to that predicted by ionic mechanisms, and was concluded to proceed by a free radical mechanism;[2] absolute rate constants for many chain propagation and termination reactions are available and the associated theories have been discussed in detail;[3] also, propagation and termination constants in free radical polymerization have been summarized and the associated theories discussed;[4] the details of 1,2- versus 1,4-addition to a conjugated diene were elucidated and discussed;[5] the relative rates of addition reactions of various radicals to olefins have been summarized and the associated theories discussed in detail;[6] and the Franck-Condon principle has been clearly defined and its significance discussed.[7]

Among early experiments which provided background data and theoretical support for the relationship between open chain structures and the corresponding cyclic structures are: (1) the mercury photosensitized decomposition of cyclohexane at 400°C which gave propylene, ethylene, methane, ethane, butadiene, propane, hydrogen, dicyclohexyl and cyclohexene.[8] (The products were rationalized on a reasonable mechanistic basis); (2) Some reactions of the cyclohexyl radical were studied and the products rationalized on a proposed mechanistic basis;[9] (3) It was shown that Ar_2-6 cyclization of

Table 10–1 Probability of Cyclization in 1,6-Dienes

Concentration of monomer mole/l	P_1/P_2	Cyclization %
7.43	0.34	25
1.00	2.50	71
0.10	25.00	96
0.01	250.00	99.6
0.001	2500.00	ca. 100.00

4-phenylbutyl radicals is not reversible up to 200°C;[10] and (4) The concept of I-strain was discussed in detail, particularly as it applies to cyclic systems including transitions from $sp^3 \rightarrow sp^2$ hybridization (or the reverse process).[11]

A statistical calculation for the probability of cyclopolymerization has been carried out, and the results were compared with experimental observations.[2-484] (See Structure 2–55 for model.) Under normal polymerization conditions, the intramolecular step leading to cyclopolymerization is not favored on a purely statistical basis, compared to the competitive intermolecular step before cyclization which would lead only to polymer containing residual unsaturated pendant groups, and eventually branching and/or crosslinking (Table 10–1). However, experimental results showed that essentially quantitative cyclization occurred under conditions of monomer concentration ≈ 8 mole/l, in contrast to a concentration = < 0.10 mole/l calculated to lead to >95% cyclization (Table 10–2). Obviously, other factors not taken into account by the probability approach are important. It was noted in this paper that the most favorable conformation for formation of a six-membered ring from a 1,6-diene requires one of the terminal hydrogen atoms to lie in the nodal plane directly between the C_2 and C_7 atoms. How-

Table 10–2 Extent of Cyclization in Polymerization of 1,6-Heptadienes

Monomer	Concentration of monomer mole/l	Soluble polymer	Estimated cyclization %	Reference
Diallyl quaternary ammonium salts	>5.0	100	96–100	12
Diallyl silanes	1.2–2.3	92–100	>95	13, 2-309
1,6-Heptadiene	1.9–6.2	100	90–96	7
1,6-Heptadiyne	3.1	10	100	14
N,N-Diallyl-melamine	(Solid state)	100	99	15
Diallyl-phosphine oxides	Bulk (melt)	100	100	16
Acrylic anhydride	~8.0	100	98–100	17, 18

ever, the other conformation which meets the π-π interaction requirement would be that with the closest approach between the C_2 and C_6 atoms, which would lead to a five-membered ring.

Among the early small-molecule radical-initiated cyclizations was that of synthesis of ethyl 1-cyano-2-methylcyclohexane carboxylate by free radical cyclization of $CH_3CH{=}CH(CH_2)_3CH(CN)CO_2C_2H_5$ in refluxing benzene in presence of benzoyl peroxide.[12]

Ethyl 2-cyano-6-heptenoate in presence of cupric salts cyclized to give mixtures of ethyl 1-cyano-2-(chloromethyl)cyclopentanecarboxylate, its methyl analog, and ethyl-1-cyano-2-methylenecyclopentanecarboxylate.[13]

Although cyclopolymerization of acrylic anhydride had been studied earlier, it was shown that cyclopolymerization also occurred by radiation initiation in toluene at 30°C.[2-309] The cyclization ratio, k_c/k_i was shown to be 11.1 mole/L and $E_c - E_i$ was 2.3 kcal/mole, in reasonable agreement with earlier reported results.

In a study of copolymerization of symmetrical nonconjugated dienes with vinyl monomers, a copolymer composition equation was developed and reactivity ratios determined.[7-6] By applying the composition equation, values of reactivity ratios for 12 sets of monomer pairs were obtained and found to be the best values with a permissible range of errors.

In one of a series of papers on synthetic linear polymers, a comparison of known specific limiting values for linear and cyclic homologous series has been made.[14] It was shown that an equation for polymer homologous compounds: $Q_{sp} = a/M = b$, where Q_{sp} = any specific property which can be derived from the additive one, M = molecular weight, a is a constant implicitly characteristic of the end group , and b is a constant explicitly characteristic of the end group, can be extended to cyclopolymer-homologous compounds with certain limitations.

Kinetic studies of polymerization of 1,2-dimethyl-5-vinyl-pyridine methyl sulphate led to the suggestion that propagation of aggregates, which may exist in solution close to saturation, could account for the observations.[15] This suggestion may also apply to cyclopolymerization of ionic monomers.

The He photoelectron spectra of α,ω-diolefins containing from 5–9 C atoms were determined. The π-level splittings were 0.34, 0.46 (or 0.31), 0.41, 0.21 and 0.14 eV, respectively.[2-522] The observed fine structures of the first two bands suggested that there was crossing of the π-levels. Extended Hueckel MO calculations on different conformations of 1,4-pentadiene and 1-5-hexadiene indicated a through-space dominated interaction throughout the series.

The ESR spectra of diallylamines and related compounds which had been treated with hydroxyl or amino radicals were assigned to five-membered ring radicals as the major radical species present; however, the dimethallylamine series gave both five- and six-membered cyclic radicals.[2-177]

The potential surfaces of the addition of methyl radicals to ethylene, acetylene, propene, and allene have been calculated, using MINDO-3.[16] The transition states have unusual structures, being reactant-like except for the attacking methyl radical which has a product-like pyramidal geometry.

Cyclopolymers having [3.3]paracyclophane repeat units were obtained by cationic initiation of 1,3-bis(4-vinylphenyl)propane but not by other initiators.[2-380] These results strongly imply, consistent with earlier proposals, that stronger through-space electronic interactions occur in cationic initiated polymerizations than in radical and anionic systems (10–2).

The role of difunctional olefinic monomers in vinyl polymerization, the theories of cyclopolymerization and cyclocopolymerization, and the structures of the complex polymers obtained from these dienes were emphasized in a review in 1982, with 99 references.[1-178]

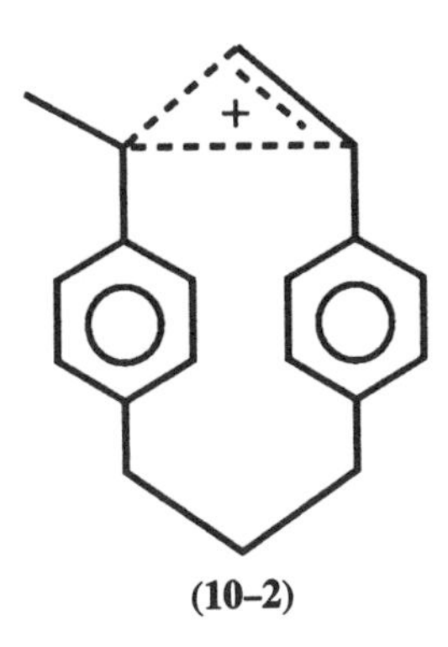

(10–2)

KINETIC STUDIES OF CYCLOPOLYMERIZATION

Although the literature dealing with cyclopolymerization shows that research in this field has generally been approached in a synthetic manner, a significant amount of material on the kinetics and driving force of the reaction has appeared. The interpretation of the work in these areas is very difficult because the results are neither conclusive nor in complete agreement with each other. The mechanism and kinetics of cyclopolymerization have recently been discussed in a review paper.[1-18]

In a study of the driving force for a reaction, it is necessary to understand the important rate and energy factors which control the reaction. In the cyclopolymerization reaction any diene unit which fails to cyclize will leave a pendant double bond. Accurate determinations of the residual unsaturation as a function of concentration and temperature have been made for certain cyclopolymers. Combining the data obtained with kinetic equations, the important relationships of the ratio of rates and the difference in energy for intramolecular and intermolecular propagations may be obtained.

The kinetic relationships between intramolecular and intermolecular propagation[2-247] for the free radical polymerization of acrylic anhydride have been derived. The rate of intermolecular propagation R_1 is given by Equation 10–1.

$$R_1 = k_i[M_1\cdot]2[M] \qquad (10\text{-}1)$$

[M] is given in moles of diene and therefore must be doubled to account for all double bonds present. Similarly, for the intramolecular propagation rate R_c,

$$R_c = k_c[M_1\cdot] \qquad (10\text{-}2)$$

the ratio of R_i to R_c results in the expression

$$R_i/R_c = 2k_i[M]/k_c \qquad (10\text{-}3)$$

The equation which relates the experimentally determined fraction of cyclic units f_c to the rates of R_i and R_c is

$$f_c = R_c/R_i + R_c \tag{10-4a}$$

or

$$1/f_c = 1 + R_i/R_c \tag{10-4b}$$

which on substitution for R_i/R_c from Equation 10–3 gives

$$1/f_c = 1 + 2k_i[M]/k_c \tag{10-5}$$

A plot of $1/f_c$ versus [M] results in a straight line with an intercept at infinite dilution corresponding to complete cyclization and a slope with a value of $2k_i/k_c$. By substitution of Equation 10–3 into the Arrhenius equation, a relationship is obtained which allows the calculation of the energy difference between an intramolecular and intermolecular propagation step. It must be assumed, however, that the degrees of polymerization are high enough that the differences in possible termination steps are negligible.

$$R_i/R_c = 2[M]A_i e^{-Ei/RT}/A_c e^{-Ec/RT} \tag{10-6a}$$

$$\log_{10}(R_i/R_c) = \log_{10}(2[M]A_i/A_c) + E_c - E_i/2.303RT \tag{10-6b}$$

The difference in energy $(E_c - E_i)$ can be calculated from the slope of a plot of $\log_{10}(R_i/R_c)$ versus 1000/T for a constant monomer concentration.

In order for the kinetic equations to be valid, the monomer concentration must be assumed to be the same before and after polymerization and any unit which fails to cyclize must not enter another polymer chain. Thus, the conversion of monomer to polymer must be as low as possible. Experimentally it was found that conversions of 0–3% were satisfactory. A similar equation was independently derived for a study of the free radical polymerization of o-divinylbenzene.[2-20] Since the assumption of a kinetic steady state may not necessarily hold in ionic polymerizations, these authors derived the same equation for the cationic polymerization of o-divinylbenzene by a statistical treatment without using the kinetic steady state assumption.[2-22]

In a study of acrylic anhydride[2-247] residual unsaturation was measured by infrared spectrometry and bromimetry and k_c/k_i = 5.9 mole/l at 45°C in cyclohexanone was obtained. The difference in the activation energies $(E_c - E_i)$ was 2.4 kcal/mole while the value of A_c/A_i was found to be 167 mole/l.[2-249] It was later reported that k_c/k_i = 11.1 mole/l for the radiation-induced polymerization of acrylic anhydride[2-249] in toluene at 30°C. The value of $E_c - E_i$ was 2.3 kcal/mole for the solution polymerization in toluene but 4.7 kcal/mole for bulk polymerization. By use of infrared spectrometry,[2-20] experimental results consistent with k_c/k_i values of 2.1, 2.3, 2.8, 3.3, and 4.0 mole/l at 20°, 30°, 50°, 70°, and 90°C, respectively, were obtained for the free radical polymerization of o-divinylbenzene in benzene. $E_c - E_i$ was 1.9 kcal/mole and cyclization was favored by a frequency factor of 10 at the monomer concentration, where the probability of finding a double bond of another molecule around the uncyclized radical is the same as that of finding the intramolecular double bond. By use of previously published data[2-368] on the radical polymerization of diallyl phthalate, a k_c/k_i value of about 0.3 kcal/mole for $E_c - E_i$ was obtained.[17] Values of k_c/k_i of 4.0 mole/l for ethylene glycol dimethacrylate and 3.4 mole/l for diethylene glycol dimethacrylate in bulk polymerization at 70°C using radical initiators have been reported.[18] A value of k_c/k_i = 130 mole/l for the radical polymerization of divinyl formal in benzene at 50°C[19] has been reported. A value of $E_c - E_i = 2.6$ kcal/mole was reported. It has been determined that $k_c/k_i = 2.4$ mole/l for the radical polymerization of methacrylic anhydride in dimethylformamide at 0°C.[20] An energy difference in $(E_c - E_i)$ of 2.6 kcal/mole was obtained and the ratio A_c/A_i was found to be 256 mole/l. A value of $k_c/k_i = 45$ mole/l for methacrylic anhydride in cyclohexanone at 37°C using radical initiators was reported also.[2-251]

These results indicate that the intramolecular propagation step requires greater energy than the intermolecular step. However, the rate of cyclization is considerably larger than that for intermolecular propagation. The values for the ratio A_c/A_i indicate a high steric factor favoring cyclization.

In the cationic polymerization of o-divinylbenzene, there was a large variation of k_c/k_i (0.1 to 5 mole/1) with various Friedel-Crafts catalysts.[2-22] The value of k_c/k_i also decreased with decrease in the dielectric constant of the solvent. A model of the propagating species was proposed to account for these effects.[2-23] In anionic polymerization[2-24] the reaction conditions which gave higher k_c/k_i values also gave greater rates of propagation in styrene polymerization. It was concluded that k_c/k_i increased with the reactivity of the growing anion. Table 10-3 compares the activation parameters in several systems for the polymerization of o-divinylbenzene. The activation parameters

Table 10-3 Activation Parameters[a] in Cyclopolymerization of Typical Monomers[b]

Monomer	Solvent	T°C	k_c/k_i mole/l	$E_c - E_i$ kcal/mole	A_c mo
Acrylic anhydride (1)	Cyclohexanone	45	5.9	2.4	16
o-Divinyl benzene (1)	Benzene	50	2.8	1.9	10
o-Divinyl benzene (2a)	Toluene	−78	0.1-5	5.3	2
o-Divinyl benzene (2b)	Methylene chloride	−78	—	6.2	5×
o-Divinyl benzene (3a)	Tetrahydro-furan	—	2.2	15	
o-Divinyl benzene (3b)	Tetrahydro-furan	—	2.8	66	
Methacrylic anhydride (1)	Cyclohexanone	37	46	—	—
Methacrylic anhydride (1)	Dimethyl formamide	0	2.4	2.6	25
Methacrylic anhydride (1)	Dimethyl formamide	80	6.2	—	—
Methacrylic anhydride (1)	Dimethyl formamide	—	11.4	—	—
Divinyl formal (1)	Benzene	50	200	2.6	—
Vinyl-*trans*-cinnamate (1)	Benzene	70	13.3	~0	—

[a] Calculated from the equation $t_c = k_c/k_i = A_c/A_i \exp[(-E_c + E_i)/RT]$.

[b] A variety of cationic species was studied. (1) Radical; (2) Cationic, (a) $BF_3O(C_2H_5)_2$, (b) $(C_6H_5)_3CBF_4$; (3) Anionic, (a) Sodium naphthalene, (b) Sodium naphthalene with $NaB(C_6H_5)_4$.

in anionic and radical cyclopolymerizations are quite alike but are rather different from those in cationic polymerizations. It was concluded that this contrast may imply that the nature of the propagating species is described by its reactivity (tightness if it is an ion-pair) alone in anionic and radical systems but that the nature of the propagating species in cationic polymerization is influenced by its tightness as well as geometrical factors such as the conformation of the growing chain end and the positions of the counteranion and the solvent molecules.

The kinetic study of the free radical polymerization of symmetrical nonconjugated diolefins was extended to include the overall rate expressions for cyclopolymerization.[17] Expressions giving the rate of loss of double bonds in terms of the initial monomer and initiator concentrations and the associated rate constants were developed. These expressions, summarized in Table 10–4, can yield information on the mechanism of termination or on k_c/k_i, if the termination mechanism is known or can be assumed. The total apparent activation energy was obtained by differentiation of the rate expression with respect to temperature.[1-18]

The rate of reaction for methacrylic anhydride in dimethylformamide at 40°C depended on $[M]^{3/2}$ $[I]^{1/2}$.[2-245] A k_c/k_i value of 11.4 mole/l was estimated.[1-18] In cyclohexanone at 37°C, a dependence on the initiator concentration of $[I]^{0.58}$ and $[I]^{0.66}$ for monomer concentrations of 0.656 and 3.282 mole/l, respectively, was reported. An order of reaction of 1, with respect to monomer concentration, up to about 3 mole/l was found[2-251]. The changing orders of reaction were attributed to the increasing viscosity of the solution, the progressive heterogeneity of the reaction medium, as well as some trapping of radicals. It has been estimated that k_c/k_i equals 12 mole/l.[1-18] For acrylic anhydride in cyclohexanone at 35°C: (a) a very high rate of polymerization, (b) an order of reaction with respect to the initiator between 0.65 and 1, and (c) a complex dependence on the monomer concentration were found.[2-243] An interpretation similar to that for the polymerization of methacrylic anhydride was suggested. An overall rate equation, $R_p = k[I]^{3/4}[M]^{3/2-2}$ for divinyl formal[19] in benzene at 50°C was obtained. It was observed that the overall rate of reaction of diallylcyanamide[21] in benzene or m-cresol at 70°C was first order with respect to concentration of initiator and either first or second order with respect to monomer concentration, depending on the solvent used. The initiator was α,α'-azobisisobutyronitrile (AIBN) in all the above cases. It was found that the rate of polymerization of diallyldimethylsilane[1-117] in bulk at 120°, 130°, and 140°C was proportional to the first power of the di-t-butyl peroxide initiator concentration.

Table 10–4 Rate and Activation Energy Expression for Cyclopolymerization

Termination reaction	Rate expression $-d[DB]/dt=$	Activation energy expression E_T = (where $E_c \approx E_i$)
$M_1 + M_1 k_{11}$	$(2k_c + k_i[M_1])(fk_d[I]/k_{11})^{1/2}$	$E_d/2 + E_{i(c)} - E_i/2$
$M_2 + M_2 k_{22}$	$(2k_c + k_i[M_1])(k_i'[M_1]/k_c)(fk_d[I]/k_{22})^{1/2}$	$E_d/2 + E_{i(c)} - E_{21} - E_c - E_{22}/2$
$M_1 + M_2 k_{12}$	$(2k_c + k_i[M_1])(k_a k_i'[M_1][I]/k_{12}k_c)^{1/2}$	$E_d/2 + E_{i(c)} - E_{21}/2 - E_c/2 - E_{12}/2$
$M_1 + M_1 k_1$	$(2k_c + k_i[M_1])(2fk_d[I]/k_i[M_1])$	$E_d + E_{i(c)} - E_1$
$M_2 + M_1 k_2$	$(2k_c + k_i[M_1])(2fk_d[I]k_i/k_2k_c)$	$E_d + E_{i(c)} - E_{21} - E_c - E_2$

It has been reported[2-249] that radiation induced bulk polymerization of acrylic anhydride at 13°C was proportional to the 0.55 power of the dose rate. This result indicated that the polymerization proceeded by a radical mechanism.

It was found that the activation energy for methacrylic anhydride[2-245] polymerization ($E_T - E_{d/2}$) was 8.0 kcal/mole. A comparison was made with ($E_T - E_{d/2}$) for methacrylic acid which was found to be 9.8 kcal/mole. It was concluded that there was little difference between normal vinyl polymerization energetics and those for cyclopolymerization. However, as indicated, it had been found[20] that the intramolecular propagation step required 2.6 kcal/mole more energy than the intermolecular step. It was suggested that a comparison between methyl methacrylate ($E_T - E_{d/2}$) = 5.0 kcal/mole)[22] and methacrylic anhydride would be more reasonable. Addition of the energy necessary for an intramolecular propagation step would give a value of 7.6 kcal/mole, quite close to the experimental value. The difference between inter-intermolecular polymerization and intra-intermolecular polymerization would be readily apparent from the latter comparison.

The radical initiated polymerization of allylsilanes[1-117] has been studied. It was concluded that the total activation energy for the polymerization of diallyldimethylsilane was about 9 kcal/mole of double bond less than that for allyltrimethylsilane. These results suggested an energetically more favorable pathway from monomer to polymer in the diallyl derivative than in the corresponding monoallyl derivative. An electronic interaction[1-17] either in the ground state or the activated state of the molecule, would be expected to provide such a pathway. These data[1-117] were, for a long time, accepted as proof that the double bond reacts more easily. However, these data were re-examined[1-18] and it was found that the conclusions were erroneous. The data actually suggested that the activation energy for the cyclization reaction was greater than the activation energy for the intermolecular reaction. This conclusion is, however, unacceptable on the basis of the other experimental results. These results show that the activation energy for the intermolecular reaction is less than that for the cyclization reaction, that is $E_i < E_c$. No cases have been reported for which $E_c < E_i$.

It has been shown [3-70] that the products of the free radical polymerization of *N*-allylmethacrylamide contained both allylic and methacrylic pendant double bonds. This suggested that in the intramolecular chain propagation reaction δ-lactam rings were formed, giving rise to two different cycloradicals probably having different reactivity. In general, for an unsymmetrical nonconjugated diene, A - - - B, intra- and intermolecular propagation reactions having rate constants $k_1 - k_{10}$ shown below would be involved, assuming that conversion to polymer is low enough to avoid branching:

Intramolecular chain propagation reactions (cyclization):

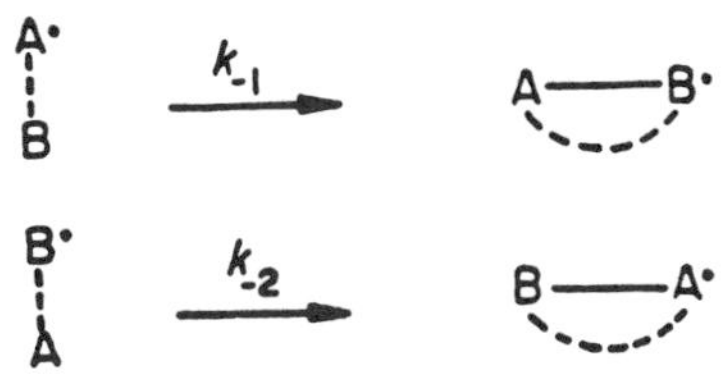

A•---B $\xrightarrow{k_{-1}}$ A—B• (cyclic)

B•---A $\xrightarrow{k_{-2}}$ B—A• (cyclic)

Intermolecular chain propagation reactions:

$$\text{B—A}^{\bullet}\ (\text{cyclic}) + \text{A}\text{---}\text{B} \xrightarrow{k_{-3}} \text{A}^{\bullet}\text{---}\text{B} \qquad \text{A}^{\bullet}\text{---}\text{B} + \text{A}\text{---}\text{B} \xrightarrow{k_{-7}} \text{A}^{\bullet}\text{---}\text{B}$$

$$\text{B—A}^{\bullet}\ (\text{cyclic}) + \text{B}\text{---}\text{A} \xrightarrow{k_{-4}} \text{B}^{\bullet}\text{---}\text{A} \qquad \text{A}^{\bullet}\text{---}\text{B} + \text{B}\text{---}\text{A} \xrightarrow{k_{-8}} \text{B}^{\bullet}\text{---}\text{A}$$

$$\text{A—B}^{\bullet}\ (\text{cyclic}) + \text{A}\text{---}\text{B} \xrightarrow{k_{-5}} \text{A}^{\bullet}\text{---}\text{B} \qquad \text{B}^{\bullet}\text{---}\text{A} + \text{A}\text{---}\text{B} \xrightarrow{k_{-9}} \text{A}^{\bullet}\text{---}\text{B}$$

$$\text{A—B}^{\bullet}\ (\text{cyclic}) + \text{B}\text{---}\text{A} \xrightarrow{k_{-6}} \text{B}^{\bullet}\text{---}\text{A} \qquad \text{B}^{\bullet}\text{---}\text{A} + \text{B}\text{---}\text{A} \xrightarrow{k_{-10}} \text{B}^{\bullet}\text{---}\text{A}$$

Assuming steady state conditions, some general equations relating cyclopolymer composition, in terms of the possible structural units, to the concentration of the monomer were derived.[1-144,23] The mole fractions f_A, f_B, f_{Ac}, and f_{Bc} of the units shown in Equation 10–9 are related by Equation 10–7:

$$\begin{aligned} f_A{:}f_B{:}f_A,f_{Ac}{:}f_{Bc} &= [\mathrm{M}](k_8[\mathrm{M}] + k_1k_6/(k_5 + k_6))(k_9 + k_{10}) \\ &= [\mathrm{M}](k_9[\mathrm{M}] + k_2k_3/(k_3 + k_4))(k_7 + k_8) \\ &= (k_8[\mathrm{M}] + k_1k_6/(k_5 + k_6))k_2 \\ &= (k_9[\mathrm{M}] + k_2k_3/(k_3 + k_4))k_1 \end{aligned} \tag{10-7}$$

from Equation 10–7:

$$f_A/f_B = \alpha([\mathrm{M}] + \beta)/([\mathrm{M}] + \gamma) \tag{10-8a}$$

and

$$f_1/f_c = \mathrm{a}([\mathrm{M}]^2 + \mathrm{b}[\mathrm{M}]) + ([\mathrm{M}] + \mathrm{c}) \tag{10-8b}$$

where $f_1/f_c = (f_A + f_B)/(f_{Ac} + f_{Bc})$ is the ratio between the mole fractions of the linear and cyclic units formed from A - - - B and

$$\alpha = (1 + k_{10}/k_9)/(1 + k_7/k_8)$$

$$\beta = k_1k_6/k_8(k_5 + k_6)$$

$$\gamma = k_2k_3/k_9(k_3 + k_4)$$

and

$$\mathrm{a} = (k_7k_9 + k_8k_{10} + 2k_8k_9)/(k_lk_9 + k_2k_8)$$

$$\mathrm{b} = [k_2k_3(k_7 + k_8)/(k_3 + k_4) + k_1k_6(k_9 + k_{10})/(k_5 + k_6)]/(k_lk_9 + k_2k_8)$$

$$\mathrm{c} = [k_lk_2/k_3/(k_3 + k_4) + k_6/(k_5 + k_6)]/(k_lk_9 + k_2k_8)]$$

—B— ⋮ A (f_A) —A— ⋮ B (f_B)

—B—A— (f_{Ac}) —A—B— (f_{Bc}) (10-9)

~~~A· ⋮ B (10-10)

~~~B· ⋮ A (10-11)

The value of the ratio between the velocity constant of the intramolecular chain propagation reaction and the sum of the velocity constants of the intermolecular chain propagation reactions for the growing radicals Equations 10–10 and 10–11 gives a measure of the ability of these two chain radicals to undergo cyclization, as expressed by the cyclization ratios, $r_{cA\cdot}$ Equation 10–12 and $r_{cB\cdot}$ Equation 10–13.

$$r_{cA\cdot} = k_1/(k_7 + k_8) \qquad (10\text{–}12)$$

$$r_{cB\cdot} = k_2/(k_9 + k_{10}) \qquad (10\text{–}13)$$

Introducing cyclization ratios into Equation 10–7 one obtains Equation 10–14:

$$f_1/f_c = (1 + f_A/f_B)[M]/(1 + f_{Ac}/f_{Bc})r_{cA\cdot} \qquad (10\text{–}14a)$$

and

$$f_A/f_B = (r_{cA\cdot}/r_{cB\cdot})(f_{Ac}/f_{Bc}) \qquad (10\text{–}14b)$$

which allow the calculation of $r_{cA\cdot}$ and $r_{cB\cdot}$ from the experimental data of an analysis of the polymers from A - - -B. If A and B are equal, namely a symmetrical nonconjugated diene, taking f_l as the mole fraction of linear units and f_c as the mole fraction of cyclic units, it has been shown that Equation 10–8 reduces to Equation 10–15, which is easily reduced

$$f_1/f_c = 2[M]k/k_c \qquad (10\text{–}15)$$

to the equations derived earlier.[2-247,2-20] In cases where the double bonds differ significantly in reactivity, it is possible to neglect any initiation or intermolecular propagation reactions involving the less reactive double bond. This led to the simple expression Equation 10–16 which has been applied to the polymerization of vinyl trans-cinnamate[3-110] and allyl ethenesulfonate.[3-121]

$$1/f_c = 1 + k_i[M]/k_c \qquad (10\text{–}16)$$

From IR, UV and bromimetry measurements on poly (vinyl *trans*cinnamate),[3-110] it has been reported that k_c/k_i = 13.3 mole/l at 70°C in benzene and that $E_c - E_i$ was about zero. At low monomer concentrations the polymerization was first order with respect to monomer concentration and 0.5 order with respect to initiator (α,α'-azobisisobutyronitrile) concentration. Above 3 mole/l, the reaction was independent of monomer concentration and first order with respect to the initiator. A value of k_c/k_i = 45 mole/l for the radical polymerization of allyl ethenesulfonate[3-121] at 45°C in benzene was obtained. Under the same conditions, allyl ethanesulfonate did not polymerize and propyl ethenesulfonate polymerized at nearly the same rate as allyl ethenesulfonate. Copolymerization of allyl ethanesulfonate and propyl ethenesulfonate yielded a polymer consisting of more than 90% of the latter. It was suggested that the high degree of cyclization in poly(allyl ethenesulfonate), 85% for [M] = 0.29 mole/l, corresponded to an enhanced reactivity of the allyl group in the monomer compared to the allyl group in allyl ethanesulfonate.

The free radical polymerization of *N*-allylmethacrylamide (a),[3-71] *N*-allylacrylamide (b),[3-72,3-73] allyl methacrylate (c),[3-111] allyl acrylate (d),[3-112] and vinyl *trans*-crotonate (e)[3-113] has been investigated. For these monomers, only one type of repeating unit was formed and Equation 10–17 holds:

$$f_c[M]/f_B = r_{cA\cdot} + r_{cB\cdot}(f_A/f_B \qquad (10\text{–}17)$$

where a through d contain allyl, e contains vinyl, a and c contain methacrylic, b and d contain acrylic and e contains crotonic double bonds, respectively. The values of the coefficients of Equations 10–8, 10–12, and 10–13 obtained from the experimental data are given in Table 10–5. With the exception of the vinyl crotonate polymerization, the values of $r_{cA\cdot}$ and $r_{cB\cdot}$ show the low tendency towards cyclization of the linear polymer radicals involved in the polymer forming processes. It is also evident that ring forma-

Table 10–5 Data from Cyclopolymerization. Studies of (a) *N*-Allylmethacrylamids, (b) *N*-Allylacrylamide, (c) Allyl Methacrylate, (d) Allyl Acrylate, and (e) Vinyl-*trans*-crotonate

Reacting species	Monomer				
	(a)	(b)	(c)	(d)	(e)
$r_{cA\cdot}$ (mole/l)	0.022	0.013	0.011	0.015	0.00...
$r_{cB\cdot}$ (mole/l)	0.26	0.54	0.16	0.68	7.57
α	7.70	5.80	8.91	4.95	1.00
β (mole/l)	0.031	0.052	0.010	0.002	0.00
δ (mole/1)	0.73	0.90	0.32	1.76	0.80
a (1/mole)	4.20	2.10	6.78	1.68	0.25
b	0.47	0.37	0.28	0.50	0.10
c (mole/1)	0.0386	0.0555	0.0124	0.0098	0.00

tion is easier through the intramolecular reaction of a growing radical with an allylic double bond as shown by the greater values of r_{Cb}. as compared with r_{Ca}.

It has been shown[23] that for [M] = ∞ the products of the polymerization of A - - - B would consist of linear repeating units only in the ratio Equation 10–18.

$$f_A/f_B = a = (1 + k_{10}/k_9)/(1 + k_7/k_8) \qquad (10\text{–}18)$$

Equation 10–18 corresponds to the copolymerization equation of two monomers in equimolecular concentration one containing only the A type double bond and one containing only the B type. Equation 10–18 shows the formal analogy between cyclopolymerization and copolymerization. However, in copolymerization the chain growth is characterized by competition between bimolecular reactions, whereas, in cyclopolymerization there is competition between monomolecular and bimolecular reactions. Hence, cyclopolymer composition depends on the concentration of the monomer while copolymer composition depends upon the relative monomer concentration and is independent of dilution.

The values of α in Table 10–5 for the polymers from (b) and (d) are in agreement with values calculated using reactivity ratios given in the literature for copolymers obtained from equimolar mixtures of only acrylic and only allylic monomers. On the other hand, for the polymers from (a) and (c), the values of α are about 60–80% lower than the ratio between methacrylic and allylic repeating units calculated for equimolar copolymers using reactivity ratios reported in the literature.

In the free radical cyclopolymerization of nonconjugated dienes, the inter- and intramolecular chain propagation reactions must be slow to allow suitable orientation of the reacting groups in order to have the formation of a large fraction of cyclic structural units.[3-112] The results from the polymerization of vinyl *trans*-crotonate were in agreement with this hypothesis. The highly reactive vinyl radical was consumed by intermolecular reactions before the pendant crotonic double bonds could assume the proper orientation for cyclization; thus r_{cA}. was zero. The reactions between the rather stable crotonyl radical and two double bonds of low reactivity were slow enough to allow a considerable amount of cyclization (87% at [M] = 0.17 mole/1) to occur. The value of α was in fairly good agreement with what was expected from copolymerization data.

The mechanism suggested for the polymerization of vinyl *trans*-crotonate[3-113] was completely different from that reported earlier[3-109] as well as that suggested for the radical polymerization of vinyl *trans*-cinnamate.[3-110] According to the latter mechanism, vinyl double bonds were responsible for the intermolecular chain propagation, in contrast to copolymerization data, whereas the cinnamyl double bonds only took part in the cyclization reactions.

The composition relationships for the copolymerization of symmetrical nonconjugated dienes with vinyl monomers[1-18] have been derived. The results of experimental investigations indicated that in some cases, the diene showed a greater reactivity than analogous monoolefins, and in other cases, the reactivities were essentially the same.

The kinetic relations between monomer concentration and composition of polymers from 1,4-dienes have been determined in terms of monomeric and structural units, with consideration of all cyclization reactions that may occur during chain growth, and these applied to the results of the analytical determination of residual unsaturation in products from free radical polymerization of divinyl ether.[24,25] Chain growth occurred exclusively by cyclic radical addition to monomer. Both five- and six-membered rings were formed in the cyclization steps. Cyclization ratios r_1 and r_2, describing the cyclization tendency of radicals -$CH_2CH(OCH{:}CH_2)CH_2\dot{C}HOCH{:}CH_2$ and (10–3) or (10–4) (to give bicyclic structural units (10–5) or (10–6), respectively), were calculated to be 100 mole/l and 4.2 mole/l respectively for the polymerization of divinyl ether in benzene at 70° in the presence of AIBN. Equations allowing calculation of the mole fraction of monocyclic, bicyclic, and linear structures showed that most of the residual vinyl groups in poly(divinyl ether) were pendant to monocyclic rather than linear units.

The cyclopolymerization of diallyl carbonate[2-288] has been studied kinetically and the results considered mechanistically. The rate of polymerization, R_p, was not proportional to the square root or the first power of the initiator concentration (I), but $R_p/(I)^{1/2}$ and $(I)^{1/2}$ bore a linear relation. The rate of polymerization, the residual unsaturation, and the degree of polymerization decreased with a decrease in monomer concentration. The relation between the rate of polymerization over the degree of polymerization $R_p/P_{n,o}$ and the monomer concentration (M), was also linear, as in $R_p/(I)^{1/2}$ and $(I)^{1/2}$. The ratio of the rate constant of the unimolecular cyclization reaction to that of the bimolecular propagation reaction of the uncyclized radical K_c was estimated as 4.0 moles/l from the dependence of the residual unsaturation on the monomer concentration.

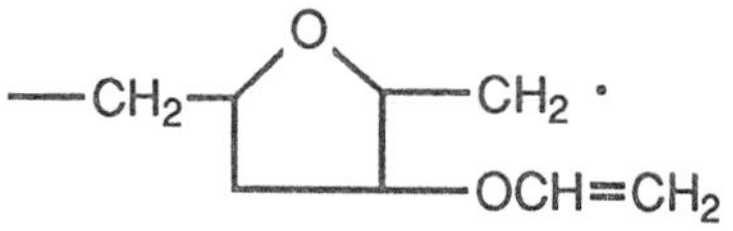

(10–3)

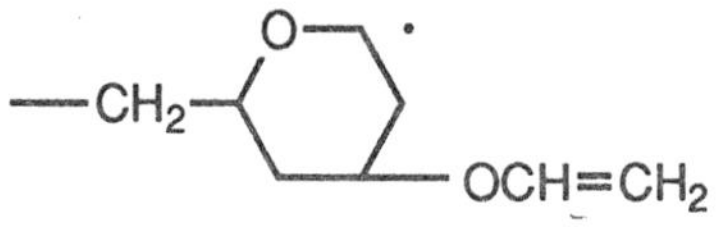

(10–4)

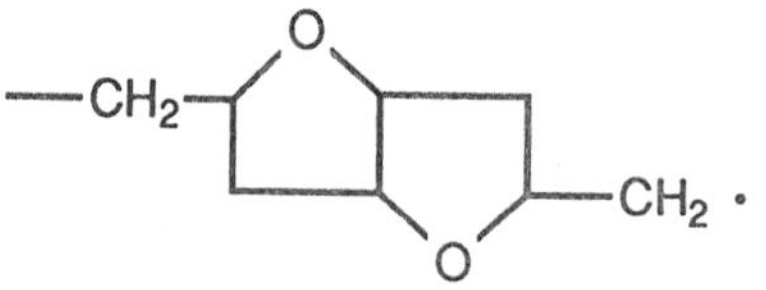

(10–5)

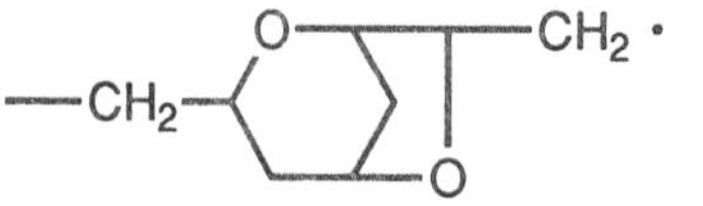

(10–6)

In a study of kinetics of benzoyl peroxide-initiated triallylidene sorbitol[2-379] homopolymerization showed that the polymerization rate R_p, initiator concentration (I), and monomer concentration (M), were related by the equation $[(I)/R_p - a]^{-1} = A(M) + B$, where a, A, and B are constants and that the degree of polymerization depended on (M) but not (I). Kinetics of benzoyl peroxide-initiated triallylidene sorbitol and tricrotylidene sorbitol copolymerizations with styrene and acrylonitrile showed that the copolymerization rates decreased with increasing triallylidene sorbitol and tricrotylidene sorbitol concentrations. The monomer reactivity ratios and Q and e values for the copolymerizations and the ratio of the unimolecular cyclization rate to the total rate of bimolecular propagation and uncyclized radical chain transfer for the homopolymerization were determined.

Many recently published kinetic studies related to cyclopolymerization have concentrated on the commercially important monomer, DADMAC and its copolymerization characteristics. The polymerization of DADMAC initiated by $(NH_4)_2S_2O_8$ followed kinetics of 2.3 order in the monomer and 0.47 order in the initiator, with overall activation energy 15.4 kcal/mol. Comparison of the experimental kinetic equation for polymerization of DADMAC with the theoretical equations for polymerization of unconjugated, symmetrical dienes suggested that the polymerization of the monomer occurred with partial cyclization and chain termination by recombination of cyclized radicals. The k and a constants of the Mark-Houwink equation for poly(DADMAC) (molecular weight 13,500–22,500; 25°; solutions in 1N NaCl) were determined (12.6×10^{14}, 0.51, respectively).[2-196] New results on cyclopolymerization kinetics of the same monomer have been published. During the cyclopolymerization of pure monomers using O-free water as solvent and $K_2S_2O_8$ as initiator, the rate of polymerization was 0.75 order in $K_2S_2O_8$ concentration, second order in DADMAC concentration, 0.5 order in DADMAC cation concentrations, and 0.5 order in Cl^- concentration.[26]

Studies of the kinetics of radical polymerization of DADMAC have been continued. The rate of free-radical initiation, determined from integral evaluation of the time dependence of conversion and inhibition, was of the 1.5 and second order in initiator and monomer, respectively. The overall rate of polymerization was proportional to the 0.75 and third power of initiator and monomer, respectively, owing to the effect of the rate of initiation.[2-204] A study of the kinetics of radical polymerization of this monomer in aqueous water-methanol, and methanol solutions, with or without the presence of low-molecular-weight electrolytes, revealed that the rate of polymerization sharply and nonlinearly increased with increasing initial concentration of monomer.[27] The mechanism of the initiation and termination reactions with persulfate as initiator has been studied. In aqueous solution, primary radicals are formed simultaneously by activated decomposition of ion pairs ($S_2O_8^{-2}$/monomer associations), forming sulfate ion radicals, and redox reaction of $S_2O_8^{-2}$ with Cl^- ion pairs with monomer, forming sulfate anion radicals and $Cl\cdot$. The partial reactions were shown by quantitative analysis to proceed at equal rates. $Cl\cdot$ formed in the second reaction acts as both initiator and terminator, the additional chain termination being regarded as transfer to monomer in formal kinetic terms.[2-134]

The kinetics and mechanism of the radical cyclopolymerization of DADMAC has been surveyed. The kinetic analysis, taking into consideration nearly complete cyclization, a linear increase of $k_pk_t^{0.5}$ with [M], and different mechanism of initiation depending on the nature of the initiator, led to rate equations which fit the experimental data well. Initiation with $S_2O_8^{-2}$ had the following peculiarities: formation of primary radical by redox reaction with chloride ions and interaction with the monomer cation, additional termination by chlorine atoms, and an experimental chain transfer constant to monomer which included transfer to monomer and termination by chlorine radicals.[4-44]

Kinetic Studies Related to Cyclopolymerization

A number of studies dealing with topics which relate to or involve cyclization reactions have been undertaken. Among these are ring-chain competition kinetics, theoretical calculations related to small and large rings, diunsaturated monomer relation to gelation, statistical probability of ring closure, intramolecular reactions and cyclization dynamics. The direct relationship of these subjects to the theory of cyclization is apparent.

Ring-chain competition kinetics in a branched polymerization reaction has been studied.[28] A kinetic scheme was developed which accurately described the kinetics of a simple branched polymerization including cyclization reactions.

Ring-chain competition kinetics in linear polymerization reactions have been studied.[29] The spanning-tree model for f-functional polycondensation processes was evaluated against new exact computations of ring-chain kinetics; for a linear copolymerization of alternating bifunctional units, the model gave excellent results and required less computation than the exact method.

Orbital symmetry conservation of chemical reactions has been considered and the reaction mechanism of cyclization of linear polyenes and the reverse reaction have been reviewed with 14 references.[30]

Exact mathematical expressions have been derived for determining the distribution function of the end-to-end vector of a random-coil polymer chain in which the chain made a definite number of turns around a fixed point (2-dimensional case) or a fixed straight line (3-dimensional case).[31] The ring-closure probability and mean end-to-end distance were also calculated.

The reaction rates and the limit of infinitely high intrinsic weight constant for the diffusion-controlled ring closure reaction of polymers exactly based on a harmonic spring model have been calculated.[32] A physical interpretation for the results based on the theory of a free particle system was discussed.

A mathematical analysis of diffusion-controlled ring closure reactions of a polymer chain based on the general theory showed that the reaction rate was affected strongly by the short-time behavior of the segmental motion.[33] A simple intuitive interpretation was given for the results.

An approximate mathematical procedure has been developed for the description of crosslinking radical polymerization of a diunsaturated monomer, with independent reactivities of double bonds based on cascade treatment and spanning-tree approximation for a kinetically controlled chain reaction valid up to the gel point.[34] The cascade formalism was explained in detail, and methods of calculating cyclization probabilities for monodisperse primary chains were discussed.

The probabilities of formation of the individual units (pendant double bond, engagement in crosslinks, and cyclization) have been calculated for polydisperse primary chains resulting from radical polymerization of a diunsaturated monomer, using a treatment involving cascade formalism.[35] The method when compared with data for diallyl phthalate polymerization somewhat underestimates cyclization.

The statistical probability of cyclization and crosslinking during polymerization for linear and branched polymer chains[36] has been considered.

A computer simulation of the intramolecular reaction of polymers has been carried out.[37] The intramolecular relation of flexible polymer chains was simulated, and the apparent first order reaction rate was obtained for various values of the degree of polymerization and the size of the reaction region.

Enzyme-simulated selective organic synthesis-successive cyclizations of polyolefins among other reactions have been discussed with 40 references.[38]

Cyclization dynamics and thermodynamics of end-to-end cyclization of polystyrene in a Θ solvent have been studied.[3-185] The cyclization of pyrene-terminated polystyrene as 10^{-6} M cyclohexane solutions at the Θ temperature (34.5°) was studied by steady-state and fluorescence decay measurements. At polydispersity $\leqslant 1.1$, fluorescence decay was exponential over 6 lifetimes for molecular weight $10^4 - 10^5$, and non exponential for shorter chains. The data fit a mechanism involving a locally excited state and an intramolecular excimer, from which a rate constant for encounter-controlled, end-to-end cyclization could be calculated. The equilibrium constant for cyclization could be calculated for shorter chains. The rate and equilibrium constants decreased with increasing d.p.

In a study of kinetics of intramolecular crosslinking and conformational properties of crosslinked chains, a kinetic model for the intramolecular crosslinking of polymers was developed and discussed.[39]

Kinetic Investigations Applied to Five- or Six-Membered Cyclic Structures

The major emphasis in mechanistic and kinetic studies dealing with cyclopolymerization has been directed toward forming highly probable ring sizes. Consequently, a diverse group of monomers have been considered, and surprisingly, good agreement among the various approaches and conclusions is apparent.

The rate constants for radical initiated polymerization of methacrylonitrile were determined. The rate constants for propagation (k_p), termination, (k_t), transfer (k_c), energies of activation (E) and temperature independent factors (A) were as follows at 25°: $k_p = 26$, $k_t = 2.1 \times 10^7$, $k_f = 5.4 \times 10^{-3}$, $E_p = 11.5$, $E_t = 5.0$, $E_f = 18.8$, $A_p = 6 \times 10^9$, $A_c = 9 \times 10^{10}$, $A_f = 3 \times 10^{11}$.[40] A comparison between these parameters and those for 2,6-dicyano-1,6-heptadiene has provided valuable insight into the mechanistic details of cyclopolymerization.

In a study of the reactions of diolefins at high temperatures, the kinetics of the cyclization of 3,7-dimethyl-1,6-octadiene were studied.[3-24] Radical initiation of polymerization yielded 51% polymer and 14% cyclization to 1,2-dimethyl-3-isopropenylcyclopentane. Although the structure of the polymer was not discussed, it was postulated that the intramolecular cyclization to the isomer did not occur via a radical mechanism.

The copolymerization of nonconjugated diolefins with vinyl monomers involves departures from conventional vinyl copolymerization kinetics and the standard forms of the binary or ternary composition equations are inadequate.[7-150] These departures include intramolecular cyclization reactions and generally the formation of some pendant double bonds. A general treatment of the copolymerization of symmetrical nonconjugated diolefins with vinyl monomers gives closed form relationships which take into account such reactions and their consequences. The general relationships may be approximated by a new series of composition equations which take into account the formation of both cyclized and noncyclized units in the chain and are simple enough for experimental use. The latter expressions are:

$$dm_1{:}dM_2{:}dM_4{::}$$

$$[M_1]([M_1] + \alpha_c + \alpha_4[M_4]){:}\ -\ [M_1]([M_1] + \alpha_4[M_4]){:}$$

$$[M_4]([M_1] + [M_4]/\delta_1)\{\alpha_4 + \alpha_c/([M_1]/\beta_4 + [M_4])\}$$

where: $\alpha_c = k_c/k_{11}$, $\alpha_4 = k_{14}/k_{11}$, $\delta_1 = k_{41}/k_{44}$, and $\beta_4 = k_{24}/k_{21}$, and $[M_1]$, $[M_2]$, and $[M_4]$ refer to the concentrations of diolefin double bonds, pendant double bonds, and comonomer, respectively.

A study has been made of copolymerization of monovinyl monomers with divinyl benzene.[7-151] The composition relations were derived both for the general case, including crosslinking reactions, and for the approximate case where the concentration of pendant unsaturation may be assumed unaffected by crosslinking reactions. The divinyl monomer must be incapable of undergoing short-range cyclization, or alternating intra-intermolecular propagation. The general case composition equation has a similarity to that for ternary polymerization of three vinyl monomers. The approximate case closely resembles the usual binary copolymerization relationship.

In a paper on cyclopolymerization, a study of the temperature effect on the polymerization mechanism of methacrylic anhydride was reported[2-308] for the monomolecular and bimolecular mechanisms of cyclopolymerization, $-d[M]/d[m] = A + B/[M]$, where [M] is the monomer concentration, [m] is the pendant double bond concentration and A and B are constants. This equation was verified by polymerizing methacrylic anhydride under different temperatures. In all cases, $-d[M]/d[m]$ gave a linear relation with $1/[M]$ as expected, and the intercepts of the plots obtained were 2.9, 2.15, and 1.7 at 60°, 40°, and 25°, respectively. The values of the ratios of the bimolecular rate to the monomolecular rate were 2.7 [M], 2.0 [M], and 1.4 [M] at 60°, 40° and 25°, respectively, and thus decrease as temperature decreases. The energy of activation of bimolecular cyclopolymerization was 3.7 kcal/mole higher than that of the monomolecular reaction.

Cyclopolymerization of the diallyl esters of oxalic, malonic, succinic, adipic, and sebacic acids has been studied kinetically and rate data as well as overall activation energies of polymerization were reported.[3-203]

Oxidative cyclization of 5-methyl-*trans*-1,5,9-decatriene was accomplished in presence of mercury salts to give mixtures containing perhydronapththalene and *cis*- and *trans*-3-butyl-4-methylcyclohexanone.[4-44]

Investigations of Kinetics of Large Ring Formation in Vinyl Polymerization

The probability of formation of large rings during crosslinking attempts using tetrafunctional vinyl monomers, as well as the intense interest during the early development of cyclopolymerization in ring size has stimulated quite a bit of activity in the study of large rings. Conformational effects come into play, as well as the early history of large ring formation in the small molecular organic chemistry, and these factors have had considerable influence on the approach to the problem as well as to the nature of the monomers studied.

The reaction kinetics of polymerization reactions involving cyclization processes was studied theoretically, and formulas were derived with which the degree of cyclization could be calculated from experimentally derived data.[1-15] It was shown that diallyl phthalate, under carefully controlled conditions, could yield a polymer that was 81% cyclized.

Diallidenepentaerythritol was polymerized via free radical initiators to give polymer yields of less than 10% in most cases.[2-423] The structure of the polymers was studied as well as the polymerization process, and it was concluded that both degradative chain transfer and cyclization occurred. The ratio of the cyclization velocity to the sum of the velocities of chain propagation and chain transfer was calculated to be 0.276.

It has been proposed that a secondary mechanism involving copolymerization of the CN bonds of acrylonitrile with vinyl chloride is more feasible than that of a transfer reaction involving the α-H atom of an antepenultimate unit of acrylonitrile for explaining the kinetic deviations from the Lewis-Mayo theory for the radical polymerization of the two monomers.[41] This secondary mechanism produces a cyclic imine radical which causes coloration of the polymer and formation of new -C:N-radicals with low reactivity vs. propagation reactions.

The rate of polymerization of diallyl phthalate by γ-irradiation was shown to be proportional to the radiation dose rate, with a net activation energy of polymerization of 5.02 kcal/mole.[42] The radical yield after irradiation, as measured by loss of diphenyl picryl hydrazyl, was 3.12. The ratio of degradative chain-transfer constant, k_u, to the polymerization constant was 0.065, and a ratio of nondegradative chain-transfer constant to k_u was 0.032. Vapor pressure osmometry indicated a degree of polymerization of 8–10, and IR spectra indicated intramolecular cyclization during polymerization.

Cyclopolymerization kinetics of *N,N'*-methylenebisacrylamide initiated by $Mn(OAc)_2 \cdot 2H_2O$ with either ethylene glycol or mercaptoethanol in aqueous acetic acid at 35–50°C was investigated.[43] Rates of polymerization and manganic ion disappearance were determined, and rate equations were derived from the observed dependence.

A kinetic study of photosensitized cyclopolymerization of *N,N'*-methylenebisacrylamide, using uranyl ion, was carried out in an effort to establish the mechanism.[2-479] Evidence for cyclization of the radical prior to propagation was obtained.

Kinetic Investigations of Cyclopolymerization of Diallyldimethylammonium Chloride (DADMAC) and Related Monomers

The application of poly(DADMAC) in a variety of industries has prompted numerous investigations of this monomer and its polymerization. Although the original proposal included the assumption that the more predictable six-membered ring was formed, it has now been shown that the ring size is almost exclusively the five-membered ring. This and related monomers appear to be the only type from which polymers formed do not consist of mixtures of ring sizes. However, no careful study of the various experimental parameters in the case of this monomer has been undertaken, as in the case of acrylic and methacrylic anhydride systems, to determine whether significant fractions of the six-membered ring can be formed.

A continuous polymerization process for dialkyldiallylammonium salts has been described.[2-193] The process involves use of 30–70% solutions of the monomer in presence of buffer salts to maintain the pH at 6.7–10.3 so that product yields were greater than 90%.

The kinetics of radical polymerization of DADMAC in presence of ammonium persulfate at 35°C was studied.[44] The reaction was 0.8 order in initiator and 2.9 order in monomer concentration at conversions $< 10\%$ and monomer concentration $> 1.5M$. The relative transfer constant to monomer was 2.5×10^{-3}. The value of $(7.25 \pm 0.9) \times 10^{-5}$ was calculated for the overall rate constant, k_1. Addition of NaCl increased the overall polymerization rate and decreased the order of the reaction in monomer. The dependence of the rate constant for chain growth and initiation on monomer concentration was related to the high reaction order in monomer. The results show that the course of the polymerization deviates considerably from the ideal model of radical polymerization.

A kinetic scheme and radical balance equation was derived for radical polymerization of DADMAC at conversions $>10\%$.[45] The mechanism explains qualitatively the reaction orders in monomer and initiators, and suggested that the initiation rate depends on monomer concentration.

Kinetic analysis of the cyclopolymerization of DADMAC, taking into consideration nearly complete cyclization, a linear increase of $k_p/k_t^{0.5}$ with [M] and different mechanisms of initiation depending on the nature of the initiator, led to rate equations which fit the experimental data well.[46] Persulfate anion as initiator has peculiarities such as formation of primary radicals by redox reaction with chloride ions and interaction with the monomer cation, additional termination by chlorine atoms, and an experimental chain transfer constant to monomer which includes transfer to monomer and termination by chlorine radicals.

Synthesis of the vesicle-forming quaternary ammonium salts, allyldidodecylmethylammonium bromide and diallyldidodecylammonium bromide was reported, and topochemical polymerization of the monomers was accomplished by γ-ray irradiation.[2-213] Only polymerized vesicles which resulted from the diallyl monomer retained the structure of the monomer vesicles and also exhibited higher stability.

The kinetics of radical polymerization of *N,N*-dimethyldiallylammonium chloride was studied in aqueous solutions of 2–5 mol/L at 60° and low monomer conversion values.[2-215] The reaction order was 0.5 with respect to initiator and showed no degradative chain transfer to monomer. The initial viscosity of monomer solutions had a significant effect on the rate constant of bimolecular chain termination.

A kinetic study of homopolymerization of DADMAC (M_1) and its copolymerization with acrylamide (M_2) in inverse emulsion in a stirred tank reactor has been published.[47] The anomalously high order with respect to monomer concentration could be explained by combining the monomer concentration equilibrium between micelles and aqueous phase, and the square dependence of the propagation rate with respect to monomer concentration. A model was developed to predict the conversion-time curve of homopolymerization of DADMAC. Monomer reactivity ratios were determined in the absence of added salt and in presence of ammonium chloride. The values obtained were (no salt): $r_1 = 0.049 \pm 0.01$; $r_2 = 7.54 \pm 0.9$; $r_1 = 0.273 \pm 0.03$; $r_2 = 7.15 \pm 0.85$; (0.748 mol/l NH_4Cl).

The kinetics of polymerization of aqueous solutions of diallyldimethylammonium chloride in inverse emulsion was studied and the mechanism discussed.[2-228] It was concluded that the mechanism conformed to the Smith-Ewart mechanism. The influences of ion strength and partition equilibrium on the rate of polymerization were studied also, and are considered to be the causes of the high order with respect to the monomer concentration.

The dependence of molecular weight of DADMAC on conversion was studied and a model was proposed for the calculation of the conversion dependence of molecular weight under consideration of gel effect and chain transfer reactions to monomer and polymer.[48] The influence of DADMAC concentration on the ionic conductivity of the solid polymer was studied. TGA showed the relative amount of loss at 22–861°, and DTA gave a glass temperature, Tg of 8°.

A copolymerization study of diallyldimethylammonium chloride with acrylamide and quaternized dimethylaminoethyl methacrylate was carried out in solution and reverse microsuspension. The copolymer reactivity ratios were determined.[7-71]

MICROSTRUCTURE OF POLYMERS

The original mechanism proposed to explain the formation of linear, saturated polymers from diallyl quaternary ammonium salts assumed that six-membered ring cyclic units would be produced. This assumption is consistent with "head-to-tail" enchainment in radical initiated polymerization of vinyl monomers, an arrangement which has been unquestionably established as the major mode of enchainment in a wide variety of vinyl polymers. To confirm the proposed structure, representative polymers were degraded, by methods which could cleave the cyclic units in the chain, and the expected products were obtained. However, the method used was not designed to distinguish between five- and six-membered rings.

This was a natural assumption in view of the thermodynamic stability of six-membered rings and the fact that the intermediates involved should be secondary radicals arising from a head-to-tail chain propagation reaction. Six-membered cyclic structures for the polymers from 1,6-heptadiene and related compounds were also proposed on the evidence that partial dehydrogenation of the polymers gave products which were characteristic of *meta*-substituted aromatic rings. Also, the exclusive formation of six-membered cyclic monomeric products in the free-radical addition reactions of 1,6-dienes with various addenda was reported later.[1-36, 2-298]

As a result of these studies, the structures of most new cyclopolymers from 1,6-dienes were assigned six-membered ring units. Since initiation and propagation steps could occur by addition to either the *head* or *tail* of the double bonds, there exists the possibility for formation of five-, six- or even seven-membered ring structures.[1-154, 1-19]

However, it had been recognized quite early that less-favored ring sizes may be formed in cyclopolymerization and that the less-stable radical may predominate in the cyclization.[1-77] It was shown that the less-stable radical predominated in the cyclization step during polymerization of allyl and methallyl crotonates (10–7). Although attack at C2 may be somewhat more favored sterically than at C3, it was reasoned that attack at C3 should be favored because of the resonance-stabilized radical formed at C2. Further evidence for formation of the less-favored ring structure and/or predominance of the less energetically favored intermediate radical in controlling the course of the cyclopolymerization process has been obtained. These results have been summarized in a recent review.[1-171]

Justification of this preference for five-membered ring formation has been based on both electronic[1-17] and steric[49, 2-484] factors. A suggestion that there may be an electronic interaction between the initially formed radical and the neighboring double bond

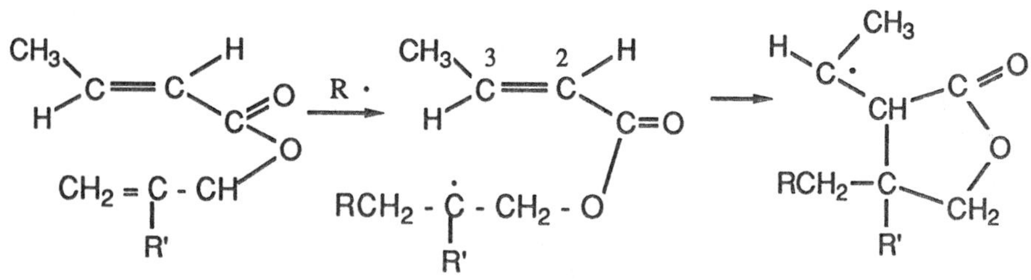

R' = H, CH_3

(10–7)

of the diene has received considerable attention.[1-17] The formation of methylcyclopentane from the reaction of 4-hexenyl mercaptan with triethyl phosphite was explained[49] as arising from the attack of the radical at the more accessible end of the double bond, the process being irreversible. It has been noted that, for approach of the radical to the double bond with the p orbitals in a common plane, formation of five-membered rings would be less sterically hindered.[2-484] One of the terminal hydrogens lies in the nodal plane directly between the radical carbon and the carbon on the terminal end of the double bond, thus hindering six-membered ring formation. No such steric interference exists for approach of the radical to the other end of the double bond, leading to five-membered rings. However, it was shown as early as 1963 that introduction of electron accepting groups onto the nitrogen atom of diallylamine enhanced its tendency to polymerize.[50]

Additional evidence for the existence of structures other than a six-membered ring came from a cyclopolymerization study of divinyl acetal.[1-20, 2-264, 2-268, 2-516] Polymers containing both five- and six-membered rings were obtained.[2-267] Hydrolysis of the cyclopolymer followed by analysis of the derived poly(vinyl alcohol) gave both 1,2- and 1,3-glycol structures. This technique was used to determine the ring content of a large number of poly(divinyl acetals). The results gave 5:6 ring size ratios of 23:77.[1-134] The 1,2-glycol content in poly(vinyl alcohol) derived from divinyl carbonate was also determined in another study and values equivalent to 6–23% of five-membered rings were found.[2-277]

Other types of cyclopolymerization reactions were also found to yield five-membered rings. A polymer with a relatively high content of five-membered cyclic anhydride units was obtained by polymerization of acrylic anhydride at 115°C in xylene.[2-243] Also, differences were observed between cyclopolymers from dimethacrylamides and polymers produced by deamination of polymethacrylamides.[2-79] It was concluded that the cyclopolymerization of dimethacrylamides gave predominantly five-membered ring units. These results have been confirmed by other investigations.[51, 2-150]

N-Phenyldimethacrylamide was found to produce polymer by γ-radiation containing 90% five-membered ring and 10% six-membered ring.[51] In contrast, it had been assumed earlier that *N*-methyldimethacrylamide yielded polymer which consisted exclusively of six-membered rings.[2-80] Other examples of cyclopolymerization of 1,6-dienes to polymers containing five-membered rings along with six-membered rings have been reported.[1-156] An explanation based upon solvent polarity has been offered[2-81] to account for ring size in cyclopolymerization of dimethyacrylamides, an increase in polarity favoring the five-membered ring. Recent evidence has been presented[2-82] in support of the five-membered cyclized radical as the intermediate in the rate-controlling step in free radical cyclopolymerization of *N*-(n-propyl)-dimethacrylamide.

The question of the ring size produced in radical addition reactions was reconsidered after it was shown that predominantly five-membered rings (94%) were obtained in the thermal decomposition of 6-heptenoyl peroxide.[1-122] It had been shown as early as 1939 that cyclohexane is more stable than methyl-cyclopentane at the temperature of the peroxide decomposition.[52] Also, in 1959, the rate of decomposition of cyclohexane-formylperoxide was shown to decompose 34 times as fast as cyclopentaneacetyl peroxide, a result which indicates that the cyclohexyl radical is more stable than the cyclopentyl radical.[53] Later, and quite significantly, it was shown that the 5-hexenyl radical gave only cyclopentylmethyl radical at −35°C, and that the reaction was irreversible up to 0°C.[54]

It was also found that in the cyclization of 1,6-heptadiene with bromotrichloromethane and other addenda, five-membered ring compounds were obtained,[53, 2-53, 56, 2-56, 57] results which contradicted earlier work. This study was extended to include diallyl ether, ethyl diallylacetate and diallylcyanamide. In each case, predominantly five-membered ring products were found. It was also observed that five-membered ring products from the cyclization of divinyl formal were obtained.[2-27]

Five membered rings resulted from radical cyclization reactions of 6-bromo-1-hexene and other alkenyl halides.[58-60] In the cyclization of allyl ethenesulphonate with butyl mercaptan[3-168] only the five-membered ring sultone and sulphonium salt were observed. Previously, in the cyclopolymerization of this monomer, only six-membered rings had been proposed.[3-121]

It has become clear that the structures of cyclopolymers or cyclocopolymers could not be assumed to contain only the thermodynamically most stable rings. In recent years, it has been realized that a detailed analysis of the microstructure of the polymer chain is necessary before the mechanism of the cyclization and the properties of the polymers can be understood. There have been three approaches to this problem:

(a) A study of intramolecular addition reactions of alkenyl radicals;
(b) detection or trapping of low molecular weight products from addition reactions of non-conjugated dienes;
(c) direct investigation of polymers using modern spectroscopic methods.

The choice for ring closure is illustrated in Structure 10–8. A free radical generated within a structure in which there is an accessible unsaturated center could lead to endo-

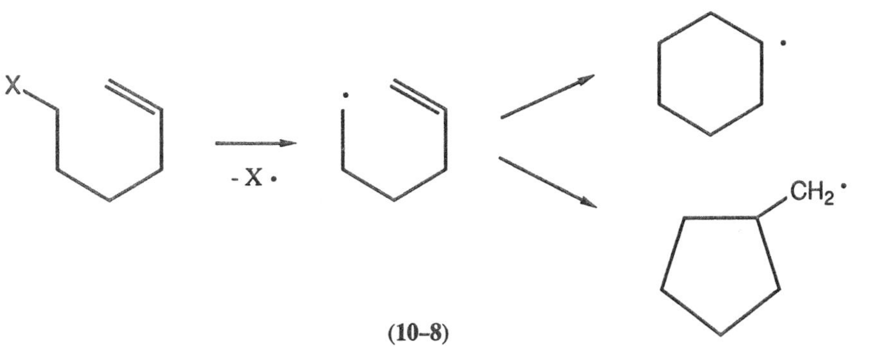

X
−X•
CH_2•

(10–8)

or exo- cyclization. Clearly, this is a model situation for the cyclopolymerization reaction.

A large number of reactions dealing with small molecular cyclization has been investigated, and several reviews have been published.[61–65] In a study of cyclization of 1-substituted 4-hexenyl radicals,[62–64] it was shown that as substitution in the 1-position by radical stabilizing groups (one or two CN or $C(O)OC_2H_5$ groups) was increased, the mixture of cyclized products changed from nearly pure cyclopentane to nearly pure cyclohexane derivatives. Higher temperatures favored cyclohexane formation. Also, it was shown that the cyclization reactions were reversible when C1 was disubstituted, although it had been shown[1–122] earlier that in the case of the primary hexenyl radical, the cyclization step is irreversible. On this basis, it was proposed that the cyclopentane product is the kinetically preferred product while the cyclohexane product is preferred via thermodynamic control, with the energy of activation for cyclization being higher for the cyclohexane derivatives.

The majority of radical cyclizations reported have been accomplished by permitting alkyl radicals to react with an unsubstituted double bond, thus producing the intermediate, non-resonance-stabilized primary or secondary radical.[11] As stabilization of the generated radical increased, the ratio of six-membered ring increased. A later investigation has shown that when the radical generated after cyclization of the six-membered ring can be stabilized by resonance, the ratio of six-membered to five-membered ring is markedly increased.[66] As shown in Structure 10–9 2-methallyloxyethyl and 2-(2-phenylallyloxy)ethyl radicals were generated by use of tri-(n-butyl) stannane[67] in presence of AIBN.[68] The results of this study strongly indicated that radical stabilization results in greater selectivity and thus an increase in the more thermodynamically stable six-membered ring. The cyclization rate constant ratios at 40°C and the derived

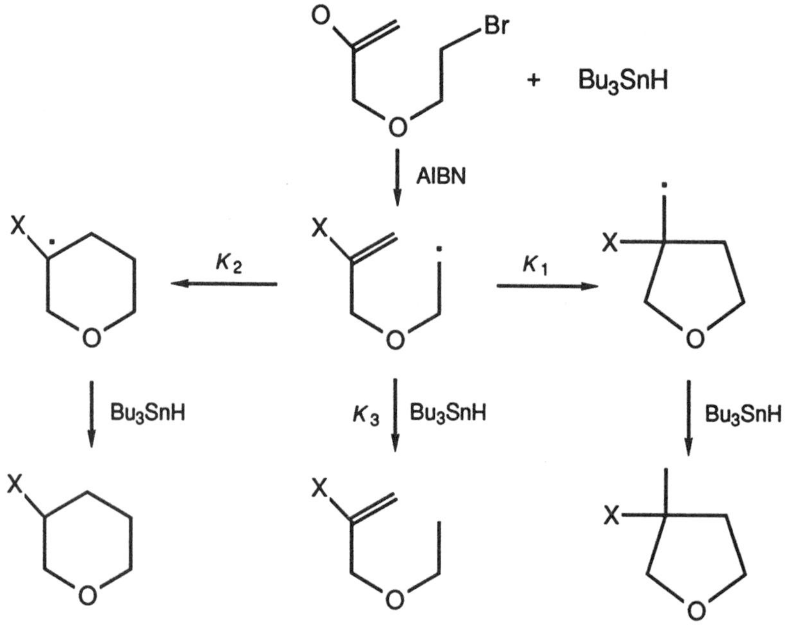

Arrhenius parameters showed that the radical-stabilizing effect of the methyl group was not sufficient to prevent the predominance of the tetrahydrofuran ring. The k_1/k_2 ratio at this temperature was 43; however, the corresponding ratio for the phenyl case was only 0.55. The Arrhenius parameter ratios, A_2/A_1, were 0.35 $\pm$0.08 for the methyl case and 2.8 $\pm$0.2 for the phenyl case (See Structure 10–21).

The ratio of five-membered to six-membered ring formed from 2-methallyloxyethyl radical was 30:1 but this ratio dropped to 1:2 in the 2-(2-phenylallyloxy)ethyl case. This remarkable difference was attributed to increased resonance stabilization of the six-membered cyclic tertiary radical.

The rate constants for cyclization at 40°C for the 1-hexenyl radical studied earlier[58]; and for the 4-(1-cyclohexenyl)butyl radical[59] were determined by comparing the reported rate constant ratios with the rate constant for hydrogen abstraction from $(C_4H_9)_3SnH$ by the 1-hexyl radical at 25°C. The values of k_c obtained were 1×10^5 sec^{-1} for the 4-(1-cyclohexenyl)butyl radical. These constants were also compared with the rate constant for addition of an ethyl radical to 1-heptene at 40°C, 1×10^3 $M^{-1}sec^{-1}$. The "effective double bond concentration" for the intramolecular cyclization of such radicals was estimated to be about 40–100 M. Such a comparison with the cyclization rate constant for the 2-methallyloxethyl radical yields a value of 62 M for the "effective double bond concentration" in this case. For the 2-(2-phenylallyloxy)ethyl radical the value would be at least 150 M, but a more accurate basis for comparison would be with the rate constant for addition of a primary radical to an α-alkyl styrene, which should have a value greater than 1×10^3 $M^{-1}sec^{-1}$ at 40°C.

This enhanced "effective double bond concentration" for reaction of the initial radical with the double bond to give a cyclic radical, coupled with the increased steric hindrance to approach of another monomer molecule, should be sufficient to explain why 1,6-dienes undergo polymerization to polymers composed almost entirely of cyclic units. Propagation by reaction of the initial radical with another molecule of monomer, a bimolecular reaction, simply cannot compete with the unimolecular cyclization processes in most cases.

The activation energy for the hydrogen abstraction reaction of a primary alkyl radical from tributyltin hydride has been calculated[68] to be between 6.8 and 8.2 kcal/mole using rate constants reported in the literature.

Some generalizations concerning the influence of steric and stereoelectronic factors on radical reactions of this type have been presented. The following guidelines were suggested:

(a) Intramolecular addition under kinetic control in lower alkenyl radicals and related structures occurs preferentially in the exo-mode;
(b) Substituents on an olefinic bond disfavor homolytic addition at the substituted position;
(c) 1,5-Ring closures of substituted hex-5-enyl and related radicals are stereoselective.

Examples[3-209, 2-515] of guideline (a) include hexenyl radicals and variously substituted hexenyl radicals and the 2-allyloxyethyl radical.[58, 69] Reactions of either 6-bromo-6-phenyl-1-hexene or *trans*-2-phenylcyclopentylmethyl bromide with tributylstanname yielded 6-phenyl-1-hexene, *trans*-1-methyl-2-phenylcyclopentane and phenylcyclohexane showing that the three corresponding radicals are interconvertible under the conditions used.[70] Direct rearrangement of the 2-phenylcyclopentyl methyl radical to 3-phenylcyclohexyl radical without ring opening was indicated. In these cases, the reactions are irreversible and the major products are five-membered ring structures arising from *exo*-cyclization. However, in other cases where the initial radical was stabilized by neighboring groups, the reactions were reversible and under thermodynamic control which led to the six-membered ring product by *endo*-cyclization. *Exo*-ring closure of radicals of this type is the kinetically favored process because the stereoelectronic requirements of the transition state can be most readily satisfied. The addition of a free radical to an alkene will occur most readily when the orbitals bearing the three electrons which are redistributed remain in the same plane throughout the reaction.[16, 1-146]

The 5-substituted hex-5-enyl radicals illustrate guideline (b). These have been found to undergo mainly six-membered ring closure[2-515, 71] because the rate of cyclization to five-membered rings is greatly retarded due to the presence of the substituent. However, conflicting results have been observed for the 5-methylhex-5-enyl[3-43] and 2-methallyloxyethyl radicals[66] where five-membered rings were still found to predominate. With the 2-(2-phenylallyloxy)ethyl radical[66] six-membered rings do predominate, but here the intermediate radical is considerably stabilized by resonance.

Cyclization of 1- or 3-substituted hex-5-enyl and related radicals occurred regiospecifically in the *exo*-mode and yielded mainly *cis*-disubstituted cyclic products, whereas 2- or 4-substituted systems gave mainly *trans*-products,[3-215, 3-210, 3-150] illustrating guideline (c).

In a study of stereoselectivity of ring closure of substituted hex-5-enyl radicals, the stereoselectivity and kinetics of ring closure of 2-substituted hex-5-enyl radicals in which the substituent was methyl or vinyl, as well as the 3- and 4-methyl-5-hexenyl radicals, were examined.[3-215] All of the radicals underwent regiospecific 1,5-ring closure with a rate constant greater than that for hex-6-enyl radicals. The 3-substituted radicals give *cis*-products preferentially. The 2- and 4-substituted radicals give mainly *trans* products. The results were rationalized in terms of conformational effects in the chair-like transition state (See Structure 3–27).

The formation of the *cis*-isomer from 1-substituted hexenyl radicals has been attributed to orbital symmetry control.[3-210] The ring closure of 2-, 3-, or 4-substituted hexenyl radicals was related to the conformational preference of the transition states.[3-215]

In these models of the cyclopolymerization reaction, significant information concerning the cyclization of radicals has been obtained. The question is whether similar products are obtained from radical addition reactions of non-conjugated dienes. E.S.R. spectroscopy has been used to study the radical addition reactions of non-conjugated dienes. Hydroxyl, amino, methyl or phenyl radicals were allowed to react with diallylmalonic acid (and its derivatives), diallyl ether, diallylamine and variously substituted diallylamines in aqueous solutions. These results were interpreted as evidence that the cyclic radicals present in highest concentration when 1,6-dienes undergo reaction were five-membered ring radicals.

The sensitivity of this cyclization to steric and resonance effects was shown by investigations on a number of substituted diallylamines. The radicals from *N,N*-bis(2-isopropylallyl)methylamine and NH_3^+ were six-membered ring radicals. This is in accord with other results on cyclization of radicals[2-515] and shows that a bulky substituent on the internal position of the double bond disfavors attack at that position.

The products from the reactions of diallylamine derivatives with radicals have been investigated.[2-104, 2-524, 2-105, 2-106] Conditions were chosen such that chain propagation to high molecular weight did not occur and the low molecular weight products were isolated and characterized by mass and ^{13}C NMR spectroscopy.[2-178, 72] *N,N*-Diallylmethylamine and derivatives (10–10) have been studied in most detail. Identified among the basic products were perhydroisoindol-5-ones (10–11), 3-azabicyclo[3.3]nonan-6-imines (10–12), or a mixture of these compounds. In certain cases, Structures 10–13 and 10–14 were found. The bicyclic adducts (10–11 and 10–12)

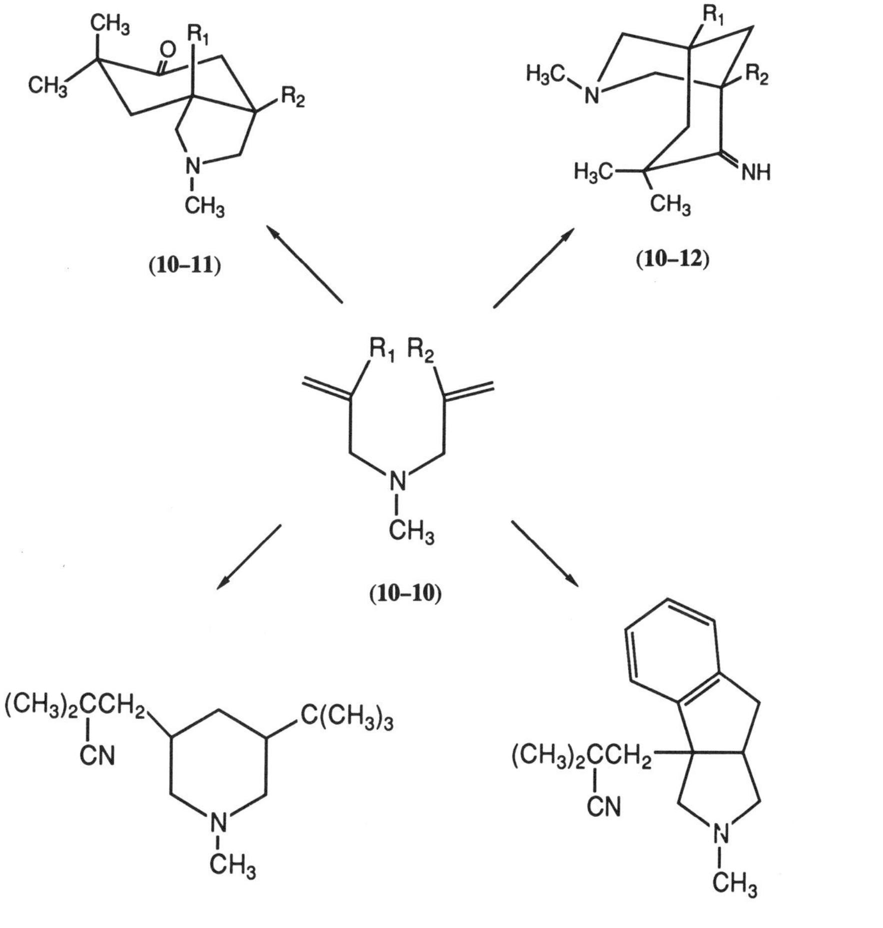

CH_3 CH_3 O R_1 R_2 N CH_3

(10–11)

R_1 H_3C N R_2 H_3C NH CH_3

(10–12)

R_1 R_2 N CH_3

(10–10)

$(CH_3)_2CCH_2$ CN $C(CH_3)_3$ N CH_3

(10–13)

$(CH_3)_2CCH_2$ CN N CH_3

(10–14)

are the result of a *backbiting* reaction between the first-formed alkyl radical and the nitrile group, to form an imine which hydrolyzes to a carbonyl function (10–15).

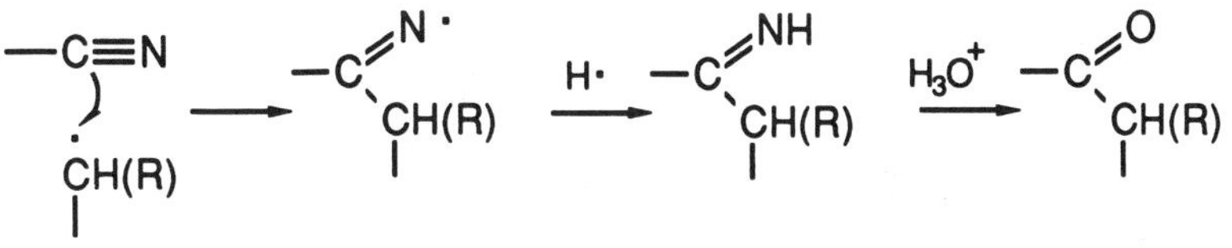

—C≡N · CH(R) ⟶ —C=N· CH(R) $\xrightarrow{H\cdot}$ —C=NH CH(R) $\xrightarrow{H_3O^+}$ —C=O CH(R)

(10–15)

The following conclusions were drawn:

(a) In these systems, the initial attack occurs on a terminal carbon of the allyl groups;
(b) Bulky alkyl substituents on one allyl group result in a preference for the unsubstituted (least hindered) group;
(c) Intramolecular cyclization invariably leads to a five-membered ring structure;
(d) Where steric hindrance occurs, six-membered rings are found.

Earlier studies of the radical addition of perfluoroalkyl iodides to dienes have been extended to include a series of *N*-substituted diallylamines.[2-64, 2-71, 2-166] By spectroscopic and chemical methods, it was shown that five-membered ring adducts were formed.

The stereochemistry of the product mixture was not discussed. However, in the reactions of diallyl ether with 1-iodo-*F*-butane, 1-iodo-*F*-hexane and 1-iodo-*F*-octane it was the *cis*-isomer which predominated, and pure samples of these isomers were isolated and characterized by ^{13}C NMR. This agrees with earlier work on diallylcyanamide but conflicts with data on 1,6-heptadiene in which it was reported that it was the *trans*-isomer which was formed most rapidly. Other workers[2-63] have observed that free-radical addition of 2-methylpropanoic acid to 1,6-heptadiene results in cyclization to the five-membered ring adduct with a ratio of 2:1 in favor of the *trans*-isomer. It has been pointed out that[73] the cyclization of hex-5-enyl radicals is not always stereoselective towards the *cis*-isomer when the substituent at C-1 is bulky.

The products of free-radical cyclization reactions (cyclotelomers) provide a valuable insight to the possible structures of analogous cyclopolymers. However, concern has been expressed[1-19, 2-27, 3-168] that cyclopolymerization and cyclotelomerization may not necessarily follow the same mechanism. For example, poly(divinylformal) yields cyclopolymers which have been shown to contain up to 30% five-membered ring units; yet in the cyclotelomers, five-membered rings predominate.[2-27] It follows that direct analysis of cyclopolymers is the most desirable approach to determination of their structures.

Pulsed Fourier transform ^{13}C NMR spectroscopy has been used most effectively to determine the structures of polymers of diallylamines[2-516, 72, 1-169, 2-107, 74] A comparison of the spectra with those of a series of pyrrolidine and piperidine model compounds showed that the cyclic units in these polymers were five-membered rings. The existence of both *cis*- and *trans*-units (ratio 5:1) was revealed from the spectra.[74] The spectra of other *n*-substituted diallylamines showed that the cyclization reaction was influenced by substituents at the β-position of the allyl groups[2-516, 2-108, 2-109, 2-111, 75] and could yield five- or six-membered rings according to the substitution pattern.

In the absence of appreciable steric hindrance, intramolecular cyclization proceeded to the five-membered ring. Where steric interactions are significant, six-membered rings were also observed. Changes in temperature affected the ratio of five-to six-membered ring structures in *N,N*-dimethallylmethylamine.

NMR spectroscopy has proved invaluable to investigation of the structures of cyclopolymers and cyclocopolymers. ^{31}P spectroscopy has been used to distinguish and quantitatively determine phosphorus atoms in five- and six-membered ring environments in the cyclopolymers from divinylphosphonates[2-350]. Using a series of phosphonates as model compounds, it was shown by direct analysis of the polymers that poly(divinylphenylphosphonate) contained five- and six-membered ring repeating units in amounts dependent upon the method of polymerization, whereas poly(divinylmethylphosphonate) only contained six-membered rings.

The [3.3.0]-fused bicyclic (five-membered ring) system was obtained by cyclopolymerization of *cis, cis*-1,5-cyclooctadiene with maleic anhydride.[76] The monomer ratio in the copolymer was 1:1 rather than 1:2 which would be required had the copolymerization occurred in an analogous manner to that proposed for the corresponding sulfur dioxide copolymer. A [3.3.2]-bicyclic ring system would be required in the 1:2 monomer ratio case.

An analogy can be drawn between five-membered ring formation in cyclopolymerization of 1,6-dienes and head-to-head polymerization in an ordinary vinyl monomer.

By means of IR, ^{1}H NMR, and ^{13}C NMR, spectroscopy, it has been shown[2-323] that the ratio of five- to six-membered ring formation in cyclopolymerization of acrylic anhydride at 50°C varied with the dipole moment (μ) of the solvent from 5% five-membered ring at $\mu = 0$ to 30% at $\mu = 4$. A pronounced temperature effect was also observed. At 90°C, the five-membered ring content was 20% at $\mu = 0$ and increased to 75% at $\mu = 4$. Cyclopolymerization in γ-butyrolactone at 120°C resulted in 90% five-membered ring. Under all conditions described, methacrylic anhydride led only to six-membered ring structures.

The five-membered ring content of poly(acrylic anhydride) has also been shown[2-324] to vary with both conversion at constant monomer concentration and with monomer concentration at constant conversion. For example, in benzene, the five-membered ring content increased from 5% at 40% conversion to 10% at 90% conversion; however, in γ-butyrolactone, the five-membered ring content increased from 28% at 15% conversion to 75% at 50% conversion, a much greater change than in the nonpolar medium.

On the basis of the previously described studies on acrylic anhydride, it was postulated[77] that, if methacrylic anhydride were polymerized above the ceiling temperature of its monoene counterpart, methacrylic isobutyric anhydride, the poly(methacrylic anhydride) should consist largely of five-membered ring anhydride structures. The ceiling temperature of methyl methacrylate is 164°C in bulk and decreases with dilution. Previous studies[2-323] had shown that cyclic poly(methacrylic anhydride) consisted exclusively of six-membered rings under the low temperature conditions. As postulated, polymerization of methacrylic anhydride at 160°C in benzonitrile was found to consist of 53% five-membered rings, and polymerization at 180°C gave almost exclusively the five-membered ring structure. Furthermore, in accord with the postulate, five-membered ring formation was found to increase with increasing temperature and dilution.

Radiation-induced cyclopolymerization of *N*-substituted dimethacrylamides has been studied extensively in the liquid, supercooled-liquid, glassy, and crystalline states.[2-81, 2-82, 2-83] In most of these cases, cyclopolymerization occurred exclusively, predominantly to five-membered rings, except in the glassy state, where no polymerization was observed. The ratio of six-membered rings was higher in the crystalline state than in the liquid states.

A single-crystal x-ray diffraction and solid state polymerization study[2-201] of *N*-(p-bromophenyl)dimethacrylamide (10–16) led to the conclusion that the monomer crystallizes in a conformation that favors the intramolecular over the intermolecular polymerization reaction. The polymer consisted exclusively of cyclopolymer having 60% five-membered and 40% six-membered rings. The ratio of five- to six-membered ring

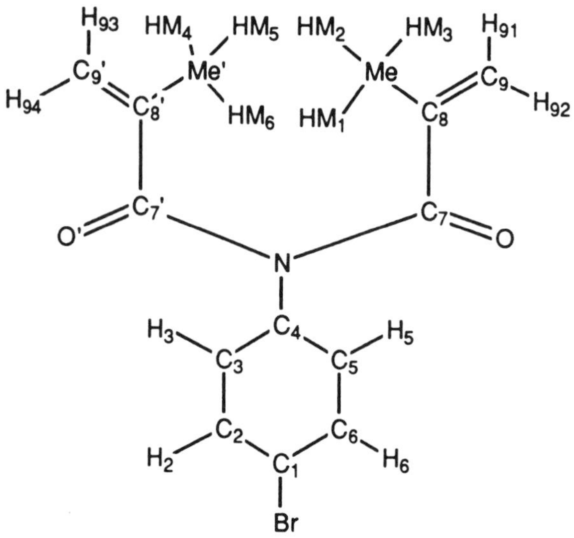

Numbering scheme used in the x-ray analysis of N-(p-bromophenyl) dimethacrylamide.

(10–16)

could not be predicted from the atomic positions in the crystal, however. Intramolecular distances (10–17) from C9 to C8′ and C9′ were 3.940 Å and 4.002 Å, respectively, whereas the closest intermolecular reactive site was nearly 0.5 Å further away (4.489 Å to C8). The closest intramolecular route for C8 was with C8′ which was 4.122 Å away. The other intramolecular site, C9′, was 4.453 Å removed, which was almost as far away as the nearest intermolecular site (C9 at 4.489 Å). Atom C8′ was fairly close to its two possible intramolecular reactors, which were C8 at 4.122 Å and C9 at 3.940 Å. The closest intermolecular reactor, C9′, was 0.6 Å further away; the closest intramolecular reacting site from C9′ was C9, which was 4.002 Å away (See Table 10–6).

Table 10–6 Pertinent Intramolecular and Intermolecular Distances (Angstroms) for Atoms in the Solid-State Polymerization of *N*-(p-Bromophenyl)dimethacrylamide

Intramolecular reactors			Closest intermolecular reactor	
Atom	Atom	Distance	Atom	Distance
C9	C8′	3.940	C8	4.489
	C9′	4.002		
C8	C8′	4.122	C9	4.489
	C9′	4.453		
C8′	C8	4.122	C9′	4.753
	C9	3.940		
C9′	C8	4.453	C8′	4.753
	C9	4.002		

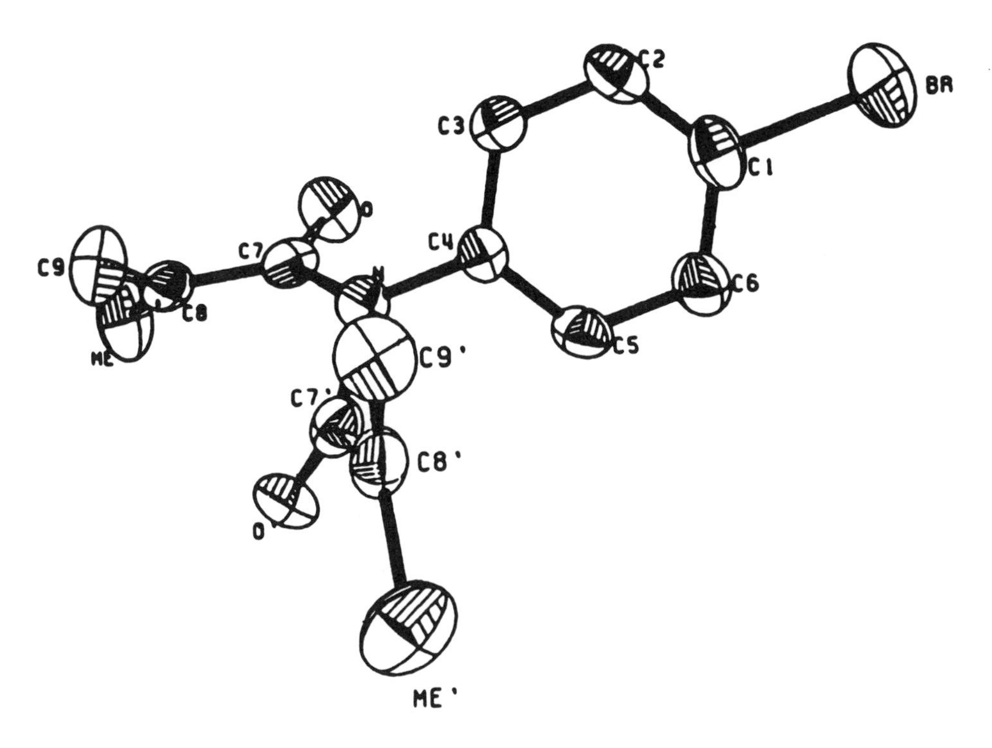

An ORTEP drawing of N-(p-bromophenyl) dimethacrylamide showing the atomic numbering and thermal ellipsoids.

(10–17)

The second intramolecular possibility, C8, was 0.4 Å further away, and the nearest potential intermolecular route was even further separated −C8′ at 4.753 Å. All of the carbons in the double bonds had at least one and, in some cases, two potential intramolecular reactive sites that were approximately 4 Å away. All other potential reactors (both intramolecular and intermolecular were at least 0.4–0.7 Å further away. Radical-initiated solution polymerization of the monomer at 58°C yielded cyclopolymer containing 80% five-membered and 20% six-membered ring. At 114°C, the cyclopolymer contained only 42% five-membered ring.

An investigation into the relative stabilities of the various conformations available to the monomer, *N*-(p-bromo-phenyl)dimethacrylamide, was made using a Perturbative Configuration Interaction method with Localized Orbitals (PCILO).[2-187] Conformers with a *cis-trans* and *trans-trans* orientation of the imide group were found that were fairly low in energy. Some energy lowering was observed for conformers in which the pi clouds of the monomer's two carbon-carbon double bonds could interact through space. One particularly low energy conformer was found by this investigation in which the two β-carbons of the double bonds were fairly close in space. If the monomer existed predominantly in this conformation in solution or had easy access to it by a rotation around a single bond, radical initiation of that conformer and a subsequent fast cyclization should result in the formation of five-membered rings in the backbone of the polymer.

Evidence has also been obtained that certain of the diallyamines and quaternary ammonium salts also yield cyclopolymer consisting largely of five-membered rings. On the basis of a ^{13}C NMR study of poly(*N,N*-diallyl-*N*-methylamine),[2-105, 2-106] it was concluded that five-membered rings were formed; however, when the allyl groups were substituted in the 2-position, mixtures of both five- and six-membered rings were formed. Also, it has now been shown by ^{13}C NMR and model compound studies that poly(diallyldimethylammonium chloride)[2-516] consists predominantly of five-membered rings linked mainly in a 3,4-*cis* configuration.

The microstructure of the cyclopolymers from divinyl formal, acetaldehyde divinyl acetal, and acetone divinyl acetal has been investigated using ^{13}C NMR spectroscopy.[2-285] By using a variety of model compounds, and from these, calculating the chemical shifts for carbon atoms in a range of possible structures for the cyclopolymers, it was shown that the polymers contained two types of structural unit. All the polymers contained the *cis*-4,-5-disubstituted-1,3-dioxolane ring as the predominant unit in the main chain with this structure being found also as a pendant group formed by an intramolecular chain-transfer mechanism. In poly(acetaldehyde divinyl acetal) the 2-methyl group may be *syn* or *anti*; the observed chemical shifts agreed remarkably well with those estimated for the *cis-syn* form. These results differ from those obtained by chemical methods on similar polymers.

Electron spin resonance studies of solid-state polymerization of irradiated *N*-(p-bromophenyl)dimethacrylamide and *sym*-dimethacryloyldimethylhydrazine have been published. The intermediate radicals formed in the cyclopolymerization of both monomers were studied. An initiation radical could be observed in each case.[2-401] A similar study of the cyclopolymerizability of *N,N′*-dimethacryloylmethylhydrazine and solvent effects on its polymerization has been published. Cyclopolymerizability of the monomer was explained by the difference in the reactivities of the two methacryloyl groups. Solvents involved not only the polymerization rate of *N,N′*-dimethacryloylmethylhydrazine but also the structure of the resulting poly(*N,N′*-dimethacryloylmethylhydrazine) to a small extent. The polymerization of *sym*-dimethacryloyldimethylhydrazine was not affected by solvents.[2-402]

In a continuing study of head-to-head polymers, *N*-phenyldimethacrylamide was cyclopolymerized to poly(*N*-phenyldimethacrylamide) containing 94% five-membered rings, but more importantly containing 6% six-membered rings in the cyclopolymer.[2-490]

To clarify the possibility of preparing a polymer that contained head-to-head (HH) methyl methacrylate units, radical cyclopolymerization of o-dimethacryloyloxybenzene was investigated. The intramolecular propagation proceeded mainly by a head-to-tail (HT) mechanism.[2-390] Preparation, thermal properties, and reaction of HH cyclopolymers by polymerization of *N*-substituted dimethyacrylamide derivatives to soluble HH cyclopolymers have been studied.[2-391] A recent study included preparation and characterization of poly(methyl acrylate) and poly(methylmethacrylate) consisting of HH and HT units through cyclopolymerization of acrylic and methacrylic anhydrides. The content of HH units of these HH/HT polymers was determined by ^{1}H NMR and ^{13}C NMR spectra.[2-326]

Consideration of the concept of I-strain[11, 1-155] in the cyclic structures formed during cyclopolymerization may shed some light on the comparison between five- and six-membered ring systems and their relative ease of formation. In five-membered ring systems, when hybridization at any one carbon is changed from sp^3 to sp^2, I-strain is

Table 10–7 Observations on I-strain in Five- and Six-membered Ring Systems

	Relative K_{eq}	Relative rate
Compound	Cyanohydrin formation	$NaBH_4$ reduction
Acetone	1.00	1.00
Cyclopentanone	3.33	15.4
Cyclohexanone	70	355

decreased. The transition from cyclopentane to cyclopentanone would reduce I-strain. In the six-membered ring, change of hybridization from sp^3 to sp^2 increases I-strain. Indicative of these internal strains are the relative rates of reduction of the two cyclic ketones with sodium borohydride, and the relative values of the equilibrium constants for cyanohydrin formation, as shown in Table 10–7.

Consideration of the fact that five-membered ring formation in the dimethacrylamide requires introduction of sp^2 hybridization at two carbons which should reduce I-strain relative to cyclopentane, whereas sp^2 hybridization at two carbons in the six-membered ring should increase I-strain relative to cyclohexane may offer a partial explanation for the five-membered ring preference in these cases. Similar comparisons have been referred to earlier.[3-157]

On the other hand, the relative differences in the calculated heats of formation[78,79] of the cyclic anhydrides and their open-chain dicarboxylic acids indicate that the six-membered ring should be favored over the five-membered ring from these considerations. These results are shown in Table 10–8.

These observations are consistent with the fact that the five-membered ring is favored at higher temperatures in cyclopolymerization of both acrylic and methacrylic anhydrides.[2-323,2-324,77]

The presence of five-membered ring in the polymer prepared by radical polymerization of methacrylic anhydride in benzonitrile at 160°C was confirmed by IR and ^{13}C NMR spectra. Formation of five-membered rings in preference to six-membered ring also increased with decreasing monomer concentration.[77]

Table 10–8 Heats of Formation of Some Acids and Their Anhydrides

Compound	Calculated heat of formation Kcal/mole	Difference (Acid-anhydride)
Propionic acid	(2×)112.1	
Propionic anhydride	161.6	82.6
Maleic acid	188.2	
Maleic anhydride	112.0	76.2
Succinic acid	225.6	
Succinic anhydride	144.7	80.9
Glutaric acid	230.6	
Glutaric anhydride	158.9	71.7

It is now well established that the ring size formed in free-radical cyclization reactions of 5-hexenyl radicals is influenced according to whether the reaction is rate or equilibrium controlled.[2-387] Five-membered ring formation in cyclopolymerization of 1,6-dienes is analogous to head-to-head polymerization in an ordinary vinyl monomer. It has been shown[2-387] by extrapolation that diallyl phthalate cyclizes via head-to-head propagation (HH) only to the extent of 15% under conditions favoring total cyclization, with this value decreasing linearly to 0% HH under conditions which favor zero cyclization. A non-linear pattern was observed, however, with diallyl *cis*-cyclohexane-1,2-dicarboxylate. This monomer showed a decrease in HH from 15% under conditions of 20% cyclization to 6% HH at 55% cyclization, followed by an increase in HH to 20% under conditions of 100% cyclization. Thus, it is obvious that more than one factor is operating to control HH propagation, which in turn controls ring sizes in cyclopolymerization. The difference, however, in the possible ring sizes in these monomers is between ten and eleven members, a range in which small differences are not so significant as between five- and six-membered ring systems. With many alkenyl radicals, the formation of a five-membered ring is kinetically favored over formation of a six-membered ring.[73] However, with 5-substituted 5-hexenyl radicals, the formation of a six-membered ring can be increased due to either the steric effect of the substituent causing a retardation in the rate of 1,5-cyclization[2-515] or a resonance-stabilizing effect that favors 1,6-cyclization.[66] 1,5-Ring closures of substituted 5-hexenyl and related radicals are stereoselective: 1- or 3-substituted systems afford mainly *cis*-disubstituted products, whereas 2- or 4-substituted systems give mainly *trans*-products.[73] Studies of the cyclopolymerization of variously substituted diallylamines suggest that these monomers also exhibit the characteristics of kinetic vs. thermodynamic control.[1-161]

The stereo-electronic requirements of the transition state for radical addition reactions has been proposed to explain the kinetic preference for the formation of five-membered ring products.[1-22, 1-173] This concept demands maximum overlap of the half-filled p-orbital with the vacant π orbital of the double bond. Approach of the radical must be along a vertical line from one of the carbon atoms of the double bond with the orbitals holding the three electrons being in the same plane throughout the reaction. Such conditions can be met by 1,5-cyclization of 5-hexenyl radicals but not by 1,6-cyclization. It has also been suggested that the predominance of *cis* addition in some cyclopolymerizations may be the result of hyperconjugative maxima of the p-orbitals with the alkyl CH_σ and σ^* orbitals, leading to a delocalized orbital of similar symmetry to the π^* orbital, which results in an attractive interaction between the alkyl substituent and the double bond (10–18). This model for the cyclization reaction can explain the increasing preference for cyclopolymers with six-membered rings as bulky 2-substituents are introduced to one or both allyl groups of *N,N*-diallylamines.

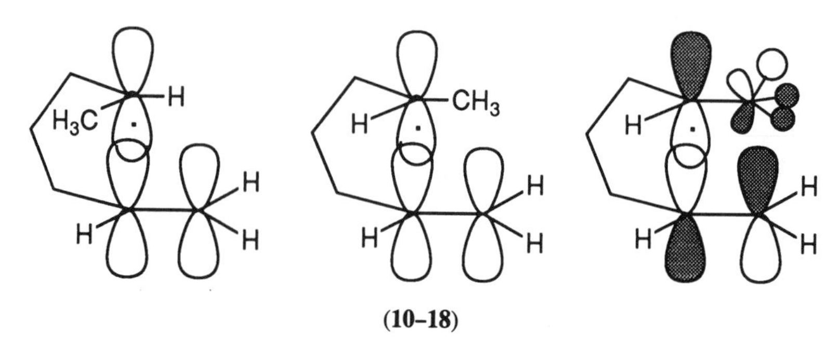

(10–18)

Other workers have provided support for this hypothesis to explain the cyclization mechanism by examination of molecular models,[2-484, 2-523] calculations of geometrical probability factors,[2-484, 80] and molecular orbital evaluations.[16, 75, 81] Calculations of probability factors indicated that steric effects are important in determining whether cyclization of β-substituted diallylmethylammonium ions results in five- or six-membered rings.[80] Evidence was obtained for the direction of intramolecular attack. The ring size in cyclopolymers will be influenced by the nature of the cyclization reaction; it is therefore important to understand the energetics and kinetics of model reactions. However, there are indications that correlation between models and polymers is unsatisfactory.[1-161] For example, the cyclization of *N,N*-di(2-ethylallyl)methylamine produced a six-membered ring, low molecular weight product exclusively, but the cyclopolymer contained 40% of a five-membered ring structure.[2-205] These results may reflect the relative abilities of the intermediate radicals to undergo propagation and/or termination reactions. In addition, the factors controlling the cyclization reactions may be influenced by the growing chain, such that polymer micro-structure is related to molecular weight. However, further studies are needed along these lines.

The cyclolinear structure of polymers of *N,N*-diallyl-*N,N*-dialkylammonium halides has been further studied. The nature of the *N*-alkyl substituents, the anion of the quaternary ammonium salt, and the solvent nature affect the structure and conformation of cyclolinear poly(*N,N*-diallyl-*N,N*-dialkylammonium halides). An increase in monomer concentration from 0.25M to 3M in preparation of poly(DADMAC) and poly(*N,N*-diethyl-*N,N*-diallylammonium chloride) in aqueous solutions and of poly(DADMAC) in methanol solutions led to no changes in the polymer structure.[2-121]

Justification of this preference for five-membered ring formation has been based on both electronic[1-155] and steric[1-122, 11] factors. A suggestion that there may be an electronic interaction[1-155] between the initially formed radical and the neighboring double bond of the diene has received considerable attention.[2-484] The formation of methylcyclopentane from the reaction of 5-hexenyl mercaptan with triethyl phosphite was explained[49] as arising from the attack of the radical at the more accessible end of the double bond, the process being irreversible. It was noted[2-484] that for approach of the radical to the double bond, with the p-orbitals in a common plane, that formation of five-membered rings would be less sterically hindered. One of the terminal hydrogens lies in the nodal plane directly between the radical carbon and the carbon on the terminal end of the double bond, thus hindering six-membered ring formation. No such steric interference exists for approach of the radical to the other end of the double bond, leading to five-membered rings.

Considerable work dealing with small molecule cyclizations has been done, and much of this work has been summarized in an earlier review article.[2-523] In a study of cyclization of 1-substituted 4-hexenyl radicals, it was shown that as substitution in the 1-position by radical stabilizing groups (one or two -CN or $-C(O)OC_2H_5$ groups) was increased, the mixture of cyclized products changed from nearly pure cyclopentane to nearly pure cyclohexane derivatives. Higher temperatures favored cyclohexane formation. Also, it was shown that the cyclization reactions were reversible when C1 was disubstituted. It had been shown earlier, however, that in the case of the primary hexenyl radical, the cyclization step is irreversible.[1-122] On this basis, it was proposed that the cyclopentane product is the kinetically preferred product, whereas the cyclohexane product is preferred via thermodynamic control, with the energy of activation for cyclization being higher for the cyclohexane derivatives.

The majority of the radical cyclizations reported have been accomplished by permitting alkyl radicals to react with a mono-substituted double bond, thus producing the intermediate, non-resonance stabilized primary or secondary radical.[2-527] As stabilization of the generated radical increased, the ratio of six-membered ring increased. Also, deuteration studies showed that in the cyclization of 1-naphthylbutyl radicals two different mechanisms were involved. These results have been used as a mechanistic tool.[2-523]

The present understanding of cyclopolymerization would indicate that the structures of cyclopolymers are highly dependent on the nature of the monomers and the reaction conditions. Predictions of structures based upon thermodynamic control of the cyclization reaction can be misleading since kinetic products are often found. An important driving force for the cyclization reaction may be an intramolecular interaction, particularly in some cases, where a charge-transfer interaction can occur, or is induced, and in cationic-initiated reactions. However, the predominance of cyclization over intermolecular propagation can be ascribed to a smaller decrease in activation entropy, which compensates for the unfavorable difference in activation energies. Further studies on the kinetics and energetics of the cyclization reaction during polymerization are necessary before a completely acceptable mechanistic explanation is forthcoming.

Theoretical Considerations Related to Cyclization and Cyclopolymerization

The frequency factor in cyclization reactions has been discussed in detail.[82] Also, orbital theory considerations supported preference for five-membered ring formation in cyclizations.[83]

The proposal that a secondary propagating radical should be predominant over a primary one is consistent with "head-to-tail" enchainment in radical-initiated polymerization of vinyl monomers, an arrangement which has been unquestionably established as the major mode of enchainment in a wide variety of vinyl monomers.[2-519]

An explanation for the effects of substituents on ring size in cyclopolymerization was presented, based on frontier orbital theory. The theory works well in explaining the ring size in cyclopolymerization except when steric hindrance is predominant in controlling the course of cyclization.[2-486]

It has been reported that the overall activation energy in free radical initiated polymerization of diallyldimethylsilane was determined to be 9 kcal/mole lower than for allyltrimethylsilane.[84]. However, as pointed out earlier, these results may have been misinterpreted.[1-18]

Small Molecule Cyclization as Model Systems for Cyclopolymerization

The ease of formation and stability vs. ring size has been discussed in detail. It was pointed out that because the bond angle of a planar cyclopentane ring is 108°, very close to the tetrahedral carbon angle, that strain in this ring should be minimal.[1-123] However, the puckered ring structures, chair, boat, etc., of cyclohexane were not considered in this context; only planar rings were considered.

As further evidence for the proposal that cyclopolymerization may be explained satisfactorily *via* stabilization through non-conjugated chromophoric interactions between the olefinic bonds of the monomer, the results that the double bond of 5-hexen-1-yl p-nitrobenzenesulfonate exerted a weak nucleophilic acceleration for solvolysis in acetic acid leading to partial formation of cyclohexyl acetate were cited.[1-121] The influence from the double bond depended considerably on the geometry of the molecule.

Solvolysis of 5-hexenyl p-nitrobenzenesulfonate in 98% formic acid containing sodium formate proceeds with participation of the olefinic bond at a rate which is about twice that of the formolysis of the hexyl ester.[85] The product, after hydrolysis, consisted of 68% cyclohexanol and 26% hexenol. Formolysis of the p-nitrobenzenesulfonates of 4-pentenol and 6-hexenol proceeded with negligible double bond participation to give mainly the products of direct substitution.

The dienol underwent facile stereoselective cyclization on treatment with formic acid (10–19) to give *syn*-$\Delta^{1,9}$-6-octalol (10–19b).[86] The mechanistic pathway of the cyclization was discussed.

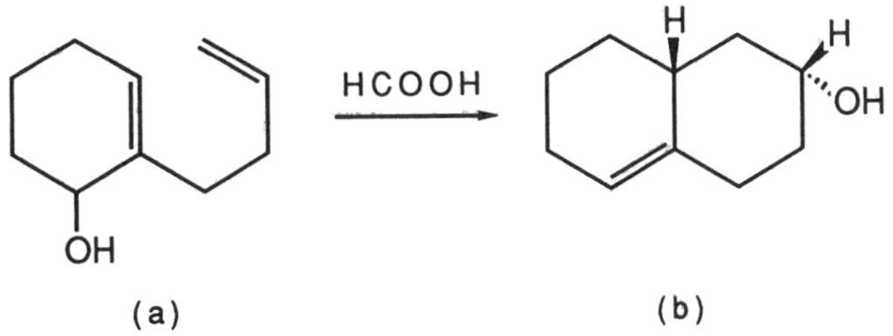

HCOOH

(a) (b)

(10–19)

The kinetics for the cyclization of radicals $CH_2{=}CH(CH_2)_nCH_2\cdot$ (n = 3, 4, 5) showed a preference for formation of cycloalkylcarbinyl radicals rather than cycloalkyl radicals because of steric demand in the transition state.[87] Also, cyclization of radicals $CH_2{=}CH(CH_2)_3CHR\cdot$ (R = H, C_3H_7), $CH_2{=}CHCH_2OCH_2CHMe\cdot$, and $CH_2{=}CH9CH_2)_4CHMe\cdot$ preferentially gave *cis* compounds, probably because of orbital symmetry considerations. The effect of methyl groups on β-scission of cyclobutylcarbinyl radicals was due to steric effects rather than conjugative effects. No evidence of homoconjugative activation of double bonds, i.e., in norbornadiene, toward free radical attack was observed.

It has been shown that bicyclic systems (See Structure 3–28) such as those shown can be formed by radical cyclization of appropriate radicals.[3-217] The rates and stereochemistry of the ring closures were determined and rationalized.

A theoretical study was undertaken of ring closure of a variety of alkenyl, arylalkenyl, alkenylvinyl and similar radicals.[3-218] The method involved the application of MM2 force-field calculations to model transition structures for which the dimensions of the arrays of reactive centers have been obtained by MNDO-UHF techniques. The results which generally accord with guidelines based on stereochemical considerations, show excellent qualitative and satisfactory quantitative agreement with experimental data. The method was successfully applied to complex systems including ring closure of alkylperoxy radicals, and formation of the triquinane system by three consecutive cyclizations.

A comparison of the dynamical behavior of closed ring and chain molecules in dilute solutions has been made.[88] The normal Fourier modes of a closed ring polymer constitute a complete basis for decoupling the hydrodynamic equations. The method was found to be exact for ring polymers and approximately correct for linear polymer chains. Ring size was not considered.

Evidence was obtained and arguments presented in support of the hypothesis that 2-Δ^3-butenylcyclohexyl cation (a) and related systems undergo ring closure stereoselectively to form preferentially the *trans*-decalin ring system (b).[89] In accord with this view, the cation, a, was not involved in the acid-catalyzed cyclization of Δ^3-butenylcyclohexene which, in contrast, gave mainly *cis*-decalin derivatives, and was therefore regarded as a concerted protonation cyclization (10–20).

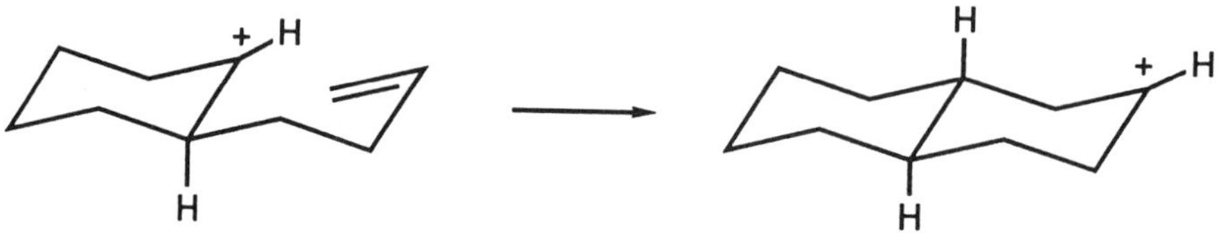

(a) (b)

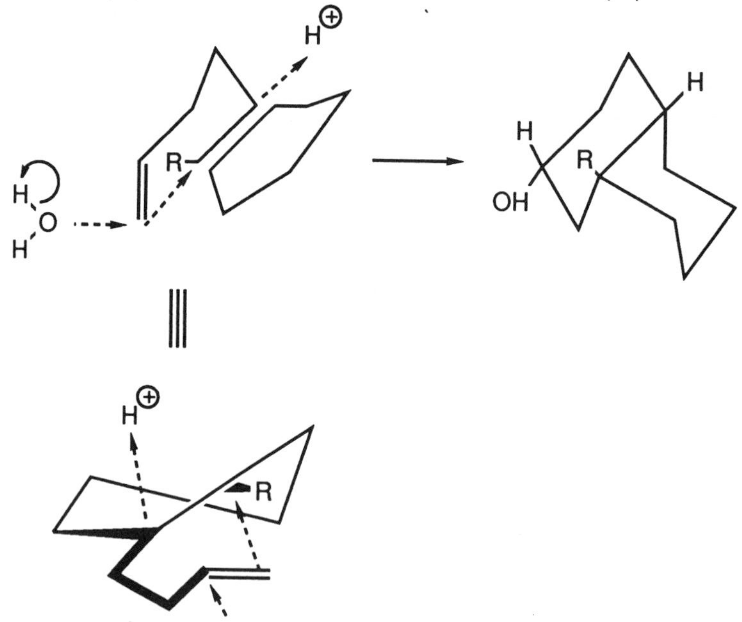

(10–20)

Investigations Dealing with Five- or Six-Membered Ring Formation in Cyclopolymerization

The acid-catalyzed cyclizations of *trans, trans*- and *cis,cis*-2,6-octadiene were investigated.[90] The stereochemistry of the products shows that the cyclization process is concerted with proton attack and follows the stereoelectronic predictions made for terpene biosynthesis. A small fraction of the time the acquisition of the nucleophile is clearly concerted with the preceding steps and leads to the theoretically expected product.

Polymerization of *N*-substituted dimethacrylamides resulted in cyclic polymers containing predominantly five-membered rings, while deamination of the corresponding polymethacrylamide formed with repeating six-membered cyclic units.[2-146]

The structures of the poly(vinyl alcohol) derived from poly(divinyl *n*-butyral) and from poly(vinyl formate) were compared with ordinary poly(vinyl alcohol). That derived from the poly(divinyl *n*-butyral) was different from the ordinary sample; however, that derived from poly(vinyl formate) showed nearly the same spectra as the ordinary atactic polymers.[2-310]

A sequence distribution study of methyl methacrylate-styrene copolymers derived from methacrylic anhydride-styrene copolymers has been published.[7-97]

Cationic polymerization of methallyl vinyl ether with boron trifluoride-diethyl ether complex at −78°C in toluene gave isotactic poly(methallyl vinyl ether).[91] The molecular weight decreased with increasing temperature. Only a viscous liquid of lower tacticity was obtained at 0°C. Crosslinking did not occur but did occur to some extent when n-heptane was used as solvent. No reference to possible cyclopolymerization was made. Five- or six-membered rings would be predicted depending on whether head-to-tail or head-to-head cyclization had occurred.

Divinyl esters of various dibasic acids have been cyclopolymerized and the polymers studied.[2-292] Hydrolysis of the esters to poly(vinyl alcohol) gave useful information for interpretation of the propagation mechanism.

Generation of o-substituted aryl radicals and identification of the cyclized products showed that 5,6- or 6,7-olefinic bonds with respect to the radical center underwent 1,5- or 1,6-cyclization, respectively, to give the thermodynamically less stable products.[92]

It has been shown that the rate of cyclopolymerization of *N*-propyldimethacrylamide, at 60°C in the presence of azobisisobutyronitrile, is less than that of benzyldimethacrylamide.[2-171] The monomer reactivity ratios for copolymerization of the propyl derivative with methyl methacrylate, styrene, acrylonitrile, and vinyl acetate were 0.25 and 3.18, 0.20 and 0.80, 0.60 and 0.40, and 7.0 and 0.02, respectively.

Cyclopolymerization of *N*-allyl-*N*-methylmethacrylamide was investigated and it was shown that radical initiation over the -78 to 120°C range gave polymers with 88–93% cyclic units consisting mainly of five-membered rings with some six-membered rings, as well as pendant methacryl groups.[3-79]

Radical polymerization of divinyl ether yielded partially cyclized polymers which contained a five-membered ring monocyclic unit and a dioxabicyclo[3.3.0]octane unit in a 1:1 ratio.[4-65]

A comparison of the observed with the theoretical extent of acetalization at the gel point of poly(vinyl alcohol) has shown that the contribution of intramolecular acetalization to the overall reaction is too large to be neglected and that two types of intramolecular crosslinking (both ineffective for gelation) were operative, one producing small rings near the chain ends and the other producing large rings.[93] A transition from gel to solubility took place at a temperature where equilibrium was reached between acetalization and deacetalization. From known rate constants for intramolecular and intermolecular acetalization, as well as that for deacetalization, the extent of reaction at the solubility-gel transition temperature was determined.

Variations in polymerization conditions (e.g., solvent, initiator type) did not affect the structure of poly(DADMAC) which gave exclusively five-membered rings with *cis*-

trans ratio of 6:1.[94] Temperature effects apparently were not studied. Predictions that the more thermodynamically favored six-membered ring should be formed to some extent at low temperature appear to be reasonable.

A study of polymerization of *N,N*-dimethyl-3,4-dimethylene-pyrrolidinium bromide, both in presence and absence of initiators has been published.[2-197] In aqueous solutions of 10–50% monomer, polymerization at 60° showed that conversion time decreased with increased monomer concentration; the molecular weight of the polymer increased with monomer concentration.

The factors influencing intramolecular addition in radical cyclopolymerization of diallyl and dimethallyl dicarboxylates, acrylic and methacrylic anhydrides, and allyl α-substituted acrylates were investigated and discussed from the standpoint of thermodynamics.[2-388] The selectivity of reaction mode of intramolecular head-to-head or head-to-tail addition was given particular consideration.

A study of cyclopolymerization of divinyl formal confirmed that the polymer varied with polymerization conditions.[2-325] The polymer contained the *cis*-dioxolane ring as the major structure, along with the *trans* ring and the branched structure. The unsaturated unit and the six-membered unit were not formed under any conditions.

Cyclopolymerization of 4-(diallylamino)pyridine hydrochloride via free radical initiation gave water-soluble polymers.[2-218] Ring size was not determined.

Radical polymerization of acrylic propionic anhydride as a monoene counterpart of acrylic anhydride was investigated in order to obtain supporting evidence for a proposed mechanism based on polymerization equilibrium for the cyclopolymerization of acrylic anhydride. The results suggested increased significance of the depropagation reaction, equilibrated with propagation, under polymerization conditions favorable to five-membered ring formation during cyclopolymerization.[95]

The temperature dependency of the cyclization constant in radical cyclopolymerization of nonconjugated dienes has been evaluated under a variety of conditions.[96] In studying the monomers, acrylic and methacrylic anhydrides, no linear relationship between K_c and $1/T$ was observed. Cyclization was accelerated at elevated temperatures. The K_c values also increased with decreased monomer concentration and increased solvent polarity. These increasing dependencies of K_c are ascribed to the increased significance of depropagation.

Poly(*N*-phenyl-3,4-dimethylenepyrrolidine) containing pyrroline rings was prepared and oxidized to polymer containing pyrrole rings.[97]

Radical cyclopolymerization of diallylsilanes and triallylsilanes has been reinvestigated.[98] Diallyldimethylsilane and diallylmethylphenylsilane gave polymers which were soluble, and very little evidence of residual unsaturation was obtained, consistent with earlier reported results. Polymer ring sizes were studied with the aid of model compounds. It was concluded that these monomers undergo only head-to-tail intramolecular cyclization to form a six-membered ring in the repeat unit.

In a continuing study of the structural chemistry of polymerizable monomers, the effects of the crystal and molecular structure of *N*-methyldimethacrylamide on the initial stage of solid-state cyclopolymerization were studied.[2-242] The shortest spacing between vinyl carbon atoms was 2.908 Å; the planes of the two intramolecular vinyl groups assumed a dihedral angle of 40.6°. The location of the vinyl carbon atoms allowed an intramolecular closure in the initiation process, i.e., tail-to-tail cyclization as observed.

Investigations Concerned with Large Ring Formation in Cyclopolymerization

In a temperature dependence study of radical cyclopolymerization of o-divinylbenzene, it was shown that a seven-membered ring propagating radical was the predominant radical observed.[99] Earlier work supported five-membered ring formation.

The cyclopolymerization of o-vinylphenyl acrylate with a conventional radical initiator and with alkylaluminum chlorides gave a polymer containing residual acrylic and vinyl double bonds.[3-105] The polymerization proceeded through head-to-tail additions with formation of a cyclic seven-membered ring, the sequence of addition being random for the radical polymerization and alternating for polymerization in the presence of the Al compounds. The latter cyclopolymerization was explained by a molecular-complex mechanism.

A high proportion of head-to-head addition was observed in the radical polymerization of allyl acetate, ranging from 7.5% at 40°C to 19% at 180°C.[100] Head-to-head addition in a vinyl monomer is analogous to five-membered ring formation in cyclization of a 1,6-diene.

Cyclopolymerization of diallyl phthalate produced a product consisting exclusively of an 11-membered ring and not the 10-membered ring expected.[2-386] The lack of intramolecular head-to-head addition in the radical copolymerization was attributed to the fact that the occurrence of head-to-head addition, i.e., the formation of 10-membered rings, requires the sacrifice of resonance stabilization between CO double bonds and aromatic benzene rings because of the loss of coplanarity.

Polymers having large rings and displaying significant phase-transfer catalytic activity were obtained from oligomeric poly(oxyethylene) divinyl ethers.[101] The ring sizes were not determined, nor was head-to-head vs. head-to-tail ring closure postulated. However, ring size predictions would lead to cyclic structures having more than twelve atoms.

Large ring-containing polyethers were produced by cationic cyclopolymerization of α,ω-divinyl ethers of oligooxyethylenes.[102] The divinyl ether monomers were prepared by treating oligomeric polyethylene glycol ($n = 2$–5) with acetylene in presence of potassium metal as catalyst. An unusual chain shortening occurred in the conversion of the polyethylene glycols to the vinyl ethers. This probably resulted from an equilibration of the ether glycol in presence of the catalyst and scrambling of molecular chain lengths. Ring sizes would be expected to be twelve-membered or larger.

Based on the extensive effort that has been expended on ring size studies in cyclopolymerization, some general conclusions can be drawn. The preference for five-membered over six-membered rings is now well established in cyclization of 5-hexenyl radicals as well as in cyclopolymerization of many 1,6-heptadienes. It is now known that ring size is influenced by whether the reaction is rate- or equilibrium controlled. The stereoelectronic requirements of the transition state for radical addition reactions (10–18) have been proposed to explain the kinetic preference for the formation of five-membered ring products. Other factors which are obviously important and influential are the steric effects and many efforts have been made to offer consistent and satisfactory explanations for these observations. Steric effects have been evaluated through examination of molecular models, calculations of geometrical probability factors, and molecular orbital calculations. The ring size in cyclopolymers will vary according to the energetics and kinetics. It is, therefore, important to understand these parameters for model systems. It has been shown that in certain systems the ring size can be controlled from almost exclusively five-membered to almost exclusively six-membered rings by controlling the experimental conditions. Factors which promote each ring size have been identified and can be controlled. A comparison between resonance stabilized and non-resonance stabilized intermediate radicals (10–21) has shown that the former situation favors the thermodynamically more stable ring over the kinetically favored but less stable ring.

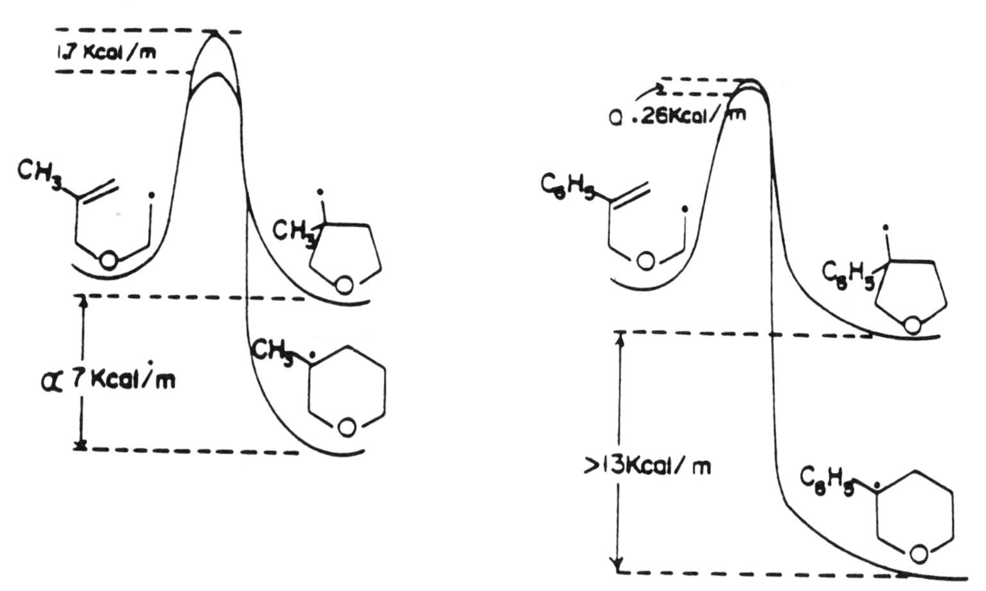

Probable reaction pathways for cyclization of radicals having methyl and phenyl stabilizing groups, respectively.

(10–21)

INTERPRETATIONS OF MECHANISTIC PROPOSALS

A satisfactory theory for cyclopolymerization must adequately account for the following observations: (a) almost all nonconjugated dienes and other similar polymerizable structures studied to date undergo cyclopolymerization in preference to cross-linking or other modes of propagation in contrast to earlier and well-accepted theories of crosslinking;[2-1] (b) such dienes invariably undergo polymerization at a higher rate than their respective monoene counterparts; (c) the cyclization step has invariably been shown to be of higher activation energy than propagation without cyclization; (d) the cyclization step often leads to propagation via the less thermodynamically stable reaction intermediate, leading to the less thermodynamically stable ring structure.

Numerous examples where the rates of cyclopolymerization have been shown to be significantly greater than those for homopolymerization of their corresponding monoolefinic counterparts have been published.[1-120] Relative rates up to 590 have been reported.[3-157]

The experimental evidence indicates that, generally, in cyclopolymerization the intramolecular propagation step is highly favored over the intermolecular step and that the reactivity of diene monomers is greater than that of corresponding monounsaturated compounds.[103] Various theories have been proposed to explain the selectivity of the intramolecular propagation step in the reaction but none have received widespread acceptance.

The Electronic Interaction Theory—Bathochromic Shifts in Ultraviolet Spectra of 1,6-Dienes

It has been proposed[1-17] that an electronic interaction may exist between the unconjugated ethylene double bonds or between the reactive species generated by initiator attack and the intramolecular double bond. Such an interaction would perhaps have a stabilizing influence on the excited state of the molecule, thus providing an energetically favorable pathway from diene to cyclic product. A formal analogy to conjugated dienes,[1-155] in which an electronic interaction between the adjacent double bonds is generally accepted to account for the bathochromic shift in their UV spectra relative to their monoolefinic counterparts (10–22), and a "through-space" electronic interaction in 1,5-and 1,6-dienes has been made.[1-155] Such interactions should be observable as bathochromic shifts in the ultraviolet spectra of these dienes. Certain abnormalities[1-118] in the UV spectra of some monomers have been observed. For example, the *cis*-and *trans*-

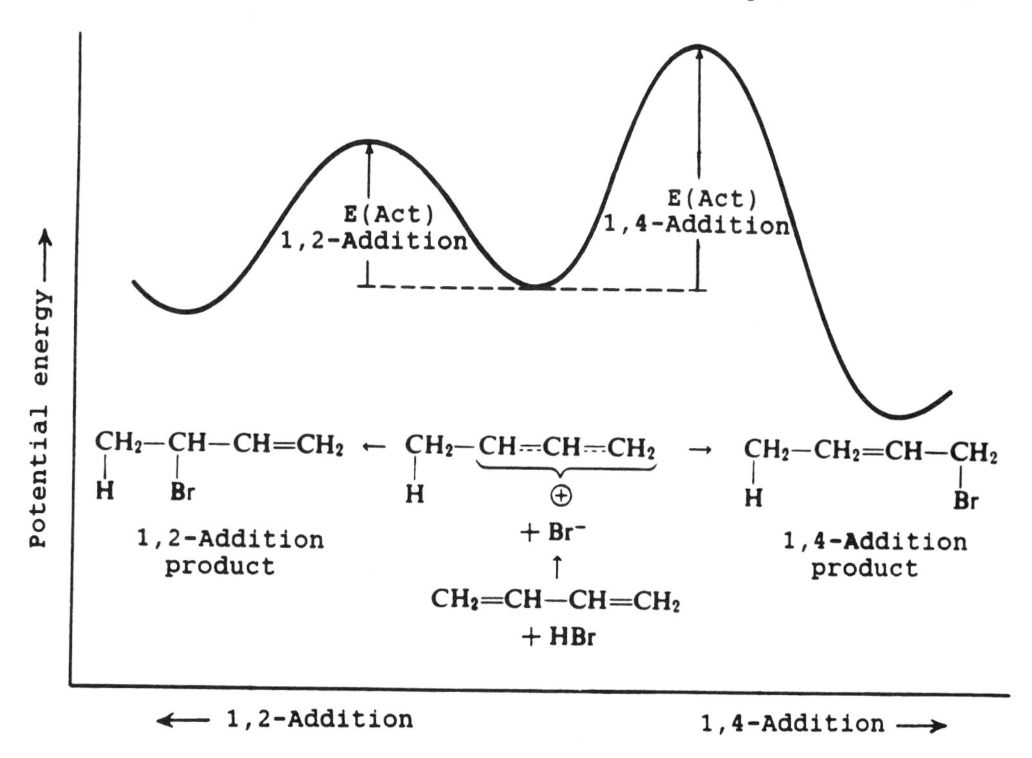

Potential energy changes during ionic addition of HBr to 1,3-butadiene: 1,2-vs. 1,4-addition.

(10–22)

isomers of 1,3,8-nonatriene exhibited bathochromic shifts of 3.5 and 5.0 nm respectively from the predicted properties based upon the mono-olefinic counterparts. These data were interpreted in terms of an excited state interaction; however, data were presented[2-51] on the UV spectra of certain dienes and tetraenes which permitted an explanation on the basis of interaction in both the ground and the excited states. Thus, the polymerization would involve attack of an initiating species on a molecule which is already in a favorable "cyclic" conformation because of a ground state interaction between the double bonds (10–23). Analogous predictions could also be made on the basis of the Franck-Condon principle.[7] This would explain the great tendency for nonconjugated dienes to cyclopolymerize.

(10–23)

The three orbitals for a 3-atom structure consist of one bonding and two degenerate antibonding orbitals (Table 10–9). Electrons in antibonding orbitals destabilize the system and raise its energy.[104]

For the three-centered structure, stability should decrease as follows: carbocation > radical > anion. If this three-centered structure is considered a transition state, then the energy of activation should increase in going from the cation to the radical to the anion. The anion (c) (Table 10–9) has been classified as antiaromatic.[105, 106]

In the carbocation case an electronic interaction as postulated would favor cyclization over intermolecular propagation by lowering the energy of activation of the cyclization reaction. There should be less stabilization in the radical case and none would be expected in the anion case. The carbocation (a) has been shown to be stable, and similar ions have been observed in NMR.[107]

The ESR spectrum of the 5-hexenyl radical has been studied.[54] At −78°C the 5-hexenyl radical was observed, and upon warming to −35°C the cyclopentylmethyl radical was detected. At intermediate temperatures both radicals were observed. A nonclassical radical was not observed. The cyclization rate constant for the previous reac-

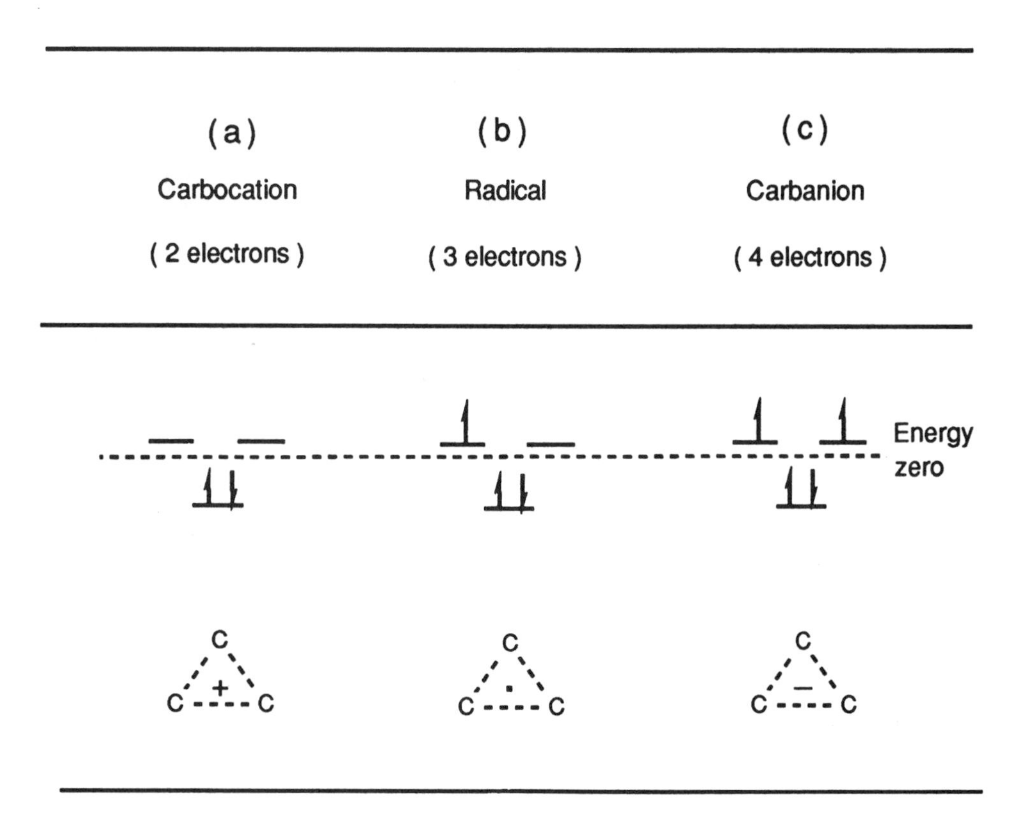

tion has been reported to be 1×10^5/sec. The addition to heptene-1 by ethyl radicals in the gas phase yielded a rate constant of $1 \times 10^3\ M^{-1}$/sec.[2-520, 108] By use of these data, an "effective" double-bond concentration of 100 M for intramolecular cyclization could be calculated.

The ring closures of Δ^3-cyclopentenyl ethyl carbocation, radical, and anion have been investigated.[68] Both the cation and radical underwent facile ring closure while the anion did not close.

The explanation for these results was that a tricentric transition state could be assumed for the radical or cationic systems—three and two electrons, respectively. In the four-electron system, or the anionic case, a linear transition state was postulated for the three centers involved. This is sterically impossible and, therefore, no reaction occurred.

It seems questionable that this is the only possible explanation for the failure of ring closure in the case of the Δ^3 cyclopentenyl ethyl carbanion. The activation energy for cyclization may merely be too high—whether in a tricentric or linear transition state. Supporting this view is the fact that no cyclopolymerizations have been reported for unsubstituted diene systems using anionic initiators. For example, 1,6-heptadiene has not been cyclopolymerized anionically, although numerous studies have been conducted using metal-coordination initiation. 2,6-Diphenylheptadiene-1,6, however, undergoes anionic cyclopolymerization under mild conditions.[1-38] The 2,6-diphenylheptadiene-1,6 system also undergoes cyclopolymerization with radical and cationic systems. The products obtained by all three initiating systems were identical.

It is interesting to note that all confirmed anionic cyclopolymerizations seem to involve systems in which the resulting carbanions are stabilized by adjacent phenyl, carbonyl, cyano, or other functionality. These groups delocalize electron density, making the double bonds of the diene more susceptible to attack—consistent with the electronic mechanism.

It has been observed[1-118] that the absorption maxima for methacrylic anhydride were shifted bathochromically from ethyl, allyl, and methallyl methacrylate. The bathochromic shift was in the direction expected if an excited state interaction of the nonconjugated double bonds existed. The wavelength maximum and extinction coefficient for methacrylic anhydride were considerably different from that obtained later.[20] However, the later results still indicated a bathochromic shift between methacrylic anhydride and its esters.

An e value of $\cong 1.2$ has been reported for methacrylic anhydride by using ^{14}C reagents to study its radical initiated homopolymerization and its copolymerization with styrene, methyl methacrylate, and methyl acrylate.[7-92] Methacrylic anhydride was less reactive than methyl methacrylate towards the benzoyloxy radical and this lower reactivity was correlated with the large e values. There was a large number of cyclized methacrylic anhydride units in its homopolymers and copolymers with styrene and the number of cyclized methacrylic anhydride units in the homopolymers prepared from low concentrations of methacrylic anhydride was unexpectedly high.

The homopolymer of methacrylic anhydride appears to possess a certain degree of stereoregularity. In a study of models of stereoregular polymers, it was shown that spectra of the telomonomer and telodimer of vinyl acetate are different from that of the polymer. It was also shown that poly(methacrylic acid) derived *via* hydrolysis of cyclopoly(methacrylic anhydride) is slightly crystalline. Poly(divinylformal), prepared by photoinitiated polymerization, also gave an x-ray pattern showing partial crystallinity.[109]

In support of the earlier proposed mechanism for cyclopolymerization, further evidence for non-conjugated chromophoric interactions was presented. Certain tetraenes, e.g. 1,3,6,8-nonatetraene, in comparison with 1,7-nonadiene, produced a bathochromic shift in the ultraviolet spectrum of 10.6 nm, in addition to two additional maxima over a 35 nm range not shown by the diene.[1-119]

A study of methacrylic propionic anhydride and vinyl methacrylate, which also exhibited bathochromic shifts, indicated that the shift was due to added electronic stabilization through resonance of the nonbonded electrons on the acyl oxygen with the unsaturated unit attached to it. However, a more recent study[110] found no evidence of shifts in the UV spectra of certain di-unsaturated esters and ethers. The data for methallyl methacrylate showed essentially no bathochromic shifts. However, a UV study of the series—allyltrimethylsilane, diallyldimethylsilane, triallylmethylsilane, and tetraallylsilane—indicated a progressive bathochromic shift with increasing number of allyl groups. This was interpreted in terms of an interspacial interaction between the double bonds of the allyl groups.[111,112] Moreover, the radical initiated polymerization of allylsilanes led to the conclusion that the total activation energy for polymerization of diallyldimethylsilane was about 9 kcal/mole, less than that for triallylmethylsilane.[84] However, the validity of these data has been questioned.[1-18]

It has been suggested that ground state interaction is not a significant factor in the cyclopolymerization of *cis*-1,3-diisocyanatocyclohexane.[113] In 1,3-disubstituted cyclohexanes, the diaxial conformation of the *cis*-isomer is, energetically, the least-favored conformation because intramolecular repulsive interactions are greater in this conformation than in the di-equatorial one. For example, in *cis*-1,3-dimethylcyclohexane at room temperature, less than one molecule in 10,000 occupies the diaxial conformation. An interspacial electronic interaction would be expected to decrease the energy of the diaxial conformation, thus increasing the proportion of this conformer in the equilibrium mixture. For the process, $\Delta G°$ value could be considered as due to two NCO-H interactions (0.5 kcal/mole),[114–117] one (NCO-NCO) repulsive interaction and one (NCO-NCO) stabilizing interaction, if it exists. Values of known 1,3-diaxial interactions range from 1.6 (CH_3-OH) to 3.7 (CH_3-CH_3) kcal/mole. A value as large as 3.7 kcal/mole would make the $\Delta G°$ value for the process (10–24) equal to 4.2 kcal/

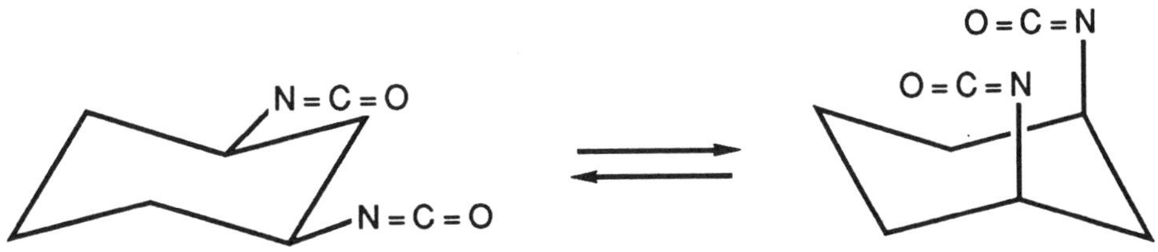

(10–24)

mole in the absence of any stabilizing interaction. A stabilizing interaction as significant as 2.5 kcal/mole would reduce it to 1.7 kcal/mole (5% diaxial conformer—a detectable proportion). The (NCO-NCO) repulsive interaction may not be as large as (CH_3-CH_3); if it were 2.5 cal/mole a stabilizing interaction of 1.3 kcal/mole could be comfortably detected. From NMR data on *cis*- and *trans*-1,3-diisocyanatocyclohexane, the conformational equilibrium in the *cis*-isomer was determined. The results indicated that the only detectable conformation of *cis*-1,3-diisocyanatocyclohexane was that in which the isocyanato groups were in equatorial positions. This was taken as evidence that a ground state interaction between the isocyanato groups in this monomer, which readily cyclopolymerizes, is not a significant factor in the cyclopolymerization mechanism.

From wave mechanical methods the predicted delocalization energy for bicycloheptadiene[118] in the ground state was zero. Heats of hydrogenation studies also gave no indication of homoconjugative stabilization of bicycloheptadiene.[119]

The fact that these monomers, which are structurally favorable for interaction to occur, do not exhibit interaction in the ground state, makes it appear unlikely that such an interaction is important in cyclopolymerization.

However, a photoelectron spectral study of α,ω-diolefins has provided evidence that there is crossing of the π-level splittings in support of a conformation of 1,5-hexadiene in which overlap of the π-bonds occurs, not parallel, but at 90° with respect to each other. Also, an extended Hueckel MO calculation on different conformations of 1,5-hexadiene indicated a through-space dominated interaction.[2-522] Also, in the case of methacrylic anhydride, the bathochromic shifts could be explained alternatively by resonance stabilization of the non-bonded electrons on the anhydride oxygen with the carbonyl groups.[2-317]

The photoelectron spectra of tetravinyl tin were reported and assigned on the basis of comparison with those of other group IVB tetravinyl derivatives, and by correlation of the electronic structures of ethylene and SnH_4.[120]

There is a reasonable basis for assuming that an interaction can exist between monomer double bonds in the excited state and between the ion or radical and the pendant double bond. Bicycloheptadiene, which shows no interaction in the ground state, has been described as possessing excited state π-electron interactions of the type shown in Structure 10–25.[118] The bond order between carbon 1 and 2 (3 and 4) is 1.5 and

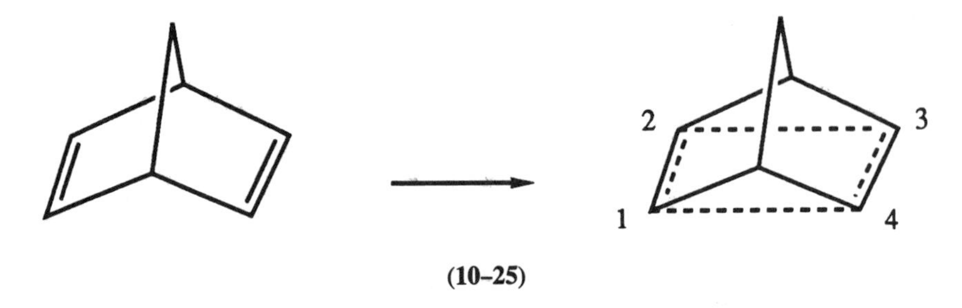

(10–25)

between carbons 1 and 4 (2 and 3) is 0.12. Overlap of π orbitals in the photoexcited state, producing unexpected absorption maxima, has also been observed with other compounds.[121–123] Chemical abnormalities appear in the reactivity of bicycloheptadienes and other compounds toward electrophilic,[124, 125] free radical[126, 127] and thermal additions,[128] and towards irradiation.[129, 130] More recently evidence has been provided for the resonance stabilization of carbanions.[131] While π-orbital interactions certainly occur with both polar and photochemical activation, the results of free radical experiments require the existence of at least two intermediate radicals.[127] However, they do not rule out the possibility that one of these may be "nonclassical" or that a third, presumably "nonclassical," radical exists in addition to the two "classical" ones.

The cyclization of unsaturated cyclic radicals has been the subject of a number of recent investigations[1-122, 49, 55-58, 62, 132, 7-87] and is of interest since the direction of ring closure varies strikingly with changes in radical structure. With highly substituted radicals the formation of cyclohexyl derivatives predominates while, somewhat surprisingly, the simple 5-hexenyl radical cyclizes almost exclusively to the cyclopentylmethyl radical, and five-membered ring products are also preferred in many cases. Although the available evidence does not allow a satisfactory explanation of the results, it seems that they could be interpreted by the existence of a "nonclassical" radical, the choice of product being determined by the steric requirements of the system.

For those open-chain systems,[2-51, 111] however, which show bathochromic shifts in their UV spectra, it must be assumed in accordance with the Franck-Condon principle[7] that the molecule is preoriented in a conformation favorable for the electronic transition to occur. However, many monomers undergo cyclopolymerization in which there is no evidence for an excited state interaction. Therefore, use of the electronic interaction as the only explanation for the preference for cyclization is not justified. It is quite likely that the additional energy requirement for cyclization comes about as the result of the introduction of angular and torsional strain during ring formation, as well as other steric factors.

Statistical Consideration in Cyclopolymerization

The probability of cyclopolymerization, assuming a freely rotating model, has been calculated.[2-484] The results indicate that from statistical considerations alone, 1,6-diene monomer concentrations must be reduced to 0.10 mole/l in order to attain 95% cyclization, whereas in many invesigations, monomer concentrations of five to eight molar have been reported to lead to essentially quantitative cyclization.[1-9, 1-28] Consistent with these results and their implications are the results of calculations which assume pseudo-second order reaction kinetics for the intermediate radical with its intramolecular double bond in order to determine the "apparent" double bond concentration required to account for the observed rates. This value is >20 moles/l for cyclopolymerization of 2,6-dicyano-1,6-heptadiene,[3-157] and has been calculated to be 40–100 moles/l for small molecule cyclizations, and the absolute unimolecular rate constant to be 10^5 sec^{-1}.[7] The absolute rate constant for addition of ethyl radicals to alkenes = 1×10^3. However, there is evidence that a unimolecular process is generally favored by the frequency factor over a bimolecular process in the same system by a factor of about 100.[133]

The general kinetic equations have been derived for radical initiated cyclopolymerization of symmetrical 1,4-dienes, e.g. divinyl ether, which occurs by the cyclization of two monomer molecules, forming six-membered monocyclic or [3.3.1]-bicyclic units.[133] However, five-membered monocyclic units are also possible as are [2.2.2]-bicyclic units, formed by head-to-head addition.

The failure of the statistical treatment[2-484] to explain the high degree of cyclization in cyclopolymerization is not paralleled by condensation polymerization. Competition between chain polymerization and ring formation is readily accounted for in bifunctional monomers with less than five and more than six members through application of statistical probability and ring stability considerations.[1-123] However, these factors do not completely account for the large tendency of monomers capable of forming five- and six-membered rings to cyclize, as in the case of cyclopolymerization.

Relative Activation Energy Requirements for the Two Steps

In a free radical cyclopolymerization study of vinyl-*trans*-cinnamate, the energy of activation for the cyclization step was found to be slightly lower than or equal to that for the intermolecular step.[3-110] The ratio for the cyclization rate constant to the vinyl propagation rate constant, k_c/k_i, was found to be 13 mole/l. The meaning of this ratio is that at that concentration of monomer, the rates of cyclization and intermolecular propagation will be equal. However, in all other cases studied, the energy of activation of the cyclization step is higher than that for the intermolecular step. This ratio was reported to be 46 mole/l for methacrylic anhydride and 5.9 mole/l for acrylic anhydride.[2-251]

In a similar study of allyl ethenesulfonate, k_c/k_i was found to be 45 mole/l at 45°C in benzene.[3-121] Under the same conditions, allyl ethanesulfonate did not polymerize and propyl ethenesulfonate polymerized at about the same rate as allyl ethenesulfonate. Copolymerization of allyl ethanesulfonate and propyl ethenesulfonate yielded a polymer consisting of more than 90% of the latter. The high degree of cyclization in poly(allyl ethenesulfonate), 85% for [M] = 0.29 mole/l, corresponded to an enhanced reactivity of the allyl group in the monomer compared to the allyl group in allyl ethanesulfonate.

In unsymmetrical monomers, where the double bonds differ significantly in reactivity, it was possible to neglect initiation or intermolecular propagation reactions involving the less reactive double bond. This permitted a simplification of the kinetic expressions which describe these systems, and such have been applied to vinyl *trans*-cinnamate.[3-131] allyl ethenesulfonate,[3-121] *N*-allylmethacrylamide,[3-71] *N*-allylacrylamide,[3-72] allyl methacrylate,[3-111] allyl acrylate,[3-133] and vinyl *trans*-crotonate.[3-134] For these monomers, only one type of repeating unit is formed, and the simplified kinetic expressions could be used. On this basis, it has been suggested[3-112] that in free radical cyclopolymerization of non-conjugated dienes the intermolecular chain propagation reactions must be slow to allow suitable orientation of the intramolecular double bond in order to have the formation of a large fraction of cyclic structural units.

In a cationic polymerization study of o-divinylbenzene[2-22] there was a large variation of k_c/k_i (0.12 to 5 mole/l) with various Friedel-Crafts catalysts. The value of k_c/k_i also decreased with decrease in the dielectric constant of the solvent. In an anionic polymerization study of the same monomer,[2-24] reaction conditions which gave higher k_c/k_i values also gave higher rates of propagation in styrene polymerization. It was concluded the k_c/k_i increased with the reactivity of the growing chain.

Small Molecular Radical Cyclizations

The results of monomeric radical cyclizations also appear to support the electronic interaction theory in that the less thermodynamically favored cyclopentane adduct, with the apparent development of an intermediate primary radical, appears to predominate in many of these studies.[55, 57] The results have been explained on the basis of a "non-classical" radical, consistent with the electronic interaction theory under discussion. The chain transfer step could lead to five- or six-membered ring depending upon the steric requirements of the diene and the incoming chain transfer agent, an explanation which avoids the necessity of developing the unstable primary radical. Recent data on interconversion studies of 2-phenylcyclopentylmethyl and 3-phenylcyclohexyl radicals may support such an intermediate,[70] although studies on the cyclization of 4-(1-cyclohexenyl) butyl radical are interpreted as failing to support such an intermediate.[59]

Cation Initiated Cyclopolymerization

There is little question, however, that cationic initiated intramolecular cyclizations[85, 86, 89, 90, 3-207, 1-121] and polymerizations[74, 2-4] are in accord with an electronic interaction theory. The monomeric cyclization studies show from the stereochemistry of the products that concerted ring closures occur in some cases while in others an allylic cation is apparently present as an intermediate. Cationic initiated cyclopolymerization of 2,6-diphenyl-1,6-heptadienes[74, 2-4] led to high conversion to cyclopolymer having less than 5% residual unsaturation.

Metal alkyl coordination initiated cyclopolymerizations have been reported which support the electronic interaction theory. Such polymerization attempts on 2,2-disubstituted alkenes have been notably unsuccessful. However, 2,5-dimethyl-1,5-hexadiene was converted to a reasonably high polymer containing very little residual unsaturation by this technique while 2-methyl-1-pentene was converted only to a pentamer under the same conditions.[1-52] This remarkable result was attributed to a "driving force" inherent in the cyclopolymerization process.

1,3-Bis(4-vinylphenyl)propane can be polymerized, using cationic initiators, to a cyclopolymer having [3.3]paracyclophane units in the chain.[2-380] Cyclopolymers are not produced using radical or anionic initiators. As an explanation, an interaction was proposed between the styryl type cation and the intra-molecular styryl group (See Structure 10–2), which stabilizes the transition state leading to cyclopolymer.

The technique of complexation with alkylaluminium chlorides, which has been used for the preparation of alternating copolymers has been employed to cyclopolymerize a number of unsymmetrical monomers.[3-100, 3-104] The ability of *o*-allylphenyl acrylate, 2-(*o*-allylphenoxy)ethyl acrylate, 4-(*o*-allylphenoxy)butyl acrylate and *o*-vinylphenyl acrylate to cyclopolymerize was increased when alkylaluminium chlorides were present. These results provide evidence for an intramolecular reaction between the double bond of these monomers, one of which is strongly electron accepting due to the complexing agent, and the other electron donating (See Structure 10–2).

If electronic interactions were the only explanation for cyclization this would suggest a low-energy pathway to cyclopolymer. A study of the cyclopolymerization of methacrylic anhydride[2-317] has shown that the intramolecular step has a higher energy of activation (10.9 kJ mol^{-1}) than the intermolecular step. Similar values for acrylic anhydride (9.41 kJ mol^{-1}),[134] methacrylic anhydride (9.45 kJ mol^{-1})[58] and *o*-divinylbenzene (10.29 kJ mol^{-1})[2-26] have been determined.[134]

Relative Rates of Competing Steps and Degree of Polymerization

Two quite unique and characteristic traits of the cyclopolymerization process which must be recognized in consideration of mechanistic explanations are the facts that the intramolecular step is highly favored over the intermolecular step[1-52, 1-56] and that the degree of polymerization is higher than that for the corresponding mono-unsaturated derivatives.[1-52, 1-38, 1-56, 1-8, 5-1] As illustrative examples, neither α-methylstyrene nor allyltrimethylammonium salts undergo radical initiated polymerization very well while their diene counterparts, 2,6-diphenyl-1,6-heptadiene[1-38, 2-4] and diallyldimethylammonium salts, respectively, yield high polymers under radical conditions.

Enhanced Rate of Cyclopolymerization

A third aspect of cyclopolymerization to be considered is the fact that the rate of cyclopolymerization is significantly greater than that of homopolymerization of the mono-olefinic counterpart of the diene.[3-110, 2-251, 2-243, 19] Relative rates up to 590 have been reported[3-157] in a comparison between 2-cyano-1-heptene and 2,6-dicyano-1,6-heptadiene; however, methacrylonitrile is reported to polymerize at about the same rate as the diene.[40] Although in a telomerization study of allyl ethyl ether and diallyl ether with bromotrichloromethane, it was reported that allyl ethyl ether was slightly more reactive than diallyl ether,[4-30] suggesting that the second double bond did not assist in the addition to the first one, even though the overall product was cyclized. These results, however, are inconsistent with those of most other similar experiments.

Low Volume Shrinkage in Cyclopolymerization

Polymerization of methacrylic anhydride has been shown to be accompanied by a volume shrinkage of only 13.7 ml/mole of unsaturation as compared to 22–23 ml/mole for methyl, ethyl, and n-propyl methacrylates.[2-246] The low volume shrinkage was interpreted to indicate a preorientation of the methacrylic anhydride in a cyclic conformation, thus requiring little volume change on polymerization. This preorientation was also suggested to be the driving-force for cyclization of the molecule. However, as has been pointed out in a later paper,[2-317] low volume shrinkage also may be associated with polymers containing cyclic units in or near the chain as the volume shrinkages of phenyl, benzyl, and cyclohexyl methacrylates are all very close to 15 ml/mole.[2-518]

Preorientation in Solid-State Polymerization

In an effort to establish whether preorientation of the diene molecule in a conformation favorable for cyclopolymerization occurs, several solid-state polymerization studies of characteristic 1,6-dienes have been undertaken. Irradiation of *N,N*-diallylmelamine with γ-rays gave a soluble polymer[1-18] that contained negligible unsaturation. *N*-Phenyldimethacrylamide[2-150, 51] and *N*-methyldimethacrylamide[51] have been polymerized in the solid-state by use of suitable irradiation. By use of X-radiation at a dose rate of 0.2 Mrad/hr[2-150] at a temperature 5° below the monomer melting point for 100 min. 10% conversion to a polymer having $[\eta] = 0.26$ dl/g was observed. The polymer was found to contain predominantly five-membered rings. *N*-Methyldimethacrylamide[51] *via* γ-ray initiated solid-state polymerization led to cyclopolymer containing 70% five-membered rings and 30% six-membered rings. *N*-Phenyldimethacrylamide[51] led to cyclopolymer containing 90% five-membered rings and 10% six-membered rings. Dimethacrylamide[51] led to a polymer containing only six-membered rings. *N*-Methyl- and *N*-phenyl-*N*-isobutyrylmethacrylamide did not undergo radical initiated polymerization as did the dimethacrylamides[51] A single-crystal x-ray diffraction and solid-state polymerization study[66] of *N*-(p-bromophenyl)dimethylacrylamide (See Structures 10–16, and 10–17) led to the conclusion that the monomer crystallizes in a conformation that favors the intramolecular over the intermolecular polymerization reaction. The polymer consisted of exclusively cyclopolymer having 60% five-membered and 40% six-membered rings. The ratio of five- to six-membered ring could not be predicted from the atomic positions in the crystal, however. Radical-initiated solution polymerization of the monomer at 58°C yielded cyclopolymer containing 80% five-membered and 20% six-membered ring. At 114°C, the cyclopolymer contained only 42% five-membered ring. The facile polymerization of these monomers in solid-state supports the preorientation theory, particularly in the crystal; however, these investigations offered no satisfactory explanation to account for the distribution of five- and six-membered rings in the polymers.

Further considerations of the analogy between 1,6-non-conjugated dienes and conjugated dienes the assumption that was made that the activation energy of cyclization must be lowered to explain cyclopolymerization may be open to question, particularly in the case of cationic-initiated cyclopolymerization. Addition of hydrogen bromide to 1,3-butadiene[2] yields both the 1,2- and 1,4-addition products, and the proportions in which they are obtained are considerably affected by the temperature of the reaction. Reaction at −78°C produces a mixture containing about 20% of the 1,4-product and about 80% of the 1,2-product. However, reaction at higher temperature (about 40°C) yields a mixture of about 80% of the 1,4-product and about 20% of the 1,2-product. Also, the fact that either compound is converted to the same mixture as obtained by reaction at the higher temperature when equilibrated at that temperature led to the conclusion that the mixture was the result of an equilibrium between the two. Since the 1,4-adduct predominates (80:20) at the elevated temperature, it is the more stable isomer. Since the reaction at −78°C leads to a preponderance of the 1,2-adduct (80:20), it forms faster than the 1,4-product. The proportions in which the products are formed remain the same regardless of the temperature of the reaction, but as the temperature is raised there is faster conversion of the initially formed products to the equilibrium mixture. This is a classical example of kinetic control of product ratio at the lower temperature and thermodynamic control of product ratio at the higher temperature, and can be represented by the energy diagram shown in Structure 10–22. These conclusions have been justified in terms of molecular orbital theory.[5]

Therefore, it is not inconsistent that the activation energy for the 1,6-addition step in cyclopolymerization, the cyclization step which, by analogy is related to 1,4-addition in 1,3-butadiene, is higher than that for the 1,2-addition step, the intermolecular propagation step, which, by analogy to the butadiene case, corresponds to 1,2-addition. In fact, it has now been established that cyclopolymers are reversible above their ceiling temperatures and five- and six-membered ring content is equilibrated. If, for example, in the case of 2,6-diphenyl-1,6-heptadiene, the following "non-classical" carbocation (10–26) contributes to stabilization of the product of initial attack on monomer by initia-

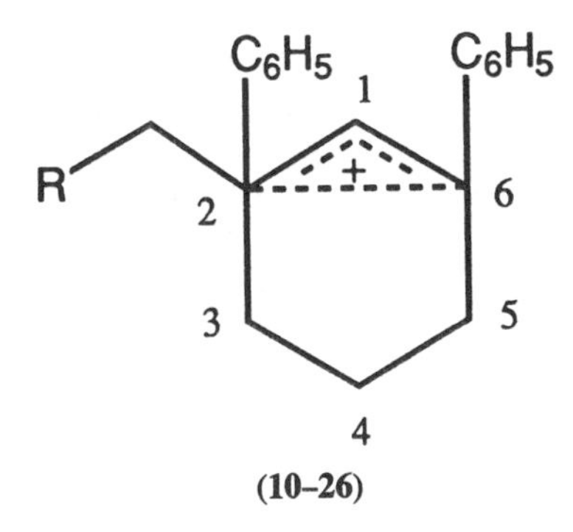

(10–26)

tor, the products of the next step of the propagation could conceivably be governed by the relative stabilities of the products. Attack from above the plane by another monomer molecule could occur with approximately equal probability at C-2 as at C-6 (or at C-1, to produce the five-membered ring). Furthermore, formation of the product *via* attack at C-2 would probably require less energy than that resulting from attack at C-6, owing to the angular and torsional strain introduced in ring formation. On the other hand, the cyclized product should be favored from an enthalpy standpoint since two π-systems are converted to σ-systems by attack at C-6 whereas only one π-system is consumed by attack at C-2.

The above analogy can be considered valid only for cationic initiated systems since there is no evidence available in the 1,3-butadiene case that free radical addition of HBr forms the 1,2-addition product faster than the 1,4-product is formed.

Further spectroscopic studies[2-322, 3-23, 3-138] have failed to produce evidence for interspacial interactions in monomers which cyclopolymerize. Kinetic evidence has been obtained[3-211, 3-215] which shows that the reactivity of each double bond in the radical (10–27) is about five times greater than the double bond in the 5-hexenyl radical under similar conditions. At first, this was attributed to a homoconjugative interaction which stabilized conformations favorable for intramolecular reaction, and lowered the energy of the orbital involved in formation of the new bond. However, it was found that the 3-propyl-5-hexenyl radical also cyclized more rapidly than the 5-hexenyl radical and thus the enhanced reactivity of Structure 10–27 cannot be attributed solely to homoconjugation.

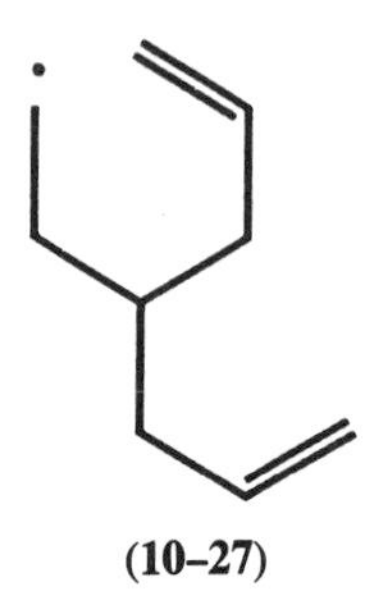

(10–27)

The Steric Control Theory

A number of investigations point towards a thermodynamic explanation for cyclopolymerization. It was proposed as early as 1967 that steric factors alone control cyclopolymerization.[1–18] The nature of an unsaturated acyclic radical may, even in the absence of electronic effects, favor the cyclization reaction over linear propagation. It has been noted[1–18] that a pendant group on the β carbon atom of a monovinyl monomer has a significant effect on the absolute rate constants for propagation and termination. The values for methyl acrylate and butyl acrylate are 720 versus 14 and 2.2×10^6 versus 9×10^3 l/mole/s for propagation and termination respectively. Thus the rate of intermolecular reaction of an initiated monomer molecule would be expected to be lower than that of a similar monomer with less than five atoms in the pendant group. However, the pendant double bond would frequently be presented to the reactive species in a conformation which is favorable for reaction. Further, the cyclic radical formed would be less sterically hindered than its noncyclic counterpart, thus facilitating the approach and reaction of the next monomer unit. This effect may be important in explaining the high degrees of polymerization obtained through the cyclopolymerization of otherwise poorly polymerizable structures.

However, reported values of the propagation rate constants for methyl acrylate at 25°C vary from 580 to 1580 while the corresponding termination rate constants vary from 6.5×10^6 to 55×10^6. Also, the propagation rate constants reported for n-butyl acrylate at the same temperature vary from 13 to 2100 while the corresponding termination rate constants vary from 1.8×10^4 to 3.3×10^8. Because of the pronounced autoacceleration experienced with methyl acrylate and the accompanying necessity for extrapolating measurements to a conversion of less than one percent, the data on this monomer are less accurate than for most other monomers investigated.[3] A comparison between methyl methacrylate and n-butyl methacrylate would have been more realistic and when these comparisons are made, the above argument is not well supported. The propagation and termination rate constants of methyl methacrylate at 30°C, determined by the rotating sector method are 143 and 12.2×10^6, respectively. Values for n-butyl methacrylate at the same temperature and determined by the same method are 369 and 10.2×10^6, respectively.[4] Thus, it appears that from this comparison the length of the pendant group on the β carbon has little effect on the absolute rate constants for propagation and termination. Also, while *N*-methyldimethacrylamide readily undergoes free-radical initiated cyclopolymerization, *N*-methylisobutyrylmethacrylamide yields no polymer under identical conditions.

The decrease in entropy for a cyclization step would perhaps be expected to be smaller than that for addition of a new monomer unit. Only rotational motion will be lost in cyclization; however, on adding a new monomer molecule to the chain the loss of translational and rotational degrees of freedom will result. Therefore, as far as entropy is important, cyclization would be favored over intermolecular propagation.

A mechanism advocating a complete steric control of cyclopolymerization would require $E_c - E_I$ to be zero, since the double bonds are considered to be identical throughout and the cyclization and intermolecular activation energies would be the same.[1-18] Such a mechanism is not consistent with the fact that $E_c - E_I$ is, experimentally, most often greater than zero.

The kinetic relationship between intramolecular and intermolecular propagation for free radical polymerization of acrylic anhydride has been derived.[2-247] If accurate determinations of the residual unsaturation as a function of concentration and temperature can be made for the polymer, these data, when used in conjunction with the kinetic equations, permit the important relationships of the ratio of rates and the difference in energy of activation requirements for the intramolecular and intermolecular propagations to be determined. The ratio k_c/k_i, defined as the ratio of the cyclization rate constant to the intermolecular propagation rate constant for acrylic anhydride, was found to be 5.9 mole/l. at 45°C in cyclohexanone. The difference in the activation energies $(E_c - E_i)$ was 2.4 kcal/mole while the value for the ratio of the frequency factors for the two steps, A_c/A_i, was found to be 167 mole/l.

For the radiation induced polymerization of acrylic anhydride in toluene at 30°C, the value of $k_c/k_i = 11.1$ mole/l. and $E_c - E_i$ was 2.3 kcal/mole.[2-249]

THE MECHANISM OF CYCLOPOLYMERIZATION

The polarity of the polymerization medium in cyclopolymerization of *N*-substituted dimethacrylimides has a marked effect on the structure of the cyclopolymer.[2-81] Increasing solvent polarity favors an increase in the ratio of five- to six-membered ring. The explanation offered is based upon an IR spectroscopic study which supported a planar conformation of the imide groups of acrylic imides with a transoid orientation in nonpolar solvents. In polar solvents, conformations with cisoid orientation could be detected. In dimethacrylimides, where steric factors have a marked effect on the imide group conformation, a nonplanar conformation was observed. Evidence has been presented that in polymers and copolymers of divinyl carbonate[2-278] considerable five-membered ring structure is produced.

For free radical polymerization of *o*-divinylbenzene in benzene, k_c/k_i values of 2.1, 2.3, 2.8, 3.3, and 4.0 mole/l at 20, 30, 50, 70, and 90°C, respectively, were found, and $E_c - E_i$ was 1.9 kcal/mole; cyclization was favored by a frequency factor of 10 at the monomer concentration where the probability of finding a double bond of another molecule around the uncyclized radical is the same as that of finding the intramolecular double bond.[2-20]

Similar studies on divinylformal by radical polymerization in benzene at 50°C indicate the k_c/k_i ratio to be 200 moles/l and $E_c - E_i$ to be 2.6 kcal/mole.[19] For methacrylic anhydride in cyclohexanone at 37°C with radical initiators, k_c/k_i was found to be 46 mole/l.[2-251] However, a radical polymerization study of the same monomer in dimethylformamide at 0°C, yielded a value of k_c/k_i of 2.4 mole/l; $E_c - E_i$ was found to be 2.6 kcal/mole and A_c/A_i was 256 mole/l.[2-317] The value of k_c/k_i was 6.2 mole/l. at 80°C. The marked difference in the k_c/k_i ratio from that reported in cyclohexanone was attributed to a solvent effect. A previous study of methacrylic anhydride[2-245] in dimethylformamide led to an estimate of 11.4 mole/l for k_c/k_i.

For larger ring sizes, the available data appear to support the conclusions that the most important factors involved in cyclization are statistical probability and thermodynamic stability. Calculations[17] based upon published data[2-368] on radical polymerization of diallylphthalate requiring an eleven-membered ring, gave a value of k_c/k_i of 4.25 moles and a value of 0.3 kcal/mole for $E_c - E_i$. Similar data for k_c/k_i for ethyleneglycol dimethacrylate and diethyleneglycol dimethacrylate were found to be 4.0 mole/l and 3.4 mole/l, respectively.[18] The cyclic structures resulting from cyclization of these monomers would be expected to contain nine members and twelve members, respectively.

Certain mechanistic aspects of the cyclopolymerization of several bifunctional monomers[2-287] have been discussed. The cyclization constant r_c was defined on the basis of the assumption that intramolecular cyclization and intermolecular propagation of uncyclized growing species are competitive. The variation of r_c with polymerization conditions resulted in the following conclusions in polymerizations of *o*-divinylbenzene: (a) the intramolecular cyclizations had higher activation energies than the intermolecular propagation in all of the radical, cationic, and anionic polymerizations; (b) in cationic polymerization with Friedel-Crafts catalysts, the extent of cyclization was higher when stronger Lewis acids were used as catalyst. A propagation scheme was proposed that took into consideration the catalyst effect, the solvent effect, and the common ion effect; (c) in the anionic polymerization, both the free anion and the ion pair were involved in propagation as already proposed for the living polymerization of styrene. Telomerization in $CHCl_3$, was carried out in order to obtain direct evidence for cyclopolymerization and to elucidate the cyclic structure. The structures of telomers derived from diallyl ether, divinyl ether, and divinyl formal indicated the presence of the propagating CH_2 radical contrary to initial expectations.

A comparison of the activation parameters in cyclopolymerization of *o*-divinylbenzene, using different initiating systems has been shown in Table 10–3. These results are in accord with the respective driving forces which would be predicted for the relative interactions of neighboring double bonds with cation, free radical, and anion centers.[5-8] Apparently inconsistent with these concepts, however, is the fact that 2,6-diphenyl-1,6-heptadiene undergoes essentially complete cyclization at moderate monomer concentrations by an anionic initiation.[1-38,2-4] However, the phenyl substituted carbanion is subject to resonance stabilization.

Cyclopolymerization has been extended to the polymerization of many dialdehydes and diepoxides. As shown typically in the case of phthalaldehyde, the cyclization of dialdehydes was apparently a concerted or intermediate type reaction rather than a competitive reaction such as observed for divinylbenzene.

A comparison between the most probable propagation mechanisms of cyclopolymerization of *o*-divinylbenzene, *cis*–1,2-divinylcyclohexane, *o*-phthalaldehyde, *cis* cyclohexene-4,5-dicarboxaldehyde and *o*-vinylbenzaldehyde by cationic initiators has been made.[2-27] *o*-Divinylbenzene was proposed to propagate by a stepwise process and an intermediate, while the divinyl cyclohexane and the aldehydes were proposed to proceed by a concerted process. *o*-Vinylbenzaldehyde showed the existence of both reaction schemes, depending upon the type of cation produced.

The mechanism of polymerization of divinylbutyral has been studied.[2-275,2-276] It was found that this monomer possesses high radical reactivity and an easily abstracted hydrogen atom. In cyclic free radical polymerization, initiation occurred via hydrogen abstractions, resulting in multifunctional growth including isomerization (to ester and ketal) and crosslinking. The polymerization rate was proportional to [initiator concentration]$^{0.5}$ but poly(divinyl butyral) molecular weight was not. Chain transfer constants were very high, and a poly-functional propagation terminated by recombination was proposed.

It has been shown that free radical initiation of *o*-allylstyrene[3-2] leads to cyclic polymer possessing six-membered rings. In cationic polymerization, the allyl group isomerizes to β-methylvinyl and the polymer contains a considerable quantity of five-membered ring.

The probable mechanism of polymerization of 1,5- and 1,6-nonconjugated dienes[3-75] has been reviewed, as well as the structure and properties of the cyclic polymers formed. The radical polymerizations of *N*-allylmethacrylamide and *N*-allylacrylamide in methanol, and allyl acrylate, or allyl methacrylate in benzene at 60°, were discussed specifically.

An *e* value of $\cong 1.2$ for methacrylic anhydride[7-92] has been determined by using ^{14}C reagents to study the benzoyl peroxide-initiated polymerization of methacrylic anhydride and its copolymerization with styrene, methyl methacrylate, and methyl acrylate. Methacrylic anhydride was less reactive than methyl methacrylate towards the benzoyloxy radical, and this lower reactivity was correlated with the large *e* value. There was a large number of cyclized methacrylic anhydride units in its homopolymers and copolymers with styrene and the number of cyclized methacrylic anhydride units in the homopolymers prepared from low concentrations of methacrylic anhydride was unexpectedly high. Monomer reactivity ratios for the copolymers of methacrylic anhydride with styrene, methyl methacrylate, and methyl acrylate were given.

Divinyl carbonate or ethyl vinyl carbonate were copolymerized with *p*-chlorostyrene and vinyl acetate.[7-89] Alfrey-Price $Q = 0.035$ or 0.0025 and $e = 0.23$ or -0.26 for divinyl carbonate or ethyl vinyl carbonate, respectively. These values suggest that copolymerization reactivities of divinyl carbonate and ethyl vinyl carbonate are similar and are close to that of vinyl acetate ($Q = 0.026$, $e = -0.22$). The cyclization reaction had little effect on the copolymerization parameters. The extent of cyclization of divinyl carbonate was lower in divinyl carbonate-*p*-chlorostyrene than in divinyl carbonate-vinyl acetate copolymers.

A method for repression of cyclopolymerization in the γ-radiation-initiated polymerization of 1-vinyluracil has been reported.[135] Amorphous poly(1-vinyluracil) of varying syndiotacticity was prepared with complete repression of cyclopolymerization through γ-radiation at low temperatures, at high monomer concentrations, and in the solid state, and by polymerization of negatively charged monomers and those containing bulky substituents. Extent of cyclopolymerization decreased from 20% to 0% in solution with temperature reduction from 0° to −78°. At 0°, high concentrations of negatively charged 1-vinyluracil eliminated cyclopolymerization. In the solid state, polymerization occurred at 40° without any cyclization. At very high doses, there was some crosslinking. 6-Methyl-l-vinyluracil, when polymerized in solution at 25°, also gave no cyclopolymer.

Solvent effects in radical polymerization of vinyl cinnamate have been studied.[3-155] The cyclopolymerization of vinyl cinnamate was carried out in various solvents, and the content of the cyclized unit in the polymer was determined by IR spectroscopy. The solvent effect on the cyclization was insignificant, but the overall polymerization rate in homogeneous media was dependent on the solvent polarity.

The IR absorption spectra of certain "4-substituted" 1,6-heptadienes were correlated with their polymerization characteristics.[136] Included were *N,N*-diallylamides, diallylcyanamide, diallyl(tricyanovinyl)amine, diallylammonium bromide, diallylmethylphosphine oxide, diallylphenylphosphine oxide, diallylmethylphenylphosphonium bromide, and dimethyl-4-penteneamine. The frequencies of the in-plane deformation vibration of the terminal methylene groups varied from 1411 to 1437 cm^{-1}. The higher frequencies were attributed to the electron withdrawal effect of the substituent in the 4-position. Those compounds having high absorption frequencies also undergo polymerization more readily than those in the lower frequency range. These correlations were presented as substantiating evidence for an intramolecular interaction in the monomers which can in part explain the facile but highly improbable cyclopolymerization mechanism.

It was shown that in the cationic cyclopolymerization of 1,1′-divinylferrocene[2-342] with Lewis acids and alkylaluminum catalysts, the cyclization constant, r_c (the ratio of rate constants of the intramolecular cyclization and the intermolecular propagation) at O° in CH_2Cl_2, was approximately 1 mole/l with boron trifluoride-diethyl etherate, stannic chloride, and titanium tetrachloride, but increased to 1.8–2.8 mole/l with the aluminum-containing catalysts aluminum trichloride, ethylaluminum dichloride, and diethylaluminum chloride. The r_c value (determined by applying the equation previously used for cyclopolymerizations of *o*-divinylbenzene) also increased (from 0.672 to 1.49 moles/l) with increasing polarity of the polymerization solvent. The apparent activation energy of the cationic intramolecular cyclization was 1.2–1.5 kcal/mole, closer to the values of the anionic and radical cyclopolymerizations of *o*-divinylbenzene (2.2 and 1.9 kcal/mole). The intramolecular cyclization of the growing carbocation pair became more favorable as the tightness of the ion pair decreased, because of the increase in reactivity.

In a review article, evidence was presented based upon the dilution effect to further support the participation of charge-transfer (donor-acceptor) complexes in cyclocopolymerization.[1-21]

It has been shown that irradiation (solid state)- and free-radical-initiated polymerization of dimethacrylamide, *N*-methyldimethacrylamide, and *N*-phenyldimethacrylamide[137] give cyclopolymers containing only five- and six-membered rings as determined by IR and NMR spectra. These results support a preorientation of the two methacrylic double bonds in the solid-state. For example, poly (*N*-methyldimethacrylamide) contained 10% piperidinedione units and 90 % pyrrolidinedione units.

The UV spectra of a number of dimethacrylamides[138] have been studied and the results compared with spectra of *N*-isobutyrylmethacrylamides in an effort to determine any possible relationship between the spectral characteristics, e.g., intramolecular electronic interactions and the cyclopolymerization characteristics of the dimethacrylamides. No definite relationships were observed.

The effect of the stability of the cyclized radical on the rate of cyclopolymerization[3-157,139] has been evaluated. The enhanced rate of cyclopolymerization of divinyl monomers over that of the corresponding monovinyl monomers was considered. The relative polymerization rates of the following monovinyl-divinyl monomer pairs were determined: methyl methacrylate-methacrylic anhydride; 2-cyano-1-heptene-2,6-dicyano-1,6-heptadiene; and 2-phenyl-1-heptene-2,6-diphenyl-1,6-heptadiene. Also studied were 2-phenylallyl methacrylate and 2-phenylallyl 2-carbethoxyallyl ether. An attempt was made to prepare methacrylic isobutyric anhydride as the monovinyl monomer to compare with methacrylic anhydride, but its inherent instability precluded isolation in a pure state. The overall rate of the divinyl monomer was considerably greater than that of the corresponding monovinyl monomer, the ratios varying from 2 to 10. From the rate data obtained, it was possible to estimate the effective concentration of the intramolecular double bond with respect to the radical, and values greater than 20 M at 50° were obtained with 2,6-dicyano-1,6-heptadiene. This value compares favorably with a concentration of 21.8 M for liquid ethylene at −102°. A concerted cyclization step which requires considerable preorientation prior to reaction is indicated. This favorable preorientation may be related to an electronic interaction between the developing radical cite and the intramolecular double bond.

Solid state polymerization studies of certain 1,6-diene monomers[51] were carried out in order to compare the structure of the polymers with those obtained by free radical initiated solution polymerization, in an effort to draw conclusions about the most probable conformation of the monomer in the crystal. Polymerization characteristics of these dienes were compared with suitable monoolefinic counterparts. Studied were dimethacrylamide and *N*-isobutyrylmethacrylamide, and their *N*-methyl and *N*-phenyl derivatives. Solution polymerization of all monomers and solid state polymerization of solid monomers (*N*-methyl and *N*-phenyl-*N*-isobutyryl-methacrylamide were liquids which did not polymerize) led to polymers of comparable structure in each case. Dimethacrylamide yielded a six-membered ring polymer. *N*-Methyldimethacrylamide gave a predominantly five-membered ring polymer but up to 30% six-membered rings. *N*-Phenyldimethacrylamide gave a polymer containing 90% five-membered rings. The formation of comparable cyclic polymers by both methods means that the 1,6-diene molecules are present in the crystal in a conformation favorable for cyclopolymerization.

The temperature independent factor of cyclization parameters has been evaluated.[134] The close agreement between this factor calculated from the entropy change in the intramolecular cyclization reactions and in the intermolecular additions of cyclizable chain radicals onto monomer molecules and the exponential values for the free radical copolymerization of acrylic anhydride and divinyl ether indicated that the high fractions of cyclic structural units in cyclopolymers from symmetrical unconjugated dienes can be thermodynamically accounted for by an entropic effect largely exceeding the energetic one. The entropy decrease was consistently smaller in the intramolecular than in the intermolecular reactions.

The composition of cyclopolymers in terms of structural units has been determined.[140] A method based on the analytical evaluation of residual unsaturation and on the knowledge of the cyclization ratios was presented for the determination of the cyclopolymer composition. The application of this method to the free radical polymerization products of vinyl *trans*-crotonate and of divinyl ether was described.

The effect of pressures up to 4000 atm. on the free radical cyclopolymerization of acrylic anhydride in solution has been investigated.[141] Both the molecular weight and degree of cyclization of the polymer were increased by pressure. The rate of polymerization at first decreased with increasing pressure, but above 2500 atm a normal acceleration occurred. Evidence was also presented for a bimolecular termination in the cyclopolymerization of diallylcyanamide[142] under pressure. The acceleration of the cyclopolymerization of this monomer in the presence of tert-butyl perbenzoate with pressure increases suggested that the termination at 3000 atm. is bimolecular and supports the theory that termination occurs by an intramolecular degradative chain transfer reaction at one atmosphere.

It is obvious that the structures of cyclopolymers and cyclocopolymers are highly dependent on the nature of the monomers and the reaction conditions. Predictions of structures based upon thermodynamic control of the cyclization reaction can be misleading, since kinetic products are often found. An important driving force for the cyclization reaction may be an intramolecular interaction, particularly in some cases, where a charge-transfer interaction can occur or is induced, and in cationic-initiated reactions. However, the predominance of cyclization over intermolecular propagation can be ascribed to a smaller decrease in activation entropy which compensates for the unfavorable difference in activation energies.[134,2-26,2-318] Therefore, the Gibbs free energy of activation would be smaller for cyclopolymerization than for intermolecular propagation.

It has been proposed that steric factors control and promote cyclopolymerization.[1-18] A significant effect of the size of the pendant group on the β-carbon of a monovinyl monomer on the rate constants for propagation and termination was observed. The k_p and k_t values of methyl acrylate and n-butyl acrylate were compared and the n-butyl group was considered to be responsible for a k_p value ratio of 50 in favor of the methyl ester. Also, k_t values favored the methyl ester by a factor of 220. However, as was pointed out earlier, the reported values for these systems vary widely, and some uncertainty exists which leaves some question about the validity of this comparison. The presence of the pendant group was interpreted to shield the radical to the intermolecular reaction. The cyclized radical would have less steric hindrance than the uncyclized radical and therefore would react faster. k_c/k_p should be greater than unity. Both k_c and k_p should be affected by the length of the pendant group containing a terminal double bond. As the length increases, the absolute value of k_c should fall due to fewer configurations of the pendant group, which would lead to cyclization.

The value of k_p should also decrease because greater steric bulk shields the radical more and thus retards intermolecular propagation. Variation in the k_c/k_p would not be as drastic as would be the case in which k_c changed. It should be noted that only k_c would change in the electronic interaction model.

The value of $E_c - E_p$ should be close to zero for the steric control mechanism since the double bonds are considered identical throughout the reaction. This mechanism would not require E_c to be lower than E_p in cyclopolymerization. Since $k = A \exp -(\Delta E_a/RT)$ and E_c and E_p are proposed to be equal, the only factor favoring cyclization over propagation would be the pre-exponential term A for a given concentration of diene. Consequently, as was emphasized earlier, the steric control mechanism alone, cannot account for the experimental facts.

Measurements of the degree of residual unsaturation in cyclopolymers as a function of monomer concentration can yield values for k_c/k_p. Several measurements have been made but these shed little light on the mechanism of cyclopolymerization because only a single system with one initiator was usually studied.

Data on the variation in unsaturation of polymer as a function of concentration for the polymerization of diallyl phthalate to low conversions have been published.[2-368] An interesting dependence on the number of cyclized and uncyclized units at different monomer concentrations was observed. At [M] = 8.68 moles/l, 44% of the groups were cyclized. At [M] = 6.40 moles/l, 50%; [M] = 4.22, 57%; and at [M] = 2.50, 68%. The above data were used to obtain a value of k_c/k_p = 4.25 moles/l.[17] The data gave a linear plot and the intercept was 1.25.

Values of k_c/k_p of 5.9 moles/l for acrylic anhydride,[2-243] 13 moles/l for poly(vinyl-*trans*-cinnamate),[3-110] 50 moles/l for poly(methacrylic anhydride)[7-84] and 130 moles/l for poly(divinyl acetal)[19] have been reported.

Polymerizations of *o*-divinylbenzene using radical,[2-20] anionic,[2-24] and cationic[2-22,2-23] initiators have been studied and k_c/k_p values were obtained for these polymerizations. IR, NMR, and viscometric studies were used to determine residual unsaturation. In free-radical polymerization, k_c/k_p varied from 2 to 4 moles/l. In anionic systems, k_c/k_p values were generally below 1.0 mole/l, while in cationic polymerization k_c/k_p varied widely from 0.1 to 5.0 moles/l depending upon polymerization conditions. No discussion of the intercepts was made in the three systems.

The two mechanisms that have been considered in this volume are the electronic-interaction mechanism and the steric-control mechanism. Both mechanisms have major weaknesses. The major weakness of the electronic-interaction mechanism is that the activation energy of cyclization is equal to or greater than that for linear propagation. The steric-control mechanism does not explain why $E_c - E_p$ is most often greater than zero and why the reactivity of the diolefinic monomer is most often greater than the reactivity of closely related monoolefins. Another weakness of the steric-control mechanism is that it would require very rapid rotation of the pendant group to exclude the intermolecular propagation. At low temperature and high monomer concentration, cyclopolymerization is observed. Based on probability calculations, the steric mechanism alone seems insufficient to explain the results.[2-484]

In an extension of the steric control hypothesis, the proposal was advanced that highly cyclized polymers may be obtained from monomers whose monofunctional counterparts do not homopolymerize. This has been successfully applied to a number of systems, including *N*-substituted dimethacrylamides.[2-453, 2-82 to 2-84, 3-79, 3-80, 2-174] Thus, *N*-isobutyryl-*N*-propylmethacrylamide did not homopolymerize. However, complete cyclization occurred on homopolymerization or copolymerization of *N*-propyldimethacrylamide. It was proposed that the loss of internal rotation degrees of freedom is less effective in decreasing entropy than a loss of translational and rotational degrees of freedom and the activation entropies favoring cyclization. Earlier studies had established the validity of such conclusions; for example, it had been known since the early history of cyclopolymerization that allyltrimethylammonium chloride does not polymerize or polymerizes very slowly whereas the corresponding diallyl monomer,

DADMAC, polymerizes readily to high polymers. Other examples to illustrate these phenomena have been cited.

On the basis of the present evidence it appears that, in one case of cyclopolymerization or another, all of the factors considered such as statistical probability, thermodynamic stability, entropy effects, non-conjugated interactions, and polymerization conditions are important in understanding the mechanism of cyclopolymerization.

FURTHER MECHANISTIC STUDIES

Allyl monomers are classically known for their degradative chain transfer characteristics. The ability of the diallyl quaternary ammonium monomers to form high polymers apparently without interference from degradative chain transfer appears to be an exception. An explanation for the lack of degradational chain transfer on the monomer during radical polymerization of *N,N*-diallyl-*N,N*-dialkylammonium halides, has been offered (10–28).

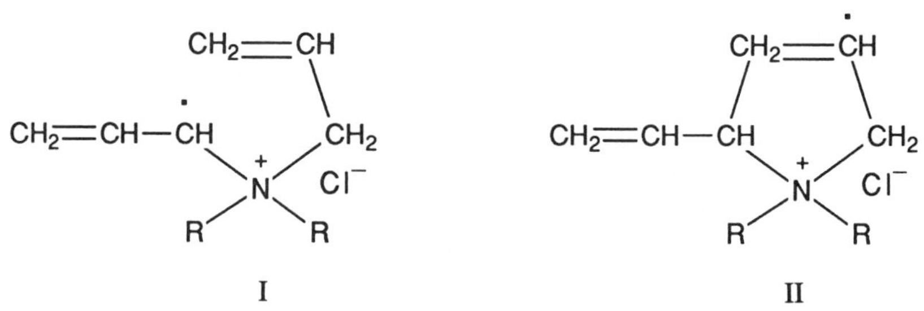

I II

(10–28)

Formation of the intermediate I (R - methyl, ethyl) in chain transfer on the monomer in radical polymerization of *N,N*-diallyl-*N,N*-diethylammonium chloride and DADMAC, and its rearrangement to the radical II, was postulated as an explanation of experimental data indicating formation in the chain transfer of a radical more active than the allyl radical (e.g., I prior to rearrangement) and participation of this active radical in chain propagation to give a linear polymer, rather than a branched or crosslinked polymer that would be formed if I acted as a chain initiator. The presence of one, rather than two, terminal bonds, and of a methyl terminal group, as indicated by the ^{1}H and ^{13}C NMR spectroscopy of the isolated polymers, supported the postulated mechanism.[143]

MINDO/3 and ^{13}C NMR results have been used to predict the free radical polymerizability of allyl monomers. Polymerizable monoallyl and diallyl monomers exhibited high negative eigenvalues for the C-H bond α to the allyl group and small $\Delta\delta$ values (distance between the β- and γ-carbon NMR peaks) as correlated by MINDO/3 calculations and ^{13}C NMR results, respectively. Monomers with low negative eigenvalues (corresponding to weak α-C-H bonds) underwent degradative chain transfer instead of polymerization. Monomers with intermediate eigenvalues or $\Delta\delta$ values polymerized with difficulty.[2–487]

Matrix polymerization of methacrylic and acrylic acid *via* a radical mechanism in presence of polyethylene glycol as the polymer matrix has been reviewed with 49 references.[144]

A new model was derived which described sterically interacting polymers in solution, a finite fraction of which may form closed rings.[145] The osmotic pressure and the single chain end-to-end distances were obtained in the case of an athermal solvent. The effect of ring formation was discussed.

A computer simulation using the upper-lower bound and scaling methods was employed to investigate the influence of cyclization on the extent of gelation of telechelic linear prepolymers with tetrafunctional crosslinking agents at the critical point.[2-491]

The cyclopolymerizability of *N*-allyl-*N*-methylacrylamide was investigated and compared with that of *N*-allyl-*N*-methylmethacrylamide which had been shown to yield a highly cyclized polymer. The structural study showed that the acrylamide monomer gave a high retention of the allyl group and little cyclization. Only the acryloyl group underwent propagation. The different relative rates of the competing propagation reactions in the two system were responsible for the major differences in behavior.

Extended cooperative electronic effects were observed in poly((E,E)-[6.2]paracyclophane-1,5-diene) (10–29) by use of fluorescence spectroscopy.[146] The spectra were interpreted as the extended excimer fluorescence from multiples of electronically interacting repeat units of the polymer. Similar results were obtained for a structurally related polymer containing [3.3]paracyclophane rings.

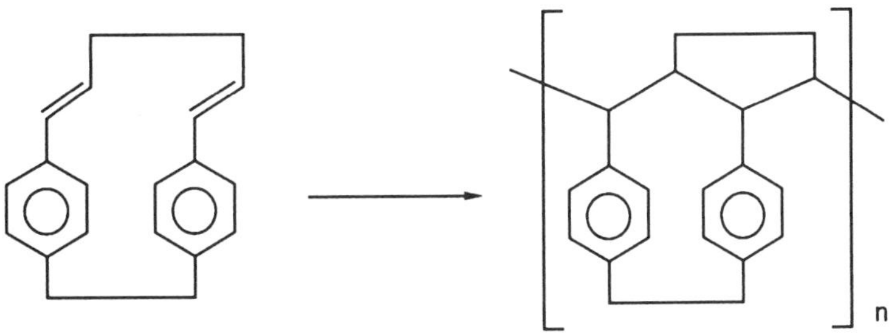

(10–29)

The radical polymerization of methacrylicisobutyric anhydride was discussed in detail in terms of the dependence of the rate of polymerization on monomer concentration, polymerization temperature, and solvent polarity compared to methyl methacrylate.[147] The increased significance of unimolecular depropagation in contrast to bimolecular propagation was clearly demonstrated with decreased monomer concentration, higher polymerization temperature, and increased solvent polarity as reflected in a reduced rate of polymerization compared to an ordinary vinyl polymerization. These results suggested a much lowered ceiling temperature of the mixed anhydride compared to methacrylates.

In a study of the features of intramolecular cyclization in radical polymerization of tetrafunctional monomers, using the Monte-Carlo method, the results indicated that two mechanisms of intramolecular cyclization are possible in the pregel stage.[148] Cyclization occurred by reaction of two radicals (chain termination) or by reaction of a radical with a double bond (intramolecular chain propagation). Cyclization could be regulated by proper selection of reaction conditions.

The effect of intramolecular cyclization on the curing region was discussed on the basis of the classical curing region theory.[149] The theory was confirmed with experimental data.

Aqueous solutions of diallyldimethylammonium fluoride were polymerized in presence of free-radical initiation to yield high molecular weight polymers, with intrinsic viscosity of 0.57 dl/g.[2-180] The increased molecular weight over poly(DADMAC) is not unexpected since chloride ion has been shown to function as a chain transfer agent in presence of certain radical initiators.

Poly(DADMAC) was obtained by sodium persulfate-initiated polymerization of the monomer in the presence of disodium versenate, and the effects of the concentration of initiator and disodium versenate, as well as the temperature and pH of the medium on yield and intrinsic viscosity of the polymer were determined.[2-120]

The molecular weight distribution for poly(diallyldimethylammonium chloride) was determined by fractional precipitation from methanol-dioxane.[2-531] Distribution curves prepared by various means showed no substantial differences in the high-molecular-weight regions. Reproducibility was good at intrinsic viscosity > 0.3 dl/g.

Some general conclusions can be drawn based on the extensive effort that has been devoted to mechanistic studies on cyclopolymerization. Present understanding would indicate that the structures of cyclopolymers and cyclocopolymers are highly dependent on the nature of the monomers and the reaction conditions. Predictions of structures based upon thermodynamic control of the cyclization reaction can be misleading, since kinetic products are often found. An important driving force for the cyclization reaction may be an intramolecular interaction, particularly in some cases, where a charge-transfer (donor-acceptor) interaction can occur, or is induced, as well as in cationic-initiated reactions. However, the predominance of cyclization over intermolecular propagation can be ascribed to a smaller decrease in activation entropy which compensates for the unfavorable difference in activation energies. Consequently, the Gibbs free energy of activation would favor cyclopolymerization over the intermolecular propagation.

REFERENCES

1. Szwarc, M., Wiley Interscience, N. Y., Publishers, 1968; CA 70, B29542j (1969)
2. Kharasch, M. S., Margolis, F. T. and Mayo, F. R., J. Org. Chem., 1937, 1, 393.
3. Flory, P. J., Principles of Polymer Chemistry, Cornell 1953, 158.
4. Brandrup, J. and Immergut, E. H., Polymer Handbook, 1966, 11; CA 64, 16009g (1966).
5. Pilar, F. L., J. Chem. Phys., 1958, 29, 1119.
6. Pryor, W. A., Free Radicals, McGraw Hill, NY, 1966, 221.
7. Moore, W. J., Physical Chem. 3rd. Ed., Prentice-Hall, 1962, 598.
8. Arai, S., Sato, S. and Shida, S., J. Chem. Phys., 1960, 33, 1277.
9. Gordon, A. S. and Smith, S. R., J. Phys. Chem., 1962, 66, 521.
10. Julia, M. and Malassine, B., Tetrahedron, 1974, 30, 695; CA 81, 77151e (1974).
11. Eliel, E. L., Stereochemistry of Carbon Compounds, 1962, 265, (269).
12. Julia, M. and Maumy, M., Org. Syn., 1973, 53, 1865; CA 81, 135504k (1974).
13. Julia, M. and Barreau, M., C. R. Hebd. Seances Acad. Sci., Ser. C, 1975, 280, 957; CA 83, 27660r (1975).
14. Geczy, I., J. Polym. Sci.(C), 1967, 16, 2991; CA 68, 30304y (1968).
15. Kabanov, V. A., Patrikeeva, T. L., Kargina, O. V. and Kargin, V. A., J. Polym. Sci. C, 1968, 23, 357; CA 69, 52529y (1968).
16. Dewar, M. J. S. and Olivella, S., J. Am. Chem. Soc., 1978, 100, 5290.
17. Gibbs, W. E., J. Polym. Sci. A, 1964, 2, 4815; CA 62, 1748f (1965).
18. Aso, C., Mem. Fac. Eng. Kyushu Univ., 1963, 22, 119; CA 59, 4045f (1963).

19. Minoura, Y. and Mitoh, M., J. Polym. Sci. A, 1965, 3, 2149; CA 63, 5749h (1965).
20. Gray, T. F., Ph.D thesis, Univ. of Florida, 1964, CA 63, 13423b (1965).
21. Uno, K.,Tsuruoka, K. and Iwakura, Y., J. Polym. Sci. A1, 1968, 6, 85; CA 68, 69656q (1968).
22. Matheson, M. S., Auer, E. E., Bevilacqua, E. B. and Hart, E. J., J. Am. Chem. Soc., 1949, 71, 497; CA 43, 4934f (1949).
23. Trossarelli, L., Guaita, M. and Priola, A., Ricerca Sci.35, Ser. 2(11-A), 1965, 8, 379; CA 64, 5272e (1966).
24. Guaita, M., Camino, G. and Trossarelli, L., Makromol. Chem. , 1970, 131, 237; CA 72, 79535w (1970).
25. Guaita, M., Camino, G. and Trossarelli, L., Kinet. Mech.Polyreactions, IUPAC Pre.3, 1969, 3 (Prepr.) 203; CA 75, 49669n (1971).
26. Reinisch, G., Jaeger, W., Hahn, M. and Wandrey, C., Macromol. Symp. (Proc. IUPAC), 1982, 28, 83; CA 99, 158900x (1983).
27. Topchiev, D. A. and Nazhmetdinova, G. T., Vysokomol. Soedin., Ser. A, 1983, 25, 636; CA 98, 179981u (1983).
28. Temple, W. B., Makromol. Chem., 1972, 160, 277; CA 78, 4599k (1972).
29. Gordon, M. and Temple, W. H., Makromol. Chem., 1972, 160, 263; CA 78, 4598j (1973).
30. Fueno, T., Kagaku (Kyoto), 1973, 28, 157; CA 81, 3204c (1974).
31. Saito, N. and Chen, Y. D., J. Chem. Phys., 1973, 59, 3701; CA 80, 37584w (1974).
32. Sunagawa, S. and Doi, M., Polym. J., 1975, 7, 604; CA 84, 18011y (1976).
33. Doi, M., Chem Phys., 1975, 9, 455; CA 83, 147842y (1975),
34. Dusek, K. and Ilavsku, M., J. Polym. Sci., Polym. Symp., 1975, 53, 57; CA 84, 122414t (1976).
35. Dusek, K. and Ilavsky, M., J. Polym. Sci., Polym. Symp., 1975, 53, 75; CA 84, 122415u (1976).
36. Erukhimovich, I. Y., Irzhak, V. I. and Rostiashvili, V. G., Vysokomol. Soedin., Ser. B, 1976, 18, 486; CA 85, 109245q (1976).
37. Sakate, M. and Doi, M., Polym. J., 1976, 8, 409; CA 85, 178171z (1976).
38. Yoshida, Z., Kagaku No Ryoiki, Zokan, 1976, 113, 1; CA 86, 42532f (1977).
39. Plate, N. A., Noa, O. V., Romantsova, I. I. and Taran, Yu. A., ACS Symp. Ser., 1980, 121, 25; CA 93, 72408g (1980),
40. Grassie, N. and Vance, E., Trans. Faraday Soc., 1956, 52, 727; CA 51, 758g (1957).
41. Graillat, C., Guillot, J. and Guyot, A., J. Macromol. Sci. (Chem.), A, 1974, 8, 1099; CA 82, 4646f (1975).
42. Divakar, D. S. and Rao, K. N., J. Polym. Sci., Polym. Chem. Ed., 1975, 13, 295; CA 82, 156828e (1975).
43. Rathnasababapathy, S., Marisami, N. and Manickam, S. P., J. Macromol. Sci. (Chem.), A, 1988, 25, 97; CA 108, 38479z (1988).
44. Wandrey, C., Jaeger, W. and Reinisch, G., Acta Polym., 1981, 32, 197; CA 95, 7858f (1981).
45. Wandrey, C., Jaeger, W. and Reinisch, G., Acta Polym., 1981, 32, 257; CA 95, 43752a (1981).
46. Jaeger, W., Hahn, M., Wandrey, Ch., Seehaus, F. and Reinisch, G., J. Macromol. Sci. (Chem.), A, 1984, 21, 593; CA 100, 192371z (1984).
47. Huang, P. C., Singh, P. and Reichert, K. H., Polym. React. Eng.: Emuls. Polym., 1986, 2, 125; CA 106, 176919c (1987).
48. Huang, P. C. and Reichert, K. H., Angew.Makromol. Chem., 1989, 165, 1; CA 110, 193723j (1989).
49. Walling, C. and Pearson, M. S. , J. Am. Chem. Soc., 1964, 86, 2262; CA 61, 2926f (1964).

50. Matsoyan, S. G., Pogosyan, G. M., Dzhagalyan, A. O., and Mushegyan, A. V., Vysokomol. Soedin., 1963, 5, 854; CA 59, 7655c (1963).
51. Butler, G. B. and Myers, G. R., J. Macromol. Sci. (Chem), A, 1971, 5, 135; CA 73, 120936p (1970).
52. Glasebrook, A. L. and Tonell, W. G., J. Am. Chem. Soc., 1939, 61, 1717; CA 33, 6809(6) (1939).
53. Hart, H. and Wyman, P. D., J. Amer. Chem. Soc., 1959, 81, 4891; CA 54, 20899e (1960).
54. Kochi, J. K. and Krusic, P. J., J. Am. Chem. Soc., 1969, 91, 3940.
55. Brace, N. O., J. Am. Chem. Soc., 1964, 86, 523; CA 60, 7923g (1964).
56. Brace, N. O., J. Org. Chem., 1967, 32, 2711; CA 67, 90421y (1967).
57. Cadogan, J. I. G., Hey, D. H. and Hock, A. O. S., Chemy. Ind., 1964, 753; CA 61, 4208h (1964).
58. Walling, C.,Cooley, J. H.,Ponaras, A. A. and Racah, E. J., J. Am. Chem. Soc., 1966, 88, 5361; CA 66, 94482w (1967).
59. Struble, D. L., Beckwith, A. L. J. and Gream, G. E., Tetrahedron Lett., 1968, 3701.
60. Beckwith, A. L. J. and Gara, W. B., J. Amer. Chem. Soc., 1969, 91, 5691; CA 71, 112136w (1969).
61. Walling, C., Molecular Rearrangements, de Mayo P.Ed. 1963, 1, 407.
62. Julia, M., Pure Appl. Chem., 1967, 15, 167; CA 68, 86436k (1968).
63. Julia, M., Record Chem. Prog., 1964, 25, 1.
64. Julia, M., Acc. Chem. Res., 1971, 4, 386.
65. Wilt, J. W., Free Radicals, Kochi, J. K.(Ed),NY Wiley 1973, 1, 333.
66. Smith, T. W. and Butler, G. B., J. Org. Chem., 1978, 43, 6.
67. Kuivila, H. G. and Beumel, Jr., O.F., J. Amer. Chem. Soc., 1961, 83, 1246; CA 55, 13347i (1961).
68. Wilt, J., Massie, S. and Dabek, J., J. Org. Chem., 1970, 35, 2803; CA 73, 65783j (1970).
69. Smith, T. W., PhD. Dissertation, Univ. Fla., 1972 (June).
70. Walling, C. and Cioffari, A., J. Am. Chem. Soc., 1972, 94, 6064.
71. Julia, M., Descoins, C., Baillarge, M., Jacquet, B., Uguen, D. and Groeger, F. A., Tetrahedron, 1975, 31, 1737; CA 83, 146797a (1975).
72. Hawthorne, J. H., Johns, S. R., and Willing, R. I., Austr. J. Chem., 1976, 29, 315; CA 85, 4676c (1976).
73. Beckwith, A. L. J., Easton, C. J. and Serelis, A. K., J. Chem. Soc., Chem. Commun., 1980, 482.
74. Hawthorne, J. H., Johns, S. R., Solomon, D. H. and Willing, R. I., Austr. J. Chem., 1979, 32, 1155.
75. Ottenbrite, R. M. and Shillady, D. D., IUPAC, E. J. Goethals, Ed. Pergamon P., 1980, 143; CA 94, 122020t (1981).
76. Dowbenko, R. and Chang, W.-H., J. Polym. Sci. B, 1964, 2, 469; CA 60, 13325b (1964).
77. Matsumoto, A., Kitamura, T., Oiwa, M. and Butler, G. B., Makromol. Chem., Rapid Commun., 1981, 2, 683; CA 96, 7162v (1982).
78. McGraw-Hill Book Co., Inc., New York, N. Y., International Critical Tables, V, 1929, 162.
79. Chemical Rubber Co., Handbook of Chem. and Phys., 1968, 49th Ed., 184.
80. Hamaan, S. D., Pompe, A., Solomon, D. H., and Sperling, T. H., Austr. J. Chem., 1976, 29, 1975; CA 86, 42769p (1977).
81. Ottenbrite, R. M., Polym. Preprints, ACS, Div. Polym. Chem. 1981, 22, 42.
82. Frost, A. A. and Pearson, R. G., Kinetics & Mechanism, Wiley, 1961, 297.
83. Julia, M. and Maumy, M., Bull. Soc. Chim. Fr., 1966, 434.
84. Hrivik, A. and Mikulasova, D., Chem. Zvesti., 1958, 12, 32.

85. Johnson, W. S., Bailey, D. M., Owyang, R., Bell, R. A., Jaques, B. and Crandall, J. K., J. Am. Chem. Soc., 1964, 86, 1959.
86. Johnson, W. S., Lunn, W. H. and Fitzo, K., J. Am. Chem. Soc., 1964, 86, 1972.
87. Beckwith, A. L. J., Colloq. Int. C.N.R.S., 1977, (Pub. 1978), 278, 373; CA 91, 192441t (1979).
88. Allegra, G., Ganazzoli, F. and Ullman, R., J. Chem. Phys., 1986, 84, 2350; CA 104, 149857h (1986).
89. Johnson, W. S., Gray, S. I., Crandall, J. K. and Bailey, D. M., J. Am. Chem. Soc., 1964, 86, 1966.
90. Ulery, H. K. and Richards, J. H., J. Amer. Chem. Soc., 1964, 86, 3113.
91. Yuki, H., Hatada, K., Emura, T. and Nagata, K., Bull. Chem. Soc. Jap., 1971, 44, 537; CA 74, 126113k (1971).
92. Beckwith, A. L. J. and Gara, W. B., J. Chem. Soc., Perkin Trans. 2, 1975, 6, 593; CA 83, 27243g (1975).
93. Iwata, H. and Ikada, Y., Bull. Inst. Chem. Res. Kyoto Univ., 1979, 57, 184; CA 91, 124184m (1979).
94. Wandrey, C., Jaeger, W., Reinisch, G., Hahn, M., Engelhardt,G., Jancke, H and Ballschuh, D., Acta Polym., 1981, 32, 177; CA 94, 157343c (1981).
95. Matsumoto, A., Funamoto. H., Oiwa, M. and Butler, G. B., J. Polym. Sci., Polym. Chem. Ed., 1986, 24, 2937; CA 106, 19095p (1987).
96. Matsumoto, A., Terada, T. and Oiwa, M., J. Polym. Sci.: Part A: Polym. Chem., 1987, 25, 775; CA 107, 59511y (1987).
97. Ottenbrite, R. M. and Chen, H., ACS Symp. Ser.(Chem. React. Polym.) Eng. 1988, 364, 127; CA 108, 132387q (1988).
98. Saigo, K., Tateishi, K. and Adachi, H., J. Polym. Sci., Part A, Pol. Chem., 1988, 26, 2085; CA 109, 190903m (1988).
99. Camino, G., Casorati, E., Chiantore, O., Costa, L., Guaita, M., and Trossarelli, L., Conv. Ital. Sci. Macromol., [Atti], 1977, 3, 67; CA 89, 24889k (1978).
100. Matsumoto, A., Iwanami, K. and Oiwa, M., J. Polym. Sci., Polym. Lett. Ed., 1980, 18, 211; CA 92, 164360t (1980).
101. Mathias, L. J., Canterberry, J. B. and South, M., Proc. IUPAC, I.U.P.A.C. Macromol. Symp., 1982, 28, 212; CA 99, 140493g (1983).
102. Mathias, L. J., Canterberry, J. B. and South, M., J. Polym. Sci., Polym. Lett. Ed., 1982, 20, 473; CA 97, 182982 (1982).
103. U.S. Dept. Com. Office Tech. Serv., P.B. Rept., 1959, 145, 435.
104. Pryor, W. A., McGraw-Hill, New York, Publishers, 1966, 266.
105. Breslow, R., Chem. Eng. News, 1965, (July 28) 43(26), 90.
106. Dewar, M. J. S., Advan, Chem. Phys., 1965, 8, 65.
107. Winstein, S., Quart. Rev., 1969, 1969, 141.
108. James, D. G. and Ogawa, T., Can. J. Chem., 1965, 43, 640.
109. Spencer, R., Burdick, M. and Frankovitch, K., US Dep. Com., OTS, PB Rept., 144 900, 1959, 24; CA 56, 7483g (1962).
110. Baucom, K. B., Ph.D. Dissertation, Univ. of Florida, 1971; CA 77, 35003v (1972).
111. Butler, G. B. and Iachia, B., J. Macromol. Sci. (Chem.), 1969, 3, 803; CA 71, 30793u (1969).
112. Butler, G. B., Polym. Preprints, 1967, 8, 35; CA 66, 105216r (1967).
113. Corfield, G. C. and Crawshaw, A., J. Polym. Sci. A1, 1969, 7, 1179; CA 71, 49091k (1969).
114. Corfield, G. C., Crawshaw, A. and Thomas, W. A., Chem. Communs., 1967, 1044; CA 68, 48882v (1968).
115. Corfield, G. C. and Crawshaw, A., J. Chem. Soc. B, 1969, 495; CA 71, 29918g (1969).

116. Jensen, F. R., Bushweller, C. H. and Beck, B. H., J. Amer. Chem. Soc., 1969, 91, 344; CA 70, 56999v (1969).
117. Herlinger, H. and Naegele, W., Tetrahedron Lett., 1968, 4383; CA 69, 86137j (1968).
118. Wilcox, C. F., Winstein, S. and McMillan, W. G., J. Am. Chem. Soc., 1960, 82, 5450; CA 55, 19821b (1961).
119. Turner, R. B., Meador, W. R. and Winkler, R. E., J. Am. Chem. Soc., 1957, 79, 4116; CA 52, 1190g (1958).
120. Novak, I. Cvitas, T. and Klasinc, L., J. Organomet. Chem., 1981, 220, 145; CA 95, 219477f (1981).
121. Jones, E. R. H., Mansfield G. H. and Whiting, M. C., J. Chem. Soc., 1956, 4073; CA 51, 4967e (1957).
122. Bowden, K. and Jones, E. R. H., J. Chem. Soc., 1946, 52; CA 40, 2790(8) (1946).
123. Lunt, J. C. and Sondheimer, F., J. Chem. Soc., 1950, 3361; CA 45, 6591c (1951).
124. Winstein, S. and Shatavsky, M., Chemy. Ind., 1956, 56; CA 50, 13761h (1956).
125. Winstein, S. and Shatavsky, M., J. Am. Chem. Soc., 1956, 78, 592; CA 50, 12908a (1956).
126. Cristol, S. J., Brindell, G. D. and Reeder, J. A., J. Am. Chem. Soc., 1958, 80, 635; CA 52, 10983c (1958).
127. Kuivila, H. G., Accts. Chem. Res., 1968, 1, 299; CA 70, 11727p (1969).
128. Ullman, E. F., Chemy. Ind., 1958, 1173; CA 53, 11271h (1959).
129. Cristol, S. J. and Snell, R. L., J. Am. Chem. Soc., 1954, 76, 5000; CA 49, 9566c (1955).
130. Cristol, S. J. and Snell, R. L., J. Am. Chem. Soc., 1958, 79, 1950; CA 52, 16244g (1958).
131. Miller, B., J. Am. Chem. Soc., 1969, 91, 751; CA 70, 77401j (1969).
132. Aso, C., Kunitake, T. and Tsutsumi, F., Kogyo-kwagaku zasshi, 1967, 70, 2043; CA 68, 114987m (1968).
133. Barton, J. M., J. Polym. Sci. B, 1966, 4, 513; CA 65, 10671h (1966).
134. Guaita, M., Makromol. Chem., 1972, 157, 111; CA 77, 102326t (1972).
135. Kaye, H. and Chang, S. H., Macromolecules, 1972, 5 (4), 397; CA 77, 127299z (1972).
136. Butler, G. B. and Miller, W. L., J. Macromol. Sci. (Chem), 1969, 3, 1493; CA 71, 124983h (1969).
137. Butler, G. B. and Myers, G. R., Kinet. Mech. Polyreact. IUPAC, Preprints 1969, 4, 153; CA 75, 64402p (1971).
138. Myers, G. R., Ph.D. Dissertation, Univ. of Florida, 1969, CA 74, 100422k (1971).
139. Butler, G. B., Kimura, S. and Baucom, K. B., ACS Polym. Preprnt. Div. Polym. Chem., 1970, 11 (1), 48; CA 76, 25654y (1972).
140. Guaita, M., Camino, G. and Trossarelli, L., J. Macromol. Sci. (Chem), 1971, 5, 1941; CA 73, 131377h, (1970).
141. Higgins, J. P. J. and Weale, K. E., J. Polym. Sci. A1, 1970, 8, 1705; CA 73, 35798a (1970).
142. Higgins, J. P. J. and Weale, K. E., J. Polym. Sci. B, 1969, 7, 153; CA 70, 115596m (1969).
143. Kabanov, V. A., Topchiev, D. A. and Nazhmetdinova, G. T., Vysokomol. Soedin., Ser. B, 1984, 26, 51; CA 100, 139676b (1984).
144. Ranby, B., Kem. Tidskr., 1971, 83, 32; CA 76, 4169d (1972).
145. Tanaka, F., J. Phys. Soc. Jpn., 1984, 53, 1652; CA 101, 24187w (1984).
146. Glatzhofer, D. T. and Longone, D. T., J. Polym. Sci., Chem., 1986, 24, 947; CA 105, 61221f (1986).
147. Matsumoto, A., Funamoto, H., Oiwa, M. and Butler, G. B., J. Macromol. Sci. (Chem.), A, 1987, 24, 183; CA 106, 120260p (1987).

148. Romantsova, I. I., Pavlova, O. V., Kireeva, S. M. and Sivergin, Yu. M., Vysokomol. Soedin., Ser. A, 1989, 31, 2618; CA 112, 179933n (1990).
149. Zhang, W., Zheng, F. and Tang, X., Jilin Daxue Ziran Kexue Xuebao, (2), 1989, 109; CA 112, 180006u (1990).

11

Mechanism of Cyclocopolymerization

Cyclocopolymerization has been used to describe the specific examples of the broader area of cyclopolymerization[1-156] in which at least two different monomers contribute to formation of the cyclic structure during the propagation steps. The first example of cyclocopolymerization observed was DIVEMA or "Pyran Copolymer" (See Structure 1–5). In an effort to produce a highly crosslinked, high capacity, carboxyl-containing ion exchange polymer for potential use in the treatment of sodium edema as it applies to high blood pressure, divinyl ether (DVE) was selected as a crosslinking agent for maleic anhydride (MA). A quantitative yield of copolymer was obtained. In an attempt to open the anhydride ring with aqueous sodium hydroxide to convert the polymer to its carboxylate anion form, which was predicted to be insoluble, surprisingly, it was found to be completely soluble in the aqueous alkaline medium. The presence of cyclic ether groups in the copolymer chain was confirmed by chemical evidence which involved cleavage by hydriodic acid and incorporation of iodine in the copolymer. This result prompted a series of structural and mechanistic investigations to explain these unpredictable results, particularly when considered in light of the earlier well-established principle regarding gelation and crosslinking.[1-23]

A discussion of the mechanism of cyclocopolymerization has been published;[1-158] many of the details of which will not be repeated here. Even though the DIVEMA copolymerization was observed to yield soluble, noncrosslinking copolymer[1-87, 1-88, 1-17, 1-16] before the postulation of the alternating intra-intermolecular chain propagation (now known commonly as "cyclopolymerization") to explain the failure of 1,6-dienes to yield crosslinked polymers,[1-156] it was only after unquestionable evidence had been obtained for the cyclopolymerization mechanism[1-9] that a plausible explanation for this most unusual behavior in the copolymerization of 1,4-dienes with alkenes was forthcoming. A satisfactory mechanism for this behavior must account for (a) the failure of the system to crosslink in accordance with the widely accepted (at the time) theory,[1-23] (b) the essential absence of carbon-carbon double bond content in the copolymer; (c) polymer composition equivalent to a diene-olefin molar ratio of 1:2, and (d) essentially quantitative conversion of monomers to copolymer when the monomer molar ratio was 1:2.

Although copolymers of certain 1,6-dienes with a number of well known olefinic monomers have now been reported,[1-35] a fundamental assumption in copolymerizations of this type was that the intramolecular cyclization step involved only the 1,6-diene, and all members of the cyclic structure were contributed by this comonomer. The DIVEMA copolymerization represented the first example of a cyclopolymerization in which both comonomers were involved in the cyclization step and the cyclic structure contained members contributed by each comonomer. The mechanism proposed assumed that one comonomer, the 1,4-diene, contributed four members to the developing cyclic structure while the other comonomer, the alkene, contributed the remaining two members.

The proposed structure for the DIVEMA copolymer was supported by; (a) the elemental analysis for the copolymer obtained at high conversion was consistent; (b) the IR spectrum was essentially devoid of residual double bond absorption and contained characteristic absorption bands for cyclic anhydride and six-membered cyclic ether structures; (c) the copolymer composition of 1:2 molar ratio of DVE to MA was consistent with the known reactivity ratios of these types of monomers; and (d) the presence of the cyclic ether group was confirmed by chemical evidence which involved cleavage by hydriodic acid and incorporation of iodine into the polymer, as mentioned above.

KINETIC STUDIES

A fundamental study of this type of copolymerization[1,8-11] has been carried out. A series of copolymer composition equations were derived that are consistent with the addition of monoolefin to diene radicals by a concerted bimolecular step proceeding through a cyclic transition state to produce the cyclic repeating unit. It was recognized quite early[8-14] that the above requirements could be satisfied if the monomer pair formed a charge-transfer complex that participated in the copolymerization. Considerable evidence has now been presented for charge-transfer complex participation in a variety of highly alternating cyclocopolymerizations.[8-62]

A general copolymer composition equation for the cyclocopolymerization of 1,4-dienes (M_1) and monoolefins (M_2) based on the kinetic scheme shown in Equations 11-1 through 11-9 has been derived.[8-11]

$$m_1\cdot + M_1 \xrightarrow{k_{11}} m_1\cdot \tag{11-1}$$

$$m_1\cdot + M_2 \xrightarrow{k_{12}} m_3\cdot \tag{11-2}$$

$$m_3\cdot \xrightarrow{k_1} m_c\cdot \tag{11-3}$$

$$m_3\cdot + M_1 \xrightarrow{k_{11}} m_1\cdot \tag{11-4}$$

$$m_3\cdot + M_2 \xrightarrow{k_{12}} m_2\cdot \tag{11-5}$$

$$m_c\cdot + M_1 \xrightarrow{k_{c1}} m_1\cdot \tag{11-6}$$

$$m_c\cdot + M_2 \xrightarrow{k_{c2}} m_2\cdot \tag{11-7}$$

$$m_2\cdot + M_1 \xrightarrow{k_{21}} m_1\cdot \tag{11-8}$$

$$m_2\cdot + M_2 \xrightarrow{k_{22}} m_2\cdot \tag{11-9}$$

which led to Equation 11–10:

$$n = \frac{(1+r_1x)\{1/[M_2]+(1/a)(1+x/r_3)\}}{(1/a)\{(x/r_3)+(r_2/x)+2\}+(1/[M_2])\{1+(1+r_2/x)(1+r_cx)-1\}} \quad \text{Equation (11–10)}$$

where $x = [M_1]/[M_2]$, $r_1 = k_{11}/k_{12}$, $r_2 = k_{22}/k_{21}$, $r_3 = k_{32}/k_{31}$, $r_c = k_{c1}/k_{c2}$, $a = k_c/k_{32}$.

Equation 11–10 is a differential copolymer composition equation that is applicable to the proposed scheme of cyclocopolymerization. The equation may be applied by letting $n = [m_1]/[m_2]$, the fractional ratio of monomers combined in the copolymer at low conversions.

A similar equation (Equation 11–11) can be derived relating the relative rate of addition of diene and the rate of cyclization, as follows:

$$\begin{aligned} d\{M_1]/d[M_c] &= [(K_{11} + K_{12})/K_{12}K_c](K_{31} + K_{32} + k_c \\ &= (r_1x + 1)\{([M_1]/r_3a) + ([M_2]/a) + 1\} \end{aligned} \tag{11-11}$$

Equation 11–11 applies at low conversions where $d[m_1]/d[m_c] = [m_1]/[m_c]$, the ratio of the total fraction of diene (unsaturated and cyclic) to the fraction of diene in cyclized units, in the copolymer.

If precise analytical methods are available for determining both the total fraction of diene in the copolymer and the fraction of either cyclic units or pendant vinyl groups, then by making a series of such measurements for different initial monomer feed compositions, values for r_1, r_3, and α could be obtained from Equation 11–11. Then the remaining two parameters, r_2 and r_c could be obtained from Equation 11–10.

In certain special cases Equation 11–10 may be approximated to simpler forms as in these examples:

Case 1: If $k_c >> k_{32}$ so that α is very large and cyclization is the predominant reaction of the radicals $m_3\cdot$ then Equation 11–10 becomes Equation 11–12.

$$n = (1 + r_1x)(1 + r_1x)/[r_1x = (r_2/x) + 2] \tag{11-12}$$

This is equivalent to considering the addition of monoolefin to diene radicals to be a concerted bimolecular step proceeding through a cyclic transition state and producing the cyclic repeating unit.

Case 2: If in addition there is a strong alternating tendency so that $(r_1, r_2, r_c) \rightarrow o$ then Equation 11–12 reduces in the limit to $n = 1/2$. This predicts an alternating copolymer composition of 2:1 molar in contrast to 1:1 for the similar limiting case of the classical binary copolymer composition equation.

Case 3: If the diene has a negligible tendency to add to its own radicals and $r_1 \cong r_c \cong 0$, and there is also predominant cyclization, then Equation 11–12 gives Equation 11–13.

$$n = 1/[r_2/x + 2] \tag{11-13}$$

A plot of 1/n against 1/x should be linear with a slope r_2 and an intercept 2.0.

The application of each of these special cases of the theory has been shown for experimental systems. The divinyl ether-maleic anhydride system represents an example of special Case 2 in which there is a strong alternating tendency and the resultant copolymer composition is 1:2 in DIVEMA.

The copolymerizations of 3,3-dimethyl-1,4-pentadiene with acrylonitrile (system A) and divinyl sulfone with acrylonitrile (system B) had been investigated (See Structure 8–12), azobisisobutyronitrile being used as the initiator. The polymers obtained were soluble in DMF in every case. The IR spectra of the polymers indicated there was little unsaturation. The reactivity ratio parameters were determined for the above systems. For system A, $r_1 \cong r_c \cong 0$, $r_2 = 3.31$. For system B, $r_1 = 0.364$, $r_2 = 1.94$, and $r_c = 0.067$. The intrinsic viscosities of the copolymers of divinyl sulfone with acrylonitrile indicated that divinyl sulfone acted as a chain transfer agent. The copolymers obtained in system A had intrinsic viscosities $[\eta]$ which varied from 0.71 dl/g for a copolymer that contained 0.97 mole fraction of acrylonitrile (AN) to 0.49 dl/g for one that contained 0.94 mole fraction of AN. This system was shown to conform to special Case (3) of the copolymer equation (See Equation 11–13). The copolymer obtained in system B had $[\eta]$ values which gradually increased from 0.08 dl/g to 0.64 dl/g for copolymers varying from 0.14 to 0.86 mole fraction AN. This system was shown to conform to special Case (1) of the copolymer equation referred to earlier.

The copolymerizations of divinyl ether with fumaronitrile[8–13] (C), tetracyanoethylene (D), and 4-vinylpyridine (E) were also investigated. The compositions of the copolymers were calculated from their nitrogen and unsaturation content. Over a wide range of initial monomer compositions. the mole fraction of C in the copolymers was in the range 0.55 to 0.63, and the copolymers contained only 2 to 3 percent unsaturation, including a high degree of cyclization. The composition of the copolymers of D indicated that cyclization occurred to only a small extent, since the copolymers contained rather high unsaturation content. The values of $r_1 = 0.23$ and $r_2 = 0.12$ were obtained. The mole fraction of E in the copolymers was between 0.85 and 0.998. If the assumption was made that $r_1 \cong r_c \cong 0$, and there was predominant cyclization, $r_2 = 32.0$ in this case. The difference in the composition of the copolymers was attributed to the difference between the electron density of the double bonds in C, D, and E.

For system (C), $\eta_{red.}$ (dl/g; conc. = 0.5 g/dl, DMF) varied over the range 0.12 to 0.18. Mole fraction of C in the copolymers varied from 0.63 for monomer feed of 0.92 to 0.55 for monomer feed of 0.05, indicating that some DVE units participated in the propagation *via* homopolymerization. This system was shown to conform approximately to special Case 2 of the copolymer composition equation (Equation 11–10) referred to earlier. For system (D), $\eta_{red.}$ (dl/g; conc. = 0.5 g/dl, DMF) varied over the range 0.05 to 0.07. Mole fractions of D in the copolymers varied from 0.39 for a monomer feed of 0.25 to 0.64 for a monomer feed of 0.90, indicating that cyclization was minimal in this case and that essentially alternating copolymerization occurred. Analysis of residual unsaturation was in accord with this conclusion. Reactivity ratios were determined as follows: $r_1 = 0.23$, $r_2 = 0.12$. For system (E), $\eta_{red.} = 0.30$ dl/g (conc. = 0.5 g/dl, DMF) for a copolymer containing 0.98 mole fraction E, the copolymer was obtained from a monomer feed of mole fraction E = 0.50.

Table 11–1 Relative Composition of the Copolymers ($[M_1] = [M_2] = 0.5$

Monoolefin	m_1^a	e^b
p-Chlorostyrene[c]	0.0000	−0.3
4-Vinylpyridine	0.025	−0.19
Acrylonitrile	0.251	+1.11
Fumaronitrile	0.393	>1.2
Tetracyanoethylene	0.523	>1.2

[a] m_1 is the molar fraction of divinyl ether in the copolymer.

[b] Value of e with respect to styrene.[3]

[c] Data of Ref. 2.

These results led to a comparison of the behavior of p-chlorostyrene, 4-vinylpyridine, acrylonitrile, fumaronitrile, and tetracyanoethylene in their copolymerization with divinyl ether, shown in Table 11–1.

The ease with which divinyl ether added to the monoolefins increased from 4-vinylpyridine to tetracyanoethylene. p-Chlorostyrene led only to the homopolymer.[2]

The same order of reactivity was obtained when the copolymerization rates, shown in Table 11–2 were compared.

Tetracyanoethylene reacted most rapidly with DVE and was incorporated into the copolymer to a greater extent than any other monoolefin studied, and also produced the most nearly alternating copolymer. The electron acceptor effect of the four -C≡N groups of the tetracyanoethylene decreased the electronic density of the double bond and thus increased the affinity of the divinyl ether for the tetracyanoethylene.[3]

These conclusions were compared with the values of *e*, a measure of polarity. The effect of the cyanide groups became obvious when the copolymerization rates were compared with the copolymer compositions. This effect was strengthened when the number of cyanide substituents increased, thus decreasing the electronic density of the double bond and making it more active in its reaction with the diene.

The steric effect does not seem to be very important. Otherwise, the order in the copolymerization rates would be the reverse of that observed. Therefore, the electronic effect is predominant in these systems. The relative rate and composition of the copolymers were explained on the basis of polarity; the greater the difference in polarity

Table 11–2 Relative Copolymerization Rates

Monoolefin[a]	Temp., °C	Reaction time min.	Conversion %
4-Vinylpyridine	60	80	3.68
Fumaronitrile	60	90	8.56
Tetracyano-ethylene	50	60	11.56

[a] Initial feed monomer molar fraction: $[M_1] = [M_2] = 0.5$

between the two comonomers, the greater the copolymerization rate and the greater the degree of incorporation of monoolefin in the copolymer.

The percentage of cyclization in the copolymerization of a 1,4-diene with a monoolefin may be related to the polarity difference of the two monomers. It has been observed that the percentage of cyclization decreases when the polarity difference between the two monomers increases, Table 11–3.

In a parallel study, the kinetics of the AIBN-initiated copolymerization of divinyl ether (DVE) and ethyl vinyl ether (EVE) with maleic anhydride (MA) were studied in seven different solvents.[8-51] Solvents studied were: acetone, methylene chloride, acetonitrile, benzonitrile, dimethyl formamide, acetophenone, and toluene. The yield at 100% conversion as a function of the feed composition when the total monomer concentration was kept constant gave a confirmation of the composition of these copolymers: DVE/MA = 1/2 and EVE/MA = 1/1. The study of the initial rate as a function of the feed composition made it possible to determine the relative values of the different propagation rate constants consistent with a mechanism by successive and selective additions: in the EVE-MA system, the addition of EVE was slower than the addition of MA; in the DVE-MA system, the addition of DVE was slower than the addition of the first MA molecule, but the addition of the second MA molecule was slower than the first. The study of the dependence of the monomer concentration, of the AIBN concentration, and of the efficiency of the initiator, on the rate of polymerization, showed finally that the true order of the monomer concentration was close to one, although its apparent order varied form one to two. From all the kinetic data it was observed that the mechanism of these copolymerizations could be explained without relying on the participation of the DA complex formed between the monomers. It was pointed out, however, that participation of the complex in a competing mechanism with the above could not be completely excluded.

The mechanism of DVE-MA copolymerization as proposed earlier (See Structure 1–5) consists of three steps (four steps, if the cyclization is included); therefore, a kinetic study presented complications; thus, it was decided to study also the copolymerization of ethyl vinyl ether (EVE) with MA because this system is similar to the DVE-MA system but offers a simpler kinetic treatment, only two steps being involved.

Previous studies of the complexes between vinyl ethers and different electron acceptors always showed 1:1 stoichiometry, and also demonstrated that their equilibrium constants are very small. These studies were extended to include the solvents used in the polymerization experiments. Similar results were obtained: DVE-MA and EVE-MA complexes showed 1:1 stoichiometry in all solvents.

Table 11–3 Effect of Polarity Difference of Comonomers on Cyclization in 1,4 Diene

Monoolefin[a]	Cyclization, %
Acrylonitrile[b]	100
Fumaronitrile	96.3
Tetracyanoethylene	18.4

[a] Initial monomer concentration: $[M_1] = [M_2] = 0.5$

[b] Data of Ref. 8–11

Copolymer yield as a function of monomer composition was determined for different feed compositions after different times of polymerization when the total monomer concentration was kept constant.

When the time of polymerization was long enough so that the rate became very slow, it was observed that the yield for different feed compositions corresponded exactly to the theoretical maximum of conversion of a perfect 1:1 copolymer for the EVE-MA system or a perfect 1:2 copolymer for the DVE-MA system. However, in dimethylformamide (DMF), both EVE-MA and DVE-MA copolymerizations were so slow that a maximum of 15% conversion was observed, even after 24 hours, for MA molar fraction in the feed equal to 0.1–0.2.

The study of the initial rate, as a function of the feed composition when the total monomer concentration was kept constant, should give information on the role of the complex in the mechanism of polymerization. The facts that the copolymers always showed constant compositions and that the rate of copolymerization was extremely fast compared to the rate of homopolymerization of the monomers suggested strongly the existence of specific interactions. As was pointed out earlier, two theories are generally considered. The first[8-14, 8-43] assumes that the specific interaction occurs between the monomers; a complex is formed with a reactivity higher than the monomers; the polymerization is considered as homopolymerization of the complex, and an alternating copolymer is formed. The other[8-85] involves specific interactions between a growing radical and the monomer of different polarity leading also to alternate polymerization; even if a complex formation is not excluded (a complex is evidently expected between monomers of different polarity...), its direct participation in the mechanism is considered to be open to question.

It was observed that the feed composition for which the initial rate was maximum when the total concentration was kept constant did not generally correspond to the composition of the polymers. This conclusion would support preferentially a mechanism by successive and selective additions of monomers rather than by complex participation. However, an alternative explanation to account for these results involved solvation of MA,[8-28] as was pointed out earlier. This system was examined by making the supposition that the copolymerization was only caused by the homopolymerization of the EVE-MA complex. The rate of polymerization could be expressed by equating rate to the product of a rate constant, c, the concentration of k complex, and r, the concentration of growing radicals (Equation 11-14). The concentration

$$\text{rate} = kcr \tag{11-14}$$

of complex is related to the concentration of EVE and MA (represented by a and b) by the equilibrium constant K (for a 1:1 complex)(Equation 11-15) For constant total monomer concentration,

$$K = \frac{c}{(a - c)(b - c)} \tag{11-15}$$

$a + b = \text{constant}$, but for different molar fractions of EVE and MA (M_a and M_b), c takes a maximum value when $a = b$, or $M_a = M_b = 0.5$.

The concentration, r, of growing radicals depends on the concentration of radicals initiating the polymerization per unit time multiplied by the mean lifetime of the radicals. If the variations of initial rate (for variations of M_a and M_b) are only due to variations of r, then there must be variations either of the concentration of the radicals initiating the polymerization or of the mean lifetime of the radicals. If variations in the rate are due only to variations of the concentration of radicals initiating the polymerization, it would be expected that the nitrogen percentage of the polymers (the nitrogen percentage being a direct measurement of the number of butyronitrile radicals initiating the polymerization) should be constant, since for each initiating radical, a constant number of monomers would be added. However,the data indicated that the percentage of nitrogen generally decreased when the rate increased. Therefore, variations of the rate could not be due only to variations of the number of initiating radicals. However, there is no absolute indication of the possible influence of the lifetime of the radicals on the rate of polymerization. There is only a presumption that this influence was negligible by considering the polymerization in acetophenone; in the EVE-MA system, the maximum of the rate was observed in the EVE-rich portion, while in the DVE-MA it was in the MA-rich portion. If the rate depends only on the lifetime of the radicals, this would mean that the lifetime of radicals was higher for lower MA molar fraction in the EVE-MA system, but for higher MA molar fraction in the DVE-MA system. Such phenomena are not highly probable. However, as was shown later, available evidence indicated some chain transfer had occurred in these systems.

On the basis of these considerations, it could be assumed that the concentration, r, of growing radicals was constant. The rate then would only depend on the concentration, c, of complex; and from Equation 11–15, the initial rate when $a + b = \text{constant}$, would take a maximum value when $a = b$, or $M_a = M_b = 0.5$.

It was observed that in every solvent the maximum of the initial rate was found in the EVE-rich portion, i.e., when $a > b$. For this reason it was difficult to explain the mechanism of polymerization by only the participation of the complex as the reactive species. A similar conclusion was drawn when it was observed that the dependence of total monomer concentration on the initial rate could be explained without the consideration of a complex in the mechanism. It is therefore reasonable that the copolymerization of EVE-MA, and the similar copolymerization of DVE-MA, follows preferentially a mechanism of successive and selective additions of monomers on radicals of different polarity. However, the present study does not permit the total exclusion of participation of a complex in addition to the mechanism by which simple monomer additions occur. These conclusions are in accord with those subsequently drawn and extensively discussed with reference to other systems.[8-102,8-103]

If only a mechanism by selective additions is considered, the location of the molar fraction for which the initial rate is maximum permits the evolution of the ratio of the rate constants of the two steps of the EVE-MA copolymerization. If a, b, a·, and b·, represent, respectively, the concentration of EVE, MA, EVE·, and MA·, and k_1' and k_2' the rate constants of addition of EVE and MA, the rate may be expressed by Equation 11–16:

$$\text{rate} = k_1' a b\cdot + k_2' b a\cdot \tag{11–16}$$

Since the composition of the copolymer is 1:1, the rates of addition of EVE and MA are equal (stationary conditions) (Equation 11–17):

$$k_1' a b\cdot = k_2' b a\cdot \tag{11–17}$$

If r is the total concentration of growing radicals ($r = a\cdot + b\cdot$), the following expression is obtained (Equation 11–18):

$$\text{rate} = 2r\frac{k_1'k_2'ab}{k_1'a + k_2'b} \qquad (11\text{–}18)$$

When $a + b = m =$ constant, the maximum of the rate for variations of $M_a(M_a = a/m;\ M_b = b/m)$ occurs when (Equation 11–19)

$$d\ \text{rate}/da = 0 \qquad (11\text{–}19)$$

It was shown that r is constant with the molar fractions of the components; therefore, the derivative of expression (See Structure 11–18) for variations of a is given in Equation 11–20:

$$\frac{d\ \text{rate}}{da} = (m - a)^2 - a\frac{2k_1'}{k_2'} \qquad (11\text{–}20)$$

The maximum initial rate occurs when (Equation 11–21):

$$\frac{m - a^2}{a} = \frac{k_1'}{k_2'} \qquad (11\text{–}21)$$

or when (Equation 11–22):

$$\frac{1 - M_a^2}{M_a} = \frac{k_1'}{k_2'} \qquad (11\text{–}22)$$

M_a in relation (Equation 11–22) is the value of the molar fraction of a for which the rate is maximum; since M_a is determined experimentally, it was possible to evaluate the ratio k_1'/k_2'. The results clearly showed that $k_2' > k_1'$; however, there was no clear correlation between k_1'/k_2' and the dielectric constant of the solvent, although the values ranged from 0.012 for DMF to 0.67 for toluene. The mole fraction of EVE in the monomer feed composition for which the initial rate was maximum ranged from 0.55 for toluene to 0.9 for DMF.

The addition of EVE is a slower process than the addition of MA; the radical of MA is a stable radical, i.e., less reactive, so that the addition of EVE becomes a relatively slow process. It was also observed that in toluene k_2' was comparatively smaller than k_1' than in the other solvents. It is quite possible that the strong complexation of MA with toluene prevents the addition of MA to some extent.

The DVE-MA copolymerization probably follows a mechanism similar to the EVE-MA copolymerization. Similar expressions can be derived. If a, b, $a\cdot$, $b\cdot$, and $a_c\cdot$, respectively represent the concentration of DVE, MA, DVE·, MA·, and cyclic radicals, and if k_1, k_2, and k_3 represent the rate constants of addition of DVE, and of the first and the second MA molecules, the rate expression becomes (Equation 11–23):

$$\text{rate} = k_1ab\cdot + k_2ba\cdot + k_3ba_c\cdot \qquad (11\text{–}23)$$

The total concentration of radical, r, equals $a\cdot + b\cdot + a_c\cdot$; the concentration of the radical just before the cyclization ($b_c\cdot$) can be neglected since it is generally assumed that the cyclization step is a very fast process. In the same manner as in the EVE-MA system, Equation 11-24 was obtained:

$$\text{Rate} = 3r\frac{k_1k_2k_3ab}{k_1k_2a + k_1k_3a + k_2k_3b} \qquad (11\text{-}24)$$

For $a + b = m =$ constant, the derivative is (Equation 11-25):

$$\frac{d\ \text{rate}}{da} = (m - a)^2 - a^2\frac{k_1}{k_2} + \frac{k_1}{k_3} \qquad (11\text{-}25)$$

and the maximum of the initial rate would be observed when the derivative equals 0, therefore (Equation 11-26):

$$\frac{1 - M_a^2}{M_a} = \frac{k_1}{k_2} + \frac{k_1}{k_3} \qquad (11\text{-}26)$$

It was therefore not possible to determine both ratios k_1/k_2 and k_1/k_3 with only the experimental determination of M_a. However, if one assumes that the ratio k_1'/k_2' is comparable to the ratio k_1/k_2 (it does not mean that $k_1' = k_1$ and $k_2' = k_2'$), it becomes possible to evaluate k_1/k_3. The results clearly showed that $k_2 > k_3$ (with the exception of toluene) and that $k_2 > k_1$. There was no clear correlation between k_1/k_2 and the dielectric constant of the solvent. The mole fraction of DVE in the monomer feed composition for which the initial rate was maximum ranged from 0.33 in acetophenone and benzonitrile to 0.85 in DMF. There was no obvious correlation between these values and the dielectric constants of the solvents. Except in toluene, $k_3 < k_2$, the addition of the first molecule of MA was faster than the second one; it was concluded tentatively that the radical $a\cdot$ is more reactive than the radical $a_c\cdot$. This conclusion, however, must be accepted only with caution because it was drawn after making several assumptions.

The order of the polymerization was determined by high precision measurements. The equilibrium constant of the complex formation is generally very small; the concentration of complex is smaller than the concentration of the components. Therefore Equation 11-15 becomes Equation 11-27:

$$Kab = c \qquad (11\text{-}27)$$

When the feed composition is kept constant (a/b-constant), the concentration of the complex is a function of the square of the total monomer concentration, $a + b$. If the complex is the reactive specie of the copolymerization, the rate depends also on the square of the total monomer concentration. An order of two is expected. On the other hand, if the polymerization occurs by successive additions of monomers, a first-order dependence is expected as shown in Equations 11-18 and 11-24. Therefore, the study of the rate of polymerization as a function of the total monomer concentration when the feed composition was kept constant appears to be an interesting experimental approach for further examination of the concept of participation of the complex in the mechanism of copolymerization.

The rate can be expressed by Equation 11-28:

$$\text{rate} = \text{function of } (a + b)^n \qquad (11\text{-}28)$$

The details of the results of the rate study in each solvent and for each monomer concentration showed that, in general, the rates were higher in nonpolar solvents than in polar solvents. These observations were consistent with those previously reported.[8-28] The orders n observed in each solvent were reported to range from 1.00 for acetone to 1.93 for acetonitrile; however, no clear correlation with dielectric constant of solvent was observed. In acetonitrile the experiments were carried out for two different feed compositions; n was found to be constant and is therefore not affected by the feed composition.

It should be pointed out, however, that all of the solvents used were relatively polar in comparison to toluene, and may affect the results by complexation with the monomer, MA. The results in toluene were almost always opposite to those obtained in the other solvents studied.

The order n of the total monomer concentration for DVE-MA took different values from 1 to 2, and seemed to increase with an increase in the dielectric constant of the solvent, although, as was pointed out earlier, there was no complete correlation between the values of n and the dielectric constant of the solvent. Even if it is assumed that the mechanism of polymerization is by successive addition of monomers (Order 1) in some solvents while in others the complex is involved (Order 2), the results are not consistent with this viewpoint. It was generally found that the equilibrium constant for complexation decreased with an increase of dielectric constant of the solvent. Therefore, a decrease of the participation of the complex and a decrease of the order would be expected for an increase of the dielectric constant of the solvent. However, the reverse was observed; in acetonitrile (dielectric constant =37), n = 1.93; in methylene chloride (dielectric constant = 9.08), n = 1.06; in toluene (dielectric constant = 2.43), n = 1.15. It was therefore considered that the order n was probably an apparent order and that other factors were involved. For this reason the dependence of the AIBN concentration on the rate of polymerization was determined. Similar methods were used for the accurate determination of the order m of the AIBN concentration.

The order m in different solvents varied from 0.5 to 1 and seemed also to increase with an increase of the dielectric constant of the solvent although there was no clear correlation throughout the dielectric range studied. But the most significant observation lies in the fact that the orders n and m follow exactly the same trend of variations in the different solvents (except for one point).

The efficiency of the initiator as a function of monomer concentration was determined. The number, N_R, of nitrogen atoms from AIBN initiator bound to the polymer per volume and time unit was determined. The number N_R was taken to be equal to the number of butyronitrile radicals that effectively initiated the polymerization. It was observed that N_R increased with an increase of the total monomer concentration; the higher the monomer concentration, the higher the concentration of growing radicals. The number N_R can be expressed by Equation 11–29:

$$N_R = \text{function of } (M)^p \qquad (11\text{–}29)$$

M represents the total monomer concentration.

Since the efficiency of the initiator is equal to the number N_R divided by the total number of radicals produced by the decomposition of the initiator, the efficiency is also a function of $(M)^p$. The order p was found by plotting the log of N_R vs. the log of the total monomer concentration. The values of p ranged from 0.41 in benzonitrile to 1.13 in acetonitrile. It was pointed out that the values of p were susceptible to relatively large errors since the nitrogen percentages of the polymers were generally very low.

The classical expression of the rate of polymerization initiated by a radical initiator (AIBN) is generally given by Equation 11–30:

$$\text{rate} = \text{function of } (M)^q[f(AIBN)]^m$$

or

$$= \text{function of } (M)^q[(M)^p(AIBN)]^m$$

or

$$= \text{function of } (M)^{q+pm}(AIBN)^m \tag{11–30}$$

While n represented the apparent order of monomer concentration, q represents here the true order, and one has Equation (11–31):

$$n = q + pm \tag{11–31}$$

From the experimental determinations of n, p, and m, q was easily determined. Values for q ranged from 0.62 in toluene to 1.09 in benzonitrile.

As observed, the values of q are generally close to 1, except in toluene where q is lower. An order of 1 is generally expected for radical-initiated polymerization when initiator efficiency and other factors can be neglected. The kinetic data could thus be explained without the necessary participation of a complex, and DVE-MA copolymerization mechanism could involve specific interaction between the monomers and the radicals of different polarity leading to alternating copolymers. The deviations of q from unity were explained by considering the fact that errors of 5% on n, p, and m could lead easily to errors of 20% on q.

Additional experiments were carried out for the EVE-MA copolymerization in acetophenone and methylene chloride. The values of n were 1.90 and 0.92 respectively, while those for m were 0.77 and 0.59. The orders n and m were similar to those obtained in the DVE-MA copolymerization.

The influence of the solvent on the mechanism of termination and in the transfer reaction was also considered. It has been pointed out that m varies from 0.5 to 1 and increases with the dielectric constant of the solvent. In acetophenone and benzonitrile, m was higher than expected. An order of 0.5 indicates a termination by a coupling reaction between growing radicals, while an order of 1.0 indicates a noncoupling termination (reaction between a radical and another specie, or no termination at all). It appears therefore that there was a change in the termination mechanism with the dielectric constant of the solvent; in high dielectric constant solvents, noncoupling termination is more probable, but in low dielectric constant solvents, termination by coupling appears to predominate. In the absence of chain transfer, the number of polymeric chains formed per unit time, N_{pm}, is equal to the number of initiating radicals formed per unit time, N_R, for termination by a noncoupling mechanism or to one half of this number for termination by a coupling mechanism (Equations 11–32 and 11–33):

For noncoupling termination,

$$N_{pm} = N_R \text{ or } N_{pm}/N_R = 1 \tag{11–32}$$

For coupling termination,

$$N_{pm} = 1/2N_R \text{ or } 2N_{pm}/N_R = 1 \tag{11–33}$$

The number N_{pm} was determined from the rate of polymerization, the molecular weight measurements, and Avogadro's number. It was observed that $N_{pm} > N_R$ (it follows that

$2N_{pm} >> N_R$) (an exception was observed in acetonitrile). Therefore, it seems that transfer reactions occur by an increase in the number of polymeric molecules relative to the number of radicals initiating the polymerization.

The extent of the transfer can be evaluated by considering the ratio to be N_{pm}/N_R when the order is $m = 1$ as in benzonitrile, and the ratio to be $2N_{pm}/N_R$ when the order is $m = 0.5$ as in methylene chloride. It was observed that the transfers occurred more often in solvents of low dielectric constant than in solvents of high dielectric constant. Moreover, it was observed that the transfers increased with an increase of the total monomer concentration (except in DMF).

All of these observations suggest the following picture for the transfer and termination processes. A growing radical is solvated by monomer or solvent molecules, but solvation by monomer molecules is relatively favored in solvents of low dielectric constant. Transfer to monomer occurs easier and termination by coupling predominates. In solvents of high dielectric constant, less transfer to monomer occurs and noncoupling termination predominates due to the high solvation of the radicals by solvent molecules.

It is evident that this picture represents only a trend. Finally, that the order m was greater than expected in acetophenone and benzonitrile can be explained by considering that these solvents possess some electron-acceptor character and are capable of a high degree of solvation; noncoupling termination now becomes predominant.

The phase of the polymerization was also considered. With polymerization, it has been observed that the rate of polymerization with precipitation of the polymer during polymerization is generally faster than the rate of polymerization without precipitation. However, in some solvents the phase depends on the total monomer concentration; no specific change of the rate was observed when the polymerization became heterogeneous. In the EVE-MA case, the phase of polymerization was homogeneous in all solvents except CH_2Cl_2 and toluene. In the DVE-MA case, the phase was homogeneous only in DMF, was heterogeneous in CH_2Cl_2, toluene and acetonitrile, but changed from homogeneous to heterogeneous in acetophenone, acetone, and benzonitrile depending on total monomer concentration. The plot of log of rate versus log of monomer concentration remained linear regardless of the phase.

Spontaneous copolymerization of DVE, an electron-donor 1,4-diene, and divinyl sulfone, an electron-acceptor 1,4-diene has been reported to yield a copolymer postulated to contain a 3-methylene-[3.3.1]-bicyclononane repeating unit as shown in Structure 11-1:[8-29] When divinyl ether and divinyl sulfone were mixed in a 1:1 molar ratio

Donor Acceptor

(11-1)

in dioxane, a spontaneous reaction occurred, and a temperature decrease of 5°C was noted. The solution immediately became cloudy, and a precipitate formed. The copolymer was soluble in DMF and had a melting point of 115 to 120°C and $[\eta] = 0.19$ dl/g (DMF). Based upon elemental, IR and NMR analyses, and the known structure of homopolymers of both divinyl ether and divinyl sulfone, a mechanism involving a complex between the two monomers was proposed. It was pointed out that the endothermic heat of mixing could well be an entropy effect signifying a large degree of reorientation of the monomer molecules to a preferred conformation. To establish that initiation was not occurring as the result of the presence of oxygen or other peroxidic impurity, the reaction was repeated under conditions that permitted thorough degassing. The reaction was still spontaneous and was noted to proceed even at −180°C.

In an unusual type of cyclocopolymerization, DVE has been shown to undergo copolymerization with benzaldehyde (BA) utilizing the cationic initiator, $BF_3O(C_2H_5)_2$.[7–184] BA is not homopolymerizable, yet is known to copolymerize with styrene and other monomers by a cationic mechanism. Copolymerization of DVE and BA proceeded smoothly at −78°C when the mole fraction of BA was high to yield copolymers of 7000 to 9000 molecular weight. When the mole fraction of BA in the monomer feed was 0.89 the mole fraction of BA in the copolymer was 0.66 corresponding to the structure shown in Structure 11–2. However, lower mole fractions of BA in the monomer feed generally led to a lower BA ratio in the copolymer, since DVE slowly homopolymerizes in contact with this initiator.

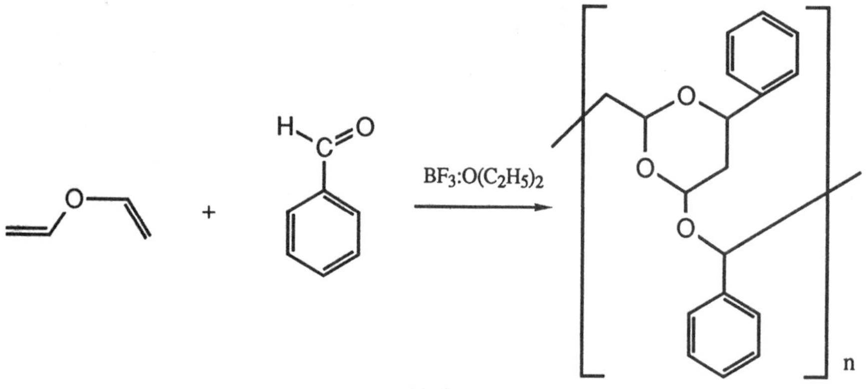

(11–2)

RING SIZE INVESTIGATIONS IN CYCLOCOPOLYMERS

As has been pointed out earlier, the first example of a cyclocopolymer was synthesized in 1951 by copolymerizing divinyl ether (DVE) with maleic anhydride (MA) as shown in Equation 8–1, the copolymer structure can best be represented as an alternating copolymer of DVE and MA in which one MA molecule participates in ring formation.

A radical attacks one of the double bonds of divinyl ether to give the secondary radical $m_1\cdot$. Since vinyl ethers copolymerize well with maleic anhydride, the next step would be the addition of $m_1\cdot$ to maleic anhydride to give radical $m_3\cdot$. If this radical then attacks the β-carbon of the remaining vinyl ether group, a six-membered ring would be formed, and the resulting radical $m_c\cdot$ would again react with maleic anhydride, giving the 1:2 copolymer. The proposed structure was supported by (a) the elemental analysis for the copolymer obtained at high conversion was consistent; (b) the IR spectrum was essentially devoid of residual double bond absorption and contained characteristic absorption bands for cyclic anhydride and six-membered cyclic ether structures; (c) the copolymer composition of 1:2 molar ratio of DVE to MA was consistent with the known reactivity ratios of these types of monomers; (d) the presence of the cyclic ether group was confirmed by chemical evidence which involved cleavage by hydriodic acid and incorporation of iodine into the polymer.

An earlier review has summarized much of the published literature dealing with the size of the ring produced in cyclocopolymerization.[10-139] The proposed six-membered ring was based upon the principle that the more stable radical of a pair of competing propagating radicals should control the course of the reaction.[3-1] Earlier work[8-22] had led to the proposal that the more thermodynamically stable six-membered ring was formed. By analogy to the case of cyclopolymerization of 1,6-dienes, it is now apparent that cyclocopolymerization could lead to the thermodynamically favored six-membered ring, to the kinetically favored five-membered ring, or to a mixture of both.

Although the structure proposed for the cyclocopolymer of divinyl ether and maleic anhydride was supported[1-16] by a broad IR absorption band at 1085–1100 cm^{-1}, attributed to the C-O stretch of a *six* membered cyclic ether, it is now recognized that these structure studies did not exclude the possibility of five-membered rings. However, at the time it was difficult to identify an unknown cyclic ether using the C-O stretch since all ethers have strong, broad absorptions in this region.

Tetrahydrofuran has an intense broad absorption centered at 1070 cm^{-1} and tetrahydropyran has numerous strong bands covering the region 1000–1100 cm^{-1}. Therefore, the presence of a polymer unit involving a five-membered cyclic ether should not be ruled out.

As was discussed earlier, there is little doubt that a 1:2 copolymer is formed; however, there are a number of alternative structures which cannot be eliminated from consideration. Thus, if radical $m_3\cdot$ were to add to the α rather than to the β-vinylic carbon, a polymer containing tetrahydrofuran rings would be formed instead of tetrahydropyran rings (Structure 11–3). Although according to radical stabilities it seems unreasonable to form a primary radical in preference to the more stable secondary radical, a considerable amount of evidence has been accumulating that the formation of five-membered

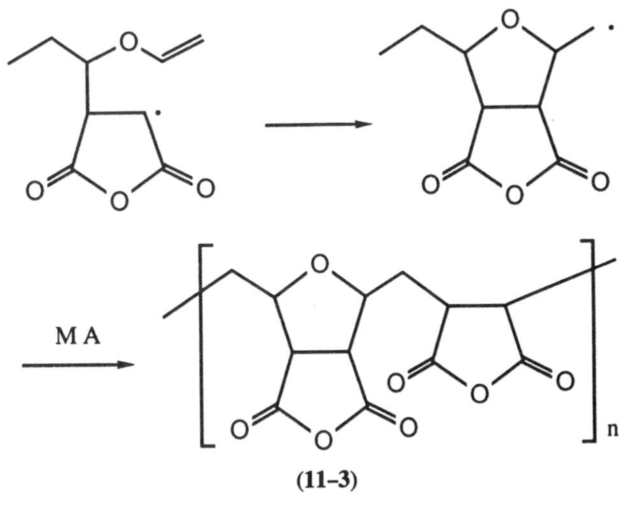

MA

(11–3)

rings is preferred over six, that is, the reactions are under kinetic rather than thermodynamic control.

However, support for the six-membered ring unit in a cyclocopolymer of divinyl ether and dichloromaleic anhydride has been claimed[8-21] on the basis of reaction of the polymer with sodium hydroxide which was followed by back titration using high-frequency oscillometry. It was suggested that only one mole of hydrogen chloride was eliminated. Further, it was proposed that only a six-membered cyclic ether unit (a) could lose hydrogen chloride and that it would be only one mole (Structure 11–4). The

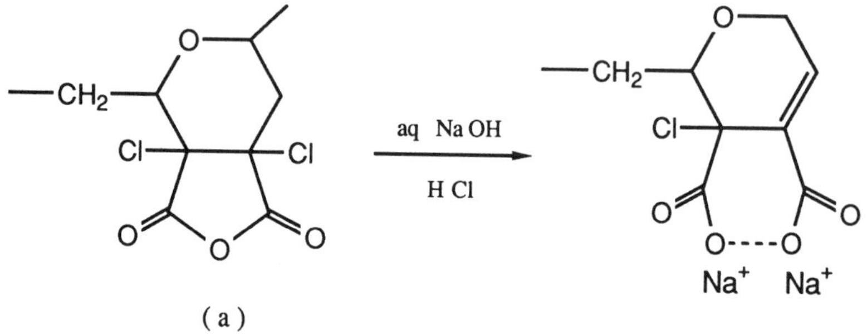

(a)

aq Na OH

H Cl

(11–4)

argument was that loss of a second mole would be extremely difficult due to the destabilizing effect of the adjacent oxygen atom on the incipient negative charge. In the five-membered ring unit [(a), Structure 11–5] both hydrogens which could be removed

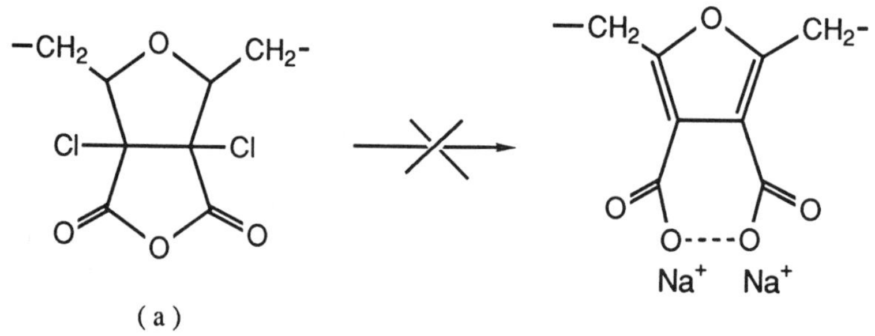

(a)

(11–5)

would be on a carbon atom adjacent to the oxygen atom. However, it had been shown that thermal decomposition of 3-oxa-6-azoniaspiro[5.5]undecane hydroxide (Structure 11–6) gave a mixture consisting of 1-(2-vinyloxyethyl)piperidine (a, 86%) and 4-(pent-

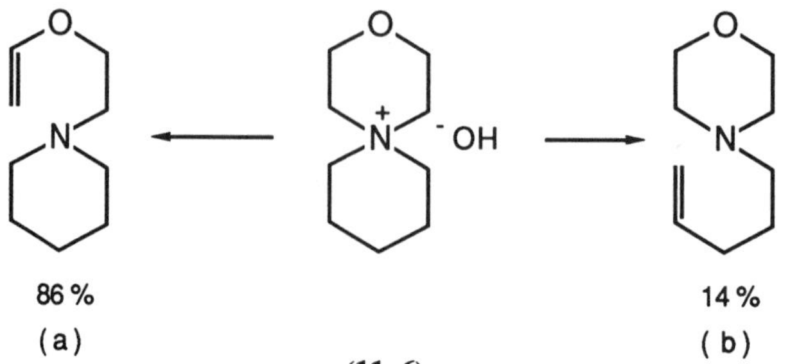

86 %
(a)

14 %
(b)

(11–6)

4-enyl)morpholine (b, 14%). Thus, elimination in the morpholine ring is appreciably easier than in the piperidine ring.[3-213] This suggests that the inductive effect of the oxygen atom causes an increased acidity of the adjacent C-H bond. Without further evidence, particularly studies on model compounds, it is not possible to rule out the existence of a five-membered ring structure in the cyclocopolymer of divinyl ether and dichloromaleic anhydride.

Another structural complication arises form the possibility of chain-branching in the polymer. Although alkyl vinyl ethers do not homopolymerize with a free-radical catalyst, divinyl ether does. At low conversions a soluble polymer can be obtained, which contains only about 20% of the expected unsaturation, so that undoubtedly cyclic polymers are being formed;[4] here too, there is little direct evidence about ring size, and polymers with both five- and six-membered rings may be formed, with vinyl ether groups pendant from the ring or the chain.[4-61] (Structure 11–7). The rate of divinyl

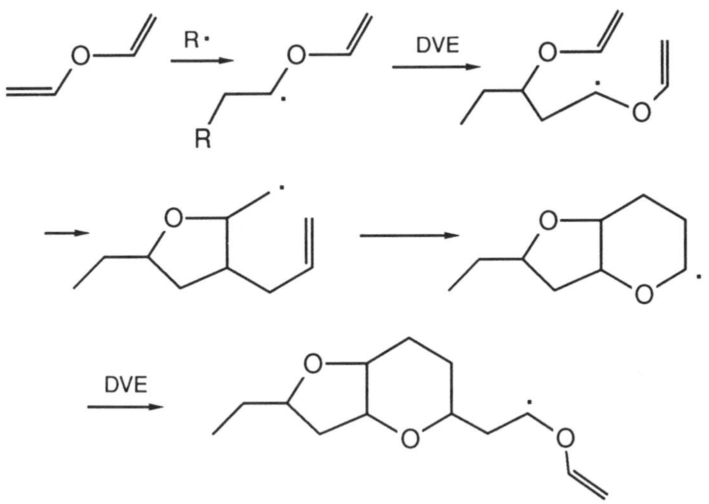

(11–7)

ether homopolymerization is only about one tenth that of its copolymerization with maleic anhydride, so the probability is quite high that some structures of this type would be formed during the polymerization. Since maleic anhydride copolymerizes well with vinyl ethers, the pendant vinyl ether groups would be branch points, leading to long-chain branching in the copolymer, and the degree of unsaturation in the final polymer should be low, as found.

A cyclopolymer of divinyl ether has also been studied using NMR spectroscopy.[4-64] Again the carbon chemical shifts for all the possible structural units and the IR sterioisomers were estimated, using model compounds as a basis for these calculations. The conclusion was that only one kind of monocyclic intermediate was formed, a five-membered ring product with *trans*-stereochemistry at the new bond (See Structure 11–7). Here again, there is a bulky substituent at the radical site and the cyclization of the initially formed radical was not stereoselective towards *cis*-stereochemistry. The monocyclic and bicyclic units were present in the polymer in approximately equal amounts.

The cyclocopolymer of divinyl ether and maleic anhydride has been studied by a number of groups in attempts to clarify its structure.

The methyl ester (DME) of DIVEMA has been used as a molecular probe to identify the structure of DIVEMA.[8-50] Solution light-scattering, gel permeation chromatography, and intrinsic viscosity measurements have shown that tetrahydrofuran at 30°C is a theta solvent for DME and that DME has a random coil conformation with possible long chain branching at higher molecular weights. Determination of the characteristic ratio of DME required identification of its molecular structure. Molecular model studies revealed that the bulky methyl ester groups cause much more steric hindrance in the generally accepted tetrahydropyran structure of DME than in an alternative tetrahydrofuran structure. This observation, together with the polymer solution measurements, indicates the latter structure is more in accord with experimental data, suggesting that both DME and the parent DIVEMA contain tetrahydrofuran in their structures. However, potentiometric titration[5, 8-2, 1-164] has shown that at pH 7, only approximately three of the four carboxyl groups are neutralized. These results have been interpreted to mean that one of the carboxyl groups in the hydrolyzed form of the copolymer is different from the other three, a structural feature that is not consistent with the tetrahydrofuran structure but is consistent with the tetrahydropyran structure.

Also, it has been suggested that pendant vinyl ether groups in units formed by the homopolymerization of divinyl ether (See Structure 11–7) could act as branch points since maleic anhydride copolymerizes readily with vinyl ethers.[1-171]

The results of a study of DIVEMA using the deuterated monomers, dideuteromaleic anhydride and tetradeuterodivinyl ether, and their respective copolymers (Structure 11–8) and (Structure 11–9), followed by a 300 MHz ^{1}H NMR spectroscopic study of the copolymer structures supported predominantly a six-membered ring structure in polymers prepared in cyclohexanone as solvent.[8-53] The spectra led to the suggestion that the *trans*-fused bicyclic unit (Structure 11–10) predominated over the *cis*-fused unit (Structure 11–11), but that both *cis*-and *trans*-chain anhydrides were present in approximately equal amounts (Structures 11–8, 9).

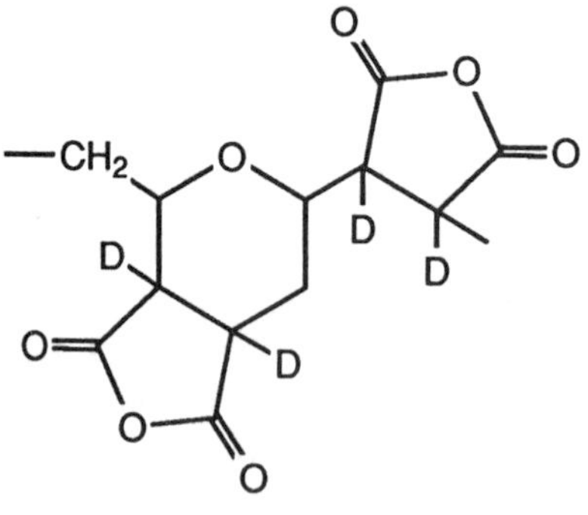

(11–8)

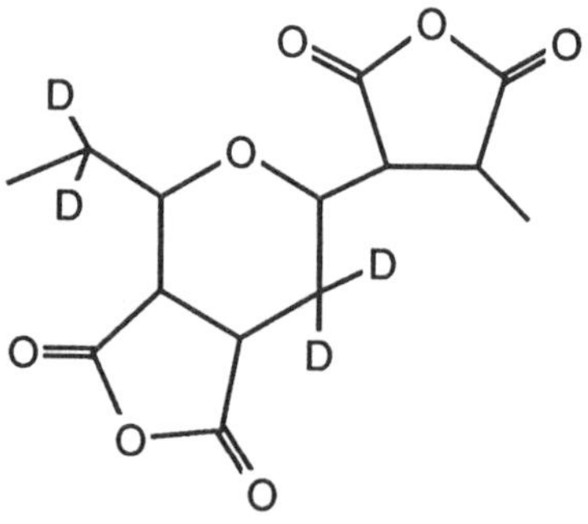

(11–9)

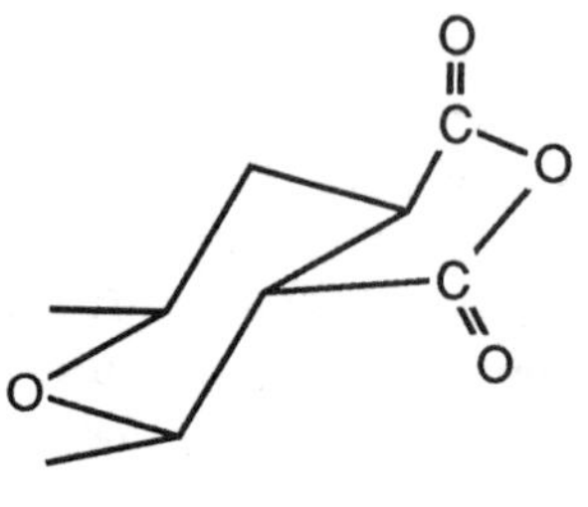

(11–10)

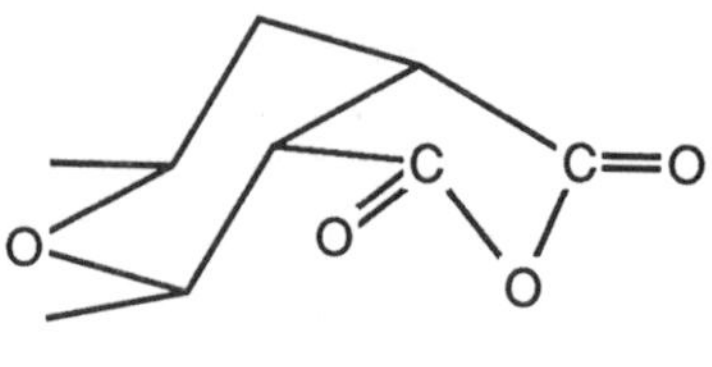

(11–11)

Differing results have been obtained from a ^{13}C NMR study of this cyclocopolymer.[8-2] The carbon chemical shifts for possible units were calculated with the aid of model compounds. The major peaks of a cyclopolymer prepared in chloroform were consistent with the presence of a symmetrical bicyclic unit having *cis*-fused rings attached to a *trans* monocyclic anhydride (Structure 11–12). A polymer produced in acetone/carbon disulphide was estimated to contain 90% six-membered and 10% five-membered rings.

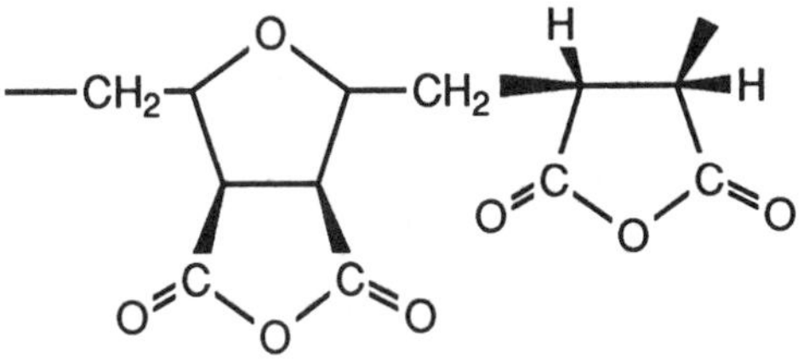

(11–12)

Further evidence has been presented, based primarily on ^{13}C NMR data, that for a copolymer prepared in toluene or benzene, both six-membered ring and five-membered ring structures are present in a ratio of 56:44.[6, 8-60] Using model compounds, all of the ^{13}C peaks in the spectrum of the copolymer were assigned and the only unique (that is, without overlapping) carbon resonance was that of the methylene carbon in the six-membered ring. 2,6-Dimethyltetrahydropyran-3,4-dicarboxylic acid, 2,5-dimethyltetrahydrofuran-3,4-dicarboxylic acid, and 2,3-dimethylsuccinic anhydride were used as model compounds. All spectra examined (including those published earlier)[8-2] contained this peak. There appears to be insufficient spectroscopic evidence available for complete stereochemical assignments to be made. Other copolymers with divinyl ether have also been studied.[8-61]

It is now clear that the original, not unreasonable, proposal of a six-membered ring as the fundamental unit of cyclopolymers and cyclocopolymers is not totally correct. In the polymerization of diallylamines, which have been thoroughly studied, there is agreement that both five- and six-membered ring structures exist in amounts which are dependent upon the structure of the monomer and the conditions of the polymerization process. In other cases, for example the cyclopolymers of divinyl acetals or the cyclocopolymers of divinyl ether and maleic anhydride, different approaches, using different samples of the polymers, have yielded conflicting results. A more systematic study of such examples by complementary methods would be necessary before relationships to monomer structures and polymerization process, or inadequacies in techniques, can be assessed. Obviously, the structures of cyclopolymers and cyclocopolymers cannot be predicted with certainty, as yet, but must be accurately determined if an understanding of their properties and the most probable mechanism of this interesting process is to be established.

Telomerization studies have been used extensively in investigations dealing with ring size and stereochemistry in cyclopolymers.[1-36, 10-57] However, there is some question about whether telomerization and cyclopolymerization would follow the same pathways.[7, 1-19]

Although cyclopolymerization and cyclocopolymerization are certainly related mechanistically, the unusual nature of the latter reaction, in which bimolecular ring closure is highly favored, requires further explanation.[1-158] The early kinetic treatment,[8-11] which suggested that the bimolecular ring closure step is concerted, led to the proposal that the propagating species is a charge-transfer complex.[8-14] Since the complex is a 1:1 molar complex, the cyclocopolymerization may involve an alternating copolymerization between the complex and a second molecule of maleic anhydride. This proposal was supported by evidence for the existence of complexes in several comonomer pairs. Solvent effects on the cyclocopolymerization of maleic anhydride and divinyl ester have provided further support to a charge-transfer complex as the active species.[8-28] However, it has been concluded[8-51] from kinetic data that the mechanism can be explained without invoking charge-transfer complex participation in the mechanism. The role of such complexes has been discussed to considerable length in Chapter 8.

Large-ring-containing cyclocopolymers have been reported. Cyclocopolymerization of the divinyl monomer 1,2-bis[(2-ethenyloxy)]ethoxy)benzene with maleic anhydride *via* radical initiation has been shown[2-464] to produce the 1:2 alternating cyclocopolymer (See Equation 8-8). The monomer pair formed a charge-transfer (or donor-acceptor) complex having an equilibrium constant (K) of 0.280, suggesting participation of the complex in the copolymerization. Also consistent with this proposal was the observation that the maximum rate of copolymerization occurred at the monomer molar ratio of 1:2.

An experimentally reproducible procedure for the alternating cyclopolymerization of maleic anhydride and divinyl ether (DIVEMA) has been published in Macromolecular Syntheses. The polymer was characterized by IR, NMR, and gel permeation chromatography.[8-62] The prolonged interest in DIVEMA is perhaps due to its pronounced biological activity. Calcium salts of DIVEMA have been shown to have much reduced toxicity. Solid calcium salts of DIVEMA were as effective as the corresponding alkali metal salts in treating tumors or viruses, and are much less toxic.[8-160] Additional examples of cyclocopolymerization of 1,4- and 1,5-dienes have been discussed in detail in Chapter 8 as well as being included in several reviews.[1-171, 1-183, 1-178]

Chloromaleic anhydride (CMA) was found to copolymerize with divinyl ether (DVE) to form soluble copolymers of 1:1 composition, devoid of residual unsaturation.[8-19] A bicyclic structure was proposed in which the polymer backbone consisted only of divinyl ether units. DVE and CMA copolymerized readily in methylene chloride at 60°C in the presence of azobisisobutyronitrile to give soluble copolymers of molecular weight ca. 11,000. Copolymer compositions were varied between 0.496 and 0.513 mole fraction CMA. The constant 1:1 composition of the soluble copolymers, and the absence of residual unsaturation led to the proposal of this structure (Structure 11-13) for the copolymers:

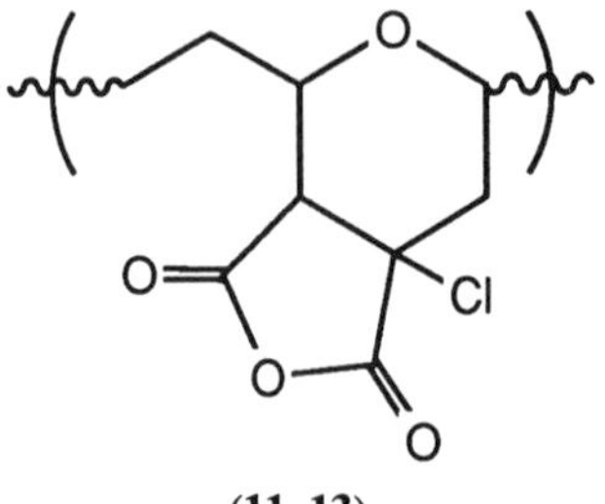

O
Cl
O
O
O

(11-13)

The ease with which the copolymers underwent dehydrohalogenation suggest that the hydrogen and chlorine atoms on the anhydride unit are in a *trans* conformation as a result of a stepwise cyclization process.

Oxidation of the hydrolyzed, dehydrohalogenated copolymers afforded the corresponding vic-diol copolymers. The absence of a significant decrease in copolymer molecular weight upon periodic acid cleavage of the vic-diol copolymers supported the proposed structure. Functional group analyses and softening points were in accord with the structures of the derived copolymers.

Studies of this system should provide additional information about the mechanism of cyclocopolymerization. Assuming the participation of a charge transfer complex in the polymerization, if σ-bond formation between the donor and acceptor portions occurs *via* a concerted process, one would expect the hydrogen and chlorine atoms on the CMA unit to be present in a *cis* configuration. On the other hand, if a stepwise process were involved, a *trans* structure might be predicted. The ease of dehydrohalogenation of the copolymers would be expected to indicate which structure is present, since *trans* eliminations are known to proceed much more readily (Structure 11–14):

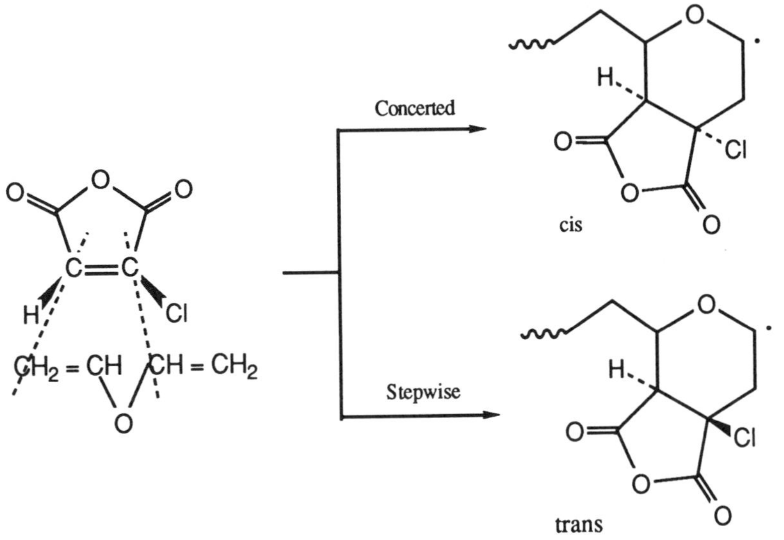

Concerted

cis

Stepwise

trans

(11–14)

Spectroscopic studies indicated the presence of a DVE-CMA charge-transfer complex in chloroform solution. The equilibrium constant for complex formation in $CHCl_3$ (25°C) was found to be 1.5×10^{-1}; $\epsilon = 5.6 \times 10^2$; $\lambda = 299$ nm. The concentration of complex, as in the DVE-MA case is therefore low.

Since CMA units were absent from the copolymer backbone, no major decrease in molecular weight upon periodic acid cleavage was expected to occur. Thus, the following series of reactions were conducted on DVE-CMA copolymer samples, and the molecular weight after each transformation determined (Structure 11–15).

The molecular weights of the derived copolymers indicated only a gradual decrease in molecular weight with each reaction step, amounting to approximately one scission per chain, indicating that only a small amount of random degradation occurred under the reaction conditions employed. For example, Str. 11–13 had M.W. = 11,000; 11–15a, 5240; 11–15b, 2270; and 11–15c, 1500. Of primary significance, however, was the absence of catastrophic decreases in molecular weight upon periodic acid cleavage of the vic-diol copolymer (11–15b). From these results it was concluded that virtually all the CMA units were present in a bicyclic structure (See Structure 11–13).

1) KOH 2) H^+ → (a) → 1) $KMnO_4$ 2) H^+

(b) → HIO_4 → (c)

(11–15)

Finally, a significant decrease in copolymer softening point in proceeding from the cyclic vic-diol copolymer (11–15b) to the linear cleaved copolymer (11–15c) would have been expected. In one experiment, a sample of (11–15b) (M_n = 1085) was found to soften at 215°, while after cleavage to (11–15c) (M_n = 1050), a softening point of 125° was observed. These results were interpreted to support Structure 11–13 for the copolymer, and to support a step-wise cyclization process in this case.

An example of cyclocopolymerization leading to tricyclic polymers has been reported.[8] Polymers obtained from the transannular reaction of sulfur monochloride (S_2Cl_2) with *trans, trans, cis*–1,5,9-cyclododecatriene and *trans,trans,trans*–1,5,9-cyclododecatriene were insoluble, highly reactive, sulfonium polymers. The products resulted from transannular additions and subsequent transannular displacements. Only incomplete structural assignment could be made for the ionic structures, but the reactions were studied by use of model compounds, for example, *trans*–2,9-dichloro-13-thiabicyclo[8.2.1]-5-tridecene. The polymers were reduced to give diepisulfides.

INTERPRETATION OF MECHANISTIC PROPOSALS FOR CYCLOCOPOLYMERIZATION

Mechanistic aspects of the cyclocopolymerization reaction have been reviewed on a number of occasions.[1-156, 1-158, 1-171] The unusual nature of this reaction, in which bimolecular ring closure is highly favored, led to the proposal[1-21] that the propagating species is a charge-transfer complex. This proposal was supported by evidence for the existence of complexes in several comonomer pairs. Solvent effects on the cyclocopolymerization of maleic anhydride and divinyl ether have been studied and further evidence produced to support a charge-transfer complex as the active species in the polymerization process.[8-28] However, it was later concluded from kinetic data that the mechanism could be explained without involving a charge-transfer complex.[8-51]

Although acrylonitrile forms a cyclocopolymer with divinyl ether, the structure is irregular due to the tendency of acrylonitrile to homopolymerize. Addition of Lewis acids, such as zinc chloride and triethylaluminium, increased the rate of polymerization and produced alternating copolymers (2:1) of acrylonitrile and divinyl ether. A charge-transfer complex was identified between divinyl ether and a complex of acrylonitrile with zinc chloride, which added support to a mechanism involving a charge-transfer complex. In a detailed study of charge-transfer complex formation and cyclocopolymerization between divinyl ether and several substituted maleic anhydrides the following conclusions were offered:[8-23]

a. a strong charge-transfer complex produces a 1:1 cyclocopolymer;
b. a sterically hindered anhydride yields a 1:1 cyclocopolymer;
c. a weak charge-transfer complex and a reactive anhydride yields a 1:2 cyclocopolymer.

Anhydrides, which fall into categories (a), (b), and (c) respectively, include chloromaleic anhydride, dimethylmaleic anhydride and maleic anhydride.

Justification for the strong support of the six-membered ring in DIVEMA was based on molecular orbital considerations.[8-53] Structural analysis supported the highly energetically favored six-membered ring copolymer as the product of the copolymerization of the DVE-MA system. During the cyclization step, the vinyloxy double bond would be attacked by a radical on the anhydride unit (Structure 11–16).

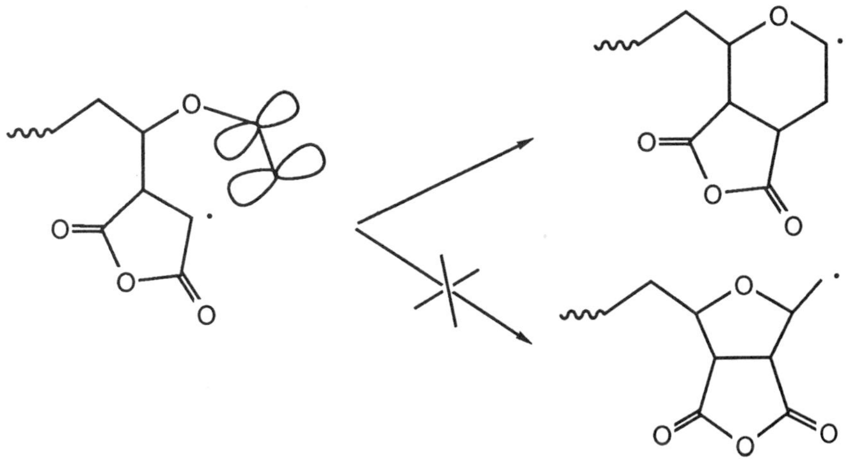

(11–16)

The process would be highly enthalpy controlled as shown by considering the closer energy gap between the highest occupied molecular orbital (HOMO) of the vinyloxy double bond and the singly occupied molecular orbital (SOMO) of the anhydride radical and that of the corresponding symmetrical diene cyclization (Structure 11–17).

The HOMO of the vinyloxy double bond is polarized to have higher orbital density on the terminal position.[9] Therefore, a fast radical addition on the terminal carbon of the double bond leads to a six-membered ring radical. A copolymer with six-membered ring structure is thus obtained.

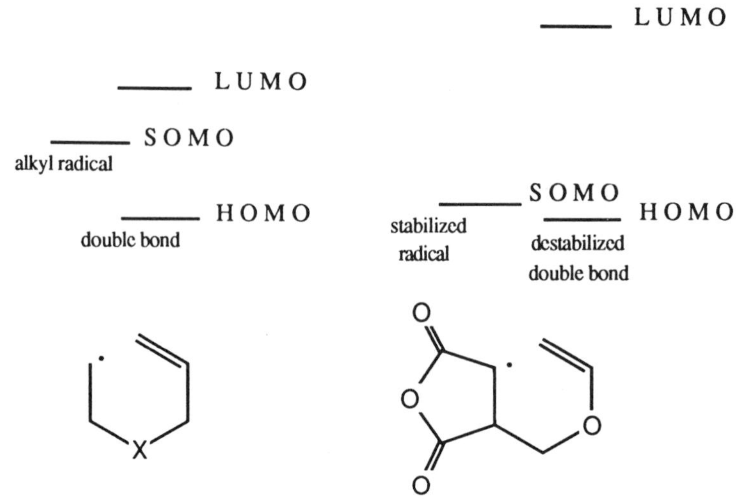

LUMO

LUMO

SOMO
alkyl radical

SOMO

HOMO
double bond

stabilized radical

HOMO
destabilized double bond

(11–17)

A charge-transfer complex has been proposed to explain the rapid cyclization. The charge-transfer complex concept can be applied here with the help of HOMO-LUMO theory to predict the ring structure of the cyclization step. The Mulliken theory of overlapping and orientation principle predicts that stabilization in the molecular complex formation should be determined essentially by the overlap of the donor HOMO and the acceptor LUMO.[1-109] By examination of IR and Raman spectra of DVE, the presence of two rotational isomers[10] was proposed. The more stable isomer has C symmetry, in which the two vinyl groups, although coplanar, are non-equivalent. The microwave spectrum was examined and assigned to the C conformer.[11] A small nonplanarity caused by H-H repulsion between the α hydrogen of the *cis*-vinyl group and the β hydrogen of the *trans*-vinyl group was found (Structure 11–18).

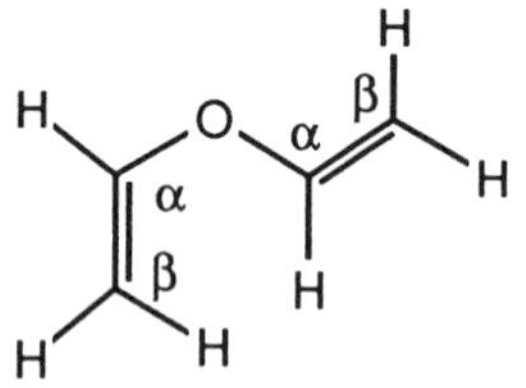

Cis - trans
(11–18)

The charge distributions in vinyl ether and vinyl methyl ether were calculated by the CNDO/2 method.[9] A large electron density was found on the terminal position (Structure 11–19). This orbital charge density actually describes the orbital density of

(11–19)

the HOMO of DVE. The LUMO of MA has been described as Structure 11–20a with higher orbital densities on the double bond carbons but antisymmetrical to the plane of symmetry of this compound.[12]

Therefore, the most stable conformation for a DVE-MA complex can be expected as shown in (11–20), based on the conformer structure of DVE and the molecular densities of both comonomers to give (11–20b) followed by cyclization.

(a) Path a R· (c) Path b R· (b)

(11–20)

When this complex is initiated by a radical, a six-membered cyclic radical would be formed, concerted (Path a) or stepwise, by an anhydride radical addition on the terminal carbon of the vinyloxy unit (Path b). This complexation reduces the energy gap between the complex and the propagating anhydride radical. Thus a radical addition on the complex should occur, and the reaction is supposed to be fast. This special complexation and/or interaction should significantly reduce the activation enthalpy for the formation of a six-membered ring. In the temperature range studied, a five-membered ring formation cannot compete at all, which explains the temperature independence on the structure of the cyclocopolymerization.

Based on this study, the following conclusions were drawn:

1. The intramolecular cyclization is favored over the intermolecular addition, because of the lower entropy change of the former process than the latter. This explains the high degree of cyclization.
2. The entropy preference cannot be explained on the basis of activation energies and statistical probability. Preorientation by the delocalization of the radical with the intramolecular double bond or the formation of complex offers a satisfactory explanation.
3. This preorientation would lead to a six-membered ring structure by a favorable energy factor based on the HOMO orbital density of DVE. For a symmetrical nonconjugated diene, the five-membered ring cyclization is favored by the entropy factor.
4. A faster rate of this cyclocopolymerization than the copolymerization of the corresponding monoolefin pairs can be explained by the closer energy of the anhydride radical to the complex.

Tetrahydronaphthoquinone and dimethyl tetrahydronaphthoquinone formed donor-acceptor complexes with divinyl ether.[8-21] Since participation of such complexes has been considered in the cyclocopolymerization of 1,4-dienes with monoolefins such as divinyl ether-maleic anhydride and divinyl ether-fumaronitrile systems, radical copolymerization of tetrahydronaphthoquinone and dimethyl tetrahydronaphthoquinone with divinyl ether was studied. It was found that these copolymers have constant 1:1 composition regardless of the comonomer ratio. The terpolymerization of divinyl ether-tetrahydronaphthoquinone-dimethyltetrahydronaphthoquinone confirmed the 1:1 donor-acceptor composition in the polymer. Spectroscopic data suggested a cyclized repeating unit in which the copolymer main chain consisted of only divinyl ether units. There was a marked difference between these copolymers and the typical cyclocopolymers, such as those of divinyl ether-maleic anhydride and divinyl ether-fumaronitrile, in which the copolymer main chains consisted of divinyl ether and the comonomer alternately. The overall composition was 1:2. These results were interpreted in terms of the steric effect of the bulky acceptor monomers and the electronic interactions between the comonomers. Competition between an acceptor monomer and the charge-transfer complex toward the cyclized divinyl ether radical in the propagation step appears to favor the charge-transfer complex.

A soluble cyclocopolymer of 1:2 composition was obtained by UV irradiation of divinyl ether and fumaronitrile in methanolic solution.[8-57] By appropriate calculations based on the results of the rate observed under different concentration conditions and with different UV filters, it was shown that both the complex formed between divinyl ether and fumaronitrile, and the noncomplexed species (divinyl ether, fumaronitrile, and methanol) were able to initiate polymerization by light excitation. A study was completed on the same polymerization in the same solvent initiated by a radical initiator. The characteristics of the polymers were the same as those of the photoinitiated polymers. The kinetic results were similar for the polymers obtained by both kinds of initiation, indicating a similar polymerization mechanism.

Acrylonitrile is known to form cyclocopolymers with divinyl ether and 3,3-dimethyl-1,4-pentadiene. Since acrylonitrile has a high tendency toward homopolymerization, the copolymers are not of regular structure. Lewis acids, such as $ZnCl_2$ and $Al(C_2H_5)_3$ were used[8–23] to increase the e-values of acrylonitrile and methacrylonitrile through complexation. Acrylonitrile, methacrylonitrile, and 2- and 4-vinylpyridine were copolymerized with divinyl ether and 1,4-pentadiene with Lewis acids. In all cases the rate of copolymerization was enhanced and the alternating tendency of the cyclopolymer increased with the amount of added Lewis acids. A 1:2 divinyl ether:acrylonitrile alternating cyclopolymer was obtained spontaneously or with AIBN and $Al(C_2H_5)_3$ in hexane. Also, 1:2 alternating cyclocopolymer was obtained in acetone by using a large amount of $ZnCl_2$. The identification of complexation between the divinyl ether and the 2:1 acrylonitrile-zinc chloride complex, and between the 1-hexene and the 2:1 acrylonitrile-zinc chloride complex may support the participation of a complex between all 1,4-dienes studied and the monoolefin-Lewis acid complexes to increase the rate and the alternating tendency.

The equilibrium constants for charge-transfer complex formation between divinyl ether and several substituted maleic anhydrides and 2-cyclopentene-1,4-dione in $CHCl_3$ were measured.[8–22] The copolymerization of divinyl ether with these monoolefins produced regular cyclocopolymers of constant 1:1 or 1:2 (divinyl ether:monoolefin) composition regardless of the comonomer ratio. Comparison of the complexation and the cyclocopolymerization led to the following conclusions: (a) A strong complex gives regular cyclocopolymer of constant 1:1 composition having a copolymer backbone made up of only 1,4-diene units; (b) when a monoolefin is unreactive (often sterically), the 1:1 cyclocopolymer is produced; (c) if complexation is weak and the monoolefin is reactive toward radicals (but not so reactive as to homopolymerize easily), a 1:2 alternating cyclocopolymer is produced. A facile and quantitative elimination of hydrogen halides with dilute aqueous NaOH solution was found. Comparison of the elimination reactions for divinyl ether-chloromaleic anhydride, divinyl ether-bromomaleic anhydride, and divinyl ether-dichloromaleic anhydride 1:1 regular alternating cyclocopolymers led to a conclusion that supports the six-membered ring structure of the repeating cyclic unit formed in cyclocopolymerization.

CONCLUSIONS

Current concepts of cyclocopolymerization would indicate that the structures of cyclocopolymers are highly dependent on the nature of the monomers and the reaction conditions. Predictions of structures based upon thermodynamic control of the cyclization reaction can be misleading, since kinetic products are often found. An important driving force for the cyclization reaction may be an intramolecular interaction, particularly in some cases, where a charge-transfer interaction can occur or is induced, and in cationic-initiated reactions. However, the predominance of cyclization over intermolecular propagation can be ascribed to a smaller decrease in activation entropy which compensates for the unfavorable difference in activation energies. Further studies on the kinetics and energetics of the cyclization reaction during polymerization are necessary before a completely acceptable mechanistic explanation is forthcoming.

Even 30 years after the early studies in this field, a considerable number of research papers and patents are published annually, particularly on the biological properties of

DIVEMA and related structures (see below). The scope of the cyclocopolymerization mechanism for the synthesis of polymers with desirable structures and properties, the microstructure of the polymers and the driving force for these reactions are areas which are still actively investigated.

A SURVEY OF THE BIOLOGICAL ACTIVITY OF DIVEMA AND ITS DERIVATIVES

The use of synthetic polymers as biologically active materials is increasing. Among those polymers receiving most attention by investigators is the alternating cyclocopolymer of divinyl ether and maleic anhydride (See Equation 8–1), commonly known as "pyran" copolymer or DIVEMA. This copolymer has been investigated extensively for a variety of biological effects since the initial observation by scientists at the National Cancer Institute that the hydrolyzed and neutralized copolymer has considerable antitumor activity along with a much reduced toxicity in comparison with other polyanions investigated. DIVEMA has also been shown to be an interferon inducer, to possess antiviral, antibacterial, antifungal, anticoagulant, and antiarthritic activity, to aid in removing polymeric plutonium from the liver, to inhibit viral RNA-dependent DNA polymerase (reverse transcriptase), and to activate macrophages in effecting the immune response of test animals.[8–48]

In a review on mechanistic aspects of synthesis of this copolymer,[1-171] the importance of molecular weight and molecular weight distribution was discussed. A biphasic response of the reticuloendothelial system to the copolymer drug had been observed, which led to the postulate that the copolymer may consist of a toxic molecular weight fraction and another fraction of lower toxicity. Other evidence for molecular weight dependence was based upon the observation that antiviral activity of the copolymer requires a higher molecular weight than antitumor activity. The latter was optimum with low molecular weight samples of narrow molecular weight distribution.

These results led to development of a method for synthesizing a copolymer of narrow molecular weight distribution by photochemically initiating a solution polymerization in acetone, using tetrahydrofuran as a chain transfer agent. This method gave samples having molecular weight distribution ranging from 1.6 to 2.6, in contrast to 3.7 to 7.7 for earlier prepared samples.

Certain of the biological properties of DIVEMA have been summarized in Chapter 8. This section provides further literature citations for the extensive efforts of numerous investigators to provide a thorough understanding of the interactions of this amazingly prolific substance with living systems, including its antitumor activity.

Reviews on Biological Activity of DIVEMA

Several reviews have been published in efforts to keep other investigators informed of progress in the various areas of biological activity of DIVEMA being studied.[13–28]

DIVEMA as an Inducer of Interferon

Soon after the discovery of interferon and its relationship to antiviral activity, DIVEMA was shown to possess the capability of inducing interferon generation in host animals. Although many of the publications cited in this section deal with interferon induction as well as the major topic of classification, a limited number of publications dealt largely with the interferon-generating aspects of the biological properties of DIVEMA.[29-53]

Effects of DIVEMA on the Reticuloendothelial System (RES), Immunological Response, and Macrophage Activity

(a) Reticuloendothelial System: DIVEMA has been investigated extensively *via* changes in the reticuloendothelial system of host animals. Although many other studies classified in this summary also deal with this system, a large number of investigations have been classified under this heading.[54-62]

(b) Immulogical Effects: The immulogical effects of DIVEMA have been classified among the various publications as an immunoadjuvant,[63-70] immunomodulator[71-78] and immunostimulator,[79-85] or its action as an immunoresponse,[86-89] or immunotherapy.[90-94] The reports are rather numerous.

(c) Macrophage Activity: A number of investigations have shown the effect of DIVEMA to specifically involve macrophage activity. Although others may involve activity of this class as well, the papers cited here have emphasized this type of activity.[95-116]

Activity of DIVEMA against Specific Viral or Tumor Systems

(a) Lewis Lung Carcinoma has been the target of antiviral investigations of DIVEMA and MVE-2 over a long period.[117-124]

(b) L1210 leukemia has been targeted for treatment by DIVEMA or MVE-2,4 and 5 in several investigations.[125-132]

(c) A new mouse lung tumor isolated prior to 1974 as a spontaneous tumor from a mouse, and designated as MLC109 (or M109) alveogenic carcinoma has been the target of the antiviral effect of DIVEMA and/or MVE-2 since its discovery.[133-140]

(d) DIVEMA and MVE-2 have been investigated to some extent as antiviral agents against MBL-2 leukemia cells.[141-146]

(e) Friend Leukemia Virus was among the earliest viruses targeted for treatment by DIVEMA.[147] Several investigations have followed this early work.[148-153]

(f) Foot and Mouth Disease Virus (FMDV) has long been a problem with certain farm animals. DIVEMA has shown promise as an effective antiviral agent against this virus.[154-159]

(g) DIVEMA has been studied extensively as an antiviral agent towards Herpes Simplex Virus, both Types 1 and 2 (HSV).[160-166]

Effects of Molecular Weight of DIVEMA on its Biological Response and Physically Modified Forms of DIVEMA (MVE)

It is a well-established fact that radically-initiated vinyl polymers consist of polymer having a distribution of molecular species. It was observed quite early in the history of the biological investigations of DIVEMA that biological responses to different samples were different.

(a) Molecular Weight and Its Effect on Biological Properties: A number of investigations have resulted in reports of these molecular weight dependent biological effects.[167-176]

(b) DIVEMA of Controlled Molecular Weight (MW) and Molecular Weight Distribution (MWD): Although many investigations have dealt with the modified form of

DIVEMA referred to as MVE (most often MVE-2) several comparative studies on a relatively large number of investigations have been concerned specifically with MVE.[177–198]

Chemically Modified Derivatives of DIVEMA

A wide variety of cyclocopolymers structurally related were reported in Chapter 8. However, a number of modified polymers based on chemical reactions have been reported and their biological effects documented.[199–212]

The Effects of DIVEMA on Foreign Substances

The lethal effects of asbestos fibers have been well documented. Studies have been designed to determine the effects of DIVEMA on the biological surface activity of asbestos particles.[213,214]

Chemically induced tumors have been studied over a lengthy period. The effects of DIVEMA on such tumors have been investigated to some degree.[215–218]

The problem of removal of hepatic plutonium by use of DIVEMA as an adjunct to other treatment has been investigated.[219–223]

Other Specific Activities of DIVEMA or MVE-2

(a) Clinical Programs: A clinical study of DIVEMA was undertaken early in the history of development of its biological response.[224–227] A second clinical study was authorized following the development of MVE-2, a modified form of DIVEMA having controlled molecular weight and molecular weight distribution.[228]

(b) Antibacterial: The antibacterial properties of DIVEMA have been discussed to some extent in Chapter 8. However, some specific reports have appeared, and it has been shown that DIVEMA increases host resistance to bacteria as well as to tumors, viruses and fungi.[229–230]

(c) Polymerase Specificity: DIVEMA is a specific inhibitor of viral DNA polymerases as well as wheat germ RNA polymerase and DNA-dependent RNA polymerase from *E. coli*.[231–237]

(d) Phagocytic Activity: Many papers have dealt with phagocytic activity; however, a few have been concerned with this activity specifically.[238,239]

(e) The Effect of DIVEMA on Natural Killer Cells: Several investigations have included results on natural killer (NK) cells. Augmentation of NK activity as well as macrophages occurred, but often conditions optimal for boosting NK activity were different from those optimal for macrophage activation.[240–246]

(f) Interaction of DIVEMA with Cell Membranes: A limited number of investigations have dealt with the interaction of DIVEMA and/or MVE-2 with cell membranes or related systems under a variety of conditions.[247–249]

Particular attention was given to the importance of divalent metal ions, especially Ca^{2+} in promoting the association of DIVEMA with cell and lipid surfaces.[250–253] DIVEMA showed a minor positive effect in adenovirus 12 oncogenesis in hamsters;[251] and caused transient tumor inhibition and some prolonged life of tumor-bearing mice.[252]

DIVEMA was shown to have heparin-like properties by changing the fibrogenogen content of human plasma so that normal clots do not form.[253] and was confirmed to be mitogenic for both B and T cells *in vivo* and B cells *in vitro*.[254]

Four different samples of DIVEMA tested showed greater activity against vaccinia virus in chick embryo cell cultures than in tissue cultures.[255]

DIVEMA prolonged the allograft survival in a study of its effect on the rejection of a neurine leukemic allograft by mice[256] as well as to elicit a migratory response by human neutrophils or monocytes when incubated with serum.[257]

(g) Control of Plant Viruses and Viroids by DIVEMA: Plant viruses and viroids can be inhibited by treating plants infected with or exposed to viruses or viroids by a virus-inhibiting treatment with DIVEMA.[258-260]

REFERENCES

1. Barton, J. M., Butler, G. B. and Chapin, E. C., Polym. Preprints, 1964, 5, 216; CA 62, 14830g (1965).
2. Allen, V. and Butler. G. B., 1965; Quoted in Ref. 1–171.
3. Alfrey, T., Bohrer, J. J. and Mark, H., "Copolymerization", Wiley Interscience, 1952, 91.
4. Aso, C., Ushio, S., and Sogabe, M., Makromol. Chem., 1967, 100, 100; CA 70, 29413t (1969).
5. Ottenbrite, R. M., Goodell, E. M. and Munson, A. E., Polymer, 1977, 18, 461; CA 87, 161438k (1977).
6. Freeman, W. J. and Breslow, D. S., Org. Coat. Plast. Chem., 1981, 44, 108; CA 98, 4941d (1983).
7. Aso, C., Kumitake, T. and Tagami, S., Prog. Polym. Sci. Japan, 1971, 1, 174.
8. Lautenschlaeger, F. K., J. Polym. Sci., Part A-1, 1971, 9, 377; CA 74, 126103g (1971).
9. Fueno, T., Kajimoto, O. and Kobayashi, M., Bull. Chem. Soc. Jpn., 1974, 46, 2316.
10. Clague, A. D. H. and Danti, A., J. Mol. Spectrosc., 1967, 22, 371.
11. Hirose, C. and Curl, R. F., J. Mol. Spectrosc., 1971, 38, 358.
12. Fukui, K., "Theory of Orient. and Stereoselectivity 1975 60.
13. Regelson, W., L'Interferon, 1970, 6, 353.
14. Grossberg, S. E., N. Engl. J. Med., 1972, 287, 122.
15. Finter, N. B., "Interferons and Interferon Inducers" NY, 1973.
16. Regelson, W., Polymer Science and Technology, 1973, 2, 161.
17. Gopalakrishnan, K., Indian J. Cancer, 1975, 12, 30.
18. Ottenbrite, R. M. and Regelson, W., Encycl. of Polymer Sci. and Tech., 1977, 2.
19. Von Hoff, D. D., Rozencweig, M., Soper, W. T., Helman, L. J., Penta, J. S., Davis, H. and L. Muggia, F. M., Cancer Treat, Rep., 1977, 61, 759.
20. Adams, D. O., Ann. Rev. Allergy, (Med. Exam. Press), 1978.
21. Levy, H. B., "Polymeric Drugs", 1978, 305.
22. Morahan, P. S., Prog. Cancer Res., 1981, 16, 185; CA 94, 167227g (1981).
23. Munson, A. E., White, K. L., Jr. and Klykken, P. C., Prog. Cancer Res. Ther , 1981, 16, 329; CA 94, 167228h (1981).
24. Regelson, W., Pharmacol. Ther., 1981, 15, 1; CA 96, 96854n (1982).
25. Breslow, D. S., ACS Symp. Ser., 1982, 195, 1; CA 97, 198584n (1982).
26. Carrano, R. A., Iuliucci, J. D., Luce, J. K., Page, J. A. and Imondi, A. R., Immunol. Ser., 1984, 25, 243; CA 103, 64204x (1985).
27. Breslow, D. S., Chemtech, 1985, 15, 302; CA 102, 220100x (1985).
28. Regelson, W., Int. Encycl. Pharmacol. Ther., 1985, 429; CA 104, 17b (1986).
29. Merigan, T. C., Ciba Found. Sym. on Interferon, London, 1967, 50.
30. Finkelstein, M. S., Bausek, G. H. and Merigan, T. C., Science, 1968, 161, 465.
31. DeClercq, E. and DeSomer, P., Proc. Soc. Exp. Biol. Med., 1969, 132, 699.
32. Minor, N. A. and Koehler, J., Abstr. V8, Bact. Proc., 1969, 150;

33. Nussenzweig, R. S., Vilcek, J. and Jahiel, R., Abstr. 3102, Fed. Proc., 1969, 28,; 815.
34. Pearson, J. W., Griffin, W. and Chirigos, M. A., Proc. Amer. Cancer Res., Abstr. 269, 1969, 10.
35. DeClercq, E., Eckstein, F. and Merigan, T. C., Ann. N.Y. Acad. Sci., 1970, 173, 444.
36. DeClercq, E. and Merigan, T. C., Arch. Intern. Med., 1970, 126, 94.
37. DeClercq, E., Newer, M. R. and Merigan, T. C., J. Clinical Invest., 1970, 49, 1565.
38. Billiau, A., Muyembe, J. J. and DeSomer, P., Nature New Biology, 1971, 232, 183.
39. Pindak, F. F., Schmidt, J. P., Giron, D. J. and Allen, P. T., Proc. Soc. Exp. Biol. Med., 1971, 138, 317.
40. Finkelstein, M. S., McWilliams, M. and Huizenga, C. G., J. Immunol., 1972, 108, 183.
41. Gazdar, A. F., Steinberg, A. D., Spahn, G. F. and Baron, S., Proc. Soc. Exp. Biol. Med., 1972, 139, 1132.
42. Kanady, M. J. and Smith, W. R., Proc. Soc. Exp. Biol. Med., 1972, 141, 794.
43. Morahan, P. S., Regelson, W. and Munson, A. E., Antimicrob. Agents Chemother., 1972, 2, 16.
44. Cahn, F. and Lubin, M., Intervirology, 1973, 1, 376.
45. Tilles, J. G. and Braun, P., Proc. Exp. Biol. Med., 1973, 144, 460.
46. Merigan, T. C., Cancer Chemother. Rep., 1974, 58, 571.
47. Rabinovitch, M., Manejias, R. E., Russo, M. and Abbey, E. E., Cell. Immunol., 1977, 29, 86.
48. Gresser, I., Maury, C., Bandu, M. T., Tovey, M. and Maunoury, M. T., Int. J. Cancer, 1978, 21, 72; CA 88, 115299p (1978).
49. Djeu, J. Y., Heinbaugh, J. A., Holden, H. T. and Herberman, R. B., J. Immunol., 1979, 122, 175; CA 90, 119575q (1979).
50. Giron, D. J., Liu, R. Y., Hemphill, F. E., Pindak, F. F. and Schmidt, J. P., Proc. Soc. Exp. Biol. Med., 1980, 163, 146; CA 92, 108978p (1980).
51. Herberman, R. B., Brunda, M. J., Cannon, G. B., Djeu, J. Y., Nunn, H. M. E., Jett, J. R., Ortaldo, J. R., Reynolds, C., Riccardi, C., and Santoni, A., Prog. Cancer Res. Ther., 1981, 16, 253; CA 95, 18406h (1981).
52. Bartocci, A., Read, E. L., Welker, R. D., Schlick, E., Papademetriou, V. and Chirigos, M. A., Cancer Res., 1982, 42, 3514; CA 97, 142968y (1982).
53. Koestler, T. P., Badger, A, M., Rieman, D. J., Greig, R. and Poste, G., Cell Immunol., 1985, 96, 113; CA 103, 194581d (1985).
54. Munson, A. E. and Wooles, W. R., The Pharmacologist, 1969, 11:174, 261.
55. Regelson, W. and Munson, A. E., Proc. Am. Assoc. Cancer Res., 1970, 11, 66.
56. Regelson, W., Munson, A. E., Wooles, W. R., Lawrence, W, Jr. and Levy, H., L'Interferon, 1970, 6, 381.
57. Hirsch, M. S., Black, P. H. and Proffitt, M. R., Fed. Proc., 1971, 30, 1852.
58. Munson, A. E. and Regelson, W., Proc. Soc. Exper. Biol. Med., 1971, 137, 553.
59. Wooles, W. R. and Munson, A. E., J. Reticuloendothel. Soc., 1971, 9, 108.
60. Drummond, D. and Munson, A., J. Reticuloendothel, Abst.No. 3, 1974, 15, 2a.
61. Kartasheva, A. L., Dubovik, B. V., Etlis, V. S., Nikitenko, N. V., Kutyreva, V. S. and Bandurko, L. N., Zh. Mikrobiol. Epidemiol. Immunobiol., 1977, 143; CA 88, 32539 (1978).
62. Formtling, R. A., Shadomy, H. J. and Kaplan, A, M., Mycopathologia, 1984, 85, 3; CA 101, 65594v (1984).
63. Regelson, W., Ann. Rep. Med. Chem., 1973, 8, 160.
64. Kaplan, A. M., Morahan, P., Baird, L., Snodgrass, M. and Regelson, W., Proc. Amer. Assoc. Cancer Res., Abst.550 1974, 15, 138.
65. Baird, L. G. and Kaplan, A. M., Cell. Immunol., 1975, 20, 167.
66. Baird, L. G., Diss. Abstr. Int. B, 1976, 37, 1615; CA 86, 331 (1977).
67. Weissman, R. M., Coffey, D. S. and Scott, W. W., Oncology, 1977, 34, 133.

68. Chirigos, M. A., Stylos, W. A., Schultz, R. M. and Fullen, J., Cancer Res., 1978, 38, 1093.
69. Drozhennikov, V. A., Kartasheva, A. L., Orlova, E. B. and Perevezentseva, O. S., Byull. Eksp. Biol. Med., 1979, 88, 678; CA 92, 158326v (1980).
70. Kartasheva, A. L., Yuferova, N. V., Drozhennikov, V. A., Orlova, E. B., Perevezentseva, O. S. and Filatov, P. P., Radiobiologiya, 1981, 21, 217; CA 95, 2702r (1981).
71. Hirsch, M. S., Black, P. H., Wood, M. L. and Monaco, A. P., J. Immununol., 1973, 111, 91.
72. Mohr, S. J. and Chirigos, M. A., Prog. in Cancer Res. and Therapy, 1977, 2, 421.
73. Papas, T. S., Chirikjian, J. G., Woods., W. A. and Chirigos, M. A., "Mod. Host Imm. Res. Treat. Ind. Neop., 1974, Proc. 28.
74. Tagliabue, A., Mantovani, A., Polentarutti, N., Vecchi, A. and Spreafico, F., J. Natl. Cancer Inst., 1977, 59, 1019.
75. Etlis, V. S., Kutyreva, V. S., Dubovik, B. V. and Bogomazov, S. D., Khim.-Farm. Zh., 1978, 12, 130; CA 89, 36602z (1978).
76. Puccetti, P. and Giampietri, A., Pharmacol. Res. Commun., 1978, 10, 489; CA 89, 190584b (1978).
77. Stringfellow, D. A., Fitzpatrick, F. A., Sun, F. F. and McGuire, J. C., Prostaglandins, 1978, 16, 901; CA 90, 132909w (1979).
78. Reczek, P. R., Res. Commun. Chem. Pathol. Pharmacol., 1981, 34, 105; CA 96, 832e (1982).
79. Pearson, J. W., Chirigos, M. A., Chaparas, S. D. and Sher, N. A., J. Natl. Cancer Inst., 1974, 52, 463.
80. Fiel, R. J., Mark, E. H. and Levine, H. I., J. Colloid Interface Sci., 1976, 55, 133.
81. Morahan, P. S., Schuller, G. B., Snodgrass, M. J. and Kaplan, A. M., J. Infect. Dis., 1976, 133, A249.
82. Kaplan, A. M., Baird, L. G. and Morahan. P. S., Progress in Cancer Res. and Therapy, 1977, 2, 461.
83. Mohr, S. J., Massicot, J. G. and Chirigos, M. A., Cancer Res., 1978, 38, 1610.
84. Stinnett, J. D., Morris, M. J. and Alexander, J. W., J. Reticuloendothel, 1979, 25, 525; CA 91, 49593p (1979).
85. Marinova, S. and Stoichkov, I., Probl. Onkol., 1980, 8, 40; CA 95, 162131g (1981).
86. Baird, L. G. and Kaplan, A. M., J. Reticuloendothel, Abstr. No. 40, 1974, 16, 20a.
87. Schultz, R. M., Woods, W. A., Mohr, S. J. and Chirigos, M. A., Cancer Res., 1976, 36, 1641.
88. Webster, G. F. and McArthur, W. P., Int. Arch. Allergy Appl. Immunol., 1981, 66, 308; CA 95, 180784b (1981).
89. Stoichkova, N. and Marinova, S., Eksp. Med. Morfol., 1985, 24, 34; CA 104, 424g (1986).
90. Mavligit, G. M., "Immunocancerology in Solid Tumors". 1976, 27.
91. Collins, A. L. and Song, C. W., Radiat. Res., 1977, 70, 688.
92. Collins, A. L. T. and Song, C. W., J. Natl. Cancer Inst., 1978, 60, 1477; CA 89, 123063t (1978).
93. Stoichkov, I., Chirigos, M., Schultz, R. M., Pavlidis, N. A. and Goldin, A., Onkologiya (Sofia), 1981, 18, 149; CA 96, 173805b (1982).
94. Milas, L., Hunter, N., Ito, H. and Turic, M., Tumour Prog. Markers, M. Eu. Can. Res.6, 1981, 241; CA 100, 114658j (1984).
95. Kaplan, A. M., Morahan, P. S. and Munson, J. A., J. Reticuloendothel, Abstr. No. 29, 1974, 15, 17a.
96. Kaplan, A. M., Morahan, P. S. and Regelson, W., J. Natl. Cancer Inst., 1974, 52, 1919.
97. Kaplan, A. M. and Morahan, P. S., J. Reticuloendothel, Abstr. No. 60, 1975, 18, 31a.

98. Schultz, R. M., Papamatheakis, J. D. and Chirigos, M. A., J. Reticuloendothel Soc., Abstr. No. 81, 1975, 18, 41a.
99. Kaplan, A. M. and Morahan, P. S., Ann. N.Y. Acad. Sci., 1976, 276, 134.
100. Morahan, P. S. and Kaplan, A. M., Int. J. Cancer, 1976, 17, 82.
101. Schultz, R. M., Chirigos, M. A. and Kennell, C. J., Proc. Am. Assoc. Cancer Res., 1976, 17, 28.
102. Meltzer, M. S. and Oppenheim, J. J., J. Immunol., 1977, 118, 77.
103. Morahan, P. S., Glasgow, L. A., Crane, J. L, Jr. and Kern, E. R., Cell. Immunol., 1977, 28, 404.
104. Morahan, P. S. and Kaplan, A. M., Prog. in Cancer Res. and Therapy, 1977, 2, 449.
105. Schultz, R. M., Chirigos, M. A., Mohr, S. J. and Woods, W. A., Prog. in Cancer Res. and Therapy, 1977, 2, 437.
106. Schultz, R. M., Papamatheakis, J. D. and Chirigos, M. A., Science., 1977, 197, 674.
107. Schultz, R. M., Papamatheakis, J. D., Luetzeler, J. and Chirigos, M. A., Cancer Res., 1977, 37, 3338.
108. Morahan, P. S. and Kaplan, A. M., Progress in Cancer Res. and Therapy, 1978, 7.
109. Regelson, W., Encyclopedia Pharmacol. Therapeutics, 1977.
110. Schultz, R. M., Pavlidis, N. A., Chirigos, M. A. and Weiss, J. F., Cell Immunol., 1978, 38, 302; CA 89, 86533y (1978).
111. Baerlin, E., Leser, H. G., Deimann, W., Till, G. and Gemsa, D., Heterog. Mononucl. Phagocytes (Workshop) 1980, 243; CA 96, 49551b (1982).
112. Baerlin, E., Leser, H. G., Deimann, W., Resch, K. and Gemsa, D., Int. Arch. Allergy Appl. Immunol., 1981, 66, 180; CA 96, 67147d (1982).
113. Giampietri, A., Puccetti, P. and Contessa, A. R., J. Immunopharmacol., 1981, 3, 251; CA 96, 135496u (1982).
114. Hamilton, T. A., Weiel, J. E. and Adams, D. O., J. Immunol., 1984, 132, 2285; CA 100, 207778p (1984).
115. White, K. L. and Anderson, A. C., Agents Actions, 1984, 15, 562; CA 102, 60398v (1985).
116. Oda, T., Maeda, H., Ueda, M., Kobayasi, T., Hirano, T. and Ohashi, S., Igaku no Ayumi., 1985, 132, 866; CA 103, 16832d (1985).
117. Morahan, P. S., Munson, J. A., Baird, L. G., Kaplan, A. M. and Regelson, W., Cancer Res., 1974, 34, 506.
118. Snodgrass, M., Kaplan, A. M. and Morahan, P., J. Reticuloendothel Soc., Abstr. No 28, 1974, 16, 11a.
119. Kaplan, A. M., Oyler, S. D., Regelson, W. and Morahan, P. S., J. Reticuloendothel, Abstr. No. 1, 1975, 18, 1a.
120. Snodgrass, M. J., Morahan, P. S. and Kaplan, A. M., J. Natl. Cancer Inst., 1975, 55, 455.
121. Stinnett, J. D., J. Reticuloendothel Soc., Abst. No. 58, 1975, 18, 30a.
122. Kaplan, A. M. and Morahan, P. S., Proc. Am. Assoc. Cancer Res., 1976, 17, 156.
123. Marks, T. A., Woodman, R. J. and Kline, I., Fed. Proc., 1976, 35, 623.
124. Kaplan, A. M., Walker, P. L. and Morahan, P. S., "Mod. Host Imm. Res. Treat. Ind. Neop." 1974, No. 28.
125. Mohr, S. J. and Chirigos, M. A., Prog. Cancer Res. Ther., 1978, 7, 415; CA 89, 74112x (1978).
126. Chirigos, M. A. and Stylos, W. A., Cancer Res., 1980, 40, 1967; CA 93, 107035m (1980).
127. Chirigos, M., Papademetriou, V., Bartocci, A. and Read, E., Adv. Imm. Proc. Int. Conf. Imm. 1st., 1980 (Publ. 1981), 217; CA 95, 126010h (1981).
128. Ramonas, L. M., Erickson, L. C., Klesse, W., Kohn, K. W. and Zaharko, D. S., Mol. Pharmacol., 1981, 19, 331; CA 94, 202589a (1981).

129. Rao, V. S. and Mitchell, M. S., J. Biol. Response Modif., 1983, 2, 67; CA 99, 256g (1983).
130. Schlick, E., Hartung, K., Piccoli, M., Bartocci, A. and Chirigos, M., Immunol. Ser., 1984, 25, 513; CA 103, 69574c 1985).
131. Zaharko, D. S. and Covey, J. M., Cancer Treat. Rep., 1984, 68, 1255; CA 102, 238h (1985).
132. Zaharko, D. S., Covey, J. M. and Muneses, C. C., JNCL, J. Natl. Cancer Inst., 1985, 74, 1319; CA 103, 64496u (1985).
133. Woodman, R. J. and Gang, M., Proc. Amer. Assoc. Cancer Res., Abst.436 1974, 15, 109.
134. Marks, T. A., Woodman, R. J., Geran, R. I., Billups, L. H. and Madison. R. M., Cancer Treat. Rep., 1977, 61, 1459.
135. Schultz, R. M., Papamatheakis, J. D., Luetzeler, J., Ruiz, P. and Chirigos, M. A., Cancer Res., 1977, 37, 358.
136. Pavlidis, N. A., Schultz, R. M., Chirigos, M. A. and Luetzeler, J., Cancer Treat. Rep., 1978, 62, 1817; Ca 90, 132815n (1979).
137. Schultz, R. M., Papamatheakis, J. D. and Chirigos, M. A., Prog. Cancer Res. Ther., 1978, 7, 459; CA 89, 53420f (1978).
138. Schrecker, A. W., Schultz, R. M. and Chirigos, M. A., J. Immunopharmacol, 1979, 1, 219; CA 91, 106483d (1979).
139. Loveless, S. E. and Munson, A. E., Cancer Res., 1981, 41, 3901; CA 95, 181064d (1981).
140. Klykken, P. C., Loveless, S. E., Morahan, P. S. and Munson, A. E., J. Immunopharmacol., 1983, 5, 31; CA 100, 17363g (1984).
141. Schultz, R. M., Papamatheakis, J. D. and Chirigos, M. A., Proc. Am. Assoc. Cancer Res., 1977, 18, 23.
142. Schultz, R. M., Papamatheakis, J. D. and Chirigos, M. A., Cell. Immunol., 1977, 29, 403.
143. Schultz, R. M., Chirigos, M. A. and Papamatheakis, J. D., Cancer Immunol. Immunotherapy, 1978, 3, 183; CA 89, 53394a (1978).
144. Bartocci, A., Papademetriou, V. and Chirigos, M. A., J. Immunopharmacol, 1980, 2, 149; CA 97, 418v (1982).
145. Schlick, E., Riffmann, R., Chirigos, M. A., Weiker, R. D. and Herberman, R. B., Cancer Res., 1985, 45, 1108; CA 102, 197617k (1985).
146. Chirigos, M. A., Saito, T., Schlick, E. and Ruffman, R., NIH, Cancer Treat. Symp., 1985, 1, 11; CA 103, 64451a (1985).
147. Regelson, W., Proc. Int. Sym. Ath. Reticulo. Italy, 1966.
148. Regelson, W. and Foltyn, O., Proc. Amer. Assoc. Cancer Res., 1966, 7, 228.
149. Schuller, G. B., Morahan, P. S. and Snodgrass, M. J., J. Reticuloendothel Soc., Abstr. No 28, 1974, 15, 16a.
150. Schuller, G. B., Morahan, P. S. and Snodgrass, M., Cancer Res., 1975, 35, 1915.
151. Roberts, P. S. and Regelson, W., Prog. Cancer Res. and Therapy, 1977, 2, 549.
152. Schuller, G. B., Diss. Abstr. Int. B., 1977, 37, 3784; CA 86, 137801 (1977).
153. Schuller, G. B. and Morahan, P. S., Cancer Res., 1977, 37, 4064.
154. Richmond, J. Y. and Campbell, C. H., Arch. ges. Virusforsch., 1972, 36, 232.
155. Sellers, R. F., Herniman, K. A. J. and Hawkins, C. W., Res. Vet. Sci., 1972, 13, 339.
156. Campbell, C. H. and Richmond, J. Y., Infect. Immun., 1973, 7, 199.
157. McVicar, J. W., Richmond, J. Y., Campbell, C. H. and Hamilton, L. D., Can. J. Comp. Med., 1973, 37, 362.
158. Campbell, C. H., Richmond, J. Y. and McKercher, P. D., Arch. ges. Virusforsch, 1974, 46, 334.
159. Richmond, J. Y. and Campbell, C. H., Infect. Immun., 1974, 10, 1029.

160. Galin, M. A. and Weissenbacher, M., J. Gen. Physiol., 1970, 56, 249.
161. Morahan, P. S. and McCord, R. S., J. Immunol., 1975, 115, 311.
162. McCord, R. S., Breinig, M. K. and Morahan, P. S., Antimicrob. Agents Chemother., 1976, 10, 28.
163. Morahan, P. S., Kern, E. R. and Glasgow, L. A., Proc. Soc. Exp. Biol. Med., 1977, 154, 615.
164. Breinig, M. C., Wright, L. L., McGeorge, M. B. and Morahan, P. S., Arch. Virology, 1978, 57, 25.
165. Thomson, T. A., Hilfenhaus, J., Moser, H. and Morohan, P. S., Infect. Immun., 1983, 41, 556; CA 99, 86277z (1983).
166. Morahan, P. S., Dempsey, W. L., Volkman, A. and Connor, J., Infect. Immun., 1986, 51, 87; CA 104, 101992w (1986).
167. Regelson, W., J. Med. (Basel), 1974, 5, 50.
168. Regelson, W., Morahan, P., Kaplan, A. M., Baird, L. G. and Munson, J. A., Excerpta Medica, Amsterdam, Neth., 1974, 97; CA 83, 71832 (1975).
169. Regelson, W., Morahan, P. and Kaplan, A., "Polyelectrolytes and their Appl." 1975, 1, 131.
170. George, J. and Chirikjian, J. G., Nucleic Acids Res., 1978, 5, 2223; CA 89, 125096e (1978).
171. Lavine, H. I., Mark, E. H. and Fiel, R. J., J. Colloid Interface Sci., 1978, 63, 242; CA 88, 131132j (1978).
172. Morahan, P. S., Barnes D. W. and Munson, A. E., Cancer Treat. Rep., 1978, 62, 1797; Ca 90, 114934w (1979).
173. Barnes, D. W., Morahan, P. S., Loveless, S. and Munson, A. E., J. Pharmacol. Exp. Ther., 1979, 208, 392; CA 90, 162210d (1979).
174. Zander, A, R., Spitzer, G., Verma, D. S., Ginzbarg, S. and Dicks, K. A., Biomed. Express, 1980, 33, 69; CA 93, 142997z (1980).
175. Ottenbrite, R. M., Munson, A. E., Klykken, P. C. and Kaplan, A. M., Org. Coat. Plast. Chem., 1981, 44, 115; CA 96, 11141w (1982).
176. Ottenbrite, R. M., J. Macromol. Sci. (Chem.), 1985, A22, 819; CA 103, 64383e (1985).
177. Dean, J. H., Luster, M. I., Boorman, G. A., Lauer, L. D., Adams, D. O., Padarthaingh, M. L., Jerrella, T. R. and Mantovani, A., Prog. Cancer Res. Ther., 1981, 16, 267; CA 94, 167885v (1981).
178. Milas, L., Hersh, E. M. and Hunter, N., Collog. Inst. Natl. Sante. Rech. Med., 1980 (Publ. 1981), 97, 147; CA 95, 144024v (1982).
179. Milas, L., Hersh, E. M. and Hunter, N., Cancer Res., 1981, 41, 2378; CA 95, 18362r (1982).
180. McCormick. D. L., Becci, P. J. and Moon, R. C., Carcinogenesis (London), 1982, 3, 1473; CA 98, 65191f (1983).
181. Adams, D. O., Johnson, W. J., Marino, P. A. and Dean, J. H., Cancer Res., 1983, 43, 3633; CA 99, 169186v (1983).
182. Khato, J., Chirogos, M. A. and Sieber, S. M., J. Immunopharmacol., 1983, 5, 65; CA 100, 17304h (1984).
183. Rosenblum, M. G., Hersh, E, M. and Loo, T. L., J. Chromatogr., 1983, 272, 200; CA 98, 119029e (1983).
184. Schultz, R. M. and Altom, M. G., J. Immunopharmacol., 1983, 5, 277; CA 100, 114670g (1984).
185. Wiltrout, R. H., Brunda, M. J., Gorelik, E., Peterson, E. S., Dunn, J. J., Leonhardt, J., Varesio, L., Reynolds, C. W. and Holden, H. T., J. Reticuloendothel. Soc., 1983, 34, 253; CA 100, 4676f (1984).
186. Baldwin, J. R., Carrano, R. A., Imondi, A. R., Iuliucci, J. D. and Hagerman, L. M., Polym. Mater. Sci. Eng., 1984, 51, 136; CA 101, 122676z (1984).

187. Piccoli, M., Saito, T. and Chirigos, M. A., Int. J. Immunopharmacol., 1984, 6, 569; CA 102, 55802t (1985).
188. Schlick, E., Hartung, K. and Chirigos, M. A., Cancer Immunol. Immunother., 1984, 18, 226; CA 102, 142852c (1985).
189. Schlick, E., Welker, R., Piccoli, M., Ruffmann, R. and Chirigos, M., Prog. Clin. Biol. Res., 1984, 161, 511; CA 101, 122685b (1984).
190. Shopp, G. M. and Munson, A. E., Prog. Clin. Biol. Res., 1984, 161, 501; CA 101, 204018a (1984).
191. Talmadge, J. E., Maluish, A. E., Collins, M., Schneider, M., Herberman, R. B., Oldham, R. K. and Wiltrout, R. H., J. Biol. Response Modif., 1984, 3, 634; CA 102, 89797k (1985).
192. Chirigos, M. A., Schlick, E. and Hartung, K., Prostaglandine Immun., 1985, 161; CA 104, 67308r (1986).
193. Ruffmann, R., Schlick, E., Tartaris, T., Gruys, E., Welker, R. D., Saito, T. and Chirigos, M. A., Med. Oncol. Tumor Pharmacother., 1985, 2, 195; CA 104, 33379z (1986).
194. Saito, T., Welker, R. D., Fukui, H., Herberman, R. B. and Chirigos, M. A., Cell Immunol., 1985, 90, 577; CA 102, 89812m (1985).
195. Schlick, E., Ruffmann, R., Hartung, K. and Chirigos, M. A., J. Immunopharmacol., 1985, 7, 141; CA 103, 16543k (1985).
196. Talmadge, J. E., Herberman, R. H., Chirigos, M. A., Maluish, A. E., Schneider, M. A., Adams, J. S., Philips, H., Thurman, G. B., Varesio, L., J., et al, Immunol., 1985, 135, 2483; CA 103, 158855n (1985).
197. Chirigos, M. A., Schlick, E., Saito, T. and Gruys, E., Methods Find. Exp. Clin. Pharmacol., 1986, 8, 27; CA 104, 199687y (1986).
198. Zhang, S. R., Salup, R. R., Urias, P. E., Twilley, T. A., Talmadge, J., Herberman, R. B. and Wiltrout, R. H., Cancer Immunol. Immunother., 1986, 21, 19; CA 104, 179871a (1986).
199. Papamatheakis, J. D., Schultz, R. M., Chirigos, M. A. and Massicot, J. D., Cancer Treatment Repts., 1978, 62, 1845.
200. Przybylski, M., Fung, W. P., Ringsdorf, H., Adamson, R. H. and Zaharko, D. S., Proc. Am. Assoc. Cancer Res., 1978, 19, Abstr. 6.
201. Fung, W-P., Przybylski, M., Ringsdorf, H. and Zaharko, D. S., J. Natl. Cancer Inst., 1979, 62, 1261; CA 91, 83051v (1979).
202. Hirano, T., Klesse, W. and Ringsdorf, H., Makromol. Chem., 1979, 180, 1125; CA 90, 204801e (1979).
203. Ringsdorf, H. and Przybylski, M., Ger. Pat. 2,830,901, 1979 (Feb. 1); CA 90, 157088r (1979).
204. Hirano, T., Ringsdorf, H. and Zaharko, D. S., Cancer Res., 1980, 40, 2263; CA 94, 197k (1981).
205. Gros, L., Lloyd, J. B., Ringsdorf, H., Schnee, R., Schorlemmer, H. U. and Stotter, H., Polym. Prepr., ACS (Div. Polym. Chem.), 1981, 22, 21; CA 98, 66587b (1983).
206. Pratten, M. K., Duncan, R., Cable, H. C., Schnee, R., Ringsdorf, H. and Lloyd, J. B., Chem.-Biol. Interact., 1981, 35, 319; CA 95, 73082z (1981).
207. Gros, L., Ringsdorf, H., Schnee, R., Lloyd, J. B., Ruede, E., Schorlemmer, H. U. and Stoetter, H., ACS Symp. Ser., 1982, 195, 83; CA 97, 121585f (1982).
208. Agency of Industrial Sciences and Technology, Jap. Pat. JP 60 67,426 [85 67,426], 1985 (Apr. 17); CA 103, 92828x (1985).
209. Agency of Industrial Sciences and Technology, Jap. Pat. JP 60 67,490 [85 67,490], 1985 (Apr. 17); CA 104, 34302f (1986).
210. Agency of Industrial Sciences and Technology, Jap. Pat. JP 60 67,493 [85 67,493], 1985 (Apr. 17); CA 103, 71625p (1985).

211. Agency of Industrial Sciences and Technology, Jap. Pat. JP 60 81,197 [85 81,197], 1985 (May 9); CA 104, 6146q (1986).
212. Zunino, F., Gambetta, R. A. and Penco, S., Belg. BE 902,344, 1985 (Sept. 2); CA 104, 130218d (1986).
213. Schnitzer, R. J., Bunescu, G. and Baden, V., Ann. N.Y. Acad. Sci., 1971, 172, 757.
214. Schnitzer, R. J., Environ. Health Perspectives, 1974, 9, 261.
215. Kripke, M. L. and Borsos, T., J. Natl. Cancer Inst., 1974, 53, 1409.
216. Harmel, R. P., Jr. and Zbar, B., J. Natl. Cancer Inst., 1975, 54, 989.
217. Elzay, R. P. and Regelson, W., J. Dent. Res., 1976, 55, 1138.
218. Gould, A. R., Miller, C. H. and Kafrawy, A. H., Arch. Oral Biol, 1981, 26, 761; CA 96, 117234c (1982).
219. Lindenbaum, A, Rosenthal, M. W., Baxter, D. W., Egan, N. E., Kalesperis, G. S., Moretti, E. S. and Russell, J. J., Div. Biol. Med. Res., Argonne Nat. Lab., 1972 (Ann. Rep.), ANL-7970, 121.
220. Baxter, D. W., Rosenthal, M. W. and Lindenbaum, A., Ann. Meeting Radiation Res. Soc., MO, 1973, Abstr. 21, Bd-10.
221. Lindenbaum, A., Rosenthal, M. W. and Guilmette, R. A., Proc. IAEA: Treat. Diag. Int. Dep. Rad., 1975, 357; CA 84, 173824 (1976).
222. Guilmette, R. A. and Lindenbaum, A., "Health Effects of Plutonium and Radium" 1976, 223.
223. Lindenbaum, A. and Rosenthal, M. W., U. S. Pat. 4,143,131, 1979 (Mar. 6); CA 90, 182449s (1979).
224. Anon., JAMA Medical News, 1967, 201, 27.
225. Dennis, L. H., Angeles, A., Baig, M. and Shnider, B. I., Amer. Soc. Clin. Pharm. Chem. Abst. Meet 1969.
226. Leavitt, T. J., Merigan, T. C. and Freeman, J. M., Amer. J. Dis. Child., 1971, 121, 43.
227. Munson, A. E., Regelson, W. and Munson, J. A., J. Tox. Appl. Pharm., 1972, 22, 299.
228. Carrano, R. A., Kinoshita, F. K., Imondi, A. R. and Iulicci, J. D., Prog. Cancer Res. Ther., 1981, 16, 345; CA 94, 185145h (1981).
229. Pindak, F. F., Appl. Microbiol., 1970, 19, 188; CA 72, 76931 (1970).
230. Baird, L. G., Kaplan, A. M. and Regelson, W., J. Retculoendothel, Abstr. No. 90, 1974, 15, 51a.
231. Chirigos, M. A. and Papas, T. S., Adv. Pharmacol. Chemother., 1975, 12, 89.
232. Chirikjian, J. G., Rye, L. and Papas, T. S., Proc. Nat. Acad. Sci. USA, 1975, 72, 1142.
233. Fiel, R. J., Musser, D. A. and Munson, B. R., J. Natl. Cancer Inst., 1976, 57, 1319.
234. Papas, T. S., Pry, T. W., Schafer, M. P. and Sonstegard, R. A., Cancer Res., 1977, 37, 3214.
235. DiCioccio, R. A. and Srivastava, B. I. S., Biochem. J., 1978, 175, 519; CA 90, 147580f (1979).
236. Greene, R. and Munson, B., Can. J. Biochem., 1980, 58, 295; CA 93, 2962j (1980).
237. O'Conner, D. L. and Kumar, S. A., Biochem. Int., 1983, 6, 489; CA 98, 194027f (1983).
238. Munson, A. E., Regelson, W. and Wooles, W. R., Abst. 12, 6th. Ann. Meet. J. Reticul. S. 1969, 632.
239. Kapila, K., Smith, C. and Rubin, A. A., J. Reticuloendothel, 1971, 9, 447.
240. Puccetti, P., Santoni, A., Riccardi, C., Holden, H. T. and Herberman, R. B., Int. J. Cancer, 1979, 24, 819; CA 92, 104422e (1980).
241. Santoni, A., Puccetti, P., Riccardi, C., Herberman, R. B. and Bonmassar, E., Int. J. Cancer, 1979, 24, 656; CA 92, 34384t (1980).
242. Santoni, A., Riccardi, C., Barlozzari, T. and Herberman, R. B., Nat. Cell-Mediated Immun. Tumors, 1980, 753; CA 93, 230820s (1980).
243. Santoni, A., Riccardi, C., Barlozzari, T. and Herberman, R. B., Int. J. Cancer, 1980, 26, 837; CA 94, 95943m (1981).

244. Chirigos, M. A., Schlick, E., Piccoli, M., Read, E., Hartung, K. and Bartocci, A., Adv. Immunopharm. 2, Proc. Int. Conf.2nd 1982, 669; CA 99, 115683m (1983).
245. Hartung, K., Schlick, E., Stevenson, H. C. and Chirigos, M. A., J. Immunopharmacol., 1983, 5, 129; CA 100, 18427m (1984).
246. Janiak, M., Budzynski, W., Gnatowski, B., Radzikowski, C., Szmigielski, S., Jeljaszewicz, J. and Pulverer, G., Immunobiology (Stuttgart), 1984, 167, 328; CA 102, 39603y (1985).
247. Boyd, P. M. and Tirrell, D. A., Polym. Prepr. ACS, Div. Polym. Chem., 1980, 21, 188; CA 96, 30133v (1982).
248. Zander, A. R., Spitzer, G., Verma, D. S., Beran, M. and Dicke, K. A., Exp. Hematol. (Copenhagen), 1980, 8, 521; CA 93, 88664y (1980).
249. Marwaha, L. K., Boyde, P. M. and Tirrell, D. A., Org. Coat. Plast. Chem., 1981, 44, 211; CA 97, 207887t (1982).
250. Marwaha, L. K. and Tirrell, D. A., ACS Symp. Ser., 1982, 186, 163; CA 97, 267v (1982).
251. Larson, V. M., Clark, W. R. and Hilleman, M. R., Proc. Soc. Exp. Biol. Med., 1969, 131, 1002.
252. Sandberg, J. and Goldin, A., Cancer Chemother. Rep., 1971, 55, 233.
253. Roberts, P. S., Regelson, W. and Kingsbury, B., J. Lab. Clinical Med., 1973, 82, 822.
254. Baird, L. G. and Kaplan, A. M., J. Reticuloendothel Soc., Abstr. No. 36, 1975, 18, 19a.
255. Ter-Pogosyan, R. A., Vartevanyan, Z. T., Kamalyan, L. A., Dubovik, B. V., Kartasheva, A. L. and Etlis, V. S., Zh. Eksp. Klin. Med., 1975, 15, 15; CA 85, 40739 (1976).
256. Mohr, S. J., Chirigos, M. A., Fuhrman, F. and Smith, G., Cancer Res., 1976, 36, 1315.;
257. Majeski, J. A. and Stinnett, J. D., J. Natl. Cancer Inst., 1977, 58, 781.
258. Breslow, D. S. and Chadwick, A. A., U.S. Pat. US 3,996,347, 1976.
259. Hercules Inc., Fr. Pat. 2,368,222, 1978 (May 19); CA 90, 116427g (1979).
260. Hercules, Inc., Isreali Pat. 50,689, 1980 (Jan. 31); CA 93, 20743t (1980).

12

Practical Significance of Cyclopolymerization

Many diverse commercial products have been developed by utilizing cyclopolymerization. The first cyclopolymer to be manufactured in commercial quantities apparently was poly(dimethyldiallylammonium chloride).[2-154] This polymer is manufactured by a number of suppliers and has been shown to possess optimum functional properties for application to electrographic paper reproduction processes.[2-90] In such processes, proper functioning of the photo-responsive coating depends on the rapid dissipation of static electrical charges by the substrate. Electroconductivity, normally expressed as resistivity, is dependent upon the water content of the substrate. Polyelectrolytes, because of their ionogenic nature, provide both a source of ions and a high degree of hygroscopicity; both properties are necessary in order to provide electroconductivity.

APPLICATIONS OF POLYMERS AND COPOLYMERS OF DIALLYLDIMETHYLAMMONIUM CHLORIDE (DADMAC)

A review and use analysis of cyclopolymerization of dialkyldiallylammonium halides has recently been published.[1-169] In addition to the above referenced use of these materials, other uses as paper additives include antistatic agents, fluorescent whiteners, paperboard reinforcement, and retention aids. In the water treatment field, the polymer is used as a flocculent and/or as a primary coagulant or coagulant aid in potable water, wastewater, coal flotation, foam flotation of metal sulfides, etc. It is also reported to be used in the zinc, tin, and lead electroplating industries, as well as in the cosmetic field, as a biocide in water, as a demulsifier of dispersed oils, and as a detergent additive.

Anion-exchange materials containing the quaternary ammonium cation in the polymeric network were synthesized as early as 1949,[1-1] However, a reinvestigation in this area[1] has shown that the earlier studied ion exchanger, *N,N,N′, N′*,-tetraallyl-*N,N′*-dimethylethylenediammonium dichloride, gave both superior rates and capacity for use in extraction of uranium.

Some of the experimental procedures for making copolymers of dimethyldiallylammonium chloride, as well as some experimental variables, and procedures for evaluating the copolymers in flocculation and sludge dewatering, have been summarized.[2-91, 2]

The preparation of acrylamide-DADMAC-diethyldiallylammonium chloride copolymer by copolymerizing the corresponding monomers in the presence of EDTA, Na salicylate, $(NH_4)_2S_2O_4$, $NaHSO_3$, and $CuSO_4$ has been reported.[7-23] The partially hydrolyzed copolymer was useful as a retention aid in papermaking.

The preparation of a polysalt by adding poly(DADMAC) to the sodium salt of poly(2-hydroxy-3-methacryloyloxypropane-1-sulfonic acid) has been reported.[7-24]

As the basis of an investigation of the binding properties of polyion complexes, the copolymer of DADMAC with sulfur dioxide was added to sodium poly(methacrylate) to form the complex.[3] Methyl orange (as well as homologous derivatives through pentyl) was studied. Hydrophobic and electrostatic interactions which accompany the binding were proven to be important.

Water-soluble copolymers of dimethyldiallylammonium chloride and methylstearyldiallylammonium chloride by radical initiation have been prepared.[7-25] The copolymers are useful as antistatic agents for textiles, electrical conductive coatings for paper, and softening agents for paper and textiles. Acrylate copolymers containing DADMAC units in the chain have been found to function as positively-charged electrophotographic toners.[4]

A comparison was made between the ferric chloride-lime and polyelectrolyte processes for chemical conditioning of sludges for vacuum filtration dewatering.[5] The requirements for lime varied with sludge type and source and increased with sludge age. Poly(DADMAC) has also been found to be useful as a suspension stabilizer in conjunction with gelation as an aid in controlling particle size in suspension copolymerization of styrene, chloromethylstyrene and divinylbenzene;[6] it has also been used in fixation of reactive dyes on textiles,[7] as well as in the development of preshampoo compositions.[8]

The copolymers of dialkyldiallylammonium salts with sulfur dioxide first reported in 1966[7-112] are manufactured commercially and have similar industrial uses and properties to the homopolymer reported above.

Application in Flocculation and Coagulation

Synthesis of cationic polyelectrolytes from acrylamide and DADMAC were prepared in aqueous solutions and their flocculating capacities were tested in aqueous suspensions of TiO_2. Acrylamide homopolymer of molecular weight $<10^6$ had virtually no flocculating capacity. DADMAC homopolymer accelerated the sedimentation of TiO_2, and this acceleration increased with increasing molecular weight as the optimum doses of the homopolymer decreased. The flocculating capacity increased and the optimum dose decreased with increasing molecular weight of the copolymers, as well, but larger doses than those of the homopolymer were required for producing the same sedimentation rates.[7-32]

Diallyldialkylammonium chlorides for use as flocculating agents were polymerized in the presence of water-soluble azo initiators to give high-molecular-weight polymers.[9] Both poly(DADMAC) and its copolymer with acrylamide have been found to be useful as filter aids and pigment disperants for dewatering pigments, by adding them to the slurry before dewatering.[10] Several cationic polymers, including poly(DADMAC), were studied as flocculants for colloidal silica and their respective optimum doses determined as a function of their respective structures and molecular weights.[11] The addition of poly(DADMAC) inhibited deposit formation in the evaporator in the concentration of black sulfite liquor to the extent of about 83% when 250 ppm of the polymer of molecular weight about 14,000 was used.[12]

Cationic copolymers of acrylamide with DADMAC are perhaps among the most important and extensively used polymer types as flocculation and coagulation agents in water and wastewater treatment. These copolymers have been extensively studied both from the standpoint of efficiency in synthesis and effective comonomer ratios in the copolymer. The reactivity ratios for copolymerization of DADMAC (M_1) and acrylamide (M_2) have been determined.[7-28] Both r_1 and r_2 depended on DADMAC concentration in the initial comonomer solution, and increased with decreasing DADMAC concentration. An increase in total monomer concentration increased r_1 but r_2 remained nearly constant. The results were explained by association of the cationic monomer and electrostatic interactions. Although r_1 and r_2 values vary as indicated, the most commonly used values appear to be: $r_1 = 0.26$; $r_2 = 6.12$. Because of the markedly different rates of reaction of the two monomers, a factor of about fifty in favor of acrylamide, the composition of the copolymers would constantly change with conversion and at total conversion of acrylamide, significant quantities of DADMAC remain unpolymerized in the final solution. Efforts to reduce this residual monomer content are referred to later. The energetics of this copolymerization are qualitatively illustrated in Figure 12–1.

The addition of activated carbon and poly(DADMAC) to a flocculant containing a mineral acid along with sodium silicate accelerated flocculation and the removal of organic impurities in purification of water.[13] The flocculating efficiency of the polymer increased in the removal of nonhydrated phosphatides from citric acid-treated soybean and sunflower oils as the hydration temperature rose from 20° to 80°.[14] Also, when the polymer was used with nonionic or anionic polyacrylamide, sedimentation of solids

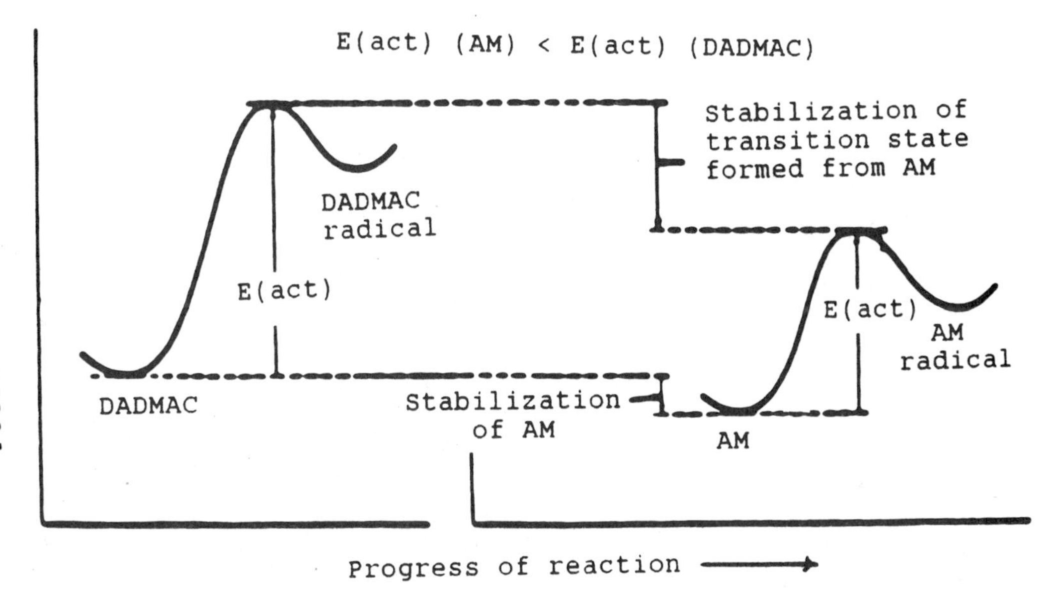

Figure 12–1 Energetics of reaction rates of acrylamide vs. DADMAC. (Plots aligned with each other for easy comparison).

from coal-containing wastewater[15] was accelerated. When this polymer, among other flocculants studied, was added at overdose levels to effluents from clarifiers, and the effluent passed through cation and anion exchange resins, deterioration of the resins was observed. The polymer was less damaging than other flocculants studied.[16]

In a study of the mechanism of flocculation of sulfate lignin with poly(DADMAC) in treatment of pulp wastewater, the optimal flocculant dose increased with decreasing molecular mass due to the decreasing adsorption capacity of the polyelectrolyte. The latter also explains an increase in the flocculation range and a slower change in the zeta-potential of the lignin particles with increasing lignin concentration in the solution at decreasing molecular mass of the polyelectrolyte.[17] The addition of the polymer containing unreacted monomer to oil-in-water emulsions from various sources significantly improved separation of the emulsions.[18]

A study of flocculation mechanism during the use of poly(DADMAC), polyethylenimine and cationic polyacrylamide has been published. The flocculation of mechanical pulp containing kaolin by polyelectrolytes, i.e., poly(DADMAC), polyethylenimine and acrylamide-2-propenyltrimethylammonium chloride copolymer was explained by changing of electric charge and bridge formation of pulp particles in the suspension. Poly(DADMAC) and acrylamide-2-propenyltrimethylammonium chloride showed an electrostatic effect and bridge formation-van der Waals forces, respectively, whereas polyethylenimine had an intermediate effect.[19]

Cationic polymers of DADMAC and copolymers of DADMAC and acrylamide were shown to be effective flocculants of suspended matter, such as montmorillonite clay from water, as early as 1964.[1-152]

A method of purification of DADMAC which involves a vacuum stripping step, a steam stripping step at a regulated pH, followed by an activated carbon treatment has been described.[2-157] Improved properties of the polymers obtained from monomer purified by this method were obtained.

Cyclic polymers and copolymers with low viscosity have been studied.[7-56] Poly(DADMAC) and diacetoneacrylamide-DADMAC copolymers with low intrinsic viscosity and monomer content and containing ring structures were manufactured using redox catalysts and glycerol chain transfer agent, for use as flocculating agents for wastewater and as coatings in the manufacture of paper insulators.

A gel permeation chromatographic process (GPC) was developed to fractionate high molecular weight polyelectrolytes, typically poly(DADMAC).[20] A copolymer of DADMAC with *N,N*-diethylacrylamide has been developed for use as a cationic flocculant.[7-65] Cationic polyelectrolytes have been used in the process of handling slack sludge from aqueous latex paint waste.[21]

Poly(DADMAC) was synthesized by a one-stage process from dimethylamine and allyl chloride to obtain monomer conversions from 85–92%, for use in flotation treatment of wastewaters from petroleum refineries.[2-116]

A DADMAC-methacrylic acid copolymer has been used in flocculation and desalination processes.[22] Poly(DADMAC) and its uses in the water treatment, papermaking, mineral processing and petroleum recovery industries were surveyed in 1982.[23] The adsorption of a DADMAC-acrylic acid copolymer on to silica particles has been studied. The extent of adsorption was dependent on pH and increased considerably in presence of sodium chloride electrolyte.[24]

A review on the synthesis, properties and uses of poly(DADMAC) in water and wastewater treatment with fifteen references was published in 1984.[25] Residual poly(DADMAC) when used as a flocculant in the food industry was determined with relative error of 0.19–6.4% by use of eosine.[2-135]

Monomer reactivity ratios were determined precisely in radical copolymerization of DADMAC with acrylamide only by stepwise calculation within limited ranges of monomer ratios. Both r_1 and r_2 depended on [DADMAC].[7-28] A spectrophotometric method for determination of microamounts of poly(DADMAC) which used flow injection techniques has been developed.[26]

Poly(DADMAC) was useful for coagulation of oily wastewater containing nonionic surfactants and emulsified oil.[27] Its relative efficiency for swimming-pool water treatment has been determined.[28] It has also been used as a component of transparent soaps.[29]

A spectroscopic method of analysis of residual poly(DADMAC) by use of azo dyes has been developed.[30] The polymer has been promoted as a drainage, dewatering and retention aid for pulp stocks, and wastewater.[31] A review with ten references on use of polyelectrolytes as flocculating agents for water and wastewater treatment has been published. Poly(DADMAC) was among those discussed.[32]

Its use in flocculation of a slurry consisting of montmorillonite, kaolinite, and mica in the presence of anionic polyelectrolytes has been studied and the results reported.[33] Poly(DADMAC) was used as a major component of a polyelectrolyte complex used for immobilizing *paracoccus denitrificans* cells and denitrifying activity.[34]

Poly(DADMAC) and related polymers of high molecular weight have been shown to have high efficiency as flocculants and coagulants in a number of other applications.[35] It has also been used as the cationic component of an agent for dewatering sewage sludge which consisted of both a cationic coagulant and an anionic coagulant in controlled ratios.[7-78]

Ampholytic copolymers and terpolymers of DADMAC were found to be superior flocculants in petroleum refineries and automobile oily wastewaters.[36] Color contamination of potable influent water resulting from colloidal organic material was reduced by addition of poly(DADMAC).[37]

Flocculation of cellulosic fibers by cationic polyacrylamides (copolymers of acrylamide and DADMAC) with different charge densities was studied. Rapid flocculation occurred independent of charge densities for the four polymers studied.[38] Poly(DADMAC) was an effective flocculant for regeneration of spent washing solutions, particularly in washing beverage containers.[39]

DADMAC-acrylamide copolymer was found to be an effective flocculant for sludge material.[40] This copolymer was also studied in an investigation to determine certain fundamental aspects of dual component retention aid systems.[41]

Organic sludge was dewatered by use of poly(DADMAC) in conjunction with other polymers or copolymers, then partially dried on a belt-press machine to give product with water content of wt. 76% in contrast to wt. 81% for a conventional coagulant.[42]

A cationic terpolymer of DADMAC, acrylamide, and *N,N*-dimethylaminoethylmethacrylate methochloride salt was effective in dewatering an organic sludge. The terpolymer gave superior results in comparison to the methacrylate homopolymer alone.[43]

Higher molecular weight poly(DADMAC) was obtained by polymerization of the monomer in presence of fluoride ion. Molecular weight increased linearly with fluoride ion concentration.[2-231] Poly(DADMAC) was found to be useful as a coagulant for removal of heavy metals from waste streams.[44]

Poly(DADMAC) was also used in the preparation of aqueous suspensions of calcium-containing fillers useful in the manufacturer of paper.[45] The preparation of water soluble copolymers of DADMAC and triallylmethylammonium chloride and their uses have been described.[46]

Poly(DADMAC), designated as VPK-402 has been found to be effective for decolorization of pulping wastewaters.[47] A polyelectrolyte complex derived from poly(DADMAC) and an anionic polymer has been shown to be effective as a flocculant of mycelia from *aspergillus terreus*, and the succeeding process of production of itaconic acid with the flocculated mycelia.[48]

Poly(DADMAC) has been used as one component of a flocculation agent for accelerated sedimentation of suspended matter from wastewater.[49] It has also been used as a component of a flocculation system for reducing turbidity from aqueous solutions.[50]

Poly(DADMAC) was shown to be an effective coagulant for turbid waters.[51] The amount of heavy metal was also reduced up to 50%. The LD_{50} values of the flocculant, poly(DADMAC) were determined in mice, rats, and guinea pigs and were found to be 1720, 3000, and 3250 mg/kg, respectively.[52]

A review of the synthesis of poly(DADMAC) and similar polymers with 41 references was published in 1989.[53]

Surface water with high turbidity was successfully flocculated with a combination of aluminum chlorohydrate and high molecular weight poly(DADMAC).[54] Comparative systems gave less satisfactory results. It has also been used effectively in the removal of surfactants from wastewaters by an ultrafiltration method combined with complexation.[55]

Several additional publications and patents have dealt with various aspects of the application of poly(DADMAC) and its copolymers to flocculation and coagulation or other aspects of water and wastewater treatment.[56–77]

A wide variety of novel uses of poly(DADMAC) and its derivatives have been published during 1990. It has been reported to improve the effectiveness of a PO_4^{3-} precipitation agent which included a trivalent metal salt, e.g., an Al or Fe salt, especially $Al_2(OH)_5Cl$.[78] It has also been used in purification of spent culture liquor in the yeast industry.[79]

The copolymer of DADMAC with sulfur dioxide has been found to be useful as coagulants and thickening agents, and for soil stabilization.[80] Copolymers of DADMAC and acrylamide of high molecular-weight having intrinsic viscosity of 18.3 dL/g and cationicity of 1.0 eq./kg. were useful as flocculants and drainage and retention aids.[81]

Poly(DADMAC) has been used to remove phospholipids from vegetable oil,[82] and to precipitate amorphous calcium phosphate from supersaturated solutions.[83] A DADMAC-acrylamide copolymer of high solids content as a microemulsion has exhibited improved performance as a flocculant and thickener.[84]

The MW and MWD of poly(DADMAC) has been determined by ultrafiltration.[85] The high-MW polymer was useful as a coagulant and flocculant for wastewater.[86] A major disadvantage in many uses of poly(DADMAC) involves the presence of unreacted monomer. A process has been described which makes it possible to remove the residual monomer from the polymer.[87]

A copolymer of DADMAC with *N*-vinylformamide has been found to be useful in its non-hydrolyzed form in treating wastewater and as a dewatering agent.[88] The DADMAC homopolymer has been used as wetting agents for polymer slurries, followed by mechanical dewatering.[89]

A toxicological study of poly(DADMAC) manufactured in Russia under the trade designation of VPK-402 has been published.[90]

Applications Based on Electroconductive Properties

In formulating electroconductive coatings having good solvent holdout and low tackiness on paper, the conductive coating consisting of poly(DADMAC)-fluorosurfactant mixture with calcium carbonate was applied to a barrier-coated papersheet, such coating comprising an ethylene-vinyl acetate copolymer, or other similar materials.[91] Water-insensitive electroconductive coatings, useful for paper in electrography, have been described in which the coatings contain copolymers of DADMAC and *N*-(hydroxymethyl)acrylamide, or copolymers of these two monomers with additional monomers.[92] Electroconductive coatings on paper with improved tack properties consist of poly(DADMAC), a binder, a pigment, and a siloxane.[93] Paper coated in this manner exhibited no coating pick-up on the rolls.

In the production of paper for electrographic reproduction, a water-soluble electroconductive coating, which would allow subsequent coatings such as ZnO or dielectric resins to be applied from an aqueous based system, was applied to a substrate as a double coating; e.g., a strong cationic polymer, poly(DADMAC), followed by a strong anionic polymer, e.g., poly(Na vinylsulfonate) was effective.[94] The polymer has also been used as an additive to improve the quality and effectiveness of electrodeposition of copper layers on conducting plates.[95] Conductive high-pressure laminates were prepared by laminating melamine-formaldehyde resin-impregnated paper with paper containing melamine resins, carbon black, and poly(DADMAC) and paper impregnated with phenolic resins.[96]

Crosslinked poly(DADMAC) when bonded to 2,6-dichlorophenolindophenol constitutes an immobilized redox network which functions as a graphite-electrode coating for electrodes used in a catalytic study of cytochrome.[97] Electrically conductive coatings for paper having nontacky surface properties, low surface-resistance and excellent solvent resistance, and useful for copying or electrographic paper, were prepared from a blend of poly(DADMAC) and poly(methacrylic acid).[98]

Ion-transport and electrical response properties of solid poly(DADMAC) plasticized with polyethylene glycol, were improved. The process gave compositions in which it could be demonstrated that the chloride ion was mobile[99] by use of a technique which employed ion-blocking platinum electrodes and ion-reversible calomel electrodes. Ionic conductance was also induced in the trifluoromethylsulfonic acid salt of poly(DADMAC) by plastization with polyethylene glycol.[100]

Electroconductive coating materials with superior heat stability, moisture resistance, and electroconductivty can be manufactured by emulsion polymerization of suitable vinyl monomers in the presence of a variety of polymeric emulsifiers included among which was poly(DADMAC).[101] This polymer has also been shown to effectively inhibit aluminum corrosion in primary cells with aluminum anodes and alkaline electrolyte. These conclusions were supported by double layer capacitance measurements.[102]

Poly(DADMAC) has been found to be useful in the reprographic paper industry as an electroconductive polymer.[103] Paper made conductive with this material exhibited surface electrical conductivity three to four orders of magnitude higher than untreated paper. It has also been used as the cationic polyelectrolyte in providing electric conduction properties to cellulose fibers in paper production.[2-513]

A theoretical model for the electric double layer of fibers with an adsorbed layer of poly(DADMAC) as cationic polyelectrolyte has been developed.[2-510] The effect of multilayer formation of this polymer on the electrical conduction behavior of carboxymethylcellulose (CMC) fiber was investigated.[2-511] The excluded volume effect and protonic transition mechanism accounted for the conduction behavior of the fiber.

A mathematical model for electric conduction properties of cellulose fiber with adsorbed layers of poly(DADMAC) and its computer simulation has been developed.[2-512] The model correlated the electric conduction with the hydrodynamic properties of the cellulosic pad and permitted estimation of specific conductance values of fiber surfaces using the hydrodynamic data obtained by the streaming current method.

Organic polyelectrolyte-based coatings for fabrication of electrographic and electrophotographic coatings were studied. Optimal results were obtained using poly(DADMAC) with different paper bases.[104]

The effect of relative humidity on the surface electrical conductivity of a cellulose substrate with a polyelectrolyte coating of poly(DADMAC) (VPK-402) was considerable. This dependence was determined by the sorption process characteristics of the hydrophilic polyelectrolyte coating.[105]

Polymers or copolymers giving conductive, flexible, tack-free coatings, useful on electrostatic copying paper included poly(DADMAC) and its copolymers.[106] Electrical conductive polymer compositions were prepared by use of a polymeric peroxydisulfate.[107] This salt was obtained from DADMAC and an excess of potassium persulfate.

Polymerization of DADMAC in inverse emulsion was studied and the dependence of molecular weight on conversion determined.[10-48] A model was proposed for the calculation of conversion dependence on molecular weight. The influence of monomer concentration during polymerization on the ionic conductance of the solid was also studied.

Additional publications or patents on the use of poly(DADMAC) or related copolymers in the areas of electroconductive polymers and related subjects have appeared in the literature.[108-130] Many have dealt with improving the surface resistivity and solvent holdout in conditioning paper for maximum electroconductive properties. The paper was coated with aqueous solutions of poly(DADMAC), sodium alginate, and tamarind seed gum. After conditioning, the treated paper exhibited good surface resistivity and solvent holdout.[122]

Polymers having a conjugated main chain are of interest as electroconductive polymers. A variety of dipropargyl derivatives have been cyclopolymerized to yield polymers of interest in this area. The cyclopolymerization of dipropargylgermanium compounds has now been accomplished, and the electrical conducting properties of the polymers reported.[131]

Although the mechanism of the electrical conductance of poly(DADMAC) and similar polymers is different from that of those polymers having conjugated main chains, the conductance properties of poly(DADMAC), when used on copying and printing papers, have been further improved.[132]

Applications in the Paper and Textile Industries

Grafted copolymers of DADMAC have been used in developing organic pigments useful in loading paper. Thus, as acrylamide-DADMAC prepolymer was grafted with styrene to give a graft copolymer which was crosslinked with glyoxal.[133]

A graft copolymer of cellulose and DADMAC was prepared by impregnating cellulose with aqueous monomer containing $(NH_4)_2S_2O_8$, removing the excess and heating the impregnated cellulose. The extent of grafting of the cationic monomer could be increased to 17.5% and monomer conversion was as high as 25%.[7-35]

Rosin emulsion sizes which show good paper-strengthening properties contain formaldehyde treated tall-oil resin fortified with fumaric acid and a copolymer of methacrylamide and DADMAC.[134] Stable, pumpable, and solvent-free colloidal polyampholyte latices useful as pigment retention and drainage aids in the manufacture of paper have been synthesized by copolymerizing DADMAC with a suitable ratio of acrylamide and acrylic acid in presence of a water-soluble free radical initiator and tetrasodium ethylenediaminetetraacetate.[135]

A composition for producing fiberboard with increased physicomechanical indexes contains poly(DADMAC) as the sizing additive.[136] Also, a composition useful for dyeing paper has been developed by grafting styrene to a copolymer of acrylamide and DADMAC, and subsequently, crosslinking the graft copolymer with glyoxal.[137] The beatability of cellulose pulp in the paper-making process can be improved by treatment with poly(DADMAC), polysaccharides, cationic surfactants and certain metal salts.[138] High-molecular-weight, water-soluble polyammonium compounds of DADMAC, useful in the coating and preparation of paper, were prepared in high space-time yield by the continuous or batch addition of water-insoluble azo compound initiators in methanol or dimethylformamide to aqueous monomers.[2-207] The wet tensile strength of paper has been found to be vastly improved by incorporating a reaction product of glyoxal with an acrylamide-DADMAC copolymer into the pulp slurry, followed by an additional glyoxal treatment to aid in crosslinking the cellulose.[139] Poly(DADMAC) was evaluated as an alkenylsuccinic anhydride emulsification and retention aid in a paper slurry and compared in terms of improved properties with conventional alkenylsuccinic anhydride emulsions in water or cationic starch.[140]

Cotton, silk, wool, and nylon fabrics were printed with reactive dyes by impregnating the fabric with diallylamine hydrochloride polymer or diallylamine hydrochloride-DADMAC copolymer as a dye fixing agent before printing the fabric.[141] The washfastness rating of a dyed or printed fabric was more than doubled, compared with the untreated fabric, when treated with poly(DADMAC).[7] Reactive-dyed cotton fabrics treated with poly(diallylamine hydrochloride) at pH ~ 10 exhibit vastly improved washfastness over a fabric treated with the polymer at pH = 6.5.[2-202] Starching compositions for fabrics which showed good stiffness included poly(DADMAC) as a major component.[142]

Poly(N,N-diallylethylammonium phosphate) and related polymers have been shown to be useful as antistatic agents for textiles.[2-208] Textile materials with good antistatic properties at low humidity can be prepared by treating the fabric with poly(DADMAC).[143] The treatment of fabrics dyed by reactive dyes with poly(DADMAC) and related polymers gave fabrics showing improved colorfastness to washing, abrasion, chlorine, and light.[144]

The polymer of 1,1-diallylguanidine salts has been shown to be useful in improving wet-strength properties of paper as well as a flocculant for solids suspended in aqueous media.[2-175]

An acrylamide-acrylic acid-amylopectin-DADMAC graft copolymer gave values of 74.0 and 77.0% TiO_2 pigment retention when used at 0.01 and 0.03% concentration in a papermaking process, compared to 64.5 and 66.5%, respectively, for a common standard and 40.5% for a blank control (all values at 11% alum).[7-62]

Heat-resistant acrylic fibers were obtained from a copolymer of acrylonitrile with diallylamine. The copolymer was wet spun into a coagulating bath, drawn 700% in boiling water, dried, and heat-treated 10 minutes in steam at 110° to give 3-denier/filament fibers with distortion ratio at 100° in the wet state 6.4%, compared with 25.1% for fibers spun from acrylonitrile-methyl acrylate copolymer.[7-111]

A paper fiber additive containing polyacrylamide blended with glyoxal and poly(DADMAC) has been described as a cationic regulator. The addition of glyoxal-crosslinked polyacrylamide and poly(DADMAC) to cellulose pulp gave paper with improved dry and wet strength.[145]

An acrylamide-DADMAC copolymer was chlorinated to partially convert the amide groups to *N*-chloroamide. The resulting copolymer was found to improve the wet strength properties of paper.[146] A highly branched DADMAC-triallylmethylammonium chloride copolymer was effective in a process for solids retention and accelerated dewatering in paper manufacturer.[2-227]

Poly(DADMAC) could also be used at 0.1% concentration in corrugated paperboard manufacture. The advantages were: (1) dewatering time was decreased from 6.0 to 4.8 seconds; (2) solids content and porosity of web increased; (3) oil penetration of sheet was accelerated, and (4) machine speed could be increased by 3–4%, and the chemical oxygen demand of the effluent was reduced.[147]

The colloid titration method was used to determine the amount of poly(DADMAC) adsorbed on linters and carboxymethylcellulose fibers from dilute aqueous solutions.[2-509]

Poly(DADMAC) has been used as a heat and mass exchanger. Application of the polymer to the surface of wool fibers resulted in improvements in the transfer of water vapor.[148] High molecular-weight poly(DADMAC) was found to be useful in coatings and paper manufacture.[2-122]

Washfast antibacterial acrylic fibers were prepared by finishing the spun fibers with emulsions containing poly(DADMAC) among other compounds.[149]

A dye fastness improver for a reactive dye-treated textile fabric was prepared from a copolymer of an allylamine salt and a diallylamine salt.[150]

Cellulose pulp slurry was treated with an acrylamide-DADMAC copolymer along with other components to impart temporary wet strength to the paper.[151]

A method of reducing the amount of colloidal pitch particles in aqueous pulps employed poly(DADMAC) as an additive on to the fiber.[152] The polymer was used as a water-in-oil emulsion which could be inverted to produce a polymer solution rapidly.

Polymers of diallylamine hydrochloride, when applied to reactive-dyed fabrics, imparted improved washfastness, lightfastness and chlorine fastness to the fabric.[153] Also, copolymers of diallylamines and triallylamines were useful as color fastness improving agents. When treated with these copolymers, dyed cotton knit goods showed good colorfastness against washing, light, irradiation, and chlorine containing water treatment.[154]

Polymers of diallylamines have been found to be effective as color fastness improving agents, especially for reactive-dyed fabrics.[155] Copolymers of diallylamine and methyldiallylamine were also found to improve dye fastness of reactive-dyed textiles.[7-37]

In a process for production of high-wood containing, coated papers from thermomechanical pulp, poly(DADMAC) was used to suppress slime formation.[156] A hydrolyzed acrylamide-DADMAC copolymer was found to be an effective deinking agent for wastepaper.[157] Poly(DADMAC) was used as an effective pretreatment agent for kaolin filler in filler loading of intaglio printing paper.[158]

In a process for manufacture of acrylic fiber with high water-retention properties, poly(DADMAC) was used as a component of the chemical treatment which contributed to improved water-retention in the fiber.[159]

Poly(DADMAC) and a copolymer of diallylamine hydrochloride and sulfur dioxide were found to be useful as fixation agents for dyeing of cellulosic fibers with reactive dyes.[160] A mixture of the homopolymers of diallyldimethylammonium chloride and diallylamine hydrochloride were used in a 70:30 ratio to aid in dyeing of cellulosic fibers.[161]

A DADMAC-hydroxyethyl cellulose graft copolymer was shown to be effective as dye enhancers for fabric treatment, effecting improved dye yield, levelness and fastness, and could be applied before, during, or after dyeing, optionally in conjunction with durable press resins, antistatic and soilproofing agents, and water repellents.[162]

Pigment dispersions with adjustable cationic properties for brush-on paper coatings were manufactured by coating pigment particles in aqueous suspension with poly(DADMAC) as a dispersion agent for the coated particles.[163] A copolymer of DADMAC and *N*-methylol acrylamide was used to treat cellulose materials dyed with direct dyes to improve color fastness.[164]

In a method for manufacture of antistatic paper, useful for packaging and cushioning, the preferred antistatic polymer was poly(DADMAC).[165] Effective clarification of white pulping liquor was attained by use of poly(DADMAC) in conjunction with an anionic polyacrylamide in paper manufacture.[166]

DADMAC monomer was used as a component of a monomer mixture which was grafted to corn starch. The grafted polysaccharide was further treated to give a solution for use as wet-strength additives in paper manufacture.[167]

Diallylamine derivative copolymers were found to be effective as a component of coatings with good adhesion, corrosion resistance, weldability, and press workability in a method for increasing paper strength.[168]

Poly(DADMAC) was used as a major component of an agent for improving fastness of reactive-dyed fabrics.[169] The dry strength of paper and cardboard prepared with starch was increased by pre-heating the starch with poly(DADMAC).[170] Poly(DADMAC) was found to be an effective component of an aqueous solution used in an antimicrobial treatment of textile material.[171]

A water-soluble cationic poly(diallylamine)-epichlorohydrin resin was heated with a carboxymethyl hydroxyethyl cellulose solution in aqueous sodium hydroxide to yield crosslinked structures. The products were useful as ion exchangers.[2-234]

Poly(DADMAC) was found to be an effective conditioner for pulp mill wastewaters.[172] The use of poly(DADMAC) as a retention aid significantly increased the first-pass and system retention of suspended solids and the paper quality in paper manufacture.[173]

Poly(DADMAC) was used effectively in conjunction with enzymatically-degraded starch to impart high dry strength properties in the manufacture of paper and cardboard.[174] It was also used together with two cationic polyacrylamides in a study to determine the adsorption kinetics of polyelectrolytes on the cellulose pulp fibers in turbulent flow.[175]

Triallylmethylammonium chloride was used as a component of a monomer mixture to form water-in-oil dispersions for use in thickening of aqueous solutions and textile printing pastes.[176]

Further studies which have increased the number of diverse applications of poly(DADMAC) and its copolymers in the paper and textile fields have appeared in the literature.[177–200]

Further development in the uses of poly(DADMAC) and its derivatives have been reported during 1990. A novel improvement in printing involves the development of ink-jet recording. Poly(DADMAC) has been reported by numerous investigations to be a useful adjunct to this development. As a component of an ink-jet recording material, it has been reported to contribute to superior ink-fixing properties.[201]

Antistatic properties of textiles, resistant to laundering, have been improved by impregnating the material with a composition including poly(DADMAC).[202]

The molecular weight and cationicity of polymers as flocculants are critical properties. A recent process for the production of high-molecular-weight copolymers of DADMAC and acrylamide has been reported. Copolymers having intrinsic viscosities as high as 16 dL/g in 4% NaCl at 25°C have been reported.[203]

Poly(diallylamines) have also been reported to contribute to improvements in ink-jet printed copies.[204] Poly(DADMAC) has been used to improve the properties of composite fiberboards manufactured from cellulosic fibers, perlite, and mineral wool.[205] It has also been used with lanolin to improve the properties of detergents useful in shampoos,[206] and when crosslinked with methacrylamide compounds, to improve the wet-strength properties of paper.[207]

Diallyl derivatives of aziridinium salts have been used in a process for dyeing of textiles with pigments.[208] Poly(DADMAC) in conjunction with poly(vinyl alcohol) has been used to improve the quality of sheets for ink-jet printing.[209] Unhydrolyzed DADMAC copolymers with *N*-vinylformamide have been claimed to produce paper with improved properties when used in the papermaking process.[210]

A novel use of poly(DADMAC) has apparently surfaced in 1990. It has been found to be effective in modification of the technological properties of ligno-cellulosic fibers in the manufacture of fiber-reinforced cement slabs.[211] It has also been shown to give paper of improved bursting strength when used for pretreating the stock in paper manufacture.[212]

Cosmetic Applications and Hair Treatment

An approach to developing compositions for straightening or resetting hair under strongly alkaline conditions uses poly(DADMAC), along with an aqueous mineral oil dispersion and a fatty alcohol nonionic emulsifier.[213] A preshampoo type hair treatment composition which produces excellent hair-conditioning effects includes poly(DADMAC) as one of the major components.[214]

Oxidizing solutions for hair wave setting, which prevent hair damage and maintain waves for a prolonged period include as one important component poly(DADMAC).[215] This polymer, in conjunction with certain quaternary ammonium salts, e.g., distearyldimethylammonium chloride and methyl cellulose, produce hair rinses stable at 40° and −5°, which when used, were found to produce soft, manageable hair.[216]

Poly(DADMAC) or DADMAC-acrylamide copolymers have been found to be effective components of hair rinses.[217] The homopolymer has been used in formulating a pre-treatment solution for permanent waving of new hair growth.[218] A hydrophilic polymer, useful as a cationic gel base for use in direct hair dyes, was prepared by copolymerizing diallyl fumarate or diallyl maleate with diallyldimethylammonium salts.[219] An effective hair conditioner has been formulated which includes, as a significant component, polymers or copolymers of DADMAC.[220] Shampoos, hair rinses and other hair preparations which have good consistency and are stable against temperature changes contain poly(DADMAC) as a significant component.[5-37]

Semitransparent, stable gel-like shaving creams with good spreadability and foaming properties were formulated by use of a polysalt of poly(DADMAC) with a suitable anionic polymer, e.g., alginic acid.[221]

Further uses of poly(DADMAC) in formulating hair conditioners and shampoo compositions have been reported.[222, 223] The compositions gave better combing property and luster to the hair.

An ultra-mild skin cleansing bar has been formulated with use of a copolymer of DADMAC with acrylamide as an ingredient. The soap had good bar firmness and lathering properties.[224]

Fine hair conditioning effects have been claimed from shampoos formulated by use of poly(DADMAC) as the cationic component. The formulation was stable after storing at 45°C for thirty days.[225]

In addition to use of a copolymer of DADMAC and acrylamide in shampoo compositions, a sulfonic acid-type anionic surfactant was also added.[226] The result was a composition which gave creamy, fine, and smooth foams, without affecting the hair.

A graft copolymer, prepared by polymerizing DADMAC monomer on to hydroxyethyl cellulose, was used as a major component of hair-conditioning formulations which were effective as rinses, lotions, shampoos, or fixative compositions.[227]

Mild skin-cleansing soap bars with low moisture content were formulated by use of a copolymer of acrylamide with DADMAC monomer. In a standard test, the bars exhibited a firmness penetrometer value of 2.7.[228]

Polysalts derived from poly(DADMAC) and an anionic surfactant have been used as a major component of cosmetic pearlescent liquid soap.[229]

Hair preparations, such as shampoos and hair rinses, contained imidazolinium compounds and poly(DADMAC) among the components.[230] The preparations were stable against temperature changes and had good consistency.

Transparent soaps which contained poly(DADMAC) exhibited improved swelling degree, formability, and solidification point, as well as good swelling and cracking resistance in comparison to the product without the cationic polymer.[231]

A deodorant composition which contained poly(DADMAC) or a copolymer of DADMAC with acrylamide as well as an antimicrobial agent, gave 33% greater antimicrobial activity in a standard test than a similar formulation without the polymer or copolymer.[232]

Poly(DADMAC) has been used in antiperspirant compositions to increase the residual amount of the antiperspirant on the skin, and to enhance the antiperspirant's effect, making it possible to reduce the amount of the antiperspirant applied.[233]

Hair preparations have been described which contain a cationic polymer, for example, poly(DADMAC) and quaternary silk polypeptides, derived from silk protein.[234]

Thickening agents useful for pharmaceutical formulations and cosmetics were prepared by use of a cationic graft copolymer prepared by grafting DADMAC to cellulose or cellulose derivatives and an anionic carboxy polymer, among other components.[235] A gel for treatment of psoriasis was prepared from this formulation.

A hair treating composition which contained a copolymer of DADMAC and acrylamide gave a 16% better performance in a standard test than a similar composition without the copolymer.[236]

Other hair-cosmetic compositions which contained both a cationic copolymer of DADMAC and acrylamide or the cationic poly(DADMAC) and a cyclodextrine gave better performance in a standard test than either formulation in which only one of the above components was present.[237]

Hair dyeing preparations which contain polymers or copolymers of DADMAC and acidic dyes gave good dyeing ability, developed fast colors, were shampoo-resistant and were not skin irritants.[238]

Further applications of poly(DADMAC) and its copolymers in the areas of cosmetics and hair treatment have been published in the recent literature.[239-254]

Poly(DADMAC) and its derivatives have found many additional applications in cosmetics, particularly in hair conditioning. New publications in 1990 in this area include use of poly(DADMAC) as a component of hair conditioners,[255-258] and in liquid soap compositions,[259] and to prevent long range crystallization of fatty acid derivatives in liquid compositions useful as hair tonics.[260]

Biological, Medical, and Food Applications

Poly(DADMAC) has been shown to be four to five times more active than cholestyramine in binding bile acids in cholesterol biosynthesis in the rat from ^{14}C-acetate.[261] The polymer has also been found to be more effective than alum and a variety of other cationic polymers in removing viruses from water during drinking water purification by coagulation and flocculation.[262] It is also useful for preserving and/or disinfecting contact lenses.[263]

A terpolymer was prepared from *N*-carbomethoxymethyl-*N*-methyl-*N,N*-diallylammonium chloride, *N*-carbomethoxymethyl-*N*-dodecyl-*N,N*-diallylammonium chloride and *N*-dodecyl-*N*-methyl-*N,N*-diallylammonium chloride. The polymers were prepared as antimicrobials for contact lens sterilization solutions or for hydrophilic ointments for treatment of bacterial lesion skin infections such as burns or ulcers.[264]

Poly(DADMAC) has been approved by the U.S. Food and Drug administration for use at controlled levels as a flocculent in the clarification of water used to manufacture paper and paperboard for contact with food products.[265] The copolymer of DADMAC and acrylamide has been approved by the U.S. Food and Drug Administration for use at controlled levels on the finished product as a drainage and retention aid in the manufacture of paper and paperboard for contact with food products.[266] DADMAC-acrylamide-potassium acrylate terpolymer has been approved by the U.S. Food and Drug Administration for use at $\leqslant 0.05\%$ by weight of finished product as a retention/drainage aid in the manufacture of paper and paperboard for food contact.[267]

The homopolymer of DADMAC has been found to be effective in removing polyphenols and other undesirable impurities from hopped wort to improve the colloidal stability of beverages derived from this source.[268] The total phenol content was reduced by 20% and the anthocyanogen content by 21%. It has also been shown to be effective for obtaining sheep red cell monolayers on glass in assays of antibody-dependent (K-) and natural killer-cells in the blood of patients with asthma.[269] A study of the effect of poly(DADMAC) on the esterolysis of 8-acetoxyquinoline has shown that hydrolysis occurred and that the polyion did not affect the reaction rate.[270]

Poly(DADMAC) has been found to be effective as a flocculant in clarification of brandy.[271] The flocculant removed metal ions such as iron, copper and calcium, and reduced the content of acids, esters and aldehydes. Antibacterial acrylic fibers with excellent washfastness can be manufactured by finishing the spun fibers with emulsions containing poly(DADMAC) as a significant component, followed by heat-treatment.[149] This polymer has also been used to form polyelectrolyte complexes with sodium cellulose sulfate which constitute semipermeable membranes for use as microcapsules containing hormone-producing tissue or cells mixed with a culture medium.[272] The treatment of the inner wall of the container for blood serum separation with the alternating 1:1 copolymer of DADMAC and sulfur dioxide facilitates serum separation from heparinized blood.[273] Heparinized blood in the container was allowed to stand for coagulation and then centrifuged to obtain the separated serum. The immunostimulating activity of unnatural polyelectrolytes, such as polyacrylic anhydride, capable of forming covalent conjugates with corpuscular and protein antigens has been studied. Polyacrylic acid anhydride mixed with bovine serum albumin forms a covalent bond, where polyacrylic acid forms an unstable electrostatic complex. Both polyacrylic acid and polyacrylic acid anhydride can bind to erythrocytes. The immunostimulatory action of polyacrylic acid depended somewhat on the degree of polymerization, but it did not depend on whether antigen and polyelectrolyte were added separately or together. On the other hand, formation of antibody-producing cells in spleens of mice injected with polyacrylic acid anhydride markedly increased with the increased number of anhydride groups and was higher when the antigen-polyion covalent complex was injected, as opposed to injecting them separately.[274]

An analytical material for determining hydrogen peroxide, organic peroxides and hydrogen peroxide-forming material mixtures has been developed. A test strip consists of a transparent carrier layer and a porous gelatin layer containing poly(DADMAC), a p-phenylenediamine derivative as indicator, and a naphthol or pyrazolone derivative as coupler. The strips are useful for glucose determination in body fluids.[275]

A new encapsulation technique involves the formation of a polyelectrolyte complex membrane from a polyanion and poly(DADMAC) which can be used for long-term culturing of pancreatic islets.[276] A polyelectrolyte complex derived from poly(DADMAC) and sodium cellulose sulfate has been used to prepare capsules which consisted of a liquid core and a semipermeable membrane.[277]

A stable microbiocidal aqueous dispersion comprising methylene bisthiocyanate and a self-inverting emulsion of a copolyampholyte containing DADMAC was suitable for inhibiting the growth of slime-forming bacteria, fungi, and algae.[7-29]

A copolymer of DADMAC with acrylic or methacrylic acid has been used as a bactericide for water-oil base cutting fluids.[7-33] The effect of polyelectrolyte flocculants on microorganisms in receiving streams has been studied. Tests of polyelectrolytic toxicity to aquatic microorganisms demonstrated that poly(DADMAC) did not inhibit the cumulative oxygen uptake of a culture of mixed aquatic microorganisms, as measured with an electrolyte respirometer. Although oxygen uptake curves showed that the polymer could not support microbial activity by the mixed aquatic microbe, a Pseudomonas species was isolated from the St. Lawrence River which could biodegrade and assimilate the polyelectrolyte. This bacterial species was also able to utilize the monomer as a source of both carbon and nitrogen.[7-34]

Algicidal compositions which contained the alternating copolymer of DADMAC monomer and sulfur dioxide along with dimethyl-, allyl-, and diallylamine, inhibited the total growth of *Mougeotia genuflexa* and *Spirogyra crassa* in mixed culture much better than other available algicides.[278]

Antideterioration agents for carboxymethylcellulose fibers which contained DADMAC were also useful as water retainers for farmland soils and as road sealants.[279]

A container for blood testing was prepared using a formulation which included the alternating copolymer of DADMAC and sulfur dioxide.[280]

Biocidal agents with toxicity against *Rhizoctonia solani* and *Xanthomonas malvacearum* were formulated from complexes with poly(DADMAC).[281]

A formulation for use in polyethylene tubes, as containers for conducting blood tests, was prepared by use of the alternating copolymer of sulfur dioxide and DADMAC along with a blood coagulation promoter.[282]

A modified formulation for use in glass test tubes as containers for blood testing contained antifibrinolysis agents as well as the alternating copolymer of DADMAC monomer and sulfur dioxide.[283]

Copolymers of methylbenzyl-, methyl-p-nitrobenzyl- and DADMAC with sulfur dioxide were evaluated as catalysts in hydrolysis of dinitrophenyl phosphate dianions.[284] The acceleration factor was greatest for the methyl-p-nitrobenzyl copolymer, this factor being as high as 60.

The salt linkage formation of poly(DADMAC) with the acidic groups in the polyanion complex between human carboxyhemoglobin and potassium (vinyl alcohol) sulfate was studied by colloid titration. At pH 7.7, the carboxy group of the hemoglobin stoichiometrically formed a linkage with the polymer. However, at pH 12, the polymer also formed a linkage with sulfate anion formed by cleavage of the complex.[2-502]

Quaternary ammonium salts of monodisperse conidine oligomers with antiheparin activity have been studied. The salts were prepared by polymerizing conidine, and are isomeric with the six-membered ring structure originally proposed for poly(DADMAC).[285]

The following generalized formula for the quaternary ammonium salts of monodisperse conidine oligomers with antiheparin activity was provided (Structure 12–1, n = 5–40, X - Cl, Br).[286]

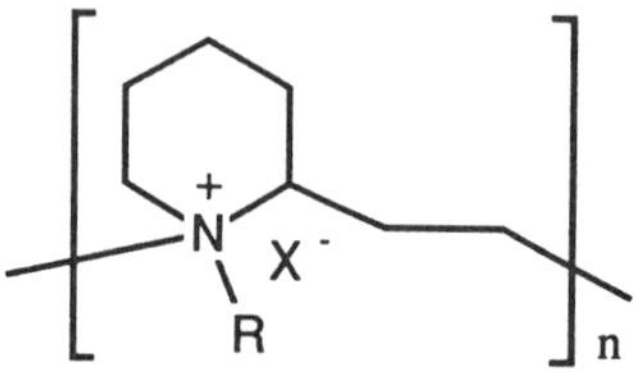

(12–1)

In a study of the role of leukocyte factors and cationic polyelectrolytes in phagocytosis of groups A *streptococci* and *Candida albicans* by neutrophils, macrophages, fibroblasts and epithelial cells, it was suggested that cationic polyelectrolytes, an example of which is poly(DADMAC), may prove important as agents capable of enhancing the penetration into cells of both viable and nonviable particles, genetic materials and drugs.[287]

Cyclopolymerization as a means of constructing polymers with defined arrangement of functional groups in the chain has been investigated.[2-366] Water-soluble acrylamide copolymers of various molecular weights containing salicylaldehyde and lysine functions were prepared from the cyclopolymerizable monomer (*N*″-5-methacryloylamino-salicyclidene-*N*′-methacryloyl-(s)-lysinato) (pyridine) copper II, and acrylamide, followed by destruction of the copper complex by use of the sodium salt of EDTA. At pH>6 the formation of "internal" aldimine occurs in these samples. The degree of assembly of "internal" Schiff base is considerably higher than in the model low-molecular-weight system and does not depend on the molecular weight of the polyacrylamide, the polymer concentration in buffer solution, or the presence of a crosslinking agent. Based on these data, it was concluded that in the copolymerization of (*N*″-5-methacryloylaminosalicylidene-*N*′-methacryloyl-(S)-lysinato)(pyridine) copper(II) with acrylamide, methacryloyl moieties, bonded with the amino groups of salicylaldehyde and lysine, to form macrocycles, the cyclopolymerization rate being much higher than the rate of crosslinking. After the removal of Cu ions from the polymer, the lysine and salicylaldehyde moieties remained attached to the same polymer chain in the immediate vicinity of each other, thus favoring the formation of "internal" aldimine. The materials are of interest as antitumor agents, enzyme models and in blood technology.

Poly(DADMAC), as an interface for retarding the back-reaction in photoinduced electron transfer reactions between zinc tetrabis(4-*N*-methylpyridinium)-porphyrin perchlorate and propyl viologensulfonate decreased the back electron transfer rate by a factor of two without affecting other bimolecular reaction rates.[288]

Poly(DADMAC) was used as a component of an agent for microbiocidal disinfection which totally inhibited the growth of *Chlorella vulgaris*, in vitro[289]

A graft copolymer of cellulose and DADMAC was shown to possess hemostatic properties, which improved when the extent of grafting increased.[290] Treatment of the copolymer with 2% heparin showed that only 0.35–0.4% was bound to heparin.

Vesicle-forming quaternary ammonium salts, allyldidodecylmethylammonium bromide and diallyldidodecylammonium bromide, were polymerized by x-irradiation.[2-213] The structure of the monomer vesicles was retained only in the polymerized vesicles of the diallyl monomer.

Poly(DADMAC) has been approved by USFDA as a pigment dispersing agent in limited quantities in coatings for paper and paperboard for food contact.[291]

Polyallylamine has been used as a component of a preservative composition for food.[292] In a standard test, the composition prevented mildew and putrefaction up to 90 days.

An antithrombogenic material was developed using polyallylamine as a component which could be made into a film useful in making blood bags.[293]

Polymers capable of reversibly stretching and shrinking were developed from an acidic polymeric electrolyte and a basic polymeric electrolyte in which polyallylamine represented the basic component.[294] The materials are useful as chemical actuators of artificial limbs, prostheses, and robots.

Wastewater and process waters from microbiological protein manufacture was purified for reuse by use of a flocculation composition of which poly(DADMAC) was a major component.[295]

Further studies on polymerization of vesicular properties of allyl and diallyl based monomeric and polymeric quaternary ammonium salts showed that those derived from the diallyl monomer showed long-term stability up to two years.[2-230]

Vesicles formed from diallyldodecylammonium bromide by ultrasound and polymerized by free radical initiation or irradiation were stable and freeze fracture electron micrographs showed that the products were single-compartment dilayer membrane vesicles.[2-233]

Poly(DADMAC) was used in preparation of uncontaminated encapsulated cells of *Yarrorvia lipolvtica* in cellulose sulfate. It was possible to fill the prepared capsules and have their surface free of cells.[296]

A variety of additional publications or patents have appeared in the literature dealing with novel applications of poly(DADMAC) or its copolymers in the areas of the biological, medical or food industries.[297-309]

Poly(DADMAC) has been reported to be effective as a component of synergistic algicides, useful in cooling towers, which have low toxicity to nontarget organisms.[310] It has also been used in clarifying sugar liquors by removing floc precipitates consisting mainly of calcium phosphate.[311] Further use of DADMAC derivatives, in this case the copolymer of DADMAC with sulfur dioxide, as a component of synergistic algicides which controlled *Chlorella vulgaris* at 10 ppm, vs. 40 ppm for a control has been reported.[312]

An interesting and novel application appears to be the use of poly(DADMAC) as a component of an improved method of rat liver microsome encapsulation.[313]

Applications in Coal, Minerals, and Glass

Poly(DADMAC) has been found to be effective in the flotation process of potassium ores.[314] The application of polyelectrolytes and coagulants, including poly(DADMAC) in the dissolved and induced air flotation and vacuum flotation processes used in mining operations, along with future trends have been discussed.[1-174] Aqueous solutions or gels of poly(DADMAC) have been used effectively as fluid-loss additives, fracturing fluids and permeability-increasing fluids in stimulation of reservoir rocks in petroleum and natural gas wells.[315] The compression strength of frozen granulated coal aggregate was reduced by almost 50% if the granulates were sprayed before freezing with a solution containing poly(DADMAC) among other components.[316]

Treatment of glass and glass fibers with a solution containing polyethylene oxide and poly(DADMAC) results in production of workable and hydrophilic glass fiber strands.[317] The polymer has also been used effectively to treat glass and glass fibers to produce chopped strands of glass and glass fibers having excellent water retentivity and dispersibility which are useful as a substitute for asbestos.[318] It has also been shown to be a significant component of a solution for treatment of bundled glass fiber filaments to give chopped glass fiber strands having excellent workability and hydrophilic properties.[319] The viscosity of aqueous coal slurries containing 40–60% coal fines was reduced by almost 50% by addition of 1–10 lbs/ton of the polymer.[320]

Acrylonitrile was polymerized in the presence of (12–2a) sodium alcoholate to cyanoethylated polyacrylonitrile (a) which was hydrogenated to give poly(2-piperidylidene-methylene) (12–2b) useful as a flocculation agent.[321]

Poly(DADMAC), when present in an aqueous drilling fluid, was particularly effective in preventing clay swelling and/or fines migration.[322]

The removal of hydrogen sulfide from a fluid stream, for example from geothermal steam, involves a method which employs poly(DADMAC) among other similar materials, as a soluble cationic polymeric catalyst.[323] In a standard test, approximately 100%

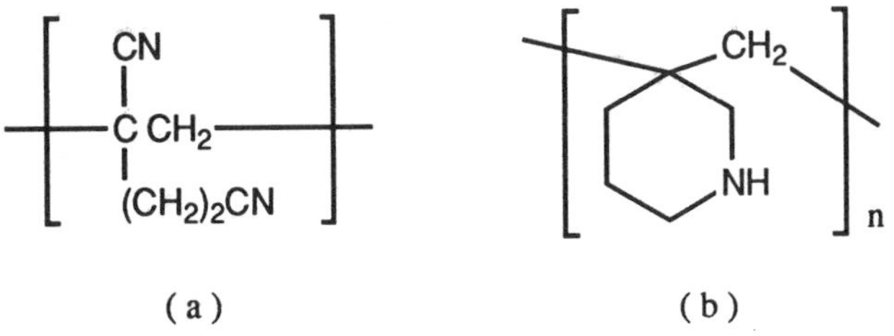

(a) (b)

(12–2)

increase in performance was observed with the cationic catalyst over that without the catalyst.

DADMAC-acrylamide copolymers were used as improved dewatering acids for mineral processing.[324] Slurries from powdered ores or coal were treated for clarification as well as dewatering and in a comparative test, it was shown that residual turbidity was reduced by ≈200% over another similar mixture of polymer components.

Weakly basic poly(diallylamine) resins were tested for gold adsorption, and the results showed that these resins have high capacity and good selectivity for gold over most of the pH range and in many different types of solutions, including contaminated wastes.[325]

Zeta-potential measurements on poly(DADMAC)-modified glass fibers showed that the zeta-potential of poly(DADMAC)-modified fibers in a humid atmosphere was −49.9 mV as compared to a value of −88.6 mV for an unmodified fiber.[326]

The applications of poly(DADMAC) and its copolymers in the mining, minerals and glass industries have increased rapidly. Accordingly, the number of published reports on these novel applications has increased.[327–341]

Poly(DADMAC) has been used as a component of a process of concentration of uranium from ocean water by a combination of ultrafiltration with complexation.[342] It has also been claimed as a component in a process for flocculation of coal fines from aqueous suspension.[343] Aqueous solutions of poly(DADMAC) have been used to treat glass fibers in a process for manufacture of non-agglomerating, well-dispersible glass fibers and glass-fiber rovings.[344]

Membrane Applications

Poly(DADMAC) has been used effectively as the polycation in formation of polyelectrolyte complex membranes. The filtration properties of such membranes could be varied by controlling the ratio of two anionic components.[345] Ion-selective improved semipermeable membranes useful in improved separation of molecular mixtures have been prepared by coating a preformed laminate with a terpolymer of diallylamine hydrochloride, sulfur dioxide, and DADMAC, followed by crosslinking through the amine function with 5-(chlorosulfonyl)-isophthaloyl chloride.[346]

Membrane filtration of fermentation media and other biological material was found to be accelerated by flocculating solid substances with poly(DADMAC) before filtration.[347]

Membranes of poly(DADMAC) with high swelling ability and electrical conductivity have been formed from solutions which contained aprotic solvents.[7–76] The electrical properties of the membranes in nonaqueous media depended on the polymer and nature of the media. The increase in donor properties of the solvent decreased the ion exchangeable sorption of salts and led to an increase of the membrane activity. Electrical conductivity and transference numbers were measured in water-dimethylsulfoxide mixtures.

In fabrication of nitrate-selective electrodes with polymer membranes containing immobilized sensors, covalently bound quaternary ammonium salts were used as sensor groups.[348] Triallylbutylammonium bromide gave superior results. After polymerization, the crosslink density was 1.7×10^{-5} mole cm^{-3}.

Photosensitive Langmuir-Blodget membranes, useful for photochromic optical memories or optical switching devices, were manufactured by a process which utilized poly(DADMAC) as a major component.[349]

Applications as Soil Treatment Aids

Poly(DADMAC) has been effectively used as a component in a sequential process for stabilization of clay-containing earth formations.[350] A simulated test well comprising sand, fine sediments, and clay was calibrated with 500 ml. standard brine and treated with 4% aqueous poly(DADMAC) (Mol. Wt.: 37,000), 300, standard brine 500, fresh water 500, 15% aqueous HCl 400, and fresh water 500 ml. Results indicated that the formation had greater stability than after treatment in the absence of poly(DADMAC). The polymer has also been used as a significant component of a solution for increasing the content of aggregates in alkaline soils.[351]

Poly(DADMAC) has been used as a component of a preservative for agricultural products and their processing.[352] The preservative prevented microbial growth in a simulated test of food products in contact with *Leuconostoc dextranium, Baccillus polymyxa* and soil, which assured a high, broad spectrum organism count, for 168 days at ambient temperatures.

Hydrophilic cationic polymers with comonomers (e.g. allyltrimethylammonium chloride) crosslinked at intervals of 50–100 monomer units, with chloride ion as counter ion of the DADMAC units in the chain were found to be effective in increasing the water-holding capacity of sandy soils.[353] Nutrient storage of the soil was also improved.

A process of lowering serum cholesterol involved orally administering a crosslinked poly(DADMAC).[354] Crosslinking was accomplished by use of small amounts of epichlorohydrin and 1-chloro-2,3-dihydroxypropane. The residue was 90% insoluble in water.

Poly(DADMAC) has been claimed as a suitable cationic polyelectrolyte for enhanced adhesion of agrochemicals to leaf and soil surfaces.[355]

Miscellaneous Applications

Poly(DADMAC) has been found to function as a superior mordant in mordanting of filter dyes in photographic materials.[356] Polymers or copolymers of DADMAC have been used effectively in light-sensitive photographic silver halide materials having improved antistatic characteristics when the polymer or copolymer was incorporated into a antihalation back layer, interlayer or top layer.[7-175] The removal of dust from gases in microbiological processes is accomplished by spray scrubbing with water containing poly(DADMAC) as a surfactant.[357]

The homopolymer has also been used to prevent dust formation, blow-off, and wetting of loose materials by applying the polymer as a surfactant along with a binder, e.g., poly(acrylic acid), to the surface of the loose material as a coating.[358] Ink-jet recording sheets, having complete water resistance and excellent lightfastness, are obtained by coating substrates with a mixture of poly(DADMAC) and poly(ethylenimine).[359]

Emulsions containing a saponified DADMAC-vinyl acetate copolymer or allyltrimethylammonium chloride-vinyl acetate copolymer were prepared and mixed with thickeners, freeze-thaw stabilizers, rustproofing agents, preservatives, defoamers, emulsifiers, or dispersants, and water.[7-30] Polyampholytes and their salts were prepared from anionic, cationic, and some nonionic monomers and used as scale and corrosion inhibitors, in drilling fluids, and in waterflood oil recovery. A typical example of such a polyampholyte is an acrylic acid-DADMAC-*N*,*N*-dimethylacrylamide terpolymer (70:20:10).[7-31]

A high quality ink-jet recording sheet was prepared by coating paper with a composition consisting mainly of an acrylamide-DADMAC copolymer. The prepared sheet provides water resistant images and high resolving power.[360]

The adsorption of poly(DADMAC) on carboxymethylated cellulose fiber at different pH values was measured, and the process shown to be an ion-exchange reaction.[361]

Copolymers of DADMAC and diallyldimethylammonium dihydrogenphosphate of molecular weight 30,000–300,000 have been found to be useful as inhibitors of salt deposition in oil recovery.[362]

Recording liquids for jet printing and writing which produce fast-drying, sharp images are made up of water-soluble dyes and poly(DADMAC) as the major components.[363] The liquid was stable, and formed prints with good drying properties and light resistance.

Polymeric conditioners for improved electroless metalization of nonmetallic surfaces, which produce uniform metal coatings with good adhesion comprise poly(DADMAC) of molecular weight 15,000.[364] A glass-fiber epoxy resin composite was treated with this conditioner to give improved copper coatings.

Adhesives containing poly(DADMAC) were useful for reversible bonding of packaging materials such as paperboard.[365]

The copolymer of DADMAC and sulfur dioxide was used as a macroion in a study of surface tension of synthetic polyelectrolytic solutions at the air-water interface.[366] Further studies on interfacial phenomena, using the colloid titration method for determining the charge characteristics of a variety of materials used in the papermaking industry, including poly(DADMAC), have been carried out.[367]

Poly(DADMAC) was used as the cationic component of a series of interpolymer complexes formed with polycarboxylic acids. Specific conductance of the complexes was determined.[368] Stable polyelectrolyte complexes of different compositions were prepared by reaction of poly(DADMAC) with hexachloroplatinate ion in aqueous solution.[369]

Poly(DADMAC) has been characterized in its applications in bulk films and on electrode surfaces.[370] Poly(DADMAC) was used as the cationic component for generating an improved coprecipitated hydrogel for use in pressure tolerant gas-diffusion electrodes.[371]

An acrylamide-DADMAC copolymer was effective as a component of an agent for detackification of spray paint.[372]

Poly(DADMAC) has been used in a study of binding of polyelectrolytes to oppositely charged ionic micelles at critical micelle surface charge densities.[373]

Poly(DADMAC) has been used as a component for felt composites with good elongation and stiffness in contact with vinyl flooring plasticizers.[374]

The water resistance of shaped articles, formed by removing water from an aqueous dispersion of vermiculite lamellae, by incorporating a polymer in the dispersion, was improved by adding as a component poly(DADMAC).[375] Poly(DADMAC) has also been found to be effective as an antistatic base in a photographic element support.[376] Copolymers of DADMAC have been used in a composition for improving the adhesion of glues for labels.[377]

It has been shown that the binding of 2-p-toluidinylnaphthalene-6-sulfonate, useful as a fluorescent probe to detect hydrophobic environments of biopolymers, to the copolymer of DADMAC with sulfur dioxide which occurs through the sulfonate and quaternary ammonium groups, is exothermic and involves a positive entropy gain.[378] The contribution of the entropy term to the free energy change increases with increasing hydrophobicity of the polymers.

A cationic monomer derived from allyl bromide and *N,N*-dimethyltetradecylamine was claimed to copolymerize, in presence of a radical initiator, with acrylamide to produce a copolymer which effectively enhanced the viscosity of brine solutions at different salt levels.[379]

A synergistic antifouling composition contains dialkyldiallylammonium chloride-sulfur dioxide copolymer and/or dialkyldiallylammonium chloride polymer and *N*-(2-hydroxyalkyl)aminomethanol.[380] The control of algae and microorganisms by this preparation in a test system was demonstrated.

Poly(DADMAC) was used as a component of a suspension polymerization medium for vinyl polymerization to form controlled-size beads for ion exchange use.[7-63]

Heat regenerable, amphoteric ion exchangers having salt uptake 1.32 mg/g and exchange acid capacity 3.58 meq/g, and exchange base capacity 2.64 meq/g consisted of a hydrolyzed terpolymer of divinyl benzene, ethyl acrylate and triallylamine.[7-59]

Poly(DADMAC) was used to form stoichiometric complexes with oppositely charged polymers as the basis for a study of their adsorption properties.[381] Conductance and turbidity measurements were carried out.

Acids thickened with branched emulsion or suspension polymers of DADMAC have been developed as oil-well drilling and fracturing fluids for stimulation of production.[382] Triallylmethylammonium chloride was used in limited quantities as a branching comonomer.

An α, ω-nonconjugated diene, ethylene glycol dimethacrylate, functioned as a tetrafunctional monomer in radiation crosslinking of poly(vinyl chloride), although some loss of crosslinking ability was observed by cyclization of the bis-methacrylate.[7-175]

An analytical procedure which is becoming more useful in the polyelectrolyte field is referred to as polyelectrolytic titrations.[383] This technique was used effectively to continuously control the papermaking process with regard to deviations of charge conditions of the pulp suspension.

Poly(DADMAC) was used as the cationic standard in the polyelectrolytic titration method for determination of concentrations of a variety of polymers.[384] The end point of the titration was traced with high sensitivity using a streaming current detector.

Poly(DADMAC) was used as the polymeric cation in the colloid titration method for determining residual concentrations of carrageenan and celluloid sulfate.[385]

Poly(diallyldimethylammonium peroxydisulfate) was synthesized for use as a catalyst for radical polymerization.[386]

Studies in the interaction of poly(DADMAC) with ferro- and ferricyanide anions indicate that the binding of these ions with the polymer proceeded by an anionic mechanism.[2-495]

Phase diagrams were obtained for a strong cation, poly(DADMAC), with a nonionic surfactant (Triton X-100) and an ionic surfactant (SDS) in presence of NaCl.[2-497] Phase separation depended on the ionic strength and the mole fraction of anionic surfactant and was relatively insensitive to polymer or total surfactant concentrations. Turbidimetric titration curves were obtained.

Emulsions were prepared from a saponified copolymer of DADMAC-vinyl acetate.[7-30] These emulsions were useful in thickeners, freeze-thaw stabilizers, rustproofing agents, preservatives, defoamers, emulsifiers or dispersants.

Polyampholytes of acrylic acid-DADMAC-*N,N*-dimethylacrylamide were useful as scale and corrosion inhibitors, in drilling fluids and in waterflood oil recovery.[7-31]

A composition for stabilizing clays in cementing of oil and gas boreholes was developed which included poly(DADMAC) as a major component.[387] The respective flow rates in a simulated study were 112.0, 156.5, 119.6, and 133.6% compared to 17.4, 1.8, 0, and 0% for acrylamide copolymer with 2-acrylamido-2-methylpropanesulfonic acid.

Poly(DADMAC) or a copolymer of DADMAC with vinyl pyrrollidone were employed as antistatic agents in light-sensitive silver halide materials.[388]

The particle size of polyelectrolyte complexes formed between poly(DADMAC) with sodium cellulose sulfate in water increased with increasing NaCl concentration.[2-499]

Poly(DADMAC) has been used as a component of a drilling fluid for oil and gas well technology.[389] In connection with waterflood and surfactant-flood petroleum recovery, partially hydrolyzed acrylamide-bis-methacrylimide copolymers were found to have better thermal stability.[390]

Crosslinked poly(DADMAC) has been used as a major component of a temporary blocking drilling fluid for closing pressure storage chambers in petroleum recovery.[391]

In a study of the effect of different stabilization methods on the chemical composition and quality of wine, it was shown that the highest wine quality was obtained by treatment with poly(DADMAC) which reduced the levels of protein nitrogen, polyphenols and iron from 18.4, 450.3, and 12.3 mg/L, to 8.7, 358.1, and 4.7 mg/L, respectively.[392]

Poly(DADMAC) was used as the standard reagent for the cationic polymer used in polyelectrolyte titration.[393] Use of this technique in paper chemistry was discussed.

Terpolymers of DADMAC, acrylic acid and 2-acrylamido-1-methylpropylsulfonic acid have been shown to be useful as scale and corrosion inhibitors.[394]

A review of recent results on studies of the structure of polyelectrolyte complexes, including those with poly(DADMAC) was published in 1985 with 17 references.[395]

Organic acids, useful in the food and pharmaceutical industries were manufactured by microorganisms fixed in a semipermeable membrane or capsule of polyelectrolyte complexes.[396] *Yarrowia lipolytica E.H. 59* was fixed on a complex of poly(DADMAC) and cellulose sulfate, and resulted in a yield of 0.7 g of citric acid/g of glucose.

A terpolymer of DADMAC, acrylamide and vinylphosphonic acid was used as an acid-soluble viscosifier for fracturing-acidizing of petroleum and gas wells.[397] The electrochemistry of electrodes was studied using poly(DADMAC) as supporting electrolyte.[398]

The interaction of a globular protein, bovine serum albumin, with polyelectrolytes, poly(DADMAC) and poly-*N*-ethyl-4-vinylpyridinium bromide was studied by determining the kinetics of exchange of the protein between the polyelectrolytes. The exchange was quite rapid ($<$ 10s).[399]

Whole cells from *Paracoccus denitrificans* were immobilized with a polyelectrolyte complex of which poly(DADMAC) was a component.[400] The immobilized cells exhibited activity at pH 4, at which the free cells lost their activity.

Poly(DADMAC) has been used in the quenching bath to decrease crack formation and increase efficiency of quenching in the mining machine construction industry.[401] It has also been used as a component of an enzyme carrier. Lipase from *Candida cylindracea* could be immobilized to yield 92 units of lipase/g of this carrier.[402]

Poly(DADMAC) can be determined spectroscopically by use of the reagent Solochrome Violet RS.[403] The sensitivity of the method was excellent. The polymer of mol. wt. 5,000–100,000 was used to prevent algae growth in swimming pools.[404] It has also been used to quench fluorescence and excimer formation of sodium pyrene-3-sulfonate.[405]

Poly(DADMAC), (polyelectrolyte PKB-1) was manufactured after the chemical stability of various construction and engineering materials in its presence were tested under manufacturing conditions.[406]

Poly(DADMAC) was used to improve print sharpness in the ink-jet printing of fabrics.[407] DADMAC was used as a component of a monomer mixture to produce polymers which were self-crosslinkable in the presence of base catalysts.[408]

Surface charges on liposomes made from purified phospholipids were determined by colloidal microtitration using poly(DADMAC).[409] Poly(DADMAC) was also used as a component of a system to remove trihalomethane compounds from water.[410]

Polyelectrolyte-surfactant complexes prepared by substituting the chloride ions of poly(DADMAC) with alkyl sulfonates, were characterized by a variety of methods.[411]

Turbidimetric titrations and quasielastic light scattering studies were used to determine the size distribution of complexes formed between poly(DADMAC) and bovine serum albumin.[412] Equilibrium binding of mixed micelles to poly(DADMAC) was studied by turbidimetric titration, quasielastic light scattering and ultrafiltration.[413]

Poly(DADMAC) was used as a component of an insoluble ionomeric ionically conductive hydrophilic hydrogel for use in pressure-tolerant gas-diffusion electrodes.[414]

Poly(DADMAC) has been used as a component of an ionomeric ionically conductive gas-impermeable layer in pressure-tolerant gas-diffusion electrodes.[415] The layer comprises the cationic polymer which has been crosslinked in situ by exposure to the required amount of γ-radiation. Poly(DADMAC) was also used as a component of a system used with ionomer membranes in pressure-tolerant gas-diffusion electrodes.[416]

The dust generated during handling, transportation and stockpiling coal can be suppressed by use of a formulated aqueous solution including poly(DADMAC) as one of the components.[417]

Gamma irradiation of a film of a solution of DADMAC gave 98% conversion to polymer at 1 Mrad.[2-229] The importance of high conversion is obvious since it has generally been difficult to attain high conversions in solution because of early solidification of the concentrated monomer solution.

The residual monomer in commercial poly(DADMAC) was irradiated by γ-radiation to form insoluble networks that could be immobilized on electrode surfaces.[418]

Poly(DADMAC) was used as the cationic component of a polyelectrolyte complex-coimmobilized *Nitrosomonac luropaea* and *Paracoccus denitrificans* cells to effect simultaneous occurrence of nitrification and denitrification under oxygen gradient.[419]

A copolymer of DADMAC with sulfur dioxide was used as an immobilization matrix for complexing certain organic compounds for isomerization.[420]

The copolymer of DADMAC with sulfur dioxide was studied as a binding agent for the azo dye, chrome violet K.[421] Cobaltic ions enhanced the binding rate; in contrast, cupric, nickel and zinc ions had little or no effect on the binding.

Poly(DADMAC) was used to improve the adhesion of colloidal particles to a glass surface when subjected to flow.[422]

The process for manufacture of 1,2-dichloropropane by chlorination of propane gave increased purity in the product when poly(DADMAC) was used as catalyst.[423]

Poly(DADMAC) was used as an additive to improve the quality of vegetable oils during refining.[424] It was also used as a component of a water-resistant binder with good adhesive properties.[425] The other component was 8–12% hydrolyzed polyacrylonitrile.

Symplexes (polysalts) were obtained by mixing the alternating copolymer of maleic acid and allyamine with cationic polyelectrolytes, e.g. poly(DADMAC). The products exhibited polyampholyte character.[7-82]

A polysalt was formed between poly(DADMAC) and the alternating copolymer between maleic acid and methyldiallylamine.[7-81] Copolymers of maleic anhydride and methyldiallylamine exhibited amphoteric character.[7-77]

Adsorption isotherms of poly(DADMAC) on fractioned spruce groundwood and on cellulose and lignin surfaces were determined at different pH values at 25 °C.[426]

Heterogeneous catalysts containing complexed copper were obtained by mixing a copolymer of DADMAC and sulfur dioxide with copper sulfate.[7-125]

Poly(DADMAC) has been used as a component of a composition for recovering oil from formation, to improve the rheological characteristics and petroleum-displacing capacity of the composition.[427]

The cupric copper complexation capacities of aquatic organic matter were determined by streaming current detection, a method which involved titration of humic anions with poly(DADMAC).[428]

Sealing in recirculated water is prevented by adding effective quantities of a copolymer of DADMAC with varying ratios of acrylic acid.[429]

A variety of unusual or nonclassical uses of poly(DADMAC) and its copolymers have been discovered. Among these are to increase the speed of photographic emulsions;[430] in suspension vinyl polymerization to uniform polymer beads;[431] as an additive for alkaline plating baths for zinc;[432] to improve rechargeable electrochemical cells;[433] as a corrosion inhibitor for dry cell batteries;[434] to increase the service life of secondary alkaline batteries;[435–437] as brighteners for acidic electroplating baths;[438, 439] to make tin electroplates resistant to natural whisker growth;[440] to improve the action of fluorescent whiteners in the paper industry;[441] as a component of a detergent mixture for foam baths;[442] to reduce the electric charge on ammonium nitrate explosives;[443] to prevent salt deposition in petroleum production equipment;[444] to improve the molecular weight of aromatic polyamides in two-phase systems;[445] to improve the solvent dispersibility of pigment compositions;[446,447] to improve the antistatic and heat sealable properties of biaxially oriented film;[448] to improve electric razor preshave compositions;[449] to improve antistatic coating compositions;[450] to improve stimulation of subterranean formations of petroleum or natural gas wells;[451] to aid in emulsion breaking in fat- or oil-in-water emulsions;[452] to increase the curing rate of adhesives;[453] to improve the properties of fillers for filled plastics and rubbers;[454] to improve the extraction and purification of L-glutamic acid;[455] as an antistatic agent in a process for preparing nonelectroscopic toners with a liquid slip agent;[456] to improve detackification of paint spray operation wastes;[457] to improve production of granulated phenol-formaldehyde resols;[458] as a monomer in pearl polymerization;[459] as an electrolyte to improve the charge-discharge efficiency of secondary lithium batteries;[460] as a component of a pigment dispersion for use in brushing-on colors[461] and in producing high-quality ink-jet printing.[462]

The solubilization of the hydrophobic dye, solvent Orange 2, in polyelectrolyte-micelle complexes has been studied. Poly(DADMAC) was used as the polyelectrolyte, and mixed anionic and nonionic micelles were utilized.[463] This polymer has also been used as a component in ion-exchange electrokinetic chromatography with polymer ions for the separation of isomeric ions having identical electrophoretic mobilities.[464]

Poly(DADMAC) has been used as the polycation in a continuing investigation of the stoichiometry of polysalts and of the preferential bonding from polymer mixtures.[465]

In a study of polyampholytes based on maleic acid and allylamines, methyldiallylamine and DADMAC were used as monomers.[466] The copolymers of the above monomer pairs were soluble over the entire pH range.

In a continuing study of association in poly(DADMAC)-micelle complexes, it was found that a higher order association in the complexes occurred.[467] An increase in micelle dimensions appeared to lead to a transition from intrapolymer-complexes, in which several micelles were bound by a single polymer chain, to interpolymer association, in which many polymer chains were involved in a higher order aggregate.

Water-soluble lignin graft polymers useful in petroleum recovery have been claimed by grafting DADMAC and acrylamide to lignin by use of free radical initiation.[468]

DADMAC and diallylamine have been used as monomers for synthesizing vinyl grafted lignite fluid loss additives, useful in drilling and cementing oil wells.[469] Wastewater containing suspended solids and petroleum products can be recovered by treatment with poly(DADMAC) as a flocculating agent, followed by separation of the residues after settling.[470]

In a study of protein-polyelectrolyte complexing selectivity, in which poly(DADMAC) was complexed with globular protein, it was shown by turbidimetric titration that at the phase boundary, phase separation may be a consequence of the saturation of the protein-binding sites on the polycation.[471]

Poly(DADMAC) has been used to broaden the working range of a zinc-plating electrode.[472] In a 1990 Russian paper, the development of new products from the polyelectrolyte, poly(DADMAC), designated as VPK-402, was discussed.[473] Also, poly(DADMAC), crosslinked with methylenebisacrylamide, has been used as a sorbent.[474]

Further applications of poly(DADMAC) in oil well technology includes the development of compositions for stimulating, drilling or completion fluids. The polymer was shown to be effective as a viscosifier because of its resistance to hydrolysis.[475]

APPLICATIONS OF OTHER CYCLOPOLYMERS OR CYCLOCOPOLYMERS

The potential for cyclopolymers to compete with their classical counterparts is obvious. The limiting features, however, generally relate to added monomer cost or other lack of accessibility. Consequently, the advantages for the cyclopolymers or the corresponding copolymers must arise from some property inherent in the cyclic structure and its effect on the polymer properties. Some obvious effects are an increase in chain stiffness, increased thermal stability, and possibly stronger interchain interactions. Other possible effects could be associated with the different chemical or biological effects of the specific functional groups present in the cyclopolymer but not present in the classical counterpart. For example, the cyclic *N*-methylmethacrylimide polymers may be contrasted with poly(*N,N*-dimethylacrylamide). Many comparative studies of this nature have been made. It has been shown that increased thermal stability and chain stiffness are inherent contributions of the cyclic unit in the chain in a number of cases. This section describes a variety of applications in which these novel properties play a role as well as a number of other potential applications inherent in the chemical or biological effects of the cyclic unit.

Novel Applications in Water and Wastewater Treatment

The major developments in this area appear to be directly related to the early observation that poly(DADMAC) was effective at low concentrations as a flocculation and coagulation agent. Extensive commercial development has occurred with this polymer. The major applications in this field have been discussed above.

Predictably, copolymers of diallamine and its derivatives, including DADMAC, with sulfur dioxide should be useful in this area and such is the case.[7-112] Also, copolymers of diallylglycinonitrile with acrylic acid are effective as scale formation inhibitors in boiler water.[7-27] Specifically, DADMAC-SO_2 copolymers synthesized by photoinitiation, have found application in coagulation and flocculation.[7-117, 7-118, 7-119] *N*-Methyldiallylamine copolymers have found application as water-soluble thermosetting polyampholytes.[476] Poly(diallylamines) have also found use as components of polymeric emulsions.[477]

Applications as Ion Exchangers (Sirotherm Process)

A novel development in ion exchange technology involved the concept that a mixed bed of cationic and anionic polymers could be thermally regenerated at elevated temperatures to effectively reverse the salt uptake cycle which had occurred at lower temperature. This process has been referred to as the "Sirotherm process" and was developed in Australia with the objective of economically recovering useable water from brackish water. This section includes a discussion of these developments as well as a limited number of other developments in ion exchange uses of cyclopolymers.

The first example of a strongly basic ion exchanger was the result of the early discovery of cyclopolymerization in 1949.[1-1, 478, 7-66] Many important developments have occurred during the intervening years. Developments in connection with the research program leading to the Sirotherm process have shown that poly(triallylamine) or its copolymers are the most effective cationic components of the needed mixed bed system.[479, 1-160, 1-168, 2-112, 4-27, 2-94] The complete details of the Sirotherm process, including the demineralization plant, were published in a series of papers in 1965–66.[480–484] 1,6-Bis-(diallylaminohexane) was used as a crosslinker for the allylamine polymer system.[2-112] The practical applications of the poly(diallylamine) developed during the Sirotherm program were discussed in a 1976 review.[1-160]

Highly crosslinked polyquaternary ammonium monomers were developed rather early, e.g. hexaallylethylenediammonium salts and hexaallylhexamethylenediammonium salts as well as the corresponding dimethyltetraallyl derivatives were synthesized as early as 1952.[485, 2-138, 486] A later reinvestigation of cyclopolymerization led to the development of a strongly basic ion exchange system that was reported to have superior rate of exchange and ion exchange capacity in the extraction of uranium from ores (Structure 12–3).[1]

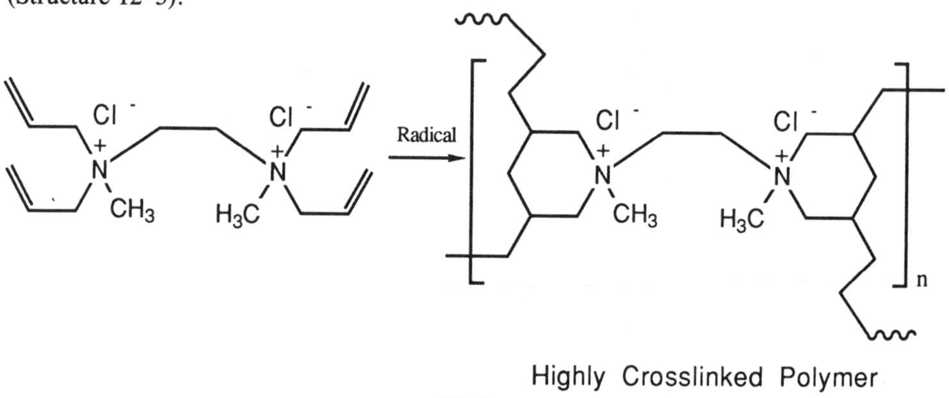

Radical

Highly Crosslinked Polymer

(12–3)

Poly(triallylamine) or its copolymers have been further described,[487] and modifications have been discovered which significantly increase its rate of ion exchange,[488] as well as its performance as a composite of mixed bed systems.[489]

Ion-exchange electrokinetic chromatography has been developed for the separation of analyte ions having identical electrophoretic mobilities in capillary electrophoresis. The separation principle is based on the differential ion-pair formation of the analyte ion with a polymer ion. Poly(DADMAC) has been employed effectively as the cationic polymer for use in this novel technique.[490]

Cyclopolymers as Catalysts

Cyclopolymerization of α, ω-dienes capable of leading to large rings have been studied and described in Chapters 2 and 3. However, monomers designed to lead to crown ether type polymers have been synthesized and their cyclopolymerization studied.[491] Divinyl ethers of tri-, tetra-, and pentaethylene glycol have been cationically polymerized to polymers useful in separating metal ions.[2-472, 2-476] Also, biomimetic catalysts based on cyclopolymers of 4-diallylaminopyridine have been reported.[492, 2-240]

A gel-polymer matrix based on diallyl monomers has been used as the basis for studying diffusion processes of fluorine- and silicon methacrylates into the matrix.[493]

Polymeric catalysts for esterolysis and esterification reactions have been developed by cyclopolymerization of 4-(diallylamino)pyridine. Copolymers with neutral, anionic, and cationic monomers have also been investigated as catalysts.[494] Their catalytic effectiveness was discussed.

Cyclobutene ring opening has been shown to be a useful reaction for the synthesis of well-defined double stranded molecules, including polymers (See Structure 9–4).[495]

Applications Related to Resist Technology

Cyclopolymers of phthaldehyde, first reported in 1967,[5-55] have recently been shown to have potential as resists. The polymer has a ceiling temperature of −43°C and decomposes thermally to regenerate monomer in 70% yield, and is volatilized by a photochemical reaction.[5-77] A copolymer of allyl acrylate with styrene has been used to make negative electron beam resists with high gel rigidity.[3-153] Allyl methacrylate has been used as a component of a coating for lithographic plates which gave clear images.[496] Also, copolymers of allyl methacrylate and propargyl methacrylate have been used to provide very high speed as electron beam and x-ray resists.[7-188]

Tetraallyloxyethane has been used as the preferred termonomer with methyl methacrylate and potassium acrylate to produce, by terpolymerization, the antistatic layer, one of the two layers which make up the backing for photographic film support.[497]

Cyclopolymerization of bis(fluoroalkyl) dipropargyl malonate derivatives in presence of $MoCl_5$- and WCl_6-based catalysts led to soluble polymers which presumably contained a conjugated main chain.[498] The mechanical, thermal and electrical properties of the polymers were determined.[498]

Applications in the Elastomer, Paper, and Textile Industries

1,6-Diene hydrocarbons were cyclopolymerized by coordination catalysts as early as 1963, and shown to have elastomer properties.[1-52, 2-3] Later, copolymers of 1,5-hexadiene and ethylene were prepared and their elastomeric properties studied.[7-8] During the development of ethylene-propylene-diene (EPDM) elastomers, 1,4-hexadiene was studied as the diene used to provide a vulcanization site. However, it was found that considerable potential residual unsaturation was lost by cyclization of a fraction of the diene.[3-59] As early as 1960, cyclopolymers of 1,5-hexadiene were extensively studied and this investigation added more definitive evidence for the nature of the repeating unit in poly-1,5-hexadiene.[1-53] Every polymer prepared was crystalline, had a density exceeding 1.0 g/cc and a melting point in excess of 100°C, and was very flexible. Poly-1,5-hexadiene had a tensile strength of 5400 psi, a melting point of 139°C, and a density of 1.122 g/cc, and yet was very flexible, having an apparent modulus of elasticity at -50°C of 200,000 psi.

The x-ray fiber repeat distance of poly-1,5-hexadiene was found to be 4.80 Å, which corresponded to a chain structure consisting of a linear, zigzag array of 1-methylene-3-cyclopentyl groups. Although a repeat distance of 4.80 Å was calculated, if the methylene substituents in the 1,3-positions of the cyclopentane ring were placed *trans* on a planar ring, this system did not take into account known facts about 1,3-disubstituted cyclopentanes. *cis*-1,3-Dimethylcyclopentane is known to be more stable than the *trans* isomer by about 0.5 kcal/mole, and this is most likely the result of ring puckering, a proved phenomenon in cyclopentane and its derivatives. If the methylene substituents of poly-1,5-hexadiene were placed *cis* and the number-2 carbon atom of the ring puckered toward the methylene substituents to a perpendicular distance of 0.46 Å above the plane of planar cyclopentane, the distance between the methylene groups (fiber repeat distance) would be 4.80 Å. This system (See Structure 2–7) would[1-12] account for all the known facts about cyclopentane and poly-1,5-hexadiene.

Crystalline copolymers of ethylene and 1,5-hexadiene over the 15–93 mole % ethylene compositional range have been prepared and studied.[1-60, 7-11] The density of the copolymers ranged almost linearly from 0.931 g/cc for pure polyethylene to 1.08 for pure poly(1,5-hexadiene). The effect of copolymer composition was a reduction of the apparent modulus of elasticity at −50°C as the diene monomer content increased, these modulia passing through a minimum at about 50 mole % diene.

Fiber-forming properties superior to those of polyacrylonitrile were determined for a copolymer of 2,6-dicyano-1,6-heptadiene and acrylonitrile.[1-26] A copolymer containing 15% of the diene monomer was spun and drawn fourfold to strong filaments having a wet initial modulus of 10 g/denier at 90°C and a recovery from 3% elongation in 50°C water of 56%. Typical values for a similar fiber from polyacrylonitrile are 3 g/denier and 40%, respectively.

Applications of poly(DADMAC) in the paper and textile industries have been discussed at length above. However, other cyclopolymers or the corresponding copolymers have found potential uses in these industries. Polymethacrylimide has been studied as a component of both fibers and films.[2-167] Dyeable fibers have been obtained by blending polydiallylamine with polypropylene before spinning.[2-149] Copolymers of diallylamines have found uses as dye fastness improvers.[7-37]

Water-absorbent resins were prepared by copolymerizing sodium acrylate with *N*-methylenebisacrylamide.[499] The polymers were obtained as sperical shaped beads of average particle size diameter of 100–600 μm, and of narrow size distribution.

Thermal Stability of Cyclopolymers

A comparison study of the thermal stability of the cyclocopolymer of *cis,cis*-1,5-cyclooctadiene and sulfur dioxide with a conventional copolymer of an alkene and sulfur dioxide has been carried out.[1-91, 4-41] The copolymer was obtained in film-and fiber-forming molecular weights, had no residual unsaturation, and contained two moles of sulfur dioxide per mole of cyclooctadiene (See Structure 1–19). The copolymer decomposed at temperatures above 250° with gas evolution. The copolymer was prepared in 98% yield (based on cyclooctadiene) and had an inherent viscosity (ln $\eta_{rel/c}$) of 2.2 at 0.5 g per 100 ml of dimethyl sulfoxide solution.

The thermal stability of this bicyclic copolymer was studied.[4-42] Degradation followed first-order kinetics at 220°C or above. Random cleavage of the copolymer chain occurred. Pyrolysis of the copolymer up to 75% loss of weight yielded the monomers as cracking products in the molar ratio of the copolymer composition. The activation energy E_a for the degradation process was 41 kcal/mole with a frequency factor of $5.2 \times 10^{13} s^{-1}$. In comparison, E_a for the propylene-SO_2 copolymer is 32 kcal/mole. This copolymer decomposes approximately 100 times faster at 229° than the bicyclic copolymer does at 230°. The bicyclic sulfone structure is inherently more stable than the linear α,β-sulfone structure of the olefin-SO_2 copolymers.

Poly(1,5-hexadiene)[1-52] was reinvestigated and some of its practical properties were determined. It was shown to contain approximately 5% residual unsaturation. Since not all of the monomer units incorporated into the chain were cyclized, the polymer more closely resembled a copolymer than the homopolymer, and could be vulcanized through the residual unsaturation.[2-16] The density of the polymer exceeded 1.0 g/cc, which is higher than any aliphatic poly-olefin. It had about the same tensile strength and melting point as high density polyethylene but was about as flexible as low density polyethylene. The thermal stability of the polymer has been studied, and consistent with comparisons made between cyclopolymers and their monoalkene counterparts, this polymer exhibited much greater thermal stability than polyethylene (See Figure 12–2).[2-3]

Thermally stable poly(diallylarylphosphine oxides) were prepared both by direct polymerization of the corresponding monomers and by thermal degradation of appropriate poly(diallylphosphonium salts), e.g., poly(diallyldiphenylphosphonium chloride), to give the phosphine oxide polymer. The latter method gave much higher molecular weights than the former method.[1-37]

Antitumor and other Medical Applications

The alternating cyclocopolymer of divinyl ether and maleic anhydride, DIVEMA, has been extensively investigated for its antitumor properties, and has been discussed earlier (Chapter 8). Similar structures, as well as other unrelated cyclopolymers, have also been found to possess antitumor or other biological properties.

The alternating cyclocopolymer of divinyl ether with citraconic anhydride has been studied for its antitumor properties.[8-32, 8-66] The effects of molecular weight and molecular weight distribution of DIVEMA on its biological properties have been evaluated.[8-2, 8-48] The extensive biological activity studies of DIVEMA were reviewed in 1980 with

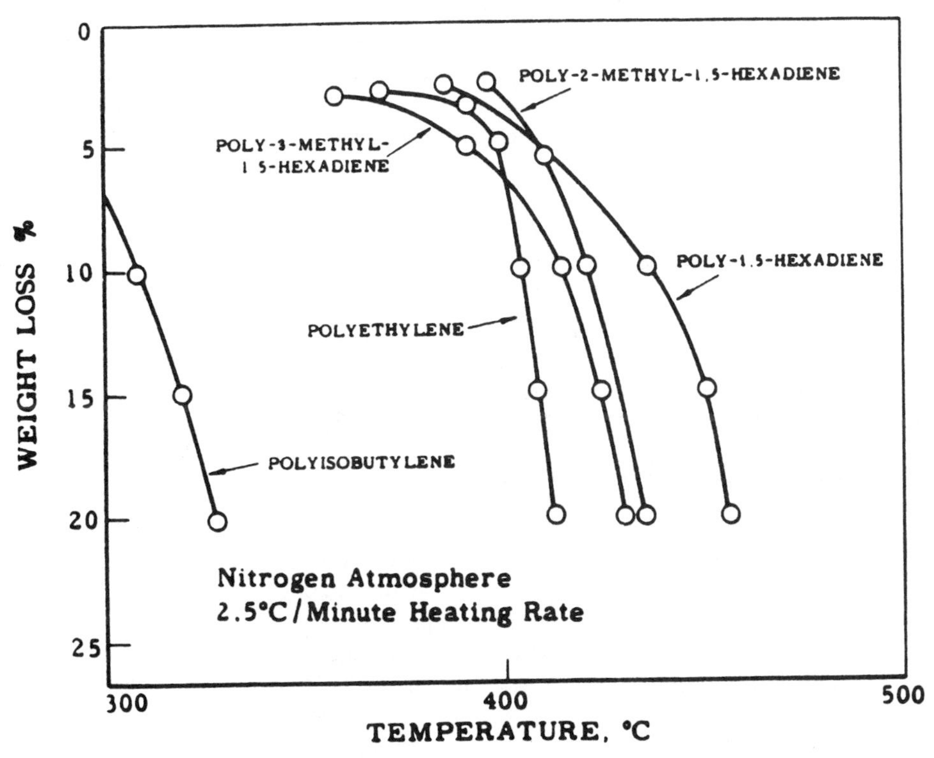

Figure 12–2 Thermogravimetric curves of polyisobutylene, polyethylene and poly(1,5-dienes).

162 references,[1-171] and the antitumor properties of a group of closely related cyclocopolymers were summarized in 1985.[8-161]. In another study DIVEMA was derivatized with the antitumor drugs, adriamycin and daunomycin, and the derived products were evaluated for their biological activity.[500] The fine structure of DIVEMA is still somewhat uncertain, and further studies have been published dealing with ring size in this polymer.[8-33, 8-60]

A variety of other medical applications of cyclopolymers have been evaluated. A copolymer of triallyl cyanurate with vinyl acetate was found useful as an absorbent for low-density lipoprotein removal from blood plasma.[501] Poly(DADMAC) was found to be useful as the cationic polymer in forming microcapsules with polysalt walls.[502] Poly(DADMAC) has also been found useful in coil-globule conformational transition studies of adsorption of human serum albumin.[503] A polymer of the diglycidyl ether of 1,4-butanediol has been used as an affinity chromatographic substrate for separation of blood factor VIII complex.[504]

DIVEMA has been modified[505] by covalently bonding methotrexate, *N*-[4-(*N*-methyl-2,4-diamino-6-pteridinyl-methylamino)-benzoyl] glutamic acid, a widely used antitumor agent and a folic acid antagonist, to several copolymers of various molecular weights. This copolymer was selected as a potential carrier for this agent, because of its established antitumor and immunestimulating activity. Bonding to DIVEMA was accomplished without crosslinking by nucleophilic addition reactions of the pteridinylamino groups under mild conditions. The polymeric derivatives were purified by membrane filtration, solvent extraction, and reprecipitation and were characterized by thin-layer chromatography, infrared, nuclear magnetic resonance, ultraviolet and mass spectrometry, and elemental analysis. Copolymers having initial number average molecular weights ranging from 4,440 to 24,000 were used. Using molar ratios of the DIVEMA repeating unit of the copolymer to methotrexate ranging from 1.9 to 3.5, and varying the reaction time from 24 to 96 hours, methotrexate residues ranging from 3.0 to 42.6 mole percent were introduced into the copolymers. The linkage to the polymer occurred at the 2- or 4-amino groups of the pteridine ring.

A methotrexate substituted derivative of DIVEMA was evaluated (NIH screening test no. NSC 282447) for dihydrofolate reductase inhibitory activity *in vitro* and antitumor activity *in vivo*. The preliminary pharmacological studies showed inhibition of dihydrofolate reductase and cytotoxicity to L-1210 lymphoid leukemia cells *in vitro* similar to free methotrexate.

DIVEMA has recently been shown to be an inhibitor of topoisomerase I from human spleen.[506] The inhibition was specific. Topoisomerase II was not inhibited.

Other Areas of Application of Cyclopolymers

Cyclopolymers or the corresponding copolymers have found applications as chelators, in chemical association, in composites, as crosslinking agents, in electrode technology, in food application, as herbicides, in oil well technology, in optical applications, including synthesis of optically active polymers, and in printing.

Poly(diallylamines) have been reported to be efficient as chelators when appropriately substituted on nitrogen.[2-110] A variety of studies have been conducted in which chemical association, or ionic binding had occurred. Polysalts are readily formed between polycations and polyanions. Other investigations have been concerned with use of cyclopolymers as suspension stabilizers,[6] and as binding agents for certain dyes in analyses.[507] Allyl methacrylate has been used to improve impact resistance in poly(vinyl chloride) composites.[508] Dimethacrylates have been studied to determine the relationship between the cyclization tendency and crosslinking capacity.[7-177] Poly(DADMAC) has been used in electrode technology.[509] Allyl methacrylate has been approved by the Federal Drug Administration as a minor component of packaging material for contact with food products.[510] Polymeric herbicides have been developed by converting the herbicide, metribuzin, to its bis-acryloyl derivative followed by cyclopolymerization.[2-214, 2-220] Although poly(DADMAC) has been used extensively in oil well technology, other cyclopolymers have seen only limited use. However, a terpolymer of diallylacetamide has been found to have application to this industry.[7-36] Optically active polymers have been synthesized from the divinylacetal of R(+)-3,7-dimethyloctanal.[2-307] Also, diallyl fumarate has been used in preparation of optical lenses with high refractive index, low shrinkage, and good heat resistance.[511] Vinyl methacrylate has been found useful in manufacture of wetting-free lithographic printing plates.[512] α, ω-Divinylpolydimethylsiloxane was also used in the composition.

The solution properties of a fluorocarbon-containing hydrophobically associating copolymer were determined.[513] The copolymers were obtained by copolymerization of acrylamide with 2-(*N*-ethylperfluorooctanesulfonamido)ethyl acrylate or 2-octanesulfonamido-(*N*-ethylperfluoroethyl methacrylate).

Polyelectrolyte-enhanced ultrafiltration (PEUF) is a method of removing ions from water.[514] In PEUF, a polyelectrolyte of opposite charge to the target ion is added to the solution and the target ion binds to the polymer. In this paper, poly(styrene sulfonate) was used as the polyanion. However, classical concentration polarization behavior is observed in PEUF but is not severe in the operating range of practical importance. No report for use of poly(DADMAC) as a polycation in PEUF was found but would be a logical choice.

Radical cyclopolymerization of ethyl vinyl citraconate led to synthesis of polymers containing 4-membered rings.[515] Some of the properties of the polymers were studied.

Characterization of the cyclopolymer of methyl α-hydroxymethylacrylate using nutation NMR spectroscopy has been reported.[516] The polymer was prepared from dilabeled monomer, and the method depended on the proposition that directly bonded ^{13}C spin pairs, detectable by the NMR experiment, could only be formed in the case of five-membered ring formation. The results supported a cyclopolymer consisting primarily of six-membered rings.

Epoxide monomers, capable of cyclopolymerization, and their polymers were prepared.[517] The monomers were prepared by epoxidation of bis-allyl ethers and were obtained in high yield. Epoxidation of allyl ether was not reported.

Cyclopolymerization of dimethyldipropargylgermanium and diphenyldipropargylgermanium using $MoCl_5$- and WCl_6-based catalysts have been reported.[518] Poly(diphenyl-dipropargylgermanium) was soluble, had a number-average MW of 12,000, and had a polyene structure with recurring cyclic units. The electrical conductive properties of the polymers were examined.

The solid-state cross-polymerization reactions of poly(α, ω-alkyldiyne) macromonomers were studied.[419] The results showed that a broad distribution of conjugated lengths was present in the partially cross-polymerized macromonomers.

REFERENCES

1. Clingman, A., Parrish, J. and Stevenson, K., J. Appl. Chem., 1963, 13, 1; CA 58, 7398c (1963).
2. Flock, H. G. and Rausch, E. G., Water-soluble Polymers, N. M. Bikales, 1973, 2, 21; CA 81, 111018d (1974).
3. Takagichi, T., Kozuku, H. and Kuroki, N., J. Polym. Sci., Polym. Chem. Ed., 1981, 19, 3237.
4. Williams, M. W. and Auclair, C. J., U.S. Pat. 4,299,898, 1982; CA 96, 172138z (1982).
5. Christensen, G. L., and Wavro, S. G., Ind. Waste, Proc. Mid-Atl. Conf. 13th., 1981, 404; CA 97, 28129a (1982).
6. Balakrishnam, T. and Ford, W. T., J. Appl. Polym. Sci., 1982, 27, 133; CA 96, 86051j (1982).
7. Nippon Senka Kogyo Co., Ltd., Jpn. Kokai Tokkyo Koho, Jap. Pat. 56 128,382, 1981 (Oct. 7); CA 96, 21247h (1982).
8. Kao Soap Co., Ltd., Jap. Pat. 81 120,613, 1981 (Sept. 22); CA 96, 11509f (1982).
9. Nippon Kayaku Co., Ltd., Jap. Pat. 81 18,611, 1981 (Feb. 21); CA 95, 43953s (1981).
10. Sharpe, A. J., Falcione, R. J. and Boothe, J. E., Eur. Pat. Appl. EP 45,562, 1982 (Feb. 10); CA 96, 182447p (1982).
11. Yorke, M. A., Proc. IUPAC, Macromol. Symp., 28th, 1982, 888; CA 99, 140755u (1983).
12. Yorke, M. A., U.S. Pat. 4,357,207, 1982 (Nov. 2); CA 98, 18345z (1983).
13. Kul'skii, L.A., Pilipenko, A.T., Znamenskaya, M.V., Zul'figarov, O.S., Kruglitskii, N.N. Shmidt, B.B., et al. USSR SU 1,114,625, 1984 (Sept. 23); CA 102, 84216k (1985).
14. Paronyan, V.K., Askinazi, A.I., Gubman, I.I., Gaponenko, V.G., Zamskaya, R.S., Kalasheva, N. Baklanova, V.A., et al, Maslo-Zhir. Prom-st., 1986, 1, 11; CA 105, 8291c (1986).
15. Schoenherr, D., Froemling, U., Munick, H., Jaeger, W., Wandrey, C., Hahn, M. and Seehaus, F., Ger. (East) DD 217,193, 1985 (Jan. 9); CA 104, 39232t (1986).
16. May, L. M. and Carlson, W. M., Proc.-Int. Water Conf., Eng. Soc. W. Pa. 1985, 46, 235; CA 105, 11743g (1986).
17. Chernoberezhskii, A. Y. and Semenov, V. P., Zh. Prikl. Khim. (Leningrad), 1985, 58, 2279; CA 104, 55677x (1986).
18. Hofinger, M., Hille, M. and Boehm, R., Ger. Offen. DE 3,446,489, 1986 (July 3); CA 105,117419w (1986).
19. Nicke, R., Borchers, B. and Tappe, M., Zellst. Pap. (Leipzig), 1985, 34, 215; CA 104, 170355c (1986).
20. Butler, G. B., U.S. Pat. US 3,962,206, 1976 (June 8); CA 85, 7900r (1976).
21. Sherwood, J. C. and Brunbeck, R. T., U.S. Pat. 4,312,759, 1982 (Jan. 26); CA 96, 109682g (1982).
22. Topchiev, D. A., Kaptsov, N. N., Gudkova, L. A., Kabanov, V. A. Martynenko, A. I., Trushin, B. N. and Parkhamovich, E, S., USSR, SU 910,664, 1982 (Mar. 7); CA 97, 73635y (1982).
23. Skeist Lab., Inc., Prospectus for Multiple-Client Study, 1982, 1.
24. Williams, P. A., Harrop, R. and Robb, I. D., J. Colloid Interface Sci., 1984, 102, 548; CA 102, 62794v (1985).
25. Wandrey, C., Jaeger, W., Starke, W. and Wotzke, J., Wasserwirtsch.-Wassertech., WWT, 1984, 34, 185; CA 102, 171981z (1985).
26. Toei, K., Zaitsu, T. and Igarashi, C., Anal. Chem. Acta, 1985, 174, 369; CA 104, 10259w (1986).
27. Horiuchi, M., Watanabe, M. and Goto, Y., Jap. Pat. JP 60,238,193 [85,238,193], 1985 (Nov. 27); CA 104, 115484u (1986).

28. Jessen, H. J., Acta Hydrochim. Hydrobiol., 1985, 13, 235; CA 102, 209051m (1985).
29. Ignatenko, V. G., Narymskaya, R. A., Chernoberezhskii, A. Yu. and Stepanova, I. A., Bum. Prom-st., 1986, 20; CA 106, 86446s (1987).
30. Kirie, K., Zaitsu, T. and Igarashi, C., Jap. Pat. 61 86,656 [86 86,656], 1986 (May 2); CA 106, 112916s (1987).
31. Baron, Jr., J. J., Nowakowski, T., Farrington, T. A. and Mahn, F. R., Eur. Pat. EP 204,404, 1986 (Dec. 10); CA 107, 9165t (1987).
32. Philipp, B., Jaeger, W., Gohlke,U., Wandrey, C., Hahn, M. Dietrich, K., Reinisch, G. and Koetz, J., et al, Pap. Puu, 1986, 68, 419; CA 105, 139038k (1986).
33. Karasev, K. I. and Startseva, E. P., Izv. Vyssh. Uchebn. Zaved., Khim. Tek., 1986, 29, 67; CA 106, 104413k (1987).
34. Pominville, D. A., U.S. Pat. 637,824, 1987 (Jan. 20); CA 106, 125430d (1987).
35. Farrar, D. and Langley, J., Eur. Pat. EP 228,868, 1987 (Jul. 15); CA 107, 199140b (1987).
36. Bhattacharyya, B. R., Srivatsa, S. R. and Dwyer, M. L., U.S. Pat. US 4,715,962, 1987 (Dec. 29); CA 108, 118362W (1988).
37. Walterick, Jr., G. C., U.S. Pat. 4,668,404, 1987 (May 26); CA 107, 102393a (1987).
38. Waagberg, L. and Lindstroem, T., Nord. Pulp Pap. Res. J., 1987, 2, 152; CA 108, 114466k (1988).
39. Reineke, W., Ger. (East) DD 249,464, 1987 (Sept. 9); CA 108, 100658n (1988).
40. Pech, R., Fr. Demande FR 2,589,145, 1987 (Apr. 30); CA 107, 242052p (1988).
41. Waagberg, L. and Lindstroem, T., Nord. Pulp Pap. Res. J., 1987, 2, 49; CA 107, 200721a (1987).
42. Sato, H., Goda, S., Koike, S. and Kurita, R., Jap. Pat. JP 63,240,999 [88,240,999], 1988 (Oct. 6); CA 110, 218519n (1989).
43. Yagiyu, T., Takai, Y. and Noda, K., Jap. Pat. JP 63,242,309 [88,242,309], 1988 (Oct. 7); CA 110, 218517k (1989).
44. Spence, M. D., Kozaruk, J. M., Melvin, M. and Gardocki, S. M., U.S. Pat. 4,758,353, 1988 (Jul. 19); CA 109, 115555w (1988).
45. Bown, R. and Pownall, P. G., GB 2,200,104, 1988 (Jul. 27); CA 109, 212638h (1988)
46. Jaeger, W., Wandrey, C., Hahn, M., Ballschuh, D., Ohme, R., Staeck, R. and Biering, H., Eup. Pat. EP 264,710, 1988 (Apr. 27); CA 109, 129853d (1988).
47. Chernoberezhskii, A. Yu., Korovin, L. K., Parkhamovich, E. S. and Ledashchev, V. V., Bum. Prom-st, 1988, 6; CA 109, 196484x (1988).
48. Kokufuta, E., Suzuki, H. and Nakamura, I., J. Ferment. Technol., 1988, 66, 433; CA 109, 188749x (1988).
49. Karasev, K. I., Myazin, V. P. and Lavrov, A. Yu., USSR SU 1,393,801, 1988 (May 7); CA 109, 60947t (1988).
50. Hassick, D. E. and Miknevich, J. P., U.S. Pat. US 4,746,457, 1988 (May 24); CA 109, 79457e (1988).
51. Dwyer, M. L., U.S. Pat. US 4,769,155, 1988 (Sep. 6); CA 109, 215720q (1988).
52. Vitvitskaya, B. R., Korolev, A. A., Skachkova, I. N., Savonicheva, G. A., Sergeev, S. G., and Nilova, O. L., Gig. Sanit., 1988, 66; CA 109, 1964y (1988).
53. Jaeger, W., Gohike, U., Hahn, M., Wandrey, C. and Dietrich, K., Acta Polym., 1989, 40, 161; CA 110, 136201p (1989).
54. Hassick, D. E. and Miknevich, J. P., U.S. Pat. US 4,800,039, 1989 (Jan. 24); CA 110, 218741d (1989).
55. Kochergin, N. V., Besterekov, U. B., Kamshibaev, A. A. and Abdiev, K. Zh., Khim. Prom-st. (Moscow), (9), 1989, 683; CA 112, 164349c (1990).
56. Slagel, R. C., Ger. Pat. DE 2,057,578, 1971 (June 3); CA 75, 77536k (1971).
57. Flock, H. G., Jr. and Hoover, M. F., GB Pat. 1,287,489, 1972 (August 31); CA 77, 168494c (1972).

58. Rausch, E. G. and Muia, R. A., Ger. Pat. DE 2,311,222, 1973 (Sept. 13); CA 80, 19344p (1974).
59. Herpers, F. J., Jr. and Untiedt, D. I., Ger. Pat. DE 2,426,691, 1974 (Dec. 19); CA 82, 158158k (1975).
60. Nakayama, M., Jap. Pat. 50,091,148, 1975 (July 21); CA 84, 21829b (1976).
61. Hurlock, J. R., Ballweber, E. G. and Connelly, L. J., U. S. Pat. 3,920,599, 1975 (Nov. 18); CA 84, 45246k (1976).
62. Tanaka, T., Oka, J., Saito, K. and Nakajima, K., Jap. Pat. 50,060,434, 1975 (May.24); CA 84, 154053r (1976).
63. Walker, J. L. and Flock, H. G., Jr., Can. Pat. CA 985,598, 1976 (March 16); CA 85, 182239v (1976).
64. Morgan, J. E. and Boothe, J. E., U. S. Pat. 3,968,037, 1976 (July 6); CA 86, 21560y (1977).
65. Herpers, F. J., JR. and Untiedt, D. I., U. S. Pat. 4,014,808, 1977 (March 29); CA 87, 169572v (1977).
66. Welcher, R. P., Rabinowitz, R. and Cibulskas, A. S., U. S. Pat. 4092467, 1978 (May 30); CA 89, 90460q (1978).
67. Tasaki, T. and Kawamata, N., Jap. Pat. 53,093,655, 1978 (Aug. 16); CA 89, 185581u (1978).
68. Goodman, R. M., U. S. Pat. 4,166,040, 1979 (Aug. 28); CA 92, 64495x (1980).
69. Paul, S. N., Waller, J. E. and Cairns, J. E., Ger. Pat. DE 2,942,111, 1980 (Apr. 30); CA 93, 209779a (1980).
70. Nitto Boseki Co., Ltd., Jap. Pat. 57,092,012, 1982 (June 8); CA 97, 216932f (1982).
71. Rey, P. R., U. S. Pat. US 4,430,248, 1984 (Feb. 7); CA 100, 123499k.
72. Huang, Shu Jen W., U. S. Pat. US 4,450,092, 1984 (May 22); CA 101, 59958a (1984)
73. Neigel, D., Eur. Pat. EP 103,698, 1984 (Mar. 28); CA 101, 73258d (1984).
74. Hoerder, M., Finke, N. and Jobsky, K., Ger. (East) Pat. DD 227,950, 1985 (Oct. 2); CA 104, 173922w (1986).
75. Roe, W. J. and Malito, J. T., U. S. Pat. US 4,578,255, 1986 (March 25); CA 104, 228166k (1986).
76. Samakaev, R. Ku., Dytyuk, L. T., Akhmetov, V. N., Shmatkov, A. V. and Soldatov, P. Ya., USSR Pat. SU 1,370,095, 1988 (Jan. 30); CA 108, 156228q (1988).
77. Hayashi, I. and Muto, T., Jap. Pat. JP 63,286,405, 1988 (Nov. 24); CA 111, 8016r (1989).
78. Albert, A., Fed. Rep. Ger. Pat. DE 3,825,281, 1990 (Feb. 1); CA 112, 124586w (1990).
79. Makarov, V. L., Khol'kin, Yu. I. and Kind, V. B., Gidroliz. Lesokhim. Prom-st., 1990, 2, 8; CA 112, 215119n (1990).
80. Shamaeva, Z. G., Vorob'eva, A. I., Kozlov, V. G. and Leplyanin, G. V., Plast. Massy, 1990, 5, 15; CA 113, 60236t (1990).
81. Gartner, H. A., Eur. Pat. Appl. EP 363,024, 1990 (Apr. 11); CA 113, 98266r, (1990).
82. Kolganova, I. V., Shnaider, M. A., Topchiev, D. A., Klyachko, Yu. A. and Paronyan, V. Kh., Izv. Vyssh. Uchebn. Zaved. Pishch. Tek., 1989, 5, 30; CA 113, 170692w (1990).
83. Amjad, Z., Colloids Surf., 1990, 48, 95; CA 113, 198785c (1990).
84. Kozakiewicz, J. J. and Duplaise, D. L., J. Appl. Polym. Sci., 1990, 41, 2349; CA 113, 213214m (1990).
85. Timofeeva, G. I., Pavlova, S. A., Wandrey, C., Jaeger, W., Hahn, M., Linow, K. J. and Goernitz, E., Acta Polym., 1990, 41, 479; CA 113, 232417w (1990).
86. Waldmann, J. J., U. S. Pat. 4,891,422, 1990 (Jan 2); CA 114, 29583a (1991).
87. Moench, D. and Hartmann, H. Ger. Offen DE 3,908,803, 1990, (Sept. 20); CA 114, 43777y (1991).
88. Moench, D., Hartmann, H. and Buechner, K. H., Ger. Offen. DE 3,909,005, 1990 (Sept. 20); CA 114, 48976x (1991).
89. Imai, K., Eur. Pat. Appl. EP 398,738, 1990 (Nov. 22); CA 114, 103903y (1991).

90. Chernoberezhskii, A. Yu., Korovin, L. K. and Beim, A. M., Bum. Prom-st., 1990, 9, 14; CA 114,128353g (1991).
91. Windhager, R. H. and Hwang, M. H., U.S. Pat. 4,259,411, 1981 (Mar. 31); CA, 95, 44908t (1981).
92. Sinkovitz, G. D. and Dixon, K. W., U.S. Pat. 4,316,943, 1982 (Feb. 23); CA 96, 182932t (1982).
93. Windhager, R. H., Eur. Pat. Appl. EP 39,564, 1981 (Nov. 11); CA 96, 37194z (1982).
94. Sharpe, A. J., Windhager, R. H. and Zierden, T. W., Eur. Pat. Appl. EP 60,666, 1982 (Sept. 22); CA 98, 25517k (1983).
95. Wuensche, E., Uhlmann, S., Ohme, R., Rusche, J. and Ballschuh, D., Ger. (East) DD 159,268, 1983 (Mar. 2); CA 99, 60925j (1983).
96. Berbeco, G. R., PCT Int. Appl. WO 84 02,881, 1984 (Aug. 2); CA 101, 172700y (1984).
97. De Castro, E. S., Heineman, W. R. and Mark, J. E., Polym. Prepr. (ACS, Div. Polym. Chem.), 1984, 25 (1), 191; CA 101, 30056u (1984).
98. Daiichi Kogyo Seiyaku Co. Ltd., Jap. Pat. JP 59,126,453 [84,126,453], 1984 (Jul. 21); CA 101, 212831a (1984).
99. Hardy, L. C. and Shriver, D. F., Macromolecules, 1982, 17, 975; CA 100, 157301e (1984).
100. Hardy, L. C. and Shriver, D. F., J. Am. Chem. Soc., 1985, 107, 3823; CA 103, 15009d (1985).
101. Kimura, T. and Igarashi, T., Jap. Pat. 61 81,401 [86 81,401], 1986, (Apr. 25); CA 105, 154821s (1986).
102. Hirai, T., Yamaki, J., Okada, T. and Yamaji, A., Electrochim. Acta, 1985, 30, 61; CA 105, 31820h (1986).
103. Dolinski, R. and Dean, W., Chem. Tech., 1971 (May) 304; CA 75, 56713a (1971).
104. Razin, G. P. and Nekhaichuk, A. D., Nov. Issled. Khim. Teknol. Bum., 1985, 23; CA 106, 224301u (1987).
105. Akim, E. L., Golovkov, A. S., Morev, A. V. and Krymer, M. G., Khim. Dreu., 1986, 15; CA 106, 111173k (1987).
106. Sumi, H. and Hotta, H., Ger. Offen. DE 3,716,356, 1987 (Nov. 19); CA 108, 114544j (1988).
107. Eiffler, J., Eur. Pat. EP 308,109, 1989 (Mar. 22); CA 111, 215101e (1989).
108. Hoover, M. F. and Carothers, R. O., U. S. Pat. 3,490,938, 1970 (Jan. 20); CA 72, 68126n (1970).
109. Growald, B., Dahlquist, J. A. and Marlor, G. A., U. S. Pat. 3,674,711, 1972 (July 4); CA 77, 115608m (1972).
110. Sharpe, A. J., Jr., Windhager, R. H. and Hearp, K., Ger. Pat. DE 2,454,333, 1975 (May 22); CA 83, 62308d (1975).
111. Calgon Corp., Neth. Pat. NL 7,414,247, 1975 (May 21); CA 84, 172153c (1976).
112. Ott, R. J. and Franke, H. G., U. S. Pat. 4,002,475, 1977 (Jan. 11); CA 86, 197952s (1977).
113. Schoeller, F. Jr., Belg. Pat. BE 843,485, 1976 (Oct.18); CA 87, 60786s (1977).
114. Sharpe, A. J., Jr., Windhager, R. H. and Hearp, K. S., U. S. Pat. 4,040,984, 1977 (Aug. 9); CA 87, 119476w (1977).
115. Jansma, R. H. and Albrecht, W. E., U. S. Pat. 4,084,034, 1978 (Apr. 11); CA 89, 26375p (1978).
116. Calgon Corp., Fr. Pat. FR 2,362,426, 1978 (March 17); CA 89, 188968m (1978).
117. Hwang, M. H., U. S. Pat. 4,132,674, 1979 (Jan. 2); CA 90, 106018k (1979).
118. Hayama, K., Aoi, H. and Ito, I., Jap. Pat. 54,065,754, 1979 (May 26); CA 91, 124448a (1979).
119. Calgon Corp., Jap. Pat. 55,066,944, 1980 (May 20); CA 93, 206436a (1980).
120. Windhager, R. H. and Hwang, Mei H., Eur. Pat. EP 12,517, 1980 (June 25); CA 94, 123255k (1981).

121. Hwang, Mei H., U. S. Pat. 4,282,118, 1981 (Aug. 4); CA 95, 178656t (1981).
122. Yin, R. I., Eur. Pat. EP 44695, 1982 (Jan. 27); CA 96, 164283c (1982).
123. Savit, J., Brit. Pat. GB 2,090,198, 1982 (July 7); CA 97, 118218h (1982).
124. Rutherford, S. L. and Savit, J., Fr. Pat. FR 2,496,287, 1982 (June 18); CA 97, 191278g (1982).
125. Stulgis, S., Chauser, M. G., Vilcinskiene, I., Daraceuniene, J., Ropite, A. Gabrielyan, S. M., Mardoyan, M. K., Cherkashin, M. I., Sidaravicius, I. and Daugela, R., USSR Pat. SU 978,096, 1982 (Nov. 30); CA 98, 98812n (1983).
126. Yin, R. I., U. S. Pat. 4,373,011, 1983 (Feb. 8); CA 98, 145194z (1983).
127. Bigelow, H. E., Can. Pat. CA 1,140,332, 1983 (Feb. 1); CA 98, 170373q (1983).
128. Jujo Paper Co., Ltd., Jap. Pat. JP 59,020,696, 1984 (Feb. 2); CA 100, 211975t (1984).
129. Eklund, N. and Huang, J., U. S. Pat. US 4,483,913, 1984 (Nov. 20); CA 102, 87691x (1985).
130. Orlov, I. G., Sidaravicius, D., Topchiev, D. A., Cherkashin, M. I., Kabanov, V. A., Valsiuniene, V., Vapsinaskaite, I., Koshelev, K. K. and Pavlenko, N. E. et al., USSR Pat. SU 1,416,927, 1988 (Aug. 15); CA 110, 104953u (1989).
131. Butera, R. J., Simic-Glavaski, B. and Lando, J. B., Macromolecules, 1990, 23, 211.
132. Komiya, K., Beppu, K. and Kanai, S., Japan Pat. JP 02,214,706 (90,214,706), 1990 (Aug. 27); CA 114, 63590k (1991).
133. Hercules, Inc., Belg. Pat. 881,397, 1980 (Jul. 29); CA 94, 193965r (1981).
134. DIC Hercules, Inc. Jap. Pat. JP 81 169,898, 1981 (Dec. 26); CA 96, 183108r (1982).
135. Iovine, C. P., and Ray-Chaudhuri, D. K., U.S. Pat. US 4,305,860, 1981 (Dec. 15); CA 96, 106052r (1982).
136. Sukhaya, T.V., Snopkov, V.B., Martsul, V.N., Topchiev, D.A., Kabanov, V.A., Kaptsov, N., Gavrilenko, L.P., et al, USSR SU 881,100, 1981 (Nov. 15); CA 96, 70697p (1982).
137. Spence, G. G. and Maslanka, W. W., Braz. Pedido PI BR 80 00,267, 1981 (Aug. 4); CA 96, 87112s (1982).
138. Loth, F., Fanter, C., Dautzenberg, H. and Dietrich, K., Ger. (East) DD 208,159, 1984 (Mar. 28); CA 101, 112687q (1984).
139. American Cyanamid Co., Jap. Pat. JP 58,156,098 [83,156,098], 1983 (Sept. 16); CA 100, 176752u (1984).
140. Rende, D. S. and Breslin, M. D., Eur. Pat. Appl. EP 151,994, 1985 (Aug. 21); CA 103, 179872f (1985).
141. Nippon Senka Kogyo Co., Ltd., Jap. Pat. 82,11,288, 1982 (Jan. 20); CA 96, 219207p (1982).
142. Kao Soap Co., Ltd., Jap. Pat. 59 88,978 [84 88,978], 1984 (May 23); CA 101, 193616r (1984).
143. Daiichi Kogyo Seiyaku Co., Ltd., Jap. Pat. JP 59,216,982 [84,216,982], 1984 (Dec.7); CA 102, 168283t (1985).
144. Daiichi Kogyo Seiyaku Co.,Ltd., Jap. Pat. JP 6071,786 [8571,786], 1985 (Apr. 23); CA 103, 124944a (1985).
145. Ballweber, E. G., Jansma, R. H. and Phillips, K. G., U.S. Pat. 4,217,425, 1980 (Aug. 12); CA 93, 169989k (1980).
146. Jarovitsky, P. A., Yun-Lung, F., Dexter, R. W. and Huang, S. Y., Eur. Pat. EP 245,702, 1987 (Nov. 19); CA 108, 114546m (1988).
147. Nicke, R., Zellst. Pap. (Leipzig), 1982, 31, 19; CA 97, 40564w (1982).
148. Leeder, J. D., Watt, I. C. and Banks, P. J., J. Macromol. Sci. (Chem), A, 1982, 17, 327; CA 96, 70041v (1982).
149. Kanebo, Ltd., Kanebo Synth. Fibers, Ltd., Jap. Pat. 59,163,427 [84,163,427], 1984 (Sept. 14); CA 102, 8213s (1985).

150. Iwata, M., Saka, T., Yamato, H. and Kondo, N., Jap. Pat. JP 61,203,110 [86,203,110], 1986 (Sep. 9); CA 106, 68695t (1987).
151. Guerro, G. J., Proverb, R. J. and Tarvin, R. F., Eur. Pat. EP 133,699, 1985 (Mar. 6); CA 103, 24013c (1985).
152. Molnar, M. J., Can. CA 1,194,254, 1985 (Oct. 1); CA 104, 20917y (1986).
153. Iwata, M., Saka, T., Yamato, H. and Kondo, N., Jap. Pat. JP 61 203,155 [86 203,155], 1986 (Sep. 9); CA 106, 68684p (1987).
154. Iwata, M., Saka, T., Ikeda, T. and Kondo, N., Jap. Pat. JP 61,130,317 [86,130,317], 1986 (Jun. 18); CA 106, 50824z (1987).
155. Iwata, M., Saka, T., Ikeda, T. and Kondo, N., Jap. Pat. JP 61,130,318 [86,130,318], 1986 (Jun. 18); CA 107, 40556v (1987).
156. Arheilger, D. and Von Medvey, I., Wochenbl. Papierfabr., 1986, 114, 958; CA 106, 121639f (1987).
157. Tefft, E. R., Eur. Pat. Appl. 172,684, 1986 (Feb. 26); CA 104, 226594z (1986).
158. Wilengowski, H., Jokiel, M. and Farnach, R., Zellst. Pap. (Leipzig), 1987, 36, 21; CA 107, 98459g (1987).
159. Lehmann, A., Tretner, W., Irlbacher,, G., Mai, J., Hille, C., Gliemann, M., Bergmann, H. D. and Peter, E., Ger. (East) DD 250,148, 1987 (Sep. 30); CA 109, 151329j (1988).
160. Fukunishi, A., Tsunekawa, T., Okamoto, M. and Kasuya, K., Jap. Pat. JP 62,282,083 [87,282,083], 1987 (Dec. 7); CA 108, 114157k (1988).
161. Fukunishi,A., Tsunekawa, T., Okamoto, M. and Kasuya, K., Jap. Pat. JP 2,257,481 [87,257,481], 1987 (Nov. 10); CA 108, 77082j (1988).
162. Tambor, M., Cope, J. L. and Jerome, J. L., U.S. Pat. 4,737,156, 1988 (Apr. 12); CA 109, 112013g (1988).
163. Hofmann, H. P. and Weigl, J., Ger. Offen. DE 3,707,221, 1988 (Sep. 15); CA 110, 175395z (1989).
164. Daniher, F. A. and Aspland, J. R., U.S. Pat. US 4,735,628, 1988 (Apr. 5); CA 109, 75137c (1988).
165. Armington, S. E., Halperin, S. A., Pickett, G. E. and Metz, B. A., PCT Int. Appl. WO 88 01,910, 1988 (Mar. 24); CA 109, 24417m (1988).
166. Tsmyg, N. G., Muzychenko, M. P., Selyuzhitskii, M. I. and Korostelev, S. I., Bum. Prom-st., 1988, 4; CA 110, 175327d (1989).
167. Tsai, J. J. H., Jobe, P. and Billmers, R. L., Eur. Pat. EP 283,824, 1988 (Sep. 28); CA 110, 77921y (1989).
168. Matsuda, N. and Kondo, N., EP 282,081, 1988 (Sep. 14), CA 110, 59784b (1989).
169. Komiya, K., Kanai, S. and Beppu, K., Jap. Pat. 63,264,985 [88,264,985], 1988 (Nov. 1); CA 110, 214695u (1989).
170. Degen, H. J., Pfohl, S., Weberndoefer, V., Rehmer, G., Kroener, M. and Stange, A., Ger. Offen. DE 3,706,525, 1988 (Sep. 8); CA 110, 59782z (1989).
171. Korchagin, M.V., Polyakova, L. A., Krichevskii, G. E., Kazakeviciute, G., Vainshel'boim, A. L., Chauser, M. G., Fedorovskaya, R. F., Ryazantseva, T. B., Penenzhik, M A., et al, USSR SU 1,509,462, 1989 (Sep. 23); CA 112, 181344q (1990).
172. Rose, G. R., U.S. Pat. 4,851,128, 1989 (Jul. 25); CA 112, 25137u (1990).
173. Westman, L., Grundmark, H. and Petersson, J., Nord. Pulp Pap. Res. J., 1989, 4,3, 113; CA 111, 176540a (1989).
174. Stange, A., Degen, H. J., Auhorn, W., Weberndoerfer, V., Kroener, M.and Hartmann, H., Ger. Offen. DE 3,724,646, 1989 (Feb. 2); CA 111, 80224z (1989).
175. Falk, M., Oedberg, L., Waagberg, L. and Risinger, G., Colloids Surf., 1989, 40, 115; CA 111, 216180s (1989).
176. Dahmen, K. and Kuester, E., Ger. Offen. DE 3,730,781, 1989 (Mar. 23); CA 111, 196677d (1989).

177. Negi, Y., Harada, T., Ishitsuka, O. and Ono, H., Jap. Pat. 44,008,729, 1969 (April 23); CA 72, 4296k (1970).
178. Harada, T., Jap. Pat. 47,017,918, 1972 (May 24); CA 77, 153851t (1972).
179. Nishikaji, T., Kamata, O. and Mori, K., Jap. Pat. 52,148,211, 1977 (Dec.9); CA 88, 172196b (1978).
180. Komoto, Y., Isome, Y. and Kusunose, T., Jap. Pat. 53,035,018, 1978 (Apr. 1); CA 89, 60965r (1978).
181. Kishioka, H., Jap. Pat. 53,070,178, 1978 (June 22); CA 89, 164877e (1978).
182. Reuss, P. J. and Weigl, J., U. S. Pat. 4,210,488, 1980 (July 1); CA 93, 116250f (1980).
183. Ballweber, E. G., Jansma, R. H. and Phillips, K. G., U. S. Pat. 4,233,411, 1980 (Nov. 11); CA 94, 32556t (1981).
184. Svitel'skii, V. P., Mil'shtein, A. D., Gritsulyak, V. N., Dzygun, V. A., Fedotovskii, L. B., Zernov, A. M., Trushin, B. N., Parkhamovich, E. S. and Topchiev, D. A., USSR Pat. SU 896,137, 1982 (Jan. 7); CA 96, 183099p (1982).
185. Daigle, B. and D'Errico, M. J., Eur. Pat. EP 58,621, 1982 (Aug. 25); CA 97, 184264w (1982).
186. Valendo, P. F., Tsmyg, N. G., Kabanov, V. A., Topchiev, D. A., Varfolomeev, D. F., Kirillov, T. S., Schmidt, B. B. and Kartashevskii, A. I., USSR Pat. SU 1,044,709, 1983 (Sept. 30); CA 99, 214407f (1983).
187. Arnhold, S., Gergele, H., Jacobasch, H. J., Thiel, B. and Grosse, I., Ger. (East) Pat. DD 201458, 1983 (July 20); CA 99, 214500f (1983).
188. Whitfield, J. M. and Cosper, D. R., U. S. Pat. US 4,432,834, 1984 (Feb. 21); CA 100, 193906q (1984).
189. Nikka Chemical Industry Co., Ltd., Jap. Pat. JP 59,036,788, 1984 (Feb. 29); CA 101, 39771f (1984).
190. Mitsubishi Paper Mills, Ltd., Jap. Pat. JP 60,059,195, 1985 (April 5); CA 103, 89263d (1985).
191. Flesher, P. and Farrar, D., Eur. Pat. EP 170,394, 1986 (Feb. 5); CA 104, 208358w (1986).
192. Hoshino, E., Nakae, T. and Murata, M., Jap. Pat. JP 61,047,800, 1986 (March 8); CA 105, 135985g (1986).
193. Rende, D. S. and Breslin, M. D., U. S. Pat. US 4,657,946, 1987 (Apr. 14); CA 107, 200761p (1987).
194. Okuma, S., Yamagishi, K., Hara, M., Suzuki, K. and Yamamoto, T., Jap. Pat. JP 62,265,328, 1987 (Nov. 18); CA 108, 151820s (1988).
195. Pratt, R. J., Slepetys, R. A., Nemeh, S. and Willis, M. J., Eur. Pat. EP 245,553, 1987 (Nov. 19); CA 108, 188738x (1988).
196. Farrar, D., Langley, J. and Allen, A., Eur. Pat. EP 262,945, 1988 (Apr. 6); CA 109, 38956y (1988).
197. Tsmyg, N. G., Muzychenko, M. P., Rokhlov, L. A., Selyuzhitskii, M. I., Korostelev, S. I., Topchiev, D. A. and Shurupov, E. V., USSR Pat. SU 1,375,706, 1988 (Feb. 23); CA 109, 56877c (1988).
198. Bauer, W., Keil, K. H., Nagl, G., Kaiser, M. and Steinbach, J., Ger. Offen. Pat. DE 3,703,293, 1988 (Aug. 18); CA 110, 40439w (1989).
199. Nemeh, S., Sennett, P. and Slepetys, R. A., U. S. Pat. US 4,772,332, 1988 (Sept. 20); CA 110, 175404b (1989).
200. Hahn, E., Henning, G., Mielke, M., Degen, H. J. and Pfohl, S., Fed. Rep. Ger. EP 309,908, 1989 (April 5); CA 111, 99250b (1989).
201. Kotaki, Y., Mori, T., Higuma, M. and Sato, H., Europ. Pat. Appl. EP 350,257, 1990 (Jan. 10); CA 112, 208060h (1990).
202. Zemaitaitis, A., Pacauskaite, A., Kavaliunas, R. and Brazauskas, V., USSR Pat. SU 1,544,853, 1990 (Feb. 23); CA 113, 25517p (1990).

203. Gartner, H. A., Eur. Pat. Appl. EP 359,509, 1990 (Mar. 21); CA 113, 41539c (1990).
204. Fukunishi, A. and Okamoto, M. Japan Pat. JP 02 01,358 (90 01,358), 1990 (Jan. 5); CA 113, 49925w (1990).
205. Felegi, J., Jr. and Kehrer, K. P., PCT Int. Appl. WO 90 06,342, 1990 (June 14); CA 113, 99724a (1990).
206. Niimi, A. and Watanabe, K., Japan Pat. JP 02,103,291 (90,103,291), 1990 (Apr. 16); CA 113, 103212b (1990).
207. Moench, D., Stange, A., Liebe, J., Hartmann, H., Merger, F. and Schwartz, M., Eur. Pat. Appl. EP 364,798, 1990 (Apr. 25); CA 113, 173389p (1990).
208. Martini, T., Karsunky, U., Sternberger, K. and Keil, K. H., Ger. Offen. DE 3,844,194, 1990 (July 5); CA 113, 174033e (1990).
209. Kojima, Y. and Omori, T., Eur. Pat. Appl. EP 379,964, 1990 (Aug. 1); CA 114, 45184b (1991).
210. Moench, D., Hartmann, H., Freudenberg, E. and Stange, A., Ger. Offen DE 3,909,004, 1990 (Sept. 27); CA 114, 64619p (1991).
211. Kuehne, G. Baustoffindustrie, 1990, 33, 91; CA 114, 87519v (1991).
212. Svitel'skii, V.P., Mil'shtein, A.D., Polevaya, V.I., Zhukotskaya, L.I., Dzygun, V.A. and Shurupov, E.V. USSR SU 1,585,420, 1990 (Aug. 15); CA 114, 104629a (1991).
213. Johnson Products Co., Inc., Fr. Demande 2,457,685, 1980 (Dec. 26); CA 95, 67816a (1981).
214. Matsunaga, K., Tsushima, R. and Okumura, T., Eur. Pat. Appl. EP 53,448, 1982 (June 9); CA 97, 133354t (1982).
215. Kao Soap Co., Ltd., Jap. Pat. 57 212,111 {82 212,111], 1982 (Dec. 27); CA 98, 166760w (1983).
216. Kao Soap Co., Ltd., Jap. Pat. 82,128,619, 1982 (Aug. 10), CA 97, 222748s (1983).
217. Kao Soap Co., Ltd., Jap. Pat. 82,126,409, 1982 (Aug. 6), CA 97, 222747r (1983).
218. Hoch, D., Wajaroff, T. and Konrad, E., Ger. Offen. DE 3,301,515, 1984 (July 19); CA 101, 197932p (1984).
219. Doerfel, K., Ballschuh, D., Rusche, J. and Ohme, R., Ger.(East) Pat. DD 217,987, 1985 (Jan. 30); CA 103, 146972k (1985).
220. Shiseido Co., Ltd., Jap. Pat. 59 184,115 [84 184,115], 1984 (Oct. 19); CA 102, 50715g (1985).
221. Su, D. T. T., Ger. Offen. DE 3,445.749, 1985 (July 4), CA 103, 146985s (1985).
222. Yoshioka, K. and Kamimura, Y., Jap. Pat. JP 63 57,512 [88 57,512], 1988 (Mar. 12); CA 110, 121006e (1989).
223. Shinjo, Z., Jap. Pat. 01 128,914 [89 128,914], 1989 (May 22); CA 111, 219097f (1989).
224. Winkler, W. M., Seaman, S. A. and Visscher, M. O., Eur. Pat. Appl. EP 308,190, 1989 (Mar. 22); CA 111, 239321c (1989).
225. Maruyama, Y., Jap. Pat. 01 26,510 [89 26,510], 1989 (Jan. 27); CA 111, 102518c (1989).
226. Ikeuchi, T., Jap. Pat. JP 63,313,716 [88,313,716], 1988 (Dec, 21); CA 110, 160219w (1989).
227. Iovine, C. P. and Nowak, F. A., Jr., U.S. Pat. US 4,803,071, 1989 (Feb. 7); CA 111, 120612h (1989).
228. Jordan, N. W., Winklee, W. M., Seaman, S. A. and McGuffey, H. W., Eur. Pat. EP 308,189, 1989 (Mar. 22); CA 111, 239322d (1989).
229. Boothe, J. E., Morse, L. D. and Klein, W. L., Eur. Pat. EP 266,111, 1988 (May 4); CA 110, 179261a (1989).
230. Shiseido Co., Ltd., Jap. Pat. 59 176,204 [84 176.204], 1984 (Oct.5); CA 102, 50711c (1985).
231. Matsumoto, S., Mori, K. and Ito, H., Jap. Pat. 61 190,597 [86 190,597], 1986 (Aug. 25); CA 106, 86727j (1987).

232. Klein, W. L. and Sykes, A. R., EP 200,548, 1986 (Nov. 5); CA 106, 107774q (1987).
233. Klein, W. L. and Sykes, A. R., EP 222,580, 1987 (May 20), CA 108, 209990r (1988).
234. Yoshioka, K. and Kamimura, Y., Jap. Pat. JP 62 178,510 [87 178,510], 1987 (Aug. 5); CA 108, 26821s (1988).
235. Gollier, J. F., Dubief, C. and Mondet, J., Ger. Offen. DE 3,716,381, 1987 (Nov. 19); CA 109, 27612g (1988).
236. Homan, G. R. and Cornwall, S. M., Eur. Pat. EP 219,830, 1987 (Apr. 29); CA 107, 183322b (1987).
237. Iwao, S. and Kuwana, H., Eur. Pat. EP 246,090, 1987 (Nov. 19); CA 108,81809u (1988).
238. Fukunishi, A., Tsunekawa, T. and Kawai, M., Jap. Pat. JP 01 279,820 [89 279,820], 1989 (Nov. 10); CA 112, 124930d (1990).
239. Sokol, P. E., U. S. Pat. 3,986,825, 1976 (Oct. 19); CA 86, 60406x (1977).
240. Sokol, P. E., U. S. Pat. 4,027,008, 1977 (May 31); CA 87, 90592t (1977).
241. Hsiung, Du. Y. and Mueller, W. H., U. S. Pat. 4175572, 1979 (Nov. 27); CA 92, 135133r (1980),
242. Shiseido Co., Ltd., Jap. Pat. JP 58,167,700, 1983 (Oct. 3); CA 100, 87682d (1984).
243. Shiseido Co., Ltd., Jap. Pat. JP 58 138,799, 1983 (Aug. 17); CA 100, 126720m (1984).
244. Walsh, M. F., Brit. Pat. GB 2,122,898, 1984 (Jan. 25); CA 101, 12011w (1984).
245. Lion Corp., Jap. Pat. JP 60 001,116, 1985 (Jan. 7); CA 102, 190848c (1985).
246. Puri, A. K., UK Pat. WO 8,600,013, 1986 (Jan. 3); CA 104, 155714a (1986).
247. Yanagida, T. and Murotani, I., Jap. Pat. JP 60 221,493, 1985 (Nov. 6); CA 104, 226755c (1986).
248. Fukuda, T. and Ogawa, M., Jap. Pat. JP 61 007,399, 1986 (Jan. 14); CA 105, 81227r (1986).
249. Miyazawz, K., Ohata, Y. and Ogawa, M., Jap. Pat. JP 62 004,799, 1987 (Jan. 10); CA 107, 42116a (1987).
250. Miyazawa, K., Ohata, Y. and Ogawa, M., Jap. Pat. JP 62 018,500, 1987 (Jan. 27); CA 107, 79981e (1987).
251. Dawson, G. G. and Williams, M. K., Eur. Pat. EP 222,525, 1987 (May 20); CA 107, 98637p (1987).
252. Dawson, G. G. and Williams, M. K., Uk Pat. GB 2,182,343, 1987 (May 13); CA 107, 98639r (1987).
253. Hefford, R. J. W. and Murray, A. M., Eup. Pat. EP 257,807, 1988 (March 2); CA 109, 155956j (1988).
254. Ogoshi, M., Miyazawa, K. and Uehara, K., Jap. Pat. JP 63 054,313, 1988 (March 8); CA 110, 121005d (1989).
255. Scandel, J., Brit.UK Pat. Appl. GB 2,220,216, 1990 (Jan. 4); CA 113, 29116e (1990).
256. Yoshioka, K. and Kamimura, Y., Japan Pat. JP 02 45,409 (90 45,409), 1990 (Feb. 15); CA 113, 197655y (1990).
257. Yoshioka, K. and Kamimura, Y., Japan Pat. JP 02 45,410 (90 45,410), 1990 (Feb. 15); CA 113, 197656z (1990).
258. Goetz, H., Hartmann, P., Koehler, J. and Lang, G., Ger. Offen. DE 3,826,369, 1990 (Feb. 8); CA 113, 217778w (1990).
259. Takada, Y., Yuizono, M., Kumagai, M and Muramatsu, K., Japan Pat. JP 02 142,900 (90 142,900), 1990 (May 31); CA 114, 45521j (1991).
260. Nishida, Y., Japan Pat. JP 02 166,161 (90 166,161), 1990 (June 26); CA 114, 88411r (1991).
261. Gilfillan, J. L. and Huff, J. W., Res. Commun. Chem. Pathol. Pharmacol., 1981, 33, 373; CA 95, 197363m (1981).
262. Malek, B., George, D. B. and Filip, D. S., J. Am. Water Works Assoc., 1981, 73, 164; CA 95, 30124x (1981).

263. Smith, F. X. and Riedhammer, T. M., Eur. Pat. Appl. EP 53451, 1982 (June 9); CA 97, 98406n (1982).
264. Howes, J. G. B., Eur. Pat. Appl. EP 55,048, 1982 (June 30); CA 97, 188331p (1982).
265. U.S. Food and Drug Adm., Fed. Regist., 1982 (Oct. 1), 47 (191), 43365; CA 97, 180371f (1982).
266. U.S. Food and Drug Adm., Fed. Regist., 1982 (Aug. 31), 47 (169), 38274; CA 97, 143241t (1982).
267. U.S. Food and Drug Adm., Fed. Regist., 1982, 47 (181), 41103; CA 97, 196977u (1982).
268. Mueka, O., Pruegel, M., Peuckert, W., Rusche, J., Ballschuh, D. and Ohme, R., Ger. (East) DD 155,326, 1982 (June 2); CA 97, 196832t (1982).
269. Brysin, V. G., Zaurov, D. D., Gudkova, L. A. and Topchiev, D. A., Lab. Delo, 1984, 2, 75; CA 100, 172754y (1984).
270. Arcelli, A. and Concilio, C., J. Chem. Soc., Perkin Trans. 2, 1983, 9, 1327; CA 100, 22200a (1984).
271. Shnaider, M.A., Kamenskaya, E.V., Korshunova, M.L., Sirbiladze, A.L. and Topchiev, D.A., Izv. Vyssh. Uchebn. Zaved. Pishch Teknol 1983, 6, 32; CA 100, 137290x (1984).
272. Loth, F., Dautzenberg, H., Hahn, H. J., Jahr, H., Mehlis, B. & Fechner, K., Ger. (East) DD 217,821, 1985 (Jan. 23); CA 103, 92897u (1985).
273. Sekisui Chemical Co., Ltd., Jap. Pat. 60 06,865 [85 06,865], 1985 (Jan. 14); CA 102, 145699u (1985).
274. Khaitov, R. M., Mustafaev, M. I., Norimov, A. S. and Abramenko, T. V., Immunologiya (Moscow), 1986, 2, 22; CA 105, 77190f (1986).
275. Plaschnik, D., Knabe, G., Kallies, K., Kretzschmar, E., Pfuetzner, L. and Huenniger, H., Ger. (East) DD 224,951, 1985 (Jul. 17); CA 104, 145088k (1986).
276. Braun, K., Besch, W., Lucke, S. and Hahn, H. J., Exp. Clin. Endocrinol., 1986, 87, 313; CA 105, 213976w (1986).
277. Dautzenberg, H., Loth, F., Fechner, K., Mehlis, B. and Pommerening, K., Makromol. Chem., Suppl., 1985, 9, 203; CA 104, 39647g (1986).
278. Yajima, M., Yosui to Haisui, 1989, 31, 203; CA 111, 19375h (1989).
279. Emori, S., Sato,H., Sudo, K. and Ekusa, K., Jap. Pat. JP 63 21,975 [88 21,975], 1988 (Jan. 29); CA 110, 191755x (1989).
280. Anraku, H., Jap. Pat. JP 63,275,954 [88,275,954], 1988 (Nov. 14); CA 111, 21151s (1989).
281. Abdullaev, O. G., Tsoi, O. G., Kalendareva, T. I., Rashidova, S. Sh. and Shtil'man, M. I., Vysokomol. Soedin., Ser. B, 1989, 31, 271; CA 111, 210485y (1989).
282. Anraku, H., Jap. Pat. JP 63 83,670 [88 83,670], 1988 (Apr. 14); CA 111, 3732m (1989),
283. Anraku, H., Jap. Pat. JP 63 82,361 [88 82,361], 1988 (Apr. 13); CA 111, 3731k (1989),
284. Ueda, T., Harada, S. and Ise, N., Polymer J., 1972, 3, 476; CA 77, 152788r (1972).
285. Nekrasov, A.V., Berestetskaya, T.Z., Efimov, V.S., Chernova, O.V. and El'tsefon, B.S., USSR SU 891,703, 1981 (Dec. 3); CA 96, 200381u (1982).
286. Nekrasov, A.V., Berestetskaya, T.Z., Efimov, V.S., Chernova, O.V. and El'tsefon, B.S., USSR SU 891,703, 1981 (Dec. 23); CA 98, 17220t (1983).
287. Ginsburg, I., Sela, M.N., Morag, A., Ravid, Z., Duchan, Z., Ferne, M., Thomas, P.P., Davies, P., et al. Inflammation (NY), 1981, 5, 289; CA 97, 67142g (1982).
288. Otvos, J. W., Casti, T. E. and Calvin, M., Sci. Pap. Inst. Phys. Chem. Res. (Jpn.), 1984, 78, 129; CA 103, 169705q (1985).
289. Reidmann, W. D., Banasiak, L., Brunner, G., Kochmann, W., Lyr, H., Naumann, J., Pfeiffer, H. D., Reissmueller, H. and Schwotzer, H., Ger. (East) DD 215,690, 1984 (Nov. 21); CA 103, 100408c (1985).
290. Shkurnikova, I. S., Vasil'chenko, E. A., Penenzhik, M. A., Pismannik, K. D.,Topchiev, D. A. and Virnik, A. D., Khim. Drev., 1985, 30; CA 102, 154751g (1985).

291. U. S. Food and Drug Administration, Fed. Regist., 1987 (May 11), 52, 17553; CA 107, 38210j (1987).
292. Motoyama, S., Ohno, S., Umeda, S., Ikema, R., Inohana, S.and Hoshino, H., EP 229,380, 1987 (Jul. 22); CA 107, 153160e (1987).
293. Yahagase, A., Kawachi, Y., Murashige, Y. and Fukahori, N., Jap. Pat. JP 62 53,666 [87 53,666], 1987 (Mar. 9); CA 108, 192806k (1988).
294. Suzuki, M., Jap. Pat. 62,115,064 [87,115,064], 1987 (May 26); CA 108, 23066p (1988).
295. Unger, R., Vollhardt, D., Lerche, K. H., Kretzschmar, G., Hallensleben, S. and Klaus, A., Ger. (East) DD 249,378, 1987 (Sep. 9); CA 108, 100659p (1988).
296. Foerster, M., Schellenberger, A., Doepfer, K. P., Mansfeld, J., Dautzenberg, H., Kluge, H. and Roembach, J., Ger. (East) Pat. DD 272,868, 1989 (Oct. 25); CA 112, 213100a (1990).
297. Hoover. M. F., U. S. Pat. 3,539,684, 1970 (Nov. 10); CA 74, 31069m (1971).
298. Shigematsu, T., Shibahara, T., Nakashima, T. and Teraoka, T., Jap. Pat. 54 157,826, 1979 (Dec. 13); CA 92, 158963a (1980).
299. Lyr, H., Ballschuh, D., Banasiak, L., Gruendel, E., Mueller, A., Ohme, R., Otto, D. and Rusche, J., Ger. (East) Pat. DD 145490, 1980 (Dec, 17); CA 95, 37133h (1981).
300. Dautzenberg, H., Loth, F., Pommerening, K., Linow, K.J. and Bartsch, D., Ger. (East) Pat. DD 160,393, 1983 (July 27); CA 100, 157805x (1984).
301. Smith, A. L., U. S. Pat. US 4,462,914, 1984 (July 31); CA 101, 157454c (1984).
302. Riedmann, W. D., Banasiak, L., Brunner, G., Kochmann, W., Lyr, H., Naumann, J.,Pfeiffer, H. D., Reissmueller, H. and Schwotzer, H., Ger. (East) Pat. DD 215690, 1984 (Nov. 21); CA 103, 100408c (1985).
303. Reidmann, W. D., Ballschuh, D., Banasiak, L., Bluemke, R., Kochmann, W., Lyr, H., Ohme, R., Rusche, J. and Thust, U., Ger. (East) Pat. DD 161,130, 1985 (Feb. 20); CA 103, 122028f (1985).
304. Brysin, V. G., Omerov, M. M., Zaurov, D. D. and Topchiev, D. A., USSR Pat. SU 1,182,397, 1985 (Sept. 30); CA 104, 65418c (1986).
305. Paronyan, V. Kh., Shnaider, M. A., Topchiev, D. A., Askinazi, A. I., Schmidt, A. A., Kasparov, G. N., Klachko, Y. A., Grubman, I. I. and Kolchanova, I. V., USSR Pat. SU 1,180,384, 1986 (July 7); CA 105, 207834g (1986).
306. Dautzenberg, H., Loth, F., Pommerening, K., Linow, K. J. and Bartsch, D., Ger. Dem. Pat. CH 659,591, 1987 (Feb. 13); CA 107, 223273v (1987).
307. Mishchenko, S. V., Topchiev, D. A., Kaptsov, N. N., Merkushov, M. V. and Zhirnov, A. I., USSR Pat. SU 1,407,475, 1988 (July 7); CA 109, 209930s (1988).
308. Anraku, H., Jap. Pat. JP 63082362, 1988 (April 13); CA 110, 228133w (1989).
309. Thompson, J. A., U. S. Pat. US 4830776, 1989 (May 16); CA 111, 112015a (1989).
310. Morita, H., Arai, N. and Sekidera, Y., Japan Pat. 02 56,404 (90 56,404), 1990 (Feb. 26); CA 113, 206717z (1990).
311. Bell, L. and Craig, T., Canad. Pat. 1,272,191, 1990 (July 31); CA 114, 45357k (1991).
312. Arai, N. and Morita, H., Japan Pat. JP 02,207007 (90,207,007), 1990 (Aug. 16); CA 114, 77022p (1991).
313. Stange, J., Dummler, W., Bruegmann, E., Falkenberg, D., Siegmund, E., Ernst, B. and Dautzenberg, H., Z. Med. Laboratoriumsdiagn., 1990, 31, 366; CA 114, 88609m (1991).
314. Kovyneva, R.M., Markin, A.D., Aleksandrovich, K.M., Trushin, B.N. and Parkhamovich, E. S., Vestai Akad Navuk BSSR, Ser. Khim. Navuk 1981, 6, 107; CA 97, 26809y (1982).
315. Pacholke, G., Rettig, D., Eins, I., Shul'ta, G., Foerster, M., Ballschuh, D., Rusche, J. Markert, H., et al, Ger. (East) DD 159,657, 1983, (Mar. 23); CA 99, 90848d (1983).
316. Roe, W. J., U.S. Pat. 4,426,409, 1984 (Jan. 17); CA 100, 104804q (1984).
317. Nippon Valqua Industries, Ltd., Jap. Pat. 59 50,054, 1984 (Mar. 22); CA 101, 976442y (1984).

318. Nippon Valqua Industries, Ltd., Jap. Pat. 59 50,052, 1984 (Mar. 22); CA 101, 96445b (1984).
319. Nippon Valque Industries,Ltd., Jap. Pat. 59 50,053, 1984 (Mar. 22); CA 101 96444a (1984).
320. Rey, P. A., U.S. Pat. 4,536,186, 1985 (Aug. 20); CA 103, 163307q (1985).
321. Popovici, A., Cercet. Miniere, 1968, 11, 393; CA 74, 64809f (1971).
322. Hollenbeak, K. H. and Brown, P. S. Jr., Eur. Pat. EP 251,558, 1988 (Jan. 7); CA 108, 153399k (1988).
323. Lampton, R. D., Jr. and Hopkins, T. M., U.S. Pat. 4,614,644, 1986 (Sep. 30); CA 106, 69694d (1987).
324. Richardson, P. F. and Bhattacharyya, B. R., U.S. Pat. 4,673,511, 1987 (Jun. 16); CA 107, 158894g (1987).
325. Hodgkin, J. H. and Eibl, R., React. Polym. Ion Exch., Sorbents, 1988, 9, 285; CA 110, 61496c (1989).
326. Forkel, K., Masthoff, R., Klatt, B., Ballschuh, D. and Ohme, R., Z. Chem., 1989, 29, 345; CA 112, 24528k (1990).
327. Kirwin, R. C., Hart, W. L. and Antonetti, J. M., Ger. Pat. DE 2,259,009, 1973 (June 14); CA 79, 80956k (1973).
328. Hart, W. L., Antonetti, J. M. and Henricks, J. M., FR 2,175,174, 1973 (Nov. 23); CA 80, 85941g (1974).
329. Antonetti, J. M. and Snow, G. F., U. S. Pat. 4,141,691, 1979 (Feb. 27); CA 91, 7279k (1979).
330. Braun, D. B. and Fan, Y. L., Eur. Pat. EP 55,489, 1982 (July 7); CA 98, 38285h (1983).
331. Schoenherr, D., Froemling, U., Munick, H., Jaeger, W., Wandrey, C., Hahn, M. and Seehaus, F., Ger. (East) Pat. DD 203,305, 1983 (Oct. 19); CA 100, 108683x (1984).
332. Nippon Valqua Industries, Ltd., Jap. Pat. JP 59 050,054, 1984 (March 22); CA 101, 96442y (1984).
333. Nippon Valqua Industries, Ltd., Jap. Pat. JP 59 050,053, 1984 (March 22); CA 101, 96444a (1984).
334. Grimmecke, H. D., Gabert, A., Kupper, H., Typelt, H. and Fiebig, H., Ger. (East) Pat. DD 225,708, 1985 (Aug. 7); CA 104, 126145k (1986).
335. Kelly, K. P. and Looney, J. R., UK Pat. GB 2,167,468, 1986 (May 29); CA 105, 136809q (1986).
336. Andreson, B. A., Kabanov, V. A., Topchiev, D. A., Zezin, A. B., Bochkarev, G. P., Vtyaganov, I. V. and Shakhmaev, Z, M., USSR Pat. SU 1,252,329, 1986 (August 23); CA 105, 211519a (1986).
337. Lampton, R. D., Jr. and Hopkins, T. M. II, U. S. Pat. US 4,614,644, 1986 (Sept. 30); CA 106, 69694d (1987).
338. Garland, E. K., Eur. Pat. EP 211,338, 1987 (Feb. 25); CA 106, 158837v (1987).
339. Lampton, R. D., Jr. and Hopkins, T. M. III, U. S. Pat. US 4,629,608, 1986 (Dec. 16); CA 106, 216394m (1987).
340. Nemeh, S. and Slepetys, R. A., Eur. Pat. EP 260,945, 1988 (March 23); CA 109, 212619c (1988).
341. Ergin, Yu. V., Fazlutdinov, K. S., Kostrova, L.I., Khatmullin, F. G. and Tukaev, R. A., USSR Pat. SU 1,511,375, 1989 (Sept. 30); CA 112, 23436s (1990).
342. Petrochenkova, E. V., Dytnerskii, Yu. I., Volchek, K. A., Tokareva, G. I. and Topchiev, D. A., Khim. Tekhnol. Vody, 1990, 12, 171; CA 113, 9782r (1990).
343. Vasconcellos, S. R. and Luong, P. T., U. S. Pat. 4,906,386, 1990 (Mar. 6); CA 113, 26758e (1990).
344. Forkel, K., Ebert, R., Hoebbel, D., Ballschuh, D. and Ohme, R., Ger. (East) Pat. DD 282,678, 1990 (Sept. 19); CA 114, 127723r (1991).

345. Groebe, V., Luu, T.H., Philipp, B. and Schwarz, H. H., Acta Polym., 1981, 32, 488; CA 95, 188247x (1981).
346. Teijin, Ltd., Jpn. Pat. 81 10,531, 1981 (Feb. 3); CA 95, 63379n (1981).
347. Teich, W. and Zirkler, W., Ger. (East) DD 214,063, 1984 (Oct. 3); CA 102, 147537p (1985).
348. Ebdon, L., Braven, J. and Frampton, N. C., Analyst (London), 1990, 115, 189; CA 112, 209936k (1990).
349. Fujihira, M., Japan Pat. JP 02,102,253 (90,102,253), 1990 (Apr. 13); CA 113, 173700h (1990).
350. Halliburton Co., Brit. Pat. 1,590,345, 1981 (June 3); CA 95, 206518v (1981).
351. Kabanov, V.A., Topchiev, D.A., Kaptsov, N.N., Zezin, A.B., Sulga, G., Mozheiko, L.N., Rekners, F., Gudkova, L. A. and Nashmetdinova, G. T., USSR Pat. SU 865,887, 1981 (Sept. 23), 109; CA 96, 51344e (1982).
352. Reidmann, W. D., Ballschuh, D., Banasiak, L., Bluemke, R., Kochmann, W., Lyr, H., Ohme, R., Rusche, J. and Thust, U., Ger. (East) DD 161,130, 1985 (Feb. 20); CA 103, 122028f (1985).
353. Benkenstein, H., Kullmann, A., Pagel, H., Krueger, W., Ohme, R. and Ballschuh, D., Wiss. Z. Humb.U. Berlin, Math. Nat.Reihe 1987, 36, 330; CA 107, 153486r (1987).
354. Buntin, G. A. and Scheve, B. J., U.S. Pat. US 4,759,923, 1988 (Jul. 26); CA 110, 185962w (1989).
355. Chamberlain, P. and Langley, J. G., Eur. Pat. Appl. EP 365,279, 1990 (Apr. 25); CA 113, 2087j (1990).
356. Meisel, U., Plaschnick, D., Luu, T.H., Wandrey, C., Bach, G., Jaeger, W., Hahn, M. and Phillip, B., et al, Ger. (East) DD 153,002, 1981 (Dec. 16); CA 97, 118184u (1982).
357. Muratkozieva, N.A., Russkikh, A.N., Skvortsova, M.M., Chegodaev, F.N. and Shmakova, T.V., USSR SU 1,095,961, 1984 (June 7); CA 101, 113126t (1984).
358. Pokhodenko, N. T., Shmidt, B. B., Petrumina, O. A., Gafner, V. V., Gumerov, F. Z., Topchiev, D.A., Kabanov, V. A., Zezin, A. B. and Papisov, I. M., USSR SU 1,250,569, 1986 (Aug. 15); CA 105, 213630d, (1986).
359. Jujo Paper Mfg. Co., Ltd., Jap. Pat. JP 60,109,894 [85,109,894], 1985 (Jun. 15); CA 103, 224474r (1986).
360. Tokita, M., Kobayashi, A. and Yasuda, K., Jap. Pat. 01 75,281 [89 75,281], 1989 (Mar 20); CA 111, 105922r (1989).
361. Waagberg, L., Oedberg, L. and Glad-Nordmark, G., Nord. Pulp Pap. Res. J., 1989, 4, 71; CA 111, 176513u (1989).
362. Topchiev, D. A., Murzabekova, T. G., Kolganova, I. V., Nechaeva, A. V., Davydova, S. L., Krut'ko, E. B. and Kabanov, V. A., USSR SU 1,409,635, 1988 (July 15); CA 110, 25512j (1989).
363. Takimoto, H. and Yoneyama, T., Jap. Pat. 01 81,868 [89 81,868], 1989 (Mar. 28); CA 111, 235272q (1989).
364. Tschang, C. J., Rehmer, G., Winkler, E., Gotsmann, G. and Glaser, K., Ger. Offen. DE 3,743,742, 1989 (July 6); CA 111, 175641x (1989).
365. Richter, H. U., Ihle, G., Weisbach, V., Schilling, G. and Welzel, H. P., Ger. (East) DD 261,800, 1988 (Nov. 9); CA 111, 24694r (1989).
366. Okubo, T., J. Colloid Interface Sci., 1988, 125, 386; CA 110, 13901j (1989).
367. Onabe, F., Osawa, I., Usuda, M. and Kadoya, T., Mokuzai Gakkaishi, 1984, 30, 839; CA 102, 26627j (1985).
368. Leca, M., Olteanu, M. and Ionescu-Bujor, I., Rev. Roum. Chim., 1989, 34, 811; CA 111, 115887x (1989).
369. Kudaibergenov, S. E., El'chibaeva, Z. S., Jaeger, W., Reinisch, G. and Bekturov, E. A., Izv. Akad. Nauk. Kaz. SSR, Ser. Khim., 1989, 50; CA 111, 233823w (1989).
370. Huber, E. W., Diss. Abstr. Int. B, 1989, 49, 3148; CA 111, 224253h (1989).

371. Gordon, A. Z., Yeager, E. B., Tryk, D. S. and Hossain, M. S., PCT Int. Appl. WO 88 06,646, 1988 (Sep. 7); CA 110, 138678y (1989).
372. Faust, C. A., Eur. Pat. EP 293,129, 1988 (Nov. 30); CA 110, 140953c (1989).
373. Dubin, P. L., The, S. S., McQuigg, D. W., Chew, C. H. and Gan, L. M., Langmuir, 1989, 5, 89; CA 110, 45399t (1989).
374. Camisa, J. D., U.S. Pat. US 4,810,329, 1989 (Mar. 7); CA 111, 197102z (1989).
375. Brungardt, C. L. and Rush, P. K., Eur. Pat. EP 321,688, 1989 (June 28); CA 111, 179839r (1989).
376. Besio, M. and Valsecchi, A., Eur. Pat. EP 320,692, 1989 (Jun. 21); CA 111, 184085w (1989).
377. Adler, E., Welzel, H. P., Ohme, R. and Ballschuh, D., Ger. (East) Pat. DD 261,801, 1988 (Nov. 9); CA 111, 155391n (1989).
378. Takagishi, T., Naoi, Y. and Kuroki, N., J. Polym. Sci., (Chem.), 1979, 17, 1953; CA 91, 108436q (1979).
379. Peiffer, D. G., U.S. Pat. 4,853,447, 1989 (Aug. 1); CA 112, 57001m (1990).
380. Yajima, M., Jap. Pat. 01,294,602 [89,294,602], 1989 (Nov. 28); CA 113, 36373q (1990).
381. Kurakawa, Y., Shirakawa, N., Terada, M. and Yui, N., J. Appl. Polym. Sci., 1980, 25, 1645; CA 93, 186911w (1980).
382. Dixon, K. W., U.S. Pat. US 4,225,445, 1980 (Sept. 30); CA 93, 242396w (1980).
383. Schempp, W. and Tran, H. T., Wochenbl. Papierfabr., 1981, 109, 728; CA 95, 221563z (1982).
384. Schempp, W. and Tran, H. T., Wochenbl. Papierfabr., 1981, 109, 731; CA 95, 221563z (1982).
385. Toei, K. and Zaitsu, T., Bunseki Kagaku, 1982, 31, 543; CA 97, 207390a (1982).
386. Ballschuh, D., Rusche, J. and Ohme, R., Ger. (East) DD 158,783, 1983 (Feb. 2); CA 99, 38975d (1983).
387. Borchardt, J. and Smith, C., Ger. Offen. DE 3,213,799, 1982 (Nov. 4); CA 98, 94650k (1983).
388. Plaschnick, D, Kuhrt, A., Meisel, U., Bley, M., Jaeger, W. Wandrey, C. and Linow, K. J., Ger. (East) DD 200,395, 1983 (May 4); CA 99, 184959t (1983).
389. Andreson, B.A., Topchiev, D.A., Kabanov, V.A., Bochkarev G.P., Varfolomeev, D.F., Schmidt, B.B., Einkeeve, E.K. and Sharipov, A.U., USSR Pat. SU 1,129,215, 1984 (Dec. 15); CA 102, 151942j (1985).
390. Martin, F. D., Hatch, M. J., Abouelezz, M. and Okley, J. C., Polym. Mater. Sci. Eng., 1984, 51, 688; CA 101, 213571c (1984).
391. Goerisch, K., Hammer, G., Kappler, W., Metze, J., Przyborowski, H., Ohme, R., Ballschuh, D. and Rusche, J., Ger (East) Pat. DD 213,942, 1984 (Sept. 26); CA 102, 134769r (1985).
392. Klyachko, Yu. A., Shargorodskii, V.B. and Shnaider, M. A., Vinodel. Vinograd. SSSR, 1984, 7, 51; CA 102, 60703r (1985).
393. Krause, T., Schempp, W. and Hess, P., Dtsch. Papierwirtsch, 1984, 127; CA 102, 26650 (1985).
394. Matz, G. F., U.S. Pat. 4,536,292, 1985 (Aug. 20); CA 103, 200673d (1985).
395. Philipp, B., Linow, K. J. and Dautzenberg, H., Wiss. Z. Tech.Hoch.C. Schorlemmer L-M 1985, 27, 295; CA 103, 178830d (1985).
396. Schellenberger, A., Doepfer, K. P., Straube, G. and Dautzenberg, H., Ger. (East) DD 242,055, 1987 (Jan. 14); CA 107, 152919x (1987).
397. Dawson, J. C., U.S. Pat. US 4,604,218, 1986 (Aug. 5); CA 105, 194262s (1986).
398. Elliott, C. M., Redepenning, J. G. and Balk, E. M., J. Electroanal. Chem. Interfac. Electroc 1986, 213, 203; CA 106, 40310q (1987).
399. Izumrudov, V. A., Zezin, A. B. and Kabanov, V. A., Dokl. Akad. Nauk SSSR, 1986, 291, 1150; CA 107, 92146f (1987).

400. Kokufuta, E., Shimohashi, M. and Nakamura, I., J. Ferment. Technol., 1986, 64, 533; CA 106, 125617v (1987).
401. Kovaleva, M. P., Kats, A. A., Rozin, M. M., Kudryavtseva, N. Y., Markus, E. Y., Kryuchkov, V. V., Amburg, L. A. and Parkhamovich, E. S., USSR SU 1,257,102, 1986 (Sep. 15); CA 106, 36875k (1987).
402. Keil, K. H., Wullbrandt, D., Keller, R. and Engelhardt, F., Ger. Offen. DE 3,537,259, 1987 (Apr. 23); CA 107, 116135e (1987).
403. Croitoru, V. and Ciocodei, F. G., Bul. Inst. Politeh Bucuresti, Ser. Chim. 1987, 49, 49; CA 109, 129957r (1988).
404. Biering, H., Wagner, D., Pfeiffer, H. D., Naumann, J., Thust, U., Ballschuh, D., Bech, R., Ohme, R. and Rusche, J., Ger.(East) Pat. DD 246,680, 1987 (Jun. 17); CA 108, 81760w (1988).
405. Becker, H. G. O., Schuetz, R., Kuzmin, M. G., Sadovskii, N. A. and Soboleva, I. V., J. Prakt. Chem., 1987, 329, 87; CA 107, 197352y (1987).
406. Danilov, I. N., Ishmaeva, A. M. and Kuznetsov, V. A., Khim. Tekhnol. Topl. Masel, 1987, 13; CA 106, 197200b (1987).
407. Handa, N., Yoshida, Y. and Masuda, Y., Jap. Pat. JP 62,170,591, 1987 (Jul. 27); CA 108, 39573n (1988).
408. National Starch and Chemical Corp., Jap. Pat. 62 36,409 [87 36,409], 1987 (Feb. 17); CA 107, 97294n (1987).
409. Noda, Y., Bunseki kagaku, 1987, 36, 403; CA 107, 193454k (1987).
410. Walterick, Jr., G. C. and Fillipo, B. K., Eur. Pat. Appl. EP 211,527, 1987 (Feb. 25); CA 107, 64563w (1987).
411. Hoffmann, H., Seoud, O. El., Huber, G. and Baecher, R., Proc. Int. Meet. Polym. Sci. Technol., 1988, 317; CA 109, 231942u (1989).
412. Dubin, P. L. and Murrell, J. M., Macromolecules, 1988, 21, 2291; CA 109, 38607k (1988).
413. Dubin, P. L., Rigsbee, D. R., Gan, L. M. and Fallon, M. A., Macromolecules, 1988, 21, 2555; CA 109, 74361j (1988).
414. Gordon, A. Z., Yeager, E. B., Tryk, D. S. and Hossain, M. S., PCT Int. Appl. WO 88 06,645, 1988 (Sep. 7); CA 109, 213625p (1988).
415. Gordon, A. Z., Yeager, E. B., Tryk, D. S. and Hossain, M. S., PCT Int. Appl. WO 88 06,644, 1988 (Sep. 7); CA 109, 213626q (1988).
416. Gordon, A. Z., Yeager, E. B., Tryk, D. S. and Hossain, M. S., PCT Int. Appl. WO 88 06,642, 1988 (Sep. 7); CA 109, 213627r (1988).
417. Howells, M., AU 568,721, 1988 (Jan. 7); CA 109, 25179x (1988).
418. Huber, E. W. and Heineman, W. R., Anal. Chem., 1988, 60, 2467; CA 109, 178844k (1988).
419. Kokufuta, E., Shimohashi, M. and Nakamura, I, Biotechnol. Bioeng., 1988, 31, 382; CA 108, 172881c (1988).
420. Miki, S., Maruyama, T., Ohno, T., Tohma, T., Toyama, S. and Yoshida, Z., Chem. Lett., 1988, 861; CA 109, 171408n (1988).
421. Takagishi, T. and Matsui, N., Chem. Express, 1988, 3, 395; CA 109, 192136n (1988).
422. Varennes, S. and Van de Ven, T. G. M., Colloids Surf., 1988, 33, 63; CA 109, 198091r (1988).
423. Rysaev, U. Sh., Rasulev, Z. G., Abdrashitov, Ya M., Potapov, A. M., Rysaev, F. Sh. and Kolesnikov, I. M., USSR SU 1,407,927, 1988 (Jul. 7); CA 109, 233180t (1989).
424. Kornena, E. P., Tarabaricheva, L. A., Levchenko, S. A. and Kushchenko, L. P., USSR SU 1,386,642, 1988 (Apr.7); CA 109, 229197m (1989).
425. Artamonova, N. D., Baranovskii, V. Yu., Belen'kii, P. G., Boiko, A. Yu., Bolyachevskaya,K. I., Vinogradov, V. V., Zezin, A. B., Kabanov, V. A., Kalyuzhnaya,

R. I., et al, USSR SU 1,507,771, 1989 (Sept. 15); CA 112, 78762a (1990).
426. Laatikainen, M., J. Colloid Interface Sci., 1989, 132, 451; CA 111, 176473f (1989).
427. Ergin, Yu V., Kostrova, L.I., Khatmullin, F. G. and Tukaev, R. A., USSR Pat. SU 1,523,655, 1989 (Nov. 23); CA 112, 101921n (1990).
428. Weis, M., Valera, F. S. and Frimmel. F. H., Z. Wasser Abwasser Forsch, 1989, 22, 253; CA 112, 106168g (1990).
429. Amjad, Z. and Masler, W. F., U.S. Pat. US 4,885,097, 1989 (Dec. 5); CA 112, 223096t (1990).
430. Wood, H. W., Ger. Pat. DE 1,929,741, 1970 (Jan. 2); CA 72, 84894s (1970).
431. Hamann, H. C. and Clemens, D. H., U. S. Pat. 3,728,318, 1973 (April 17); CA 79, 6262q (1973).
432. Fukuda, A. and Igarashi, T., Jap. Pat. 49,070,832, 1974 (July 9); CA 82, 36689p (1975).
433. Ohsawa, K., Hirasa, K. and Kinoshita, H., Ger. Pat. DE 2,404,418, 1974 (Aug. 8); CA 82, 46292t (1975).
434. Davis, S. M., U. S. Pat. 3,877,993, 1975 (April 15); CA 83, 118589r (1975).
435. Ohsawa, K., Jap. Pat. 50,042,345, 1975 (April 17); CA 83, 182044m (1975).
436. Ohsawa, K., Jap. Pat. 50,054,843, 1975 (May 14), CA 83, 166874b (1975).
437. Ohsawa, K., Jap. Pat. 50,077,839, 1975 (June 25); CA 84, 92654b (1976).
438. Saito, K., Takasugi, M. and Oka, J., Jap. Pat. 51,023,439, 1976 (Feb. 25); CA 85, 70072d (1976).
439. Ohsawa, K., Jap. Pat. 51,032,434, 1976 (March 19); CA 85, 101265a (1976).
440. Ohsawa, K. and Koizumi, T., Jap. Pat. 52,075,621, 1977 (June 24); CA 88, 56477v (1978).
441. Mobil Oil A.-G. in Deutschland, Ger. Pat. DE 2,628,571, 1977 (Dec. 29); CA 88, 91317q (1978).
442. Hewitt, G. T., Ger. Pat. DE 2,750,777, 1978 (May 18); CA 89, 26458t (1978).
443. Nippon Kayaku Co., Ltd., Jap. Pat. 57,034,094, 1982 (Feb. 24); CA 97, 25957h (1982).
444. Topchiev, D. A., Kaptsov, N. N., Gudkova, L. A., Kabanov, V. A., Dytyuk, L. T., Samakaev, R. K. and Trushin, B. N., USSR SU 903,563, 1982 (Feb. 7); CA 97, 26177r (1982).
445. Ulrich, H. H., Reinisch, G., Gohlke, U., Leibnitz, E., Goebel, K. H. and Dietrich, K., Ger. (East) Pat. DD 160,112, 1983 (May 4); CA 99, 123161e (1983).
446. Shiseido Co., Ltd., Jap. Pat. JP 58,213,055, 1983 (Dec. 10); CA 100, 122844g (1984).
447. Shiseido Co., Ltd., Jap. Pat. JP 58,213,054, 1983 (Dec. 10); CA 100, 122845h (1984).
448. Touhsaent, R. E. and Steiner, R. H., Eur. Pat. EP 102,209, 1984 (Mar. 7); CA 100, 211823s (1984).
449. Scodari, N. F., U. S. Pat. US 4,457,912, 1984 (July 3); CA 101, 116592c (1984).
450. Savit, J., U. S. Pat. US 4,454,058, 1984 (June 12); CA 101, 132717c (1984).
451. Almond, S. W. and Scott, E., U. S. Pat. US 4,487,867, 1984 (Dec. 11); CA 102, 98131r (1985).
452. Brandenburg, W., Evers, J., Arndt, K. D., Klose, W., Gille, G., Borner, G. and Popp, P.,, Ger. (East) Pat. DD 226,481, 1985 (Aug. 28); CA 104, 170810x (1986).
453. Dubikovskaya, L. V., Kryuchkov, V. V., Kozlova, G. I., Fedoseenkova, L.I., Amburg, L. A., Parkhamovich, E. S., Spirin, G. V. and Malyuta, V. G., USSR Pat. SU 1,227,642, 1986 (April 30); CA 105, 98889f (1986).
454. Renger, P., Janowski, F., Wolf, F., Matzen, D., Haeussler, L. and Cotta, F., Ger. (East) Pat. DD 232,291, 1986 (Jan. 22); CA 105, 209962w (1986).
455. Ghiocel, R. R., Carpov, A., Maxim, S., Jocus, M., Crihalmeanu, A., Szabo, L. and Pop, D., Rom. Pat. RO 88,879, 1986 (Feb. 28); CA 106, 31394t (1987).
456. Ruskin, M. K., U. S. Pat. US 4,661,439, 1987 (April 28); CA 107, 106302t (1987).
457. Kaiser, H. J., U. S. Pat. US 4,759,855, 1988 (July 26); CA 109, 115558z (1988).

458. Ena, A. B., Dubikovskaya, L. V., Paramonova, N. P., Petrova, V. Ku., Parkhamovich, E. S., Nekhorosheva, T. A. and Topchiev, D. A., USSR Pat. SU 1426979, 1988 (Sept. 30); CA 110, 96496e (1989).
459. Rehmner, G., Niessner, M., Heide, W., Hartmann, H. and Peters, K. C., Fed. Rep. Ger. Pat. DE 3,709,921, 1988 (Oct. 6); CA 110, 155039h (1989).
460. Hirai, T., Yamaki, J., Tobishima, S., Arakawa, M. and Yoshimatsu, I., Jap. Pat. JP 01,030,179, 1989 (Feb. 1); CA 110, 216345d (1989).
461. Weigl, J., Hofmann, H. P. and Von Raven, A., Fed. Rep. Ger. Pat. DE 3,730,833, 1989 (March 23); CA 111, 41687h (1989).
462. Ariga, T., Murakami, K., Shimada, M., Nagai, K. and Kamimura, H., Jap. Pat. JP 63,299,971, 1988 (Dec, 7); CA 111, 136200h (1989).
463. Sudbeck, E. A., Dubin, P. L., Curran, M. E. and Skelton, J., J. Colloid Interface Sci., 1991, 142, 512; CA 114, 129842c (1991).
464. Terabe, S. and Isemura, T., Anal. Chem., 1990, 62, 650; CA 112, 151026v (1990).
465. Dautzenberg, H., Linow, K. J. and Rother, G., Acta Polym., 1990, 41, 98; CA 112, 159370y (1990).
466. Khamzamulina, R. E., Kudaibergenov, S. E., Bekturov, E. A., Ketz, I., Filipp, B. and Han, M., Izv. Akad. Nauk Kaz. SSR, Ser. Khim., 1990, 1, 45; CA 113, 41754u (1990).
467. Dubin, P. L., Vea, M. E. Y., Fallon, M. A. The, S. S., Rigsbee, D. R. and Gan, L. M., Langmuir, 1990, 6, 1422; CA 113, 85273r (1990).
468. Meister, J. J., U. S. Pat. 4,931,527, 1990, June 5; CA 113, 134453k (1990).
469. Huddleston, D. A. and Williamson, C. D., U. S. Pat. 4,938,803, 1990 (July 3); CA 113, 155632h (1990).
470. Topchiev, D.A., Martynenko, A.I., Kabanova, E.Yu., Varyushina, G.P., Kuznetsov, O.Yu. & Kirsh, Yu. E., USSR SU 1,578,083, 1990 (July 15); CA 113, 196166q (1990).
471. Strege, M. A., Dubin, P. L., West, J. S. and Flinta, C. D., ACS Symp. Ser., 1990, 419, 158; CA 114, 2441r (1991).
472. Blinov, V. M., Gnedenkov, L. Yu., Trofimenko, V. V., Loshkarev, Yu. M. and Livshits, A. B., USSR Pat. SU 1,581,781, 1990 (July 30); CA 114 52048p (1991).
473. Krishtal, N. F., Poselenov, I. A., Nakhimovich,. M. L. and Sankov, B. G., Ref. Zh., Khim., 1990, Abstr. 17R2038; CA 114, 68848k (1991).
474. Dragalova, E. K., Topchiev, D. A. and Galushko, T. V., USSR SU 1,595,851, 1990 (Sept. 30); CA 114, 83194n (1991).
475. Farrar, D., PCT Int. Appl. WO 90 14,403, 1990 (Nov. 29); CA 114, 85131p (1991)
476. Hahn, M., Jaeger, W. and Schicktanz, A., Ger. (East) Pat. DD 274,432, 1989, (Dec. 20); CA 113, 24692e (1990).
477. Ichihara, A., Onishi, H. and Niike, H., Jap. Pat. JP 01 34,431 [89 34,431], 1989 (Feb. 3); CA 112, 99496g (1990).
478. Butler, G. B. and Bunch, R. L., U.S. Pat. US 2,687,382, 1954 (Aug. 24); CA 49, 1244a (1955).
479. Bolto, B. A., McNeill, R., Macpherson, A. S., Siudak, R., Weiss, D. E. and Willis, D., Aust. J. Chem., 1968, 21, 2703; CA 70, 6830n (1969).
480. Weiss, D. E., Bolto, B.A., McNeill, R., Macpherson, A. S., Siudak, R., Swinton, E. A.and Willis, D., J. Inst. Engrs., Australia, 1965, 37, 193; CA 63, 14536h (1965).
481. Weiss, D. E., Bolto, B. A., McNeill, R., Macpherson, A. S., Siudak, R., Swinton, E. A. Willis, D., Aust. J. Chem., 1966, 19, 561; CA 65, 4045c (1966).
482. Weiss, D. E., Bolto, B. A., McNeill, R., Macpherson, A. S., Siudak, R., Swinton, E. A. & Willis, D., Aust. J. Chem., 1966, 19, 589; CA 65, 4045e (1966).
483. Weiss, D. E., Bolto, B. A., McNeill, R., Macpherson, A. S., Siudak, R., Swinton, E. A. & Willis, D., Aust. J. Chem., 1966, 19, 765; CA 65, 4045g (1966).

484. Weiss, D. E., Bolto, B. A., McNeill, R., Macpherson, A. S., Siudak, R., Swinton, E. A. & Willis, D., Aust. J. Chem., 1966, 19, 791; CA 65, 4046a (1966).
485. Butler, G. B. and Bunch, R. L., J. Amer. Chem. Soc., 1952, 74, 3453; CA 48, 3251a (1954).
486. Butler, G. B. and Goette, R. L., U.S. Pat. US 2,946,757, 1960 (July 26).
487. Commonwealth Scientific and Indust. Research Organization, Jap. Pat. JP 79 83,092, 1979 (Jul. 2); CA 91, 124403g (1979).
488. Bolto, B A. and Siudak, R. V., J. Polym. Sci., Symp. 55, 1976, 55, 87; CA 86, 91188g (1977).
489. Bolto, B. A., Jackson, M. B. and Siudak, R. V., J. Polym. Sci., Symposium 55, 1976, 55, 95; CA 86, 156628x (1977).
490. Terabe, S. and Isemura, T., J. Chromatogr., 1990, 515, 667; CA 113, 183996y (1990).
491. Mathias, L. J. and Canterberry, J. B., Polym. Prepr., ACS, Polym. Div., 1982, 23(1), 171; CA 100, 23043p (1984).
492. Mathias, L. J. and Cei, G., Polym. Mater. Sci. Eng., 1988, 58, 244; CA 109, 151857e (1988).
493. Kandelaki, S.A., Gogoberishvili, K.M., Lekishvili, N.G., Dzhikiya, O.D., Chagulov, V.S., and Khananashvili, L.M., Soobshch. Akad. Nauk Gruz. SSR, 1989, 134, 553; CA 112, 119876m (1990).
494. Cei, G. and Mathias, L. J., Macromolecules, 1990, 23, 4127; CA 113, 131335n (1990).
495. Godt, A. and Schlueter, A. D., Chem. Ber., 1991, 124, 149; CA 114, 102865a (1991).
496. Azuma, T., Koizumi, S. and Kita, N., Jap. Pat. JP 01 40,943 [89 40,943], 1989 (Feb. 13); CA 111, 48181x (1989).
497. Berendsen, J., Vandenbruaene, R. and Vaes, J., Res. Discl., 1990, 312, 312; CA 113, 14599r (1990).
498. Han, S. H., Kim, U. Y., Kang, Y. S. and Choi, S. K., Macromolecules, 1991, 24, 973; CA 114, 102889m (1991).
499. Nagasuna, K., Namba, T., Miyake, K., Kimura, K. and Shimomura, T., Eur. Pat. Appl. EP 349,241, 1990, Jan. 3; CA 113, 60613p (1990).
500. Hirano, T., Ohashi, S., Morimoto, S., Tsuda, K., Kobayashi, T. and Tsukagoshi, S., Makromol. Chem., 1986, 187, 2815; CA 106, 38380u (1987).
501. Kuroda, T. and Yamawaki, N., Jpn. Pat. JP 01,315,338 [89,315,338], 1989, (Dec. 20); CA 113, 29347f (1990).
502. Schwab, G. and Brandon, R. L., U.S. Pat. 4,857,406, 1987 (Apr. 10); CA 112, 88379v (1990).
503. Kudaibergenov, S. E. and Bekturov, E. A., Vysokomol. Soedin., Ser. A, 1989, 31, 2614; CA 112, 235966q (1990).
504. Pepper, D. S., Eur. Pat. EP 209,251, 1987 (Jan. 21); CA 106, 192311d (1987).
505. Przybylski, M., Fell, E. and Ringsdorf, H., Makromol. Chem., 1978, 179, 1719.
506. Tang, B., Wang, L. and Xu, X., J. Bioact. Compat. Polym., 1990, 5, 65; CA 113, 73567b (1990).
507. Bel'gibaeva, Z. K., Jaeger, W., Khamzamulina, R. E., Reinisch, G. and Bekturov, E. A., Izu. Akad. Nauk Kaz. SSR, Ser. Khim., 1987, 52; CA 108, 22401g (1988).
508. Sasaki, I., Yamamoto, A. and Yanagase, A., Eur. Pat. Appl. EP 326,041, 1989 (Aug. 2); CA 112, 37387p (1990).
509. Decastro, E. S., Smith, D. A., Mark, J. E. and Heineman, W. R., J. Electroanal. Chem. Interfac. Electroc 1982, 138, 197; CA 97, 135620a (1982).
510. Anon., Fed. Regist., 1978, Nov. 24, 43(227) 54927; CA 90, 70678j (1979).
511. Murata, Y., Koinuma, Y., Amaya, N., Otsu, T. and Nisimura, M, U.S. Pat. 4,855,374, 1989 (Aug. 8); CA 112, 57438j (1990).

512. Kita, N., Jap. Pat. JP 01,223,473, 1989 (Aug. 29); CA 112, 88366p (1980).
513. Zhang, Y. X., Da, A. H., Hogen-Esch, T. E. and Butler, G. B., Polym. Ltrs., 1990, 28, 213; CA 113, 41764x (1990).
514. Scamehorn, J. F., Christian, S. D., Tucker, E. E. and Tan, B. I., Colloids Surf., 1990, 49, 259; CA 114, 68769k (1991).
515. Takase, I., Hirose, K. and Aida, H., Kobunshi Ronbunshu, 1990, 47 953; CA 114, 102920q (1991).
516. Mathias, L. J., Colletti, R. F. and Bielecki, A., J. Am. Chem. Soc., 1991, 113, 1550; CA 114, 103120r (1991).
517. Wolleb, H., Eur. Pat. Appl. EP 370,946, 1990, May 30; CA 114, 103490m (1991).
518. Cho, O. K., Kim, Y. H., Choi, K. Y. and Choi, S. K., Macromolecules, 1990, 23, 12; CA 112, 36517u (1990).
519. Butera, R. J., Simic-Glavaski, B. and Lando, J. B., Macromolecules, 1990, 23, 199.

Index

A

B

D

E

F

H

I

L

M

N

O

P

Q

R

S

V

W